APPLIED
MOLECULAR
BIOTECHNOLOGY

*The Next Generation of
Genetic Engineering*

APPLIED MOLECULAR BIOTECHNOLOGY

The Next Generation of Genetic Engineering

Edited by

Muhammad Sarwar Khan
Iqrar Ahmad Khan
Debmalya Barh

CRC Press
Taylor & Francis Group
Boca Raton London New York

CRC Press is an imprint of the
Taylor & Francis Group, an **informa** business

CRC Press
Taylor & Francis Group
6000 Broken Sound Parkway NW, Suite 300
Boca Raton, FL 33487-2742

First issued in paperback 2019

ISBN-13: 978-1-4987-1481-5 (hbk)
ISBN-13: 978-0-367-87247-2 (pbk)

Library of Congress Cataloging-in-Publication Data

Names: Khan, Muhammad Sarwar, editor. | Khan, Iqrar A. (Iqrar Ahmad), editor. | Barh, Debmalya, editor.
Title: Applied molecular biotechnology : the next generation of genetic engineering / editors, Muhammad Sarwar Khan, Iqrar Ahmad Khan, and Debmalya Barh.
Description: Boca Raton : CRC Press/Taylor & Francis, 2015. | Includes bibliographical references and index.
Identifiers: LCCN 2015039023 | ISBN 9781498714815 (alk. paper)
Subjects: | MESH: Genetic Engineering--trends. | Biotechnology--methods. | Metagenomics--methods.
Classification: LCC QH442 | NLM QU 550.5.G47 | DDC 616.02/7--dc23
LC record available at http://lccn.loc.gov/2015039023

Visit the Taylor & Francis Web site at
http://www.taylorandfrancis.com

and the CRC Press Web site at
http://www.crcpress.com

Contents

Foreword

During the last few decades several fundamental discoveries in life sciences have given rise to modern biotechnology, which is essentially based on breakthroughs in molecular biology. It is currently one of the fastest growing areas of science and thus, this century has rightly been termed the "Century of Biology," hoping that such advances in life sciences will yield changes more momentous than electricity and computers.

The tremendous advances in biotechnology have also had a profound effect not only on agriculture but also on medicine and the environment. About 20% of the world's pharmaceuticals are produced by using biotechnological processes and it has been estimated that about 50% of all pharmaceuticals will be produced in this manner by the year 2020. The most well-known products are insulin and many other biologicals, including interferon, other cytokines, and several antibiotics. With the developments in DNA sequencing, "personalized medicine" is on the anvil.

In the case of agriculture, the gains due to high-yielding varieties as a result of the green revolution in the 1960s have tapered off due to the increasing cost of inputs and scarcity of water in many countries. In order to make agriculture sustainable, modern biotechnology has played a significant role by developing pest-resistant and drought-tolerant crop varieties. As a result, there has been a 10-fold increase in the area under transgenic crops in the world since 1996 when it was 1.7 million ha, which has increased to 1785 million ha in 2014. This has come about by utilizing a series of molecular biology technologies related to genomics and new tools of bioinformatics.

During the last decade there have been several new discoveries in molecular biology that have rapidly found their way to applications. The whole new world of "omics" comprising genomics, proteomics, transcriptiomics, and metabolomics has revolutionized the way translational research is carried out. In addition, strides in computational, structural, and organelle biology have opened new vistas for developing useful products.

This book, *Applied Molecular Biotechnology: The Next Generation of Genetic Engineering*, is timely and very much needed by our younger generation of researchers. It comprises well-documented chapters on all aspects of new and emerging technologies related to plants, animals, industry, and the environment. I would like to compliment the editors for compiling such a comprehensive book covering all important aspects of the technology. I am sure it will be well received by the readers.

Professor Dr. Kauser Abdulla Malik, HI, SI, TI
Distinguished National Professor
Dean for Postgraduate Studies
Forman Christian College (A Chartered University)
Lahore

Preface

The twenty-first century is an era of technology and its applications. In the recent past, several path-breaking innovations in the field of life sciences have enabled us to resolve a number of serious global issues using cutting-edge technologies. This book provides important revolutionary molecular biology technique-based next-generation genetic engineering toward finding solutions to our needs in the fields of plant, animal, industrial, and environmental biotechnology.

The book, *Applied Molecular Biotechnology: The Next Generation of Genetic Engineering*, consists of 25 chapters subdivided into 3 sections. Chapter 1 by Dr. Malviya and colleagues provides an overview of omics-based latest tools and approaches used in modern biotechnology. Section I (Plant biotechnology) begins with Chapter 2 where Dr. Agarwal's group has demonstrated in detail how the various molecular biology technologies can be used to develop transgenic plants. In Chapter 3, Dr. Khan's team elaborated on how these transgenic plants can fulfill our ever-growing demand for food and other plant-derived products. Drs. Yagi and Shiina have given a detailed account of the chloroplast gene expression system and its applications in Chapter 4. Chapter 5 is dedicated to organelle biotechnology. In this chapter, mitochondrial omics have been discussed by Dr. Waqar Hameed. Dr. Mehboob-ur-Rahman and colleagues discuss various approaches and applications of plant functional genomics in the next chapter (Chapter 6). In Chapter 7, Dr. Jha et al. describe the current status and future prospects of whole genome resequencing toward crop improvement. Plant–microbe and plant–insect interactions are key phenomena in plant molecular biotechnology where plant protection and productivity are concerned. A detailed account on these aspects is presented by Dr. Ahmad's group in Chapter 8. In Chapter 9, Dr. de Sousa and colleagues have given brief, but highly useful, content in "Biotechnology for improved crop productivity and quality." The last chapter (Chapter 10) by Drs. Shafiq and Khan provides an overview of the methods that unveil the epigenetic code.

Section II (Animal biotechnology) consists of six chapters (Chapters 11 through 16). Chapter 11 by Dr. Shafique et al. deals with various animal models used in biomedical research. The most recent trend in medical genomics, that is, pharmacogenomics, is discussed in Chapter 12 by Drs. Chatterjee and Lo, titled "Variations in our genome: From disease to individualized cure." Chapter 13 by Dr. Dwivedi's group discusses how modern biotechnological approaches are used in the molecular diagnosis of various diseases such as cervical cancer, obesity, and diabetes. Chapters 14 and 15 are dedicated to these two highly important topics. While the screening and diagnosis techniques of cervical cancer are discussed in Chapter 14 by Dr. Nawaz and colleagues, Chapter 15 deals with the biological aspects of diabetes and obesity and are documented by Drs. Awan and Najam. The last chapter (Chapter 16) under Section II deals with human tissue banking and its role in biomedical research, by Drs. Mian and Ashankyty.

Section III combines industrial and environmental biotechnology. The general aspects of microbial biotechnology written by Drs. Aguilera and Aguilera-Gómez are included in Chapter 17. Dr. Mubin's group discusses the molecular aspects of viral diseases in Chapter 18. Chapter 19 by Dr. Ullah et al. deals with the production of industrial commodities using viral biotechnology. Chapter 20 deals with an important topic "Cell free biosystems" by Drs. Sharma and Khurana. The biotechnological uses of magnetic nanoparticles are covered by Dr. Majeed and colleagues in Chapter 21. Various applications of biotechnology in food and chemical industries are discussed in Chapter 22 by Dr. Bokhari and coauthors. In Chapter 23, Dr. Mohan's team describes how modern biotechnology can be applied to conserve our ecosystem. In this section, a special topic on marine biotechnology is also included. Dr. Rastogi and colleagues in Chapter 24 have provided a comprehensive account on various anticancer drugs from marine resources. The final chapter (Chapter 25) of this book deals with biofuel genomics by Dr. Khan and colleagues where genome-scale plant genetic engineering is described to develop transgenic plants for optimum biofuel production.

Since biotechnology itself is an interdisciplinary subject, it is difficult to cover every aspect of the subject in a single book. In this book, we have tried our best to provide an overview of the latest trends of application of molecular biology techniques in plant, animal, industrial, and environmental biotechnology. We do hope this book will be a worthwhile resource to students and researchers in the field of molecular biotechnology. We welcome your suggestions to improve the next edition of the book.

Muhammad Sarwar Khan, PhD
Iqrar Ahmad Khan, PhD
Debmalya Barh, PhD

Editors

Muhammad Sarwar Khan is a highly regarded molecular biologist who earned a PhD from the University of Cambridge. He was awarded the Rockefeller Foundation Fellowship under the Rice Biotechnology Program for Developing Countries to carry out research on plastid transformation at Waksman Institute of Microbiology, Rutgers, The State University of New Jersey, which was published in *Nature Biotechnology*. Dr. Khan was a founding head of the Biotech Interdisciplinary Division at the National Institute for Biotechnology and Genetic Engineering (NIBGE), and is currently serving as the director of the Center of Agricultural Biochemistry and Biotechnology (CABB), University of Agriculture, Faisalabad, Pakistan.

Dr. Khan has been honored with prestigious and befitting awards, including the President's Medal for Technology, Gold Medal in Agriculture by the Pakistan Academy of Sciences, Performance Gold Medal by the Pakistan Atomic Energy Commission, and a Biotechnologist Award by the National Commission of Biotechnology. He is a fellow of the Cambridge Commonwealth Society, the Cambridge Philosophical Society, and the Rockefeller Foundation.

Dr. Khan has supervised about 80 PhD and MPhil students and researchers who are serving at national and international levels in various research institutes and universities. He has published extensively in high-impact journals including *Nature* and *Nature Biotechnology*, and is author of a number of books and book chapters. Dr. Khan has made immense contributions in the field of chloroplast genetic engineering and is a pioneer in expressing oxygen-loving green fluorescent protein (GFP) from jellyfish in chloroplasts—plant organelles with reduced environment. He also pioneered plastid transformation in rice and sugarcane, recalcitrant plant species. As far as translational research is concerned, Dr. Khan has developed borer-resistant transgenic sugarcane plants with no toxin residues in the juice. His current research interests include development of edible-marker-carrying transgenics and cost-effective therapeutics and edible vaccines.

 **Iqrar Ahmad Khan** has had a long career in education and agriculture earning a PhD from the University of California, Riverside. He is currently serving as vice chancellor of the University of Agriculture, Faisalabad (UAF), Pakistan (since 2008). Dr. Khan has supervised more than 100 graduate students and researchers, established a Center of Agricultural Biotechnology and cofounded the Deutscher Akademischer Austauschdienst, German Academic Exchange Service (DAAD)-sponsored International Center for Decent Work and Development (ICDD), USAID-funded Center of Advanced Studies in Agriculture and Food Security, a French Learning Center and the Chinese Confucius Institute. He has organized numerous international conferences and established academic linkages across the continents. He has released a potato variety (PARS-70), pioneered research on breeding seedless Kinnow, and discovered new botanical varieties of wheat. Dr. Khan initiated an internationally acclaimed program to solve a devastating problem called Witches' Broom Disease of Lime in Oman. He is currently leading international projects to combat citrus greening disease and mango sudden death. He has published more than 270 articles, 5 books, and several book chapters.

Dr. Khan has a diplomatic skill that has attracted international partnerships and academic linkages (Afghanistan, Australia, South Korea, China, Germany, France, Malaysia, Indonesia, Turkey, Iran, India, Oman, Canada, UK, and USA). He has managed collaborative research projects sponsored by national and international agencies. As vice chancellor, he has revamped UAF academic, research, and outreach programs. He added new academic disciplines to narrow the knowledge gaps in microbiology, biotechnology, environmental sciences, food and nutrition, climate change, engineering, rural development, and education. UAF has achieved top ranks within the national system as well as in the Quacquarelli Symonds (QS) and National Taiwan University (NTU) rankings. Dr. Khan has a special knack for problem-solving research. He has set up an incubation center for the commercialization of knowledge. Exhibitions and business plan competitions have been made biannual features. A range of new quality assurance mechanisms have been added and special initiatives taken to narrow the gender gap.

Dr. Khan is a fellow of Pakistan Academy of Sciences and member of several professional societies and associations. In recognition of his outstanding contributions in the area of agriculture and food security he was honored with a civil award, *Sitara-e-Imtiaz*, by the government of Pakistan and very recently with *Ordre des Palmes Académiques* (with the grade of Officer) by the French Government for his exceptional role as an educator.

Debmalya Barh, MSc, MTech, MPhil, PhD, PGDM, is the founder of the Institute of Integrative Omics and Applied Biotechnology (IIOAB), India, a virtual global platform for multidisciplinary research and advocacy. He is a biotechnologist and has decades of experience in integrative omics applied to translational research. Dr. Barh has written more than 150 publications and is a globally recognized editor for editing omics-related cutting-edge reference books for top-notch international publishers. He also serves as a reviewer for several professional international journals of global repute. Owing to his significant contributions in promoting R&D globally using unique strategies, in the year 2010 he was recognized by *Who's Who in the World* and in 2014 he was entered in the *Limca Book of Records*—the Indian equivalent to the *Guinness Book of World Records*.

Contributors

Sandhya Agarwal
Metahelix Life Sciences Ltd
Bengaluru, Karnataka, India

Margarita Aguilera
Department of Microbiology
University of Granada
Granada, Spain

Jesús Manuel Aguilera-Gómez
Department of Sculpture
University of Granada
Granada, Spain

Jam Nazeer Ahmad
Department of Entomology
Integrated Genomic, Cellular,
    Developmental and Biotechnology
    Laboratory
University of Agriculture
Faisalabad, Pakistan

Niaz Ahmad
Agricultural Biotechnology Division
National Institute for Biotechnology
    and Genetic Engineering
Faisalabad, Pakistan

Samina Jam Nazeer Ahmad
Department of Botany
University of Agriculture
Faisalabad, Pakistan

Yog Raj Ahuja
Department of Genetics and Molecular
    Medicine
Vasavi Medical and Research Centre
Hyderabad, Telangana, India

Ibraheem Ashankyty
The Molecular Diagnostics and
    Personalised Therapeutics Unit
University of Ha'il
Ha'il, Kingdom of Saudi Arabia

Fazli Rabbi Awan
Diabetes and Cardio-Metabolic
    Disorders Lab
Health Biotechnology Division
National Institute for Biotechnology and
    Genetic Engineering
Faisalabad, Pakistan

Vasco Ariston de Carvalho Azevedo
Laboratory of Cellular and Molecular
    Genetics
Federal University of Minas Gerais
Belo Horizonte, Minas Gerais, Brazil

Debmalya Barh
Institute of Integrative Omics
    and Applied Biotechnology
    (IIOAB)
Purba Medinipur, West Bengal, India

Syed Ali Imran Bokhari
Department of Bioinformatics and
    Biotechnology
International Islamic University
Islamabad, Pakistan

Steven J. Burgess
Department of Plant Sciences
University of Cambridge
Cambridge, United Kingdom

Hugh J. Byrne
Focas Research Institute
Dublin Institute of Technology
Dublin, Ireland

Bishwanath Chatterjee
Department of Developmental Biology
University of Pittsburgh School of
    Medicine
Pittsburgh, Pennsylvania

Gaurav Chikara
Department of Pharmacology
All India Institute of Medical Sciences
Jodhpur, Rajasthan, India

and

King George Medical University
Lucknow, Uttar Pradesh, India

Cassiana Severiano de Sousa
Laboratory of Cellular and Molecular
    Genetics
Federal University of Minas Gerais
Belo Horizonte, Minas Gerais, Brazil

Shailendra Dwivedi
Department of Biochemistry
All India Institute of Medical Sciences
Jodhpur, Rajasthan, India

and

Department of Pharmacology and
    Therapeutics
King George Medical University
Lucknow, Uttar Pradesh, India

Sandrine Eveillard
INRA, UMR Biologie du Fruit et Pathologie
Villenave d'Ornon, France

Apul Goel
Department of Urology
King George Medical University
Lucknow, Uttar Pradesh, India

Alka Grover
Amity Institute of Biotechnology
Amity University
Uttar Pradesh, India

Alvina Gul
Atta-ur-Rahman School of Applied
    Biosciences
National University of Sciences and
    Technology
Islamabad, Pakistan

Maryyam Gul
Plant Genomics and Molecular Breeding
    Lab
National Institute for Biotechnology and
    Genetic Engineering
Faisalabad, Pakistan

Muhammad Waqar Hameed
Dr. Panjwani Center for Molecular
    Medicine and Drug Research
International Center for Chemical and
    Biological Sciences
University of Karachi
Karachi, Pakistan

Muhammad Asif Hanif
Department of Chemistry
University of Agriculture
Faisalabad, Pakistan

Syed Shah Hassan
Laboratory of Cellular and Molecular
    Genetics
Federal University of Minas Gerais
Belo Horizonte, Minas Gerais, Brazil

Sehrish Ijaz
Centre of Agricultural Biochemistry and
    Biotechnology
University of Agriculture
Faisalabad, Pakistan

Rintu Jha
Crop Improvement Division
Indian Institute of Pulses Research (IIPR)
Kanpur, Uttar Pradesh, India

Uday Chand Jha
Crop Improvement Division
Indian Institute of Pulses Research (IIPR)
Kanpur, Uttar Pradesh, India

Faiz Ahmad Joyia
Centre of Agricultural Biochemistry and
    Biotechnology
University of Agriculture
Faisalabad, Pakistan

Abdul Rehman Khan
Department of Environmental Sciences
COMSATS Institute of Information
    Technology
Abbottabad, Pakistan

Azka Khan
Atta-ur-Rahman School of Applied
    Biosciences
National University of Sciences and
    Technology
Islamabad, Pakistan

Muhammad Sarwar Khan
Centre of Agricultural Biochemistry and
    Biotechnology
University of Agriculture
Faisalabad, Pakistan

Sanjay Khattri
Department of Pharmacology and
    Therapeutics
King George Medical University
Lucknow, Uttar Pradesh, India

SM Paul Khurana
Amity Institute of Biotechnology
Amity University
Gurgaon, Haryana, India

Cecilia W. Lo
Department of Developmental Biology
University of Pittsburgh School of Medicine
Pittsburgh, Pennsylvania

Fiona M. Lyng
DIT Centre for Radiation and
    Environmental Science (RESC)
Focas Research Institute
Dublin Institute of Technology
Dublin, Ireland

Muhammad Irfan Majeed
Department of Chemistry
University of Agriculture
Faisalabad, Pakistan

Neha Malviya
Department of Biotechnology
D.D.U. Gorakhpur University
Gorakhpur, Uttar Pradesh, India

Muhammad Aamer Mehmood
Department of Bioinformatics and
    Biotechnology
GC University Faisalabad
Faisalabad, Pakistan

Maria Andréia Corrêa Mendonça
Goiano Federal Institute
Rio Verde, Goiás, Brazi

Shahid Mian
The Molecular Diagnostics and
    Personalised Therapeutics Unit
University of Ha'il
Ha'il, Kingdom of Saudi Arabia

Sanjeev Misra
Department of Surgical Oncology
King George Medical University
Lucknow, Uttar Pradesh, India

and

All India Institute of Medical Sciences
Jodhpur, Rajasthan, India

Vasavi Mohan
Department of Genetics and Molecular
    Medicine
Vasavi Medical and Research Centre
Hyderabad, Telangana, India

Mohammed Khaliq Mohiuddin
Department of Genetics and Molecular
    Medicine
Vasavi Medical and Research Centre
Hyderabad, Telangana, India

Muhammad Mubin
Centre of Agricultural Biochemistry and
    Biotechnology
University of Agriculture
Faisalabad, Pakistan

Ghulam Mustafa
Centre of Agricultural Biochemistry and
    Biotechnology
University of Agriculture
Faisalabad, Pakistan

Syeda Sadia Najam
Diabetes and Cardio-Metabolic
    Disorders Lab
Health Biotechnology Division
National Institute for Biotechnology and
    Genetic Engineering
Faisalabad, Pakistan

Haq Nawaz
Department of Chemistry
University of Agriculture
Faisalabad, Pakistan

Muhammad Shah Nawaz-ul-Rehman
Centre of Agricultural Biochemistry and
    Biotechnology
University of Agriculture
Faisalabad, Pakistan

Shahid Nazir
Biotechnology Research Institute
Ayub Agricultural Research Institute
Faisalabad, Pakistan

Aimen Niaz
Atta-ur-Rahman School of Applied
    Biosciences
National University of Sciences and
    Technology
Islamabad, Pakistan

Rajeev Kumar Pandey
Department of Dermatology and
    Allergic Diseases
University of Ulm
Ulm, Germany

Kamlesh Kumar Pant
Department of Pharmacology and
    Therapeutics
King George Medical University
Lucknow, Uttar Pradesh, India

Puneet Pareek
Department of Radiotherapy
All India Institute of Medical Sciences
Jodhpur, Rajasthan, India

Swarup K. Parida
National Institute of Plant Genome
    Research (NIPGR)
JNU Campus
New Delhi, India

Mehboob-ur-Rahman
Plant Genomics and Molecular Breeding Lab
National Institute for Biotechnology
    and Genetic Engineering
Faisalabad, Pakistan

Zainab Rahmat
Plant Genomics and Molecular
    Breeding Lab
National Institute for Biotechnology and
    Genetic Engineering (NIBGE)
Faisalabad, Pakistan

Nosheen Rashid
Faisalabad Institute of Research Science
    and Technology (FIRST)
Faisalabad, Pakistan

Amit Rastogi
Department of Molecular and Cellular
    Engineering
Sam Higginbottom Institute of
    Agricultural, Technology and Sciences
Allahabad, Uttar Pradesh, India

Sameen Ruqia
Atta-ur-Rahman School of Applied
    Biosciences
National University of Sciences and
    Technology
Islamabad, Pakistan

Saurabh Samdariya
Department of Radiotherapy
All India Institute of Medical Sciences
Jodhpur, Rajasthan, India

Sarfraz Shafiq
Department of Environmental Sciences
COMSATS Institute of Information
    Technology
Abbottabad, Pakistan

and

Center for Plant Biology
School of Life Sciences
Tsinghua University
and
Tsinghua-Peking Center for Life
    Sciences
Beijing, China

Adeena Shafique
Atta-ur-Rahman School of Applied
    Biosciences
National University of Sciences and
    Technology
Islamabad, Pakistan

Sara Shakir
Centre of Agricultural Biochemistry and
    Biotechnology
University of Agriculture
Faisalabad, Pakistan

Manju Sharma
Amity Institute of Biotechnology
Amity University
Gurgaon, Haryana, India

Praveen Sharma
Department of Biochemistry
All India Institute of Medical Sciences
Jodhpur, Rajasthan, India

Takashi Shiina
Graduate School of Life and
    Environmental Science
Kyoto Prefectural University
Kyoto, Japan

Narendra Pratap Singh
Crop Improvement Division
Indian Institute of Pulses
    Research (IIPR)
Kanpur, Uttar Pradesh, India

Bien Tan
School of Chemistry and Chemical
    Engineering
Huazhong University of Science and
    Technology
Wuhan, China

Aiman Tanveer
Department of Biotechnology
D.D.U. Gorakhpur University
Gorakhpur, Uttar Pradesh, India

Hafeez Ullah
PMAS Arid Agriculture University
Rawalpindi, Pakistan

Kinza Waqar
Atta-ur-Rahman School of Applied
    Biosciences
National University of Sciences and
    Technology
Islamabad, Pakistan

Dinesh Yadav
Department of Biotechnology
D.D.U. Gorakhpur University
Gorakhpur, Uttar Pradesh, India

Sangeeta Yadav
Department of Biotechnology
D.D.U. Gorakhpur University
Gorakhpur, Uttar Pradesh, India

Yusuke Yagi
Faculty of Agriculture
Kyushu University
Fukuoka, Japan

Yusuf Zafar
Department of Technical
    Cooperation
IAEA, Vienna International Centre
Vienna, Austria

chapter one

Emerging tools and approaches to biotechnology in the omics era

Neha Malviya, Aiman Tanveer, Sangeeta Yadav, and Dinesh Yadav

Contents

Abstract

Omics reflects totality and is generally referred for study of biomolecules influencing the structural and functional aspects of an organism. Further, for convenience the science of omics are classified into several branches such as genomics, proteomics, lipidomics, transcriptomics, cytomics, etc. Genomics is a well-known field of omics, which includes the study of total genetic content of an organism and often encompasses several other branches such as cognitive, comparative, functional, personal genomics, epigenomics, and metagenomics. The study of transcriptome termed as transcriptomics is an emerging field that covers the total set of transcripts in an organism along with the set of all ribonucleic acid (RNA) molecules. Along with the deoxyribonucleic acid and RNA, the proteins are the key players associated with the maintenance of cellular functions. Proteomics is applied to the exploration of protein structure and functionality. The nonproteinaceous repertoires of the cell metabolites are the intermediate products of metabolism. Its study, metabolomics is a rapidly evolving new discipline having potential implications. Besides the above-mentioned branches of omics, cytomics and lipidomics are also an indispensable element of the omics study. There are several basic and advanced techniques for the exploration of these branches and continuous efforts are being made to utilize these tools in different fields of scientific research. Elucidation and utilization of these applications is a hallmark for global research.

Keywords: Omics, Genomics, Proteomics, Transcriptomics, Metabolomics.

Introduction

Omics term is used with suffix -ome to address the study of respective field in totality such as genomics for genome, proteomics for proteome, metabolomics for metabolome, and many more. Omics reflects the use of diverse technologies to gain an insight into the complexity of biomolecules influencing the structure and function of organisms. The omics-driven research has led to some understanding of complex regulatory networks that controls gene expression, protein modification, and metabolite composition. Transcriptomics, metabolomics, bioinformatics, and high-throughput DNA sequencing led to the deciphering of diverse regulatory networks in different systems resulting in enhanced understanding for its potential applications in clinical diagnosis, prognosis, and therapeutic purposes.

The relationship between different branches of omics is shown in Figure 1.1.

General classifications of omics

Genomics

The outcome of innovations in sequencing technologies led to deciphering of whole genome sequences of different organisms and the branch of omics popularly known as

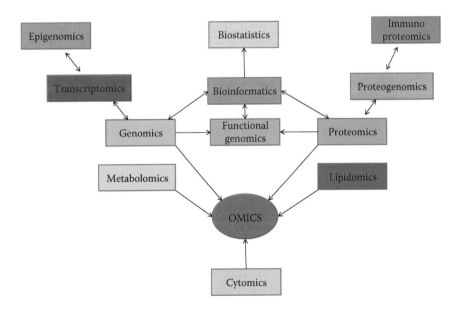

Figure 1.1 Pictorial representation showing a relationship among different branches of omics.

genomics came into existence. The genomics reflects the study of the genome in totality and is further classified into several subbranches as listed below.

Comparative genomics

The major principal underlying comparative genomics is that the common feature between the two organisms is encoded by the conserved DNA sequence. Conversely, divergence between the species is defined by the sequences that encode proteins or ribonucleic acids (RNAs). The comparative genomics provides an insight into the evolutionary aspects of organisms compared based on the sequences at whole genome level. It helps in discriminating conserved sequences from divergent ones. Further, comparative genomics can also be useful to understand the variability in terms of functional DNA segments, such as coding exons, noncoding RNAs, and also some gene regulatory regions. The genome sequences are compared by aligning them to score the match or mismatch between them. Various softwares and algorithms have been developed for the alignment of several genome sequences simultaneously and elucidate genome evolution and function.

Functional genomics

Functional genomics is applied to test and extend hypotheses that emerge from the analysis of sequence data. Elucidating the functions of identified genes of the sequenced genome of an organism is the sole purpose of functional genomics. While sequencing projects yield preliminary results, functional genomics focuses on the functional aspects such as regulation of gene expression, functions and interaction of different genes, etc. Functional genomics means genome-wide analysis through high-throughput methods. Hence it provides an overview of the biological information encoded by the organism's genome. The encyclopedia of DNA elements (ENCODE) project is a much-anticipated project, which aims to recognize all the functional elements of genomic DNA both in coding and noncoding regions.

Metagenomics

Metagenomics is emerging as an important discipline to access the biocatalytic potential of unculturable microorganisms. Despite very rich microbial diversity of the range of a million species per 1 g of soil, very few microorganisms can be cultured under *in vitro* conditions. With the advances made in the field of metagenomics, DNA can be extracted from environmental samples from which genomic library can be prepared. This library can be further explored by screening the clones for biological activity to identify clones possessing desired characteristics. A number of biocatalysts such as laccase, xylanase, endoglucanase, exoglucanase, and lipase have been recently identified from metagenomic libraries.

Epigenomics

Epigenetics refers to the external modification of DNA. It alters the physical structure of DNA without altering the DNA sequence. The DNA methylations, that is, the addition of a methyl group, or a "chemical cap," to part of the DNA molecule and histone modification are examples of epigenetic changes. Epigenetic changes can be carried over to the following generation if the modifications occur in sperm or egg cells. But most of these epigenetic changes get corrected during reprogramming of fertilized eggs. Cellular differentiation is also an example of epigenetic change in eukaryotic biology. Epigenetic mechanisms are influenced by many other factors such as prenatal development and in childhood, environmental influence, drugs, aging, diet, etc.

Personal genomics

The completion of human genome project has provided valuable information regarding variations of the human genome. Single nucleotide polymorphism (SNP), copy number variation, and complex structural variations can be typed with the help of sequencing data. An ambitious personal genome project (PGP) has been initiated to truly understand the genesis of most complex human traits—from deadly diseases to the talents and other features that makes every individual unique. PGP is widely supported by the nonprofit *PersonalGenomes.org*, which works to publicize genomic technology and knowledge at a global level. This might be useful for disease management and understanding of human health. It also deals with the ethical, legal, and social issues (ELSI) related to personal genomics.

Cognitive genomics

The brain is an important organ of an organism that helps to deal with the complex, information-rich environment. The blueprint for the brain is contained in the genetic material, that is, DNA of an organism. Brain development and cognitive or behavioral variability among individuals is a complicated process that results from genetic attribute of the person as well as their interactions with the environment. In cognitive genomics, cognitive function of the genes and also the noncoding sequences of an organism's genome related to health and activity of the brain are being studied. Genomic locations, allele frequencies, and precise DNA variations are analyzed in cognitive genomics. Cognitive genomics have immense potential for investigating the genetic reasons for neurodegenerative and mental disorders such as Down syndrome, Autism, and Alzheimer's disease.

Transcriptomics

Till date several genome sequencing projects have been completed and efforts are now being made to decipher the functional roles of different identified genes, their role in different cellular processes, genes regulation, genes and gene product interaction, and expression level of genes in various cell types. Transcription being the primary step in gene regulation processes, the information about the transcript levels is a prerequisite for understanding gene regulatory networks. The functional elucidation of the identified genes in totality is a subject matter of transcriptomics. It deals with the study of the complete set of RNAs/transcriptomes encoded by the genome of a cell or organism at a specific time and under a specific set of conditions. The techniques that are frequently used for genome-wide analysis for gene expression are complementary DNA (cDNA) microarrays and protein microarrays, cDNA–amplified fragment length polymorphism (AFLP), and serial analysis of gene expression (SAGE).

Proteomics

Proteomics is a comprehensive study of proteins in totality identified in a cell, organ, or organism at a particular time. The complexity of diverse physiological processes and biological structures hinders the applicability of proteomics, though the advent of recent proteomic techniques enables large-scale, high-throughput analyses, identification, and functional study of the proteome. For convenience, the proteomics can further be studied in different subbranches such as structural genomics, immunoproteomics, proteogenomics, nutriproteomics, etc.

Structural genomics

It aims to decipher the 3D structure of all proteins encoded by a particular genome using either experimental tools or *in silico* tools or sometimes both. In structural genomics, the structure of the total number of identified protein of particular genome is determined while in traditional structural prediction, structure of only one particular protein is determined. Through the availability of full genome sequences of number of organisms, structural prediction can be done by using both the experimental and modeling approaches, as well as previously known protein structures. The sequence-structure–function relationship provides an opportunity to analyze the putative functions of the identified proteins of an organism under purview of structural genomics.

Immunoproteomics

Immunoproteomics is the study of proteins solely associated with immune response with the aid of diverse techniques and approaches. Immunoproteomics encompasses a rapidly growing collection of techniques for identifying and measuring antigenic peptides or proteins. The approaches include gel- and array-based, mass spectrometry, DNA-, and bioinformatics-based techniques. Immunoproteomics is purposely used for understanding of disease, its progression, vaccine preparation, and biomarkers.

Proteogenomics

Proteogenomics is the study that uses proteomic information, mainly derived from mass spectrometry, to improve gene annotations. It is a field of junction of the genomics and proteomics. Previously, the genomics and proteomics studies were done independently. In genomics studies, large-scale annotation was done for identification of genes and its corresponding protein sequences, after sequencing of the genome. The proteomics aims to elucidate the protein expression observed in different tissues under specific conditions along with an insight into various posttranslational modifications. In proteogenomics there is amalgamation of both genomics as well as proteomics for elucidating the gene structures.

Nutriproteomics

The study of proteins of nutritional values of an organism can be referred as nutriproteomics. It can be defined as the interaction of nutrients with the proteins by studying the effect of nutrients on protein synthesis, interaction of nutrients with proteins, and modulation of protein–protein interaction through nutrients.

Metabolism

The diverse chemical reactions leading to growth and development of an organism is often referred to as metabolism and includes both anabolic and catabolic reactions.

Metabolomics

It is a branch of omics associated with study of metabolites of an organism in totality. This advancing field of science has much importance in pharmacological studies, functional genomics, toxicology, drug discovery, nutrition, cancer, and diabetes. As we know metabolites are the end result of all regulatory complex processes present in the cells hence metabolic changes represent reporters of alterations in the body in response to a drug or a disease.

Metabonomics

Metabonomics reflects the quantitative estimation of the metabolite of a particular organism and also includes the study of factors both exogenous and endogenous influencing the change in metabolite concentration. The exogenous factors such as environmental factors, xenobiotics, and endogenous factors such as physiology and development are predominantly considered in metabonomics. Like genomics, transcriptomics, and proteomics, metabonomics has immense potential in the discovery and development of new medicines.

Lipidomics

Lipidomics is the study of network of lipids and their interacting protein partners in organs, cells, and organelles. The lipidomic analysis is mainly done by mass spectrometry, commonly preceded by separation by liquid chromatography or gas chromatography.

Cytomics

Cytomics is the study of cytomes or the molecular single-cell phenotypes study resulting from genotype.

Pharmacogenomics

Pharmaconomics is the study of a complex genetic basis of interpatient variability in response to drug therapy. This subbranch of omics is an interesting field for pharmaceutical industries, clinicians, academicians, and patients as well. Using pharmacogenomics, the biopharmaceutical industries can improve the drug developmental process more rapidly and safely.

Miscellaneous

This includes

1. Nutritional genomics aims to study the relationship between human genome, nutrition, and health.
2. Toxicogenomics aims to study gene and protein activity in response to toxic substances.
3. Psychogenomics aims to understand the biological substrate of normal behavior and also understand the diseased conditions to unravel the behavioral abnormalities at genomic scale.
4. Stem cell genomics aims to understand human biology and disease states, which further progress toward clinical translation.

Basic tools and techniques used in science of omics

Electrophoretic techniques

Electrophoresis is the movement of molecules under the influence of uniform electric field. The separation of DNA, RNA, and proteins can be easily analyzed by gel electrophoresis on the basis of their size and charge. The gel which is a polymer is chosen on the basis

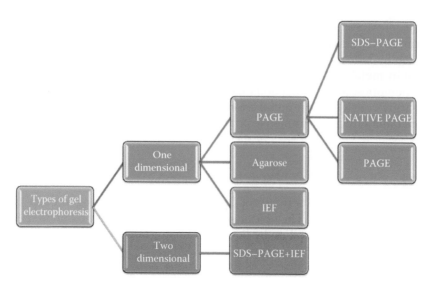

Figure 1.2 Types of gel electrophoresis.

of the specific weight of the target to be analyzed. When the target molecule is protein or small nucleic acids, polyacrylamide gel is used. If the target molecule is larger nucleic acids, then agarose gel is preferred. Broadly, there are two types of gel electrophoresis designated as one dimensional and two dimensional (2D) (Figure 1.2).

One-dimensional gel electrophoresis

Polyacrylamide gel electrophoresis
The uniform pore size provided by the polyacrylamide gel is utilized for separating proteins/DNA. The concentrations of acrylamide and bisacrylamide control the pore size of the gel.

Sodium dodecyl sulfate–polyacrylamide gel electrophoresis   It is the most widely used electrophoretic technique, which separates proteins primarily by mass.

NATIVE PAGE   It separates protein according to their charge/mass ratio.

Pulse field gel electrophoresis   DNA greater than ~40 kb length cannot be easily separated by applying constant electrical field. This problem is solved by pulsed field gel electrophoresis (PFGE) in which electric field is switched periodically between two different directions.

Agarose gel electrophoresis
It is frequently used for qualitative and quantitative estimation of nucleic acid, that is, DNA and RNA.

Isoelectric focusing
It separates the proteins in a pH gradient based on their isoelectric point (pI).

2D gel electrophoresis

It is the combination of sodium dodecyl sulfate–polyacrylamide gel electrophoresis (SDS–PAGE)/isoelectric focusing (IEF): first dimension is generally IEF. Second dimension is generally SDS–PAGE. Proteins having same molecular weight or same pI are also resolved.

Polymerase chain reaction

It can be simply defined as *in vitro* amplification of DNA in a sequential manner resulting in thousands to millions of its copies. This technique was developed by Kary Mullis in 1983. Polymerase chain reaction (PCR) consists of repetitive cycles of heating and cooling associated with events of DNA denaturation, annealing, and extension carried out at particular temperature for certain duration, which need extensive optimization depending on the template DNA and primers synthesized. The PCR utilizes the ability of DNA polymerase to synthesize new strand of DNA complementary to the target region. In general, DNA polymerase adds a nucleotide only when 3′-OH group preexists and hence it needs a primer to which it can add the first nucleotide. During the process, the DNA amplified serves as a template for subsequent amplification resulting in multiple copies at end. In this way, the DNA template is exponentially amplified. At last, after several cycles of PCR reaction, billions of copies of specific sequences (amplicons) are accumulated.

The following components are required for PCR reaction:

1. DNA template: It is the DNA having target sequence. Application of high temperature is required for denaturation. The quantity and quality of the template DNA is an important consideration for PCR amplification.
2. Thermostable DNA polymerase: *Taq* DNA polymerase is the most commonly used enzyme though *Pfu* DNA polymerase is often used because of its higher fidelity when copying DNA. The processivity and fidelity of the enzyme is an important consideration, which ultimately determines the strategy for subsequent cloning by specific cloning vectors.
3. Primers: A pair of synthetic oligonucleotides is a prerequisite to prime DNA synthesis and is an important component of the PCR reaction. The efficiency and specificity of the amplification is greatly influenced by the primers designed. The availability of softwares such as PRIMER-3, DNAStar, etc., in recent years has significantly contributed for proper designing of the primers. Some of the important considerations for primer designing are setting of GC content in the range of 40%–60%, provision for equal distribution of four bases, avoiding polypurine or polypyrimidine tracts or dinucleotide repeats, maintaining the length of primers in the range of 18–25 mer, avoiding complementarity between the forward and reverse primers, etc. In general, higher concentration of primers favors mispriming leading to nonspecific amplification.
4. Deoxynucleoside triphosphates (dNTPs): These are building blocks of new DNA strand. The standard PCR contains equimolar concentration of dATP, dGTP, dCTP, and dTTP and the concentration is in the range of 200–250 μM for each dNTP.
5. Buffer solution: It is generally provided with 10× concentration and comprises Tris-based buffer and salt like KCl and provides a suitable environment for optimum activity of DNA polymerase.
6. Divalent cations: Mg^{2+} is used commonly, but sometimes Mn^{2+} is used in the PCR buffer. It is an important cofactor for thermostable DNA polymerases. The concentration

of the Mg^{2+} is standardized for PCR amplification though 1.5 mM is the optimum concentration in most of the PCR reaction set up. The excess of Mg^{2+} reduces the enzyme fidelity and leads to nonspecific amplification.

Besides these ingredients, in many cases specific chemicals are also added, which are referred to as PCR enhancer and additives such as betaine, DMSO, formamide, BSA, etc., which often increases the specificity of PCR amplification resulting in enhanced yield and also minimizes the undesired products.

The basic steps involved in PCR amplification are (i) Denaturation: DNA melting by disrupting the hydrogen bonds between complementary bases often carried out at 94–98°C for 30 s. (ii) Annealing: Assists binding of primers to the template DNA and the annealing temperature needs optimization based on the Tm value of the primers synthesized. Generally, during standardization of annealing temperature for a particular set of reaction, the annealing temperature is kept at about 3–5°C below the Tm of the primers. (iii) Extension: Binding of primers to the template DNA during annealing step is followed by synthesis of new cDNA strand with the aid of dNTPs and DNA polymerase in 5' to 3' direction. Since *Taq* DNA polymerase has optimum activity at 72°C, the extension is carried out at this temperature, though variation can be done based on the complexity of the template DNA and primers used. Time of extension is variable and is optimized based on the expected size of DNA fragment to be amplified and type of DNA polymerase used. To ensure that any remaining single-stranded DNA is fully extended, the final extension step is carried out at 70–74°C for 5–15 min.

PCR-based DNA fingerprinting technique is quite popular in forensic sciences owing to the fact that even extremely small amounts of sample (DNA) can be amplified. PCR is often used in molecular diagnosis of many diseases. PCR-based molecular markers are routinely used for diversity analysis, it is often used in sequencing techniques, for mutagenesis studies, for expression studies, etc.

In recent years several alterations in the basic PCR has been attempted for diverse applications. The variants of PCR and its diverse application are shown in Table 1.1 and Figure 1.3.

Sequencing techniques

It is the process of deciphering the arrangement of A, C, G, and T nucleotide for a particular DNA molecule and came into existence in 1970 popularly known as Sanger (or dideoxy) method and the Maxam–Gilbert (chemical cleavage) method. The technological innovations in sequencing methods was attempted for enhancing the speed, accuracy, automation, read length, cost-effectiveness, etc., over the years and these developments have been marked as first-, second-, third-, and fourth-generation sequencing technologies as shown in Table 1.2. The second-, third-, and fourth-generation sequencing are also referred as next generation sequencing (NGS) methods.

Chemical degradation sequencing (Maxam and Gilbert method)
It uses specific chemicals for the cleavage of nucleotides and requires double-stranded template DNA radioactively labeled at one 5' end of the DNA. The method involves modification of a base chemically and then the DNA strand is cleaved by reaction that specially cleaved the DNA at the point where base is modified. Four reactions designated as G, A+G, C, and C+T is set up by degrading with specific chemicals followed by

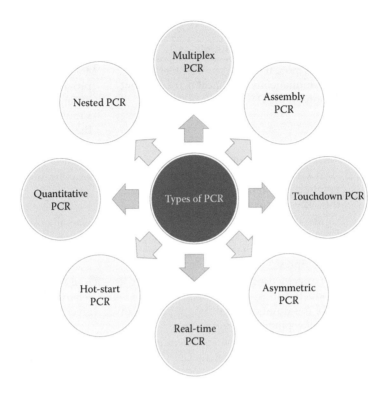

Figure 1.3 Some variants of PCR.

its analysis on polyacrylamide gel. The read length of approximately 400 bp is possible with this method. The use of toxic chemicals and difficulty in automation limits its use.

Enzymatic or chain termination method (Sanger's method)

This method involves enzymatic polymerization of DNA fragments complementary to the single-stranded template DNA in a reaction setup comprising ingredients such as dNTP, specific primer, and a modified nucleoside called terminator of dideoxynucleoside triphosphate (ddNTP) involved with chain termination. A set of four different tubes containing the appropriate amount of one of the four terminators (ddNTPs) is needed for the sequencing. Here, ^{32}P-labeled primer is preferably used for generating different fragments having the same 5′ end and these fragments are finally resolved by denaturing polyacrylamide gel electrophoresis. It requires single-stranded DNA as a template and has several advantages such as possibility for automation, increased sequence read length, no toxic chemicals required, etc. The automated DNA sequencing is an automation of Sanger's method of sequencing which has greatly enhanced the accuracy, speed, and read length. This was followed by further advancement in sequencing technologies as described in Table 1.2.

DNA and protein microarray

Microarray is a 2D array on a solid substrate that detects large amounts of biological material using multiplexed, parallel processing, and detection methods. Most commonly used is DNA microarray though protein, peptide, and carbohydrate microarrays are also gaining importance.

Table 1.1 Variants of PCR and its applications

Type of PCR	Description	Application
Multiplex PCR	Utilizes more than one set of primers meant for simultaneous amplification of different regions of the DNA.	Molecular diagnosis with provision for analyzing different markers simultaneously confirming the existence of targeted diseases.
Nested PCR	It involves two sets of primers designed in such a manner that one of the primers amplifies the shorter internal fragment of the larger fragment.	Enhances specificity of PCR amplification.
Reverse transcriptase PCR (RT-PCR)	It is associated with the amplification of RNA sequences by first converting into double-stranded cDNA using reverse transcriptase enzyme.	Generally used for expression studies as an alternative of northern hybridization.
Semiquantitative PCR	It is basically used for quantification of PCR products during the exponential phase prior to saturation stage and involves both housekeeping and target genes.	Used for determining the relative amount of cDNA in a given sample.
Real-time PCR	The simultaneous amplification and quantification of a targeted cDNA molecules.	For assessing the copy number of genes.
Asymmetric PCR	It is associated with the preferential amplification of one strand owing to the unequal concentration of primers in the reaction setup.	Helpful to generate single-stranded DNA, which can be used for sequencing and hybridization probing.
Hot-start PCR	It is basically a condition to avoid adding all the components of PCR reaction simultaneously resulting in accumulation of nonspecific PCR amplicons. The best way is to add the critical component of PCR reactions, that is, DNA polymerases only when the temperature reaches 90–95°C.	This enhances the possibility of getting the expected DNA fragment in PCR. The accumulation of nonspecific products is also minimized.
Touchdown PCR	This is basically a method of standardization of annealing temperature with primers designed from protein sequences, that is, degenerate primers. The optimization for perfect annealing temperature can be achieved by first increasing the temperature from 3°C to 5°C for few cycles above the theoretically calculated Tm of the primers, followed by reducing the temperature unless the exact annealing temperature as witnessed by the yield of the reaction.	The sole purpose is to optimize the annealing temperature for better yield.
Inverse PCR	It is a method used to allow amplification of unknown region.	It is used in chromosome walking, for cloning unknown regions of genomic sequences.

(Continued)

Table 1.1 (Continued) Variants of PCR and its applications

Type of PCR	Description	Application
Long PCR	It relies on the use of a mixture of two thermostable polymerases, one having 3′–5′ exonucleases (proofreading activity) while other lacking like use of *Taq* and *Pfu* DNA polymerases. Amplification of fragments in the range of 10–25 kb is possible by long PCR. Comparatively longer extension time, additives such as glycerol and DMSO are preferred for long PCR.	Used in physical mapping and direct cloning from genome.
In situ PCR	The PCR amplification is merged with histological localization with the aid of *in situ* hybridization technique. Several alterations like increased concentration of Mg^{2+}, increased amount of DNA polymerases is done in *in situ* PCR. Tissue preparation is an important consideration in this PCR.	It can be used for detecting cellular DNA, cDNA associated with diseased conditions.
High-fidelity PCR	Use of DNA polymerase with proofreading activity such as *Pfu, Tli,* etc.	The sole purpose is to minimize the chances of mutations resulting from the use of DNA polymerases lacking proofreading activity. It is preferred for PCR-based cloning and expression studies.
Differential display PCR	Amplify and display many cDNAs derived from mRNA of a given cell or tissue type. Uses two different types of oligonucleotides, (i) anchored antisense primers which are 10–14 mer designed complementary to poly(A) tail of mRNA and last two nucleotides of transcribed sequences, and (ii) arbitrary primer of 10 mer and often long arbitrary sense primers of 25–28 mers are also used.	The purpose is to display all mRNA of a cell. It can be used to detect differential expression of mRNAs that are expressed in low abundance.

DNA microarray

DNA microarray or DNA chip or biochip comprises millions of DNA spots onto a solid support for comprehensive genome analysis. The parallelism, miniaturization, speed, multiplexing, automation, and combinatorial synthesis are typical features of DNA microarray. Each DNA spot contains 10^{-12} moles of a specific DNA sequence, called as probe needed to hybridize cDNA. This hybridization is sensed and quantified by chemically labeled or fluorescent targets to determine relative abundance of nucleic acid sequences in the target. The following steps are associated with DNA microarray:

- Printing or deposition of high-density nucleic acid samples (cDNA or oligonucleotides) onto very precise area of the support system referred to as fabrication of chips followed by immobilization to the substrate.

Table 1.2 Overview of sequencing techniques

Method	Detail	Advantage	Disadvantage
First-generation sequencing			
Sanger and Maxim Gilbert methods	• Used either in chemical or enzymatic methods to generate a nested set of DNA fragments • Used in electrophoretic methods to separate the fragments	Highly accurate	• Requires lots of DNA (100s of ng to 1 µg) so it typically involves cloning and/or PCR • Limited throughput • High cost
Second-generation sequencing			
Roche 454 Pyrosequencing	Uses emulsion PCR to achieve clonal amplification of target sequence. The incorporation of nucleotide into the nascent DNA is manifested by luciferase by generating light.	1–5 µg DNA needed Clinical applications	400 bp length reads
Reversible terminator sequencing (Illumina)	In this technique cluster target sequence is amplified on solid surface (bridge amplification). In the sequencing reaction four types of nucleotides, each labeled by one of four fluorophores and containing a 3′ reversible terminator is utilized.	Less (<1 µg) DNA needed	75 bp length reads More false positives
ABI/SOLiD sequencing technology	Sequencing of oligonucleotides by ligation and detection (SOLiD) involves ligation through cleavable probes. Here also emulsion PCR is being used. The short read length and long run time are major limitations of this technique.	2–20 µg DNA needed	35–50 bp length reads
HeliScope sequencer (Helicos)	It is a sequencing method involving reverse terminator chemistry avoids PCR amplification and is commonly referred to as single molecule sequencing.	Enhanced accuracy of reading through homopolymers (stretches of one type of nucleotides) and also allows for RNA sequencing	The error rate is high due to noise, time consuming, 32 bp length reads
Ion torrent PGM (Life technologies)	Nanooptical DNA sequencing, during incorporation of nucleotides by DNA polymerase-based synthesis, ions are produced, which are detected by semiconductor.	Accuracy rate is high, faster, cheaper, and less than 200 bases are needed	Cannot be used for whole exome-sequencing, whole-genome sequencing, ChIP-Seq, and RNA-seq

(Continued)

Table 1.2 (Continued) Overview of sequencing techniques

Method	Detail	Advantage	Disadvantage
Third-generation sequencing			
Pacific Biosciences' real-time single molecule sequencing (PacBioRS)	Based on the properties of zero-mode waveguides. It avoids amplification of template DNA and there is a provision for real-time monitoring of the nucleotide added by an emission of light pulse. It generally results in lower read length and error rate is high	1000 bp read	Error rate of 10%–18%, high raw read error rate compared to other sequencing technologies
Combined Probe Anchor hybridization and ligation (cPAL) (Complete Genomics)	The method uses a rolling circle amplification of small DNA sequences into the so-called nanoballs. Ligation is used to determine the nucleotide sequence.	Comparatively many DNA nanoballs can be sequenced per run in cost-effective manner	Short sequencing read
Fourth-generation sequencing			
Oxford Nanopore	It combines the single molecule sequencing of third-generation sequencing technologies with nanopore technology.	High speed of sequencing, provision for whole genome scan within 15 min	Comparatively higher cost than other methods

- Hybridization of resulting microarray with a fluorescently labeled probe.
- Detection of fluorescent markers using high-resolution laser scanner.
- Gene expression pattern is analyzed based on signal emitted from each spot using digital imaging software.
- The expression pattern of two different samples can be compared.

The microarray fabrication can be performed either by *in situ* synthesis of nucleic acids or by exogenous deposition of prepared materials on solid substrates. Prefabricated microarrays can also be purchased from companies such as Affymetrix, Research Genetics, CLONTECH, Incyte Genomics, Operon Technologies, Genometrix Inc., etc.

Two commonly observed variants of DNA microarray is oligonucleotide and cDNA microarray. The commonly used methods of chip fabrication include contact printing, photolithography, pin and ring, piezoelectric printing, and bubble jet technology.

Protein microarray

Small quantities of diverse purified proteins are assorted on a solid support. The purity of proteins, possessing its native conformation and its concentration are important considerations for protein microarray. Typically, fluorescently labeled probes are used for signal generation and identification.

Other than protein and DNA microarrays several different versions of microarrays are emerging as shown in Figure 1.4.

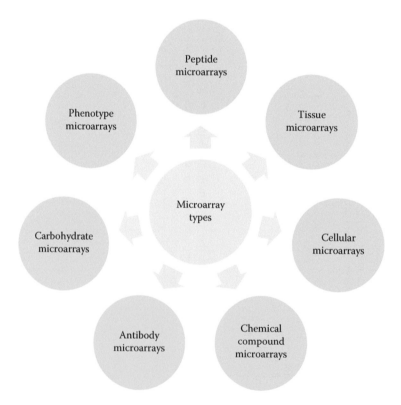

Figure 1.4 Types of microarrays.

1. *Peptide microarray*: It is a collection of small peptides attached onto solid support either glass or plastic. The peptide microarray is generally used for assessing the protein interactions, elucidating the binding and functionality with the target proteins.
2. *Tissue microarray*: In this a number of tissues, typically from different organs, are thrown together in the same block and tissue distributions of a particular antigen/protein are assessed.
3. *Cellular microarray*: Living-cell microarray allows the assessment of cellular response to different external stimuli.
4. *Chemical compound microarray*: Used for the study of the interaction between chemical compounds and biological targets.
5. *Antibody microarray*: Used for protein expression profiling and comparative analysis between different samples.
6. *Carbohydrate microarray*: Used for measurements of glycan–protein interactions and disease diagnosis.
7. *Phenotype microarray*: Used for genotype–phenotype characterization and determining most favorable conditions for different cellular activities.

Blotting techniques for the study of DNA, RNA, and proteins

Blotting techniques are techniques for identification of targeted nucleic acid or protein by immobilization onto specific support either nylon or nitrocellulose followed by detection with probes. Nucleic acids blotting techniques include blotting of nucleic acids from gels (southern hybridization, northern hybridization), dot/slot blotting, and colony/plaque blotting. This technique consists of four major steps namely, (i) resolution of protein and nucleic acid samples by electrophoretic means, (ii) transfer and immobilization on solid support (by capillary blotting, vacuum blotting, electrophoretic transfer), (iii) binding of analytical probe, and (iv) visualization of bound probe to target molecule usually by autoradiography. The fixation of the nucleic acid sample to the membrane can be achieved by several methods such as ultraviolet cross-linking, oven baking, alkali fixation, and microwave fixation. The blotting techniques designated as southern, northern, and western blotting are routinely used molecular biology technique. Southern blotting identifies DNA fragments that bind with the probes, through hybridization complementary fragments of target DNA. Northern blotting identifies messenger RNA (mRNA) after hybridization to their corresponding DNA sequences, whereas, western blotting identifies particular proteins using specific antibodies as probes.

Spectroscopic techniques

The main principle underlying spectroscopic techniques is that each and every atom and molecule absorbs and emits light at certain wavelengths. It involves the interaction between electromagnetic radiation and matter, which can be an atom, molecule, ion, or solids. This may result in absorption, emission, or scattering. Since each chemical element has its own characteristic spectrum, this nature of interaction is used to analyze matter and to interpret its physical properties. The spectroscopic technique is meant for determination of concentration or amount of a given species. The different forms of spectroscopy are discussed in the following sections.

Vibrational spectroscopy (IR and Raman)

Atoms are connected together by bonds about which they vibrate and bend. Every bond of the molecule possesses a fundamental frequency about which they vibrate. At the integral multiple of these frequencies the bonded atoms vibrate and bend.

If the frequency of the light wave matches with the bond frequency, the bond absorbs light which generally lies in the infrared (IR) region of the EM spectrum. In case of IR spectroscopy, a chemical sample is irradiated with different frequencies of IR light and corresponding absorbance at each frequency is recorded. In contrast to IR, the Raman spectroscopy uses ultraviolet (UV), VIS, or NIR as radiation source. The incident light excites the system to a higher energy level and immediately after recovering from this state, scattering reactions occur. In Rayleigh scattering, the elastically scattered light possesses energy similar to the incident light.

UV–VIS spectroscopy

The incident light in this type of spectroscopy lies in the visible and adjacent (near-UV and near-infrared [NIR]) ranges. In UV–vis spectroscopy, ultraviolet or visible light is absorbed by the molecules containing π-electrons or nonbonding electrons whereby these electrons are excited and undergo electronic transitions to higher antibonding molecular orbitals whereby the molecules undergo electronic transitions.

Nuclear magnetic resonance

It is an important technique for elucidating the structural and functional aspects of DNA and protein. The concept underlying nuclear magnetic resonance (NMR) spectroscopy is the absorption of electromagnetic radiation by atomic nuclei at radio frequencies. All electrons and some of the nuclei possess property of spin. Due to this spin, nuclei of some atoms behave as tiny magnets. Under exposure of magnetic field these nuclei occupies position at different energy levels. When the nuclei align themselves with the magnetic field which is the lowest energy state or align against the field at higher energy state. The energy needed to shift the electron from lower to higher energy state lies in the radio wave frequency. The energy released when the nucleus returns back to its lower energy state is detected. There exist different types of NMR as shown in Figure 1.5.

Structural determination of biomolecules through NMR spectroscopy Structure determination of biomolecules depends on the changes in distribution of electrons around the nucleus, which is governed by the following factors namely, (i) magnetic field around the nucleus, (ii) resonation frequency of the nucleus, and (iii) chemistry of the molecule at that atom.

Electron Spin Resonance

Electrons possess magnetic moment due to the presence of charge and spin. Bohr magnetron is used to define the magnetic moment of a free electron. When free electrons are placed in an electromagnetic field, on fulfillment of resonance conditions, energy is absorbed. This absorbance is recorded as electron spin resonance (ESR) or electron paramagnetic resonance. In case of bound electrons, ESR can be recorded for only unpaired electrons. The conformation and dynamics of the biological molecules can be easily assessed by ESR spectroscopy.

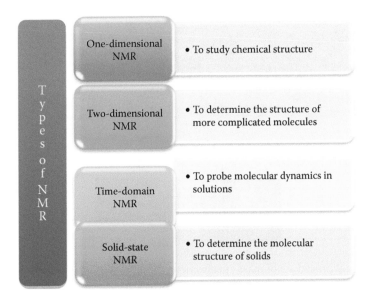

Figure 1.5 Types of NMR with a specific application.

X-ray spectroscopy

On removal of the electron from the inner shell of the electron due to excitation, there is a tendency to be replaced by another electron from an outer shell resulting in the difference in energy of the electron in the two shells emitted as an x-ray photon. X-ray released with respect to each element has its own characteristic energy corresponding to its atomic number. Hence, each element has a characteristic x-ray spectrum. This property of x-ray finds its application in the use of x-rays as a diagnostic tool for sample identification.

Electron spectroscopy

After excitation by an energy source, electrons are ejected from the inner-shell orbital of an atom. These ejected electrons are detected in electron microscopy. If electrons are ejected by x-ray as an excitation source, it is termed as electron spectroscopy for chemical analysis (ESCA). In Auger electron spectroscopy (AES), the electrons are detected, which undergo electronic transition for energy conservation.

Mass spectral techniques

Mass spectrometry is an important analytical tool for the identification and quantification of molecules in mixtures as well as to derive structural information by measuring mass-to-charge ions in gas phase. In a typical mass spectrometry procedure, the sample is first ionized leading to the rupture of sample molecules into charged entities. The generated ions are separated in relation to their mass-to-charge ratio, by putting them in electric or magnetic field. Different degrees of deflection are observed in the ions possessing different mass-to-charge ratio. The ions are detected and spectra are generated for the relative ratio of detected ions in terms of the mass-to-charge ratio. The atoms or molecules in the sample are identified by comparing the identified masses to the reference database or by characterizing the fragmentation pattern.

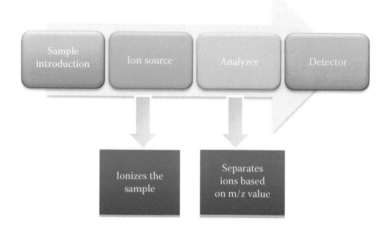

Figure 1.6 A typical procedure of mass spectrometry.

Mass spectrometers comprise three fundamental parts, namely the ionization source, the analyzer, and the detector. The nature of these components might vary based on the type of mass spectrometer. After loading the sample in mass spectrophotometer, it is vaporized and ionized by the ion source. Next, in the analyzer individual ions are deflected based on their mass and charge (m/z) under the influence of electrical and/or magnetic field. Ions deflected by analyzer hit the detectors, which commonly are electron multipliers or microchannel plates, which emit a cascade of electrons when each ion hits the detector plate. Mass spectrometers are coupled with computers with preloaded softwares that analyze the ion detector data and produces graphs that organize the detected ions by their individual m/z and relative abundance. These ions can then be processed through databases to predict the identity of the molecule based on the m/z. A typical procedure of mass spectrophotometry, types of ionization techniques, and different mass analyzers are shown in Figures 1.6 through 1.8, respectively.

Matrix-assisted laser desorption/ionization (MALDI) is a widely used technique that involves rapid photovolatilization of sample molecules, which may either be volatile or nonvolatile, embedded in a UV-absorbing matrix into gas phase as intact ions for mass spectroscopy (MS) analysis. MALDI is most commonly coupled with mass analyzer as time-of-flight (TOF) mass spectrometer. MALDI has emerged as an efficient and versatile technique for the analysis of biomolecules and large organic molecules.

Chromatographic techniques

Chromatography is an important method for separating mixture of compounds and isolate different components. It follows the principle of different rate of mobility and resistance faced by the migration and selective retardation of the solute during passage through resin particles. It comprises two phases, that is, mobile and stationary phase. Diverse chromatographic methods are present, which allow purification of different biomolecules. There are several types of chromatography such as adsorption chromatography, partition chromatography, ion-exchange chromatography, gel chromatography, and affinity chromatography.

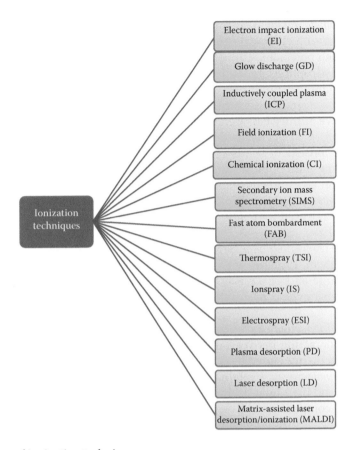

Figure 1.7 Types of ionization techniques.

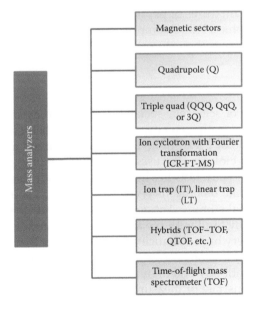

Figure 1.8 Different mass analyzers.

1. *Column chromatography*: The stationary phase is a column containing specialized beads coated with chemical functional groups. The liquid mixture is applied to the column which percolated through the matrix with separation occurring on the basis of charge, hydrophobicity, and ligand-binding properties. The eluted fractions are collected from the column.
2. *Partition chromatography*: Separation of the components of the applied sample depends on the relative affinity between the stationary and mobile phase.
3. *Size exclusion chromatography*: Separates proteins according to size. The stationary solid phase consists of beads containing pores of different dimensions. Small proteins enter the pore of beads and migrate through the column slower than the particles of larger size and consequently emerge in the later fractions.
4. *Absorption chromatography*: Proteins to be separated are loaded on the column and sequentially released and eluted by disrupting the interaction between protein and stationary phase.
5. *Ion-exchange chromatography*: Separation on the stationary phase based on the charge present on them. Proteins with a net negative charge stick on to beads possessing positive charge and *vice versa*. Proteins are released by counterions and eluted by gradually raising the concentration of monovalent ions or changing the pH of mobile phase.
6. *Hydrophobic interaction chromatography*: It separates proteins based on the hydrophobic interaction acting between the protein and the matrix coated with hydrophobic groups. Proteins are eluted by gradually decreasing the polarity of the mobile phase.
7. *Affinity chromatography*: Proteins are purified on the basis of high affinity between the protein and their ligands. Proteins are eluted by disrupting protein–ligand interaction or by competition with soluble ligand.
8. *Reversed-phase high-pressure chromatography*: Contrary to conventional chromatography, high-pressure liquid chromatography (HPLC) basically uses incompressible silica or alumina microbeads as stationary phase and pressures of few thousand psi are applied. Such type of matrix facilitates high flow rates and enhanced resolution, hence complex mixtures of peptides and lipids can be resolved through HPLC. However, reversed-phase HPLC comprises hydrophobic stationary phase of aliphatic polymers with variable carbon atoms.

Immunological techniques

Immunology is a branch of science associated with the study of the immune system of a particular organism and pathogenicity, epidemiology, diseases diagnosis, etc., are an integrated part of immunology leading to the development of immunological techniques (Figure 1.9). The forensic biology also uses several immunological techniques routinely and the basis for immunological tests lies in antigen–antibody interactions. Antibodies are basically proteins and white blood cells (WBC) are the main source of antibody production in response to external stimuli. Antigens are compounds that trigger the production of antibodies or specifically bind with the antibody in an immune system. Binding of the antibody–antigen has a very high specificity. The strength of the binding between a single binding site of an antibody and an antigen is known as affinity and the overall strength of interaction between an antibody and an antigen is termed as avidity. There are different techniques used to demonstrate or measure an immune response and also to identify or measure antigen using antibodies.

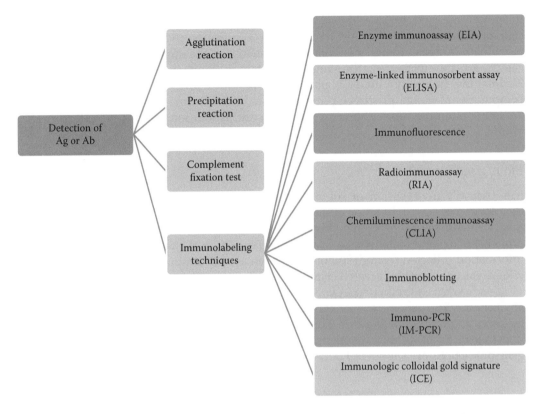

Figure 1.9 Different immunological techniques for detection of the antigen or antibody.

Lipid profiling for comprehensive analysis of lipid species within a cell or tissue

Lipids are small molecules having common physical and chemical properties. It plays a key to metabolic regulation and signaling process. Lipid profiling is a comprehensive analysis of lipid species within a cell or tissue. The two major lipid profiling methods are liquid chromatography–mass spectrometry (LC–MS) and direct infusion mass spectrometry (shotgun analysis). LC–MS-based methods typically utilize reversed-phase columns including octadecylsilane stationary phases. In shotgun analysis crude extracts are infused directly to mass spectrometer without separating previously. Lipid proofing has been applied to several studies of different diseases such as diabetes, schizophrenia, and cancer. Lipid profiling is also applied to plants and microorganisms such as yeast. Lipidomic data, with their corresponding transcriptional data and proteomic data can be used in systems biology approach for understanding the metabolic or signaling pathways of interest.

Technology integration for analysis and imaging of cellular data

Advanced microscopic imaging and analysis techniques assist the researchers to obtain insight into the individual cells, cell populations, and complex tissues and are being extensively explored for cancer to various infectious disease studies. The cytometry is frequently used for analysis of cells and cell systems by means of flow cytometry and image cytometry techniques. Flow cytometry assists the complete analysis of individual cells present in suspension and finds diverse application in immunology and hemato-pathology. Image cytometry is commonly meant for the study of individual adherent cell. Apart from measuring the same parameters as the flow cytometry it also performs 3D imaging. Image cytometry is usually carried out using automated microscopy and computational image processing and analysis. Several innovations in flow cytometry are being attempted for enhancing the speed and resolution, capturing and processing of images, etc.

Bioinformatics tools

It is basically intervention of computers in life sciences especially molecular biology. Bioinformatics is a combination of statistics, computer science, and information technology to look into biological problems. It is mainly used to analyze the data related with genes, proteins, drug, and metabolic pathway. Biological databases are a collection of biological information collected from different experimental and computational analysis including wide areas of biological research (Figure 1.10).

Gene prediction: Predicting a gene is an important step in the process of understanding any genome. The genome sequences deciphered are first subjected to bioinformatics tools for assessing the putative genes along with information about other regulatory regions known to be assisting gene functions. Some of the commonly used bioinformatics tools for gene predictions are shown in Figure 1.11.

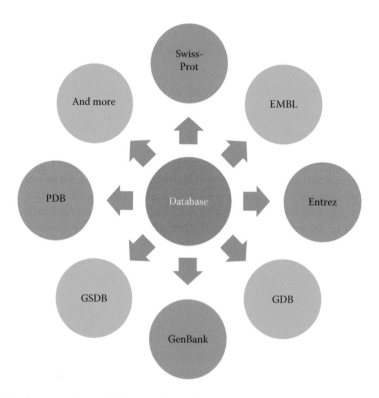

Figure 1.10 List of commonly used biological databases.

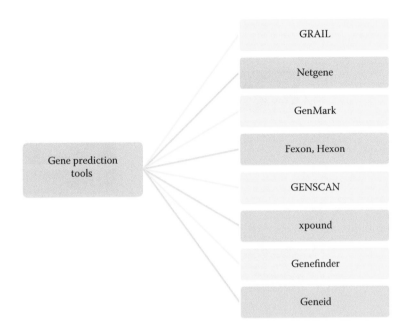

Figure 1.11 List of popularly used gene prediction tools.

Sequence alignment: Sequence alignment is a way of arranging nucleotide or amino acid sequences to identify similarity between them. Sequence similarity also defines functional, structural, or evolutionary relationships between the sequences. Commonly used bioinformatics tools for sequence alignment is shown in Figure 1.12.

Quantitative estimation through biostatistical analysis

Biostatistics is the part of statistics for designing biological experiments in the field of biology, medicines, pharmacy, agriculture, etc. Computational intervention in biostatistics has enhanced the analysis and interpretation of different types of experimental data. The exponential increase in the biological data owing to technological innovations in sequencing technologies demands development of efficient biostatistics tools. The importance of biostatistics is evident from its diverse applications in population studies, medical research, nutrition, environmental health, designing and analysis of clinical trials, analysis of genomic data, ecology, ecological forecasting, sequence analysis, system biology, and many more.

Application of science of omics in different fields

The science of omics has diverse application as discussed below.

1. *Assessment of genome structure and function*: A large volume of information can be generated through high-throughput screening, which in turn can be used for studying structural and functional aspects of genomes.

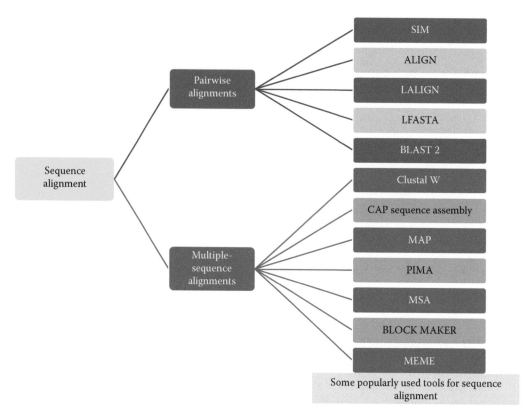

Figure 1.12 Tools for aligning sequences using pairwise and multiple-sequence alignment.

2. *Application in plant metabolomics*: Used to interpret plant growth and functions based on metabolite analysis by sophisticated tools influencing the quality of food or plant-derived medicines.
3. *Toxicogenomics for assessment of environmental pollution*: Allows large-scale screening of huge number of biological samples as potential toxicants.
4. *Therapeutic application of genomics*: Useful for gene-based diagnosis and treatment of individuals. Also helpful to manage epidemics and develop new therapies.
5. *Microbial omics and its approaches in biofuel production*: Complex microbial communities can be screened by metagenomics for targeted screening of enzymes with industrial applications in biofuel production.
6. *In crop improvement*: Genomics-based tools for crop improvements are being used with the availability of genome sequences of several crops. Several types of molecular markers have been developed associated with different agronomics traits.
7. *Evolution and diversity studies*: Useful in studying evolutionary relatedness.
8. *Analysis of RNA editing*: RNA editing is a post-transcriptional event that recodes hereditary information. RNA editing is an integral step in generating the diversity and plasticity of cellular RNA signatures.
9. *Biomarker for medical research*: Disease-related biomarkers are helpful in disease prediction, which may develop over time.

10. *Drug discovery*: Involved in the drug discovery process ranging from screening hits to its efficacy, stability, and bioavailability.
11. *Pathological studies*: Pathology is the precise study and diagnosis of disease.
12. *Clinical toxicology*: Involved in the improvement of the drug safety assessment process.

Conclusion and future perspectives   The "omics"-driven approach allows dissecting the entire biological network of genes and proteins present in a cell. The science of omics is being manifested by development of innovative techniques meant for assessing the organisms in terms of totality of biomolecules especially DNA and proteins. The comprehensive assessment of genes, proteins, lipids, metabolites, etc., are being analyzed by omics-based tools. The various types of interactions such as DNA–protein, protein–protein, cellular along with protein modifications, and metabolic pathways are also being studied in omics. The emergence of several braches of omics reveals the advancement in the technology for studying the genome, proteome, lipidome, and metabolome. Every technology has some limitations and omics too have some limitations namely reproducibility, noise, cost, etc. The integration of diverse omics-based sciences is a prerequisite for understanding the complexity of biological system and this led to emergence of system biology approach in recent years. In summary, the advent of new "omics" technologies could be an efficient tool for studying biological systems, provided it is made more accessible, high throughput, versatile, and cost-effective.

Summary

Omics deals with the study of biological molecules, their physiological function and structure in totality. The outcome of sequencing of genomes has led to the growth of genomics having diverse potential yet to be harnessed completely. The omics-based approach of science has witnessed several developments over the years with advancement in the technology such as sequencing technology. Several branches of omics have emerged, highlighting the comprehensive studies in respective branches such as proteomics, metabolomics, cytomics, metagenomics, lipidomics, transcriptomics, etc. The functional and structural genomics can be considered to be closely linked to transcriptomics and proteomics. The emerging new approach for real-time understanding of biology is "system biology," which combines the information of different field for analyzing networks and pathways that exist in biological systems. The deciphering of whole genome sequences of different organisms and technological innovations in science of omics have undoubtedly advanced our knowledge of biological system.

Take home

The omics studies are of immense use in understanding the current knowledge of genomics, proteomics, and metabolomics, which allows to access combination of genes with precision to manage various factors.

Bibliography

Barnes S, Kim H. Nutriproteomics: Identifying the molecular targets of nutritive and non-nutritive components of the diet. *Journal of Biochemistry and Molecular Biology* 2004, 37(1): 59–74.

Bickel PJ, Brown JB, Huang H, Li Q. An overview of recent developments in genomics and associated statistical methods. *Philosophical Transactions of the Royal Society A* 2009, 367: 4313–4337.

Chandramouli K, Qian PY. Proteomics: Challenges, techniques and possibilities to overcome biological sample complexity. *Human Genomics and Proteomics Volume* 2009, 1, Article ID 239204, doi:10.4061/2009/239204.

Dennis EA. Lipidomics joins the omics evolution. *Proceedings of the National Academy of Sciences* 2009, 106(7): 2089–2090.

Hardison RC. Comparative genomics. *PLoS Biology* 2003, 1(2): 156–160.

Karczewski KJ, Daneshjou R, Altman RB. Chapter 7: Pharmacogenomics. *PLoS Computational Biology* 2012, 8(12): e1002817.

Kiranmayi VS, Rao PVLNS, Bitla AR. Metabolomics—The new "omics" of health care. *The Journal of Clinical and Scientific Research* 2012, 3: 131–137.

Laakso TS, Oresic M. How to study lipidomes. *Journal of Molecular Endocrinology* 2009, 42: 185–190.

Lao YM, Jiang JG, Lu Yan. Application of metabonomic analytical techniques in the modernization and toxicology research of traditional Chinese medicine. *British Journal of Pharmacology* 2009, 157: 1128–1141.

Lay JO Jr., Borgmann S, Liyanage R, Wilkins, CL. Problems with the "omics." *Trends in Analytical Chemistry* 2006, 25(11): 1046–1056.

Lunshof JE, Bobe J, Aach J, Angrist M, Thakuria JV, Vorhaus DB, Hoehe MR, Church GM. Personal genomes in progress: From the Human Genome Project to the Personal Genome Project. *Dialogues in Clinical Neuroscience* 2010, 12(1):47–60.

Maher B. The case of the missing heritability. *Nature* 2008, 456(6): 18–21.

Mardis ER, Lunshof JE. A focus on personal genomics. *Personalized Medicine* 2009, 6(6): 603–606.

Ning MM, Lo EH. Opportunities and challenges in omics. *Translational Stroke Research* 2010, 1: 233–237.

Novik KL, Nimmrich I, Genc B, Maier S, Piepenbrock C, Olek A, Beck S. Epigenomics: Genome-wide study of methylation phenomena. *Current Issues in Molecular Biology* 2002, 4: 111–128.

Srinivasan BS, Chen J, Cheng C et al. Methods for analysis in pharmacogenomics: Lessons from the pharmacogenetics research network analysis group. *Pharmacogenomics* 2009, 10(2): 243–251.

Werner T. Next generation sequencing in functional genomics. *Briefings in Bioinformatics* 2010, 11(5): 499–511.

Wilson ID, Plumb R, Granger J, Major H, Williams R, Lenz EM. HPLC–MS-based methods for the study of metabonomics. *Journal of Chromatography B* 2005, 817: 67–76.

Web resources

http://www.nature.com/omics
http://www.niaid.nih.gov/LabsAndResources/resources/dmid/Pages/omics
http://omics.org/index.php

Further reading

Dennis EA. Lipidomics joins the omics evolution. *Proceedings of the National Academy of Sciences* 2009, 106(7): 2089–2090.

ENCODE Project Consortium. A user's guide to the encyclopedia of DNA elements (ENCODE). *PLoS Biology* 2011, 9: e1001046.

ENCODE Project Consortium et al. An integrated encyclopedia of DNA elements in the human genome. *Nature* 2012, 489: 57–74.

Hiller M. et al. Computational methods to detect conserved non-genic elements in phylogenetically isolated genomes: Application to zebrafish. *Nucleic Acids Research* 2013, 41: e151.

Hotz RL. Here's an omical tale: Scientists discover spreading suffix. *The Wall Street Journal* August 13, 2012.

Lederberg J, McCray A. "Ome Sweet" omics—A genealogical treasury of words. *The Scientist* 2001, 15(7): 8.

Lunshof JE et al. Personal genomes in progress: From the Human Genome Project to the Personal Genome Project. *Dialogues in Clinical Neuroscience* 2010, 12(1): 47–60.

Novik KL. et al. Epigenomics: Genome-wide study of methylation phenomena. *Current Issues in Molecular Biology* 2002, 4: 111–128.

Rubin GM. et al. Comparative genomics of the eukaryotes. *Science* 2000, 287: 2204–2215.

section one

Plant biotechnology

chapter two

Plant molecular biology
Tools to develop transgenics

Sandhya Agarwal, Alka Grover, and SM Paul Khurana

Contents

Abstract

Plant transgenesis is a vital tool in the field of biotechnology. It has emerged as one of the most promising advances made since the green revolution, especially for designing crops resistant to various pests, pathogens, and environmental stresses. Since the past years, crops such as soybean, cotton, corn, canola, squash, and papaya are being grown commercially for one or the other transgenic traits. The two most popular techniques for genetic transformation are *Agrobacterium tumefaciens* mediated transfer and particle bombardment. With the advent of new techniques in plant transgenesis, plant transformation vectors have also improved giving high-transformation frequencies without host range limitations. Stable transgenic plants are selected by cointroduction of a selectable marker gene, which in most cases confer resistance to antibiotics. However, once a transgenic plant is obtained, the selectable marker gene becomes dispensable and as a sound marketing strategy, new transformation techniques are emerging, which allow generation of marker-free transgenic plants. Gene silencing or ribonucleic acid (RNA) interference is another important tool of plant transgenesis, which is used frequently to understand the functioning and contribution of a target gene. Utilizing the RNA interference, a transgenic maize product showing resistance to the root worm is in pipeline for commercialization. The chapter describes different tools and techniques used in plant transgenesis with emphasis on commercial transgenic crops.

Keywords: Plant genetic transformation, *Agrobacterium tumefaciens*, Transgenic trait, Event, Marker-free transformation.

What to learn in this chapter

- Characteristics of plant transformation vectors
- How to generate a transgenic plant
- Techniques of production of marker-free plants
  - Cotransformation followed by segregation
  - Site-specific recombination
- Chloroplasts transformation technique
- Transgene detection techniques

Introduction

The fundamental aspect of plant transgenesis is the development of technology and its improvement. Plant genetic engineering is the insertion of virtually any gene of interest (GOI) into any crop species for a desired trait. Plant transgenesis is therefore, a very powerful tool bringing together useful genes from unrelated sources. The first report demonstrating the idea of expressing a foreign gene into a crop species came from transgenic tobacco (Bevan et al., 1983) when an antibiotic-resistant gene was cloned under regulatory elements of *Agrobacterium,* nopaline synthase gene, and expressed successfully. Since then, there have been continuous efforts to improve the process of genetic transformation. As for example, a whole range of vectors designed for very high and stable transgene expression are now available with specificity for particular crop species, similarly with improved tissue culture techniques, which is possible to obtain transgenic plants from recalcitrant plant species. The first report of commercial adoption of the technique came in 1994 with Calgene's delayed-ripening tomato (Flavr-Savr™). This was the first genetically modified edible crop species produced and consumed in the United States followed by consumption in Mexico, Japan, and Canada as food during 1995 and 1997.

Benefits of transgenic technology

Conventional breeding practices have limited access to gene pool from unrelated species for crop improvement, however, once a transgenic plant is created, the transgene follows typical Mendelian inheritance and is bred by using usual breeding practices. This has resulted in the creation of a number of varieties and hybrids harboring useful transgene/s. The first-generation transgenic crops exhibited enhanced input traits such as insect resistance, herbicide tolerance, and tolerance to diseases and environmental stress. These crops have resulted in benefits such as increased productivity, reduced farm costs, increased farm profits, and improved environment and health. Because of these benefits, the global cultivated area under transgenic crops increased by more than 100-fold from 1.7 million hectares in 1996 to 175 million hectares in 2013. The United States is the leading country with 70.1 million hectares under transgenic crops cultivation.

Cotton, maize, canola, and soybean are the four commercial crops that have benefited the most from transgenic technologies. Herbicide tolerance and insect resistance are the two most exploited transgenic traits. Transgenic crops harbor these traits either individually or stacked together and farmers select them depending on the requirement. In India, Bt cotton cultivation started at 50,000 hectares with commercialization in 2002 and soon it became popular due to a sharp decline in pesticide requirement and monetary benefits. With increasing hectarage every year since 2002, in 2012, 10.8-billion hectare land was cultivated with Bt cotton, which is 93% of the total cotton-cultivated land in India and is the world's largest hectarage of cotton (James, 2012). With a steep increase in adoption of Bt cotton, India had become a net importer of raw cotton till 2002–2003 and the second largest exporter of cotton. In 2013, 7.3 million small farmers planted Bt cotton in India. This gives an insight on the influence of a single technology that can modify global economics.

Production of transgenic plants

Agrobacterium: The natural plant transformation tool

Agrobacterium tumefaciens, a soil bacterium, has the natural ability to transfer genes present on its Ti plasmid or tumor-inducing plasmid to a plant in such a manner that the transferred genes establish and express precisely in the plant system. This tool created the basis of plant genetic transformation, wherein the GOI could be cloned on a plasmid designed similar to *Agrobacterium* Ti plasmid and transferred to plant cells *in vitro*. Exhaustive literature is available for successful genetic transformation in a number of plant species through *A. tumefaciens*. *Agrobacterium* thus has a very high influence on modern agricultural biotechnology. Genetic transformation mediated by this bacterium is a major tool for obtaining many commercially important transgenic crops.

The Ti plasmid harbored by wild-type *A. tumefaciens* strains range between 200 and 800 kb in size. The transfer DNA or T-DNA and virulence (*vir*) genes regions are essential features of the Ti plasmids and facilitate tumor induction in plant tissue. The tumors or crown gall disease is caused by the expression of genes coding for phytohormones such as auxins and cytokines present in the T-DNA. The T-DNA regions on native Ti plasmid are approximately 10–30 kb in size and are demarcated by T-DNA border sequences at both the ends, known as left border (LB) and right border (RB). These borders are 25 bp in length and highly homologous in sequence. They flank the T-region in a directly repeated orientation. At the time of plant transformation, the T-DNA is cut from Ti plasmid at the border sequences and exported to the plant cell.

T-DNA processing from the Ti plasmid and its subsequent transfer from *Agrobacterium* to the plant cell is due to the activity of *vir* genes. The *vir* region comprises a number of genes and each gene has a specific role in transferring the T-DNA to the plant system. The gene products nick at their LB and RB of the T-DNA region and then transfer the T-DNA into the plant cells (Gelvin, 2003). Similarly, T-DNA system is present on the hairy-root-inducing (Ri) plasmid in *Agrobacterium rhizogenes* causing rhizogenesis. Curing a particular type of plasmid and replacing with another type of tumorigenic plasmid can modify disease symptoms. For example, infection of plants with *A. tumefaciens* strain C58, which harbors the nopaline-type Ti plasmid pTiC58, develops crown gall disease, curing of this plasmid converts, and the strain to nonpathogenic. This provides information that the Ti plasmid is working as a vector to transfer the tumor-inducing genes to the infecting plant species.

Plant transformation vectors

Extensive studies are being carried out on the various components and features of Ti plasmids, to convert them into a vehicle to transfer useful genes into plants. However, because of the large size of Ti and Ri plasmids, structural analysis and modifications become difficult. The entire nucleotide sequences of several Ti and Ri plasmids have been reported and the information is useful for easier targeted replacement than before. There have been a number of reports manipulating the Ti plasmids and designing smaller-sized plasmids with very high transformation efficiencies. The Ti plasmid manipulations for plant transformation gave rise to two different types of systems involving (1) binary vectors and (2) cointegrate vectors.

Binary vectors

Described as two plasmid systems, where the *vir* and T-DNA regions of Ti plasmids could be cloned onto two separate replicons named as binary vectors and *vir* helper strain (De Framond et al., 1983; Hoekema et al., 1983). When both of these replicons are placed within the same *Agrobacterium* cell, the *vir* genes-encoded proteins act on T-DNA of the binary vector in *trans* to facilitate T-DNA processing and export to the plant cell.

The binary vectors are small-sized plasmids (8–20 kb) as compared to the Ti plasmids. In a binary vector, the T-DNA is devoid of any oncogenic genes and a multiple cloning site (MCS) is placed in between "right border" and "left border." The GOI is cloned at a unique restriction site in the multiple-cloning site. The binary vectors are broad host range vectors. Vectors contain origin of replication (oriR), which permits it to maintain in *Escherichia coli* as well as in *Agrobacterium*. Plasmid also contains selectable marker genes (SMGs) for selection in bacterial as well as in the plant system. Plant selection marker gene is placed within the T-DNA, while bacterial selection marker gene is designed outside the T-DNA region. Binary plasmid is devoid of any *vir* genes.

pBIN (Bevan, 1984), pGA (An, 1987), pPZP (Hajdukiewicz et al., 1994), and pCAM-BIA (http://www.cambia.org/daisy/cambia/585.html) series are the most studied binary

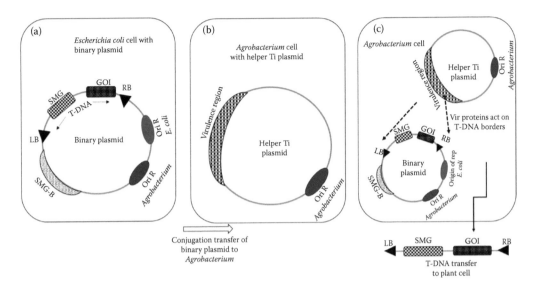

Figure 2.1 Binary vector and T-DNA transfer to plants. (a) A binary vector with T-DNA carrying GOI in *E. coli* cell. (b) *Agrobacterium* cell with helper Ti plasmid (without T-DNA). (c) *Vir* protein from helper Ti plasmid acts in *trans* on T-DNA border sequences to release and transfer T-DNA to plant cell. Abbreviations: RB—right border, LB—left border, GOI—gene of interest, SMG—selectable marker gene, SMG-B— selectable marker gene for bacteria, oriR—origin of replication.

vectors. These plasmids are either used as such or modified according to specific requirements and further used for transformation experiments.

The binary plasmid is introduced into the altered *Agrobacterium* strain via conjugation or transformation. The altered *Agrobacterium* strain harbors a disarmed Ti plasmid, which does not contain the T-DNA but only the *vir* region and is also known as *vir* helper plasmid. The altered *Agrobacterium* strain now contains two plasmids, one is the binary plasmid with GOI and the other one is the *vir* helper plasmid (Figure 2.1). The GOI present at the T-DNA of the binary vector is transferred to the plant cell by the *vir* genes *trans* acting on the binary vector T-DNA border sequences.

A number of *Agrobacterium* strains containing nononcogenic *vir* helper plasmids have been developed, such as LBA4404 (Ooms et al., 1981), GV3101 MP90 (Koncz and Schell, 1986), AGL0 (Lazo et al., 1991), and EHA105 (Hood et al., 1993). A binary system is a very convenient system that does not require microbiological specialization. Globally, this is the most preferred system.

Cointegrate vectors

The system utilizes homology-based recombination between the *E. coli* plasmid and the Ti plasmid. The homologous recombinant is also known as the "hybrid plasmid" whereas the *E. coli* plasmid is the "intermediate plasmid" and the Ti plasmid is the "acceptor plasmid."

The "intermediate plasmid" is small-sized (~5–7 kb) plasmid and harbors T-DNA without any oncogenic genes but with an MCS. A gene cassette is inserted at a unique restriction site in MCS. The vector also harbors a plant SMG within the T-DNA and an antibiotic resistance gene for bacterial selection, outside the T-DNA. An "intermediate vector" also contains a DNA fragment homologous to the "acceptor vector" and the homologous DNA fragment is cloned outside the T-DNA. This vector can replicate only in *E. coli* but not in *Agrobacterium*.

The "acceptor plasmid" is a disarmed Ti plasmid and is smaller than the native Ti plasmids (~40–60 kb). The plasmid has a broad host range and can be maintained in *E. coli* as well as in *Agrobacterium*. The "acceptor plasmid" also contains the *vir* genes, which acts in *cis* on "hybrid plasmid" and facilitates the T-DNA transfer to the plant tissue during genetic transformation. The antibiotic resistance marker gene on the "acceptor plasmid" is different from the bacterial selection marker gene present on the "intermediate plasmid," to facilitate the screening of hybrid plasmids with two different antibiotics.

The process of cointegration starts from transfer of "acceptor plasmid" to an *Agrobacterium* strain such as the LAB4404 (Figure 2.2). This is followed by the introduction of "intermediate plasmid" to the modified LBA4404 by bacterial conjugation or transformation. The cointegrate or the "hybrid plasmid" thus formed is screened by selecting *Agrobacterium* on two different antibiotics depending on the bacterial selection marker genes present on "intermediate" and "acceptor vectors." As the "intermediate plasmid" cannot replicate itself in *Agrobacterium*, the two antibiotic selection will result in screening only the "hybrid plasmid."

Zambryski et al. (1983) developed a disarmed Ti plasmid pGV3850 for homologous recombination with pBR322 plasmid. Similarly in 1996, Komari et al. designed a cointegrate system based on plasmids pSB1 and pSB11. The advantage of cointegration is the low copy number transgene inserted in transgenic plants.

Process of generating a transgenic plant

Genetic transformation is the key to generate transgenic plants with new traits for commercial applications as well as for discovery research. The various steps to generate a transgenic plant with specific GOI are as follows:

1. Designing the gene cassette for optimum expression of the transgene
2. Genetic transformation and regeneration of the transformed plant
3. Event selection

Designing the gene cassette

A gene cassette for expression in a plant species typically consists of a promoter sequence, the transgene, and a 3′ polyadenylation signal. Performance and final selection of a transgenic event is based on gene expression and generally a high transgene expression is preferred. While designing a gene cassette for high transgene expression, selection and source of the transgene play a significant role. Similarly, the choice of the promoter also contributes toward a higher gene expression. Sometimes, additional elements such as enhancers and transcriptional regulators are also used to enhance gene expression (Dai et al., 2007).

Target gene selection is the most important step toward obtaining a transgenic plant with a desired trait. For example, *cry* genes from *Bacillus thuringiensis* are the best exploited for providing protection against different insects in various plant species. The choice of *cry* gene depends on the class of target insects, such as *cry1Ab* is used for protection against Lepidopteran insects, whereas *cry34Ab1* and *cry35Ab1* are used for Dipteran insects. Similarly, for designing crops resistant to the herbicide glyphosate, *epsps* gene from the *A. tumefaciens* strain CP4 has been isolated and used to generate glyphosate-resistant maize, cotton, sugar beet, soybean, canola, and wheat (CERA, 2012). Glyphosate is one of the best-selling herbicides and till date, no other gene could provide a better herbicide

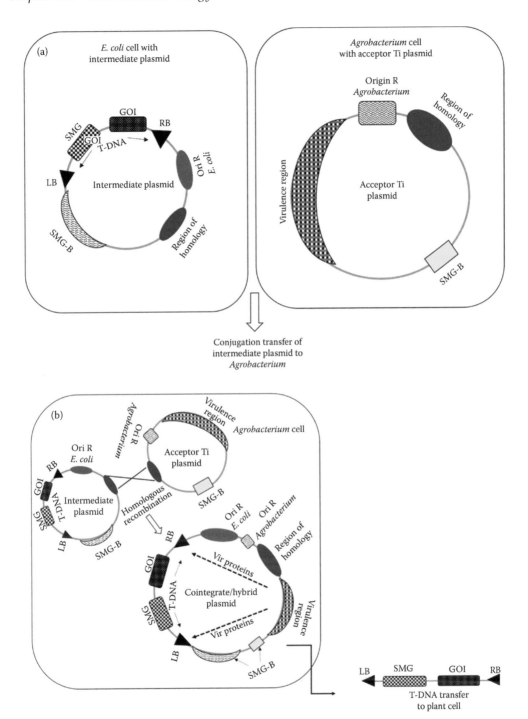

Figure 2.2 Cointegrate vector generation and T-DNA transfer to plants. (a) Intermediate and acceptor plasmids. (b) Hybrid or cointegrate plasmid generation via homologous recombination. *Vir* proteins act in *cis* on T-DNA border sequences to release and transfer T-DNA to plant cell. Abbreviations: RB—right border, LB—left border, GOI—gene of interest, SMG—selectable marker gene, SMG-B—selectable marker gene for bacteria, oriR—origin of replication.

resistance than the *cp4 epsps* gene; therefore, this gene becomes an obvious choice to design glyphosate-tolerant crops (Heck et al., 2005).

Genetic transformation permits the expression of a gene from any living organism (bacteria, yeast, birds, animals, fish, etc.) to any plant species. When the source of GOI is other than plants as, for example, bacteria, a codon optimization for the respective plant species becomes imperative. The codon optimization maximizes the gene expression by increasing the translational efficiency of the GOI. It is done by increasing the frequency of desired nucleotides according to the codon usage of highly expressed genes of that particular plant species. Plants, in general, prefer G- and C-rich codon, as compared to bacteria, where the codons are rich in A and T (http://www.kazusa.or.jp/codon). Heterologous genes with low G and C content in plants often result in a very low protein yield. The codon optimization also removes protein secondary structures and ensures proper folding, leading to a better gene expression.

A compatible combination of the transgene and promoter always results in higher gene expressions. The promoters are selected depending on the plant tissue where a maximum expression is required. Majority of commercial transgenic events harbor constitutive promoters where transgene expression is in the entire plant and not restricted to any tissue. For example, CaMV 35S from cauliflower mosaic virus and maize ubiquitin promoters are the maximum-used constitutive promoters for transformation of monocot crop species. Many times, an inducible gene expression is preferred over the constitutive one, especially regarding pathogen-inducible promoters (Mei et al., 2006). Root-specific peroxidase promoter from wheat has been used for resistance against coleopteran insects (http://cera-gmc.org/). Rice actin1 promoter is another important promoter that is often cloned along with a noncoding 5′ untranslated region (5′ UTR) and the first intron of rice actin1 gene. It has been observed that introns cloned with promoters improve the gene expression (McElroy et al., 1990). Hsp70 and Adh1 introns have been commonly used in many commercial transgenic crops (Cera, 2012).

While designing a gene cassette, the selected components are either amplified from the source genotype/s or synthetically produced. Using routine-recombinant DNA techniques, the components of a gene cassette are cloned together to create a binary or cointegrate vector for either agrotransformation or direct delivery of DNA by physical methods (Figure 2.3).

Genetic transformation and regeneration

An optimized tissue culture regeneration protocol on selective medium is prerequisite for successful genetic transformation of any plant species. Similarly, procedures resulting in higher transformation efficiency are always preferred. The transformation efficiency is the percentage of transformed explants from total inoculated explants. It is however, genotype dependent and many times, the target genotypes are not tissue culture amenable. In such cases, transformation is done with tissue culture-responsive genotypes, followed by transfer of the transgene to the desired genotype through conventional breeding.

Techniques for genetic transformation   The most-considered techniques to generate transgenic plants are

1. Biological transfer
2. Physical transfer
   a. *Biological transfer*: *Agrobacterium*-mediated transformation is the most widely used biological transfer technique. As mentioned earlier, for *Agrobacterium*-mediated

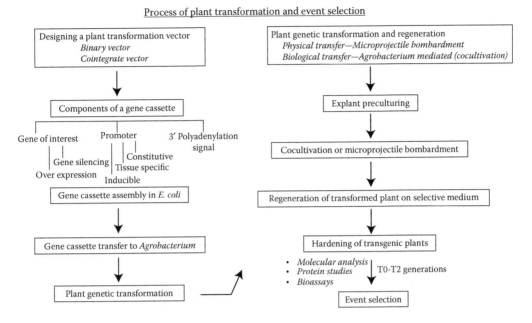

Figure 2.3 Outline of the process of plant transformation and event selection.

transformation, the T-DNA and *vir* genes are the most essential components of the Ti plasmid. To create a transgenic plant, an artificial plasmid is designed with GOI and containing all the important features of Ti plasmids while, eliminating those genes that are not involved in the transformation process. This results in smaller-sized plasmids, which are easier to handle with minimum microbiological expertise. The plasmid thus designed is mobilized to *Agrobacterium*, to facilitate the T-DNA transfer to the plant cells. The transformed plant cells are programed to regenerate into whole plantlet on the selection medium under *in vitro* conditions, resulting in a transformed plant. The *Agrobacterium*-mediated transformation is the preferred technique over the physical transfer system due to single-copy insertion and stable integration of the transgene. Unlike earlier reports, biological transfer through *Agrobacterium* is now routinely used for both dicot and monocot plant species.

b. *Physical transfer*: Engineering of important agronomic crops such as wheat, maize, and cotton through the *Agrobacterium*-mediated transformation sometimes is restricted to a few noncommercial genotypes. In such cases, biolistic transformation is the alternative technique for transformation of recalcitrant genotypes. One of the most preferred physical transformation techniques, the biolistic transformation also known as particle bombardment can be used universally for transferring DNA into any living cell. The device used for particle bombardment is known as gene gun, designed by Sanford et al. (1987). DNA to be transferred is coated on microparticles (1–4 μm) of inert metal such as gold or tungsten and accelerated to velocities about 1000–2000 sq/ft followed by transfer to explants using a burst of helium air under vacuum. Particles penetrate cell walls and membranes and enter cells in a nonlethal manner. A number of reports were published in the 1990s regarding genetic transformation of various plant species through particle bombardment (Vasil et al., 1992; Casas et al., 1993). Several transgenic events of corn commercialized in the 1990s are engineered using particle bombardment (http://cera-gmc.org/index.php?action=gm_crop_database). Jha et al. (2013) reported *Jatropha curcas* transformation with the *SbNHX1* gene using microprojectile bombardment-mediated transformation. Although the technique is a boon for transformation of recalcitrant plant species, transfer of multiple and truncated copies of the transgene reduces the usefulness of the technique. In such cases, thousands of events are generated followed by stringent event selection process to select events with single copy and intact gene cassette.

Protocol for Agrobacterium-mediated transformation   The protocol is optimized for rice cv. IR64 by Sahoo et al. (2011).

1. Seed sterilization: Surface sterilize mature seeds using—(1) 70% ethanol (v/v) for 1 min and (2) with 50% (v/v) commercial bleach for 30 min with constant shaking. Wash seeds with sterilized distilled water (8–10 times) and dry for 5 min on autoclaved Whatman No. 3 filter paper disks.
2. Embryogenic callus induction from seeds: Inoculate seeds on callus induction medium (MCI) and incubate in dark for 2 weeks at $27 \pm 1°C$. MCI composition (pH5.8): MS basal salts and vitamins, supplemented with 30 g/L maltose, 0.6 g/L L-proline, 0.3 g/L casein hydrolyzate, 0.25 mg/L 6-benzylaminopurine (BAP), 3.0 mg/L 2,4-dichlorophenoxyacetic acid (2,4-D), and gelled with 3.0 g/L phytagel.
3. Preparation of *Agrobacterium* culture: Grow *Agrobacterium* culture to OD_{600} ~1, and pellet down the cells by brief centrifugation. Resuspend the pellet in MS resuspension medium (MS basal salts and vitamins, supplemented with 68 g/L sucrose, glucose at the rate 36 g/L, 4 g/L $MgCl_2$ and 3 g/L KCl at pH 5.2) containing 150 µM acetosyringone (filter sterilized) and adjust the OD_{600} of the bacterial suspension to 0.3.
4. Cocultivation of embryogenic calli: Cocultivate 4 days subcultured embryogenic calli with *Agrobacterium* suspension and inoculate on cocultivation medium—MCCM for 48 h in dark at $27 \pm 1°C$. MCCM composition (pH 5.2): MCI medium supplemented with 10 g/L glucose and 150 µM acetosyringone (filter sterilized).
5. Selection of transgenic calli: Rinse cocultivated calli with sterile distilled water supplemented with 250 mg/L cefotaxime for 8–10 times, blot on autoclaved Whatman No. 3 filter paper disks and transfer onto MSM (first selection medium) for 12 days at $27 \pm 1°C$ in dark. For second selection, transfer the the cream-colored calli to fresh MSM medium and incubate for 10 days at $27 \pm 1°C$ in dark. Another experiment was carried out for third selection by transferring microcalli on fresh MSM medium for another 5 days at $27 \pm 1°C$ in dark. MSM composition (pH 5.8): MCI containing filter sterilized 50 mg/L hygromycin and 250 mg/L cefotaxime.
6. Regeneration from transformed calli: For the first phase of regeneration, transfer granular macrocalli to medium MSRM containing 1.0% agarose and incubate at $27 \pm 1°C$ in dark for 1 week. Transfer the dark-incubated calli onto fresh MSRM medium supplemented with 0.8% agarose and incubate at $27 \pm 1°C$ in light for next 4 days. MSRM composition (pH 5.8): MS salts, 30 g/L maltose, 0.2 mg/L naphthalene acetic acid (NAA), 2 mg/L kinetin, and filter-sterilized 30 mg/L hygromycin and 250 mg/L cefotaxime.
7. Root induction: For root induction, transfer shoots to MROM medium and maintain cultures at $27 \pm 1°C$ in light for a week. MROM composition (pH 5.8): Half- strength MS salts, 30 g/L sucrose, 3.0 g/L phytagel, and filter-sterilized 250 mg/L cefotaxime and 30 mg/L hygromycin.

Hardening of plantlets is done by regular procedure of slow acclimatization of tissue culture grown plantlets to outside environment. Once established, the plantlets are tested for the presence of GOI and further analysis is done over a few generations to identify the best event.

Event selection

Generation of hundreds of transgenic events followed by selection of the best event is the most daunting task for a successful program aiming at commercialization of transgenic event. An independent transgenic event is the plantlet, generated from a single transformed cell. The event selection criteria includes (1) presence of transgene, (2) phenotype comparable to nontransgenic line used for transformation, (3) optimum seed set, (4) single-copy T-DNA insert, (5) efficacious performance of transgenic trait in field, and (6) well-identified T-DNA insert position in transformed genotype. The selection process starts with a few hundred transgenic events and as the selection process progresses, only a few events stay at the end fulfilling all the criteria listed earlier, while the rest of the events are dropped out. The best event is thus selected for commercialization.

Types of transgene expression

Plant transformation leading to stable expression of transgene is commonly termed as transgene expression. However, in most of the transformation experiments, the gene cassette is designed to obtain a higher expression of transgene defined as overexpression. On the contrary, various experiments require downregulated expression of a particular gene termed as gene silencing. The transformation techniques for both expression systems are similar, while difference lies in designing the gene cassettes.

Overexpression of transgene

Overexpression of transgene is through the gene cassettes as described earlier, typically consisting of a strong promoter, the GOI, and a 3′ polyadenylation signal (Figure 2.4). Transcription of GOI generates ribonucleic acid (RNA) which further translates to protein. A high level of protein expression is expected from an overexpression system. There are a number of reports for transgene overexpression leading to desired characteristics such as resistance to insects, abiotic and biotic stresses, herbicides, and to understand gene functions for discovery research (Table 2.1).

Downregulation of transgene expression

Homology-dependent gene silencing or downregulation of transgene expression also known as RNA interference (RNAi) is the most studied gene-silencing technology in the recent years. To design an experiment for gene knockdown, RNAi silencing is the choice of technique due to dominant nature, whereas insertion or loss of function mutations are recessive. Most commonly, the RNAi silencing is manifested by posttranscriptional gene silencing (PTGS) leading to messenger RNA (mRNA) degradation. The construct for gene silencing is designed by an inverted repeat of the transgene separated by an intron (Figure 2.4). This results in the formation of double-stranded RNA (dsRNA), which acts as the trigger for gene silencing. The moment a dsRNA is generated in the plant system, it is cleaved into small RNAs (21–24 nucleotide length) known as small interfering RNA (siRNA) by ribonuclease III (RNase III) like endonuclease termed as Dicer. The typical

(a)

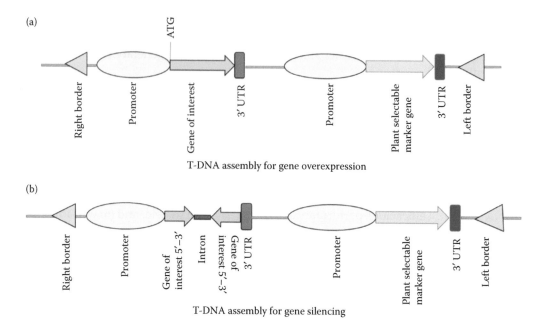

T-DNA assembly for gene overexpression

(b)

T-DNA assembly for gene silencing

Figure 2.4 Graphic representation of gene cassettes in T-DNA. (a) T-DNA representation of an overexpression vector with cassettes for GOI and SMG. (b) T-DNA representation of an RNAi vector displaying inverted repeat arrangement of GOI.

siRNA generated by Dicer degradation consists of 5′ phosphate, 3′ hydroxyl terminal followed by two single-stranded nucleotide at 3′ end. The siRNA generated by the Dicer is incorporated into multicomponent ribonuclease complex called RNA-induced silencing complex (RISC). The siRNA guides RISC to target homologous mRNA based on complementarity between the siRNA and the target mRNA. The target mRNA is cleaved by the RISC at a nucleotide across from the center of the guide siRNA strand, leading to further degradation of target mRNA. The target mRNA is no more available for translation and no protein is formed.

Though still in infancy, the RNAi technology applications fall into two categories: (1) protection of plant against viral and insect diseases and (2) manipulation of metabolic pathways. The first commercial outcome was the transgenic papaya resistant to papaya ringspot virus (http://cera-gmc.org/index.php?action=gm_crop_database&mode=ShowProd&data=55–1%2F63-1) designed with viral coat protein gene.

Screening and detection of transgenic plants

Genetic transformation is an important tool for crop improvement. Success of any transformation experiment depends on generation of a population of putative transgenic events followed by screening of transgenic events from the population. Screening and further characterization of transgenic events is done by various methods and categorized as

1. Molecular characterization: Molecular characterization involves techniques like polymerase chain reaction (PCR) and Southern blot hybridization to detect the presence of transgene, number of T-DNA inserts, and exact position of T-DNA in transgenic plant genome.

Table 2.1 Transgene downregulation and overexpression reports

Crop species	Trait	Transgene	Reference
Downregulation of gene expression			
Bean	Bean golden mosaic virus resistance	Rep(Ac1) viral gene	Aragão et al. (2013)
Cotton	Cotton boll worm resistance	Molt-regulating transcription factor gene HaHR3	Xiong et al. (2013)
Corn	Western corn root worm resistance	Snf7 homolog of western corn root worm	Ramaseshadri et al. (2013)
Cotton	Cotton bollworm resistance	P450 gene—CYP6AE14	Mao et al. (2011)
Rice	Brown plant hopper resistance	Hexose transporter NlHT1, carboxypeptidase Nlcar, and trypsin-like serine protease Nltry genes	Zha et al. (2011)
Barley	Powdery mildew resistance	Effector gene Avra10	Nowara et al. (2010)
Barley	Fusarium resistance	P450 lanosterol C-14α-demethylase (CYP51)	Koch et al. (2013)
Potato	Reduction of cold-induced sweetening	Silencing of vacuolar invertase	Zhu et al. (2014)
Transgene overexpression			
Tomato	Salt tolerance	LeNHX2 endosomal ion transporter	Huertas et al. (2013)
Maize	Drought and salt tolerance	HVA1 *Hordeum vulgaris* abundant protein	Nguyen and Sticklen (2013)
Rice	Improved nitrogen use efficiency	Alanine amino transferase	Beatty et al. (2013)
Arabidopsis	Multiple-herbicide resistance	Glutathion transferase (gstp1)	Cummins et al. (2013)
Cotton	Fusarium and Verticillium resistance	Plant defensin NaD1	Gaspar et al. (2014)
Rice	Drought tolerance	Transcription factor EDT1/HDG11	Yu et al. (2013)

2. Protein expression studies: Protein expression is an indication of proper gene functioning. Maximum studies make use of enzyme-linked immunosorbent assay (ELISA) to detect and quantify the expression of transgene protein.
3. Field performance evaluation: Efficacy of a transgenic event is the best evaluated at field level. Bioassays are done by exposing the transgenic events to those conditions for which they are designed such as pests and pathogen resistance, tolerance to salt or drought, or herbicides, etc. (Figure 2.5).

Molecular characterization

Polymerase chain reaction   Invented by Karry Mullis in 1985 (Mullis et al., 1987), the technique multiplies target DNA exponentially with the help of a pair of oligonucleotides (~20 bp) complementary to DNA molecule, one to each strand of double helix. The

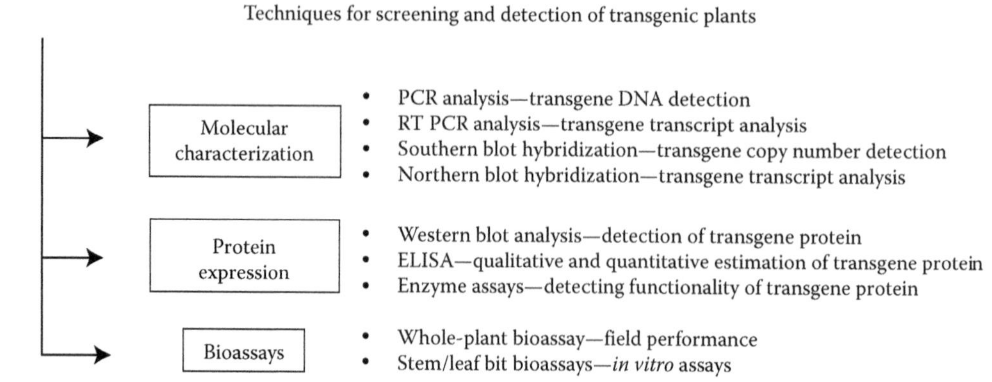

Techniques for screening and detection of transgenic plants

Molecular characterization	• PCR analysis—transgene DNA detection
	• RT PCR analysis—transgene transcript analysis
	• Southern blot hybridization—transgene copy number detection
	• Northern blot hybridization—transgene transcript analysis

Protein expression	• Western blot analysis—detection of transgene protein
	• ELISA—qualitative and quantitative estimation of transgene protein
	• Enzyme assays—detecting functionality of transgene protein

| Bioassays | • Whole-plant bioassay—field performance |
| | • Stem/leaf bit bioassays—*in vitro* assays |

Figure 2.5 Diagram explaining various levels of screening of transgenic plants.

oligonucleotides act as primers for DNA synthesis and delimit the region for amplification. Amplification is carried out by thermostable DNA polymerase I, isolated from *Thermus aquaticus* and known as *Taq* DNA polymerase.

DNA is amplified following denaturing, annealing, and extension steps repeatedly for 30–35 times. PCR is the simplest technique to detect transgenic events by using transgene specific primes (Figure 2.6).

The PCR can also be used to find out the presence/absence and quantification of transgene transcript or mRNA. This is done by RNA-dependent DNA polymerase activity of enzyme reverse transcriptase. Reverse transcriptase convert RNA molecule to cDNA or complementary DNA, which is amplified by *Taq* DNA polymerase using a routine DNA amplification reaction.

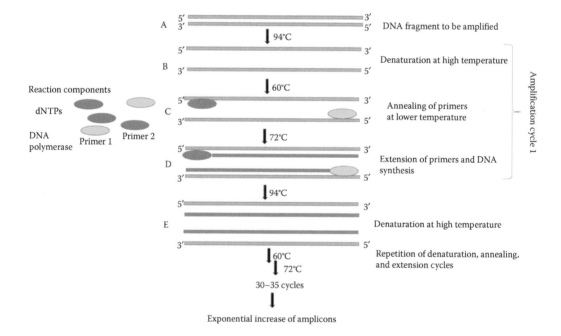

Figure 2.6 DNA synthesis in a PCR.

Southern blot analysis   Named after Edward Southern, who invented the technique in 1975, Southern blotting is used to detect specific DNA molecules from a pool of DNA by using a target specific probe. The technique is highly sensitive and routinely used to identify the transgene copy number in an event. The technique involves the following steps (Figure 2.7).

Similar to Southern blotting, Northern blotting is a technique to detect the presence of transcripts of the target DNA. In this method, an RNA sample, often the total cellular RNA, is run on a formaldehyde agarose gel. Formaldehyde unfolds and linearizes RNA by disrupting hydrogen bonds. The individual RNAs separated according to size by gel electrophoresis are transferred to nylon membrane to which the extended denatured RNAs adhere. A DNA or RNA probe is labeled and target molecule is detected following similar steps of hybridization, washings, and signal detection as mentioned earlier for Southern blotting. Major consideration for Northern blotting is the avoidance of RNases contamination to obtain the desired results.

Thermal asymmetric interlaced PCR   Developed by Liu and Whittier in 1995, TAIL PCR detects the position of transgene in the genomic DNA by identification of flanking regions. Two types of primers used in the reaction are—(1) specific primers (SPs): specific to target sequence, these primers are with higher melting temperature (58–63°C), (2) arbitrary degenerate (AD) primers: relatively smaller primers with lower melting temperature (45–48°C), these primers bind to genomic DNA. PCR amplification gives three types of products (1) primed by specific primer and AD primer (target product), (2) primed by specific primer alone (nonspecific), and (3) primed by AD primer alone (nonspecific). Reaction cycles are designed by maintaining and interlacing the annealing temperature at high and low stringency, to make sure that a target amplicon is amplified with AD primers and SPs. The product obtained in the primary or first reaction is further amplified in secondary and tertiary reaction to increase the probability of detection of right target sequence.

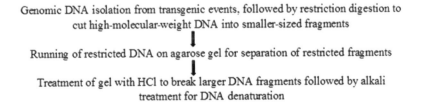

Genomic DNA isolation from transgenic events, followed by restriction digestion to cut high-molecular-weight DNA into smaller-sized fragments

↓

Running of restricted DNA on agarose gel for separation of restricted fragments

↓

Treatment of gel with HCl to break larger DNA fragments followed by alkali treatment for DNA denaturation

↓

Transfer of DNA from gel to positively charged nylon membrane through capillary action, by placing membrane directly in contact with the gel

↓

Drying of the membrane followed by hybridization with labeled probe. A probe is a DNA fragment with sequence identical to target DNA to be detected. Labelling is either radioactive or fluorescent. Probe is denatured to single-stranded DNA before hybridization

↓

Posthybridization washes and signal detection according to type of label used

Figure 2.7 Flow diagram of Southern blot hybridization.

Protein expression studies

Protein detection is a step further to molecular identification by PCR and Southern blotting. While molecular characterization provides information for the presence or absence of transgenes, the protein expression indicates the functionality of transgene. Most commonly, ELISA is the choice of technique for protein detection. ELISA involves the following steps (Figure 2.8):

1. Coating of transgene specific antibodies to microtiter plate.
2. Binding of target protein to antibodies of microtiter plate.
3. Adding secondary antibody, specific to target protein. Secondary antibody is labeled with an enzyme that catalyzes the color reaction.
4. Appropriate reagents and incubation at specific temperature allow color reaction to proceed.
5. The amount of color development is proportional to the quantity of a target protein.

Relatively faster and economical, the ELISA has emerged as a very powerful tool to screen almost all the commercially available crops. Multiplexed ELISA kits are also available for the detection of transgenic crops with stacked traits.

Field performance evaluation

Bioassays at field level are imperative to screen and detect the best event. Molecular characterization indicates the presence/absence of transgene and protein expression detects the functionality of transgene, whereas, the bioassays show the efficacy of an event in field conditions. For example, screening of plants transformed with *cry* genes is done by conducting the experiment at a hot spot for target insect. This experiment is usually done in the season when the target insects are in abundance. Similarly, screening for salt tolerance involves exposing the transgenic events to high salt containing soils. For herbicide tolerance, the events are sprayed with specific herbicide and observations are made for efficacious events. Field tests also identify plants with a phenotype comparable to the nontransgenic counterpart. Field tests are done in replications and mostly at more than one site, providing an accurate estimate on the new capabilities conferred to the plant by integrated transgene.

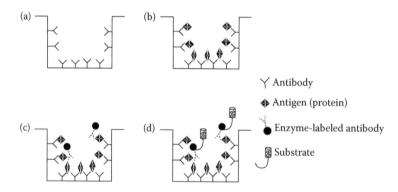

Figure 2.8 Immunodetection of a transgenic protein. (a) Transgene protein-specific antibody-coated microtiter plate. (b) Loading and binding of sample protein to antibodies. (c) Addition of enzyme-labeled secondary antibody (specific to transgene protein). (d) Color development with enzyme substrate.

Transgenic plants from laboratory to farmers field

A well-optimized tissue culture and genetic transformation system is indispensable for successful production of transgenic plants. Since the creation of the first transgenic plant, consistent efforts are being made to optimize genetic transformation for a number of plant species and crops especially for economically important crops. However, many important plant species are not tissue culture amenable. In such cases, transgenic plants are produced in the species that are tissue culture amenable and further the transgene is transferred to commercially important parental lines (for hybrid production) or varieties of that crop through routine breeding practices.

Therefore, mere production of transgenic plants in the laboratory and their establishment in the field is not sufficient, but commercialization requires production of hybrids or varieties with transgene. For example, cotton cultivar Coker is highly responsive to tissue culture and genetic transformation but is agronomically poor. In such cases, transgenic plants with particular trait are produced in Coker, followed by crossing and further backcrossing for six generations with desired parental lines or varieties (Figure 2.9).

Stable incorporation of transgene to parental lines or varieties takes 3–4 years depending on the crop species. Many times only one of the parental lines is introduced with transgene, followed by hybrid seed production. Once stabilized, the hybrids are produced, followed by seed production at commercial level.

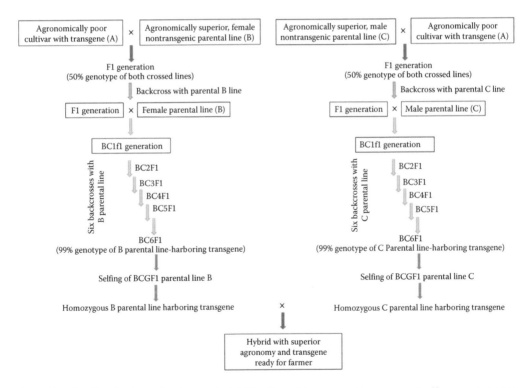

Figure 2.9 Outline for introduction and stabilization of transgene to agronomically superior genotypes and hybrid production.

Production of marker-free transgenic plants

The primary requirement of a genetic transformation system is selection of transformed cells from a population of untransformed cells. Transformed cells have the capability to generate whole plant in the presence of a selective agent. Antibiotics and herbicides are the most commonly used selective agents (Table 2.2). An SMG providing resistance to antibiotic or herbicide is often placed in the form of a gene cassette within the T-DNA region of a vector and transformed tissue is grown on the medium supplemented with selectable marker-specific selective agent. However, once the transformed plant is formed, the SMGs become dispensable (Figure 2.10).

Horizontal transfer of antibiotic-resistance genes to gut bacteria of animals and humans and vertical transfer of herbicide-resistance genes to weedy relatives are the major biosafety concerns in genetically transformed crops harboring SMGs (Dale et al., 2002). Elimination of SMG also enables transgene stacking by retransforming the

Table 2.2 SMGs used for selection of transformed plant tissue

Gene	Gene product	Selectable agent	Source	Reference
neo, nptII	Neomycin phosphotransferase	Kanamycin, neomycin, geneticin (G418)	*E. coli* Tn5	Fraley et al. (1983)
hph (aphIV)	Hygromycin phosphotransferase	Hygromycin B	*E. coli*	Waldron et al. (1985)
SPT	Streptomycin phosphotransferase	Spectinomycin, streptomycin	*E. coli* Tn5	Maliga et al. (1988)
Ble	Bleomycin	Bleomycin resistance	*E. coli* Tn5	Hille et al. (1986)
cat	Chloramphenicol	Chloramphenicol acetyl transferase	*E. coli* Tn5	De Block et al. (1984)
cp4 epsps	5-Enolpyruvyl shikimate 3-phosphate synthase	Glyphosate	*A. tumefaciens*	Barry et al. (1992)
pat, bar	Phosphinothricin acetyl transferase	Phosphinothricin	*Streptomyces hygroscopicus*	De Block et al. (1989)

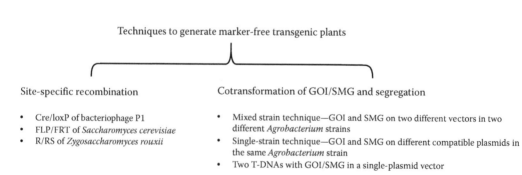

Figure 2.10 Outline of techniques to generate marker-free transgenic plants.

transformed plants with an additional GOI. There are two basic approaches to produce marker-free plants:

1. Cotransformation and further segregation of GOI and SMG
2. Site-specific recombination to eliminate SMG

Cotransformation followed by segregation

Cotransformation involves cloning of GOI and SMG into two separate transformation vectors followed by transformation. The transformation process is carried out as usual and transformed plants are selected on the medium supplemented with a selection agent specific to SMG. Two types of transgenic plants grow on the selective medium, first-category plants harbor both the cassettes of GOI and SMG and the second-category plants express only SMG cassette. Plants from the first category are allowed to set T_0 seeds. The T1 generation obtained from the T_0 seeds is segregating and both the T-DNAs segregate independently in Mendelian fashion resulting in progenies carrying only the GOI and not the SMG.

Different strategies used in the cotransformation system are (1) mixed strain technique—GOI and SMG are carried on two different vectors in two different *Agrobacterium* strains (Komari et al., 1996), (2) single-strain technique—GOI and SMG are carried on different compatible plasmids in the same *Agrobacterium* strain (Sripriya et al., 2008), and (3) two T-DNAs in a single-plasmid vector (Komari et al., 1996; Miller et al., 2002). The cotransformation system is simple and produces marker-free transgenic plants in a very clean manner, that is, segregated plants for the GOI do not contain any other DNA fragments related to marker gene cassette or residual recombination sites as is the case for recombinase system.

Two very important considerations for a cotransformation system are a high efficiency of cotransformation and high frequency of unlinked integration of two T-DNAs. According to Komari et al. (1996), the single-strain technique results in higher efficiency of cotransformation as compared to mixed strain technique. With regard to unlinked cotransformations, an octopine-type *Agrobacterium* strain LBA4404 may provide a better option as compared to nopaline-type C58 strain of *Agrobacterium* (De Block and Debrouwer, 1991). Sripriya et al. (2011) placed a transgene on a 12-copy binary vector and increased cotransformation frequency up to 86%, followed by segregation frequency of the transgene up to 40% in T1-segregating generation.

Site-specific recombination

Site-specific recombination is a process of DNA breakage and reunion at specific positions. The three most commonly used site-specific recombination systems are—(1) Cre/loxP of bacteriophage P1 (Dale and Ow, 1991; Sreekala et al., 2005), (2) FLP/FRT of *Saccharomyces cerevisiae* (Davies et al., 1999), and (3) R/RS of *Zygosaccharomyces rouxii* (Sugita et al., 2000). All these are two-component systems requiring a recombinase enzyme and specific recognition sites for recombinase action. Depending on the system, the recombinase is encoded by *cre* gene in case of Cre/LoxP, *flp* gene for FLP/FRT system, and *r* gene for R/RS system of recombination. The recombinase thus produced acts in *trans* to catalyze recombination between two short, very specific DNA sequences denoted as LoxP, FRT, and RS for three systems, respectively.

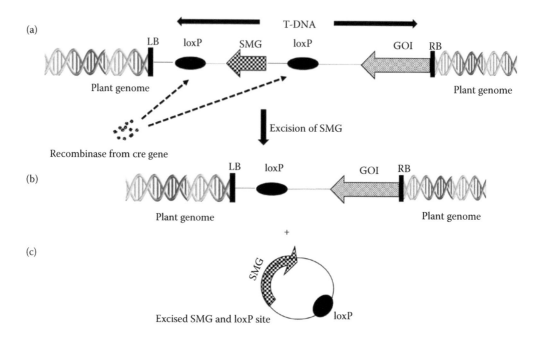

Figure 2.11 Cre/LoxP-mediated SMG excision. (a) Recombinase produced from *cre* gene acts on loxP sites and excise SMG. (b) The resulting T-DNA without marker gene cassette. (c) Excised marker gene and one loxP site circularizes and degrades. Abbreviations: LB—left border, RB—right border, GOI—gene of interest, SMG—selectable marker gene.

Out of three recombination systems, the Cre/LoxP recombination is the most widely used to generate marker-free transgenic plants. The system involves expression of *cre* gene, which could be expressed as transiently, inducibly, developmentally, or engineered as autoexcision. The cre protein (38.5 kDa) encoded by *cre* gene acts on loxP sites, a 34-bp region, flanking the marker gene cassette in the T-DNA. As explained in Figure 2.11, the recombination between directly repeated loxP sites results in excision and circularization of marker gene cassette along with one of the loxP sites. The excised circular marker gene cassette is lost naturally in the cell. Commonly, the transgenic plants with GOI and marker gene are crossed with the plants expressing *cre* gene. The marker gene is excised in the T_1 generation followed by segregation and elimination of the *cre* gene in subsequent generations.

Site-specific recombination is successfully employed to generate marker-free transgenic plants in many crops as reviewed by Tuteja et al. (2012), however; segregation of the *cre* gene over generations is time consuming and laborious. The other problem associated with these inducible or developmentally engineered systems is a poor promoter induction, which may result in incomplete transgene excision, and/or off-effects such as leaky expression or unintended excision.

Chloroplast transformation: Transplastomic plants

Chloroplasts, evolved from cyanobacteria, are the plastids containing chlorophyll and are found in green parts of the plants. Depending on the plant species, the number of plastids in each cell is variable and hundreds of chloroplasts may be present in a single cell. Each plastid contains multiple copies of its own genome resulting in a very high ploidy level. In case of *Arabidopsis thaliana*, leaf mesophyll cell contains about 120 chloroplasts and these harbor approximately 1000–1700 copies of the 154-kb plastid genome (Zoschke et al., 2007). Plastid genome is a circular molecule of double-stranded DNA. Because of the presence of thousands of copies of chloroplast DNA, it typically comprises as much as 10%–20% of the total cellular DNA content (Bendich, 1987).

Advantages of chloroplast transformation

With high polyploidy of plastid genome, an extraordinary higher accumulation of recombinant protein can be obtained in chloroplasts (Maliga and Bock, 2011). Thus, the plastid transformation provides a valuable alternative tool for genetic transformation and leads to higher expression level that the nuclear transformation lacks. The plants produced by chloroplast transformation are termed as *transplastomic*.

Another advantage of chloroplast transformation is the expression of multiple genes as a polycistronic unit, which makes it an attractive system over nuclear transformation. Plastid transformation also provides a strong biological containment as the transfer of plastids is through the maternal parent and transmission of plastids through pollen is very rare (Svab and Maliga, 2007) since pollen does not contain chloroplasts. Absence of transgene flow by outcrossing makes transplastomic plants safer as compared to the plants with nuclear transgenes. Another advantage of chloroplast transformation is the fixed integration of the transgene in the plastid genome or plastome. Transgene integration into the plastome is homology-based recombination between the target regions of the transformation vector and the plastid genome (Maliga and Bock, 2011). Chloroplast transformation eliminates the concerns of position effect, frequently observed in nuclear transgenic lines (Daniell, 2002). Homologous recombination also avoids inactivation and unpredictable rearrangements of transgene as well as host genes. Furthermore, chloroplast transformation accompanied by higher expression does not show any transgene silencing.

Chloroplast transformation process

Plastid transformation was first reported in green alga, *Chlamydomonas reinhardtii* (Boynton et al., 1988) followed by tobacco transformation (Svab et al., 1990) and thereafter transformation reports came from more than a dozen plant species including both monocotyledonous and dicotyledonous crops (Muhammad, 2012).

Plastid transformation vectors: Transformation of plastid genome involves a series of steps such as development of a transformation vector, an efficient DNA delivery system, and well-optimized plant regeneration protocol from explant. Incorporation of target gene in the plastid genome (pt DNA) is based on the homologous recombination between the vector DNA and plastid DNA. A typical vector for chloroplast transformation harbors a chloroplast promoter, 5′ UTR sequences and GOI, marker gene, and 3′ UTR along with the plastid genome homologous DNA sequences flanking at both the ends of the whole cassette, which facilitates homologous recombination. The most common promoter used for chloroplast transformation is from plastid rRNA (rrn) operon (Prrn), which is used in fusion with 5′ UTR or 5′ translation control region (5′ TCR) that includes the 5′ UTR and N terminus of the coding region. Maliga (2002) indicated the importance of 5′ UTR/5′ TCR, which can lead to a varying amount of protein accumulation as high as 10,000-fold based on the choice of translation control signals. Immediately downstream of stop codon, plastid 3′ UTRs are placed. Most commonly used 3′ UTRs are usually derived from plastid genes such as *rps16, rbcL, psbA,* and *rpl32* 3′. Spectinomycin/streptomycin is the well-optimized choice for antibiotic marker for plastid transformation.

Location of the insertion site in the plastid genome has significant effect on the level of protein accumulation. Insertion of a transgene in repeated region and regions with heavily transcribed operon in ptDNA yields higher protein levels (Maliga and Bock, 2011).

Gene gun or the biolistic process is the most successful one for delivering transformation vector to chloroplasts. Gold or tungsten microparticles (1.0–4.0 μm) are coated with vector DNA for targeted insertion by homologous recombination. Only few of the hundreds of ptDNA copies incorporate transforming DNA. This is followed by growing cells on the selective tissue culture medium, which results in enrichment of transformed ptDNA copies. Antibiotic selection results in gradual sorting process of plastid and the plastid genome (ptDNA) yielding genetically stable homoplastomic cells, which harbor only transformed ptDNA copies. Regeneration of these homoplastomic cells produces genetically stable plants (Figure 2.12).

Great potential of chloroplast transformation technology is evident from the reports in which enhanced gene expression is observed with different cry proteins (Chakrabarti et al., 2006; Liu et al., 2008) and various genes for herbicide resistance (Shimizu et al., 2008). De Cosa et al. (2001) reported that genetically engineered tobacco plants accumulated *cry2Aa2* as high as 46% of the total protein.

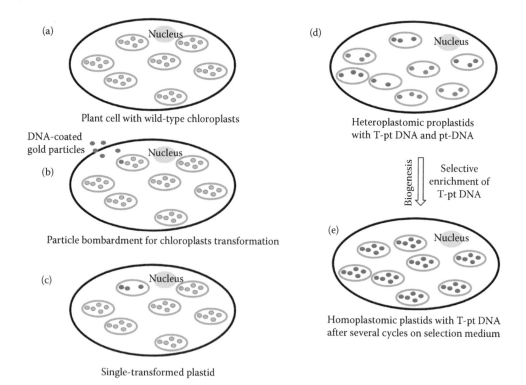

Figure 2.12 T-ptDNA sorting and generation of homoplastomic plants. (a) Plant cell with wild-type chloroplasts and ptDNA. (b) and (c) Particle bombardment and transformation of a single ptDNA (red colored). (d) Heteroplastomic proplastids with T-ptDNA and ptDNA in the same cell. (e) Homoplastomic plastids with T-ptDNA after many cycles on a selective medium.

There is enormous capacity to use chloroplasts as bioreactors for molecular farming, as each cell contains a high copy number of chloroplasts (Maliga and Bock, 2011). High level of expression in transplastomic plants makes them a suitable system for industrial production of biopharmaceuticals such as antibodies, vaccines, antimicrobials, and also for industrial enzymes. While working for production of endolysins, an antimicrobial compound, Oey et al. (2009a) observed, transplastomic protein accumulation as high as 70% of total soluble protein. This high level of expression eliminates the need to purify protein to homogeneity and where possible, the crude protein can be used as such, thus reducing the total production cost of protein. However, obtaining genetically uniform plants from the transformed plastid genomes is still a bottleneck that has to be addressed to observe the maximum potential of the technology.

Status and future of transgenic crops

The genetic transformation technology has brought together the new genes or regulatory elements in totally unrelated crop species for the benefit of mankind. For example, the crops designed for insect resistance, harbor and express genes from bacteria. This combination is not available by traditional breeding practices. Therefore, much studies are expected from transgenic crop species related to the safety of the inserted gene for food, feed, and environment. Commercial production of a transgenic crop species is approved only after it meets all the biosafety aspects. These approvals are given by government agencies, set up exclusively for safety assessment of transgenic crops. In India, the regulatory authority is Genetic Engineering Approval Committee (GEAC) constituted by Ministry of Forests and Environment and Department of Science and Technology. Many subcommittees are there under GEAC for regulating and monitoring research related to the transgenic plants, animals, and microbes. In the United States, United States Department of Agriculture (USDA), Environmental Protection Agency (EPA), and Food and Drug Administration (FDA) work in association for safety assessment of transgenic crops before it releases into the environment. The United States is the first country to adopt food and feed from the transgenic crops. Currently, there are 27 countries growing the transgenic crops with the United States, Brazil, Argentina, India, and Canada at the top five positions with heavy acarage under genetically transformed crops.

Cotton expressing the *cry* genes for resistance against ball worms is the only crop, deregulated in India (http://igmoris.nic.in/). Many crops are there in pipeline such as maize for insect resistance and herbicide tolerance, cotton for herbicide tolerance, rice for drought and salinity, bacterial leaf blight and biofortification, brinjal for fruit and stem borer, wheat for heat tolerance, fungal diseases, etc. In China, deregulated crops include cotton, papaya, sweet pepper, poplar, and tomato, whereas in the United States, the list is still longer with cotton, canola, papaya, maize, soybean, sugarbeet, alfalfa, and squash.

With the improvement of technology and success of the first-generation transgenic crops, focus is shifted to second-generation transgenic crops with value-added traits such as nutritional enhancement for food and feed. Similarly, third-generation transgenic crops are producing products beyond food and fiber such as pharmaceuticals and improved processing of bio-based fuels. The transgenic crops have the potential for substantial contribution for crop productivity and addressing global food requirement. With the introduction of new traits and crops for wider geographies, many developing countries are expected to adopt the technology in the coming years.

In India, after a gap of 2 years, recently the regulatory committees have given approval for field testing of a few transgenic events under "crop-specific dedicated monitoring protocols system." With the ever-increasing demand for food, reduction of availability of cultivable land, and fluctuating climatic conditions, adoption of technologies has become almost imperative. A faster implementation of policies for deregulation of transgenic crops is therefore, the need of the hour.

References

An, G., 1987. Binary Ti vectors for plant transformation and promoter analysis. *Methods in Enzymology.* 153, 292–305.

Aragão, F.J.L., Nogueira, E.O.P.L., Tinoco, M.L.P., Faria, J.C., 2013. Molecular characterization of the first commercial transgenic common bean immune to the bean golden mosaic virus. *Journal of Biotechnology* 166, 42–50.

Barry, G., Kishore, G., Padgette, S., Talor, M., Kolacz, K., Weldon, M., Re, D., Eichholtz, D., Fincher, K., Hallas, L., 1992. Inhibitors of amino acid biosynthesis: Strategies for imparting glyphosate tolerance to plants. In: *Biosynthesis and Molecular Regulation of Amino Acids in Plants.* American Society of Plant Physiology, Rockville, MD. pp. 139–145.

Beatty, P.H., Carroll, R.T., Shrawat, A.K., Guevara, D., Good, A.G., 2013. Physiological analysis of nitrogen-efficient rice overexpressing alanine aminotransferase under different N regimes. *Botany* 91, 866–883.

Bendich, A.J., 1987. Why do chloroplasts and mitochondria contain so many copies of their genome? *Bioessays* 6, 279–282.

Bevan, M.W., 1984. Binary *Agrobacterium* vectors for plant transformation. *Nucleic Acids Research* 12, 8711–8721.

Bevan, M.W., Flavell, R.B., Chilton, M.-D., 1983. A chimaeric antibiotic resistance gene as a selectable marker for plant cell transformation. *Nature* 304, 184–187.

Boynton, J., Gillham, N., Harris, E., Hosler, J., Johnson, A., Jones, A., Randolph-Anderson, B. et al., 1988. Chloroplast transformation in *Chlamydomonas* with high velocity microprojectiles. *Science* 240, 1534–1538.

Casas, A.M., Kononowicz, A.K., Zehr, U.B., Tomes, D.T., Axtell, J.D., Butler, L.G., Bressan, R.A., Hasegawa, P.M., 1993. Transgenic sorghum plants via microprojectile bombardment. *Proceedings of the National Academy of Science* 90, 11212–11216.

CERA, 2012. *GM Crop Database.* Centre for Environmental Risk Assessment, Washington DC.

Chakrabarti, S., Lutz, K., Lertwiriyawong, B., Svab, Z., Maliga, P., 2006. Expression of the cry9Aa2 Bt gene in tobacco chloroplasts confers resistance to potato tuber moth. *Transgenic Research* 15, 481–488.

Cummins, I., Wortley, D.J., Sabbadin, F., He, Z., Coxon, C.R., Straker, H.E., Sellars, J.D. et al., 2013. Key role for a glutathione transferase in multiple-herbicide resistance in grass weeds. *Proceedings of the National Academy of Sciences* 110, 5812–5817.

Dai, X., Xu, Y., Ma, Q., Xu, W., Wang, T., Xue, Y., Chong, K., 2007. Overexpression of an R1R2R3 MYB gene, OsMYB3R-2, increases tolerance to freezing, drought, and salt stress in transgenic *Arabidopsis*. *Plant Physiology* 143, 1739–1751.

Dale, E.C., Ow, D.W., 1991. Gene transfer with subsequent removal of the selection gene from the host genome. *Proceedings of the National Academy of Science* 88, 10558–10562.

Dale, P.J., Clarke, B., Fontes, E.M.G., 2002. Potential for the environmental impact of transgenic crops. *Nature Biotechnology* 20, 567–574.

Daniell, H., 2002. Milestones in chloroplast genetic engineering: An environmentally friendly era in biotechnology. *Trends in Plant Science* 7, 84–91.

Davies, G.J., Kilby, N.J., Riou-Khamlichi, C., Murray, J.A.H., 1999. Somatic and germinal inheritance of an FLP-mediated deletion in transgenic tobacco. *Journal of Experimental Botany* 50, 1447–1456.

De Block, M., De Brouwer, D., Tenning, P., 1989. Transformation of *Brassica napus* and *Brassica oleracea* using *Agrobacterium tumefaciens* and the expression of the bar and neo genes in the transgenic plants. *Plant Physiology* 91, 694–701.

De Block, L.M., Herrera-Estrella, I., Van Montagu, M., Schell, J., Zambryski, P., 1984. Expression of foreign genes in regenerated plants and in their progeny. *EMBO Journal* 3, 1681–1689.

De Framond, A.J., Barton, K.A., Chilton, M.-D., 1983. Mini-Ti: A new vector strategy for plant genetic engineering. *Nature Biotechnology* 1, 262–269.

De Cosa, B., Moar, W., Lee, S.B., Miller, M., Daniell, H., 2001. Overexpression of the Bt cry2Aa2 operon in chloroplasts leads to formation of insecticidal crystals. *Nature Biotechnology* 19, 71–74.

Fraley, R.T., Rogers, S.G., Horsch, R.B., Sanders, P.R., Flick, J.S., Adams, S.P., Bittner, M.L. et al., 1983. Expression of bacterial genes in plant cells. *Proceedings of the National Academy of Sciences United States* 80, 4803–4807.

Gaspar, Y.M., McKenna, J.A., McGinness, B.S., Hinch, J., Poon, S., Connelly, A.A., Anderson, M.A., Heath, R.L., 2014. Field resistance to *Fusarium oxysporum* and *Verticillium dahliae* in transgenic cotton expressing the plant defensin NaD1. *Journal of Experimental Botany* 65, 1541–1550.

Gelvin, S.B., 2003. *Agrobacterium*-mediated plant transformation: The biology behind the "Gene-Jockeying" tool. *Microbiology and Molecular Biology Reviews* 67, 16–37.

Hajdukiewicz, P., Svab, Z., Maliga, P., 1994. The small, versatile pPZP family of *Agrobacterium* binary vectors for plant transformation. *Plant Molecular Biology* 25, 989–994.

Heck, G.R., Armstrong, C.L., Astwood, J.D., Behr, C.F., Bookout, J.T., Brown, S.M., Cavato, T.A. et al., 2005. Development and characterization of a CP4 EPSPS-based, glyphosate-tolerant corn event. *Crop Science* 45, 329.

Hille, J., Verheggen, F., Roelvink, P., Franssen, H., Kammen, A., Zabel, P., 1986. Bleomycin resistance: A new dominant selectable marker for plant cell transformation. *Plant Molecular Biology* 7, 171–176.

Hoekema, A., Hirsch, P.R., Hooykaas, P.J.J., Schilperoort, R.A., 1983. A binary plant vector strategy based on separation of vir- and T-region of the *Agrobacterium tumefaciens* Ti-plasmid. *Nature* 303, 179–180.

Hood, E., Gelvin, S., Melchers, L., Hoekema, A., 1993. New *Agrobacterium* helper plasmids for gene transfer to plants. *Transgenic Research* 2, 208–218.

Huertas, R., Rubio, L., Cagnac, O., García-Sánchez, M.J., Alché, J.D.D., Venema, K., FernáNdez, J.A., RodríGuez-Rosales, M.P., 2013. The K+/H+ antiporter LeNHX2 increases salt tolerance by improving K+ homeostasis in transgenic tomato. *Plant, Cell and Environment* 36, 2135–2149.

James, C., 2012. *Global Status of Commercialized Biotech/GM Crops*. ISAAA, Ithaca, NY.

Jha, B., Mishra, A., Jha, A., Joshi, M., 2013. Developing transgenic *Jatropha* using the SbNHX1 gene from an extreme halophyte for cultivation in saline wasteland. *PLoS ONE* 8, e71136.

Koch, A., Kumar, N., Weber, L., Keller, H., Imani, J., Kogel, K.-H., 2013. Host-induced gene silencing of cytochrome P450 lanosterol C14-demethylase-encoding genes confers strong resistance to *Fusarium* species. *Proceedings of the National Academy of Sciences* 110, 19324–19329.

Komari, T., Hiei, Y., Saito, Y., Murai, N., Kumashiro, T., 1996. Vectors carrying two separate T-DNAs for co-transformation of higher plants mediated by *Agrobacterium tumefaciens* and segregation of transformants free from selection markers. *Plant Journal* 10 (1), 165–174.

Koncz, C., Schell, J., 1986. The promoter of TL-DNA gene 5 controls the tissue-specific expression of chimaeric genes carried by a novel type of *Agrobacterium* binary vector. *Molecular and General Genetics* 204, 383–396.

Lazo, G.R., Stein, P.A., Ludwig, R.A., 1991. A DNA transformation-competent *Arabidopsis* genomic library in *Agrobacterium*. *Biotechnology (NY)* 9, 963–967.

Liu, C.-W., Lin, C.-C., Yiu, J.-C., Chen, J.W., Tseng, M.-J., 2008. Expression of a *Bacillus thuringiensis* toxin (cry1Ab) gene in cabbage (*Brassica oleracea* L. var. capitata L.) chloroplasts confers high insecticidal efficacy against *Plutella xylostella*. *Theoretical and Applied Genetics* 117, 829–829.

Liu, Y.G., Whittier, R.F., 1995. Thermal asymmetric interlaced PCR: Automatable amplification and sequencing of insert end fragments from P1 and YAC clones for chromosome walking. *Genomics* 25, 674–681.

Maliga, P., 2002. Engineering the plastid genome of higher plants. *Current Opinion in Plant Biology* 5, 164–172.

Maliga, P., Bock, R., 2011. Plastid biotechnology: Food, fuel, and medicine for the 21st century. *Plant Physiology* 155, 1501–1510.

Maliga, P., Svab, Z., Harper, E.C., Jones, J.D., 1988. Improved expression of streptomycin resistance in plants due to a deletion in the streptomycin phosphotransferase coding sequence. *Molecular and General Genetics* 214, 456–459.

Mao, Y.-B., Tao, X.-Y., Xue, X.-Y., Wang, L.-J., Chen, X.-Y., 2011. Cotton plants expressing CYP6AE14 double-stranded RNA show enhanced resistance to bollworms. *Transgenic Research* 20, 665–673.

McElroy, D., Zhang, W., Cao, J., Wu, R., 1990. Isolation of an efficient actin promoter for use in rice transformation. *The Plant Cell* 2, 163–171.

Mei, C., Qi, M., Sheng, G., Yang, Y., 2006. Inducible overexpression of a rice allene oxide synthase gene increases the endogenous jasmonic acid level, PR gene expression, and host resistance to fungal infection. *Molecular Plant–Microbe Interactions* 19, 1127–1137.

Miller, M., Tagliani, L., Wang, N., Berka, B., Bidney, D., Zhao, Z.-Y., 2002. High efficiency transgene segregation in co-transformed maize plants using an *Agrobacterium tumefaciens* 2 T-DNA binary system. *Transgenic Research* 11, 381–396.

Muhammad, S.K., 2012. Plastid genome engineering in plants: Present status and future trends. *Molecular Plant Breeding* 3, 91–102.

Mullis, K.B., Erlich, H.A., Arnheim, N., Horn, G.T., Saiki, R.K., Scharf, S.J., 1987. Process for amplifying, detecting, and/or cloning nucleic acid sequences. Google Patents.

Nguyen, T.X., Sticklen, M., 2013. Barley HVA1 gene confers drought and salt tolerance in transgenic maize (*Zea mays* L.). *Advances in Crop Science and Technology* 1, 1–8.

Nowara, D., Gay, A., Lacomme, C., Shaw, J., Ridout, C., Douchkov, D., Hensel, G., Kumlehn, J., Schweizer, P., 2010. HIGS: Host-induced gene silencing in the obligate biotrophic fungal pathogen *Blumeria graminis*. *The Plant Cell* 22, 3130–3141.

Oey, M., Lohse, M., Kreikemeyer, B., Bock, R., 2009a. Exhaustion of the chloroplast protein synthesis capacity by massive expression of a highly stable protein antibiotic. *Plant Journal* 57, 436–445.

Ooms, G., Hooykaas, P.J.J., Moolenaar, G., Schilperoort, R.A., 1981. Crown gall plant tumors of abnormal morphology, induced by *Agrobacterium tumefaciens* carrying mutated octopine Ti plasmids; analysis of T-DNA functions. *Gene* 14, 33–50.

Ramaseshadri, P., Segers, G., Flannagan, R., Wiggins, E., Clinton, W., Ilagan, O., McNulty, B., Clark, T., Bolognesi, R., 2013. Physiological and cellular responses caused by RNAi-mediated suppression of Snf7 orthologue in western corn rootworm (*Diabrotica virgifera*) larvae. *PLoS ONE* 8, e54270.

Sahoo, K.K., Tripathi, A.K., Pareek, A., Sopory, S.K., Singla-Pareek, S.L., 2011a. An improved protocol for efficient transformation and regeneration of diverse indica rice cultivars. *Plant Methods* 7, 49.

Sanford, J.C., Klein, T.M., Wolf, E.D., Allen, N., 1987. Delivery of substances into cells and tissues using a particle bombardment process. *Particulate Science and Technology* 5, 27–37.

Shimizu, M., Goto, M., Hanai, M., Shimizu, T., Izawa, N., Kanamoto, H., Tomizawa, K.-I., Yokota, A., Kobayashi, H., 2008. Selectable tolerance to herbicides by mutated acetolactate synthase genes integrated into the chloroplast genome of tobacco. *Plant Physiology* 147, 1976–1983.

Southern, E.M., 1975. Detection of specific sequences among DNA fragments separated by gel electrophoresis. *Journal of Molecular Biology* 98, 503–517.

Sreekala, C., Wu, L., Gu, K., Wang, D., Tian, D., Yin, Z., 2005. Excision of a selectable marker in transgenic rice (*Oryza sativa* L.) using a chemically regulated Cre/loxP system. *Plant Cell Reports* 24, 86–94.

Sripriya, R., Raghupathy, V., Veluthambi, K., 2008. Generation of selectable marker-free sheath blight resistant transgenic rice plants by efficient co-transformation of a co-integrate vector T-DNA and a binary vector T-DNA in one *Agrobacterium tumefaciens* strain. *Plant Cell Reports* 27, 1635–1644.

Sripriya, R., Sangeetha, M., Parameswari, C., Veluthambi, B., Veluthambi, K., 2011. Improved *Agrobacterium*-mediated co-transformation and selectable marker elimination in transgenic rice by using a high copy number pBin19-derived binary vector. *Plant Science* 180, 766–774.

Sugita, K., Matsunaga, E., Kasahara, T., Ebinuma, H., 2000. Transgene stacking in plants in the absence of sexual crossing. *Molecular Breeding* 6, 529–536.

Svab, Z., Hajdukiewicz, P., Maliga, P., 1990. Stable transformation of plastids in higher plants. *Proceedings of the National Academy of Sciences* 87, 8526–8530.

Svab, Z., Maliga, P., 2007. Exceptional transmission of plastids and mitochondria from the transplastomic pollen parent and its impact on transgene containment. *Proceedings of the National Academy of Sciences* 104, 7003–7008.

Tuteja, N., Verma, S., Sahoo, R.K., Raveendar, S., Reddy, I.B.L., 2012. Recent advances in development of marker-free transgenic plants: Regulation and biosafety concern. *Journal of Biosciences* 37, 167–197.

Vasil, V., Castillo, A.M., Fromm, M.E., Vasil, I.K., 1992. Herbicide resistant fertile transgenic wheat plants obtained by microprojectile bombardment of regenerable embryogenic callus. *Nature Biotechnology* 10, 667–674.

Waldron, C., Murphy, E.B., Roberts, J.L., Gustafson, G.D., Armour, S.L., Malcolm, S.K., 1985. Resistance to hygromycin B: A new marker for plant transformation studies. *Plant Molecular Biology* 5, 103–108.

Xiong, Y., Zeng, H., Zhang, Y., Xu, D., Qiu, D., 2013. Silencing the HaHR3 gene by transgenic plant-mediated RNAi to disrupt *Helicoverpa armigera* development. *International Journal of Biological Sciences* 9, 370–381.

Yu, L., Chen, X., Wang, Z., Wang, S., Wang, Y., Zhu, Q., Li, S., Xiang, C., 2013. *Arabidopsis* enhanced drought Tolerance1/HOMEODOMAIN GLABROUS11 confers drought tolerance in transgenic rice without yield penalty. *Plant Physiology* 162, 1378–1391.

Zambryski, P., Joos, H., Genetello, C., Leemans, J., Montagu, M.V., Schell, J., 1983. Ti plasmid vector for the introduction of DNA into plant cells without alteration of their normal regeneration capacity. *EMBO Journal* 2, 2143–2150.

Zha, W., Peng, X., Chen, R., Du, B., Zhu, L., He, G., 2011. Knockdown of midgut genes by dsRNA-transgenic plant-mediated RNA interference in the Hemipteran insect *Nilaparvata lugens*. *PLoS ONE* 6, e20504.

Zhu, X., Richael, C., Chamberlain, P., Busse, J.S., Bussan, A.J., Jiang, J., Bethke, P.C., 2014. Vacuolar invertase gene silencing in potato (*Solanum tuberosum* L.) improves processing quality by decreasing the frequency of sugar-end defects. *PLoS ONE* 9, e93381.

Zoschke, R., Liere, K., Börner, T., 2007. From seedling to mature plant: *Arabidopsis* plastidial genome copy number, RNA accumulation and transcription are differentially regulated during leaf development: Plastome copy number in *Arabidopsis* leaf development. *The Plant Journal* 50, 710–722.

Web resources

https://www.isaaa.org/
http://cls.casa.colostate.edu/
http://www.gmo-free-regions.org/
http://www.pbslearningmedia.org/resource/tdc02.sci.life,gen.lp_bioengfood/bioengineered-foods/

Further reading

Banerji, D. 2013. *Plant Genetic Engineering—Elements*. Studium Press (India), New Delhi.

Chittaranjan, K., Timothy, C.H. (Eds.) 2008. *Compendium of Transgenic Crop Plants, 10 Volume Set*. Wiley-Blackwell, London, United Kingdom.

Cummings, C.H. 2008. *Uncertain Peril: Genetic Engineering and the Future of Seeds*. Beacon Press, Boston, MA.

Keith, T.A. 2002. *Genetically Modified Crops: Assessing Safety*. Taylor & Francis, New York.

Kole, C., Michler, C., Abbott, A.G., Hall, T.C. (Eds.) 2010. *Transgenic Crop Plants Volume 1: Principles and Development*. Springer, New York.

Kole, C., Michler, C., Abbott, A.G., Hall, T.C. (Eds.) 2010. *Transgenic Crop Plants, Volume 2: Utilization and Biosafety.* Springer, New York.

Liang, G.H., Daniel, Z. 2004. *Genetically Modified Crops: Their Development, Uses, and Risks.* Skinner Taylor & Francis, Oxford, UK.

Nigel, G.H. 2011. *Genetically Modified Crops.* Imperial College Press, London, UK.

Robert, P. 2010. *Food Politics: What Everyone Needs to Know.* Oxford University Press, New York.

chapter three

Plant molecular biotechnology
Applications of transgenics

**Muhammad Sarwar Khan, Ghulam Mustafa,
Shahid Nazir, and Faiz Ahmad Joyia**

Contents

Abstract

Molecular biotechnology entails deoxyribonucleic acid manipulations of living organisms, including microscopic creatures, plants, and animals by employing various customary and emerging approaches such as omics, molecular biology, molecular genetics, molecular pathology, microbiology, nanotechnology, forensic botany, molecular virology, and their biotechnological applications in the area of agriculture, health, and industry. Of these, genetic manipulation of plants is generally accomplished through a toolbox that allows engineering genes and by relocating the engineered genes into other organisms to develop processes and/or produce useful products. Hence, gene/s can be transferred from one organism to another through bypassing sexual means of reproduction across kingdoms; contributing significantly toward improvement of agronomic, horticultural, and forest plants. But, when talking about developing transgenic plants, containment of transgenes is one of the main concerns of the consumers; hence, chloroplast transformation comes in place since chloroplasts are predominantly maternally inherited to progenies. This chapter summarizes applications of plant molecular biotechnology in the multidisciplinary perspective.

Keywords: Transgenic technology, Agronomic traits, Biofortification, Metabolic engineering, Biopharming, Third generation biotechnology.

Plant biotechnology generation after generation

Agricultural biotechnology is progressing day by day and there are three generations of this technology in the fields, with numerous genetically modified plant products and by-products already an invasive part of the marketplace (James, 2014). Nevertheless, a sustainable regulatory framework is far from being implemented in developing countries such as Pakistan where biosafety regulations and guidelines were approved by the federal government back in 2005 but the devolution of ministries after the 18th amendment in the law has confused the regulatory bodies again who have to run the business, without realizing how much this industry and resultantly the economy of the country has to face.

The outcome of the first-generation plant biotechnology remained very successful where transgenic plants harboring traits of insect–pest and herbicide resistance were developed. Many Asian countries including China, India, Indonesia, Malaysia, Philippines, Thailand, Pakistan, and Vietnam are giving high priority to plant biotechnology research, hoping to improve productivity, farmers' livelihoods, and meeting food security demands. Many of these countries focus either on food or high commercial value crops to reduce the use of pesticides and herbicides (Penn, 2000). People of developing countries including India and Pakistan have benefitted directly from the products of biotechnology. A success story of agricultural biotechnology is the cultivation of Bt cotton, without going into details of how the transgenic cotton was developed, prior to the approval of the biosafety rules and guidelines in Pakistan (Tabassum et al., 2012). Within a few years, Bt cotton replaced almost 80% of the area earlier occupied by the cotton developed using conventional breeding tools. Nowadays, Bt corn, sugarcane, and other crops are being investigated for commercial release.

From a technical perspective, Bt cotton was developed by incorporating *cry1Ac* gene from the bacterium *Bacillus thuringiensis* (Bt). Initially, the two genes encoding cry proteins were identified and isolated by Monsanto in the 1980s. Of these two, the *cry1Ac* was successfully introduced into the cotton genome under CaMV promoter and a nopaline synthase (nos) terminator, the regulatory sequences for regulating the expression of the transgene. The expression cassette was introduced into the genome using a gall-forming bacterium (an *Agrobacterium*), which was then allowed to naturally invade cotton plant tissue in a tissue culture environment, inserting the deoxyribonucleic acid (DNA) containing the *cry1Ac* gene into the cotton tissue (Perlak et al., 1990). The recipient cells were then subjected to various rounds of selection and regeneration to recover a homozygous shoot. The tested shoots were multiplied and toxin levels were observed and bioassays were performed to assess the feeding larvae mortality levels. The selected plants were subjected to conventional breeding means to develop crosses with other varieties to produce a range of cry protein-carrying varieties, which were marketed with the trademark "Bollgard®." Contrary to India, where the Government of India accorded approval for the release of

Bt cotton though few states in 2002, Pakistan did not commercially adopt Bt cotton by 2010. However, the cultivation of first-generation (*cry1Ac*) Bt cotton, unapproved and unregulated, increased rapidly after 2005.

As far as the second generation of agricultural biotechnology is concerned, better taste, more nutritional contents, and longer life were the main items on the wish list of plant biotechnologists (Jefferson-Moore and Traxler, 2005). Calgene's Flavr Savr tomatoes, potato with higher solids (Kramer and Redenbaugh, 1994) are the success stories with merits and demerits. Flavr Savr tomatoes were developed in 1994 through antisense approach. Using the strategy, the engineered tomatoes were no longer able to produce polygalacturonase, an enzyme involved in fruit softening. The research was initiated with the aim to enable tomatoes to remain on the vine to ripen for a long shelf life and to develop their full flavor. The Flavr Savr tomatoes were not carrying an alien gene; rather, a gene normally present in the tomato was blocked so that a normal protein involved in ripening was not produced giving the tomato a longer shelf life. Other crops developed through genetic manipulations were Monsanto's Roundup Ready soybeans, Syngenta's StayRipe banana, and potatoes. The Roundup Ready soybean was developed by incorporating 5-enolpyruvylshikimate-3-phosphate synthase (EPSPS) from the CP4 strain of the bacteria, *Agrobacterium tumefaciens* under the control of CaMV promoter, a chloroplast transit peptide from *Petunia hybrida*, and an nos as a transcriptional termination element from *Agrobacterium*. Later on, the *cry1Ac* was stacked into the same Roundup Ready soybean lines (Stefanova et al., 2015), but liked by the consumers.

Development of transgenic plants for cost-effective and plentiful availability of industrial products, for example, enzymes, biosensors, plastics, cosmetics, and pharmaceutical products such as vaccines, antibodies, and therapeutic proteins are the main promises of the third-generation biotechnology. Success stories and their development will be discussed in the section "Plant transformation for cost-effective therapeutics and vaccines."

Role of plant tissue culture in plant biotechnology

Plant tissue culture comprises a set of *in vitro* procedures that are of pivotal importance in plant biotechnology. A competent, consistent, and reproducible *in vitro* regeneration protocol allowing the regeneration of a complete plant from a single cell is indispensable because there are two decisive issues to be mastered for a successful plant transformation event: (i) foreign DNA should be efficiently transferred into the plant cells; (ii) the transformed cells should be able to regenerate into a complete fertile plant. In most cases, recovery of transgenic plants becomes complicated because cells acquiescent for gene transfer may not be appropriate for regeneration *in vitro* (Yookongkaew et al., 2007). Hence, the development of a simple, efficient, and reproducible regeneration system is a prerequisite for any transformation approach including nuclear as well as plastid transformation.

To recover a transgenic plant from a single cell, it has to undergo successive dedifferentiation and redifferentiation phases. During the process of dedifferentiation, plant cells pass through repeated cell divisions producing a mass of undifferentiated cells known as callus. The process is termed as callogenesis. These undifferentiated cells are then cultured onto various combinations of plant growth regulators (PGRs) especially the auxins and cytokinins. Such variations in PGRs induce the process of redifferentiation causing regeneration of new plants from calli. Ideally, each and every cell of callus should be able to produce at least one complete plant. Hence, screening of transgenic plants essentially depends on the proficiency to transform the regenerable cells, which will be subsequently regenerated to recover an entire plant. In most cases, transformation efficiency is contingent to the ability of the plant cells to form complete plants through the *in vitro* regeneration process. So, the transformation efficiency would be tremendously enhanced if regeneration processes is efficient (Subroto et al., 2014).

Furthermore, *in vitro* techniques can be employed to induce genetic variation, which is inevitable, to improve vigor status of planted material and a desirable germplasm available for breeding programs. Although, *in vitro* regeneration protocols have been devised for most of the crop plants, yet continued optimization is still vital for various plants including monocots, woody, and recalcitrant species. Tissue culture techniques for various plant parts including protoplasts, anthers, microspores, ovules, and embryos have been accomplished in order to induce genetic variants for breeding purposes. Cell culture techniques have also been employed to induce somaclonal and gametoclonal variations for crop improvement. Moreover, cell and meristem culture can be successfully used for pathogen eradication from planting materials and thereby intensely increase the yield of established cultivars. Various commercial laboratories are producing millions of plants as a commercial commodity to be sold in the market. From the literature review, it is obvious that the importance of tissue culture can be expected to have an ever-increasing impact on crop improvement (Brown and Thorpe, 1995; Harriman et al., 2006).

Genetic transformation being a less-efficient process requires the application of tissue culture procedures for the screening and purification of transgenic clones (Birch, 1997). Once a transgenic event is successful, it becomes necessary to propagate the transgenic clones. Micropropagation is the most promising *in vitro* technique by virtue of its proficiency, reproducibility, and minimization of somaclonal variations in this regard (Zeng et al., 2010) especially for vegetatively propagated plants. An efficient *in vitro* methodology for purification of transgenic sugarcane clones and their mass multiplication has been devised (Mustafa and Khan, 2012a). The study comprised sugarcane regeneration from *in vitro*-grown plants to purify and multiply transgenic clones without subjecting them to field conditions and harvesting the mature cane. This may be considered as one of the best ever published reports as far as number of clones produced per explant is concerned. The technique was employed to purify transgenic sugarcane plants carrying Bt gene. Another approach with potential utility of clonal purification for transgenic rice was devised by Joyia and Khan (2012).

Manipulation of genes and regulatory sequences

Selectable marker genes and/or easily scorable reporter genes being used in plant transformation events are meant to provide a selective benefit only to the transformed cell, permitting their quick and better development by killing the nontransformed ones. Hence, these selectable marker genes are introduced into the genome along with the gene(s) of interest. The selective agents are employed at the preliminary stages of transformation for better and early selection of transgenic cells (Sawahel, 1994). The worth of a particular resistance marker depends on its mode of action and the way it is taken up or transported in plant tissues (Wilmink and Dons, 1993). During regeneration steps, the nontransformed cells die or are bleached out on the selective medium.

Among the selectable marker genes applied to plant transformation, the most popular markers are bacterial genes that confer resistance to various antibiotics including kanamycin, spectinomycin, streptomycin, and hygromycin or to the herbicide-like glufosinate. As various concerns are being raised by regulatory authorities and consumers, scientists have been encouraged to discover innovative selection strategies. Though such concerns proved to be theoretical (Flavell et al., 1992), yet the development of environmentally friendly selectable markers especially of plant origin is attaining consideration (Mustafa and Khan, 2012b).

Gene expression (selectable marker gene or gene of interest) is controlled by essential regulatory sequences. The two types of discrete *cis*-regulatory sequences, that is, promoter and enhancers have been studied. The promoters are located close to the transcription initiation site and perform in a position-dependent manner whereas the enhancers can be located far from the promoter region and regulate the gene expression in a position- and orientation-independent manner. These *cis*-acting sequences regulate the gene expression predominately by altering the rate of transcription, ribonucleic acid (RNA) processing, stability of RNA, and efficiency of ribosome to bind and translate the transcript (Wittkopp and Kalay, 2012). Hence, the appropriate *cis*-acting sequences should be cloned as (5′-UTRs) regions of plant transformation vectors (Eibl et al., 1999). The level of foreign protein accumulation is directed by the 5′-UTR ligated upstream of the open reading frame (ORF) encoding gene of interest. The most commonly used 5′-UTR and 3′-UTR are 35SP/35ST for nuclear transformation and PpsbA/TpsbA for chloroplast transformation (Zoubenko et al., 1994; Millan et al., 2003; Watson et al., 2004; Daniell et al., 2005; Kittiwongwattana et al., 2007).

On the contrary, *trans*-regulatory elements are located far from the genes they are regulating. More precisely, such regulatory elements are DNA sequences that encode transcription factors involved in the regulation of distant genes (Jones et al., 1988). Such gene-encoding various transcription factors are being used to develop the transgenic plants tolerant to a variety of abiotic stresses.

Plant transformation techniques

A range of transformation techniques have been employed for deliberate transgene delivery into plant's nuclear as well as plastid genomes. *Agrobacterium*-mediated protocols are available specifically for nuclear transformation; however, polyethylene glycol (PEG)-mediated, microinjection and the biolistic process using gene gun can be employed for both the plastid and nuclear transformation with a varying degree of success.

A. tumefaciens is a soil bacterium responsible for crown gall disease in dicots (Bevan and Chilton, 1982). The protocols of *Agrobacterium*-mediated plants transformation employing *Agrobacterium* in its original form, proved to be inefficient because of difficulty in manipulating Ti plasmid, limited host range, and necessity of wounded target tissue (Finer, 2010). Thanks to more than 30 years of intense research that the introduction of binary vector systems has simplified the procedure (Job, 2002). The host range limitations have also been resolved largely.

DNA delivery system that is being widely employed for nuclear as well as plastid transformation is the bombardment of DNA-coated gold or tungsten particle using gene gun. It is species or genotype independent. Moreover, it has no biological constraints or host limitations (Wang et al., 2009a). Another reason for its comprehensive appliance is its ability to promote cotransformation of the same cell with more than one transgene simultaneously (Altpeter et al., 2005). This transformation procedure is executed using a device known as the PDS-1000/He, a product of BioRad United States. Recently, its alternatives are being marketed by some Chinese manufacturers. It employs high pressure (upto 3000 psi) of helium gas to accelerate the DNA-coated gold or tungsten microparticles also known as microcarriers. Gold particles are preferred due to the chemical inertness of the pure gold and smooth texture; hence it will not cause toxicity to the bombarded cells (Huang and Chen, 2011). Rupture disks are employed to build He pressure according to disk capacity. When the disk ruptures, high pressure is released hitting the second disk (macrocarrier) on which the gold or tungsten particles coated with the DNA are spotted, pushing it to the stopping screen, which stops the macrocarrier; however, metal particles pass through it finally hitting the target tissues under vacuum. This procedure facilitates the delivery of the DNA coated on microcarriers into the plant cells. The delivered DNA ultimately reaches and integrates into the nuclear, mitochondrial, or chloroplast genome. Gene gun is the method of choice for the plastid transformation in plants and mitochondrial transformation in plants as well as animals (Maliga and Small, 2007).

The plastids are double-membrane-bounded cell organelles whose membranes exhibit restricted permeability for macromolecules including nucleic acids. Kohler et al. (1997) described plastid fusions and subsequent exchange of DNA; however, no mechanism has been reported governing the movement of macromolecules such as nucleic acids. The presence of plastid DNA sequences in nucleus and mitochondria is the evidence of the transmission of genetic material between the three (genetic) compartments of higher plant cells as a significant evolutionary process (Schuster and Brennicke, 1988, 1994; Joyce and Gray, 1989). Hence, the occurrence of a natural mechanism for passage of nucleic acids through the lipid bilayers enclosing the plastid organelle is evident.

Untill 1993, the double membrane of chloroplasts posed to be an impenetrable barrier obstructing the endeavors of plastome transformation when Golds et al. for the first time reported the successful plastid transformation using PEG. The underlying mechanism for DNA passage from cytoplasm into the chloroplast stroma remained unclear. In this technique, protoplasts are treated with a solution containing various ions, PEG, and DNA causing variations in the plasma membrane permeability, hence permitting the DNA molecules to pass through the double membrane. Transgene integration into the plastome is precise and predetermined by virtue of the homologous sequences in transformation vectors with 100% homology to a specific target area in wild-type (untransformed) plastome. During cell division, random distribution of plastids among progeny cells creates a heteroplastomic cell population until, after several rounds of selection, absolute segregation creates homoplastomic cells having only transformed plastome copies (Khan and Maliga, 1999). This method of plastid transformation is cost-effective; however, it essentially entails expertise in protoplast isolation, culture, and above all, an efficient and reproducible *in vitro* regeneration system making the species regenerable from protoplasts is a prerequisite.

Applications of biotechnology

The current world population is expected to rise up to 10 billion by the year 2050; however, agriculture production is not increasing at a similar pace, resulting in increased malnutrition and starvation. Green revolution resulted in approximately 1000% increase in food production and has least chances of further increment. The father of green revolution "Norman E. Borlaug" was also of the view that only biotechnology has the potential to meet the increasing demand of agricultural production by ameliorating environmental concerns (Borlaug, 2005). Since the advent of transgenic technology, it has successfully been employed to address biotic, abiotic stresses, and to enhance nutritional value (Baisakh et al., 2006) of cereal crops. Owing to the diversified impact of genetically modified (GM) crops, they were grown on an area of 181.5 million hectares. A global meta-analysis of 147 studies in the past 20 years has revealed that the adoption of GM technology has not only increased crop yield by 22% but has also reduced the use of chemical pesticide by 37% resulting in increased farmer's profit to a great extent.

Plant transformation for biotic and abiotic stress tolerance

Biotic and abiotic stresses are serious devastating agents affecting crop yield to a great extent. Transgenic technology offers enormous possibilities to address these stresses and

Table 3.1 Transgenic crops with improved agronomic traits

Sr #	Crop	Transformation event	Gene	Trait	References
1	Soybean	GTS-40-3-2	EPSPS	Herbicide resistance	Harrigan et al. (2007)
2	Maize	Bt 176	*cry1Ab*, bar	Insect resistance, herbicide resistance	Dutton et al. (2003)
3	Maize	MON810	*cry1Ab*	Insect resistance	Nguyen and Jehle (2007)
4	Canola	GT 73	cp4 EPSPS, gox	Herbicide resistance	White (1989)
5	Cotton	MON 1445/1698	cp4 EPSPS	Herbicide resistance	James (2006)
6	Cotton	MON 531	*cry1Ac*	Insect resistance	Perlak et al. (1990)
7	Sugar beet	GTSB77	cp4EPSPS, gox	Herbicide resistance	Mannerlof et al. (1997)
8	Papaya	55-1/63-1	Coat protein PRSV	Virus resistance	Gonsalves (1998)
9	Tomato	CGN89564-2	Amino glycoside 3' phosphotransferase 2	Flavr Savr	Kramer and Redenbaugh (1994)

virtually there is no place on earth where the term "transgenic plant" is not known. The first genetically engineered crops were soybean and maize, commercialized in the United States in 1996. Since then, transgenic crops have undergone great adaptation worldwide and now contribute to more than 40% of the total area under cultivation (Table 3.1). Although nuclear transformation has contributed a lot, still there are certain limitations that can be addressed through transplastomic technology. The resistance-conscious valuable agronomic traits are insect resistance and herbicide resistance, which strictly demand high accumulation levels of resistance-conferring proteins (Bock and Khan, 2004: Daniell et al., 2002). Owing to the ability of chloroplasts to accommodate high foreign protein content, it is among the fore-most-investigated traits for plastome engineering as well (Khan, 2012).

Insect resistance

Insect pests are serious disastrous agents limiting crop yield to a great extent. Plants have inbuilt defense mechanisms in the form of nonprotein antimetabolites, that is, alkaloids and protein antimetabolites, that is, α-amylase inhibitors and proteinase inhibitors. Breeders have attempted to introduce certain novel resistance genes into the desired plants but face certain limitations as they cannot be introduced across the species. Genetic engineering has made it possible to introduce a transgene from bacteria, yeast, fungi, or a plant. Bt-derived cry genes-encoding toxins provide a rich source of insecticidal proteins to control insect pests and had been used as a biological pesticide for more than 50 years (Qaim and Zilberman, 2003). They had also been used for plant transformation to engineer resistance against insect pests (Vaeck et al., 1987). More than 400 genes-encoding endotoxins have been identified from a wide range of Bt (Crickmore et al., 2007). Many of these, that is, *cry1Ab, cry1Ac, cry1Aa, cry1Ba, cry2Aa, cry3A, cry1Ca, cry1H, cry6A, cry9C,* and *cry1F,* have been employed in transgenic technology to develop transgenic plants. Each of the cry protein produces a particular endotoxin having distinctive insecticidal spectrum. Some of the crystal toxins are specific to and affect larvae of lepidopteran insects, whereas others are toxic to coleopteran or dipteran insect pests. Initially, transgene expression level wild-type Bt toxins were lower in transformed plants compared with other heterologous genes. Modifications in the coding sequence of the cry genes for appropriate codon usage of the target plant species resulted in increase in expression level to a great extent (Van der Salm et al., 1994), sufficient enough to cause mortality of target pests in the field. The first result of Bt-engineered transgenic plants was published in 1987 (Vaeck et al., 1987). Since then, several Bt-engineered transgenic crops including corn, cotton, tomato, canola, potato, chickpea, and eggplant have been developed and commercialized in the United States, Argentina, Canada, India, and Australia (Ferry et al., 2006). They have received more acceptance owing to their biosafety for human beings because cry endotoxins are active only in alkaline conditions (insect gut) but not under acidic conditions (human gut). To-date Monsanto has been the pioneer organization as far as the development and commercialization of Bt-engineered crops is concerned. They have patented various single-gene transformation events in maize (MON 810 and 863) as well as in cotton (MON 531). Several other transformation events having stacked two or more genes for insect resistance have also been developed and are in the process of approval.

One of the serious concerns about Bt-engineered crops is the suboptimal expression of toxins which may lead to increased risks of resistance development against these cry proteins (Daniell, 2000). Owing to the prokaryotic origin of cry gene, its nuclear expression is prone to certain complications regarding codon usage, affecting transgene expression. The cry genes are AT rich as compared with plant nuclear genes; therefore, they could not be expressed efficiently in the eukaryotic system of the nucleus resulting in decreased transgene expression owing to premature transcription termination, instability of messenger ribonucleic acid (mRNA), aberrant mRNA splicing, and inept codon usage. Native Bt mRNA also appeared to be degraded under the nuclear expression system (De Rocher et al., 1998). Certain synthetic Bt genes with an increased guanine–cytosine (GC) content have been reported to be expressed in the nucleus from an undetectable 0.2% to 0.3% of the total soluble cellular protein. This has resulted in the failure of Bt cotton in controlling *Heliothines armigera* in Australia and certain other bollworms in Texas, where it was cultivated on an area of ~20,000 acres (Hilder and Boulter, 1999). Thus, transgenic crops with a suboptimal expression of the transgene are not fully protected from the insect–pest attack and require repeated doses of pesticides to be applied to minimize yield losses. Resistance development against Bt toxin may be avoided or delayed by (i) sowing Bt-engineered crops along with regular non-Bt crops (refugia crop). This would help to avoid development of homogenous-resistant insects as Bt toxins are controlled by mutant, recessive alleles; hence, the resistant insects have more chances of mating with the normal susceptible insects feeding on refugia crop (Bates et al., 2005). The resultant progenies of the insects will not be resistant to Bt toxins. Planting refuges is obligatory but farmers sometimes neglect this regulation and so (ii) another option is to avoid or slow down resistance development by stacking insecticidal genes with different modes of action. Multiple genes will help to avoid or delay resistance development in insects due to broad-spectrum modes of action (Keresa, 2002) but (iii) the most appropriate solution to this is to express transgenes in plastids. Chloroplasts, owing to have prokaryotic transcriptional, translational machinery address this problem requiring no sequence manipulation for codon usage and giving out a higher expression of foreign proteins. Several Bt proteins (*cry2Aa2, cry9Aa2, cry1Ia5,* and *cry1Ac*) have been expressed in the plastid genome resulting in hyperexpression of the transgene. The first example representing the enormous potential of the transplastomic technology came from the expression of the *cry1A(c)* insecticidal protein (McBride et al., 1995).

Native Bt-coding regions were expressed up to 3%–5% of the total soluble protein (20- to 30-fold higher than current commercial nuclear transgenic plants) with stable mRNA that resulted in 100% mortality when bioassayed with tobacco budworm (*Heliothis virescens*), cotton bollworm (*Helicoverpa zea*), and the beet armyworm (*Spodoptera exigua*) (Kota et al., 1999; Maliga and Bock, 2011). In addition, most of the insect larvae feed on green plant tissues that are rich in chloroplast; thus, attacking (invading) insect pests will be exposed to the highest dose of insecticidal toxins expressed in the chloroplasts. High-level accumulation of insecticidal toxin in transplastomic plants was achieved (10%–20%) in tobacco (Chakrabarti et al., 2006) and cabbage (Liu et al., 2007), and this could be further enhanced up to 45% of the total soluble protein when expressed in the form of an operon encoding for chaperonin protein in addition to *cry2Aa2* (De Cosa et al., 2001). These chaperonin proteins not only assist the correct folding of the insecticidal toxin but also promote protein crystallization inside plastids. Thus, gene pyramiding (Bt with non-Bt insect proteins) using transplatomic technology offers a valuable tool for resistance management by overcoming it or at least, significantly delaying the broad-spectrum Bt-resistance development in the field crops (Kota et al., 1999). Although most of these studies were conducted in the model plant (*Nicotiana tobaccum*), the recent developments in insect-resistant transplastomic brassica, soybean, and eggplant offer potentiality of the technology to be transferred to valuable food crops.

In addition to cry proteins, other insecticidal resistance genes have also been exploited to engineer crop plants for insect resistance. These include vip genes from Bt, cholesterol oxidase from *Streptomyces*, protease inhibitors from plants, animals, and microbes, neuropeptides are neurotransmitters (proctolin) from insects, avidin from chicken egg, proteinase inhibitors, α-amylase inhibitors, and lectins from higher plants. Hence, multiple genes with a diversified mode of action can prove to be of great worth for the broad-spectrum control of plant insects.

Disease resistance

In addition to insects, plant pathogenic bacteria, fungi, viruses, and even nematodes cause serious diseases of plants throughout the world. These phytopathogens pose drastic effects on agricultural and native plants resulting in reduced productivity, nutritional value, and overall quality of the produced biomass. Transgenic technology has explored various pathogenesis-related (PR) genes to combat viral, bacterial, and fungal diseases. Coat protein-mediated resistance has been induced in plants to avoid viral attack. Similarly, transgenic plants expressing a defective-movement protein of tobacco mosaic virus or protease of soybean mosaic virus has been found to be resistant to viral infection. Systemic acquired resistance (SAR) is an effective strategy adopted by the plants to avoid phytopathogens, which is accompanied by the induced expression of PR genes including chitinases (ChiCs), glucanases, chitosanases, and by the elevated accumulation of salicylic acid (SA). The SA was overexpressed in tobacco by two bacterial genes involved in the conversion of chorismate into SA resulting in elevated tolerance level of the plants to pathogens. Since fungal pathogens are one of the serious yield limiting agents and may cause 70% drop-off in the crop yield. Overexpression of antifungal proteins have proved to be of great worth to control mycoparasites. Arrieta et al. (1996) reported coexpression of class I β-1,3-glucanase and thaumatin-like proteins and determined a low degree of fungal infection. Overexpression of snakin-1 gene resulted in enhanced resistance to *Rhizoctonia solani* and *Erwinia carotovora* (Natalia et al., 2008). ChiC isolated from *Streptomyces griseus* demonstrated enhanced resistance against *Alternaria solani* (Khan et al., 2008) whereas expression of Trichoderma-derived ChiC and glucanase enzymes demonstrated improved resistance to *R. solani* (Esfahani et al., 2010). Expression of *Nicotiana tabacum* AP24 osmotine, *Phyllomedusa sauvagii* dermaseptin and *Gallus gallus* lysozyme resulted in resistance development against bacterial and fungal pathogens (Rivero et al., 2012). Ribosome-inactivating protein (*rip30*) and ChiC (*chiA*) genes showed enhanced resistance to *R. solani* in a greenhouse assay (M'hamdi et al., 2012). A wild eggplant (*Solanum torvum*) derived *StoVe1* gene resulted in enhanced resistance to *Verticillium dahliae* infection (Liu et al., 2012). Five novel thionin genes, from *Brassicaceae* species were also used to develop gray mold (*Botrytis cinerea*) resistance in engineered crop plants (Hoshikawa et al., 2012). Hence, a variety of genes (from mycoparasitic fungi as well as plants) have been explored to control fungal pathogens to a great success through nuclear tranformation. Plastome engineering also offers an attractive option for the development of plant varieties resistant to the aforementioned serious phytopathogens. The antimicrobial peptide MSI-99 was expressed in tobacco chloroplast genome to attain high-level expression of transgene (De Gray et al., 2001). MSI-99 is an analog of a naturally occurring peptide (magainin 2), found in the skin of an African frog. Its amino acid sequence was modified to enhance lytic abilities. The AT content of MSI-99 was 51.3%, comparable to tobacco plastome that has 61% AT content (Shimada and Sugiura, 1991). Antimicrobial activity also appeared to be increased with increase in homotransplasmicity resulting in 88% (T1) and 96% (T2) inhibition of growth against *Pseudomonas syringae* (a major plant pathogen). Expression of MSI-99 in tobacco chloroplasts appeared to improve resistance against rice blast fungus. Crude protein extracts from the transgenic plants were evaluated for antimicrobial activity under *in vitro* and *in vivo* conditions and it displayed significant suppressive effects against rice blast isolates (Wang et al., 2015). Plastome-targeted expression of two bacterial genes (isochorismate synthase [ICS] and isochorismate pyruvate lyase [IPL]) resulted in 500–1000-fold increased accumulation of SA and SA glucoside compared to control plants, thus giving out enhanced resistance against infectious phytopathogens. This expression did not affect the plant phenotype but showed resistance to viral and fungal infections. For this, precursors of ICS (coding sequence of *Escherichia coli* entC gene) and IPL (coding sequence from *Pseudomonas fluorescens* pmsB) were cloned with and without plastid-targeting sequences. The developed transplastomic tobacco plants were highly resistant to *Oidium lycopersicon* fungi as compared with wild-type plants (Verberne et al., 2000). Similarly, overexpression of antifungal proteins (ChiCs, proteases, chitosanases, glucanases, and kinases) may further strengthen phytopathogen control program. Another strategy used chloroplast-targeted expression of *P. syringae*-derived *argK* genes, which encodes for a toxin-resistant enzyme ornithine carbamoyltransferase (ROCT). The *P. syringae* produces phaseolotoxin, which acts as a chlorosis-inducing toxin by inhibiting ornithine carbamoyltransferase (OCT) (Templeton et al., 1985). The toxin-resistant gene was fused with pea rbcS followed by *Agrobacterium*-mediated transformation. The resultant transgenic ROCT plants showed 83%–100% resistance to phaseolotoxin as compared with nontransformed OCT plants (0%–22%).

Herbicide resistance
Weeds are one of the major problems encountered in crop management. They compete not only for water and nutrients but also decrease farm yield and production. Given the harmful economic implications of poor weed management, it is hardly surprising that production of herbicides constitutes the majority of the agrichemical industry. Nonselective herbicides are always desirable for the control of weeds compared with selective herbicides as they may kill only one type of vegetation. Further, broad-spectrum herbicides, that is, glufosinate and glyphosate display low levels of toxicity, and to date, weeds have shown

minimal resistance to their repeated applications as well. They are more biodegradable as compared with selective herbicides. Considering this, nonselective herbicide-resistance genes have preferably been used to develop transgenic plants, which have the ability to withstand herbicide spray.

Herbicide resistance is such a desirable trait that in 2003, 73% of the genetically engineered crops were herbicide resistant, whereas 18% composed of insect-resistant crops and 8% composed of both herbicide- and insect-resistant crops. Various herbicide-resistance genes, that is, *bar* (Glufosinate), *aroA* (Glyphosate), *BXN* (Bromoxynil), *DHPS* (Sulfonamides), and *ALS* (Sulfonylurea) have been used to develop herbicide resistance. Although herbicide resistance is the dominant trait, yet it has certain serious concerns if expressed in nucleus, particularly in case of cross-pollinated crops where transgene escape may even lead to the development of superweeds. Further, suboptimal expression of transgenes may lead to increased risk of resistance development against the target herbicide (Hilder and Boulter, 1999).

Chloroplast genetic engineering is of great value to address these resistance-conscious traits as well as avoids transgene contamination as it is maternally inherited. Engineering herbicide resistance is of dual use as it not only encodes for a valuable trait (herbicide resistance) but also the same herbicide may be used for the selection of the transformation event. Owing to lethality of herbicide selection system, a sufficient proportion of transplastomes require to be established prior to select transformation events on herbicides. That is why a primary round of selection is made with antibiotics followed by herbicides (used for the removal of antibiotic-resistance gene as well), to effectively select transformants (Ye et al., 2003). Hence, direct selection for herbicide resistance does not work in chloroplast transformation. Various herbicide-resistant genes have been used to develop chloroplast-selectable markers. A *psbA*-mutant gene of *Chlamydomonas* appeared to tolerate metribuzin herbicide, used for the selection of transformants (Przibilla et al., 1991). Desired mutants of *psbA*, having the potential to tolerate other classes of herbicides can also be developed to engineer crop plants using plastid transformation technology as well (Day and Goldschmidt-Clermont, 2011). Sulfometuron methyl (SMM) is another herbicide that targets acetohydroxyacid synthase (AHAS). The said gene is encoded by the nucleus in most of the plants but is reported to be encoded by the plastome in Porphyridium (a unicellular red algae). Therefore, mutant AHAS can be used as a dominant selectable marker to engineer plastid genome of Porphyridium (Lapidot et al., 2002). The herbicide sulcotrione is tolerant to the enzyme 4-hydroxyphenylpyruvate dioxygenase (HPPD) which is involved in the biosynthsis of quinones and vitamin E. Overexpression of barley-derived HPPD in tobacco chloroplast showed tolerance to sulcotrione (Falk et al., 2005). Similarly, a gene from *P. fluorescens* provided tolerance to isoxaflutole, when expressed in tobacco and in soybean plastome (Dufourmantel et al., 2007). These herbicide-tolerant genes can be used as secondary-selectable markers after the excision of primary-selectable markers (antibiotics). Herbicide tolerance can be engineered in plants either by the introduction of herbicide-resistant proteins or by the modification of herbicide metabolism and compartmentation. Increasing metabolism may be more attractive strategy as the phytotoxic chemicals (herbicides) are altered neither by interfering with plant primary metabolism nor by residing inside the plant. So far, herbicide metabolism has been engineered by microorganism-derived genes as well as certain plant-derived genes (Duke, 1996). Certain multigene families have been explored to develop herbicide-resistant transgenic plants, that is, cytochrome P450 mono-oxygenases (Werck-Reichhart et al., 2000), glutathione S-transferases (McGonigle et al., 2000), and glycosyl transferases (Brazier et al., 2002) and resistance to Bialaphos/Liberty (Iamtham and Day, 2000; Lutz et al., 2001), glyphosate

(Ye et al., 2001), isoxaflutole (Dufourmantel et al., 2007), and sulfonyl-urea herbicides (Shimizu et al., 2008).

EPSPS (5-enol-pyruvyl shikimate-3-phosphate synthase) is a protein of choice to develop resistance against broad-spectrum herbicide Glyphosate. Bacteria are the prime natural source of EPSPS enzymes. Certain mutants of the gene have been developed to encode for increased resistance using nuclear transformation system. Prompt transplastomic expression of the EPSPS requires codon optimization, in accordance to plastid codon usage where A or T as third codon position is strongly preferred. Regulatory sequences, too, determine the expression level to a great extent. The EPSPS has been found to be accumulated up to 10% of total soluble protein of plants but a problem with the EPSPS gene expression in chloroplast is that chloroplast localization of the enzyme requires 250-fold higher enzyme expression to give a similar level of resistance as that of nuclear expression. Thus, persistant expression of this broad-spectrum herbicide-resistant gene in chloroplast requires to be investigated. Petunia-derived EPSPS gene resulted in glyphosate resistance when expressed in tobacco chloroplast, using *aadA*-based selection system (Daniell et al., 1998). Further, they evaluated maternal inheritance of EPSPS gene, proving worth of the transplastomic technology that it eliminates gene escape, therefore no superweeds. Bacterial *bar* gene encodes for phosphinothricin acetyl transferase (PAT), which poses resistance to phosphinothricin (PPT, Basta, or glufosinate) (Iamtham and Day, 2000; Lutz et al., 2001). It has appeared to be a nice substitute in chloroplast transformation system where the transgene was expressed up to 7% of total soluble protein, giving out field-level tolerance to phosphinothricin (Maliga and Bock, 2011).

Abiotic stress tolerance

Genetic engineering has undoubtedly opened a new avenue to overcome crop losses due to various abiotic stresses prevalent in the agricultural ecosystems. Nevertheless, abiotic stresses remain the greatest constraint to crop production. They have estimated to cause up to 60% reduction in crop yield, worldwide (Acquaah, 2007). During the last two decades, a large number of crop plants have been engineered with genes conferring abiotic stress-tolerance traits (Bhatnagar-Mathur et al., 2008). Plants have evolved different adaptive responses to neutralize the negative effects of these stresses. A common response to such stresses is accumulation of various small nontoxic organic molecules involved in osmoregulation or stabilization of protein complexes and membranes at the cellular level (Hincha and Hagemann, 2004).

Several plant-derived stress-induced genes have been overexpressed in transgenic plants including a gene encoding for hybrid proline-rich protein from the pigeon pea, a dehydration-responsive element binding (DREB)/cold-binding factor (CBF) from wheat, TSRF1, an ethylene-responsive factor (ERF) transcription factor from tomato and sugarcane, ERF, and to combat abiotic stresses (Trujillo et al., 2008). Zhang et al. (2006) used *G. frondosa* and TSase gene to attain increased yield under drought conditions. A heterologous gene (P5CS) derived by a stress-inducible promoter (AIPC) was expressed to explore the role of proline as a proficient osmoprotectant under drought conditions (Molinari et al., 2007). An upstream regulatory sequence of *Arabidopsis thaliana prd29A* resulted in enhanced resistance to abiotic stresses (Wu et al., 2008). Similarly, Kumar et al. (2014) reported a genetic improvement for drought and salinity stress tolerance by expressing *Arabidopsis*-derived vacuolar pyrophosphatase (avp1) gene. The overexpression of genes-encoding LEA proteins has also proved to improve stress tolerance in transgenic plants. Expression of the barley gene *HVA1* in wheat (Sivamani et al., 2000) and rice (Xu et al., 1996) conferred increased drought and salinity tolerance to engineered plants. HVA1 protein expression helped in better performance of transgenic rice plants by protecting cell membrane from injury under drought stress (Babu et al., 2004). The mulberry plants engineered with barley HVA1 gene showed better cellular membrane stability, less photooxidative damage, improved photosynthetic yield, and better water use efficiency compared with nontransgenic plants exposed to drought and salinity stress. Results also indicated that the production of HVA1 proteins helps in better performance of transgenic plants by promoting membrane stability of plasma membrane as well as chloroplastic membrane (Lal et al., 2008). Similarly, maize plants engineered with barley HVA1 gene showed increased tolerance to salts (100–300 mM NaCl), survived for 15 days of complete drought, and depicted higher leaf relative water content (RWC) as compared with nontransformed plants (Nguyen and Sticklen, 2013). Constitutive expression of the cold-regulated *COR15a* gene of *A. thaliana* has also shown a significant increase in the survival of isolated protoplasts frozen over the range of $-4.5°C$ to $-7°C$ temperature.

Transplastomic technology has successfully been employed to address these stresses, particularly drought, salt, cold, frost, and high temperature (Wang et al., 2009b; George et al., 2010). Among various strategies opted to combat these stresses is overexpression of osmolytes, that is, glycinebetaine. Plastids are organs of great value to engineer majority of the key enzymes involved in plant stress metabolism as plastids are the site of function of these osmolytes. Betain aldehyde dehydrogenase (BADH) is an enzyme, responsible for converting betaine aldehyde into betain. Spinach- and sugarbeet-derived Badh gene(s) were targeted to tobacco plastome. The enzyme was successfully produced in plastids and conferred resistance to betaine aldehyde. Similarly, spinach-derived choline mono-oxygenase (CMO) (an enzyme involved in the conversion of choline into betaine aldehyde) was expressed in tobacco. The resultant enzyme was transported to plastids but these plants were able to accumulate a very low level of betaine (Nuccio et al., 1998), probably owing to the absence of engineered BADH activity in chloroplast. This suggested that BADH as well as CMO pathways should be introduced and localized into transplastomes as the presence of both BADH and CMO is essential for the synthesis of betaine in chloroplast.

Transplastomic carrot plants exhibited high levels of salt tolerance when targeted with Badh gene using carrot-specific chloroplast vectors. Bombarded calli were selected on variable concentrations of spectinomycin and betaine aldehyde. The calli surviving on the antibiotics were screened and tested for transgene integration into the plastome by polymerase chain reaction and Southern blot analyses. Putative transgenic calli cultures were maintained under *in vitro* conditions for regeneration. Plants were then shifted to soil in pots to induce taproot system. The resultant transplastomic plants had 74.8% recombinant (BADH) enzyme activity. The transformed carrot cells were able to survive NaCl upto 400 mM, whereas severe growth retardation was observed in untransformed plants when grown in 200-mM NaCl (Kumar et al., 2004). Increased biosynthesis of glycine betaine has also appeared to improve abiotic stress tolerance. *Beta vulgaris*-derived choline mono-oxygenase (BvCMO) was expressed in tobacco plastome under the control of ribosomal RNA operon promoter and a synthetic T7 gene G10 leader sequence. The *trans*-protein was observed to be accumulated in roots, leaves, seeds, and the resultant transplastomic plants showed increased photosynthetic efficiency when grown on 150-mM NaCl (Zhang et al., 2008).

Trehalose is a primary-storage carbohydrate that helps to tolerate abiotic stresses by protecting both the biological membranes and proteins (Leslies et al., 1995). It has been well documented that trehalose biosynthesis may alter stress tolerance in crop plants as well (Goddijn and van Dun, 1999; Lee et al., 2012). Owing to the involvement of trehalose in drought tolerance, efforts have been made to improve dessication tolerance by transplastomic technology. Yeast *trehalose phosphate synthase* (*TPS1*) was expressed in tobacco plastome to enable plants to cope with water shortage. Transcript analysis revealed that transplastomic expression was 169 times more as compared with nuclear-expressed transgene. Both the nuclear and chloroplast-transgenic plants showed enzyme activity but trehalose accumulation was 15–20-fold higher in plastid-engineered plants and plants were able to grow even on 6% PEG. Further, thylakoid membranes of engineered chloroplasts depicted high integrity by retaining the chlorophyll and surviving dehydration with normal growth. The nucleus-engineered plants showed stunted growth whereas control-untransformed plants were not able to survive dehydration. Therefore, the transplastomic technology offers an effective strategy to engineer crop plants for abiotic stresses with no pleiotropic effects or gene escape to wild relatives (Lee et al., 2003).

Plant transformation for nutrient fortification

More than 50% of the world population is facing malnutrition (Christou and Twyman, 2004). Developing countries, in particular, rely on cereals (the only staple food) for nutrition. Milled cereal grains are poor source of essential amino acids and other nutrients. Biofortification provides dietary supplements to tackle malnutrition problems (White and Broadley, 2005). A rapid way to enhance (fortify) the nutritional value of food crops is through transgenic technology. In addition to the pioneering work in rice (Beyer, 2010; Potrykus, 2010), certain essential amino acids (Wakasa et al., 2006), long-chain fatty acids (Wu et al., 2005), vitamins (Van Eenennaam et al., 2003), and minerals (Drakakaki et al., 2005) have been expressed to provide an adequate source of nutrients for human consumption (Farre et al., 2010).

Genetic manipulation of plants to overexpress the ferritin protein for improved iron content has been done over the years. Ferritin is an iron-storage protein that can be derived from many different sources including bacteria, fungi, plants, and animals. Soybean ferritin was introduced in rice and resulted in a threefold increase in iron content (Goto et al., 1999). Lucca et al. (2001, 2002) engineered rice with ferritin from *Phaseolus vulgaris* and phytase from *Aspergillus fumigatus* under the control of endosperm-specific promoter. The resultant transgenic plants depicted a twofold increase in iron and a 130-fold increment in phytase activity in their seeds. Recombinant expression of the soybean ferritin and *Aspergillus niger* phytase for increased iron bioavailability has been achieved in wheat (Brinch-Pedersen et al., 2000), maize (Drakakaki et al., 2005), and rice (Qu et al., 2005). Kobayashi et al. (2013) explored two new iron regulators in rice (OsHRZ1 and OsHRZ2), whose expression appeared to be induced under Fe deficiency. OsHRZ-knockdown plants showed iron efficiency as well as accumulation with an enhanced expression of iron utilization-related genes, revealing that they are negative regulators of iron deficiency responses. Thus, exploring insights of iron accumulation in cereal crop would be of great worth to engineer crop plants for iron fortification. Landini et al. (2012) demonstrated that iodine content can be increased in *A. thaliana* either by easing its uptake through the overexpression of human sodium iodide symporter (NIS) or by the reduction of its volatilization, knockout HOL-1 (a halide methyl transferase).

Similarly, efforts have been made to engineer β-carotene and ascorbic acid pathways (Wurbs et al., 2007; Apel and Bock, 2009) for increased bioavailability of vitamins. Diretto et al. (2006) silenced the first dedicated step in the β-epsilon branch of carotenoid biosynthesis and confirmed tuber-specific silencing of Lcy-e. Resultant tubers depicted a significant increase in β-carotene level (up to 14-fold). Total tuber carotenoids appeared to be increased up to 2.5-fold. They also observed that expression of many other genes involved in this pathway was modified. An enzyme involved in the synthesis of α-tocopherol and γ-tocopherol methyltransferase was used to increase the vitamin E activity of *A. thaliana* seed oil (Shintani and DellaPenna, 1998).

Transplastomic technology has also been used to engineer β-carotene pathway in tomato. Bacteria-derived *crtY* gene was expressed in tomato plastome under the control of *atpl* promoter by particle bombardment and resulted in the conversion of lycopene to β-carotene. The amount of β-carotene was increased from 6.91 µg/g fresh weight (in wild type) to 28.6 µg/g fresh weight in transplastomic fruits, although the total carotenoid content was decreased by >10% (Wurbs et al., 2007). Daffodil lycopene β-cyclase was expressed in tomato plastids under the control of the ribosomal ribonucleic acid (rRNA) operon promoter for the conversion of lycopene (a major storage carotenoid of the tomato fruit) to provitamin A. β-carotene (provitamin A) level was increased to 95 µg/g fresh weight in tomato fruits as compared with 19 µg/g fresh weight in wild-type untransformed fruits, whereas total carotenoids were increased from 76.67 µg/g fresh weight (in wild-type plants) to 115 µg/g fresh weight in engineered tomato plants (Apel and Bock, 2009).

Furthermore, to use chloroplast transformation to produce α-tocopherol-rich vegetable, tocopherol cyclase-overexpressing transplastomic lettuce plants (pLTC) were generated. The total tocopherol levels and vitamin E activity increased in the pLTC plants compared with the wild-type lettuce plants. Lu et al. (2013) studied the impact of the three plastid-localized enzymes involved in tocopherol biosynthesis pathway in chromoplasts and chloroplasts. They engineered a tocopherol metabolic pathway for enhanced expression of vitamin E, using the transplastomic technology. The plastid-expressed operon resulted in 10-fold increase in total tocopherol content as compared with nuclear expression of histidine-containing phosphotransfer (HPT) in *A. thaliana*, which resulted in up to 4.4-fold increase in leaf tocopherol (Collakova and DellaPenna, 2003). In addition, they also found that tocopherol metabolism affects the pathways involved in the biosynthesis of photosynthetic pigments and also have a positive impact on the photosynthetic efficiency of plants as well. Thus, transplastomic approach is useful for improving the levels and composition of tocopherol (Yabuta et al., 2012).

Plant transformation for metabolic engineering

Plant metabolic engineering has made great strides during the last two decades, with notable success stories. Engineering metabolic pathways in plants will not only help them to fulfill nutritive needs but also increase their photosynthetic efficiency. Biofuel feed stock may also be engineered for increased fermentable saccharides by incorporating self-destructing lignin. Hence, metabolic engineering involves valuable alteration of metabolic pathways prevailing in plants in order to achieve better understanding as well as to use cellular pathways for energy transduction, and supramolecular assembly. Metabolic engineering in plants may be categorized into (a) primary metabolic pathways, that is, carbohydrates, amino acids, lipids, and (b) secondary metabolic pathways, that is, alkaloids, flavonoids, terpenoids, quinones, and lignins (Gomez-Galera et al., 2007). Engineering these pathways may enhance the production of a large number of compounds including energy-rich foods, vitamins, different pharmaceuticals, and the compounds of industrial importance.

Metabolic engineering approaches have been employed to increase biomass oil content and develop novel bluish color in plants (Katsumoto et al., 2007; Wood, 2014). Leaf oil content was increased either by ectopic expression of certain transcription factors involved in the regulation of seed development and maturation process (Santos Mendoza et al., 2005), overexpression of acetyl-CoA carboxyalse, or through the downregulation of genes involved in metabolic pathways that compete for the available carbon (Sanjaya et al., 2011). In *A. thaliana,* disruption of sugar-dependant1 (a lipase-related pathway) also resulted in enhanced TAG content in nonseed tissues (Kelly et al., 2013). Coexpression of diacylglycerol acyltransferase (DGAT1), oleosin genes, and WRINKLED1 transcription factor (WRI1) resulted in accumulation of TAG in tobacco leaves up to 15%. This resulted in increment in both oil contents as well a fatty acid synthesis (Reynolds et al., 2015). Expression of the viola F3'5'H gene in rose resulted in the accumulation of a high percentage of delphinidin (95%) and a novel bluish flower color was developed. They proposed that in addition to viola F3'5'H gene, overexpression of Iris x hollandica dihydroflavonol 4-reductase (DFR) and downregulation of (endogenous) DFR gene results in more exclusive accumulation of delphinidin, irrespective of the host plant. Hyperaccumulation of delphinidin in rose petals resulted in blue flowers, which could never be achieved through hybridization breeding. Further, the transgene integration was stable and exclusive accumulation of delphinidin was inherited to the next progenies (Katsumoto et al., 2007).

Engineering photosynthetic as well as respiratory pathways will not only ensure the increasing demand of food supply but will also increase feed stock provision to produce green energy. Rubisco is the key enzyme involved in catalyzing the first key step of CO_2 fixation as part of the Calvin cycle (Spreitzer and Salvucci, 2002). Its ability to use oxygen as a substrate in photorespiration as well as low turnover rate makes it notoriously inefficient. This induces plants to produce more and more Rubisco (higher than any other protein), making it the most abundant protein all over the world. Alternative carbon fixation pathways have been adapted by the plants to improve their photosynthetic efficiency by reducing the oxygenase activity of Rubisco and actively concentrating on CO_2 (Caemmerer et al., 2012). Photorespiratory pathway was engineered in *Arabidopsis* to alleviate photorespiratory losses and enhance photosynthesis by the release of CO_2 in the vicinity of Rubisco (Kebeish et al., 2007). Khan (2007) proposed a novel strategy to increase biomass production of crop plants by engineering photorespiratory pathway in chloroplast. Nazir and Khan (2012) concluded that expressing pinus chlB gene in tobacco chloroplasts promote rooting and early chlorophyll pigment development.

Similarly, strategies have been proposed to degrade lignin into valuable aromatic monomers either chemically or with the help of enzymes present in microbes (white rot fungi) having ability to degrade it naturally (Bugg et al., 2011). In *A. thaliana,* expression of 4-O-methyltransferase also resulted in reduced lignin content by blocking access to lignin precursors (p-hydroxyls), required for their polymerization. No profound effect was observed on growth and scarification yield was also improved by 25% (Zhang et al., 2012). Thus, researchers are striving hard to develop ideal plants having the ability to degrade their lignin or cross-link the lignin, causing it to precipitate and get it separate easily from cellulose, releasing pure cellulose that could be more easily degraded into glucose resulting in cheaper bioethanol production.

Plant transformation for cost-effective therapeutics and vaccines

Therapeutic proteins and vaccines are the biological preparation that develops immunity to a specific disease. These proteins play a very prominent role in health improvement of humans and animals. They are the most powerful tools in the struggle against various infectious diseases and have begun presenting affectivity in avoidance and treatment against a range of pathogens (Daniell et al., 2009; Yusibov et al., 2011). Despite the successful performance of immunization plans to reduce infectious diseases, still about 15 million people die every year due to these preventable causes, most from developing countries (Anonymous, 2008). According to the World Health Organization, there is a major need for developing new cost-effective pharmaceuticals against several diseases especially for resource-poor countries (Anonymous, 2009). Therefore, more efforts are required to address the vaccination demand in these countries (Lossl and Waheed, 2011). Currently, existing methods for the commercial manufacturing of pharmaceutical proteins mainly include the use of bacteria, yeast, mammalian cell cultures, etc. These methods have their unique benefits, but their overall application is restricted by insufficient scalability, expensive production, high distribution costs, and safety issues (Yusibov and Rabindran, 2008). As a potential solution to these problems, plants as production platform have won considerable attention in the last decade (Rybicki, 2010). Plants offer the unique opportunity to be engineered as biofactories for the production of therapeutic proteins and secondary metabolites (Hassan et al., 2011) and they are considered as an alternative and attractive source for vaccine production (Wagner et al., 2004). Plant-based production of biopharmaceutical proteins is one of the most important application of biotechnology and genetic engineering in modern medicine era. The first recombinant protein of pharmaceutical interest produced in tobacco and sunflower was human growth hormone about 29 years ago (Barta et al., 1986). Now, it is a routine work and about 60 antigens and 30 biopharmaceutical/therapeutic proteins have been successfully produced using various plant species (Cardi et al., 2010). The manufacturing of recombinant pharmaceutical protein in genetically engineered cells instead of avirulent/killed strains offers unique features including superior safety, less antigenic competition, opportunity to target vaccines to precise locations, and to distinguish vaccinated animals from infected ones (Arntzen et al., 2005; Daniell et al., 2009). Conventional production of biopharmaceutical proteins using mammalian cells, bacteria, yeast, insects cells etc., contains drawbacks related to absence/modification of posttranslational alterations or potential existence of endotoxins, pyrogens, human pathogens, etc. Additionally, the use of fermentors enhances the cost and limits of manufacturing systems (Yusibov et al., 2011).

Plant-based treatments of human diseases predate the most primitive phases of recorded society. By the sixteenth century, botanical gardens offered assets of *"Materia Medica"* for educating therapeutic use, and herbal medicine increased until the seventeenth century when more scientific "pharmacological" medications were discovered (Winslow and Kroll, 1998). Later, the active code in many therapeutic plants was recognized and purified for remedial use. Even today, about one-fourth of existing prescription drugs has a botanical origin. Plant molecular farming is a promising technology that relies on recombinant protocols for plant-based production of valuable proteins and vaccines. Plants not only provide food and shelter for humans but with the progression of molecular biology techniques, they are also serving as green factories for the production of various types of valuable recombinant proteins to cure human beings from different diseases (Shanmugaraj and Ramalingam, 2014). Transgenic plants offer a safe, efficacious, and reasonably cheap source for the production of therapeutic proteins and vaccines to the poor world. Therapeutic production in plants can be achieved by the development of stable or transient expression systems. Transient system is more efficient as it requires less time with enhanced protein production, uniformity in protein accumulation, advantages of scalability, no biosafety issues, etc. (Daniell et al., 2009). Increased protein accumulation in plants ultimately reduces the downstream processing costs as well. But this mainly depends on the crop plant and tissues used for the expression of foreign proteins. Production cost may vary from 0.1 to 1.0 US$/g of antibodies production in plants. This also depicts 1000-fold reduction in manufacturing cost when the same amount of protein is expressed in cell lines (Molowa et al., 2002).

Therapeutic proteins expressed in transgenic plants

Proteins that are engineered in laboratory for pharmaceutical purpose are known as the therapeutic proteins. These proteins are used for the treatments of a wide range of human diseases including cancer, hepatitis, hemophilia, anemia, multiple sclerosis, different types of infectious diseases, etc. With the advancements in recombinant DNA technology, these proteins can be generated in various hosts such as mammalian cells, bacteria, yeast, transgenic plants, etc. Among these, mammalian cells are the most favorite choice for the manufacturing of therapeutic proteins as the activity of proteins requires some posttranslational modifications that naturally occur in mammalian cells. Monoclonal antibodies are the largest category of genetic engineering and biotechnology-derived drugs. Efficient manufacturing of these antibodies has been a typical benchmark in assessing commercial viability of an expression system for pharmaceutical protein productions. These are extremely expensive to manufacture in mammalian cells and therefore, are expensive (Kelley, 2007). Development of new platform is definitely very much important to reduce the production cost and to improve the scalability of these proteins. Transgenic plants are the alternate system for the production of antibodies that offers only the viable and large-scale production system for these antibodies (Larrick and Thomas, 2001) on an economical basis. Recombinant DNA technology and genetic transformation protocols now made it feasible that these drugs can be expressed in plant expression system in their active form. Antibodies against different types of diseases including dental caries, cholera, diarrhea, arthritis, malaria, cancers, hepatitis, influenza, etc., have been successfully engineered in the transgenic plants as well. Some of these plant-expressed pharmaceutical proteins/antibodies are currently under clinical trials as well. Important therapeutic proteins/antibodies are expressed in various crop species.

Vaccines/antigens expressed in transgenic plants

Infectious diseases cause about 50% deaths in developing nations. Vaccination is one of the most preventative means to secure from these deadly diseases. Currently used vaccines are isolated from mammalian cells, which are very expensive and not affordable to poor people living in developing countries. Alternate sources for the production of affordable and cheaper vaccines are of great importance for the developing nations worldwide. Advancements in molecular biology tools made it possible to manufacture vaccine antigens in plants in their active form and these are easily available. Plant-derived vaccines are cheaper, safer than, and as much active as those produced from mammals (Houdebine, 2009). Vaccines are the biological preparations that provide protection against a specific disease. These are usually prepared from the weakest or killed forms of microbes, their toxin, or one of their surface proteins that stimulate the body's immune system to identify the agent as a risk, obliterate it, and maintain a proof of it, so that the immune system can more easily distinguish and destroy any of these microorganisms in future. Significant efforts have been made since the idea for the vaccine production on transgenic plants by Arntzen et al. (1994). The plant-derived vaccine has been produced against different diseases, which includes vibrio cholera, rotavirus, hepatitis B virus (HBV), Norwalk virus, enterotoxigenic, rabies, respiratory syncytial virus, etc. Vaccines against human and animal diseases such as hepatitis B for human (Richter et al., 2000) and foot-and-mouth disease in animals (Wigdorovitz et al., 1999) have been expressed in transgenic plants. In the last decade, many vaccine antigens have been successfully expressed in transgenic plants. Many of these plant-produced vaccines were isolated, purified, and used as injectable vaccines. Efforts are in progress to produce an edible vaccine in transgenic plants and have successfully immunized test animals against cholera, HBV, rabies, rotavirus, etc.

The edible vaccine quantity expressed in plant is relatively low and research is under way to improve the concentration of these vaccines using different plants and tissues. These edible vaccines are being expressed and tested in potatoes, tomatoes, banana, and carrots (Thomas et al., 2002). Many therapeutic vaccine antigens are being expressed in stable or transient state in different plant tissues. Some of the stably introduced vaccine antigens in various plant species are given in Table 3.2.

Comparison of plant-based versus conventional non-plant-based expression systems
Production of recombinant pharmaceutical protein in plants offers several exceptional advantages in comparison with mammalian cells, bacteria, viruses, etc. Unlike bacteria and mammals, plants are capable of posttranslational modification known as glycosylation, required for biological activity of proteins such as antibiotics (Rogers, 2003; Goldstein and Thomas, 2004). Glycosylation of protein means the addition of certain sugar molecules to its structure causing it to fold in a specific and unique orientation required for proper activity and function. In addition to glycosylation, plants are competent in producing more complex proteins without any assistance, whereas mammalian cells required support to glycosylate during the production of complex proteins. The system needed for the production of pharmaceutical and therapeutic proteins in mammalian cells is expensive and intricate to maintain while plant production system is more attractive, less expensive, and easy to maintain (Fischer et al., 2004). Additionally, the chances of contamination during the production therapeutic protein are less in plant cells as compared with bacterial and mammalian cells. So, all these features make the plant cells an ideal and excellent alternate target for the production of therapeutic proteins, as they have the ability to glycosylate complex proteins and are cheap to grow in large amounts.

Conclusion

Plant biotechnology explores contemporary aspects of innovative research with an ultimate goal to improve quality and quantity of eatables as well as to secure human health through the production of therapeutic proteins, antibiotics, vaccines, etc. Though science is moving ahead rapidly and valuable milestones have been achieved, that is, biotic and abiotic stress-tolerant crop plants have been developed, plants have been engineered for increased nutrients as well as to express certain valuable therapeutic proteins. Although certain serious concerns are associated with the first-generation transgenic crop plants, scientists had been striving hard to fix these concerns through innovative research. Regarding antibiotic-resistance genes (used for the selection of transformation events), techniques have been devised to develop marker-free plants. As far as insect resistance is concerned, multiple genes with multiple mode of action have been employed to avoid resistance development. To avoid transgene flow or development of superweeds, transgene containment exploiting chloroplast genome is possible. Almost all concerns raised in first-generation biotechnology have been addressed and fixed in second-generation biotechnology, and so on. In addition, a regulatory system has been upgraded keeping in view the sensitivity of the risks involved in transgene development and their commercialization. Yet, the scientific community should continue research with enthusiasm to address the existing loopholes in plant biotechnology so that mankind may benefit from this technology of tremendous potential. To conclude, plant biotechnology is striding out nicely and a day will come when transgenic plants will be the only desirable entity to feed the ever-increasing population.

Table 3.2 Status of plant-based therapeutic proteins and vaccines

Sr. no	Product	Target disease	Host crop	Development stage	Reference
Therapeutic proteins					
1	Glucocerebrosidase	Gaucher disease	Carrot	Stage III	Shaaltiel et al. (2007)
2	Gastric lipase	Cystic fibrosis, pancreatitis	Maize	Stage II (marketed as analytical reagent)	Roussel et al. (2002)
3	Lactoferrin	Gastrointestinal infections	Maize	Clinical trial (marketed as analytical reagent)	Samyn-Petit et al. (2001)
4	Human intrinsic factor	Vitamin B12 deficiency	*Arabidopsis*	Stage II	Fedosov et al. (2003)
Vaccines					
5	*E. coli* heat-labile toxin	Diarrhea	Potato Maize	Clinical trial	Savarino et al. (2002)
6	Norwalk virus capsid protein	Diarrhea	Potato	Clinical trial	Tacket et al. (2000)
7	HBV surface antigen	Hepatitis B	Potato Lettuce	Clinical trial	Thanavala et al. (1995)
8	Rabies glycoprotein	Rabies	Spinach	Clinical trial	Yusibov et al. (2002)
9	Newcastle disease virus HN	Newcastle disease (poultry)	Tobacco	Approved by USDA	Mihaliak et al. (2005)
10	Personalized anti-idiotype single-chain FVs	No lymphoma *n*-Hodgkin's	Tobacco	Clinical trial	Bendandi et al. (2010)
11	H5n1 influenza Ha VLP	H5n1 "avian" influenza	Tobacco	Stage II (approved in Canada)	Shoji et al. (2009)
12	H1N1 influenza HAC1	H1N1 "swine" influenza	Tobacco	Stage I	Shoji et al. (2011)
Antibodies					
13	CaroRx	Dental caries	Tobacco	Stage II (EU approved)	De Muynck et al. (2010)
14	Anti-cD20	Non-Hodgkin's lymphoma, rheumatoid arthritis	Duckweed	Preclinical	Cox et al. (2006)
15	Anti-αccr5	HIV	Tobacco	Preclinical	Pogue et al. (2010)
16	Anti-HIV gp120	HIV	Maize, tobacco	Preclinical	De Muynck et al. (2010)
17	Anti-HBsag scFV	Hepatitis B vaccine purification	Tobacco	Marketed in Cuba	Rademacher et al. (2008)

References

Acquaah, G. 2007. *Principles of Plant Genetics and Breeding*. Blackwell, Oxford, UK.

Altpeter, F., Varshney, A., Abderhalden, O., Douchkov, D., Sautter, C., Kumlehn, J., Dudler, R. and Schweizer, P. 2005. Stable expression of a defense-related gene in wheat epidermis under transcriptional control of a novel promoter confers pathogen resistance. *Plant Molecular Biology.* 57: 271–283.

Arntzen, C.J., Mason, H.S., Shi, J., Haq, T.A., Estes, M.K. and Clements, J.D. 1994. In: Brown, F., Chanock, R.M., Ginsberg, H.S. and Lemer, R.A. (eds.), *Vaccines '94: Modern Approaches to New Vaccines including Prevention of AIDS*, Cold Spring Harbor Laboratory Press, New York, pp. 339–344.

Anonymous. 2008. *The Global Burden of Diseases 2004 Update*. World Health Organization, Switzerland.

Anonymous. 2009. *Millennium Development Goals: Progress towards the Health Related Millennium Development Goals*. World Health Organization, New York.

Apel, W. and Bock, R. 2009. Enhancement of carotenoid biosynthesis in transplastomic tomatoes by induced lycopene-to-provitamin A conversion. *Plant Physiology.* 151: 59–66.

Arntzen, C., Stanley, P. and Betty, D. 2005. Plant-derived vaccines and antibodies: Potential and limitations. *Vaccine.* 23: 1753–1756.

Arrieta, J.G., Enriquez, G.A., Suarez, V., Estevez, A., Fernandez, M.E., Menedez, C., Coego, A. et al. 1996. Transgenic potato expressing two antifungal proteins. *Biotechnologia Aplicada.* 13: 121.

Babu, C.R., Zhang, J., Blum, A., David, T.H., Wu, R. and Nguyen, H.T. 2004. *HVA1*, a LEA gene from barley confers dehydration tolerance in transgenic rice (*Oryza sativa* L.) via cell membrane protection. *Plant Science.* 166: 855–862.

Baisakh, N., Datta, K., Rai, M., Oliva, N., Tan, J., Mackill, D.J., Khush, G.S. and Datta, S.K. 2006. Marker free transgenic (MFT) golden near isogenic introgression lines (NIILs) of indica rice cv. IR64 with accumulation of provitamin A in the endosperm tissue. *Plant Biotechnol Journal.* 4: 467–475.

Barta, A., Sommergruber, K., Thompson, D., Hartmuth, K., Matzke, M.A. and Matzke, A.J. 1986. The expression of a nopaline synthase–human growth hormone chimaeric gene in transformed tobacco and sunflower callus tissue. *Plant Molecular Biology.* 6: 347–357.

Bates, S.L., Zhao, J.Z., Roush, R.T. and Shelton, A.M. 2005. Insect resistance management in GM crops: Past, present and future. *Nature Biotechnology.* 23: 57–62.

Bendandi, M., Marillonnet, S., Kandzia, R., Thieme, F., Nickstadt, A., Herz, S., Frode, R. et al. 2010. Rapid, high-yield production in plants of individualized idiotype vaccines for non-Hodgkin's lymphoma. *Annals of Oncology.* 21: 2420–2427.

Bevan, M. and Chilton, M.D. 1982. Multiple transcripts of T-DNA detected in Nopaline crown gall tumors. *Journal of Molecular Applied Genetics.* 1: 539–546.

Beyer, P. 2010. Golden rice and "golden" crops for human nutrition. *New Biotechnology.* 27: 478–481.

Bhatnagar-Mathur, P., Vadez, V. and Sharma, K.K. 2008. Transgenic approaches for abiotic stress tolerance in plants: Retrospect and prospects. *Plant Cell Reports.* 27: 411–424.

Birch, R.G. 1997. Plant transformation: Problems and strategies for practical application. *Annual Review of Plant Physiology and Plant Molecular Biology.* 48:297–326.

Bock, R. and Khan, M.S. 2004. Taming plastids for a green future. *Trends in Biotechnology.* 22: 311–318.

Borlaug, N. 2005. Biotechnology and the green revolution: Interview with Norman Borlaug. www.actionbioscience.org/biotech/borlaug.html.

Brazier, M., Cole, D.J. and Edwards, R. 2002. O-glucosyltransferase activities toward phenolic natural products and xenobiotics in wheat and herbicide-resistant and herbicide susceptible blackgrass (*Alopecurus myosuroides*). *Phytochemistry.* 59:149–156.

Brinch-Pedersen, H., Olesen, A., Rasmussen, S.K., Preben, B. and Holm, P.B. 2000. Generation of transgenic wheat (*Triticum aestivum* L.) for constitutive accumulation of an *Aspergillus phytase*. *Molecular Breeding.* 6: 195–206.

Brown, D.C.W. and Thorpe, T.A. 1995. Crop improvement through tissue culture. *World Journal of Microbiology and Biotechnology.* 11: 409–415.

Bugg, T.D., Ahmad, M., Hardiman, E.M. and Rahmanpour, R. 2011. Pathways for degradation of lignin in bacteria and fungi. *Natural Product Reports.* 28: 1883–1896.

Cardi, T., Paolo, L. and Maliga, P. 2010. Chloroplasts as expression platforms. *Expert Review of Vaccines.* 9: 893–911.

Chakrabarti, S.K., Kerry, A.L., Benjawan, L., Zora, S. and Maliga, P. 2006. Expression of the *cry9Aa2* B.t. gene in tobacco chloroplasts confers resistance to potato tuber moth. *Transgenic Research.* 15: 481–488.

Christou, P. and Twyman, R.M. 2004. The potential of genetically enhanced plants to address food insecurity. *Nutrition Research Reviews.* 17: 23–42.

Collakova, E. and DellaPenna, D. 2003. Homogentisate phytyltransferase activity is limiting for tocopherol biosynthesis in *Arabidopsis. Plant Physiology.* 131: 632–642.

Cox, K.M., Sterling, J.D., Regan, J.T., Gasdaska, J.R., Frantz, K.K., Peele, C.G., Black, A. et al. 2006. Glycan optimization of a human monoclonal antibody in the aquatic plant *Lemna minor. Nature Biotechnology.* 24: 1591–1597.

Crickmore, N., Zeigler, D.R., Schnepf, E., Van, R.J., Lereclus, D., Baum, J., Bravo, A. and Dean, D.H. 2007. Bacillus thuringiensis toxin nomenclature. http://www.lifesci.sussex.ac.uk/Home/Neil_Crickmore/Bt/

Daniell, H. 2000. Genetically modified food crops: Current concerns and solutions for next generation food crops. *Biotechnology and Genetic Engineering Reviews.* 17: 327–352.

Daniell, H., Datta, R., Varma, S., Gray, S. and Lee, S.B. 1998. Containment of herbicide resistance through genetic engineering of the chloroplast genome. *Nature Biotechnology.* 16: 345–348.

Daniell, H., Khan, M.S. and Allison, L. 2002. Milestones in chloroplast genetic engineering: An environmentally friendly era in biotechnology. *Trends in Plant Science.* 7: 84–91.

Daniell, H., Kumar, S. and Dufourmantel, N. 2005. Breakthrough in chloroplast genetic engineering of agronomically important crops. *Trends in Biotechnology.* 23: 238–245.

Daniell, H., Singh, N.D., Mason, H. and Streatfield, S.J. 2009. Plant-made vaccine antigens biopharmaceuticals. *Trends in Plant Science.* 14: 669–679.

Day, A. and Goldschmidt-Clermont, M. 2011. The chloroplast transformation toolbox: Selectable markers and marker removal. *Plant Biotechnology. J.* 9: 540–553.

De Cosa, B., Moar, W., Lee, S.B., Miller, M. and Daniell, H. 2001. Overexpression of the Bt *cry2Aa2* operon in chloroplasts leads to formation of insecticidal crystals. *Nature Biotechnology.* 19: 71–74.

De Gray, G., Rajasekaran, K., Smith, F., Sanford, J. and Daniell, H. 2001. Expression of an antimicrobial peptide via the chloroplast genome to control phytopathogenic bacteria and fungi. *Plant Physiology.* 127: 852–862.

De Muynck, B., Navarre, C. and Boutry, M. 2010. Production of antibodies in plants: Status after twenty years. *Plant Biotechnology Journal.* 8: 529–563.

De Rocher, E.J., Vargo-Gogola, T.C., Diehn, S.H. and Green, P.J. 1998. Direct evidence for rapid degradation of Bt. toxin mRNA as a cause of poor expression in plants. *Plant Physiology.* 17: 1145–1461.

Diretto, G., Tavazza, R., Welsch, R., Pizzichini, D., Mourgues, F., Papacchioli, V., Beyer, P. and Giuliano, G. 2006. Metabolic engineering of potato tuber carotenoids through tuber-specific silencing of lycopene epsilon cyclase. *BMC Plant Biology.* 26: 6–13.

Drakakaki, G., Marcel, S., Glahn, R.P., Lund, E.K., Pariagh, S., Fischer, R., Christou, P. and Stroger, E. 2005. Endosperm-specific co-expression of recombinant soybean ferritin and *Aspergillus phytase* in maize results in significant increases in the level of bioavailable iron. *Plant Molecular Biology.* 59: 869–880.

Dufourmantel, N., Dubald, M., Matringe, M., Canard, H., Garcon, F., Job, C., Kay, E. et al. 2007. Generation and characterization of soybean and marker-free tobacco plastid transformants over-expressing a bacterial 4-hydroxyphenylpyruvate dioxygenase which provides strong herbicide tolerance. *Plant Biotechnology Journal.* 5: 118–133.

Duke, S.O. 1996. *Herbicide-Resistant Crops: Agricultural, Environmental, Economic, Regulatory, and Technical Aspects.* CRC Lewis Publishers, Boca Raton, FL.

Dutton, A., Romeis, J. and Bigler, F. 2003. Assessing the risks of insect resistant transgenic plants on entomophagous arthropods: Bt-maize expressing *cry1Ab* as a case study. *Biocontrol.* 48: 611–636.

Eibl, C., Zou, Z., Beck, A., Kim, M., Mullet, J. and Koop, H.U. 1999. *In vivo* analysis of plastid psbA, rbcL and rpl32 UTR elements by chloroplast transformation: Tobacco plastid gene expression is controlled by modulation of transcript levels and translation efficiency. *The Plant Journal.* 19: 333–345.

Esfahani, K., Motallebi, M., Zamani, M.R., Sohi, H.H. and Jourabchi, E. 2010. Transformation of potato (*Solanum tuberosum* cv. Savalan) by chitinase and β-1,3-glucanase genes of mycoparasitic fungi towards improving resistance to *Rhizoctonia solani* AG-3. *Iranian Journal of Biotechnology*. 8: 73–81.

Falk, J., Brosch, M., Schafer, A., Braun, S. and Krupinska, K. 2005. Characterization of transplastomic tobacco plants with a plastid localized barley 4-hydroxyphenylpyruvate dioxygenase. *Journal of Plant Physiology*. 162: 738–742.

Farre, G., Sanahuja, G., Naqvi, S., Bai, C., Capell, T., Zhu, C. and Christou, P. 2010. Travel advice on the road to carotenoids in plants. *Plant Science*. 179: 28–48.

Fedosov, S.N., Laursen, N.B., Nexø, E., Moestrup, S.K., Petersen, T.E., Jensen, E.Ø. and Berglund, L. 2003. Human intrinsic factor expressed in the plant *Arabidopsis thaliana*. *European Journal of Biochemistry*. 270: 3362–3367.

Ferry, N., Edwards, M.G., Gatehouse, J., Capell, T., Christou, P. and Gatehouse, A.M. 2006. Transgenic plants for insect control: A forward looking scientific perspective. *Transgenic Research*. 15: 13–19.

Finer, J.J. 2010. Plant nuclear transformation. In: F. Kempken and C. Jung (eds.), *Genetic Modification of Plants, Biotechnology in Agriculture and Forestry 64*, Springer-Verlag, Berlin, Heidelberg.

Fischer, R., Stoger, E., Schillberg, S., Christou, P. and Twyman, R.M. 2004. Plant-based production of biopharmaceuticals. *Current Opinion in Plant Biology*. 7: 152–158.

Flavell, R.B., Dart, E., Fuchs, R.L. and Fraley, R.T. 1992. Selectable marker genes: Safe for plants. *Biotechnology*. 10: 141–144.

George, S., Venkataraman, G. and Parida, A. 2010. A chloroplast-localized and auxin-induced glutathione S-transferase from phreatophyte *Prosopis juliflora* confer drought tolerance on tobacco. *Journal of Plant Physiology*. 167: 311–318.

Goddijn, O.J.M. and van Dun, K. 1999. Isolation and metabolism in plants. *Trends in Plant Science*. 4: 315–319.

Golds, T., Maliga, P. and Koop, H.U. 1993. Stable plastid transformation in PEG-treated protoplasts of *Nicotiana tabacum*. *Nature Biotechnology*. 11: 95–97.

Goldstein, A.D. and Thomas, A.J. 2004. Biopharmaceuticals derived from genetically modified plants. *An International Journal of Medicine*. 97: 705–716.

Gomez-Galera, S., Pelacho, A.M., Gene, A., Capell, T. and Christou, P. 2007. The genetic manipulation of medicinal and aromatic plants. *Plant Cell Reports*. 26: 1689–1715.

Gonsalves, D. 1998. Control of papaya ringspot virus in papaya: A case study. *Annual Review of Phytopathology*. 36: 415–437.

Goto, F., Yoshihara, T., Shigemoto, N., Toki, S. and Takaiwa, F. 1999. Iron fortification of rice seed by the soybean ferritin gene. *Nature Biotechnology*. 17: 282–286.

Harrigan, G.G., Ridley, W.P., Riordan, S.G., Nemeth, M.A., Sorbet, R., Trujillo, W.A., Breeze, M.L. and Schneider, R.W. 2007. Chemical composition of glyphosate-tolerant soybean 40-3-2 grown in Europe remains equivalent with that of conventional soybean (*Glycine max* L.). *Journal of Agriculture and Food Chemistry*. 55: 6160–6168.

Harriman, R.W., Bolar, J.P. and Smith, F.D. 2006. Importance of biotechnology to the horticultural plant industry. *Journal of Crop Improvement*. 17: 1–26.

Hassan, S.W., Waheed, M.T. and Lossl, A.G. 2011. New areas of plant-made pharmaceuticals. *Expert Review of Vaccines*. 10: 151–153.

Hilder, V.A. and Boulter, D. 1999. Genetic engineering of crop plants for insect resistance—A critical review. *Crop Protection* 18: 177–191.

Hincha, D.K. and Hagemann, M. 2004. Stabilization of model membranes during drying by compatible solutes involved in the stress tolerance of plants and microorganisms. *Biochemical Journal*. 383: 277–283.

Hoshikawa, K., Ishihara, G., Takahashi, H. and Nakamura, I. 2012. Enhanced resistance to gray mold (*Botrytis cinerea*) in transgenic potato plants expressing thionin genes isolated from *Brassicaceae* species. *Plant Biotechnology*. 29: 87–93.

Houdebine, L.M. 2009. Production of pharmaceutical proteins by transgenic animals. *Comparative Immunology, Microbiology and Infectious Diseases* 32: 107–121.

Huang, P.H. and Chen, P.Y. 2011. Design of a two-stage electromagnetic impulse force circuit for gene gun. *Journal of Marine Science and Technology*. 19: 686–692.

Iamtham, S. and Day, A. 2000. Removal of antibiotic resistance genes from transgenic tobacco plastids. *Nature Biotechnology*. 18: 1172–1176.

James, C. 2006. *Global Status of Commercialized Biotech/GM Crops*. ISAAA, Ithaca, New York, USA.

James, C. 2014. *Global Status of Commercialized Biotech/GM Crops Executive Summary*. ISAAA, Ithaca, New York, USA.

Jefferson-Moore, K.Y. and Traxler, G. 2005. Second-generation GMOs: Where to from here? *AgBioForum*. 8: 143–150.

Job, D. 2002. Plant biotechnology in agriculture. *Biochimie*. 84: 1105–1110.

Jones, C.N., Rigby, P.W.J. and Ziff, E.B. 1988. Trans-acting protein factors and the regulation of eukaryotic transcription: Lessons from studies on DNA tumor viruses. *Genes and Development*. 2: 267–281.

Joyce, P.B.M. and Gray, M.W. 1989. Chloroplast-like transfer RNA genes expressed in wheat mitochondria. *Nucleic Acids Research*. 17: 5461–76.

Joyia, F.A. and Khan, M.S. 2012. Reproducible and expedient rice regeneration system using *in vitro* grown plants. *African Journal of Biotechnology*. 11: 138–144.

Katsumoto, Y., Fukuchi-Mizutani, M., Fukui, Y., Brugliera, F., Holton, T.A., Karan, M., Nakamura, N. et al. 2007. Engineering of the rose flavonoid biosynthetic pathway successfully generated blue-hued flowers accumulating delphinidin. *Plant Cell Physiology*. 48: 1589–1600.

Kebeish, R., Niessen, M., Thiruveedhi, K., Bari, R. and Hirsch, H.J. 2007. Chloroplastic photorespiratory bypass increases photosynthesis and biomass production in *Arabidopsis thaliana*. *Nature Biotechnology*. 25: 593–599.

Kelley, B. 2007. Very large scale monoclonal antibody purification: The case for conventional unit operations. *Biotechnology Progress*. 23: 995–1008.

Kelly, A.A., Van E.H., Quettier A.L., Shaw E., Menard G., Kurup S. and Eastmond P.J. 2013. The sugar-dependent1 lipase limits triacylglycerol accumulation in vegetative tissues of Arabidopsis. *Plant Physiology*. 162: 1282–1289.

Keresa, S. 2002. Genetic transformation of tobacco and potato plants for insect resistance using analogues of the squash trypsin inhibitor gene. PhD thesis. University of Zagreb, Croatia.

Khan, M.S. 2007. Engineering photorespiration in chloroplasts: A novel strategy for increasing biomass production. *Trends in Biotechnology*. 25: 437–440.

Khan, M.S. 2012. Plastid genome engineering in plants: Present status and future trends. *Molecular Plant Breeding*. 3: 91–102.

Khan, M.S. and Maliga, P. 1999. Fluorescent antibiotic resistance marker for tracking plastid transformation in higher plants. *Nature Biotechnology*. 17: 910–915.

Khan, R.S., Sjahril, R., Nakamura, I. and Mii, M. 2008. Production of transgenic potato exhibiting enhanced resistance to fungal infections and herbicide applications. *Plant Biotechnology Reports*. 2: 13–20.

Kittiwongwattana, C., Lutz, K., Clark, M. and Maliga, P. 2007. Plastid marker gene excision by the phiC31 phage site-specific recombinase. *Plant Molecular Biology*. 64: 137–143.

Kobayashi, T., Nagasaka, S., Senoura, T., Itai, R.N., Nakanishi, H. and Nishizawa, N.K. 2013. Iron-binding haemerythrin RING ubiquitin ligases regulate plant iron responses and accumulation. *Nature Communications*. 4: 2792.

Kohler, R.H., Cao, J., Zipfel, W.R., Webb, W.W. and Hanson, M.R. 1997. Exchange of protein molecules through connections between higher plant plastids. *Science*. 276: 2039–2042.

Kota, M., Daniell, H., Varma, S., Garczynski, S.F., Gould, F. and Moar, W.J. 1999. Overexpression of the *Bacillus thuringiensis* (Bt) *cry2Aa2* protein in chloroplasts confers resistance to plants against susceptible and Bt-resistant insects. *Proceedings of the National Academy of Sciences*. 96: 1840–1845.

Kramer, M.G. and Redenbaugh, K. 1994. Commercialization of tomato with an antisense polygalacturonase: The FLAVR SAVR tomato story. *Euphytica*. 79: 293–297.

Kumar, S., Dhingra, A. and Daniell, H. 2004. Plastid-expressed betaine aldehyde dehydrogenase gene in carrot cultured cells, roots, and leaves confers enhanced salt tolerance. *Plant Physiology*. 136: 2843–2854.

Kumar, T., Uzma, Khan, M.R., Abbas, Z., Ali, G.M. 2014. Genetic improvement of sugarcane for drought and salinity stress tolerance using Arabidopsis vacuolar pyrophosphatase (avp1) gene. *Molecular Biotechnology*. 56: 199–209.

Lal, S., Gulyani, V. and Khurana, P. 2008. Overexpression of *HVA1* gene from barley generates tolerance to salinity and water stress in transgenic mulberry (*Morus indica*). *Transgenic Research*. 17: 651–663.

Landini, M., Gonzali, S., Kiferle, C., Tonacchera, M., Agretti, P., Dimida, A., Vitti, P., Alpi, A., Pinchera, A. and Perata, P. 2012. Metabolic engineering of the iodine content in *Arabidopsis*. *Scientific Reports*. 2: 338.

Lapidot, M., Raveh, D., Sivan, A., Arad, S.M. and Shapira, M. 2002. Stable chloroplast transformation of the unicellular red alga *Porphyridium* species. *Plant Physiology*. 129: 7–12.

Larrick, W.J. and Thomas, W.D. 2001. Producing proteins in transgenic plants and animals. *Current Opinion in Biotechnology*. 12: 411–418.

Lee, D.H., Ryu, H., Bae, H.H. and Kang, S.G. 2012. Transgenic tobacco plants harboring the trehalose phosphate synthase TPS gene of *Escherichia coli* increased tolerance to drought stress. *Research Journal of Biotechnology*. 7: 22–26.

Lee, S.B., Kwon, H.B., Kwon, S.J., Park, S.C., Jeong, M.J., Han, S.E., Byun, M.O. and Daniel, H. 2003. Accumulation of trehalose within transgenic chloroplasts confers drought tolerance. *Molecular Breeding*. 11: 1–13.

Leslies, S.B., Israeli, E., Lighthart, B., Crowe, J.H. and Crowe, L.M. 1995. Trehalose and sucrose protect both membranes and proteins in intact bacteria during drying. *Applied and Environmental Microbiology*. 61: 3592–3597.

Lossl, A.G., Waheed, M.T. 2011. Chloroplast-derived vaccines against human diseases: achievements, challenges and scopes. *Plant Biotechnology Journal*. 9: 527–539.

Liu, C.W., Lin, C.C., Chen, J. and Tseng, M.J. 2007. Stable chloroplast transformation in cabbage (*Brassica oleracea var. capitata L.*) by particle bombardment. *Plant Cell Reports*. 26: 1733–1744.

Liu, S., Zhu, Y., Xie, C., Jue, D., Hong, Y., Chen, M., Hubdar, A.K. and Yang, Q. 2012. Transgenic potato plants expressing *StoVe1* exhibit enhanced resistance to *Verticillium dahliae*. *Plant Molecular Biology Reporter*. 30: 1032–1039.

Lu, Y., Rijzaani, H., Karcher, D., Ruf, S. and Bock, R. 2013. Efficient metabolic pathway engineering in transgenic tobacco and tomato plastids with synthetic multigene operons. *Proceedings of the National Academy of Sciences*. 110: E623–E632.

Lucca, P., Hurrell, R. and Potrykus, I. 2001. Genetic engineering approaches to improve the bioavailability and the level of iron in rice grains. *Theoretical and Applied Genetics*. 102: 392–397.

Lucca, P., Hurrell, R. and Potrykus, I. 2002. Fighting iron deficiency with iron rich rice. *Journal of the American College of Nutrition*. 21: 184S–190S.

Lutz, K.A., Knapp, J.E. and Maliga, P. 2001. Expression of bar in the plastid genome confers herbicide resistance. *Plant Physiology*. 125: 1585–1590.

Maliga, P. and Bock, R. 2011. Plastid biotechnology: Food, fuel, and medicine for the 21st century. *Plant Physiology*. 155: 1501–1510.

Maliga, P. and Small, I. 2007. Plant biotechnology: All three genomes make contributions to progress. *Current Opinion in Biotechnology*. 18: 97–99.

Mannerlof, M., Tuvesson, S., Steen, P. and Tenning, P. 1997. Transgenic sugar beet tolerance to glyphosate. *Euphytica*. 94: 83–91.

McGonigle, B., Keeler, S.J., Lau, S.M.C., Koeppe, M.K. and O'Keefe, D.P. 2000. A genomic approach to the comprehensive analysis of the glutathione S-transferase gene family in soybean and maize. *Plant Physiology*. 124: 1105–1120.

M'hamdi, M., Chikh-Rouhou, H., Boughalleb, H. and Ruiz de Galarreta, J.I. 2012. Enhanced resistance to *Rhizoctonia solani* by combined expression of chitinase and ribosome inactivating protein in transgenic potatoes (*Solanum tuberosum L.*). *Spanish Journal of Agricultural Research*. 10: 778–785.

Mihaliak, C.A., Webb, S., Miller, T., Fanton, M., Kirk, D. and Cardineau, G. 2005. Development of plant cell produced vaccines for animal health applications. *Proceedings of the 108th Annual Meeting of the United States Animal Health Association*. Greensboro, NC, 158–163.

Millan, A.F.S., Mingo-Castel, A., Miller, M. and Daniell, H. 2003. A chloroplast transgenic approach to hyper-express and purify human serum albumin, a protein highly susceptible to proteolytic degradation. *Plant Biotechnology Journal*. 1: 71–79.

Molinari, H.B.C., Marur, C.J., Daros, E., Freitas de Campos, M.K., Portela de Carvalho, J.F.R., Filho, J.C.B., Pereira, L.F.P. and Vieira, L.G.E. 2007. Evaluation of the stress-inducible production of proline in transgenic sugarcane (*Saccharum* spp.): Osmotic adjustment, chlorophyll fluorescence and oxidative stress. *Physiologia Plantar.* 130: 218–229.

Molowa, D., Shenouda, M., Meyers, A., Tublin, P. and Fein, A. 2002. The state of biologics manufacturing: Part 2. In: J.P. Morgan (ed.), *Market Analysis of the State of Manufacturing of Biologics Including mAbs.* New York, pp. 1–16.

Mustafa, G. and Khan, M.S. 2012a. Reproducible *in vitro* regeneration system for purifying sugarcane clones. *African Journal of Biotechnology.* 11: 9961–9969.

Mustafa, G. and Khan, M.S. 2012b. Prospecting the utility of antibiotics as lethal selection agents for chloroplast transformation of sugarcane. *International Journal of Agriculture and Biology.* 14: 307–310.

Natalia, I., Almasia, A.A., Bazzini, H., Hopp, E. and Vazquez-Rovere, C. 2008. Overexpression of snakin-1 gene enhances resistance to *Rhizoctonia solani* and *Erwinia carotovora* in transgenic potato plants. *Molecular Plant Pathology.* 9: 329–338.

Nazir, S. and Khan, M.S. 2012. Chloroplast-encoded *chlB* gene from *Pinus thunbergii* promotes root and chlorophyll pigment development in *Nicotiana tabaccum. Molecular Biology Reports.* 39: 10637–10646.

Nguyen, H.T. and Jehle, J.A. 2007. Quantitative analysis of the seasonal and tissue-specific expression of *cry1Ab* in transgenic maize Mon810. *Journal of Plant Diseases and Protection.* 114: 82–87.

Nguyen, T.X. and Sticklen, M. 2013. Barley HVA1 gene confers drought and salt tolerance in transgenic maize (*Zea mays* L.). *Advances in Crop Science and Technology.* 1: 1.

Nuccio, M.L., Russell, B.L., Nolte, K.D., Rathinasabapathi, B., Gage, D.A. and Hanson, A.D. 1998. The endogenous choline supply limits glycine betaine synthesis in transgenic tobacco expressing choline monooxygenase. *The Plant Journal.* 16: 487–496.

Penn, J.B. 2000. Biotechnology in the pipeline: Sparks companies' update. *Proceedings of the 2000 Beltwide Cotton Conference.* National Cotton Council, Memphis, TN.

Perlak, F.J., Deaton, R.W., Armstrong, T.A., Fuchs, R.L., Sims, S.R., Greenplate, J.T. and Fischhoff, D.A. 1990. Insect resistant cotton plants. *Bio/Technology.* 8: 939–943.

Pogue, G.P., Vojdani, F., Palmer, K.E., Hiatt, E., Hume, S. and Phelps, J. 2010. Production of pharmaceutical-grade recombinant aprotinin and a monoclonal antibody product using plant-based transient expression systems. *Plant Biotechnology Journal.* 8:638–654.

Potrykus, I. 2010. Lessons from the "Humanitarian Golden Rice" project: Regulation prevents development of public good genetically engineered crop products. *New Biotechnology.* 27: 466–472.

Przibilla, E., Heiss, S., Johanningmeier, U. and Trebst, A. 1991. Site-specific mutagenesis of the D1 subunit of photosystem II in wild-type *Chlamydomonas. Plant Cell.* 3: 169–174.

Qaim, M. and Zilberman, D. 2003. Yield effects of genetically modified crops in developing countries. *Science.* 299: 900–902.

Qu, L.Q., Yoshihara, T., Ooyama, A., Goto, F. and Takaiwa, F. 2005. Iron accumulation does not parallel the high expression level of ferritin in transgenic rice seeds. *Planta.* 222: 225–233.

Rademacher, T., Sack, M., Arcalis, E., Stadlmann, J., Balzer, S. and Altmann, F. 2008. Recombinant antibody 2G12 produced in maize endosperm efficiently neutralizes HIV-1 and contains predominantly single-GlcNAc N-glycans. *Plant Biotechnology Journal.* 6:189–201.

Reynolds, B.K., Taylor, M.C., Zhou, X.R., Vanhercke, T., Wood, C.C., Blanchard, C.L., Singh, S.P. and Petrie, J.R. 2015. Metabolic engineering of medium chain fatty acid biosynthesis in *Nicotiana benthamiana* plant leaf lipids. *Frontiers in Plant Science.* 6: 1–14.

Richter, J.L., Thanavala, Y., Arntzen, C.J. and Mason, H.S. 2000. Production of hepatitis B surface antigen in transgenic plants for oral immunization. *Nature Biotechnology.* 18: 1167–1171.

Rivero, M., Furman, N., Mencacci, N., Picca, P., Toum, L., Lentz, E., Bravo-Almonacid, F. and Mentaberry, A. 2012. Stacking of antimicrobial genes in potato transgenic plants confers increased resistance to bacterial and fungal pathogens. *Journal of Biotechnology.* 157: 334–343.

Rogers, K.K. 2003. *The Potential of Plant-Made Pharmaceuticals.* http://www.plantpharma.org/ials/index.php?id=1.

Roussel, A., Miled, N., Berti-Dupuis, L., Rivière, M., Spinelli, S., Berna, P., Gruber, V., Verger, R. and Cambillau, C. 2002. Crystal structure of the open form of dog gastric lipase in complex with a phosphonate inhibitor. *Journal of Biological Chemistry.* 277: 2266–2274.

Rybicki, E.P. 2010. Plant made vaccines for humans and animals. *Plant Biotechnology Journal.* 8: 620–637.

Samyn-Petit, B., Gruber, V., Flahaut, C., Wajda-Dubos, J.P., Farrer, S., Pons, A., Desmaizieres, G., Slomianny, M.C., Theisen, M. and Delannoy, P. 2001. N-glycosylation potential of maize: The human lactoferrin used as a model. *Glycoconjugate Journal.* 18: 519–527.

Sanjaya, Durrett, T.P., Weise, S.E. and Benning, C. 2011. Increasing the energy density of vegetative tissues by diverting carbon from starch to oil biosynthesis in transgenic Arabidopsis. *Plant Biotechnology Journal.* 9: 874–883.

Santos Mendoza, M., Dubreucq, B., Miquel, M., Caboche, M. and Lepiniec, L. 2005. LEAFY COTYLEDON 2 activation is sufficient to trigger the accumulation of oil and seed specific mRNAs in Arabidopsis leaves. *FEBS Letters.* 579: 4666–4670.

Savarino, S.J., Hall, E.R., Bassily, S., Wierzba, T.F., Youssef, F.G., Peruski, L.F. Jr., Abu-lyazeed, R. et al. 2002. Introductory evaluation of an oral, killed whole cell enterotoxigenic *Escherichia coli* plus cholera toxin B subunit vaccine in Egyptian infants. *Pediatric Infectious Disease Journal.* 21: 322–330.

Sawahel, W.A. 1994. Transgenic plants: Performance, release and containment. *World Journal of Microbiology and Biotechnology.* 10: 139–144.

Schuster, W. and Brennicke, A. 1988. Interorganellar sequence transfer: Plant mitochondrial DNA is nuclear is plastid is mitochondrial. *Plant Science (Shannon).* 54: 1–10.

Schuster, W. and Brennicke, A. 1994. The plant mitochondrial genome: Structure, information content, RNA editing and gene transfer. *Annual Review of Plant Physiology Plant Molecular Biology.* 45: 61–78.

Shaaltiel, Y., Bartfeld, D., Hashmueli, S., Baum, G., Brill-Almon, E., Galili, G., Dym, O. et al. 2007. Production of glucocerebrosidase with terminal mannose glycans for enzyme replacement therapy of Gaucher's disease using a plant cell system. *Plant Biotechnology Journal.* 5: 579–590.

Shanmugaraj, B.M. and Ramalingam, S. 2014. Plant expression platform for the production of recombinant pharmaceutical proteins. *Austin Journal of Biotechnology and Bioengineering.* 1: 4.

Shimada, H. and Sugiura, M. 1991. Fine structural features of chloroplast genome: Comparison of the sequenced chloroplast genome. *Nucleic Acid Research.* 19: 983–995.

Shimizu, M., Goto, M., Hanai, M., Shimizu, T., Izawa, N., Kanamoto, H., Tomizawa, K.I., Yokota, A. and Kobayashi, H. 2008. Selectable tolerance to herbicides by mutated acetolactate synthase genes integrated into the chloroplast genome of tobacco. *Plant Physiology.* 147: 1976–1983.

Shintani, D. and DellaPenna, D. 1998. Elevating the vitamin E content of plants through metabolic engineering. *Science.* 282: 2098–2100.

Shoji, Y., Bi, H., Musiychuk, K., Rhee, A., Horsey, A., Roy, G., Green, B. et al. 2009. Plant-derived hemagglutinin protects ferrets against challenge infection with the A/Indonesia/05/05 strain of avian influenza. *Vaccine.* 27: 1087–1092.

Shoji, Y., Chichester, J.A., Jones, M., Manceva, S.D., Damon, E., Mett, V., Musiychuk, K. et al. 2011. Plant-based rapid production of recombinant subunit hemagglutinin vaccines targeting H1N1 and H5N1 influenza. *Human Vaccine.* 7: 41–50.

Sivamani, E., Bahieldin, A., Wraith, J.M., Al-Niemi, T., Dyer, W.E., Ho, T.H.D. and Qu, R. 2000. Improved biomass productivity and water use efficiency under water deficit conditions in transgenic wheat constitutively expressing the barley *HVA1* gene. *Plant Science.* 155: 1–9.

Spreitzer, R.J. and Salvucci, M.E. 2002. Rubisco: Structure, regulatory interactions, and possibilities for a better enzyme. *Annual Review of Plant Biology.* 53: 449–475.

Stefanova, P., Angelova, G., Georgieva, T., Gotcheva, V. and Angelov, A. 2015. A novel multiplex PCR method for simultaneous detection of genetically modified soybean events. *International Journal of Current Microbiology and Applied Sciences.* 4(4): 256–268.

Subroto, A.P., Utomo, C., Darmawan, C., Hendroko, R. and Liwang, T. 2014. Tissue culture media optimization and genetic transformation of *Jatropha curcas* genotype Jatromas cotyledon explants. *Energy Procedia.* 47: 15–20.

Tabassum, S., Anwar, Z., Khattak, J.Z., Mahmood, S., Khan, F.A.R., Javed, H., Ashraf, M. and Hussain, S. 2012. The future of biotechnology in Pakistan. *Journal of Asian Scientific Research.* 2: 518–523.

Tacket, C.O., Mason, H.S., Losonsky, G., Estes, M.K., Levine, M.M. and Arntzen, C.J. 2000. Human immune responses to a novel Norwalk virus vaccine delivered in transgenic potatoes. *Journal of Infectious Diseases.* 182: 302–305.

Templeton, M.D., Mitchell, R.E., Sullivan, P.A. and Shepherd, M.G. 1985. The inactivation of ornithine transcarbamoylase by N6 (*N'*-sulpho-diaminophosphinyl)-L-ornithine. *Biochemical Journal*. 228: 347–352.

Thanavala, Y., Yang, Y.F., Lyons, P., Mason, H.S. and Arntzen, C. 1995. Immunogenicity of transgenic plant-derived hepatitis B surface antigen. *Proceedings of the National Academy of Sciences United States*. 92: 3358–3361.

Trujillo, L.E., Sotolongo, M., Menendez, C., Ochogavia, M.E., Coll, Y., Hernandez, I., Borras-Hidalgo, O., Thomma, P.B.H.J., Vera, P. and Hernandez, L. 2008. SodERF3, a novel sugarcane ethylene responsive factor (ERF), enhances salt and drought tolerance when overexpressed in tobacco plants. *Plant Cell Physiology*. 49: 512–525.

Vaeck, M., Reynaerts, A., Hoftey, H., Jansens, S., De Beuckleer, M., Dean, C., Zabeau, M., Van Montagu, M. and Leemans, J. 1987. Transgenic plants protected from insect attack. *Nature*. 327: 33–37.

Van Eenennaam, A.L., Lincoln, K., Durrett, T.P., Valentin, H.E., Shewmaker, C.K., Thorne, G.M., Jiang, J. et al. 2003. Engineering vitamin E content: From *Arabidopsis* mutant to soy oil. *Plant Cell*. 15(12): 3007–3019.

Verberne, C., Verpoorte, R., Bol, J.F., Mercado-Blanco, J. and Linthorst, H.J.M. 2000. Overproduction of salicylic acid in plants by bacterial transgenes enhances pathogen resistance. *Nature Biotechnology*. 18: 779–783.

Von Caemmerer, S.V., Quick, W.P. and Furbank, R.T. 2012. The development of C4 rice: Current progress and future challenges. *Science*. 336: 1671–1672.

Wagner, B., Fuchs, H., Adhami, F., Ma, Y., Scheiner, O. and Breineder, H. 2004. Plant virus expression systems for transient production of recombinant allergens in *Nicotiana benthamiana*. *Methods* 32: 227–234.

Wakasa, K., Hasegawa, H., Nemoto, H., Matsuda, F., Miyazawa, H., Tozawa, Y., Morino, K. et al. 2006. High-level tryptophan accumulation in seeds of transgenic rice and its limited effects on agronomic traits and seed metabolite profile. *Journal of Experimental Botany*. 57: 3069–3078.

Wang, H.H., Yin, W.B. and Hu, Z.M. 2009b. Advances in chloroplast engineering. *Journal of Genetics and Genomics*. 36(7): 387–398.

Wang, Y.P., Wei, Z.Y., Zhang, Y.Y., Lin, C.J., Zhong, X.F., Wang, Y.L., Ma, J.Y., Ma, J. and Xing, S.C. 2015. Chloroplast-expressed MSI-99 in tobacco improves disease resistance and displays inhibitory effect against rice blast fungus. *International Journal of Molecular Sciences*. 16: 4628–4641.

Wang, Z., Gerstein, M. and Snyder, M. 2009a. RNA-Seq: A revolutionary tool for transcriptomics. *Nature Reviews Genetics*. 10: 57–63.

Watson, J., Koya, V., Leppla, S.H. and Daniell, H. 2004. Expression of *Bacillus anthracis* protective antigen in transgenic chloroplasts of tobacco, a non-food/feed crop. *Vaccine*. 22: 4374–4384.

Werck-Reichhart, D., Hehn, H. and Didierjean, L. 2000. Cytochromes P450 for engineering herbicide tolerance. *Trends in Plant Sciences*. 5: 116–123.

White, F.F. 1989. Vectors for gene transfer in higher plants. In: S. Kung and C.J. Arntzen (eds.), *Plant Biotechnology*. Butterworths, Boston. 3–34.

White, P.J. and Broadley, M.R. 2005. Biofortifying crops with essential mineral elements. *Trends in Plant Sciences*. 10: 586–593.

Wigdorovitz, A., Carrilloa, C., Santosa, M.J.D., Tronoa, K., Peraltaa, A., Gomezb, M.C., Riosb, R.D. et al. 1999. Induction of a protective antibody response to foot and mouth disease in mice following oral or parenteral immunization with alfalfa transgenic plants expressing the viral structural protein VP1. *Virology* 255: 347–353.

Wilmink, A. and Dons, J.M.M. 1993. Selective agents and marker genes for use in transformation of monocotyledonous plants. *Plant Molecular Biology Reporter*. 11: 165–185.

Winslow, L.C. and Kroll, D.J. 1998. Herbs as medicines. *Archives of Internal Medicine*. 158: 2192–2199.

Wittkopp, P.J. and Kalay, G. 2012. *Cis*-regulatory elements: Molecular mechanisms and evolutionary processes underlying divergence. *Nature Reviews Genetics*. 13: 59–69.

Wood, C.C. 2014. Leafy biofactories: Producing industrial oils in non-seed biomass. *EMBO Reports*. 15: 201–202.

Wu, G., Truksa, M., Datla, N., Vrinten, P., Bauer, J., Zank, T., Cirpus, P., Heinz, E. and Qiu, X. 2005. Stepwise engineering to produce high yields of very long-chain polyunsaturated fatty acids in plants. *Nature Biotechnology*. 23(8): 1013–1017.

Wu, Y., Zhou, H., Que, Y.X., Chen, R.K. and Zhang, M.Q. 2008. Cloning and identification of promoter Prd29A and its application in sugarcane drought resistance. *Sugar Technology.* 10: 36–41.

Wurbs, D., Ruf, S. and Bock, R. 2007. Contained metabolic engineering in tomatoes by expression of carotenoid biosynthesis genes from the plastid genome. *Plant Journal.* 49: 276–288.

Xu, D., Duan, X., Wang, B., Hong, B., Ho, T.H.D. and Wu, R. 1996. Expression of a late embryogenesis abundant protein gene, *HVA1*, from barley confers tolerance to water deficit and salt stress in transgenic rice. *Plant Physiology.* 110: 249–257.

Yabuta, Y., Tanaka, H., Yoshimura, S., Suzuki, A., Tamoi, M., Maruta, T. and Shigeoka, S. 2012. Improvement of vitamin E quality and quantity in tobacco and lettuce by chloroplast genetic engineering. *Transgenic Research.* 22: 391–402.

Ye, G.N., Colburn, S.M., Xu, C.W., Hajdukiewicz, P.T.J. and Staub, J.M. 2003. Persistence of unselected transgenic DNA during a plastid transformation and segregation approach to herbicide resistance. *Plant Physiology.* 133: 402–410.

Ye, G.N., Hajdukiewicz, P.T.J., Broyles, D., Rodriguez, D., Xu, C.W., Nehra, N. and Staub, J.M. 2001. Plastid-expressed 5-enolpyruvylshikimate-3-phosphate synthase genes provide high level glyphosate tolerance in tobacco. *Plant Journal.* 25: 261–270.

Ye, X., Al-Babili, S., Klöti, A., Zhang, J., Lucca, P., Beyer, P. and Potrykus, I. 2000. Engineering the provitamin A (β-carotene) biosynthetic pathway into (carotenoid-free) rice endosperm. *Science.* 287: 303–305.

Yookongkaew, N., Srivatanakul, M. and Narangajavana, J. 2007. Development of genotype-independent regeneration system for transformation of rice (*Oryza sativa* ssp. *indica*). *Journal of Plant Research.* 120: 237–245.

Yusibov, V., Hooper, D.C., Spitsin, S.V., Fleysh, N., Kean, R.B., Mikheeva, T., Deka, D. et al. 2002. Expression in plants and immunogenicity of plant virus-based experimental rabies vaccine. *Vaccine.* 20: 3155–3164.

Yusibov, V. and Rabindran, S. 2008. Recent progress in the development of plant derived vaccines. *Expert Review of Vaccines.* 7: 1173–1183.

Yusibov, V., Streatfield, S.J. and Kushnir, N. 2011. Clinical development of plant-produced recombinant pharmaceuticals: Vaccines, antibodies and beyond. *Human Vaccines.* 7: 313–321.

Zeng, F., Qian, J., Luo, W., Zhan, Y., Xin, Y. and Yang, C. 2010. Stability of transgenes in long-term micropropagation of plants of transgenic birch (*Betula platyphylla*). *Biotechnology Letters.* 32: 151–156.

Zhang, J., Tan, W., Yang, X.H. and Zhang, H.X. 2008. Plastid-expressed choline monooxygenase gene improves salt and drought tolerance through accumulation of glycine betaine in tobacco. *Plant Cell Reports.* 27: 1113–1124.

Zhang, K., Bhuiya, M.W., Pazo, J.R., Miao, Y., Kim, H., Ralph, J. and Liu, C.J. 2012. An engineered monolignol 4-O-methyltransferase depresses lignin biosynthesis and confers novel metabolic capability in *Arabidopsis. The Plant Cell.* 24: 3135–3152.

Zhang, S.Z., Yang, B.P., Feng, C.L., Chen, R.K., Luo, J.P., Cai, W.W. and Liu, F.H. 2006. Expression of the Grifola frondosa trehalose synthase gene and improvement of drought-tolerance in sugarcane (*Saccharum officinarum* L.). *Journal of Integrative Plant Biology.* 48: 453–459.

Zoubenko, O.V., Allison, L.A., Svab, Z. and Maliga, P. 1994. Efficient targeting of foreign genes into the tobacco plastid genome. *Nucleic Acids Research.* 22: 3819–3824.

chapter four

The chloroplast gene-expression system

Yusuke Yagi and Takashi Shiina

Contents

Abstract

Chloroplasts are semiautonomous organelles that contain only limited coding information and are dependent on a large number of nucleus-encoded proteins. During plant evolution, chloroplasts lost many prokaryotic deoxyribonucleic acid (DNA)-binding proteins and transcription regulators that were present in the original endosymbiont. Thus, chloroplasts have a unique hybrid transcription system composed of the remaining prokaryotic components as well as the nucleus-encoded eukaryotic components. Recent proteome and transcriptome analyses have provided insights into chloroplast transcription systems and their evolution. In this chapter, we review the chloroplast-specific transcription systems, focusing on the multiple ribonucleic acid polymerases, eukaryotic transcription regulators in chloroplasts, unique factors involved in posttranscriptional modification in chloroplasts, chloroplast promoters, and dynamics of chloroplast nucleoids.

Keywords: Chloroplasts, Plastids, Transcription, Organelle RNA Metabolism.

Introduction

Chloroplasts are the green organelles found in almost all land plants and algae. Chloroplasts convert light energy into chemical energy by photosynthesis, which feeds all living organisms on Earth. Furthermore, the oxygen produced by photosynthesis supports modern atmospheric conditions. Chloroplasts are not responsible for photosynthesis. In plant cells, chloroplasts represent just one of the specialized forms that organelles called plastids, for "plastic" organelles, can take. As the name implies, plastids can change in form, size, and color in distinct cell types and in response to environment cues such as light conditions, chloroplasts in leaves, amyloplasts in roots, chromoplasts in fruits, etioplasts in etiolated plants, and proplastids in seeds. Plastids are also involved in multiple essential metabolic reactions, such as the production of starch, all 20 amino acids, and fatty acids. Their plasticity supports the multiple functions of plastids in the plant cells.

Chloroplasts are believed to have arisen from an endosymbiotic event between a photosynthetic cyanobacterium and a eukaryotic cell that had already acquired a mitochondrion. Although chloroplasts of modern plants and algae have retained the genome of the symbiont, the genome has drastically shrunk during intracellular evolution. Many chloroplast-encoded genes were lost or transferred to the nucleus soon after endosymbiosis. In 1986, the first chloroplast genomes were fully sequenced by the Sugiura and Oyama groups using tobacco (*Nicotiana tabacum*) and the common liverwort (*Marchantia polymorpha*), respectively (Ohyama et al. 1986; Shinozaki et al. 1986). The tobacco chloroplast genome is 150 kbp, with two 2.5-kbp inverted repeats (IRs) that are separated by single-copy regions (SSC and LSC; Figure 4.1). It contains 50 protein-coding genes involved in photosynthesis, gene expression, lipid metabolism and other processes, 30 transfer ribonucleic acid (tRNA) genes, and full sets of ribosomal ribonucleic acid (rRNA) genes. Unlike genes in cyanobacteria and other bacteria, chloroplast-encoded genes in land plants possess several introns (a single group I intron and ~20 group II introns). The chloroplast genomes of green algae have a relatively high number of group I introns, suggesting that algal introns were acquired independently of those in land plants (Steinmetz et al. 1982; Erickson et al. 1984; Rochaix et al. 1985).

In spite of their small genomes (0.15 Mbp in land plant chloroplasts versus 3 Mbp in cyanobacteria), chloroplast gene expression is regulated by more complex systems compared to the simple prokaryotic regulatory system. Chloroplast gene expression is mediated by two distinct types of RNA polymerase (RNAP) and is highly dependent on the posttranscriptional regulation, such as processing of polycistronic transcripts, intron splicing, and RNA editing. Moreover, recent RNA-seq analyses of chloroplast transcripts identified unexpected diversification of RNA molecules such as noncoding RNAs and antisense RNAs (Hotto et al. 2011; Zhelyazkova et al. 2012). However, the genes encoded in the chloroplast genomes are insufficient to regulate their complicated gene expression, and the chloroplast gene-expression machinery includes various nucleus-encoded regulatory components.

The basic chloroplast gene expression is mediated by prokaryotic machineries derived from the ancestral cyanobacterium. Similar to bacterial nucleoids, chloroplast DNA forms a

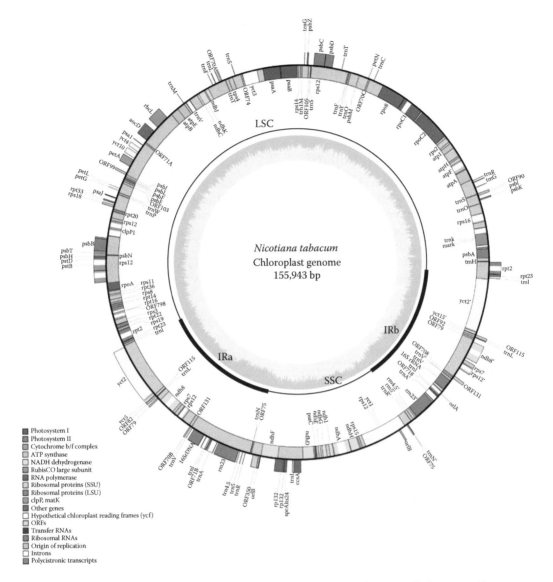

Figure 4.1 Chloroplast genome map. The chloroplast genome was drawn with the organelle genome draw tool (http://ogdraw.mpimp-golm.mpg.de/) using *N. tabacum* chloroplast DNA sequences (Genbank accession number NP_001879). The black inner circle shows the inverted repeat structures (IRa and IRb) and single-copy regions (LSC, large single copy; SSC, small single copy). The inside of the inner circle (gray zone) represents GC contents. Genes are shown on the outer circle as colored boxes. Genes transcribed clockwise and counterclockwise are depicted outside and inside the circle, respectively. Different colors represent different gene functions, as shown below the map.

chloroplast nucleoid that is packed with various proteins and RNA molecules (Sakai et al. 2004). As expected, a bacterial-type RNAP has been identified in chloroplasts. In contrast to the basic expression machinery, chloroplasts lost homologs of bacterial regulatory elements such as transcription factors and nucleoid proteins at an early stage of chloroplast evolution. Genomics and proteomics analyses in *Arabidopsis thaliana* have estimated that chloroplasts contain only ~800 proteins homologous to those of cyanobacteria among a total of 2300 chloroplast proteins (Abdallah et al. 2000). The remaining 1500 genes for chloroplast proteins may have been newly acquired from the nuclear genome of host cells after the endosymbiosis event. Indeed, higher-plant chloroplasts have acquired phage-type RNAP that resembles mitochondrial RNAP. Furthermore, the recent proteomic analyses identified many nonbacterial nucleoid components that play critical roles in chloroplast gene expression including transcription, posttranscriptional RNA processing, and translation. Here, we summarize the current knowledge regarding the chloroplast gene-expression system.

Multiple RNAPs in chloroplasts

The bacterial-type RNAP in chloroplasts, called plastid-encoded plastid RNA polymerase (PEP), shares functional similarity with bacterial multisubunit RNAP (Igloi and Kossel 1992) (Figure 4.2a). Enzyme activity of PEP is inhibited specifically by tagetoxin (Mathews and Durbin 1990; Sexton et al. 1990; Rajasekhar et al. 1991; Sakai et al. 1998) and partly by rifampicin, inhibitors of bacterial RNAP (Orozco et al. 1985; Pfannschmidt and Link 1994). Furthermore, *Escherichia coli* RNAP β and β' subunits are functionally exchangeable with recombinant PEP subunits, indicating the functional equivalence of RNAP core subunits

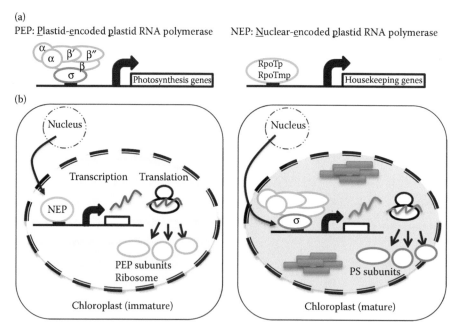

Figure 4.2 Basic transcriptional machineries in higher plants. (a) Two types of chloroplast RNAPs. Higher plants have two distinct types of chloroplast RNAP, PEP, and nucleus-encoded plastid RNA polymerase (NEP). PEP (left panel) is a bacterial-type multi-subunit RNAP composed of the core enzymatic subunits α, β, β', b" (blue), and a sigma subunit (red) that is responsible for promoter recognition. Core subunits are generally encoded in chloroplast genomes, whereas sigma subunits are encoded by nuclear genes. PEP mainly transcribes chloroplast-encoded photosynthesis genes. NEP (right panel) is a monomeric enzyme that resembles mitochondrial T7-type RNAPs. NEP is involved in the transcription of housekeeping genes such as *rpo* genes for PEP core subunits, and ribosomal protein-coding genes. (b) NEP–PEP cascade during chloroplast development. In immature chloroplasts, NEP plays a dominant role and is responsible for transcription of housekeeping genes for the chloroplast genetic machineries such as PEP and ribosomes. Subsequently, the resultant PEP transcribes photosynthesis genes.

between bacteria and chloroplasts (Severinov et al. 1996). Thus, the available evidence supports the idea that the PEP originated from the cyanobacterial endosymbiont.

Bacterial RNAPs are composed of a core Rpo complex ($\alpha\xi2$, β, and β'), which has the catalytic enzyme activity, and a sigma factor, which recognizes promoter sequences (Ishihama 2000). Unlike in other bacteria, cyanobacterial RNAP *rpoC*, which encodes the β' subunit, is split into two genes, *rpoC1* and *rpoC2* encoding the N-terminal (β') and C-terminal domain (β''), respectively. These split *rpoC* genes have been maintained in chloroplasts. Whereas all genes for chloroplast sigma factors have been transferred to the nuclear genome and genes for core subunits are typically retained in the chloroplast genome as *rpoA*, *rpoB*, *rpoC1*, and *rpoC2*. However, in some cases, *rpoA* genes have also been transferred to the nuclear genome, for example, in the nonphotosynthetic parasitic plant *Epifagus virginiana* (Morden et al. 1991), and the moss *Physcomitrella patens* (Sugiura et al. 2003). *RpoB, C1*, and *C2* are transcribed polycistronically from the *rpoB* operon, whereas *rpoA* is in an operon with several ribosomal protein-coding genes in chloroplasts.

Analysis of PEP-deficient plants such as ribosome-deficient mutants of barley (*Hordeum vulgare*), *iojap* mutants of maize (*Zea mays*; Han et al. 1992), and tobacco mutants disrupted in *rpo* genes generated by gene targeting using chloroplast transformation (Allison et al. 1996; De Santis-MacIossek et al. 1999; Kapoor and Sugiura 1999), revealed that almost all photosynthesis genes are silenced in these mutants, although a set of housekeeping genes are still active. Mapping the transcription initiation sites of these PEP-independent genes revealed that the upstream-flanking sequence of these transcripts lack typical bacterial promoter elements, such as the −35 and −10 elements that are recognized by sigma factors (Allison et al. 1996; Hajdukiewicz et al. 1997; Kapoor et al. 1997). Furthermore, the inhibitor sensitivity of this transcription activity is similar to that of phage T7 RNAP, but not to that of bacterial RNAP (Kapoor et al. 1997; Sakai et al. 1998). The chloroplast genome includes no genes coding for phage-type RNAPs, but multiple phage-type RNAP genes were identified from an *Arabidopsis* nuclear genomic library using a *Chenopodium album* mitochondrial RNAP probe (Weihe et al. 1997; Weihe and Borner 1999). Localization analysis revealed that one of these gene products is targeted to mitochondria (*RpoTm*) and one to chloroplasts (*RpoTp*) (Hedtke et al. 1999). This additional *RpoTp* RNAP in the chloroplast is termed as nuclear-encoded plastid RNA polymerase (NEP; Figure 4.2a). Fractionation of the plastid RNAP activities identified a 110-kDa single-subunit RNAP that could initiate transcription from the T7 promoter, but not from a typical σ^{70}-type *RbcL* promoter (Lerbs-Mache 1993). The expected molecular size of RpoTp is ~113 kDa, suggesting that *RpoTp* encodes the 110-kDa single-subunit RNAP identified by the biochemical approach. Overexpression of *AtRpoTp* in tobacco resulted in enhanced transcription of the sets of genes that are transcribed by NEP in PEP-deficient plants (Liere et al. 2004). Recombinant RpoTp proteins could initiate transcription from a typical T7 promoter and chloroplast NEP promoters (Liere and Link 1994). These findings demonstrate that *RpoTp* encodes a NEP enzyme and reveal a second transcription apparatus in chloroplasts.

In addition to RpoTp and RpoTm, a third phage-type RNA polymerase (RpoTmp) has been identified in dicotyledonous plants such as *Arabidopsis* and tobacco. RpoTmp is localized to both mitochondria and chloroplasts (Hedtke et al. 2000; Kobayashi et al. 2001; Hedtke et al. 2002). However, RpoTmp is not encoded in monocotyledonous plant genomes (Chang et al. 1999; Ikeda and Gray 1999; Emanuel et al. 2004). *Arabidopsis* knockout mutants of RpoTmp exhibit delayed greening, defective light-induced expression of several chloroplast mRNAs, and decreased accumulation of NEP-dependent transcripts (Baba et al. 2004). RpoTmp also functions specifically in the transcription of the *rrn* operon in proplastids and amyloplasts during seed imbibition and germination (Courtois et al. 2007).

Furthermore, double mutants of *RpoTp* and *RpoTmp* showed a severe phenotype with a strong growth defect compared to single-mutant plants (Hricova et al. 2006). These findings suggested that NEP enzymes RpoTp and RpoTmp are both active in chloroplasts of dicots, although the functional interplay between these NEPs remains largely unclear.

Green algae, such as *Chlamydomonas reinhardtii*, *Ostreococcus tauri*, and *Thalassiosira pseudonana* possess only one *RpoT* gene, which likely encodes mitochondrial RNAP (Surzycki 1969; Guertin and Bellemare 1979; Eberhard et al. 2002; Armbrust et al. 2004; Derelle et al. 2006). Similarly, the genome of the lycophyte *Selaginella moellendorffii* contains only one *RpoT* gene, the product of which targets to the mitochondria (Yin et al. 2009). On the contrary, the moss *P. patens* has three *RpoT* genes. Green fluorescent protein (GFP)-fused moss RpoTs have been shown to target the mitochondria exclusively, suggesting that the moss *RpoT* genes also encode mitochondrial RNAP (Kabeya et al. 2002; Richter et al. 2002, 2014). Moreover, phylogenetic analysis of plant *RpoT* genes suggests that NEP appeared by gene duplication of mitochondrial RNAP after the separation of angiosperms from gymnosperms.

As described earlier, there are two distinct types of RNAPs in chloroplasts of higher plants. PEP and NEP selectively transcribe distinct sets of chloroplast genes such as photosynthesis genes and housekeeping genes, respectively. Analysis of the chloroplast gene-expression pattern in Δ *rpoB* plants revealed that chloroplast genes can be categorized into three subgroups, classes I–III. Class I mainly contains photosynthesis-related genes that are transcribed by PEP, whereas class III genes (*accD* and the *rpoB* operon) are exclusively transcribed by NEP (Allison et al. 1996; Hajdukiewicz et al. 1997). Class II includes many other housekeeping genes (*clpP* and the *rrn* operon) that are transcribed by both PEP and NEP.

Expression profiles of plastid genes have been analyzed in various types of plastids. In higher plants, chloroplasts develop from proplastids and undifferentiated plastids that are found in meristematic tissues (Figure 4.2b). In early chloroplast development, NEP transcribes a group of genes-encoding PEP subunits (*rpo* genes) and components of ribosomes (*rps* and *rpl* genes) in advance of the expression of photosynthesis genes. Subsequently, PEP actively transcribes photosynthesis genes to produce the photosynthetic machinery, while NEP activity gradually declines during the greening process (Figure 4.2). It has been reported that barley *RpoTp* transcripts are highly accumulated in leaf bases containing immature chloroplasts and decrease at the tip of leaves containing mature chloroplasts (Emanuel et al. 2004). The developmental expression of NEP and PEP (the NEP–PEP cascade) may play a critical role in the regulation of chloroplast development (Figure 4.2b).

High NEP and low PEP activities have also been reported in other types of plastids such as amyloplasts in roots and chromoplasts in fruit tissues. Whole-plastid gene transcription analyses of potato (*Solanum tuberosum*) tuber amyloplasts compared with leaf chloroplasts revealed that accumulation of photosynthesis-related gene transcripts is greatly decreased (but not completely lost) in amyloplasts, whereas that of *accD*, *rpo*, *ycf1*, *ycf2*, and *ycf15* transcripts is increased (Valkov et al. 2009). Moreover, during tomato fruit ripening, a subset of NEP-dependent genes including *accD*, *trnA*, and *rpoC2* are specifically upregulated (Kahlau and Bock 2008). These results suggest that the differential function of both RNAPs is essential for plastid differentiation.

Plastid transcription

Overview of bacterial transcription regulators

Bacterial transcription is controlled at multiple steps including initiation, elongation, pausing, termination, and recycling of RNAP. At the initiation step, the holo-RNAP with different sigma factors recognizes distinct types of promoters and initiates transcription (Murakami and Darst 2003). Once the transcript reaches ~15 nucleotides, the RNAP holoenzyme releases the sigma factor and converts into an elongation complex (EC) for RNA synthesis. Transcription factors bind to promoter sequences independent of RNAP, and either activate or repress promoter activity to control the transcription initiation. Additional proteins associate with the EC (RNAP-associated proteins; Nus factors [NusA, NusB, NusG, and NusE]) (Borukhov et al. 2005; Burmann and Rosch 2011), RfaH (Svetlov et al. 2007), ribosomal protein S4 (Torres et al. 2001), Gre-factors (GreA, GreB) (Borukhov et al. 2005), Mfd (Borukhov et al. 2005), rho (Richardson 1993), and bacteriophage factors (Das 1992) during the elongation and termination steps, and modulate multiple steps of transcription, including transcription pausing, arrest, termination, or antitermination.

Chloroplast sigma factors

Chloroplast sigma factors were first detected immunologically in mustard (*Sinapis alba* L.) leaves (Bülow and Link 1988). Moreover, three chloroplast σ-like factors (SLF67, SLF52, and SLF29) have been purified biochemically from chloroplasts and etioplasts of mustard seedlings (Tiller et al. 1991; Tiller and Link 1993b). The biochemical-isolated sigma-like factors (SLFs) have distinct promoter-binding activities and ionic strength requirements, suggesting that chloroplast SLFs could confer the promoter recognition specificity of PEP, like bacterial sigma factors. The first genes-encoding chloroplast sigma factors were identified in the unicellular red alga *Cyanidium caldarium* (Liu and Troxler 1996; Tanaka et al. 1996), which contains three genes for chloroplast sigma factors (*CcaSigA*, *CcaSigB*, and *CcaSigC*). Recombinant CcaSigA protein confers transcription initiation activity to the *E. coli* RNAP core complex (Tanaka et al. 1996).

The nuclear genome of the most primitive red algae, *Cyanidoschyzon merolae*, also encodes multiple genes for sigma factors (CmeSig1–CmeSig4) (Matsuzaki et al. 2004). Among higher plants, *Arabidopsis* encodes six-sigma genes (*Sig1*–*Sig6*) (Isono et al. 1997; Tanaka et al. 1997; Kanamaru et al. 1999; Allison 2000; Fujiwara et al. 2000). Similarly, all the sequenced genomes of land plant species possess multiple sigma genes (five genes in maize, four genes in rice [*Oryza sativa*]), three genes in *Physcomitrella* (Tozawa et al. 1998; Lahiri et al. 1999; Tan and Troxler 1999; Lahiri and Allison 2000; Hara et al. 2001a,b; Ichikawa et al. 2004; Kasai et al. 2004). However, unicellular green algae including *C. reinhardtii* (Carter et al. 2004; Bohne et al. 2006) and the primitive alga *O. tauri* (Derelle et al. 2006) have only one chloroplast sigma factor.

Bacterial sigma factors form two distinct families based on their structural conservation: the σ^{70} and σ^{54} families (Wosten 1998). The σ^{70} family comprises primary σ factors (group 1) and nonessential σ factors (groups 2 and 3). Group 1 σ^{70} factors are responsible for transcription of housekeeping genes during the exponential growth phase and are essential for cell survival. Group 2 σ^{70} factors are closely related to group 1 sigma factors, but are not essential for cell viability. Group 3 σ^{70} factors are alternative sigma factors that recognize specific promoter sequences in response to environmental cues (Lonetto et al. 1992). Phylogenetic analysis of bacterial and chloroplast sigma factors indicates that chloroplast sigma factors fall into a monophyletic group (Shiina et al. 2005). All plant sigma factors are related to bacterial σ^{70} primary group 1 and group 2 factors, but not to alternative (σ^{70} group 3) or σ^{54} sigma factors. On the basis of this, it is assumed that only the primary σ^{70} factor was retained in chloroplasts, whereas other nonessential sigma factors were lost during the early chloroplast evolution. Characterization of multiple chloroplast sigma factors is described in the section "Chloroplast promoters."

PEP-associated proteins

Two types of PEP-containing preparations have been biochemically isolated in mustard and *Arabidopsis*: soluble RNA polymerase (sRNAP) and plastid transcriptionally active chromosome (pTAC) attached to chloroplast membranes (Hess and Borner 1999). *In vitro* transcription assays with these preparations revealed that transcription by sRNAP requires exogenously added template DNAs, whereas the pTAC can initiate transcription from the endogenous chloroplast DNA (Igloi and Kossel 1992; Krause et al. 2000). Both sRNAP and pTAC perform transcription of protein-coding genes, tRNAs, and rRNAs (Reiss and Link 1985; Rajasekhar et al. 1991; Krupinska and Falk 1994). Interestingly, protein compositions of highly purified sRNAP fractions are developmentally regulated (Pfannschmidt and Link 1994) (Figure 4.3a). sRNAP of etioplasts in dark-grown leaves possess only RNAP

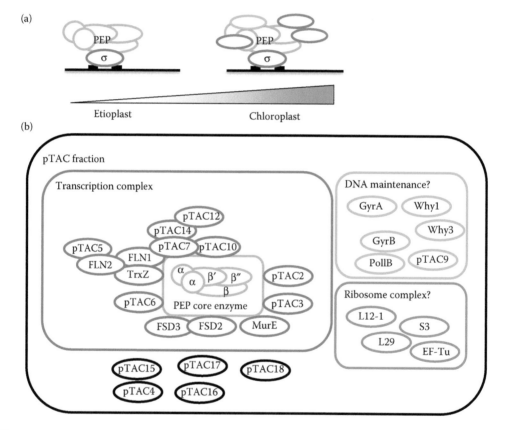

Figure 4.3 Transcription complexes in chloroplasts. (a) Developmental regulation of PEP complexes. Etioplasts have a simple and small PEP complex composed of core units (α, β, β′, and β″) and a sigma factor. During the greening process of higher plants, the PEP complex binds several additional PAPs and becomes larger (from 0.4 to 1 MDa). (b) pTAC proteins. pTAC fractions comprise PEP core enzymes (light-blue box) and many novel protein components that may directly or indirectly interact with PEP core subunits (red and purple circles, respectively). pTAC proteins form a kind of network, and pTAC7 is assumed to play a bridging role among PEP accessory proteins including pTAC10, pTAC12–pTAC14, and FLN1–FLN2–TRX-z–pTAC5 (Huang et al. 2013). The pTAC fraction also contains other proteins that are functionally categorized into several groups such as those for maintenance of chloroplast DNA and translation.

subunits similar to the *E. coli* RNAP core complex. Etioplasts convert to photosynthetically active chloroplasts under light. During the transition to chloroplasts in mustard, the RNAP develops a more complex form that is composed of not only RNAP core subunits but also at least 13 additional polypeptides (Pfannschmidt and Link 1994). It is assumed that the complex sRNAP in chloroplasts is converted from the simpler sRNAP in etioplasts by recruiting additional components (Figure 4.3a).

Proteomic analyses of pTAC fractions isolated from mature chloroplasts of *Arabidopsis* and mustard have identified 35 polypeptides including 18 novel proteins termed pTAC1–pTAC18, in addition to PEP core subunits, DNA polymerase, DNA gyrase, Fe-dependent superoxide dismutases (FeSODs), phosphofructokinase–B-type enzymes (PFKB1 and PFKB2), thioredoxin (TRX-z), and three ribosomal proteins (Pfalz et al. 2006). Analysis of domain structures in pTAC1-18 revealed that some of them have DNA- and/or RNA-binding domains, protein–protein interaction domains, or epitopes described for other cellular functions, but pTAC6, 7, 8, and 12 have no known functional domains. *Arabidopsis* knockout mutants of pTAC components generally exhibit seedling-lethal symptoms or chlorophyll-deficient phenotypes. PEP-dependent transcription is significantly reduced in the pTAC mutants, whereas NEP-dependent genes are actively transcribed. These phenotypes and chloroplast gene-expression patterns resemble those of *rpo* mutants (Allison et al. 1996; Hajdukiewicz et al. 1997), suggesting a role for pTAC proteins in the PEP transcription.

Further purification of the mustard PEP complex by heparin–sepharose chromatography and blue native two-dimensional (2-D) gel electrophoresis (Steiner et al. 2011), as well as immunoaffinity purification of the PEP complex using His-tagged α subunits produced via chloroplast transformation techniques in tobacco, identified at least 10 PEP-associated proteins (PAPs) (Suzuki et al. 2004). Recently, it has been shown that one of typical PAPs, pTAC3, associates with the PEP complex in all three steps of the transcription cycle including initiation, elongation, and termination (Yagi et al. 2012). Several studies on protein–protein interactions among PAPs have been reported (reviewed in Pfalz and Pfannschmidt 2013) (Figure 4.3b). Almost all PAP genes, except for *Trx-z*, are conserved among all land plants, but not in the green alga *Chlamydomonas*, suggesting that terrestrial plants may have acquired noncyanobacterial proteins that associate with the PEP complex during land plant evolution to regulate plastid transcription (Pfalz and Pfannschmidt 2013).

Although the primary functions of each PAP during the PEP transcription cycle remain largely unknown, it has been proposed that assembly of PAPs with the PEP complex may have a major role in checkpoints for chloroplast development (Steiner et al. 2011; Pfalz and Pfannschmidt 2013). Early in plant development, chloroplast-encoded *rpo* genes for PEP core subunits are transcribed by the NEP system to produce a basic PEP complex (PEP-B enzyme). PEP-B is responsible for the major activity in etioplasts and in immature plastids during greening. This step might be a first checkpoint in the establishment of chloroplast transcription machineries. Subsequently, PEP-B converts into PEP-A during chloroplast development via posttranslational modification by unknown modifying components. PEP-A associates with PAPs to form a larger complex. PEP-A formation is strictly dependent on light. Indeed, it has been reported that expression of *pTAC3/PAP1* genes is induced by light during the greening process (Yagi et al. 2012). Furthermore, mutants deficient in each of the PAPs generally show an abnormal chloroplast development and gene-expression patterns, suggesting that assembly of PAPs into the PEP complex may be a second checkpoint in the establishment of the plastid gene-expression machinery.

Transcription elongation and termination

Knowledge about post-initiation steps in chloroplast transcription is limited. A highly active PEP fraction isolated from chloroplasts contains only one putative transcription elongation factor, pTAC13, which has partial similarity to the prokaryotic transcription elongation factor NusG (Pfalz et al. 2006). No other PAPs have similarity to bacterial RNAP-associated proteins, suggesting that PAPs did not originate in the cyanobacterial lineage. On the contrary, Etched 1 (ET1), a candidate plastid elongation regulator, has significant similarity to the nuclear transcription elongation factor TFIIS (da Costa e Silva et al. 2004). The *ET1* gene underlies the *virescent* mutants of maize that exhibit aberrant chloroplast development in kernels and leaves. The chloroplast transcription activity is significantly reduced in the ET1-deficient mutants. Eukaryotic TFIIS is a functional counterpart of the bacterial transcription elongation factor GreB. Both GreB and TFIIS reactivate arrested RNAP to synthesize RNA efficiently. Although the molecular function of ET1 in PEP transcription remains elusive, further analysis may shed light on the regulation of PEP transcription at the elongation step in chloroplasts (Conaway et al. 2003).

In the transcription termination step, RNAP complex dissociates from the template DNA and releases the RNA transcript. This step is regulated by two mechanisms in bacteria: Rho protein-dependent termination and intrinsic sequence-dependent termination (Rho independent) (Santangelo and Artsimovitch 2011). It has been reported that the chloroplast transcription termination occurs at intrinsic bacterial-like terminators *in vitro* (Chen et al. 1995). On the contrary, most chloroplast 3′-termini are generated by 3′-to-5′ exoribonucleases (described in detail in the section "NEP regulators").

Light-dependent PEP transcription

Light plays an important role in the regulation of plastid transcription to build and maintain the photosynthetic apparatus in higher plants. Microarray analyses of transcription profiles of entire plastid chromosomes from tobacco revealed that accumulation of most photosynthesis-related gene transcripts is strongly increased by light, whereas housekeeping gene transcripts are accumulated constitutively (Nakamura et al. 2003). Furthermore, light-dependent upregulation of the overall transcriptional activity of the plastid genome has been reported in mature chloroplasts of spinach (Deng and Gruissem 1987), sorghum (*Sorghum bicolor*; Schrubar et al. 1991), barley (Baumgartner et al. 1993), wheat (*Triticum aestivum*) (Satoh et al. 1999), pea (*Pisum sativum*; Dubell and Mullet 1995), *Arabidopsis* (Chun et al. 2001; Hoffer and Christopher 1997), and tobacco (Shiina et al. 1998; Baena-Gonzalez et al. 2001). Plants sense light quality through photoreceptors: phytochromes sense red/far-red light, whereas blue light is detected by blue-light receptors including cryptochromes and phototropins (Briggs and Olney 2001). In mature leaves, light-dependent activation of the overall transcriptional activity of chloroplasts is mediated mainly by blue/UV-A light signaling via cryptochromes, whereas red light merely modulates plastid transcription, suggesting that cryptochromes play a major role in light-dependent chloroplast transcription (Thum et al. 2001a). On the contrary, the chloroplast transcription activity is reduced in *phyA* mutants exposed to the blue light and the UV-A light, suggesting a possible role for *phyA* in blue- and UV-B-light signaling to control chloroplast transcription (Chun et al. 2001). Light triggers translocation of phytochromes from the cytoplasm to subnuclear foci (phytochrome nuclear bodies) (Yamaguchi et al. 1999; Kircher et al. 2002; Chen et al. 2003). Recently, a *hemera* mutant was identified as a mutant with altered patterns of phytochrome nuclear bodies (Chen et al. 2010). Interestingly, *HEMERA* encodes pTAC12, which was previously identified as one of the pTAC components. HEMERA/pTAC12 is localized not only to the nucleus but also to chloroplast nucleoids, where it acts specifically in phytochrome signaling. The *hemera* mutant shows a seedling-lethal phenotype, impairment of all phytochrome responses, and reduced accumulation of PEP-dependent transcripts in chloroplasts. pTAC12 is therefore presumed to play a specific role in the light-dependent activation of PEP transcription through phytochrome-signaling pathways.

In addition to light signaling, photosynthetic status also regulates chloroplast transcription activity. Imbalanced stoichiometry between the two photosystems PSI and PSII causes unequal distribution of excitation energy, which leads to redox changes in the plastoquinone (PQ) pool (Pfannschmidt et al. 1999a,b). The redox state of the PQ pool inversely regulates the expression of *psaA/psaB* (encoding PSI reaction center proteins) and *psbA* (encoding a PSII reaction center protein) at the transcriptional level. The transcription of *psaA/psaB* is significantly upregulated by 560-nm light, which favors PSII over PSI, whereas *psbA* transcription is downregulated. Conversely, 650-nm light favors PSI over PSII, and downregulates *psaA/psaB* transcription, but upregulates *psbA* transcription. Transcription of *psaA/psaB* is enhanced by 2,5-dibromo-3-methyl-6-isopropyl-*p*-benzoquinone (DBMIB), which blocks electron flow from the PQ pool to PSI, but is suppressed by 3-(3,4′-dichlorophenyl)-1,1′-dimethyl urea (DCMU), which blocks electron flow from PSII to the PQ pool. On the contrary, the *psbA* transcription is suppressed by the DBMIB and enhanced by the DCMU. These facts suggest that the redox state of the chloroplast PQ pool controls the transcription of reaction center genes to maintain the balance of two photosystems (Pfannschmidt et al. 1999a,b).

Mutants deficient in chloroplast sensor kinase (CSK), a bacterial-type sensor histidine kinase, are defective in the redox control of *psaA/psaB* and *psbA* transcription, suggesting that CSK is responsible for the control of chloroplast gene expression by the chloroplast redox state (Puthiyaveetil et al. 2008). Moreover, it has been shown that *Arabidopsis* CSK is autophosphorylated and directly interacts with itself, SIG1, and plastid transcription kinase (PTK; Puthiyaveetil et al. 2013). The PTK is a eukaryotic serine/threonine protein kinase that regulates the overall transcriptional activity of chloroplasts through phosphorylation of PEP subunits including sigma factors. At an early stage of greening, the basic form of PEP and PEP-B is more phosphorylated than the PEP-A in mature chloroplasts (Tiller et al. 1991; Tiller and Link 1993a,b). The PTK activity is inversely regulated by its phosphorylation state and the redox state of glutathione (Baginsky et al. 1999). Reduced glutathione (GSH) negatively affects unphosphorylated PTK, whereas oxidized GSH (GSSG) enhances the kinase activity of the phosphorylated form. The amount of GSH in chloroplasts is increased by high-irradiance light, implicating glutathione as one of the redox signal mediators in chloroplasts (Baena-Gonzalez et al. 2001).

A candidate PTK in maize is cpCK2α, which is a homolog of cytosolic casein kinase (CK2) α subunit that bears an *N*-terminal chloroplast-targeting transit peptide (Ogrzewalla et al. 2002). *In vitro* kinase assays revealed that recombinant cpCK2α exhibits CK2 activity and that its activity is inhibited in the presence of GSH, as well as PTK. Moreover, cpCK2α is found in PEP-A complex and phosphorylates PEP components including SaSIG1 in mustard (Schweer et al. 2010). Plant sigma factors generally have multiple predicted phosphorylation sites for CK2. Complementation analyses using site-directed mutants in putative phosphorylation sites of AtSIG6 revealed that phosphorylation of SIG6 occurs at multiple sites, and affects promoter specificity (Schweer et al. 2010). AtSIG1 was also shown to be phosphorylated *in vivo* and its phosphorylation is involved in the transcriptional regulation of *psaA/psaB* and *psbA* by photosynthetic redox state through altered promoter specificity

(Shimizu et al. 2010). These results imply that cpCK2α may be involved in redox-dependent phosphorylation of PEP and sigma factors.

Several PAPs are likely involved in redox sensing, suggesting that PEP may also be regulated by PAP-mediated redox signaling. Iron superoxide dismutases (FeSODs, FSDs) are the primary antioxidant enzymes in chloroplasts (Myouga et al. 2008). Among three FSDs in *Arabidopsis* (FSD1, FSD2, and FSD3), FSD2 and FSD3 have been detected in the pTAC fraction (Pfalz et al. 2006). Yeast two hybrid and BiFC assays indicated that FSD2 and FSD3 form a heteromeric complex on chloroplast nucleoids. A *fsd2 fsd3* double mutant exhibits a severe albino phenotype and similar transcription patterns to those found in PEP-deficient plants. Overexpression of FSD2 and FSD3 results in higher tolerance to oxidative stress induced by methyl viologen than displayed by mutants deficient in those proteins, suggesting that FSD2 and FDS3 have a role in preventing reactive oxygen species (ROS) damage of the chloroplast nucleoids or PEP by scavenging the ROS. The redox sensor TRX-z has also been reported to be a PEP regulator (Arsova et al. 2010). TRX-zs modify the functions of their target proteins by reversible modification of thiol groups via disulfide reductase activity. TRX-z/PAP10 interacts with two chloroplast fructokinase-like proteins, FLN1 and FLN2 through conserved Cys motifs in the FLNs (Arsova et al. 2010; Huang et al. 2013). Furthermore, light-dependent modulation of the redox state of FLN2 is mediated by TRX-z in chloroplasts. Both FLNs have been detected in pTAC fractions. FLN1 was identified as PAP6. PEP transcription activity is regulated by light quality through the sensing of the redox state of the photosynthetic electron transport chain (as detailed earlier). Thus, it is likely that redox-related PAPs may have roles in PEP transcriptional regulation as redox sensors.

Regulation of chloroplast transcription by ppGpp

In bacteria, guanosine 3′,5′-(bis)pyrophosphate (ppGpp) acts as a second messenger for stress adaptation and modifies the promoter selectivity of RNAP through direct interaction with the β subunit of RNAP (Haugen et al. 2008). ppGpp is synthesized from GTP and ATP by RelA/SpoT-homologous enzymes (RSH). RSHs have also been identified in chloroplasts of both green algae (Kasai et al. 2002) and higher plants (van der Biezen et al. 2000; Givens et al. 2004). Moreover, a unique Ca^{2+}-activated RSH, CRSH, has been identified in angiosperms such as *Arabidopsis* (Masuda et al. 2008) and rice (Tozawa et al. 2007). It has been demonstrated that ppGpp is present in chloroplasts and that its level is markedly elevated by light and various abiotic and biotic stresses (Takahashi et al. 2004). In addition, ppGpp can inhibit chloroplast transcription (Takahashi et al. 2004) and translation (Nomura et al. 2012) *in vitro*. These facts suggest the existence of a ppGpp-signaling system for regulation of chloroplast gene expression.

NEP regulators

As mentioned earlier, two NEPs have been identified in dicots: RpoTmp and RpoTp. Mutant analyses suggest that RpoTp is the principal NEP enzyme, with a critical role in chloroplast transcription, biogenesis, and mesophyll cell proliferation (Hricova et al. 2006; Courtois et al. 2007). In contrast, RpoTmp contributes to light-induced transcription of several plastid mRNAs, and is involved in specific transcription from the PC promoter of the *rrn* operon of *Arabidopsis* during seed imbibition and germination (Baba et al. 2004; Emanuel et al. 2006; Courtois et al. 2007; Swiatecka-Hagenbruch et al. 2008). *In vitro* transcription analyses using recombinant RpoTp and RpoTmp indicated that RpoTp can accurately initiate transcription from the NEP promoters without any protein cofactor, whereas RpoTmp displays no significant promoter specificity (Kuhn et al. 2007). On the contrary, mitochondrial RNAPs in yeast and mammals require several auxiliary factors for the transcription initiation *in vivo* (Asin-Cayuela and Gustafsson 2007; Deshpande and Patel 2012). However, the NEP-associated factors involved in specific promoter recognition and transcription initiation remain largely unknown.

Although NEP enzymes are present in mature chloroplasts, the accumulation of NEP-dependent transcripts is very low, suggesting the presence of a specific repressor (or repressors) for the NEP transcription in chloroplasts. It has also been proposed that plastid tRNA^{Glu}, which is transcribed by PEP, inhibits NEP (RpoTp)-dependent transcription through direct interaction (Hanaoka et al. 2005). On the contrary, CDF2 may act as a transcription initiation factor for NEP (Bligny et al. 2000). CDF2 is required for transcription from the NEP-specific PC promoter of the spinach *rrn16* operon, which is transcribed by RpoTmp. CDF2 has two distinct forms, CDF2-A and CDF2-B. CDF2-A represses transcription initiation by PEP from the *rrn16* P1 promoter, whereas CDF2-B might bind to NEP, with the resulting complex possibly initiating transcription from the *rrn16* Pc promoter.

Recent searches for interaction partners of RpoTmp by a yeast two-hybrid screening identified two NEP interaction proteins (NIPs) (Azevedo et al. 2008). The NIPs are membrane integral proteins harboring a RING finger domain that is exposed to the stromal surface of the thylakoids. In chloroplasts, RpoTmp is tightly associated with thylakoid membranes, suggesting that NIPs anchor RpoTmp to the thylakoid membranes via their RING domain. The *rrn* transcripts from the Pc promoter are the most abundant transcripts in proplastids and develop plastids during early periods of chloroplast differentiation, and rapidly decline in mature chloroplasts. NIP accumulation shows a complementary expression pattern to *rrn* PC transcripts during seed germination. Thus, NIPs may be required for specific inhibition of RpoTmp during plastid differentiation.

Posttranscriptional regulation

In chloroplasts, RNA metabolism, including RNA processing, intron splicing, RNA degradation and stabilization, and RNA editing, play important roles in the regulation of RNA abundance (reviewed in Stern et al. 2010) (Figure 4.4). Despite the declining gene contents during chloroplast evolution, many transcripts are processed in chloroplasts. Mutants with impaired chloroplast RNA processing have been extensively studied in algae and plants. The results suggest that chloroplast RNA metabolism is essential for plant development and cell viability. In this section, we review the diversification of chloroplast RNA metabolism and the associated regulatory factors.

In chloroplasts, primary polycistronic RNA is transcribed by PEP and/or NEP, and is processed in multiple steps, including cleavage of internal polycistronic RNAs, endo- or exonucleolytic cleavages at the 5′-, and/or 3′-transcript termini and intron splicing (Figure 4.4a). For example, the *psbB* operon in *Arabidopsis* contains five protein-coding genes, *psbB*, *psbT*, *psbH*, *petB*, and *petD*. This *psbB* gene cluster generates approximately 20 processed RNAs through intercistronic processing and intron splicing. The 5′ and 3′ ends of chloroplast RNAs are generally trimmed at specific positions by RNA processing enzymes, rather than being generated by transcription initiation and termination. 5′-to-3′ exonuclease activity creates the 5′-termini of RNA. Chloroplast-localized RNaseJ, homologs of which have endonuclease and 5′-to-3′ exonuclease activity in *E. coli*, acts as a 5′-to-3′ exonuclease in chloroplasts (Stern and Gruissem 1987). Chloroplast genes generally have IR sequences at the 3′ UTR, which are predicted to form stem-loop structures (Yehudai-Resheff et al.

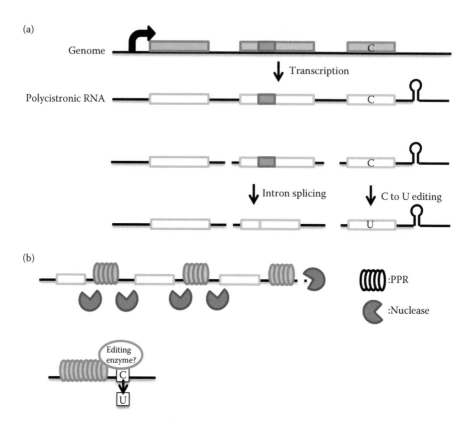

Figure 4.4 Posttranscriptional modifications of chloroplast RNA. (a) General illustration of post-transcriptional RNA modifications in chloroplasts. Several chloroplast genes form gene clusters and are under the control of a single promoter. Polycistronic transcription terminates on a stem-loop structure at the 3′ UTR. Sometimes, cleavage of primary polycistronic RNAs produces multiple mature mRNA units. Several chloroplast genes contain introns (orange boxes, exon; gray box, intron). Introns are removed from the precursor RNA via RNA splicing. Furthermore, chloroplast transcripts have 30–40 cytosine (C)-to-uridine (U) RNA-editing sites (green box). (b) The role of PPR proteins in chloroplast RNA modification. PPR proteins bind to specific positions between each mRNA unit on polycistronic RNAs to protect them from cleavage by RNases (upper diagram). PPR proteins also bind upstream of editable "C" sites on RNA (lower diagram). It is assumed that one PPR protein may be responsible for one or a few specific RNA modification events.

2001). In bacteria, secondary structures at 3′-termini are necessary to protect transcripts from degradation by the 3′-to-5′ exoribonuclease activity of polynucleotide phosphory-lase (PNPase). The 3′-termini of many chloroplast RNAs are likely protected by analogous mechanisms (Yehudai-Resheff et al. 2001). However, 3′-end mapping of chloroplast RNAs revealed that only a few transcripts retain predicted stem loops at the 3′-termini, suggest-ing the presence of other mechanisms for protecting chloroplast transcripts from endo-nucleases (Hotto et al. 2011; Zmienko et al. 2011; Zhelyazkova et al. 2012).

It has been proposed that the termini of chloroplast transcripts are covered by a "cap" to protect them from nucleolytic cleavage (Figure 4.4b). PPR10 is a sequence-specific RNA-binding protein that interacts with the processed RNA termini of the intercistronic *atpI-atpH* and *psaJ-rpl33* (Pfalz et al. 2009; Prikryl et al. 2011). RNase protection assays dem-onstrated that the association of PPR10 with target RNAs could protect the RNA from degradation by ribonucleases (Prikryl et al. 2011). This "cap" model is also supported by analysis of several other proteins, including HCF152 (Meierhoff et al. 2003), CPR1 (Barkan et al. 1994), PpPPR_38 (Hattori et al. 2007), and MRL1 (Johnson et al. 2010), which associate with intercistronic processed RNAs. Recent studies using RNA-seq in *Arabidopsis* revealed an abundance of small RNAs (16–28 nucleotides) in chloroplasts (Ruwe and Schmitz-Linneweber 2012). These small RNA sequences correspond to the 5'- and 3'-termini of chloroplast RNAs, suggesting that they may be footprints resulting from the protection of "cap" proteins against exonucleases.

PPR proteins in chloroplasts

These "cap" proteins belong to the pentatricopeptide repeat (PPR) protein family, whose members are defined by the presence of tandem arrays of a degenerate 35-amino-acid repeat termed the PPR motif (reviewed in Schmitz-Linneweber and Small 2008). The PPR gene family is eukaryote specific, and is especially expanded in angiosperms (to over 500 genes) in contrast to less than 30 PPR genes found in other eukaryotes including yeast, human, and algae (Lurin et al. 2004). Almost all PPR proteins are predicted to localize to chloroplast or mitochondria.

Computational analysis of the PPR domain in the family found that PPR proteins can be subdivided into two subgroups, the classical P-type subgroup and the PLS-type subgroup, which contain motifs of different lengths. P-type PPR proteins are characterized by tandem repeats of a degenerate 35-amino-acid motif, whereas PLS-type PPR proteins have motifs of different length, PPR-like S (short; 31 amino acids), PPR-like (long; 35 amino acids), and L2 (36 amino acids). The "cap" proteins for RNA termini are classical P-type PPR proteins, whereas the other PLS proteins are generally involved in RNA editing (see Figure 4.4b). Reverse genetics studies have revealed that mutations in individual PPR genes often cause cytoplasmic male sterility (Ding et al. 2006), retardation of plastid development (Chateigner-Boutin et al. 2008), high sensitivity to both abiotic and biotic stress (Laluk et al. 2011), or distinct types of lethal phenotypes, such as embryolethality (Sosso et al. 2012), indicating that PPR proteins play crucial roles in many different physiological processes. It is generally accepted that PPR proteins serve as RNA adapters with sequence specificity for target RNAs in organelles in a variety of posttranscriptional events including processing, splicing, protecting, and editing (Figure 4.4b). The PPR protein family in green algae contains only a dozen members, even though the chloroplast has complex RNA metabolism regulation. However, a member of the 38-amino-acid-repeat protein family, octatricopeptide repeat protein (OPR), has been identified in *Chlamydomonas* (Rahire et al. 2012), and the OPR protein family in green algae is expanded to ~70 members, including several proteins involved in specific chloroplast RNA processing.

Other chloroplast RNA-binding proteins have been identified through studies on chloroplast intron-splicing mechanisms. CRM (Till et al. 2001; Ostheimer et al. 2003), PORP (des Francs-Small et al. 2012), and APO domain (Watkins et al. 2011)-containing proteins are plant specific and are also predicted to localize to chloroplasts and/or mitochondria. These protein families bind intronic regions of specific organellar transcripts. The splicing of nuclear gene introns is catalyzed by the spliceosome, a complex of small nuclear ribonucleoproteins (snRNPs). In contrast, a different combination of splicing factors acts to promote the splicing of different subsets of introns in chloroplasts and mitochondria (de Longevialle et al. 2010).

Another noteworthy posttranscriptional event is organelle RNA editing that converts C to U, or rarely, U to C in transcripts. *Arabidopsis* contains 34 and 488 editing sites in chloroplasts and mitochondria, respectively (Shikanai 2006). The number of organellar editing sites seems to have drastically increased during the higher plant evolution. Mammalian cytosolic C to U RNA editing is mediated by two components: target RNA recognition factor and cytidine deaminases that convert a C in the RNA to U (apoB-editing catalytic subunit-1 [APOBEC-1] and APOBEC-1 complementation factor, respectively) (Teng et al. 1993; Mehta et al. 2000). A site recognition factor for chloroplast RNA editing was initially identified from a photosynthesis mutant of *Arabidopsis, chlororespiratory reduction 4* (*crr4*) (Kotera et al. 2005). The *crr4* mutant displays defective RNA editing of ndhD117166, where the *ndhD* translation initiation codon is created. CRR4 is a PPR protein belonging to the PLS subfamily and interacts with the RNA upstream of editing sites. Another genetic analysis also identified PLS-subfamily PPR proteins as specific editing site recognition factors for chloroplasts and mitochondria in higher plants and mosses (Shikanai and Fujii 2013).

Many PLS-subfamily members possess several additional motifs (E and DYW motifs) at their C-termini. The DYW motif is made up of residues that are conserved in the active site of cytidine deaminases, suggesting that DYW motifs possibly function as the catalytic domain for RNA editing (Salone et al. 2007). However, several PLS proteins involved in RNA editing do not have DYW motifs. Thus, it has now been proposed that the E/DYW motifs may recruit the editing enzymes and/or harbor the deaminase activity. In addition, multiple organellar RNA-editing factors (MORFs) have been recently identified as RNA-editing components in plant organelles (Bentolila et al. 2012; Takenaka et al. 2012). MORF proteins associate with PPR proteins and promote RNA editing. These findings indicate that RNA-editing mechanisms in plant organelles are likely more complex than the well-characterized RNA editing that occurs in mammalian species.

Chloroplast promoters

The compact genome in chloroplasts of higher plants contains a hundred genes encoding ribosomal and transfer ribonucleic acids (rRNAs and tRNAs) as well as proteins involved in photosynthesis and plastid gene transcription and translation. In contrast to bacteria, cyanobacterium-derived PEP is not sufficient to transcribe all genes in the chloroplast genomes of higher plants. As mentioned earlier, a second chloroplast RNAP NEP, is similar to T7 and related to bacteriophage-monomeric RNAPs, and is probably derived from a gene duplication of mitochondrial RNAP. PEP and NEP recognize distinct types of promoters and play critical roles in the regulation of chloroplast development.

PEP promoters

In bacteria, sigma factors are responsible for locating promoters by recognizing the two binding sites, typically located at −10 and −35 positions of the transcription initiation sites (Hawley and McClure 1983). Generally, members of the σ^{70} family of sigma factors contact two conserved promoter elements, the −10 (TATAAT) and −35 (TTGACA) elements, which are separated by approximately 17 bp (Saecker et al. 2011). Similarly, most chloroplast promoters recognized by PEP resemble the bacterial σ^{70}-type promoters with −10 and −35 consensus elements (Gatenby et al. 1981; Gruissem and Zurawski 1985; Strittmatter et al. 1985; Shiina et al. 2005) (Figure 4.5a). Recent chloroplast RNA-sequencing approaches produced a genome-wide map of transcription start sites (TSSs) in barley chloroplasts (Zhelyazkova et al. 2012). RNA-sequencing analysis of green leaves demonstrated that 89% of the mapped TSSs have a conserved −10 element (TAtaaT) three to nine nucleotides

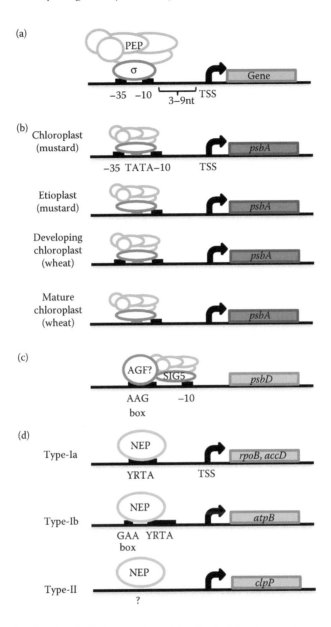

Figure 4.5 Chloroplast PEP and NEP promoters. (a) A typical PEP promoter. PEP recognizes bacterial σ[70]-type promoters, which are characterized by –10 and –35 consensus core elements spaced by 17–19 nt. The –10 elements are typically located at 3–9 nucleotides upstream of TSSs. Arrows show TSSs and the transcription direction. (b) Promoter structure of *psbA*. (c) Structure of the *psbD* LRP. *psbD* LRP lacks a functional –35 element, but has conserved AAG repeat sequences (AAG box) immediately upstream of the –35 element. Although AAG box-interacting factor (AGF, purple circle) has been detected by UV–cross-linking analysis, the gene for AGF remains unidentified. (d) NEP promoters (type-Ia, type-Ib, and type-II).

upstream (Zhelyazkova et al. 2012). These results suggest that most genes are transcribed by PEP in green leaves. On the contrary, the −35 element was mapped upstream of the −10 element in only 70% of the TSSs.

It has been previously demonstrated that some chloroplast promoters are transcribed by PEP independent of −35 elements (To et al. 1996; Satoh et al. 1997; Kim et al. 1999; Thum et al. 2001b). The *psbA* promoter is an outstandingly active PEP promoter containing a "TATA box"-like element between prokaryotic-type −35 and −10 elements. In mustard chloroplasts, both the −35 and −10 promoter elements play an essential role in the *psbA* transcription (Eisermann et al. 1990) (Figure 4.5b). Interestingly, the *psbA* transcription is dependent on the "TATA box"-like element and the −10 element, but not on the −35 element in etioplasts (Eisermann et al. 1990). On the contrary, in wheat, the −35 element is essential for the *psbA* transcription in developing chloroplasts, but not in mature chloroplasts (Satoh et al. 1999). Taken together, the lower degree of conservation of the −35 element suggests that some types of PEP can initiate transcription from promoters that lack the −35 element and/or contain other specific *cis* elements.

Higher plants have multiple sigma factors that are expected to confer promoter specificity upon the PEP core complex (Shiina et al. 2005; Lerbs-Mache 2011). By analogy with bacteria, it is expected that each sigma factor recognizes a distinct set of promoters and plays specific roles in the transcriptional regulation in response to developmental and/or environmental cues. Molecular genetic analyses revealed that SIG2 and SIG6 cooperatively regulate light-dependent chloroplast development (Hanaoka et al. 2003; Ishizaki et al. 2005). SIG2 is responsible for transcription of a group of tRNA genes, but not photosynthesis genes (Kanamaru et al. 2001). On the contrary, SIG6 targets a wide range of photosynthesis genes at an early stage of chloroplast development (Ishizaki et al. 2005). In addition, SIG3 and SIG4 specifically transcribe the *psbN* and *ndhF* genes in *Arabidopsis* (Favory et al. 2005; Zghidi et al. 2007). Recently, ChIP analysis of SIG1 revealed that it binds specifically to target promoters including those of *psaAB*, *psbBT*, *psbEFLJ*, *rbcL*, and *clpP* (Hanaoka et al. 2012). However, all these sigma factor-dependent genes are transcribed from typical σ^{70}-type promoters with −10 and −35 consensus elements. Specific promoter sequences recognized by each chloroplast sigma factor remain largely elusive except for SIG5.

SIG5 is a unique sigma factor whose expression is rapidly induced by various environmental stresses such as high osmolality, salinity, and low temperature as well as high-light stress (Tsunoyama et al. 2002; Nagashima et al. 2004). Molecular genetic analyses demonstrated that SIG5 is essential for transcription from a set of promoters, including the *psbD* light-responsive promoter (LRP) (Nagashima et al. 2004; Tsunoyama et al. 2004). The *psbD* LRP is a unique light- and stress-responsive promoter that is mapped around 900 bp upstream from the *psbD* translation initiation site (Christopher et al. 1992; Wada et al. 1994). Unlike other PEP promoters, the *psbD* LRP has a conserved upstream *cis* element termed the AAG box that is located −36 to −64 bp upstream of the TSS (Figure 4.5c). *In vitro* transcription experiments using extracts from chloroplasts demonstrated that the AAG box is essential for transcription of the *psbD* LRP, but the −35 element is not (Christopher et al. 1992; Nakahira et al. 1998). Furthermore, *in vivo* reporter assays using the chloroplast transformation technique confirmed the importance of the AAG box and the −10 element in transcription activation of the *psbD* LRP (Thum et al. 2001b). These results demonstrated that SIG5 is responsible for transcription from this unique PEP promoter, the *psbD* LRP. Interestingly, the AAG box was found in the *psbD* LRP of gymnosperms as well as angiosperms, but not in moss, suggesting a conserved role for it in the transcription of *psbD* in seed plants (Ichikawa et al. 2004).

NEP promoters

Most NEP promoters (*rpoB*, *rpoA*, and *accD*) share a core sequence, the YRTA motif (type-Ia) (Liere and Maliga 1999; Weihe and Borner 1999; Hirata et al. 2004) (Figure 4.5d). This sequence motif is conserved among higher plants, and is similar to motifs-found promoters of plant mitochondria (Binder and Brennicke 2003; Kuhn et al. 2005). A subclass of NEP promoters (type-Ib) has a GAA box upstream of the YRTA motif (Kapoor and Sugiura 1999) (Figure 4.5d). In contrast to these standard NEP promoters, type-II NEP promoters have been mapped upstream of the dicot *clpP* gene. The type-II NEP promoters lack the YRTA motif and are dependent on downstream sequences of the TSS (Weihe and Borner 1999) (Figure 4.5d). Furthermore, other non-consensus-type NEP promoters have been mapped for the *rrn* operon and certain tRNAs (reviewed by Grasser 2006).

As detailed in the section "Multiple RNAPs in chloroplasts," previous studies using PEP-deficient plants revealed that chloroplast genes can be classified into three subgroups. Briefly, class I genes-encoding photosynthesis proteins, including *psbA*, the *psbB* operon, *psbD/C*, *psbEFLJ*, *psaA/B*, and *rbcL*, are transcribed by PEP, whereas nonphotosynthetic housekeeping genes have both PEP and NEP promoters (class II). Class III comprises few genes (the *rpoB* operon, *ycf1*, *ycf2*, and *accD*) that are transcribed exclusively by NEP (Allison et al. 1996; Hajdukiewicz et al. 1997). However, the genome-wide mapping of TSSs in barley revealed that most genes including photosynthesis genes have both PEP and NEP promoters, suggesting that NEP-dependent transcripts for photosynthesis proteins could support the early stage of seedling greening (Zhelyazkova et al. 2012). Interestingly, the YRTA motif typical for type-Ia and type-Ib NEP promoters were mapped upstream of 73% of NEP-dependent TSSs. On the contrary, these analyses did not find conserved GAA boxes upstream of YRTA motifs in the barley NEP promoters. Furthermore, no conserved promoter motifs could be detected upstream of the remaining NEP-dependent TSSs lacking the YRTA motif. These results suggest that type-Ia, but not type-Ib NEP promoters play a major role in transcription by NEP in barley chloroplasts. In contrast, type-II NEP promoters, which are dependent on downstream sequences of the TSS, were identified in barley as well as in tobacco.

It is widely assumed that chloroplast-encoded genes are actively transcribed in photosynthetically active chloroplasts of leaves but that their transcription is drastically down-regulated in nongreen plastids such as proplastids and chromoplasts. In *Arabidopsis*, the transcription activity of most chloroplast-encoded genes is the highest in 7–17-day-old leaves and is significantly reduced in younger cotyledons (7-days old) and older rosette leaves (2-days old) (Zoschke et al. 2007). Examination of PEP and NEP promoter activities of *clpP* and *rrn16* revealed that PEP has a more prominent role in transcription of these genes in older leaves (Zoschke et al. 2007). On the contrary, transcription of most plastid-encoded genes is significantly downregulated during the chloroplast-to-chromoplast transition of tomato fruits. However, it should be noted that expression of *accD*, encoding an enzyme involved in fatty acid metabolism, is not greatly downregulated during the chromoplast development (Kahlau and Bock 2008). These findings suggest that the promoter activity might be regulated differentially during plastid differentiation.

Nucleoids

The number of copies of the chloroplast genome per chloroplast is not constant, but instead changes dramatically during chloroplast development. Chloroplasts in green leaves of higher plants contain many copies of the chloroplast genome, ~100 chloroplast

DNA molecules per single chloroplast, and up to 10,000 copies per cell. On the contrary, immature proplastids contain fewer DNA copies (~20 DNA molecules per single proplastid) (Miyamura et al. 1986). The plastid DNAs are wrapped by several proteins and RNA molecules, and form a specialized structure called the plastid nucleoid. Plastid nucleoids contain on average 10–20 copies of the plastid DNA and can be visualized with 4′,6-diamidino-2-phenylindole (DAPI) staining (Kuroiwa 1991). The size, shape, and distribution of plastid nucleoids vary depending on plastid type (Miyamura et al. 1986; Sato et al. 1997). Chloroplasts each contain ~20 nucleoids randomly located in the matrix of organelle. Immature proplastids of seed cells contain only one nucleoid that is located at the center of the organelle. The plastid nucleoids divide into a few small dots and redistribute to the inner envelope membranes during early chloroplast development. During the later stage of chloroplast development, nucleoids at the envelope membranes are relocated to the thylakoid membranes. Plastid nucleoid organization and dynamics are expected to be involved in the regulation of plastid function, gene expression, and differentiation. However, the underlying molecular mechanisms are unknown. In this section, we review the components of plastid nucleoids identified by several proteomics approaches and the role of nucleoids in gene expression.

In *E. coli*, packaging patterns of chromosome DNA affect gene expression and are regulated by factors including DNA topology and combinatorial interactions with nucleoid-associated proteins (NAPs) such as HU, H-NS, and FIS (reviewed by Dillon and Dorman 2010). Among bacterial NAPs, HU is one of most abundant DNA-binding proteins involved in chromosome DNA packaging. HU-like proteins (HLPs) are conserved in cyanobacteria, the red alga *Cyanidioschyzon merolae* (Kobayashi et al. 2002), and the green alga *Chlamydomonas* (Karcher et al. 2009). Knockdown analysis of a *Chlamydomonas* HLP revealed that it has roles in nucleoid maintenance and gene expression, suggesting a conservation of HU functions in the chloroplast evolution (Karcher et al. 2009). However, land plants including mosses and higher plants have lost the genes-encoding HLP-s from both chloroplast and nuclear genomes. Nevertheless, plastid nucleoids are highly organized and form a beads-on-a-string structure similar to that observed in bacterial nucleoids using atomic force microscopy.

Plant chloroplast NAPs and nucleoid-related components have been identified using several biochemical approaches. Sulfite reductase (SiR) was found to be an abundant DNA-binding protein in pea chloroplasts (Sato et al. 2001). SiR has compaction activity toward plastid DNA *in vitro* and also effectively represses transcription activity *in vitro*, suggesting that SiR regulates the global transcriptional activity of chloroplast nucleoids via changes in DNA compaction (Sekine et al. 2002). However, functional comparison of SiR proteins between different plant species (pea and maize) showed that the DNA binding and compaction activity of pea SiR is higher than that of the maize protein. Moreover, although pea SiR localizes mainly to nucleoids, no maize SiR proteins have been observed in plastid nucleoids (Sekine et al. 2007). Thus, SiR is involved in compacting plastid nucleoids, but the extent of the association of SiR with nucleoids varies among plant species.

Chloroplast nucleoids in higher plants associate with the envelope or thylakoid membranes depending on the chloroplast development. It is thought that two DNA-binding proteins, PEND and MFP1, anchor the association of nucleoids with the chloroplast membranes (Sato et al. 1998; Jeong et al. 2003). PEND contains a sequence-specific DNA-binding domain (cbZIP) and C-terminal hydrophobic domains, which are required for localization to chloroplast envelope membranes (Sato et al. 1998). On the contrary, MFP1 has been shown to localize to the thylakoid membrane, and its C-terminal domain has the DNA-binding activity (Jeong et al. 2003). Thus, it is proposed that these two anchor proteins are involved in establishing the localization pattern of nucleoids during the chloroplast development.

A fraction enriched for transcription activity from the thylakoid membrane of mature chloroplasts, termed the pTAC fraction, can be prepared. EM observation of isolated pTAC revealed chromatin-like beaded structures in which several DNA loops protrude from a central cluster region, suggesting that pTACs represent a subdomain of the chloroplast nucleoid (Yoshida et al. 1978; Briat et al. 1982). As mentioned in the section "Multiple RNAPs in chloroplasts," the pTAC fraction includes the large PEP complex composed of the Rpo core and several PAPs, proteins involved in DNA maintenance, and the other proteins with unknown function (Pfalz et al. 2006). Proteome analysis of a further-enriched pTAC fraction (TAC-II) from spinach chloroplasts identified small proteins containing a eukaryotic SWIB (SWI/SNF complex B) domain (Melonek et al. 2012). *Arabidopsis* has six small SWIB-domain-containing proteins (SWIB1-6), which are predicted to have organellar localization. Plastid nucleoid- and nucleus-localized SWIB4 protein has a histone H1 motif and can functionally complement an *E. coli* mutant lacking the histone-like nucleoid-structuring protein H-NS, indicating that SWIB4 is a candidate to be a counterpart of the bacterial NAPs. These findings seem to indicate that pTAC forms a central core of the plastid nucleoid and a transcription factory (for PEP). (Notably, NEP has been never detected in a pTAC fraction, possibly due to its less-abundant expression.)

The proteome of chloroplast nucleoids has been characterized using highly enriched nucleoid fractions of proplastids and mature chloroplasts isolated from maize leaf bases and tips, respectively (Majeran et al. 2012). The chloroplast nucleoids contain components of transcriptional machinery including PEP core Rpo and PAPs, almost all other pTAC proteins, and other proteins involved in posttranscriptional processes, such as PPR proteins, and mitochondrial transcription factor (mTERF)-domain proteins. Furthermore, 70S ribosome and ribosome assembly factors are also highly accumulated in nucleoid fractions.

Comparisons of the nucleoid proteomes of chloroplasts and proplastids has allowed the characterization of nucleoid proteins of unknown function based on their coexpression profiles (Majeran et al. 2012). These findings suggest that several posttranslational events including RNA processing, splicing and editing, and translation, occur in nucleoids, and that these processes are coregulated with transcription. Thus, proteomics analysis of the plastid nucleoids has shed light on the organization of plastid nucleoids. It has been proposed that human mitochondrial nucleoids form layered structures in which replication and transcription occur in the central core, whereas translation and complex assembly may occur in the peripheral region (Bogenhagen et al. 2008). By analogy, plastid nucleoids seem to be organized into a core domain composed of DNA maintenance and transcription machineries, and several subdomains involved in various aspects of RNA metabolism.

Evolution of plastid transcription

Plastid genomes

Chloroplasts are descendants of oxygen-producing photosynthetic cyanobacteria that were endocytosed and lived in symbiosis with primitive plant cells. The genome of the extant cyanobacterium *Synechocystis* sp. PCC6803 is 3573 kbp and contains 3317 genes (Kaneko et al. 1996). By contrast, typical chloroplasts of higher plants have smaller genomes, ranging from 120 to 170 kbp, and contain ~100 genes encoding partial sets of proteins for photosynthesis, lipid metabolism, RNAP (PEP), and ribosome components, along with several tRNA and rRNA genes (reviewed in Green 2011). Most protein-coding genes appear to have been lost or transferred to the host nuclear genome during the evolution of plants, and it is estimated that about 14%–18% of nuclear-encoded proteins are cyanobacterial in origin (Martin et al. 2002; Deusch et al. 2008). Nuclear-encoded cyanobacterium-derived proteins are synthesized on cytoplasmic ribosomes, and some of them (~800) are transported into chloroplasts. Overall, higher-plant chloroplasts contain only ~900 cyanobacterium-derived prokaryotic-type proteins. On the contrary, chloroplast proteome analyses revealed that chloroplasts contain over 3000 proteins (Abdallah et al. 2000). Therefore, chloroplast metabolism, function, and structure are largely dependent on nuclear-encoded non-bacterial-type proteins and/or hybrid components of both bacterial and eukaryotic origin.

Chloroplast sigma factors

Extant cyanobacterial genomes encode more than 100 transcription factors and six to seven sigma factors, which are involved in the developmental and environmental transcriptional regulation (Kaneko et al. 1996). The green alga *C. reinhardtii* encodes only a single σ^{70}-like factor, which functions in the chloroplast (Carter et al. 2004; Bohne et al. 2006) (Figure 4.6). The same is true of other green algae, such as *Ostreococcus* (Derelle et al. 2006). On the contrary, liverwort contains four chloroplast sigma factors, including SIG1,

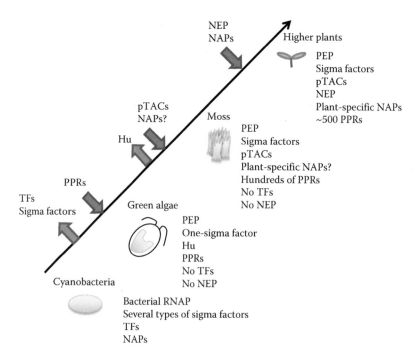

Figure 4.6 Evolution of components of the chloroplast transcription machineries. Ancient cyanobacteria have a prototype of PEP including Rpo subunits and several types of sigma factors, several transcription factors (TFs), and various NAPs. After forming an endosymbiosis, the primary chloroplast lost sigma factors, except the σ^{70} type, and all TFs. During the evolution of land plants, chloroplasts acquired more complicated transcription machineries with a variety of pTACs. In higher plants, there are multiple RNAPs (PEP, multiple sigma factors, PEP–pTAC complex, and NEP), and plant-specific NAPs. The PPR protein family appeared in green algae and expanded during land plant evolution. The increases in the number of PPR genes likely occurred in concert with the establishment of various posttranscriptional modulations.

SIG2, and SIG5 homologs (Ueda et al. 2013). The moss *P. patens* (Hara et al. 2001a,b; Rensing et al. 2008) and the early diverging vascular land plant *S. moellendorffii* (Shiina et al. 2009) also have SIG1, SIG2, and SIG5 homologs, but not SIG3, SIG4, and SIG6. As described in the section "Chloroplast sigma factors," all these sigma factors are related to bacterial σ^{70}-type factors. It is assumed that chloroplasts lost multiple sigma factors early in evolution and retained just one principal sigma factor to support the basic transcription by PEP, suggesting a decrease in demand for PEP to control gene expression in the endosymbionts. Since algal sigma factor genes share no conserved introns with higher-plant sigma factor genes (SIG1, SIG2, SIG3, SIG4, and SIG6), it is assumed that the green algal sigma factors evolved independently from those of higher plants.

However, it is likely that the number of chloroplast sigma factors subsequently expanded during the evolution of land plants from algae to meet the demand for transcriptional control of chloroplast gene expression to support cellular differentiation and responses to a fluctuating environment. It appears that SIG5 first diverged from the common ancestor of chloroplast sigma factors and evolved independently. Analyses of the intron–exon structures of plant sigma factor genes suggest that SIG1 and SIG2 likely arose from the common ancestor of higher-plant sigma factors, probably in *Streptophyta phylum*, the common ancestor of charophytes and higher plants. Subsequently, SIG3 and SIG6 likely evolved from SIG2 during the evolution of angiosperms, before the separation of monocots and dicots. On the contrary, the primitive red alga *C. merolae* retained four sigma factors that are not related to sigma factors in green algae and higher plants. These sigma factors probably evolved independently in the red algae lineage.

Evolution of transcription factors

In higher plants, neither the nuclear nor the chloroplast genome encodes the prokaryotic transcription factors, suggesting that the regulation of chloroplast transcription is mediated by eukaryotic-type transcription factors, and not by cyanobacterium-derived prokaryotic factors. It is assumed that chloroplasts have lost prokaryotic transcription factors during evolution (Figure 4.6). In fact, as mentioned earlier, higher plants have a number of host cell-derived novel transcriptional regulators and DNA-binding proteins that are involved in the regulation of chloroplast transcription. By contrast, the chloroplast genome of the primitive red alga *C. merolae* retained some transcription factors (ycf27–ycf30) that have been lost from green algae and higher plants (Minoda et al. 2010).

Evolution of NEP

One of the most important questions surrounding the evolution of plastid transcription is when plants acquired NEP (Figure 4.6). The phage-type RNAP RpoTm functions in mitochondria of most eukaryotes, including yeast, animals, and plants. It is assumed that chloroplast-localized NEP (RpoTp) was created by duplication of the gene encoding a mitochondrial counterpart. Chloroplast-localized NEP has been identified in angiosperms, including both dicots and monocots. Dicots have two types of NEP (chloroplast-localized RpoTp and dual targeting [chloroplasts and mitochondria] RpoTmp), whereas monocots have only chloroplast-targeted RpoTp. Neither the red alga *C. merolae* nor the green alga *C. reinhardtii* has chloroplast-localized NEP, suggesting that NEP was acquired during the evolution of land plants. By contrast, the moss *P. patens* contains three *RpoT* genes; one gene encodes a mitochondrial RNAP (RpoTm), whereas the other two have been demonstrated to encode potential dual-targeting transit peptides (Kabeya et al. 2002; Richter et al.

2002, 2014), suggesting the presence of RpoTmp in moss. However, the lycophyte *Selaginella* possesses only one nuclear NEP gene encoding a mitochondrial RNAP (RpoTm) (Yin et al. 2009). Thus, NEP likely plays a limited role, if any, in chloroplast transcription in lower plants such as moss and lycophytes. On the contrary, the basal angiosperm *Nuphar advena* has one plastid-localized RNAP (Yin et al. 2010). These results suggest that chloroplast NEP was acquired at a very early stage in flowering plant evolution or in gymnosperms.

References

Abdallah, F., Salamini, F. and Leister, D. 2000. A prediction of the size and evolutionary origin of the proteome of chloroplasts of *Arabidopsis*. *Trends Plant Sci*, **5**, 141–142.

Allison, L.A. 2000. The role of sigma factors in plastid transcription. *Biochimie*, **82**, 537–548.

Allison, L.A., Simon, L.D. and Maliga, P. 1996. Deletion of rpoB reveals a second distinct transcription system in plastids of higher plants. *Embo J*, **15**, 2802–2809.

Armbrust, E.V., Berges, J.A., Bowler, C., Green, B.R., Martinez, D., Putnam, N.H., Zhou, S. et al. 2004. The genome of the diatom *Thalassiosira pseudonana*: Ecology, evolution, and metabolism. *Science*, **306**, 79–86.

Arsova, B., Hoja, U., Wimmelbacher, M., Greiner, E., Ustun, S., Melzer, M., Petersen, K., Lein, W. and Bornke, F. 2010. Plastidial thioredoxin z interacts with two fructokinase-like proteins in a thiol-dependent manner: Evidence for an essential role in chloroplast development in *Arabidopsis* and *Nicotiana benthamiana*. *Plant Cell*, **22**, 1498–1515.

Asin-Cayuela, J. and Gustafsson, C.M. 2007. Mitochondrial transcription and its regulation in mammalian cells. *Trends Biochem Sci*, **32**, 111–117.

Azevedo, J., Courtois, F., Hakimi, M.A., Demarsy, E., Lagrange, T., Alcaraz, J.P., Jaiswal, P., Marechal-Drouard, L. and Lerbs-Mache, S. 2008. Intraplastidial trafficking of a phage-type RNA polymerase is mediated by a thylakoid RING-H2 protein. *Proc Natl Acad Sci USA*, **105**, 9123–9128.

Baba, K., Schmidt, J., Espinosa-Ruiz, A., Villarejo, A., Shiina, T., Gardestrom, P., Sane, A.P. and Bhalerao, R.P. 2004. Organellar gene transcription and early seedling development are affected in the rpoT;2 mutant of *Arabidopsis*. *Plant J*, **38**, 38–48.

Baena-Gonzalez, E., Baginsky, S., Mulo, P., Summer, H., Aro, E.M. and Link, G. 2001. Chloroplast transcription at different light intensities. Glutathione-mediated phosphorylation of the major RNA polymerase involved in redox-regulated organellar gene expression. *Plant Physiol*, **127**, 1044–1052.

Baginsky, S., Tiller, K., Pfannschmidt, T. and Link, G. 1999. PTK, the chloroplast RNA polymerase-associated protein kinase from mustard (*Sinapis alba*), mediates redox control of plastid *in vitro* transcription. *Plant Mol Biol*, **39**, 1013–1023.

Barkan, A., Walker, M., Nolasco, M. and Johnson, D. 1994. A nuclear mutation in maize blocks the processing and translation of several chloroplast messenger-RNAs and provides evidence for the differential translation of alternative messenger-RNA forms. *Embo J*, **13**, 3170–3181.

Baumgartner, B.J., Rapp, J.C. and Mullet, J.E. 1993. Plastid genes encoding the transcription translation apparatus are differentially transcribed early in barley (*Hordeum vulgare*) chloroplast development—Evidence for selective stabilization of Psba messenger-RNA. *Plant Physiol*, **101**, 781–791.

Bentolila, S., Heller, W.P., Sun, T., Babina, A.M., Friso, G., van Wijk, K.J. and Hanson, M.R. 2012. RIP1, a member of an *Arabidopsis* protein family, interacts with the protein RARE1 and broadly affects RNA editing. *Proc Natl Acad Sci USA*, **109**, E1453–E1461.

Binder, S. and Brennicke, A. 2003. Gene expression in plant mitochondria: Transcriptional and post-transcriptional control. *Philos Trans R Soc London Ser B, Biol Sci*, **358**, 181–188; discussion 188–189.

Bligny, M., Courtois, F., Thaminy, S., Chang, C.C., Lagrange, T., Baruah-Wolff, J., Stern, D. and Lerbs-Mache, S. 2000. Regulation of plastid rDNA transcription by interaction of CDF2 with two different RNA polymerases. *Embo J*, **19**, 1851–1860.

Bogenhagen, D.F., Rousseau, D. and Burke, S. 2008. The layered structure of human mitochondrial DNA nucleoids. *J Biol Chem*, **283**, 3665–3675.

Bohne, A.V., Irihimovitch, V., Weihe, A. and Stern, D.B. 2006. *Chlamydomonas reinhardtii* encodes a single sigma70-like factor which likely functions in chloroplast transcription. *Curr Genet*, **49**, 333–340.

Borukhov, S., Lee, J. and Laptenko, O. 2005. Bacterial transcription elongation factors: New insights into molecular mechanism of action. *Mol Microbiol*, **55**, 1315–1324.

Briat, J.F., Gigot, C., Laulhere, J.P. and Mache, R. 1982. Visualization of a spinach plastid transcriptionally active DNA-protein complex in a highly condensed structure. *Plant Physiol*, **69**, 1205–1211.

Briggs, W.R. and Olney, M.A. 2001. Photoreceptors in plant photomorphogenesis to date. Five phytochromes, two cryptochromes, one phototropin, and one superchrome. *Plant Physiol*, **125**, 85–88.

Bülow, S. and Link, G. 1988. Sigma-like activity from mustard (*Sinapis alba* L.) chloroplasts conferring DNA-binding and transcription specificity to *E. coli* core RNA polymerase. *Plant Mol Biol*, **10**, 349–357.

Burmann, B.M. and Rosch, P. 2011. The role of *E. coli* Nus-factors in transcription regulation and transcription:translation coupling: From structure to mechanism. *Transcription*, **2**, 130–134.

Carter, M.L., Smith, A.C., Kobayashi, H., Purton, S. and Herrin, D.L. 2004. Structure, circadian regulation and bioinformatic analysis of the unique sigma factor gene in *Chlamydomonas reinhardtii*. *Photosynth Res*, **82**, 339–349.

Chang, C.C., Sheen, J., Bligny, M., Niwa, Y., Lerbs-Mache, S. and Stern, D.B. 1999. Functional analysis of two maize cDNAs encoding T7-like RNA polymerases. *Plant Cell*, **11**, 911–926.

Chateigner-Boutin, A.L., Ramos-Vega, M., Guevara-Garcia, A., Andres, C., Gutierrez-Nava, M.D., Cantero, A., Delannoy, E. et al. 2008. CLB19, a pentatricopeptide repeat protein required for editing of rpoA and clpP chloroplast transcripts. *Plant J*, **56**, 590–602.

Chen, L.J., Liang, Y.J., Jeng, S.T., Orozco, E.M., Gumport, R.I., Lin, C.H. and Yang, M.T. 1995. Transcription termination at the *Escherichia coli* thra terminator by spinach chloroplast RNA polymerase *in vitro* is influenced by downstream DNA sequences. *Nucleic Acids Res*, **23**, 4690–4697.

Chen, M., Galvao, R.M., Li, M.N., Burger, B., Bugea, J., Bolado, J. and Chory, J. 2010. *Arabidopsis* HEMERA/pTAC12 initiates photomorphogenesis by phytochromes. *Cell*, **141**, U1230–U1237.

Chen, M., Schwabb, R. and Chory, J. 2003. Characterization of the requirements for localization of phytochrome B to nuclear bodies. *Proc Natl Acad Sci USA*, **100**, 14493–14498.

Christopher, D.A., Kim, M. and Mullet, J.E. 1992. A novel light-regulated promoter is conserved in cereal and dicot chloroplasts. *Plant Cell*, **4**, 785–798.

Chun, L., Kawakami, A. and Christopher, D.A. 2001. Phytochrome A mediates blue light and UV-A-dependent chloroplast gene transcription in green leaves. *Plant Physiol*, **125**, 1957–1966.

Conaway, R.C., Kong, S.E. and Conaway, J.W. 2003. TFIIS and GreB: Two like-minded transcription elongation factors with sticky fingers. *Cell*, **114**, 272–274.

Courtois, F., Merendino, L., Demarsy, E., Mache, R. and Lerbs-Mache, S. 2007. Phage-type RNA polymerase RPOTmp transcribes the rrn operon from the PC promoter at early developmental stages in *Arabidopsis*. *Plant Physiol*, **145**, 712–721.

da Costa e Silva, O., Lorbiecke, R., Garg, P., Muller, L., Wassmann, M., Lauert, P., Scanlon, M. et al. 2004. The Etched1 gene of *Zea mays* (L.) encodes a zinc ribbon protein that belongs to the transcriptionally active chromosome (TAC) of plastids and is similar to the transcription factor TFIIS. *Plant J*, **38**, 923–939.

Das, A. 1992. How the phage-lambda N-gene product suppresses transcription termination—Communication of RNA-polymerase with regulatory proteins mediated by signals in nascent RNA. *J Bacteriol*, **174**, 6711–6716.

de Longevialle, A.F., Small, I.D. and Lurin, C. 2010. Nuclearly encoded splicing factors implicated in RNA splicing in higher plant organelles. *Mol Plant*, **3**, 691–705.

Deng, X.W. and Gruissem, W. 1987. Control of plastid gene expression during development: The limited role of transcriptional regulation. *Cell*, **49**, 379–387.

Derelle, E., Ferraz, C., Rombauts, S., Rouze, P., Worden, A.Z., Robbens, S., Partensky, F. et al. 2006. Genome analysis of the smallest free-living eukaryote *Ostreococcus tauri* unveils many unique features. *Proc Natl Acad Sci USA*, **103**, 11647–11652.

De Santis-MacIossek, G., Kofer, W., Bock, A., Schoch, S., Maier, R.M., Wanner, G., Rudiger, W., Koop, H.U. and Herrmann, R.G. 1999. Targeted disruption of the plastid RNA polymerase genes rpoA, B and C1: Molecular biology, biochemistry and ultrastructure. *Plant J*, **18**, 477–489.

des Francs-Small, C.C., Kroeger, T., Zmudjak, M., Ostersetzer-Biran, O., Rahimi, N., Small, I. and Barkan, A. 2012. A PORR domain protein required for rpl2 and ccmFC intron splicing and for the biogenesis of c-type cytochromes in *Arabidopsis* mitochondria. *Plant J*, **69**, 996–1005.

Deshpande, A.P. and Patel, S.S. 2012. Mechanism of transcription initiation by the yeast mitochondrial RNA polymerase. *Biochim et Biophys Acta*, **1819**, 930–938.

Deusch, O., Landan, G., Roettger, M., Gruenheit, N., Kowallik, K.V., Allen, J.F., Martin, W. and Dagan, T. 2008. Genes of cyanobacterial origin in plant nuclear genomes point to a heterocyst-forming plastid ancestor. *Mol Biol Evol*, **25**, 748–761.

Dillon, S.C. and Dorman, C.J. 2010. Bacterial nucleoid-associated proteins, nucleoid structure and gene expression. *Nat Rev Microbiol*, **8**, 185–195.

Ding, Y.H., Liu, N.Y., Tang, Z.S., Liu, J. and Yang, W.C. 2006. *Arabidopsis* GLUTAMINE-RICH PROTEIN23 is essential for early embryogenesis and encodes a novel nuclear PPR motif protein that interacts with RNA polymerase II subunit III. *Plant Cell*, **18**, 815–830.

Dubell, A.N. and Mullet, J.E. 1995. Differential transcription of pea chloroplast genes during light-induced leaf development transcription—Continuous far-red light activates chloroplast transcription. *Plant Physiol*, **109**, 105–112.

Eberhard, S., Drapier, D. and Wollman, F.A. 2002. Searching limiting steps in the expression of chloroplast-encoded proteins: Relations between gene copy number, transcription, transcript abundance and translation rate in the chloroplast of *Chlamydomonas reinhardtii*. *Plant J*, **31**, 149–160.

Eisermann, A., Tiller, K. and Link, G. 1990. *In vitro* transcription and DNA binding characteristics of chloroplast and etioplast extracts from mustard (*Sinapis alba*) indicate differential usage of the psbA promoter. *Embo J*, **9**, 3981–3987.

Emanuel, C., von Groll, U., Muller, M., Borner, T. and Weihe, A. 2006. Development- and tissue-specific expression of the RpoT gene family of *Arabidopsis* encoding mitochondrial and plastid RNA polymerases. *Planta*, **223**, 998–1009.

Emanuel, C., Weihe, A., Graner, A., Hess, W.R. and Borner, T. 2004. Chloroplast development affects expression of phage-type RNA polymerases in barley leaves. *Plant J*, **38**, 460–472.

Erickson, J.M., Rahire, M. and Rochaix, J.D. 1984. *Chlamydomonas reinhardtii* gene for the 32000 mol. wt. protein of photosystem II contains four large introns and is located entirely within the chloroplast inverted repeat. *Embo J*, **3**, 2753–2762.

Favory, J.J., Kobayshi, M., Tanaka, K., Peltier, G., Kreis, M., Valay, J.G. and Lerbs-Mache, S. 2005. Specific function of a plastid sigma factor for ndhF gene transcription. *Nucleic Acids Res*, **33**, 5991–5999.

Fujiwara, M., Nagashima, A., Kanamaru, K., Tanaka, K. and Takahashi, H. 2000. Three new nuclear genes, sigD, sigE and sigF, encoding putative plastid RNA polymerase sigma factors in *Arabidopsis thaliana*. *FEBS Lett*, **481**, 47–52.

Gatenby, A.A., Castleton, J.A. and Saul, M.W. 1981. Expression in *E. coli* of maize and wheat chloroplast genes for large subunit of ribulose bisphosphate carboxylase. *Nature*, **291**, 117–121.

Givens, R.M., Lin, M.H., Taylor, D.J., Mechold, U., Berry, J.O. and Hernandez, V.J. 2004. Inducible expression, enzymatic activity, and origin of higher plant homologues of bacterial RelA/SpoT stress proteins in *Nicotiana tabacum*. *J Biol Chem*, **279**, 7495–7504.

Grasser, K. 2006. *Regulation of Transcription in Plants*. Oxford: Blackwell Publishing.

Green, B.R. 2011. Chloroplast genomes of photosynthetic eukaryotes. *Plant J*, **66**, 34–44.

Gruissem, W. and Zurawski, G. 1985. Analysis of promoter regions for the spinach chloroplast rbcL, atpB and psbA genes. *Embo J*, **4**, 3375–3383.

Guertin, M. and Bellemare, G. 1979. Synthesis of chloroplast ribonucleic acid in *Chlamydomonas reinhardtii* toluene-treated cells. *Eur J Biochem/FEBS*, **96**, 125–129.

Hajdukiewicz, P.T., Allison, L.A. and Maliga, P. 1997. The two RNA polymerases encoded by the nuclear and the plastid compartments transcribe distinct groups of genes in tobacco plastids. *Embo J*, **16**, 4041–4048.

Han, C.D., Coe, E.H. Jr. and Martienssen, R.A. 1992. Molecular cloning and characterization of iojap (ij), a pattern striping gene of maize. *Embo J*, **11**, 4037–4046.

Hanaoka, M., Kanamaru, K., Fujiwara, M., Takahashi, H. and Tanaka, K. 2005. Glutamyl-tRNA mediates a switch in RNA polymerase use during chloroplast biogenesis. *EMBO Rep*, **6**, 545–550.

Hanaoka, M., Kanamaru, K., Takahashi, H. and Tanaka, K. 2003. Molecular genetic analysis of chloroplast gene promoters dependent on SIG2, a nucleus-encoded sigma factor for the plastid-encoded RNA polymerase, in *Arabidopsis thaliana*. *Nucleic Acids Res*, **31**, 7090–7098.

Hanaoka, M., Kato, M., Anma, M. and Tanaka, K. 2012. SIG1, a sigma factor for the chloroplast RNA polymerase, differently associates with multiple DNA regions in the chloroplast chromosomes *in vivo*. *Int J Mol Sci*, **13**, 12182–12194.

Hara, K., Morita, M., Takahashi, R., Sugita, M., Kato, S. and Aoki, S. 2001a. Characterization of two genes, Sig1 and Sig2, encoding distinct plastid sigma factors(1) in the moss *Physcomitrella patens*: Phylogenetic relationships to plastid sigma factors in higher plants. *FEBS Lett*, **499**, 87–91.

Hara, K., Sugita, M. and Aoki, S. 2001b. Cloning and characterization of the cDNA for a plastid sigma factor from the moss *Physcomitrella patens*. *Biochim et Biophys Acta*, **1517**, 302–306.

Hattori, M., Miyake, H. and Sugita, M. 2007. A pentatricopeptide repeat protein is required for RNA processing of clpP pre-mRNA in moss chloroplasts. *J Biol Chem*, **282**, 10773–10782.

Haugen, S.P., Ross, W. and Gourse, R.L. 2008. Advances in bacterial promoter recognition and its control by factors that do not bind DNA. *Nat Rev Microbiol*, **6**, 507–519.

Hawley, D.K. and McClure, W.R. 1983. Compilation and analysis of *Escherichia coli* promoter DNA sequences. *Nucleic Acids Res*, **11**, 2237–2255.

Hedtke, B., Borner, T. and Weihe, A. 2000. One RNA polymerase serving two genomes. *EMBO Rep*, **1**, 435–440.

Hedtke, B., Legen, J., Weihe, A., Herrmann, R.G. and Borner, T. 2002. Six active phage-type RNA polymerase genes in *Nicotiana tabacum*. *Plant J*, **30**, 625–637.

Hedtke, B., Meixner, M., Gillandt, S., Richter, E., Börner, T. and Weihe, A. 1999. Green fluorescent protein as a marker to investigate targeting of organellar RNA polymerases of higher plants *in vivo*. *Plant J*, 17, 557–561.

Hess, W.R. and Borner, T. 1999. Organellar RNA polymerases of higher plants. *Int Rev Cytol— Surv Cell Biol*, **190**, 1–59.

Hirata, N., Yonekura, D., Yanagisawa, S. and Iba, K. 2004. Possible involvement of the 5′-flanking region and the 5′ UTR of plastid accD gene in NEP-dependent transcription. *Plant Cell Physiol*, **45**, 176–186.

Hoffer, P.H. and Christopher, D.A. 1997. Structure and blue-light-responsive transcription of a chloroplast psbD promoter from *Arabidopsis thaliana*. *Plant Physiol*, **115**, 213–222.

Hotto, A.M., Schmitz, R.J., Fei, Z., Ecker, J.R. and Stern, D.B. 2011. Unexpected diversity of chloroplast noncoding RNAs as revealed by deep sequencing of the *Arabidopsis* transcriptome. *G3*, **1**, 559–570.

Hricova, A., Quesada, V. and Micol, J.L. 2006. The SCABRA3 nuclear gene encodes the plastid RpoTp RNA polymerase, which is required for chloroplast biogenesis and mesophyll cell proliferation in *Arabidopsis*. *Plant Physiol*, **141**, 942–956.

Huang, C., Yu, Q.B., Lv, R.H., Yin, Q.Q., Chen, G.Y., Xu, L. and Yang, Z.N. 2013. The reduced plastid-encoded polymerase-dependent plastid gene expression leads to the delayed greening of the *Arabidopsis* fln2 mutant. *PloS One*, **8**, e73092.

Ichikawa, K., Sugita, M., Imaizumi, T., Wada, M. and Aoki, S. 2004. Differential expression on a daily basis of plastid sigma factor genes from the moss *Physcomitrella patens*. Regulatory interactions among PpSig5, the circadian clock, and blue light signaling mediated by cryptochromes. *Plant Physiol*, **136**, 4285–4298.

Igloi, G.L. and Kossel, H. 1992. The transcriptional apparatus of chloroplasts. *Crit Rev Plant Sci*, **10**, 525–558.

Ikeda, T.M. and Gray, M.W. 1999. Identification and characterization of T3/T7 bacteriophage-like RNA polymerase sequences in wheat. *Plant Mol Biol*, **40**, 567–578.

Ishihama, A. 2000. Functional modulation of *Escherichia coli* RNA polymerase. *Ann Rev Microbiol*, **54**, 499–518.

Ishizaki, Y., Tsunoyama, Y., Hatano, K., Ando, K., Kato, K., Shinmyo, A., Kobori, M., Takeba, G., Nakahira, Y. and Shiina, T. 2005. A nuclear-encoded sigma factor, *Arabidopsis* SIG6, recognizes sigma-70 type chloroplast promoters and regulates early chloroplast development in cotyledons. *Plant J*, **42**, 133–144.

Isono, K., Shimizu, M., Yoshimoto, K., Niwa, Y., Satoh, K., Yokota, A. and Kobayashi, H. 1997. Leaf-specifically expressed genes for polypeptides destined for chloroplasts with domains of sigma70 factors of bacterial RNA polymerases in *Arabidopsis thaliana*. *Proc Natl Acad Sci USA*, **94**, 14948–14953.

Jeong, S.Y., Rose, A. and Meier, I. 2003. MFP1 is a thylakoid-associated, nucleoid-binding protein with a coiled-coil structure. *Nucleic Acids Res*, **31**, 5175–5185.

Johnson, X., Wostrikoff, K., Finazzi, G., Kuras, R., Schwarz, C., Bujaldon, S., Nickelsen, J., Stern, D.B., Wollman, F.A. and Vallon, O. 2010. MRL1, a conserved pentatricopeptide repeat protein, is required for stabilization of rbcL mRNA in *Chlamydomonas* and *Arabidopsis*. *Plant Cell*, **22**, 234–248.

Kabeya, Y., Hashimoto, K. and Sato, N. 2002. Identification and characterization of two phage-type RNA polymerase cDNAs in the moss *Physcomitrella patens*: Implication of recent evolution of nuclear-encoded RNA polymerase of plastids in plants. *Plant Cell Physiol*, **43**, 245–255.

Kahlau, S. and Bock, R. 2008. Plastid transcriptomics and translatomics of tomato fruit development and chloroplast-to-chromoplast differentiation: Chromoplast gene expression largely serves the production of a single protein. *Plant Cell*, **20**, 856–874.

Kanamaru, K., Fujiwara, M., Seki, M., Katagiri, T., Nakamura, M., Mochizuki, N., Nagatani, A., Shinozaki, K., Tanaka, K. and Takahashi, H. 1999. Plastidic RNA polymerase sigma factors in *Arabidopsis*. *Plant Cell Physiol*, **40**, 832–842.

Kanamaru, K., Nagashima, A., Fujiwara, M., Shimada, H., Shirano, Y., Nakabayashi, K., Shibata, D., Tanaka, K. and Takahashi, H. 2001. An *Arabidopsis* sigma factor (SIG2)-dependent expression of plastid-encoded tRNAs in chloroplasts. *Plant Cell Physiol*, **42**, 1034–1043.

Kaneko, T., Sato, S., Kotani, H., Tanaka, A., Asamizu, E., Nakamura, Y., Miyajima, N. et al. 1996. Sequence analysis of the genome of the unicellular cyanobacterium *Synechocystis* sp. strain PCC6803. II. Sequence determination of the entire genome and assignment of potential protein-coding regions (supplement). *DNA Res*, **3**, 185–209.

Kapoor, S. and Sugiura, M. 1999. Identification of two essential sequence elements in the nonconsensus type II PatpB-290 plastid promoter by using plastid transcription extracts from cultured tobacco BY-2 cells. *Plant Cell*, **11**, 1799–1810.

Kapoor, S., Suzuki, J.Y. and Sugiura, M. 1997. Identification and functional significance of a new class of non-consensus-type plastid promoters. *Plant J*, **11**, 327–337.

Karcher, D., Koster, D., Schadach, A., Klevesath, A. and Bock, R. 2009. The *Chlamydomonas* chloroplast HLP protein is required for nucleoid organization and genome maintenance. *Mol Plant*, **2**, 1223–1232.

Kasai, K., Kawagishi-Kobayashi, M., Teraishi, M., Ito, Y., Ochi, K., Wakasa, K. and Tozawa, Y. 2004. Differential expression of three plastidial sigma factors, OsSIG1, OsSIG2A, and OsSIG2B, during leaf development in rice. *Biosci Biotechnol Biochem*, **68**, 973–977.

Kasai, K., Usami, S., Yamada, T., Endo, Y., Ochi, K. and Tozawa, Y. 2002. A RelA-SpoT homolog (Cr-RSH) identified in *Chlamydomonas reinhardtii* generates stringent factor *in vivo* and localizes to chloroplasts *in vitro*. *Nucleic Acids Res*, **30**, 4985–4992.

Kim, M., Thum, K.E., Morishige, D.T. and Mullet, J.E. 1999. Detailed architecture of the barley chloroplast psbD-psbC blue light-responsive promoter. *J Biol Chem*, **274**, 4684–4692.

Kircher, S., Gil, P., Kozma-Bognar, L., Fejes, E., Speth, V., Husselstein-Muller, T., Bauer, D., Adam, E., Schafer, E. and Nagy, F. 2002. Nucleocytoplasmic partitioning of the plant photoreceptors phytochrome A, B, C, D, and E is regulated differentially by light and exhibits a diurnal rhythm. *Plant Cell*, **14**, 1541–1555.

Kobayashi, T., Takahara, M., Miyagishima, S.Y., Kuroiwa, H., Sasaki, N., Ohta, N., Matsuzaki, M. and Kuroiwa, T. 2002. Detection and localization of a chloroplast-encoded HU-like protein that organizes chloroplast nucleoids. *Plant Cell*, **14**, 1579–1589.

Kobayashi, Y., Dokiya, Y. and Sugita, M. 2001. Dual targeting of phage-type RNA polymerase to both mitochondria and plastids is due to alternative translation initiation in single transcripts. *Biochem Biophys Res Commun*, **289**, 1106–1113.

Kotera, E., Tasaka, M. and Shikanai, T. 2005. A pentatricopeptide repeat protein is essential for RNA editing in chloroplasts. *Nature*, **433**, 326–330.

Krause, K., Maier, R.M., Kofer, W., Krupinska, K. and Herrmann, R.G. 2000. Disruption of plastid-encoded RNA polymerase genes in tobacco: Expression of only a distinct set of genes is not based on selective transcription of the plastid chromosome. *Mol Gen Genet*, **263**, 1022–1030.

Krupinska, K. and Falk, J. 1994. Changes in RNA-polymerase activity during biogenesis, maturation and senescence of barley chloroplasts—Comparative-analysis of transcripts synthesized either in run-on assays or by transcriptionally active chromosomes. *J Plant Physiol*, **143**, 298–305.

Kuhn, K., Bohne, A.V., Liere, K., Weihe, A. and Borner, T. 2007. *Arabidopsis* phage-type RNA polymerases: Accurate *in vitro* transcription of organellar genes. *Plant Cell*, **19**, 959–971.

Kuhn, K., Weihe, A. and Borner, T. 2005. Multiple promoters are a common feature of mitochondrial genes in *Arabidopsis*. *Nucleic Acids Res*, **33**, 337–346.

Kuroiwa, T. 1991. The replication, differentiation, and inheritance of plastids with emphasis on the concept of organelle nuclei. *Int Rev Cytol—Surv Cell Biol*, **128**, 1–62.

Lahiri, S.D. and Allison, L.A. 2000. Complementary expression of two plastid-localized sigma-like factors in maize. *Plant Physiol*, **123**, 883–894.

Lahiri, S.D., Yao, J., McCumbers, C. and Allison, L.A. 1999. Tissue-specific and light-dependent expression within a family of nuclear-encoded sigma-like factors from *Zea mays*. *Mol Cell Biol Res Commun*, **1**, 14–20.

Laluk, K., AbuQamar, S. and Mengiste, T. 2011. The *Arabidopsis* mitochondria-localized pentatricopeptide repeat protein PGN functions in defense against necrotrophic fungi and abiotic stress tolerance. *Plant Physiol*, **156**, 2053–2068.

Lerbs-Mache, S. 1993. The 110-kDa polypeptide of spinach plastid DNA-dependent RNA polymerase: Single-subunit enzyme or catalytic core of multimeric enzyme complexes? *Proc Natl Acad Sci USA*, **90**, 5509–5513.

Lerbs-Mache, S. 2011. Function of plastid sigma factors in higher plants: Regulation of gene expression or just preservation of constitutive transcription? *Plant Mol Biol*, **76**, 235–249.

Liere, K., Kaden, D., Maliga, P. and Borner, T. 2004. Overexpression of phage-type RNA polymerase RpoTp in tobacco demonstrates its role in chloroplast transcription by recognizing a distinct promoter type. *Nucleic Acids Res*, **32**, 1159–1165.

Liere, K. and Link, G. 1994. Structure and expression characteristics of the chloroplast DNA region containing the split gene for tRNA(Gly) (UCC) from mustard (*Sinapis alba* L.). *Curr Genet*, **26**, 557–563.

Liere, K. and Maliga, P. 1999. *In vitro* characterization of the tobacco rpoB promoter reveals a core sequence motif conserved between phage-type plastid and plant mitochondrial promoters. *Embo J*, **18**, 249–257.

Liu, B. and Troxler, R.F. 1996. Molecular characterization of a positively photoregulated nuclear gene for a chloroplast RNA polymerase sigma factor in *Cyanidium caldarium*. *Proc Natl Acad Sci USA*, **93**, 3313–3318.

Lonetto, M., Gribskov, M. and Gross, C.A. 1992. The sigma 70 family: Sequence conservation and evolutionary relationships. *J Bacteriol*, **174**, 3843–3849.

Lurin, C., Andres, C., Aubourg, S., Bellaoui, M., Bitton, F., Bruyere, C., Caboche, M. et al. 2004. Genome-wide analysis of *Arabidopsis* pentatricopeptide repeat proteins reveals their essential role in organelle biogenesis. *Plant Cell*, **16**, 2089–2103.

Majeran, W., Friso, G., Asakura, Y., Qu, X., Huang, M., Ponnala, L., Watkins, K.P., Barkan, A. and van Wijk, K.J. 2012. Nucleoid-enriched proteomes in developing plastids and chloroplasts from maize leaves: A new conceptual framework for nucleoid functions. *Plant Physiol*, **158**, 156–189.

Martin, W., Rujan, T., Richly, E., Hansen, A., Cornelsen, S., Lins, T., Leister, D., Stoebe, B., Hasegawa, M. and Penny, D. 2002. Evolutionary analysis of *Arabidopsis*, cyanobacterial, and chloroplast genomes reveals plastid phylogeny and thousands of cyanobacterial genes in the nucleus. *Proc Natl Acad Sci USA*, **99**, 12246–12251.

Masuda, S., Mizusawa, K., Narisawa, T., Tozawa, Y., Ohta, H. and Takamiya, K. 2008. The bacterial stringent response, conserved in chloroplasts, controls plant fertilization. *Plant Cell Physiol*, **49**, 135–141.

Mathews, D.E. and Durbin, R.D. 1990. Tagetitoxin inhibits RNA synthesis directed by RNA polymerases from chloroplasts and *Escherichia coli*. *J Biol Chem*, **265**, 493–498.

Matsuzaki, M., Misumi, O., Shin, I.T., Maruyama, S., Takahara, M., Miyagishima, S.Y., Mori, T. et al. 2004. Genome sequence of the ultrasmall unicellular red alga *Cyanidioschyzon merolae* 10D. *Nature*, **428**, 653–657.

Mehta, A., Kinter, M.T., Sherman, N.E. and Driscoll, D.M. 2000. Molecular cloning of apobec-1 complementation factor, a novel RNA-binding protein involved in the editing of apolipoprotein B mRNA. *Mol Cell Biol*, **20**, 1846–1854.

Meierhoff, K., Felder, S., Nakamura, T., Bechtold, N. and Schuster, G. 2003. HCF152, an *Arabidopsis* RNA binding pentatricopeptide repeat protein involved in the processing of chloroplast psbB-psbT-psbH-petB-petD RNAs. *Plant Cell*, **15**, 1480–1495.

Melonek, J., Matros, A., Trosch, M., Mock, H.P. and Krupinska, K. 2012. The core of chloroplast nucleoids contains architectural SWIB domain proteins. *Plant Cell*, **24**, 3060–3073.

Minoda, A., Weber, A.P., Tanaka, K. and Miyagishima, S.Y. 2010. Nucleus-independent control of the rubisco operon by the plastid-encoded transcription factor Ycf30 in the red alga *Cyanidioschyzon merolae*. *Plant Physiol*, **154**, 1532–1540.

Miyamura, S., Nagata, T. and Kuroiwa, T. 1986. Quantitative fluorescence microscopy on dynamic changes of plastid nucleoids during wheat development. *Protoplasma*, **133**, 66–72.

Morden, C.W., Wolfe, K.H., dePamphilis, C.W. and Palmer, J.D. 1991. Plastid translation and transcription genes in a non-photosynthetic plant: Intact, missing and pseudo genes. *Embo J*, **10**, 3281–3288.

Murakami, K.S. and Darst, S.A. 2003. Bacterial RNA polymerases: The whole story. *Curr Opin Struct Biol*, **13**, 31–39.

Myouga, F., Hosoda, C., Umezawa, T., Iizumi, H., Kuromori, T., Motohashi, R., Shono, Y., Nagata, N., Ikeuchi, M. and Shinozaki, K. 2008. A heterocomplex of iron superoxide dismutases defends chloroplast nucleoids against oxidative stress and is essential for chloroplast development in *Arabidopsis*. *Plant Cell*, **20**, 3148–3162.

Nagashima, A., Hanaoka, M., Shikanai, T., Fujiwara, M., Kanamaru, K., Takahashi, H. and Tanaka, K. 2004. The multiple-stress responsive plastid sigma factor, SIG5, directs activation of the psbD blue light-responsive promoter (BLRP) in *Arabidopsis thaliana*. *Plant Cell Physiol*, **45**, 357–368.

Nakahira, Y., Baba, K., Yoneda, A., Shiina, T. and Toyoshima, Y. 1998. Circadian-regulated transcription of the psbD light-responsive promoter in wheat chloroplasts. *Plant Physiol*, **118**, 1079–1088.

Nakamura, T., Furuhashi, Y., Hasegawa, K., Hashimoto, H., Watanabe, K., Obokata, J., Sugita, M. and Sugiura, M. 2003. Array-based analysis on tobacco plastid transcripts: Preparation of a genomic microarray containing all genes and all intergenic regions. *Plant Cell Physiol*, **44**, 861–867.

Nomura, Y., Takabayashi, T., Kuroda, H., Yukawa, Y., Sattasuk, K., Akita, M., Nozawa, A. and Tozawa, Y. 2012. ppGpp inhibits peptide elongation cycle of chloroplast translation system *in vitro*. *Plant Mol Biol*, **78**, 185–196.

Ogrzewalla, K., Piotrowski, M., Reinbothe, S. and Link, G. 2002. The plastid transcription kinase from mustard (*Sinapis alba* L.). A nuclear-encoded CK2-type chloroplast enzyme with redox-sensitive function. *Eur J Biochem/FEBS*, **269**, 3329–3337.

Ohyama, K., Fukuzawa, H., Kohchi, T., Shirai, H., Sano, T., Sano, S., Umesono, K. et al. 1986. Chloroplast gene organization deduced from complete sequence of liverwort *Marchantiapolymorpha* chloroplast DNA. *Nature*, **322**, 572–574.

Orozco, E.M., Jr., Mullet, J.E. and Chua, N.H. 1985. An *in vitro* system for accurate transcription initiation of chloroplast protein genes. *Nucleic Acids Res*, **13**, 1283–1302.

Ostheimer, G.J., Williams-Carrier, R., Belcher, S., Osborne, E., Gierke, J. and Barkan, A. 2003. Group II intron splicing factors derived by diversification of an ancient RNA-binding domain. *Embo J*, **22**, 3919–3929.

Pfalz, J., Bayraktar, O.A., Prikryl, J. and Barkan, A. 2009. Site-specific binding of a PPR protein defines and stabilizes 5′ and 3′ mRNA termini in chloroplasts. *Embo J*, **28**, 2042–2052.

Pfalz, J., Liere, K., Kandlbinder, A., Dietz, K.J. and Oelmuller, R. 2006. pTAC2, -6, and -12 are components of the transcriptionally active plastid chromosome that are required for plastid gene expression. *Plant Cell*, **18**, 176–197.

Pfalz, J. and Pfannschmidt, T. 2013. Essential nucleoid proteins in early chloroplast development. *Trends Plant Sci*, **18**, 186–194.

Pfannschmidt, T. and Link, G. 1994. Separation of two classes of plastid DNA-dependent RNA polymerases that are differentially expressed in mustard (*Sinapis alba* L.) seedlings. *Plant Mol Biol*, **25**, 69–81.

Pfannschmidt, T., Nilsson, A. and Allen, J.F. 1999a. Photosynthetic control of chloroplast gene expression. *Nature*, **397**, 625–628.

Pfannschmidt, T., Nilsson, A., Tullberg, A., Link, G. and Allen, J.F. 1999b. Direct transcriptional control of the chloroplast genes psbA and psaAB adjusts photosynthesis to light energy distribution in plants. *IUBMB Life*, **48**, 271–276.

Prikryl, J., Rojas, M., Schuster, G. and Barkan, A. 2011. Mechanism of RNA stabilization and translational activation by a pentatricopeptide repeat protein. *Proc Natl Acad Sci USA*, **108**, 415–420.

Puthiyaveetil, S., Ibrahim, I.M. and Allen, J.F. 2013. Evolutionary rewiring: A modified prokaryotic gene-regulatory pathway in chloroplasts. *Philos Trans R Soc B—Biol Sci*, **368**, 20120260.

Puthiyaveetil, S., Kavanagh, T.A., Cain, P., Sullivan, J.A., Newell, C.A., Gray, J.C., Robinson, C., van der Giezen, M., Rogers, M.B. and Allen, J.F. 2008. The ancestral symbiont sensor kinase CSK links photosynthesis with gene expression in chloroplasts. *Proc Natl Acad Sci USA*, **105**, 10061–10066.

Rahire, M., Laroche, F., Cerutti, L. and Rochaix, J.D. 2012. Identification of an OPR protein involved in the translation initiation of the PsaB subunit of photosystem I. *Plant J*, **72**, 652–661.

Rajasekhar, V.K., Sun, E., Meeker, R., Wu, B.W. and Tewari, K.K. 1991. Highly purified pea chloroplast RNA-polymerase transcribes both ribosomal-RNA and messenger-RNA genes. *Eur J Biochem/FEBS*, **195**, 215–228.

Reiss, T. and Link, G. 1985. Characterization of transcriptionally active DNA-protein complexes from chloroplasts and etioplasts of mustard (*Sinapis alba* L.). *Eur J Biochem/FEBS*, **148**, 207–212.

Rensing, S.A., Lang, D., Zimmer, A.D., Terry, A., Salamov, A., Shapiro, H., Nishiyama, T. et al. 2008. The *Physcomitrella* genome reveals evolutionary insights into the conquest of land by plants. *Science*, **319**, 64–69.

Richardson, J.P. 1993. Transcription termination. *Crit Rev Biochem Mol Biol*, **28**, 1–30.

Richter, U., Kiessling, J., Hedtke, B., Decker, E., Reski, R., Borner, T. and Weihe, A. 2002. Two RpoT genes of *Physcomitrella patens* encode phage-type RNA polymerases with dual targeting to mitochondria and plastids. *Gene*, **290**, 95–105.

Richter, U., Richter, B., Weihe, A. and Börner, T. 2014. A third mitochondrial RNA polymerase in the moss *Physcomitrella patens*. *Curr Genet*, **60**, 25–34.

Rochaix, J.D., Rahire, M. and Michel, F. 1985. The chloroplast ribosomal intron of *Chlamydomonas reinhardtii* codes for a polypeptide related to mitochondrial maturases. *Nucleic Acids Res*, **13**, 975–984.

Ruwe, H. and Schmitz-Linneweber, C. 2012. Short non-coding RNA fragments accumulating in chloroplasts: Footprints of RNA binding proteins? *Nucleic Acids Res*, **40**, 3106–3116.

Saecker, R.M., Record, M.T., Jr. and Dehaseth, P.L. 2011. Mechanism of bacterial transcription initiation: RNA polymerase—Promoter binding, isomerization to initiation-competent open complexes, and initiation of RNA synthesis. *J Mol Biol*, **412**, 754–771.

Sakai, A., Saito, C., Inada, N. and Kuroiwa, T. 1998. Transcriptional activities of the chloroplast–nuclei and proplastid–nuclei isolated from tobacco exhibit different sensitivities to tagetitoxin: Implication of the presence of distinct RNA polymerases. *Plant Cell Physiol*, **39**, 928–934.

Sakai, A., Takano, H. and Kuroiwa, T. 2004. Organelle nuclei in higher plants: Structure, composition, function, and evolution. *Int Rev Cytol*, **238**, 59–118.

Salone, V., Rudinger, M., Polsakiewicz, M., Hoffmann, B., Groth-Malonek, M., Szurek, B., Small, I., Knoop, V. and Lurin, C. 2007. A hypothesis on the identification of the editing enzyme in plant organelles. *FEBS Lett*, **581**, 4132–4138.

Santangelo, T.J. and Artsimovitch, I. 2011. Termination and antitermination: RNA polymerase runs a stop sign. *Nat Rev Microbiol*, **9**, 319–329.

Sato, N., Misumi, O., Shinada, Y., Sasaki, M. and Yoine, M. 1997. Dynamics of localization and protein composition of plastid nucleoids in light-grown pea seedlings. *Protoplasma*, **200**, 163–173.

Sato, N., Nakayama, M. and Hase, T. 2001. The 70-kDa major DNA-compacting protein of the chloroplast nucleoid is sulfite reductase. *FEBS Lett*, **487**, 347–350.

Sato, N., Ohshima, K., Watanabe, A., Ohta, N., Nishiyama, Y., Joyard, J. and Douce, R. 1998. Molecular characterization of the PEND protein, a novel bZIP protein present in the envelope membrane that is the site of nucleoid replication in developing plastids. *Plant Cell*, **10**, 859–872.

Satoh, J., Baba, K., Nakahira, Y., Shiina, T. and Toyoshima, Y. 1997. Characterization of dynamics of the psbD light-induced transcription in mature wheat chloroplasts. *Plant Mol Biol*, **33**, 267–278.

Satoh, J., Baba, K., Nakahira, Y., Tsunoyama, Y., Shiina, T. and Toyoshima, Y. 1999. Developmental stage-specific multi-subunit plastid RNA polymerases (PEP) in wheat. *Plant J*, **18**, 407–415.

Schmitz-Linneweber, C. and Small, I. 2008. Pentatricopeptide repeat proteins: A socket set for organelle gene expression. *Trends Plant Sci*, **13**, 663–670.

Schrubar, H., Wanner, G. and Westhoff, P. 1991. Transcriptional control of plastid gene-expression in greening sorghum seedlings. *Planta*, **183**, 101–111.

Schweer, J., Turkeri, H., Link, B. and Link, G. 2010. AtSIG6, a plastid sigma factor from *Arabidopsis*, reveals functional impact of cpCK2 phosphorylation. *Plant J*, **62**, 192–202.

Sekine, K., Fujiwara, M., Nakayama, M., Takao, T., Hase, T. and Sato, N. 2007. DNA binding and partial nucleoid localization of the chloroplast stromal enzyme ferredoxin: Sulfite reductase. *FEBS J*, **274**, 2054–2069.

Sekine, K., Hase, T. and Sato, N. 2002. Reversible DNA compaction by sulfite reductase regulates transcriptional activity of chloroplast nucleoids. *J Biol Chem*, **277**, 24399–24404.

Severinov, K., Mustaev, A., Kukarin, A., Muzzin, O., Bass, I., Darst, S.A. and Goldfarb, A. 1996. Structural modules of the large subunits of RNA polymerase. Introducing archaebacterial and chloroplast split sites in the beta and beta′ subunits of *Escherichia coli* RNA polymerase. *J Biol Chem*, **271**, 27969–27974.

Sexton, T.B., Christopher, D.A. and Mullet, J.E. 1990. Light-induced switch in barley psbD-psbC promoter utilization: A novel mechanism regulating chloroplast gene expression. *Embo J*, **9**, 4485–4494.

Shiina, T., Allison, L. and Maliga, P. 1998. rbcL transcript levels in tobacco plastids are independent of light: Reduced dark transcription rate is compensated by increased mRNA stability. *Plant Cell*, **10**, 1713–1722.

Shiina, T., Tsunoyama, Y., Nakahira, Y. and Khan, M.S. 2005. Plastid RNA polymerases, promoters, and transcription regulators in higher plants. *Int Rev Cytol*, **244**, 1–68.

Shiina, T., Ishizaki, Y., Yagi, Y. and Nakahira, Y. 2009. Function and evolution of plastid sigma factors. *Plant Biotechnology*, **26**, 57–66.

Shikanai, T. 2006. RNA editing in plant organelles: Machinery, physiological function and evolution. *Cell Mol Life Sci*, **63**, 698–708.

Shikanai, T. and Fujii, S. 2013. Function of PPR proteins in plastid gene expression. *RNA Biol* **10**, 1446–1456.

Shimizu, M., Kato, H., Ogawa, T., Kurachi, A., Nakagawa, Y. and Kobayashi, H. 2010. Sigma factor phosphorylation in the photosynthetic control of photosystem stoichiometry. *Proc Natl Acad Sci US A*, **107**, 10760–10764.

Shinozaki, K., Ohme, M., Tanaka, M., Wakasugi, T., Hayashida, N., Matsubayashi, T., Zaita, N. et al. 1986. The complete nucleotide sequence of the tobacco chloroplast genome: Its gene organization and expression. *Embo J*, **5**, 2043–2049.

Sosso, D., Canut, M., Gendrot, G., Dedieu, A., Chambrier, P., Barkan, A., Consonni, G. and Rogowsky, P.M. 2012. PPR8522 encodes a chloroplast-targeted pentatricopeptide repeat protein necessary for maize embryogenesis and vegetative development. *J Exp Bot*, **63**, 5843–5857.

Steiner, S., Schroter, Y., Pfalz, J. and Pfannschmidt, T. 2011. Identification of essential subunits in the plastid-encoded RNA polymerase complex reveals building blocks for proper plastid development. *Plant Physiol*, **157**, 1043–1055.

Steinmetz, A., Gubbins, E.J. and Bogorad, L. 1982. The anticodon of the maize chloroplast gene for tRNA Leu UAA is split by a large intron. *Nucleic Acids Res*, **10**, 3027–3037.

Stern, D.B., Goldschmidt-Clermont, M. and Hanson, M.R. 2010. Chloroplast RNA metabolism. *Annu Rev Plant Biol*, **61**, 125–155.

Stern, D.B. and Gruissem, W. 1987. Control of plastid gene-expression-3' inverted repeats act as messenger-RNA processing and stabilizing elements, but do not terminate transcription. *Cell*, **51**, 1145–1157.

Strittmatter, G., Gozdzicka-Jozefiak, A. and Kossel, H. 1985. Identification of an rRNA operon promoter from *Zea mays* chloroplasts which excludes the proximal tRNAValGAC from the primary transcript. *Embo J*, **4**, 599–604.

Sugiura, C., Kobayashi, Y., Aoki, S., Sugita, C. and Sugita, M. 2003. Complete chloroplast DNA sequence of the moss *Physcomitrella patens*: Evidence for the loss and relocation of rpoA from the chloroplast to the nucleus. *Nucleic Acids Res*, **31**, 5324–5331.

Surzycki, S.J. 1969. Genetic functions of the chloroplast of *Chlamydomonas reinhardtii*: Effect of rifampin on chloroplast DNA-dependent RNA polymerase. *Proc Natl Acad Sci USA*, **63**, 1327–1334.

Suzuki, J.Y., Ytterberg, A.J., Beardslee, T.A., Allison, L.A., Wijk, K.J. and Maliga, P. 2004. Affinity purification of the tobacco plastid RNA polymerase and *in vitro* reconstitution of the holoenzyme. *Plant J*, **40**, 164–172.

Svetlov, V., Belogurov, G.A., Shabrova, E., Vassylyev, D.G. and Artsimovitch, I. 2007. Allosteric control of the RNA polymerase by the elongation factor RfaH. *Nucleic Acids Res*, **35**, 5694–5705.

Swiatecka-Hagenbruch, M., Emanuel, C., Hedtke, B., Liere, K. and Borner, T. 2008. Impaired function of the phage-type RNA polymerase RpoTp in transcription of chloroplast genes is compensated by a second phage-type RNA polymerase. *Nucleic Acids Res*, **36**, 785–792.

Takahashi, K., Kasai, K. and Ochi, K. 2004. Identification of the bacterial alarmone guanosine 5'-diphosphate 3'-diphosphate (ppGpp) in plants. *Proc Natl Acad Sci USA*, **101**, 4320–4324.

Takenaka, M., Zehrmann, A., Verbitskiy, D., Kugelmann, M., Hartel, B. and Brennicke, A. 2012. Multiple organellar RNA editing factor (MORF) family proteins are required for RNA editing in mitochondria and plastids of plants. *Proc Natl Acad Sci USA*, **109**, 5104–5109.

Tan, S. and Troxler, R.F. 1999. Characterization of two chloroplast RNA polymerase sigma factors from *Zea mays*: Photoregulation and differential expression. *Proc Natl Acad Sci USA*, **96**, 5316–5321.

Tanaka, K., Oikawa, K., Ohta, N., Kuroiwa, H., Kuroiwa, T. and Takahashi, H. 1996. Nuclear encoding of a chloroplast RNA polymerase sigma subunit in a red alga. *Science*, **272**, 1932–1935.

Tanaka, K., Tozawa, Y., Mochizuki, N., Shinozaki, K., Nagatani, A., Wakasa, K. and Takahashi, H. 1997. Characterization of three cDNA species encoding plastid RNA polymerase sigma factors in *Arabidopsis thaliana*: Evidence for the sigma factor heterogeneity in higher plant plastids. *FEBS Lett*, **413**, 309–313.

Teng, B., Burant, C.F. and Davidson, N.O. 1993. Molecular cloning of an apolipoprotein B messenger RNA editing protein. *Science*, **260**, 1816–1819.

Thum, K.E., Kim, M., Christopher, D.A. and Mullet, J.E. 2001a. Cryptochrome 1, cryptochrome 2, and phytochrome A co-activate the chloroplast psbD blue light-responsive promoter. *Plant Cell*, **13**, 2747–2760.

Thum, K.E., Kim, M., Morishige, D.T., Eibl, C., Koop, H.U. and Mullet, J.E. 2001b. Analysis of barley chloroplast psbD light-responsive promoter elements in transplastomic tobacco. *Plant Mol Biol*, **47**, 353–366.

Till, B., Schmitz-Linneweber, C., Williams-Carrier, R. and Barkan, A. 2001. CRS1 is a novel group II intron splicing factor that was derived from a domain of ancient origin. *RNA—Publ RNA Soc*, **7**, 1227–1238.

Tiller, K., Eisermann, A. and Link, G. 1991. The chloroplast transcription apparatus from mustard (*Sinapis alba* L.). Evidence for three different transcription factors which resemble bacterial sigma factors. *Eur J Biochem/FEBS*, **198**, 93–99.

Tiller, K. and Link, G. 1993a. Phosphorylation and dephosphorylation affect functional-characteristics of chloroplast and etioplast transcription systems from mustard (*Sinapis alba* L). *Embo J*, **12**, 1745–1753.

Tiller, K. and Link, G. 1993b. Sigma-like transcription factors from mustard (*Sinapis alba* L.) etioplast are similar in size to, but functionally distinct from, their chloroplast counterparts. *Plant Mol Biol*, **21**, 503–513.

To, K.Y., Cheng, M.C., Suen, D.F., Mon, D.P., Chen, L.F. and Chen, S.C. 1996. Characterization of the light-responsive promoter of rice chloroplast psbD-C operon and the sequence-specific DNA binding factor. *Plant Cell Physiol*, **37**, 660–666.

Torres, M., Condon, C., Balada, J.M., Squires, C. and Squires, C.L. 2001. Ribosomal protein S4 is a transcription factor with properties remarkably similar to NusA, a protein involved in both non-ribosomal and ribosomal RNA antitermination. *Embo J*, **20**, 3811–3820.

Tozawa, Y., Nozawa, A., Kanno, T., Narisawa, T., Masuda, S., Kasai, K. and Nanamiya, H. 2007. Calcium-activated (p)ppGpp synthetase in chloroplasts of land plants. *J Biol Chem*, **282**, 35536–35545.

Tozawa, Y., Tanaka, K., Takahashi, H. and Wakasa, K. 1998. Nuclear encoding of a plastid sigma factor in rice and its tissue- and light-dependent expression. *Nucleic Acids Res*, **26**, 415–419.

Tsunoyama, Y., Ishizaki, Y., Morikawa, K., Kobori, M., Nakahira, Y., Takeba, G., Toyoshima, Y. and Shiina, T. 2004. Blue light-induced transcription of plastid-encoded psbD gene is mediated by a nuclear-encoded transcription initiation factor, AtSig5. *Proc Natl Acad Sci USA*, **101**, 3304–3309.

Tsunoyama, Y., Morikawa, K., Shiina, T. and Toyoshima, Y. 2002. Blue light specific and differential expression of a plastid sigma factor, Sig5 in *Arabidopsis thaliana*. *FEBS Lett*, **516**, 225–228.

Ueda, M., Takami, T., Peng, L., Ishizaki, K., Kohchi, T., Shikanai, T. and Nishimura, Y. 2013. Subfunctionalization of sigma factors during the evolution of land plants based on mutant analysis of liverwort (*Marchantia polymorpha* L.) MpSIG1. *Genome Biol Evol*, **5**, 1836–1848.

Valkov, V.T., Scotti, N., Kahlau, S., Maclean, D., Grillo, S., Gray, J.C., Bock, R. and Cardi, T. 2009. Genome-wide analysis of plastid gene expression in potato leaf chloroplasts and tuber amyloplasts: Transcriptional and posttranscriptional control. *Plant Physiol*, **150**, 2030–2044.

van der Biezen, E.A., Sun, J., Coleman, M.J., Bibb, M.J. and Jones, J.D. 2000. *Arabidopsis* RelA/SpoT homologs implicate (p)ppGpp in plant signaling. *Proc Natl Acad Sci USA*, **97**, 3747–3752.

Wada, T., Tunoyama, Y., Shiina, T. and Toyoshima, Y. 1994. *In vitro* analysis of light-induced transcription in the wheat psbD/C gene cluster using plastid extracts from dark-grown and short-term-illuminated seedlings. *Plant Physiol*, **104**, 1259–1267.

Watkins, K.P., Rojas, M., Friso, G., van Wijk, K.J., Meurer, J. and Barkan, A. 2011. APO1 promotes the splicing of chloroplast group II introns and harbors a plant-specific zinc-dependent RNA binding domain. *Plant Cell*, **23**, 1082–1092.

Weihe, A. and Borner, T. 1999. Transcription and the architecture of promoters in chloroplasts. *Trends Plant Sci*, **4**, 169–170.

Weihe, A., Hedtke, B. and Borner, T. 1997. Cloning and characterization of a cDNA encoding a bacteriophage-type RNA polymerase from the higher plant *Chenopodium album*. *Nucleic Acids Res*, **25**, 2319–2325.

Wosten, M.M.S.M. 1998. Eubacterial sigma-factors. *FEMS Microbiol Rev*, **22**, 127–150.

Yagi, Y., Ishizaki, Y., Nakahira, Y., Tozawa, Y. and Shiina, T. 2012. Eukaryotic-type plastid nucleoid protein pTAC3 is essential for transcription by the bacterial-type plastid RNA polymerase. *Proc Natl Acad Sci USA*, **109**, 7541–7546.

Yamaguchi, R., Nakamura, M., Mochizuki, N., Kay, S.A. and Nagatani, A. 1999. Light-dependent translocation of a phytochrome B-GFP fusion protein to the nucleus in transgenic *Arabidopsis*. *J Cell Biol*, **145**, 437–445.

Yehudai-Resheff, S., Hirsh, M. and Schuster, G. 2001. Polynucleotide phosphorylase functions as both an exonuclease and a poly(A) polymerase in spinach chloroplasts. *Mol Cell Biol*, **21**, 5408–5416.

Yin, C., Richter, U., Borner, T. and Weihe, A. 2009. Evolution of phage-type RNA polymerases in higher plants: Characterization of the single phage-type RNA polymerase gene from *Selaginella moellendorffii*. *J Mol Evol*, **68**, 528–538.

Yin, C., Richter, U., Borner, T. and Weihe, A. 2010. Evolution of plant phage-type RNA polymerases: The genome of the basal angiosperm *Nuphar advena* encodes two mitochondrial and one plastid phage-type RNA polymerases. *BMC Evol Biol*, **10**, 379.

Yoshida, Y., Laulhere, J.P., Rozier, C. and Mache, R. 1978. Visualization of folded chloroplast DNA from spinach. *Biologie Cellulaire*, **32**, 187–190.

Zghidi, W., Merendino, L., Cottet, A., Mache, R. and Lerbs-Mache, S. 2007. Nucleus-encoded plastid sigma factor SIG3 transcribes specifically the psbN gene in plastids. *Nucleic Acids Res*, **35**, 455–464.

Zhelyazkova, P., Sharma, C.M., Forstner, K.U., Liere, K., Vogel, J. and Borner, T. 2012. The primary transcriptome of barley chloroplasts: Numerous noncoding RNAs and the dominating role of the plastid-encoded RNA polymerase. *Plant Cell*, **24**, 123–136.

Zmienko, A., Guzowska-Nowowiejska, M., Urbaniak, R., Plader, W., Formanowicz, P. and Figlerowicz, M. 2011. A tiling microarray for global analysis of chloroplast genome expression in cucumber and other plants. *Plant Methods*, **7**, 29.

Zoschke, R., Liere, K. and Borner, T. 2007. From seedling to mature plant: *Arabidopsis* plastidial genome copy number, RNA accumulation and transcription are differentially regulated during leaf development. *Plant J*, **50**, 710–722.

chapter five

Molecular biology of mitochondria

Genome, transcriptome, and proteome

Muhammad Waqar Hameed

Contents

Abstract

Mitochondria are known to have evolved from free-living *a*-proteobacteria through endosymbiosis. During evolution, the endosymbiont genes were either lost or transferred to the nuclear genome. The organelle now encodes a limited subset of proteins that are chiefly components of the electron transport chain complexes, transcriptional, and translational apparatus. Gene expression in mitochondria comprises distinct transcriptional and posttranscriptional steps. Currently, researchers devote much effort to investigate the regulation of mitochondrial gene expression. Although these investigations do illustrate differences in transcriptional rates and transcript abundances of mitochondrial-encoded genes, it is still unclear whether there is regulation at the level of transcription initiation and/or there is a gene- or genome-specific regulation in mitochondria. Many nuclear genes that encode mitochondrial proteins are regulated at both transcriptional and posttranscriptional level, whereas the regulation of mitochondrial-encoded genes is less clear. Regulation of mitochondrial translation has been investigated extensively in yeast. However, regulation of plant mitochondrial genes at the translational level is much less understood. This chapter presents a summary of the advances made in mitochondrial omics and biotechnology, as well as future prospects.

Keywords: Plant mitochondria, Mitochondrial DNA, Transcription, Transcription regulation, Translation, Translation regulation.

Introduction

Mitochondria are highly dynamic and pleomorphic organelles. They have a smooth outer membrane surrounding an inner membrane that sequentially surrounds a protein-rich matrix (Gray et al., 1999a; Leaver et al., 1983; Unseld et al., 1997a). They can frequently change their size and shape and can travel long distances on the cytoskeletal tracks (Bereiter-Hahn and Voth, 1994; Boldogh et al., 2001; Nunnari et al., 1997). Their key function is the production of adenosine triphosphate (ATP) through oxidative phosphorylation (Attardi and Schatz, 1988; Kelly and Scarpulla, 2004; Saraste, 1999). However, they are also known to participate in numerous other cellular functions including ion homeostasis, intermediary metabolism, and apoptosis (Butow and Avadhani, 2004; Danial and Korsmeyer, 2004). Present-day mitochondria are believed to have evolved through endosymbiosis, where a single-cell α-proteobacteria was engulfed by a proteobacterium. The engulfed bacterium became domesticated by establishing communication networks with the host cell's nuclear genome (Dyall et al., 2004; Martin et al., 2002). During domestication, majority of the genes from the engulfed α-proteobacteria were either lost or transferred to the nuclear or to the plastid genome (Adams and Palmer, 2003; Palmer et al., 2000). The transferred genes are now expressed in the nucleus and their protein products are subsequently imported into the mitochondria (Hager and Bock, 2000). Despite these huge gene translocations, mitochondria still retain a subset of genes, which predominantly encode components of the electron transport chain complexes, transcriptional, and translational apparatus (Bullerwell and Gray, 2004; Unseld et al., 1997a).

Size of plant mitochondrial DNA

Mitochondrial DNA (mtDNA) differs considerably in size between various organisms. It ranges from a very compact ~16-kb mtDNA in humans to 30–90 kb in fungi and yeast. In plants however, it is extraordinarily large in size (ranging from ~208 kb in white mustard to ~11.3 Mb in *Silene conica*) and complex in its organization (Gualberto et al., 2014; Knoop, 2004; Palmer and Herbon, 1987a; Sloan et al., 2012a). It normally encodes 13–14 proteins for the electron transport chain complexes, an incomplete set of ribosomal proteins, three ribosomal ribonucleic acids (RNAs) (26S, 18S, and 5S), and 22 transfer ribonucleic acids (tRNAs) (Duchene et al., 2005; Usadel et al., 2005). Despite the large size, gene density in plant mitochondria is exceptionally very low. The actual coding regions represent only ~10% of the genome, with large parts taken up by open reading frames of unknown function, introns, and sequences imported from plastids and nucleus (Backert et al., 1997; Giege and Brennicke, 2001; Kubo et al., 2000; Sugiyama et al., 2005; Unseld et al., 1997a). Plant mtDNA are also enriched with repeated sequences that include tandem, short, and large repeats (Alverson et al., 2010, 2011a; Kubo and Newton, 2008; Sugiyama et al., 2005; Unseld et al., 1997a). The short repeats mediate irreversible recombination between the various mitochondrial genomic regions whereas, the large repeats (>1 kb) mediate reversible recombination between the mitochondrial genomic regions. The short repeat-mediated changes into the mtDNA are inheritable, whereas the larger repeats-mediated recombinations are noninheritable and are brought about to regulate the molecular conformation of the mtDNA (Andre et al., 1992; Lonsdale et al., 1984b; Newton et al., 2004). A high number of pairwise large repeats (nine pairs) are reported in the K-type cytoplasmic male-sterile line of wheat (Liu et al., 2011); contrarily, no repeats were found in the mtDNA of *Brassica hirta* containing the smallest plant mitochondrial genome (Palmer and Herbon, 1987a). It is now known that not only the size, and structure of plant mtDNA but also the number of pairwise large repeats vary considerably with mitochondrial haplotypes (Chang et al., 2011; Kubo and Newton, 2008).

Structural complexity of plant mtDNA

Plant mtDNA is frequently reported as a circular structure, because its sequences can be mapped as a circle (Oda et al., 1992; Palmer, 1988; Palmer and Herbon, 1987b; Sparks and Dale, 1980). However, *in vivo* studies failed to support the suggested simple circular chromosome model of plant mtDNA, first, because plants mtDNA contain repeated sequences that can frequently recombine and serve as a source for DNA rearrangements (Lonsdale et al., 1984a; Palmer and Shields, 1984). Second, the possibility to map circular mitochondrial genome does not necessarily suggest the existence of circular DNA molecules but rather it can also occur as head-to-tail concatemers and/or circularly permuted linear molecules (Bendich, 1993b). Efforts to detect predicted plant mtDNA genome-sized circular molecules largely revealed branched and linear structures (Backert and Borner, 2000; Manchekar et al., 2006a; Oldenburg and Bendich, 1996). Still the impression of circular plant mtDNA persists and the linear mtDNA observations are thought to be an artifact of DNA extraction procedures.

Currently, more than 120 complete-plant mtDNA sequences are available at PubMed (http://www.ncbi.nlm.nih.gov/genome). It has become possible due to the recent advances in the sequencing technologies enabling to gather a quick and deeper set of data. Deep sequencing of plant mtDNA has not only resulted in the identification of structural rearrangements but has also provided the frequency of variations in relative sequence abundances (Mower et al., 2012). However, the ambiguities related to the prevalence of a circular and/or linear mtDNA molecule still remains unresolved. It is because the assembly of a complete genome from the short-sequenced fragments is still a form of mapping, which retains the characteristic opacities in inferring the *in vivo* mtDNA molecular structure from the genome maps. Therefore, data from the mtDNA sequencing still need to be elaborated in a broader perspective that must integrate *in vivo* observations of the mtDNA structure. Nevertheless, results from the past sequencing and mapping techniques have supported the master circle model, whereas the latest sequencing, electrophoretic, and microscopy studies have illustrated that the genome-sized circular mtDNA molecules are rare or undetectable (Bendich, 1993a; Mower et al., 2012). In fact, the master circle model was devised for comparatively simpler mtDNA structures that can be mapped into circular molecules like that of *Brassica campestris*. Its mtDNA has only a single pair of large repeats that can potentially recombine to form two subgenomic circles (Palmer and Shields, 1984).

Given the simple origin of the master circle model, the discovery of remarkably complex plant mtDNAs (with large numbers of repeats and alternative structural acquaintances) posed additional difficulties for the master circle model. The mitochondrial genomes from lycophytes and *Silene vulgaris* contain a huge number of recombinational breakpoints that could not be resolved with any simple circular map (Grewe et al., 2009; Hecht et al., 2011; Sloan et al., 2012a,b). The most extreme example of intraspecific diversity in mtDNA sequences observed so far is *Beta vulgaris* and *S. vulgaris*, where variations in mtDNA sequences have led to a high frequency of male sterility in the natural populations (Darracq et al., 2011; Sloan et al., 2012b). Contrarily, angiosperm mtDNAs as, for example, of white mustard (*B. hirta*) (Palmer and Herbon, 1987b), grape (*Vitis vinifera*) (Goremykin et al., 2009), zucchini (*Cucurbita pepo*), mung bean (*Vigna radiata*) (Alverson et al., 2011b), duckweed (*Spirodela polyrhiza*) (Wang et al., 2012), watermelon (*Citrullus lanatus*), and zucchini (*C. pepo*) (Alverson et al., 2010) do not contain any large repetitive sequences suggesting that the repeated sequences seem not to play any role in plant mtDNA function. These angiosperm mtDNAs are special examples of the master circle model and can therefore serve as models to investigate the effects of large recombining repeats on the mtDNA structure and function.

Comparison of plant mtDNA sequences across different species has also illustrated basic differences in mtDNA structure. Cucumber (*Cucumis sativus*) mtDNA is a particular example that contains an autonomous multichromosomal genome, consisting of 1.6 Mb-, 84 kb-, and 4.5 kb-sized DNA molecules. The 1.6-Mb DNA sequence can be assembled into a master circle-like map (Alverson et al., 2011a), whereas the two smaller chromosomes appear autonomous and contain a pair of 3.6-kb repeats for recombinations. Even a more striking example of multichromosomal organization is *S. vulgaris*, whose DNA appears to be fragmented into <200-kb segments. Some of these fragments contain repeat regions whereas others lack any repeat sequences (Sloan et al., 2012b). *S. vulgaris* also contains few (1–4) autonomous mitochondrial chromosomes, which can be mapped into circular structures (Sloan et al., 2012a,b).

The mtDNA content and its structure not only differ between various plant species but it also differs between various tissues (meristematic and vegetative) of the same species. Here, the master circle is assumed to represent the inherited form whereas the subgenomic and linear molecules are believed to represent the noninheritable form of mtDNA (Woloszynska, 2010a). Addressing this prospect will necessitate detailed analysis of mtDNA structure in meristematic and vegetative tissues. Furthermore, the mtDNA structure also fluctuates with variations in the environment (Arrieta-Montiel and Mackenzie, 2011).

To untangle the mechanism(s) of plant mtDNA inheritance as a unique genome structure(s), intensive investigations are needed to cover different tissue types, various developmental stages, and environmental conditions.

Replication of plant mtDNA

The most-accepted model for the replication of animal mtDNA is the unidirectional displacement loop (D-loop) mechanism (Bogenhagen and Clayton, 2003); however, strand-coupled replication (bidirectional replication mechanism) and RITOLS (ribonucleotides are incorporated throughout the lagging strand) are also thought to carry out the replication of animal mtDNA (Bowmaker et al., 2003; Fish et al., 2004; McKinney and Oliveira, 2013). These models take the same origins of replication as a reference and only differ in the amount of RNA that is engaged in the freshly synthesized mtDNA strand. The strand-coupled replication model assumes the production of Okazaki fragments as intermediates for the synthesis of a new mtDNA strand, whereas the D-loop model assumes the incorporation of partially processed transcripts as primers into the newly synthesized mtDNA strand (McKinney and Oliveira, 2013; Reyes et al., 2013). However, plant mtDNA's distinctive structural complexity has made the D-loop mode of mtDNA replication less plausible (Arrieta-Montiel et al., 2009; Christensen, 2013; Woloszynska et al., 2012). It is still not clear whether plant mtDNA contains any definite replication origins and what the mechanism(s) that control mtDNA copy number and integrity are. Nevertheless, plant and fungi mtDNA are mostly believed to replicate by a rolling circle and recombination-dependent mechanism (Backert and Borner, 2000; Manchekar et al., 2006b; Oldenburg and Bendich, 1996, 2001).

Studies have shown that mtDNA content varies with plant development and in different plant tissues (Fujie et al., 1993; Oldenburg et al., 2013; Preuten et al., 2010; Woloszynska et al., 2012). The mtDNA content was found to be higher in root meristematic cells (Fujie et al., 1993), young leaf cells, and in the cotyledons (Oldenburg et al., 2013), whereas mtDNA levels declined in old senescent leaves (Preuten et al., 2010). Furthermore, a steady increase in the number of mitochondria and the size of cells was reported in actively growing tissues, while a decrease in both was observed in aged tissues (Preuten et al., 2010). Also, in rice egg cells, mtDNA was 10-fold higher relative to leaf protoplasts and roots (Takanashi et al., 2010). In roots, mtDNA synthesis was higher in the tip, which falls to lower levels outside this meristematic region (Fujie et al., 1993). The differences in mtDNA content (during development, with age, and organ type) suggests that mtDNA synthesis seems to maintain a full genome copy in the rapidly dividing cells, during which the transition to senescing cells/tissues/organs was transformed into substoichiometric molecules due to recombinations (Woloszynska et al., 2012). This also advocates that plant mitochondria in most cases contain less than a full genome equivalent in aged tissues. Further investigations are needed to fully understand the mechanism that is involved to maintain mtDNA in vegetative cells/tissues with no meristematic zones.

Enzymes involved in mtDNA replication

DNA polymerases

So far, the mechanisms proposed for the replication and maintenance of plant mtDNA are principally derived from yeast and mammals. In mammals, there are three nuclear-encoded DNA polymerases responsible for mtDNA replication, namely, DNA polymerase gamma, Twinkle helicase, and single-stranded DNA-binding (SSB) proteins (McKinney and Oliveira, 2013). Contrarily, in plants, only two nuclear-encoded DNA polymerases IA (Pol-IA) and IB (Pol-IB) are detected that are targeted into mitochondria and plastids (Carrie et al., 2009; Christensen et al., 2005; Elo et al., 2003b; Moriyama et al., 2011; Ono et al., 2007). These plant mtDNA polymerases display phylogenetic and functional overlap with the bacterial DNA polymerase I (PolI) (Moriyama et al., 2011; Ono et al., 2007). In *Arabidopsis*, DNA Pol-IB also functions as a DNA repair enzyme in plastids. Mutation analysis of DNA Pol-IB revealed increased susceptibility of plants to DNA-damaging agents; conversely, mutations in DNA Pol-IA showed no susceptibility to such chemical agents (Parent et al., 2011). However, DNA Pol-IA does reveal reduced seed production and slower growth rate relative to control plants (Cupp and Nielsen, 2014). This suggests that these DNA polymerases play multiple roles in the organelles and have temporal and/or spatial difference in their expression. Detailed analysis of Pol-IB mutants in *Arabidopsis* also displayed slight differences in the growth rate during germination. These mutants also showed reduced (by 30%) mtDNA content and slower respiration rates (Cupp and Nielsen, 2013). Surprisingly, Pol-IA transcript levels were enhanced (by 70%) in Pol-IB mutants suggesting that these polymerases are partially redundant. These observations further suggested that DNA Pol-IB controls mtDNA maintenance and replication, whereas Pol-IA may not (Cupp and Nielsen, 2013). The Pol-IB mutants also exhibited a significant increase in the number of smaller-sized mitochondria per cell suggesting that mutations in Pol-IB either leads to a reduction in mitochondrial fusion or an increase in mitochondrial fission, which still remains to be investigated.

DNA primase and helicase

Owing to the huge size and complexity of mitochondrial genomes in plants, both primase and helicase enzymes are expected to function in mtDNA replication. In *Arabidopsis*, an ortholog of bacteriophage T7-gp4 protein was identified that has both DNA primase and helicase properties (Diray-Arce et al., 2013). The orthologs of this bacteriophage T7-gp4 protein were found in all eukaryotes but were found to be absent in fungi. In phage, the N-terminal domain of this protein has DNA primase property, whereas the C-terminal has the helicase activity (Shutt and Gray, 2006). In metazoans, the ortholog of bacteriophage T7-gp4 is named as Twinkle proteins. Here, this protein has the DNA helicase motif (Shutt and Gray, 2006) whereas, in *Arabidopsis*, it has both motifs for primase and helicase activities (Diray-Arce et al., 2013). *Arabidopsis* also encode a truncated version of the full-length Twinkle gene that encodes a protein having only DNA primase domain (Diray-Arce et al., 2013). From this, it is conceivable that plant mtDNA houses some other proteins for the helicase function in addition to the Twinkle orthologs. It is also believed that as a substitute to the Twinkle protein, one of the mitochondria RNA polymerases (RNAPs) (Carrie and Small, 2013; Hedtke et al., 2000; Liere et al., 2011) may be capable of carrying out the DNA priming activity for mtDNA replication, as has been observed in animals (Wanrooij et al., 2008). This is just a hypothesis to consider mtRNA polymerase a helicase for the replication of mtDNA in plants but due to the high significance of this proposal, it is worthy of analysis.

Recombinase

Recombination of mtDNA is believed to require one or more DNA recombinase enzymes. In plant mitochondria, the DNA recombination is also known to play a dominant role in replication. Characterization of the mtDNA recombination in soybean (Manchekar et al., 2006b), *Arabidopsis* (Khazi et al., 2003), and turnip (Manchekar et al., 2006b) has revealed the DNA recombination structures of mtDNA. In *Arabidopsis*, the four nuclear-encoded orthologs to the bacterial RecA enzyme have been characterized. One of these enzymes is localized to the mitochondria, the second one is localized into both mitochondria and plastids, while targeting of the remaining two proteins is still unclear (Gualberto et al., 2013; Khazi et al., 2003; Wall et al., 2004). Under stress situations, in addition to RecA, another nuclear-encoded enzyme, that is, Rad51 is believed to be recruited by the mitochondrial replication machinery to facilitate mtDNA replication. In this context, Rad51 would function in the same way as has been recently reported in human cells (Sage and Knight, 2013).

SSB protein

In plants, at least two types of SSB proteins are known that are involved in the mtDNA replication, recombination, and/or repair (Edmondson et al., 2005; Zaegel et al., 2006). The first type of SSB1 proteins are localized to both mitochondria and plastids. SSB1 proteins bind to the unwound DNA molecules at the replication fork to inhibit the reannealing of DNA molecules during the replication process. SSB1 is also known to stimulate the activity of mitochondrial recombinase RecA (Cupp and Nielsen, 2014; Edmondson et al., 2005). The second type of SSB proteins are organellar single-stranded DNA binding (OSB) proteins (Zaegel et al., 2006). Mutational analysis of OSB1 and OSB4 illustrated the accumulation of aberrant mtDNA molecules into the mitochondrial that seem to have resulted from ectopic recombinations (Gualberto et al., 2014; Zaegel et al., 2006). OSB1 thus seems to function as a recombination surveillance protein to inhibit the transmission of aberrant mtDNA molecules into the new mitochondria (Zaegel et al., 2006).

Topoisomerase

Topoisomerase enzymes are needed for releasing tension induced in the double-stranded DNA molecules ahead of and behind the replication forks. In *Arabidopsis*, two dual-targeted topoisomerases known as topos I and topos II have been characterized (Carrie et al., 2009; Wall et al., 2004). Mutational analysis of topos II (that has bacterial-like DNA gyrase activity) showed an embryo-lethal phenotype (Wall et al., 2004). Topos II was found essential for the replication and transcription of mtDNA in prokaryotes. It is the only-known enzyme that bears the capacity of catalyzing the ATP-dependent DNA supercoiling (Gellert et al., 1976). Further investigations are needed to determine the exact number and types of DNA topoisomerases that function in the replication of plant mtDNA.

Fission, fusion, and segregation of mtDNA

It has been shown repeatedly that plant mitochondria can exist as heteroplasmic entities where they contain less than a full genome equivalent (Kanazawa et al., 1994; Kmiec et al., 2006; Preuten et al., 2010). The heteroplasmic status of plant mitochondria is influenced by the frequent fission and fusion processes (Arimura et al., 2004). Failure to divide after fusion of several mitochondria can result in long filamentous mitochondria. This leads to the mixing of mtDNA from fragmented mitochondrial genomes, which can ultimately combine to form a complete genome (Muise and Hauswirth, 1995; Woloszynska et al., 2006). In yeast, fusion-defective mitochondria rapidly lose their DNA whereas fusion defects in mammals mitochondria lead to dysfunction. In such cases, fusion-defective mitochondria are found to accumulate in one part of the cell as large networks, whereas other parts of the cell leave without any functional mitochondria (Westermann, 2010). Replication of plant mtDNA does not appear to be directly connected to the cell cycle because the mtDNA copy number varies with tissue types and plant development (Preuten et al., 2010). Conversely, structure and organization of mtDNA changes during the cell cycle where cage-like mitochondrial structures appear around the nucleus (Antico Arciuch et al., 2012; Segui-Simarro et al., 2008), which divides with the cell (Segui-Simarro et al., 2008). mtDNA is believed to replicate in these centralized zones (Logan, 2010) suggesting that the mtDNA replication occurs in the large and fused mitochondria (Woloszynska et al., 2012). It is also observed that mitochondria at the periphery of these centralized structure, bud out and refuse (Segui-Simarro et al., 2008). It is speculated that mtDNA segregates into these peripheral mitochondria upon fission and results in mitochondria containing less than a full genome equivalent (Preuten et al., 2010).

mtDNA recombinations

In plant mitochondria, large repeats of >1 kb are present in four forms and recombination between one pair of reciprocal forms leads to the generation of two remaining forms and *vice versa*. Since all the recombination forms appear in similar stoichiometry, the large repeats are believed to recombine frequently and reversibly to sustain interconversions between subgenomic mtDNA molecules (Zaegel et al., 2006). Short repeats of six to several hundred base pairs are not always represented by all four forms within the genome and those that are detected differ significantly in stoichiometry (Bellaoui et al., 1998; Kanazawa et al., 1994; Woloszynska et al., 2001; Woloszynska and Trojanowski, 2009). As mentioned earlier, the short repeats recombine sporadically and irreversibly, producing new and stable mtDNA arrangements. Since the plant mitochondrial genomes are rich in short-repeats recombination, it seems to significantly influence the evolution of plant mtDNA (Bellaoui et al., 1998; Grabau et al., 1992; Kanazawa and Shimamoto, 1999; Moeykens et al., 1995; Woloszynska et al., 2001; Woloszynska and Trojanowski, 2009). In *Arabidopsis* mtDNA, 22 pairs of identical repeats are present and with the exception of the two larger ones, the rest recombine infrequently (Unseld et al., 1997b). Advanced electrophoresis and electron microscopy techniques have provided further proof of mtDNA recombination in higher plants (Backert and Borner, 2000; Manchekar et al., 2006b). In soybean and *Chenopodium album*, branched DNA molecules were detected in the mitochondria (Backert and Borner, 2000). In the light of above facts, it is now widely accepted that plant mitochondria are heteroplasmic in nature, that is, along with an abundant main genome, they also contain substoichiometric DNA molecules (Backert et al., 1997; Manchekar et al., 2006b; Oldenburg and Bendich, 1996), where the main genome determines the phenotype of a plant, whereas the substoichiometric molecules remain functionally silent. This is because the substoichiometric molecules are known to accumulate 10-, 100-, or even 1000-folds less relative to the main genome (Arrieta-Montiel et al., 2001; Feng et al., 2009; Lilly et al., 2001; Woloszynska and Trojanowski, 2009).

Transcription in plant mitochondria

Mitochondria have acquired specialized components for the expression of their genome and some of these components are encoded by the mtDNA, whereas the majority is encoded by the nuclear genome. The nuclear-encoded components have acquired special mitochondrial-targeting sequences that bring these proteins into the mitochondrial compartment. Here, we describe various components of the mitochondrial transcription machinery and their role in transcription and its regulation.

Mitochondrial promoters

Plant mitochondrial promoter sequences were identified using *in vitro* capping, primer extension, mapping with nuclease S1, and sequencing techniques (Binder et al., 1991; Caoile and Stern, 1997; Dombrowski et al., 1999; Farre and Araya, 2001; Lupold et al., 1999a,b). These techniques have helped us to probe promoter motifs and transcription initiation sites in *Zea mays* (Mulligan et al., 1988a, 1991; Rapp et al., 1993), *Triticum aestivum* (Covello and Gray, 1991), *Glycine max* (Brown et al., 1991), *Oenothera berteriana* (Binder and Brennicke, 1993), *Solanum tuberosum* (Binder et al., 1994; Giese et al., 1996; Lizama et al., 1994), *Zea perennis* (Newton et al., 1995), *Pisum sativum* (Binder et al., 1995; Giese et al., 1996; Hoffmann and Binder, 2002), *Nicotiana sylvestris* (Lelandais et al., 1996), *Nicotiana tabacum* (Edqvist and Bergman, 2002), and *Sorghum* (Yan and Pring, 1997). In most of these plant species, a consensus nucleotide sequence CA/GTA motif was identified as the core element of mitochondrial promoters (Brennicke et al., 1999a; Fey and Marechal-Drouard, 1999). In dicots, this CA/GTA motif also serves as the transcription initiation site (Binder et al., 1996).

However, evaluation of promoter motifs from some other plant species such as *Oenothera* (Binder and Brennicke, 1993), *P. sativum* (Binder et al., 1995), *S. tuberosum* (Tada and Souza, 2006), and *G. max* (Brown et al., 1991) revealed an extended version of this consensus sequence (AAAATA**TCATAAGAGA**AG). This extended version consists of a conserved nonanucleotide motif (–7 to +2, bold letters) and a transcription initiation site (underlined) (Dombrowski et al., 1999). Mutational analysis of the conserved residues revealed that transcription initiation frequently extends beyond the obviously conserved consensus regions (Hoffmann and Binder, 2002). This is supported by the absence of any homology consensus promoter sequence motif in the highly transcribed 26S ribosomal RNA gene, in both monocots and dicots (Brennicke et al., 1999b). In *Arabidopsis*, in addition to the CA/GTA-type consensus motif, many unconventional promoter motifs such as RGTA and ATTA were also detected lacking any consensus sequence motif (Kuhn et al., 2005). Similarly, in monocots, the core promoter also consists of a CA/GTA motif directly upstream of the first transcribed nucleotide. In maize, analysis of *atp1* and *cox3* promoters confirmed the significance of CA/GTA motif (Caoile and Stern, 1997; Rapp et al., 1993), whereas, in sorghum, not only the conserved CA/GTA motif was detected but also an additional series of degenerate sequence motifs (AATA, CTTA, and YRTA) was identified (Yan and Pring, 1997). In most instances, the monocot consensus-type promoters also contain an AT-rich region of 6–24 nucleotides, ~10 nucleotides further upstream of the conserved motif. This AT-rich region was found necessary for the full activity of the downstream promoters (Caoile and Stern, 1997; Rapp et al., 1993; Tracy and Stern, 1995).

In addition to this, the presence of more than one promoter and multiple transcription initiation sites is also a common feature of plant mitochondria (Kuhn et al., 2005; Lupold et al., 1999b; Mulligan et al., 1988a; Tracy and Stern, 1995). In *O. berteriana*, at least 15 promoters were found active in the genome (Binder and Brennicke, 1993). In *Arabidopsis*, each of such promoters was found to have different strength (Giege et al., 2000b). Furthermore, the multiple initiation sites to transcribe a single gene were detected, as, for example, three initiation sites were observed for *cox2* and *cob* genes, whereas six were reported for *atp9* gene in maize (Mulligan et al., 1988a; Newton et al., 1995). Some plant mitochondrial promoters (like that of *cox2* gene in maize) were mapped several kilobases apart from the coding sequence (Lupold et al., 1999b), whereas in certain other cases (like *rRNA* genes), the consensus promoter sequences were not even recognized by the *in vitro* mitochondrial transcription analysis (Binder et al., 1995). The presence of multiple promoters in the plant mitochondria is suggested to ensure a faithful transcription despite mtDNA rearrangements (Kuhn et al., 2005). Maize mitochondrial *cox2* gene is an excellent example where the promoter's activity is chiefly dependent on the genomic context and reflects the significance of mtDNA intra- and intergenomic recombinations on mitochondrial gene expression (Lupold et al., 1999a).

Mitochondrial phage-type RNAPs

Plant mitochondrial transcription is executed by nuclear-encoded phage-type RNAP enzymes (Hess and Borner, 1999; Liere et al., 2011). The only-known protist that has retained eubacterial RNAP genes is the *Reclinomonas americana* (Lang et al., 1997). However, some eukaryotes (like algae and *Selaginella moellendorffii*) (Yin et al., 2009) contain only a single RNAP gene, whereas *Physcomitrella patens* and other angiosperms contain a family of RNAP genes. The autonomous duplication of mitochondrial *RpoTm* (mitochondrial localized) is believed to have provided the basis for the evolution of other two RNAPs, that is, *RpoTmp* (localized to both mitochondrial and plastids) and *RpoTp* (plastid localized) (Emanuel et al., 2004; Hedtke et al., 1997a; Swiatecka-Hagenbruch et al., 2007; Yin et al., 2010). The nuclear genome of *N. tabacum* encodes two sets of the three *RpoT* genes (six in total) (Hedtke et al., 2002). In poplar, more than one set of *RpoT* genes are present (http://genome.jgipsf.org/Poptr1 1/Poptr1 1.home.html). In *P. patens*, the two *RpoT* genes encode bifunctional RpoTmp peptides whereas the third gene encodes *Rpotm* (Kabeya and Sato, 2005; Richter et al., 2002). However, in *Nuphar advena* and *Arabidopsis thaliana*, a small family of phage-type RNAPs encodes the three *RpoT* genes that are targeted to mitochondria (*RpoTm*), plastids (*Rpotp*), or to mitochondria and plastids (*RpoTmp*) (Hedtke et al., 1997b, 2002; Yin et al., 2010).

Although these enzymes have long been known, we still lack investigations to describe the division of labor between mitochondrial *RpoTm* and *RpoTmp* in the process of transcription. In *Arabidopsis*, *RpoTm* and *RpoTmp* displayed an overlapping pattern of expression (Emmanuel et al., 2006) and were found vital for the gametogenesis and development (Tan et al., 2010). *RpoTmp* was also found to be important for the expression of plastid genes (Baba et al., 2004). Based on mutational analysis of *Rpotm* and *Rpotmp*, it was suggested that *RpoTmp* transcribes plastid and perhaps mitochondrial genes during the early stages of development whereas, *RpoTm* transcribes mitochondrial genes at the later stages of development. Furthermore, a recent study has shown *RpoTm* to be the key RNAP in plant mitochondria that can transcribe most, if not all, mitochondrial genes (Kuhn et al., 2009).

Auxiliary factors required for mtDNA transcription

In yeast, mitochondrial RNAP has lately been shown to exploit its C-terminal loop region for the recognition of promoter sequences, as observed for bacteriophage T7 RNAP (Nayak et al., 2009). The yeast RNAP acts as a single-subunit enzyme to start transcription without any accessory factor(s) (Matsunaga and Jaehning, 2004). Also, *Arabidopsis* RNAPs (*RpoTm* and *RpoTp*) were shown to recognize the promoter sequences *in vitro* on a supercoiled, but not on a linear DNA molecule without any additional cofactors (Kuhn et al., 2007). Conversely, *in vivo* transcription initiation in animals, fungi, and plant RNAPs is believed to require auxiliary factors for the precise and efficient promoter recognition. Such auxiliary factors have been identified in yeast and animals and are referred to as mitochondrial transcription factor A (mtTFA) and mitochondrial transcription factor B (mtTFB). In yeast, mtTFA stimulates transcription initiation by binding to DNA through two high-mobility group (HMG) boxes (Diffley and Stillman, 1992; Fisher et al., 1992; Parisi et al., 1993; Visacka et al., 2009; Xu and Clayton, 1992). The mtTFBs found in mitochondria are members of rRNA methyltransferases family, which are responsible to dimethylate adenosines in the small ribosomal subunit (Park et al., 2009; Richter et al., 2010; Schubot et al., 2001). In humans and mice, the two yeast mtTFB homologs (mtTFB1 and mtTFB2) were identified. The mtTFB1 protein functions as a methyltransferase, whereas mtTFB2 acts as a transcription factor (Asin-Cayuela and Gustafsson, 2007; Scarpulla, 2008). In *Arabidopsis* based on homology to mtTFA and mtTFB, a number of open-reading frames similar to methyltransferases were identified. From these, only one, that is, at5g66360 was found targeted to mitochondria. It can also carry out methylation of the conserved adenosine residues in mitochondrial rRNA (Richter et al., 2010). However, *in silico* screening for mitochondrial-targeted HMG-box-like proteins did not reveal any mtTFA homolog(s) in *Arabidopsis* (Elo et al., 2003a). Hence, it seems likely that plant mitochondria do not employ mtTFA- and mtTFB-like proteins as transcription factors.

However, mitochondrial lysates used for *in vitro* transcription analysis do indicate the presence of potential cofactors for the mitochondrial RNAP(s). In *T. aestivum*, a 69-kDa protein was detected that could initiate transcription from the *cox2* promoter *in vitro* (Ikeda and Gray, 1999). This protein shows resemblance to yeast mtTFB and is a member of pentatricopeptide repeat (PPR) proteins (Schmitz-Linneweber and Small, 2008). In *Arabidopsis*, the three homologs of this mitochondrial-localized PPR protein were identified. They were unable to interact with *RpoTm* or *RpoTmp,* however, they could interact nonspecifically with mitochondrial promoter regions (Kuhn et al., 2007). Similarly, from pea mitochondria, 43 and 32 kDa proteins were isolated that could bind to the *atp9* promoter region (Daschner et al., 2001). Other candidate cofactors of the mitochondrial RNAP(s) include *Mct* from *Z. mays* (Newton et al., 1995), *Sig2B* from maize (Beardslee et al., 2002), *Sig1,* and *Sig5* from *Arabidopsis* (Fujiwara et al., 2000). The Mct protein could bind to the mitochondrial *cox2* promoter and therefore might be involved in transcription initiation and regulation (Newton et al., 1995). The plastid-localized sigma factors were also found to localize to the mitochondria and the maize Sig2B was copurified with RpoTm (Beardslee et al., 2002), suggesting a conceivable function of these sigma factors in the regulation of mitochondrial transcription through the phage-type RNAPs.

Mitochondrial transcript maturation

In plant mitochondria, posttranscriptional changes determine the maturation of RNA and are therefore an important component of mitochondrial gene expression. It has been shown that unprocessed RNAs are nonfunctional and are rapidly degraded (Giege and Brennicke, 2001; Khvorostov et al., 2002). The posttranscriptional modifications include RNA splicing, alteration of the 5′ and 3′ ends, polyadenylation, and editing (Binder and Brennicke, 2003).

Alteration of the 5′ and 3′ ends

In higher plants, although most mitochondrial genes are transcribed to yield monocistronic products, however, bicistronic and polycistronic RNAs are also produced through the cotranscription of adjacent genes (Gray et al., 1999b). In such cases, formation of mature 5′ end is determined by the processing of the 3′ end of RNA synthesized on the adjacent gene

(Gray and Spencer, 1983; Makaroff et al., 1989; Singh and Brown, 1991). In mitochondria, the 5′ and 3′ regions of some mRNAs also contain tRNA-like structures (t-elements), which act as processing signals (Bellaoui et al., 1997). The mature 5′ end of some monocistronic RNAs also results from processing. Sequence comparison of mature rRNAs, pre-rRNAs, and several mRNAs has revealed conserved motifs at the 5′ end that possibly act as processing signals (Maloney et al., 1989; Schuster and Brennicke, 1989).

Polyadenylation

The stability of plant mitochondrial mRNAs is attributed to specific and often highly conserved inverted repeats, which form hairpins and loops at the 3′ mRNA end (Bellaoui et al., 1997; Dombrowski et al., 1997; Kuhn et al., 2001). The occurrence of several secondary structures at the 3′ end of plant mitochondrial RNAs are also known to act as processing and stability signals (Dombrowski et al., 1997). It is believed that the secondary structures present at the 3′ extremities impede the progression of 3′ to 5′ exoribonucleases. In higher plants, polyadenylation of mitochondrial RNA is shown to act as RNA degradation signal (Gagliardi and Leaver, 1999; Kuhn et al., 2001; Lupold et al., 1999c). In plant mitochondria, polyadenylation is sequence-independent and predominantly occurs in a region of a secondary-structure element at the mature 3′ end. The poly(A) tail in the mitochondrial mRNAs is short lived and a 10–19-nucleotides suffice is sufficient to increase the rate of mRNA degradation (Gagliardi and Leaver, 1999; Kuhn et al., 2001). Nucleotide addition and polyadenylation are known to occur in parallel and are alternative variants of mRNA processing at the same position (Khvorostov et al., 2002). In brief, mitochondrial RNA degradation is not only necessary to maintain the proper turnover rate of functional transcripts but is also necessary to shape the final transcriptome of plant mitochondria.

RNA splicing

In plant mitochondria, RNA splicing is a crucial step in the conversion of precursors into mature mRNAs. Mitochondrial genes contain two classes of introns, that is, group-I and group-II introns (Cho et al., 1998; Michel et al., 1989; Vaughn et al., 1995), mostly arranged in *cis*-configuration; however in some cases, *trans*-splicing is also known to produce mature and functional mRNAs (Chapdelaine and Bonen, 1991; Malek et al., 1997). In *Arabidopsis*, mitochondrial genome is interrupted altogether by 23 group-II introns and one group-I intron (Cho et al., 1998; Unseld et al., 1997b; Vaughn et al., 1995). Some genes are interrupted by more than one intron, for example, *nad7* has four introns. Mitochondrial introns at some instances encode maturases (*matR*) proteins, which are usually necessary for splicing of at least their respective intron (de Longevialle et al., 2008). Additionally, *trans*-splicing also occurs in plant mitochondria at several instances, principally where physically separated exons are flanked by partial group-II intron sequences (Chapdelaine and Bonen, 1991; Wissinger et al., 1991). In *Arabidopsis*, five *trans*-splicing events are detected in *nad1* (Wissinger et al., 1991), *nad2* (Lippok et al., 1996), and in *nad5* genes (Knoop et al., 1991). Relatively, little is known about the splicing machinery in plant mitochondria, although it is expected to be complex and is thought to contain a number of nuclear-encoded factors (Lambowitz and Zirnmerly, 2004; Lehmann and Schmidt, 2003). The mitochondrial or the nuclear-encoded maturases are believed to be involved in the organelle–intron splicing; however, this has not been demonstrated experimentally.

RNA editing

In plants, editing of mitochondrial transcripts is observed to be more intense as compared to other organisms. Its major variant is substitution of U for C, probably by deamination but substitution of C for U, has also been revealed in some species (Mulligan et al., 1999). Editing affects preferentially protein coding transcripts, however it is occasionally found in structural RNAs (Giege and Brennicke, 1999). In mRNAs, editing result in increasing the conserved similarity of the deduced protein sequences with homologous proteins in other species (Giege and Brennicke, 1999). In tRNAs, editing is believed to improve the spatial folding of these molecules to ultimately improve their functionality. The extent of editing is gene specific, and varies from one to several dozen sites per kilobase (Giege and Brennicke, 1999; Maier et al., 1996). Editing is such an important step in plant mitochondrial RNA maturation that proteins arising from nonedited mRNAs cannot become a part of the multiprotein complexes and are rapidly degraded (Giege and Brennicke, 1999, 2001). As mentioned earlier, noncoding RNA sequences (5'- and 3'-untranslated regions, introns, and intergenic spacers of cotranscribed genes) are also edited. The editing of the 5'-untranslated region improves the ribosome binding with mRNA, as observed for *O. berteriana rps14* (Maier et al., 1996). Additionally, the editing of noncoding regions may also be necessary for processing to yield functional RNA (Pring et al., 1998). It is also known that the rate of editing varies with environmental conditions, tissue, and plant age (Grosskopf and Mulligan, 1996; Howad and Kempken, 1997). RNA editing has also been analyzed with respect to its possible interaction with intron splicing events that demonstrated that editing at some sites might not be an absolute prerequisite for splicing (Binder et al., 1992; Carrillo and Bonen, 1997; Knoop et al., 1991; Lippok et al., 1994; Sutton et al., 1991; Wissinger et al., 1991; Yang and Mulligan, 1991). It is speculated that the regulation of mitochondrial gene expression in plants may have taken control of crucial editing sites such as those creating a translation initiation codon (AUG) from ACG and those that eliminate stop codons from within an mRNA molecule (Marchfelder et al., 1996; Marechal-Drouard et al., 1996).

Regulation of transcription

Gene expression is regulated at transcriptional, posttranscriptional, translational, and posttranslational levels. In the past, a number of studies were carried out to elucidate the mechanisms that govern the expression of plant mitochondrial genome. At the level of DNA, the analysis of mitochondrial transcriptional rates in maize showed rDNA to be most strongly transcribed (2–14-folds) relative to the protein-coding genes (Finnegan and Brown, 1990; Mulligan et al., 1991) where the changes in transcript levels were believed to have resulted from the difference in promoter strength (Muise and Hauswirth, 1992). Likewise, comparison of mitochondrial transcriptional rates in *Arabidopsis* and *Brassica napus* (male sterile and a fertile line) revealed species-specific changes in the transcription rates of mitochondrial genes (*nad4L, nad9, cox1, rps7, ccmB,* and *rrn5*), which was also thought to be due to differences in promoter strength (Leino et al., 2005). Increased expression of mitochondrial genes was also reported for *nad2, nad4L, nad6, nad7, cox1, cox3, atp1, atp9, ccmC, ccmFn,* and *rps7* genes after imbibition of wheat seedling (Khanam et al., 2007). A comprehensive run-on transcriptional analysis using *Arabidopsis* cell suspension cultures illustrated distinct transcriptional rates between individual genes, even for the genes-encoding components of the same multisubunit complex (Giege and Brennicke, 2001; Giege et al., 2000a). Tissue and/or cell-specific variations in the expression of mitochondrial genes have been associated with transcriptional levels in a few studies (Li et al., 1996; Topping and Leaver, 1990). In *Arabidopsis*, multiple mitochondrial promoters have been identified, however mapping studies failed to support any function of these promoters in a tissue or development specific regulation of the mitochondrial gene expression (Kuhn et al., 2005). Interestingly, the differences in transcriptional rates were found counterbalanced in the steady-state RNA pool, most likely by posttranscriptional processes and due to differences in RNA stabilities (Giege et al., 2000a; Holec et al., 2008; Mulligan et al., 1988b; Tracy and Stern, 1995).

In Brassica, nuclear background has also been shown to influence mitochondrial transcription rates and posttranscriptional steps suggesting mitochondrial *cis*-elements and *trans*-factors to be involved in these processes (Edqvist and Bergman, 2002). In *Arabidopsis* cell suspension cultures, little or no influence of *cis*- and/or *trans*-factors was noticed on mitochondrial gene expression at the level of transcription and translation in response to sugar starvation (Giege et al., 2005). However, the observed decrease in ATPase complexes could be ascribed to the nuclear-encoded components of the ATPase complex that were downregulated under sugar starvation suggesting that the precise stoichiometric proportions of different proteins were achieved posttranslationally.

In angiosperms, the presence of two mitochondrial RNAPs (*RpoTm* and *RpoTmp*) is believed to have some regulatory function. The mutants of *RpoTmp* revealed reduced transcription and transcript abundance of mitochondrial *cox1, nad2,* and *nad6* genes with a reduction in the amount of their respective complexes (I and IV). This means that in plant mitochondria, transcription regulation seems to be important for mitochondrial gene products (Kuhn et al., 2009). It is also believed that both *RpoTm* and *RpoTmp* could transcribe mitochondrial genes, with *RpoTmp* playing a dominant role in fine-tuning the expression. Additionally, mitochondrial gene copy number in *Arabidopsis* seems to vary substantially between different organs and at various developmental stages (Preuten et al., 2010). Indeed, pollens and anthers development was found to be associated with higher transcript levels, mitochondria number, and mtDNA amounts (Geddy et al., 2005; Warmke and Lee, 1978). In barley, mitochondrial transcripts and gene copy number was found to increase in senescent leaves as compared to green leaves (Hedtke et al., 1999). Furthermore, positive correlation was observed between transcripts and protein abundances in the *RpoTmp* mutant of *Arabidopsis* (Kuhn et al., 2009). Comparison of mitochondrial gene copy number and transcript levels in *B. hirta,* maize, Hauswirth, and Muise revealed a direct correlation between gene copy number and transcriptional rate; however, this appears not to be the rule as no such correlation was observed during the leaf development in *Phaseolus vulgaris* (Woloszynska, 2010b) and in *Arabidopsis* (Preuten et al., 2010). Despite the observed differences in transcriptional rates and transcript abundances, it is still unclear whether there is a regulation at the level of transcription initiation and whether there is a gene-specific regulation or rather an up- or downregulation of transcription of the complete mitochondrial genome.

Translation in plant mitochondria

Plant mitochondria also contain complete translational machinery, primarily comprising ribosomes, ribosomal RNAs, translation factors, tRNAs, and aminoacyl-tRNA synthetases. Generally, only few components of mitochondrial translational machinery are encoded by the mtDNA whereas, the majority of these components are encoded by the nuclear genome and their protein products are imported into the mitochondria.

Ribosomal proteins and ribosomal RNAs

The number of ribosomal subunits that are encoded by the mtDNA varies widely between the different plants species (Sanchez et al., 1996). Most of the ancestral ribosomal protein-coding genes were either lost or transferred to the nuclear genome, where they have acquired a promoter and a mitochondrial targeting sequence (Adams et al., 2000). In *Arabidopsis*, the mtDNA encodes a subset of ribosomal proteins including *rpl2*, *rpl5*, *rpl16*, *rsps3*, *rsps4*, *rsps7*, and *rsps12* (Unseld et al., 1997a). Among these, *rpl5* and *rpl16* were found missing in the mitochondrial genome of sugar beet, whereas *rpl13* gene was present in sugar beet and absent from the *Arabidopsis* mitochondrial genome (Kubo and Newton, 2008; Kubo et al., 2000). Plant mitochondrial genomes encode all the three ribosomal RNAs (Duchene et al., 2005; Kuhlman and Palmer, 1995).

Aminoacyl-tRNA synthetases

Aminoacyl-tRNA synthetases catalyze the addition of amino acids to their cognate tRNAs. In higher plants, the aminoacyl-tRNA synthetases are encoded by the nuclear genes and targeted to cytosol and mitochondria. They are transcribed from the same gene, however, due to initiation of translation at alternative initiation codons such as the alanyl-tRNA synthetase, valyl-, and threonyl-tRNA synthetases (Mireau et al., 1996, 2000), and they are targeted to different compartments. Also, dual targeting to chloroplasts and mitochondria has been observed for methionyl-tRNA synthetase (Menand et al., 1998), histidyl-tRNA synthetase (Akashi et al., 1998), cysteinyl-tRNA, and asparginyl-synthetases (Peeters et al., 2000). The two glycyl-tRNA synthetases (EDD1 and EDD2) were also targeted to the mitochondria and plastids, to the cytosol, and mitochondria, respectively (Duchene and Marechal-Drouard, 2001; Moschopoulos et al., 2012).

Transfer ribonucleic acids

Plant mitochondrial tRNAs are either transcribed within mitochondria from native genes and/or imported from chloroplasts or from the nucleus. As for example in *Arabidopsis*, tRNAs for six amino acids (A, V, L, T, F, and R) are not encoded by the mitochondrial genome and are therefore imported from the cytoplasm (Dietrich et al., 1996; Salinas et al., 2005; Unseld et al., 1997a). Similarly, five (A, R, I, L, and T), six (A, R, L, Y, V, C), and 11 (H, W, F, N, M, C, V, I, P, R, and S) tRNAs are imported from the cytoplasm in sugar beet, rice, and potato, respectively (Kubo et al., 2000; Marechal-Drouard et al., 1990; Notsu et al., 2002).

Translation initiation codons

In mitochondria, translation initiation codons are generally not preceded by recognizable Shine–Dalgarno-like sequences, but rather conserved sequence blocks have been described preceding several translational starts (Pring et al., 1992). Their location is persistent with a possible role in translation initiation, which cannot be experimentally confirmed due to technique limitations. Sequence analysis has shown that translation is usually, but not always, initiated with an AUG codon in higher-plant mitochondria and therefore, it remains unclear how the ribosome is guided to and anchored at the start of translation. In *Arabidopsis* GGG, AAU, and GUG are possible additional translation initiator triplets (Unseld et al., 1997a). Similarly, in *Oenothera* GUG and in radish ACG were found to be potential translation initiation sites (Bock et al., 1994; Dong et al., 1998).

mRNA surveillance

In higher-plant mitochondria, it is still ambiguous whether there exists an mRNA surveillance system that would prevent translation of aberrant transcripts. In fact, both edited and incompletely edited transcripts have been frequently detected to be associated with mitochondrial ribosomes, suggesting that mRNA surveillance mechanism might not exist in plant mitochondria (Lu and Hanson, 1994, 1996; Phreaner et al., 1996). *Trans*-factors activating translation of a particular mRNA are expected to exist in plant mitochondria and are speculated to be members of the PPR family, as described in chloroplast (Schmitz-Linneweber et al., 2005). Although the precise action of these PPR proteins has not been determined to date, however, they may be involved in the translational control mitochondrial transcripts in a gene-specific manner (Hanson and Bentolila, 2004).

Hydrophobic nature of mitochondrial proteins

In higher plants, proteins encoded by the mitochondrial genome are mostly highly hydrophobic in nature. Synthesis of such proteins in a hydrophilic environment bears the tendency to form unproductive aggregates. In the cytosol, signal recognition particles exist for the recognition of ribosomes, which synthesize hydrophobic polypeptides and target them to the membrane-embedded translocation complexes (Gilmore et al., 1982a,b). These complexes insert the hydrophobic polypeptides cotranslationally into the mitochondrial membrane. These signal recognition particles have been reported for the cytosol as well as for the plastids but no homolog of signal recognition particles was found in the mitochondria (Glick and Von Heijne, 1996). The cotranslational mode of insertion of the mitochondrial translation products was suggested mainly because of two reasons; first, electron microscopic studies revealed translation active ribosomes to be located preferentially in close proximity to the inner mitochondria membrane (Watson, 1972) and second, the biochemical fractionation experiments indicated synthesis and membrane integration of mitochondrial peptides in a synchronized manner (Liu and Spremulli, 2000; Sevarino and Poyton, 1980). Another line of evidence that supports the cotranslational insertion of mitochondrial genome-encoded proteins came from yeast, where mitochondrial ribosomes were found tightly associated with the inner mitochondrial membrane (Borst and Grivell, 1971; Bunn et al., 1970). In yeast, Oxa1 and Mba1 proteins were found to bind to the large subunit of the ribosome in such a way as to bring the polypeptide exit site of the ribosome in close proximity to the membrane insertion machinery. In this, it was the mitochondrial-encoded hydrophobic proteins that were directly inserted into the inner mitochondrial membrane (Jia et al., 2003; Ott et al., 2006; Szyrach et al., 2003).

Translational regulation

In higher-plant mitochondria, almost nothing is known about the translational regulation of mitochondrial genome. However, posttranslational processes including protein structural modifications and degradation, has been widely suggested to eliminate unwanted and superfluous protein products arising from nonstringent translation and insufficient assembly of final products. This assumption is based on the comparison of edited and unedited peptides revealing that only proteins arising from fully edited transcripts yield evolutionary conserved and functional proteins. Protein analysis in maize mitochondria using antibodies specific for edited, partially edited, and nonedited RPS12 protein revealed that partially edited proteins are indeed synthesized (Phreaner et al., 1996); however, direct sequencing of the ATP9 from wheat (Begu et al., 1990), NAD9 from potato (Grohmann et al., 1994), and ATP6 from petunia (Lu and Hanson, 1994) revealed that only fully edited proteins could become a part of their respective multi-subunit polyprotein complexes. These observations suggest that while all the mRNAs are translated indiscriminately of their editing status, selection occurs posttranslationally and nonfunctional proteins are not incorporated into the complexes and are presumably rapidly degraded. In addition to this, proteins are also targeted for degradation when they remain unassembled or appear damaged, for instance by reactive oxygen species (Giege et al., 2005; Moller and Kristensen, 2006). Mitochondrial proteins are also processed at certain instances from larger precursors; thus, protein degradation seems to be a crucial control point in plant mitochondrial genome expression as it is in the plastids (Adam et al., 1999; Figueroa-Martinez et al., 2008; Perrotta et al., 2002). Protein degradation may become particularly relevant to those proteins that are components of the multiprotein complexes and whose stability is dependent on the accumulation of other subunits in stoichiometric amounts (Binder et al., 1996).

Differences in mitochondrial-encoded protein abundances have been reported in yeast (Ibrahim and Beattie, 1976), maize (Newton and Walbot, 1985), sugarbeet (Lind et al., 1991), rice (Dai et al., 1993; Hahn and Walbot, 1989), *N. sylvestris* (De Paepe et al., 1993), and petunia (Conley and Hanson, 1994). Increased mitochondrial protein synthesis was reported in response to chloramphenicol treatment, whereas a decrease was observed in response to cycloheximide treatment and after dark and cold treatments. Furthermore, higher levels of mitochondrial proteins were reported in leaves, flowers, and shoot tissues. Gene-specific change in mitochondrial translation was revealed in yeast and was shown to be due to specific nuclear-encoded proteins corresponding to individual mRNAs (Binder et al., 1996). Contrarily, translational regulation of plant mitochondrial genes has not been observed to date. However, there is an indirect evidence for such a regulation, which is supported by the prevalence of multiple initiator codon types in plant mitochondrial genes and the participation of the 5'-untranslated leader sequences in translation initiation (Binder et al., 1996; Pring et al., 1992).

Summary and future perspectives

In short, plant mitochondria are very unique in the organization and structure of their DNA as well as in the processes of transcription and translation as compared to animals, yeast, and fungi. The complexity of plant mitochondrial genomes, including the presence of substoichiometric DNA molecules, has made the characterization of mtDNA replication very difficult. The main question that remains is the specific mechanism by which plant mtDNA is replicated. To understand this, we need to investigate plant lines with mutations in one or more replication protein genes, and their effect on mtDNA rearrangements and recombination structures by electron microscopy and other techniques. Similarly, the enzymes responsible for transcription in mitochondrial have long been known, but the cofactors needed to assist them in transcription initiation, elongation, and termination are still unknown or not fully investigated. Furthermore, mitochondrial or nuclear-encoded maturases that speculated to play an active role in the organelle–intron splicing are still unknown. Although a number of investigations have shown differences in transcriptional rates and transcript abundances in different species and in various tissues, we still do not know whether there is a regulation at the level of mitochondrial transcription initiation and whether there is gene-specific regulation or rather there is an up- or downregulation of transcription of the complete mitochondrial genome. Mitochondrial-encoded proteins are also shown to vary in response to development and environmental changes but these studies are based on one or two directional gel electrophoresis analysis and owing to the differences in the stability or life time of different proteins, the reported variations in mitochondrial protein abundance cannot directly be taken as a change translational activity. Therefore, detailed investigations involving mutants and using microarray, polysome, and western blot analysis are needed to overturn and investigate the missing components of mitochondrial transcription and translation apparatus to better understand these processes in plant mitochondria.

References

Adam, A., Endres, M., Sirrenberg, C., Lottspeich, F., Neupert, W., and Brunner, M. 1999. Tim9, a new component of the TIM22.54 translocase in mitochondria. *Embo J*, 18(2), 313–319.

Adams, K.L., Daley, D.O., Qiu, Y.L., Whelan, J., and Palmer, J.D. 2000. Repeated, recent and diverse transfers of a mitochondrial gene to the nucleus in flowering plants. *Nature*, 408(6810), 354–357.

Adams, K.L. and Palmer, J.D. 2003. Evolution of mitochondrial gene content: Gene loss and transfer to the nucleus. *Mol Phylogenet Evol*, 29(3), 380–395.

Akashi, K., Grandjean, O., and Small, I. 1998. Potential dual targeting of an *Arabidopsis* archaebacterial-like histidyl-tRNA synthetase to mitochondria and chloroplasts. *FEBS Lett*, 431(1), 39–44.

Alverson, A.J., Rice, D.W., Dickinson, S., Barry, K., and Palmer, J.D. 2011a. Origins and recombination of the bacterial-sized multichromosomal mitochondrial genome of cucumber. *Plant Cell*, 23(7), 2499–2513.

Alverson, A.J., Wei, X., Rice, D.W., Stern, D.B., Barry, K., and Palmer, J.D. 2010. Insights into the evolution of mitochondrial genome size from complete sequences of *Citrullus lanatus* and *Cucurbita pepo* (Cucurbitaceae). *Mol Biol Evol*, 27(6), 1436–1448.

Alverson, A.J., Zhuo, S., Rice, D.W., Sloan, D.B., and Palmer, J.D. 2011b. The mitochondrial genome of the legume *Vigna radiata* and the analysis of recombination across short mitochondrial repeats. *PLoS One*, 6(1), e16404. doi:10.1371/journal.pone.0016404.

Andre, C., Levy, A., and Walbot, V. 1992. Small repeated sequences and the structure of plant mitochondrial genomes. *Trends Genet*, 8(4), 128–132.

Antico Arciuch, V.G., Elguero, M.E., Poderoso, J.J., and Carreras, M.C. 2012. Mitochondrial regulation of cell cycle and proliferation. *Antioxid Redox Signal*, 16(10), 1150–1180.

Arimura, S., Yamamoto, J., Aida, G.P., Nakazono, M., and Tsutsumi, N. 2004. Frequent fusion and fission of plant mitochondria with unequal nucleoid distribution. *Proc Natl Acad Sci USA*, 101(20), 7805–7808.

Arrieta-Montiel, M., Lyznik, A., Woloszynska, M., Janska, H., Tohme, J., and Mackenzie, S. 2001. Tracing evolutionary and developmental implications of mitochondrial stoichiometric shifting in the common bean. *Genetics*, 158(2), 851–864.

Arrieta-Montiel, M.P. and Mackenzie, S.A. 2011. Plant mitochondrial genomes and recombination. In: Kempken, F., ed. *Plant Mitochondria*. New York, NY, USA: Springer-Verlag, 65–82.

Arrieta-Montiel, M.P., Shedge, V., Davila, J., Christensen, A.C., and Mackenzie, S.A. 2009. Diversity of the *Arabidopsis* mitochondrial genome occurs via nuclear-controlled recombination activity. *Genetics*, 183(4), 1261–1268.

Asin-Cayuela, J. and Gustafsson, C.M. 2007. Mitochondrial transcription and its regulation in mammalian cells. *Trends Biochem Sci*, 32(3), 111–117.

Attardi, G. and Schatz, G. 1988. Biogenesis of mitochondria. *Annu Rev Cell Biol*, 4, 289–333.

Baba, K., Schmidt, J., Espinosa-Ruiz, A., Villarejo, A., Shiina, T., Gardestrom, P., Sane, A.P., and Bhalerao, R.P. 2004. Organellar gene transcription and early seedling development are affected in the rpoT;2 mutant of *Arabidopsis*. *Plant J*, 38(1), 38–48.

Backert, S. and Borner, T. 2000. Phage T4-like intermediates of DNA replication and recombination in the mitochondria of the higher plant *Chenopodium album* (L.). *Curr Genet*, 37(5), 304–314.

Backert, S., Lurz, R., Oyarzabal, O.A., and Borner, T. 1997. High content, size and distribution of single-stranded DNA in the mitochondria of *Chenopodium album* (L.). *Plant Mol Biol*, 33(6), 1037–1050.

Beardslee, T.A., Roy-Chowdhury, S., Jaiswal, P., Buhot, L., Lerbs-Mache, S., Stern, D.B., and Allison, L.A. 2002. A nucleus-encoded maize protein with sigma factor activity accumulates in mitochondria and chloroplasts. *Plant J*, 31(2), 199–209.

Begu, D., Graves, P.V., Domec, C., Arselin, G., Litvak, S., and Araya, A. 1990. RNA editing of wheat mitochondrial ATP synthase subunit 9: Direct protein and cDNA sequencing. *Plant Cell*, 2(12), 1283–1290.

Bellaoui, M., Martin-Canadell, A., Pelletier, G., and Budar, F. 1998. Low-copy-number molecules are produced by recombination, actively maintained and can be amplified in the mitochondrial genome of *Brassicaceae*: Relationship to reversion of the male sterile phenotype in some cybrids. *Mol Gen Genet*, 257(2), 177–185.

Bellaoui, M., Pelletier, G., and Budar, F. 1997. The steady-state level of mRNA from the Ogura cytoplasmic male sterility locus in Brassica cybrids is determined post-transcriptionally by its 3′ region. *Embo J*, 16(16), 5057–5068.

Bendich, A.J. 1993a. Reaching for the ring: The study of mitochondrial genome structure. *Curr Genet*, 24(4), 279–290.

Bendich, A.J. 1993b. Reaching for the ring: The study of mitochondrial genome structure. *Curr Genet*, 24(4), 279–290.

Bereiter-Hahn, J. and Voth, M. 1994. Dynamics of mitochondria in living cells: Shape changes, dislocations, fusion, and fission of mitochondria. *Microsc Res Tech*, 27(3), 198–219.

Binder, S. and Brennicke, A. 1993. Transcription initiation sites in mitochondria of *Oenothera berteriana*. *J Biol Chem*, 268(11), 7849–7855.

Binder, S. and Brennicke, A. 2003. Gene expression in plant mitochondria: Transcriptional and post-transcriptional control. *Philos Trans R Soc Lond B Biol Sci*, 358(1429), 181–188; discussion 188–189.

Binder, S., Hatzack, F., and Brennicke, A. 1995. A novel pea mitochondrial *in vitro* transcription system recognizes homologous and heterologous mRNA and tRNA promoters. *J Biol Chem*, 270(38), 22182–22189.

Binder, S., Knoop, V., and Brennicke, A. 1991. Nucleotide sequences of the mitochondrial genes trnS(TGA) encoding tRNA(TGASer) in *Oenothera berteriana* and *Arabidopsis thaliana*. *Gene*, 102(2), 245–247.

Binder, S., Marchfelder, A., and Brennicke, A. 1996. Regulation of gene expression in plant mitochondria. *Plant Mol Biol*, 32(1–2), 303–314.

Binder, S., Marchfelder, A., Brennicke, A., and Wissinger, B. 1992. RNA editing in *trans*-splicing intron sequences of nad2 mRNAs in *Oenothera* mitochondria. *J Biol Chem*, 267(11), 7615–7623.

Binder, S., Thalheim, C., and Brennicke, A. 1994. Transcription of potato mitochondrial 26S rRNA is initiated at its mature 5′ end. *Curr Genet*, 26(5–6), 519–523.

Bock, H., Brennicke, A., and Schuster, W. 1994. Rps3 and rpl16 genes do not overlap in *Oenothera* mitochondria: GTG as a potential translation initiation codon in plant mitochondria? *Plant Mol Biol*, 24(5), 811–818.

Bogenhagen, D.F. and Clayton, D.A. 2003. The mitochondrial DNA replication bubble has not burst. *Trends Biochem Sci*, 28(7), 357–360.

Boldogh, I.R., Yang, H.C., and Pon, L.A. 2001. Mitochondrial inheritance in budding yeast. *Traffic*, 2(6), 368–374.

Borst, P. and Grivell, L.A. 1971. Mitochondrial ribosomes. *FEBS Lett*, 13(2), 73–88.

Bowmaker, M., Yang, M.Y., Yasukawa, T., Reyes, A., Jacobs, H.T., Huberman, J.A., and Holt, I.J. 2003. Mammalian mitochondrial DNA replicates bidirectionally from an initiation zone. *J Biol Chem*, 278(51), 50961–50969.

Brennicke, A., Marchfelder, A., and Binder, S. 1999a. RNA editing. *FEMS Microbiol Rev*, 23(3), 297–316.

Brennicke, A., Zabaleta, E., Dombrowski, S., Hoffmann, M., and Binder, S. 1999b. Transcription signals of mitochondrial and nuclear genes for mitochondrial proteins in dicot plants. *J Hered*, 90(3), 345–350.

Brown, G.G., Auchincloss, A.H., Covello, P.S., Gray, M.W., Menassa, R., and Singh, M. 1991. Characterization of transcription initiation sites on the soybean mitochondrial genome allows identification of a transcription-associated sequence motif. *Mol Gen Genet*, 228(3), 345–355.

Bullerwell, C.E. and Gray, M.W. 2004. Evolution of the mitochondrial genome: Protist connections to animals, fungi and plants. *Curr Opin Microbiol*, 7(5), 528–534.

Bunn, C.L., Mitchell, C.H., Lukins, H.B., and Linnane, A.W. 1970. Biogenesis of mitochondria. 18. A new class of cytoplasmically determined antibiotic resistant mutants in *Saccharomyces cerevisiae*. *Proc Natl Acad Sci USA*, 67(3), 1233–1240.

Butow, R.A. and Avadhani, N.G. 2004. Mitochondrial signaling: The retrograde response. *Mol Cell*, 14(1), 1–15.

Caoile, A.G. and Stern, D.B. 1997. A conserved core element is functionally important for maize mitochondrial promoter activity *in vitro*. *Nucleic Acids Res*, 25(20), 4055–4060.

Carrie, C., Kuhn, K., Murcha, M.W., Duncan, O., Small, I.D., O'Toole, N., and Whelan, J. 2009. Approaches to defining dual-targeted proteins in *Arabidopsis*. *Plant J*, 57(6), 1128–1139.

Carrie, C. and Small, I. 2013. A reevaluation of dual-targeting of proteins to mitochondria and chloroplasts. *Biochim Biophys Acta*, 1833(2), 253–259.

Carrillo, C. and Bonen, L. 1997. RNA editing status of nad7 intron domains in wheat mitochondria. *Nucleic Acids Res*, 25(2), 403–409.

Chang, S., Yang, T., Du, T., Huang, Y., Chen, J., Yan, J., He, J., and Guan, R. 2011. Mitochondrial genome sequencing helps show the evolutionary mechanism of mitochondrial genome formation in Brassica. *BMC Genomics*, 12, 497.

Chapdelaine, Y. and Bonen, L. 1991. The wheat mitochondrial gene for subunit I of the NADH dehydrogenase complex: A *trans*-splicing model for this gene-in-pieces. *Cell*, 65(3), 465–472.

Cho, Y., Qiu, Y.L., Kuhlman, P., and Palmer, J.D. 1998. Explosive invasion of plant mitochondria by a group I intron. *Proc Natl Acad Sci USA*, 95(24), 14244–14249.

Christensen, A.C. 2013. Plant mitochondrial genome evolution can be explained by DNA repair mechanisms. *Genome Biol Evol*, 5(6), 1079–1086.

Christensen, A.C., Lyznik, A., Mohammed, S., Elowsky, C.G., Elo, A., Yule, R., and Mackenzie, S.A. 2005. Dual-domain, dual-targeting organellar protein presequences in *Arabidopsis* can use non-AUG start codons. *Plant Cell*, 17(10), 2805–2816.

Conley, C.A. and Hanson, M.R. 1994. Tissue-specific protein expression in plant mitochondria. *Plant Cell*, 6(1), 85–91.

Covello, P.S. and Gray, M.W. 1991. Sequence analysis of wheat mitochondrial transcripts capped *in vitro*: Definitive identification of transcription initiation sites. *Curr Genet*, 20(3), 245–251.

Cupp, J.D. and Nielsen, B.L. 2013. *Arabidopsis thaliana* organellar DNA polymerase IB mutants exhibit reduced mtDNA levels with a decrease in mitochondrial area density. *Physiol Plant*, 149(1), 91–103.

Cupp, J.D. and Nielsen, B.L. 2014. Minireview: DNA replication in plant mitochondria. *Mitochondrion*, 19 (Pt B), 231–237.

Dai, H., Lo, Y.S., Charn, C.G., Ruddat, M., and Chiang, K.S. 1993. Characterization of protein synthesis by isolated rice mitochondria. *Theor Appl Genet*, 86(2–3), 312–316.

Danial, N.N. and Korsmeyer, S.J. 2004. Cell death: Critical control points. *Cell*, 116(2), 205–219.

Darracq, A., Varre, J.-S., and Touzet, P. 2011. A scenario of mitochondrial genome evolution in maize based on rearrangement events. *BMC Genomics*, 11, 233–233.

Daschner, K., Couee, I., and Binder, S. 2001. The mitochondrial isovaleryl-coenzyme a dehydrogenase of *Arabidopsis* oxidizes intermediates of leucine and valine catabolism. *Plant Physiol*, 126(2), 601–612.

de Longevialle, A.F., Hendrickson, L., Taylor, N.L., Delannoy, E., Lurin, C., Badger, M., Millar, A.H., and Small, I. 2008. The pentatricopeptide repeat gene OTP51 with two LAGLIDADG motifs is required for the *cis*-splicing of plastid ycf3 intron 2 in *Arabidopsis thaliana*. *Plant J*, 56(1), 157–168.

De Paepe, R., Forchioni, A., Chetrit, P., and Vedel, F. 1993. Specific mitochondrial proteins in pollen: Presence of an additional ATP synthase beta subunit. *Proc Natl Acad Sci USA*, 90(13), 5934–5938.

Dietrich, A., Small, I., Cosset, A., Weil, J.H., and Marechal-Drouard, L. 1996. Editing and import: Strategies for providing plant mitochondria with a complete set of functional transfer RNAs. *Biochimie*, 78(6), 518–529.

Diffley, J.F. and Stillman, B. 1992. DNA binding properties of an HMG1-related protein from yeast mitochondria. *J Biol Chem*, 267(5), 3368–3374.

Diray-Arce, J., Liu, B., Cupp, J.D., Hunt, T., and Nielsen, B.L. 2013. The *Arabidopsis* At1g30680 gene encodes a homologue to the phage T7 gp4 protein that has both DNA primase and DNA helicase activities. *BMC Plant Biol*, 13, 36.

Dombrowski, S., Brennicke, A., and Binder, S. 1997. 3'-Inverted repeats in plant mitochondrial mRNAs are processing signals rather than transcription terminators. *Embo J*, 16(16), 5069–5076.

Dombrowski, S., Hoffmann, M., Guha, C., and Binder, S. 1999. Continuous primary sequence requirements in the 18-nucleotide promoter of dicot plant mitochondria. *J Biol Chem*, 274(15), 10094–10099.

Dong, F.G., Wilson, K.G., and Makaroff, C.A. 1998. The radish (*Raphanus sativus* L.) mitochondrial cox2 gene contains an ACG at the predicted translation initiation site. *Curr Genet*, 34(2), 79–87.

Duchene, A.M., Giritch, A., Hoffmann, B., Cognat, V., Lancelin, D., Peeters, N.M., Zaepfel, M., Marechal-Drouard, L., and Small, I.D. 2005. Dual targeting is the rule for organellar amino-acyl-tRNA synthetases in *Arabidopsis thaliana*. *Proc Natl Acad Sci USA*, 102(45), 16484–16489.

Duchene, A.M. and Marechal-Drouard, L. 2001. The chloroplast-derived trnW and trnM-e genes are not expressed in *Arabidopsis* mitochondria. *Biochem Biophys Res Commun*, 285(5), 1213–1216.

Dyall, S.D., Brown, M.T., and Johnson, P.J. 2004. Ancient invasions: From endosymbionts to organ-elles. *Science*, 304(5668), 253–257.

Edmondson, A.C., Song, D., Alvarez, L.A., Wall, M.K., Almond, D., McClellan, D.A., Maxwell, A., and Nielsen, B.L. 2005. Characterization of a mitochondrially targeted single-stranded DNA-binding protein in *Arabidopsis thaliana*. *Mol Genet Genomics*, 273(2), 115–122.

Edqvist, J. and Bergman, P. 2002. Nuclear identity specifies transcriptional initiation in plant mito-chondria. *Plant Mol Biol*, 49(1), 59–68.

Elo, A., Lyznik, A., Gonzalez, D.O., Kachman, S.D., and Mackenzie, S.A. 2003a. Nuclear genes that encode mitochondrial proteins for DNA and RNA metabolism are clustered in the *Arabidopsis* genome. *Plant Cell*, 15(7), 1619–1631.

Elo, A., Lyznik, A., Gonzalez, D.O., Kachman, S.D., and Mackenzie, S.A. 2003b. Nuclear genes that encode mitochondrial proteins for DNA and RNA metabolism are clustered in the *Arabidopsis* genome. *Plant Cell*, 15(7), 1619–1631.

Emanuel, C., Weihe, A., Graner, A., Hess, W.R., and Borner, T. 2004. Chloroplast development affects expression of phage-type RNA polymerases in barley leaves. *Plant J*, 38(3), 460–472.

Emmanuel, E., Yehuda, E., Melamed-Bessudo, C., Avivi-Ragolsky, N., and Levy, A.A. 2006. The role of AtMSH2 in homologous recombination in *Arabidopsis thaliana*. *EMBO Rep*, 7(1), 100–105.

Farre, J.C. and Araya, A. 2001. Gene expression in isolated plant mitochondria: High fidelity of tran-scription, splicing and editing of a transgene product in electroporated organelles. *Nucleic Acids Res*, 29(12), 2484–2491.

Feng, X., Kaur, A.P., Mackenzie, S.A., and Dweikat, I.M. 2009. Substoichiometric shifting in the fertil-ity reversion of cytoplasmic male sterile pearl millet. *Theor Appl Genet*, 118(7), 1361–1370.

Fey, J. and Marechal-Drouard, L. 1999. Compilation and analysis of plant mitochondrial promoter sequences: An illustration of a divergent evolution between monocot and dicot mitochondria. *Biochem Biophys Res Commun*, 256(2), 409–414.

Figueroa-Martinez, F., Funes, S., Franzen, L.G., and Gonzalez-Halphen, D. 2008. Reconstructing the mitochondrial protein import machinery of *Chlamydomonas reinhardtii*. *Genetics*, 179(1), 149–155.

Finnegan, P.M. and Brown, G.G. 1990. Transcriptional and post-transcriptional regulation of RNA levels in maize mitochondria. *Plant Cell*, 2(1), 71–83.

Fish, J., Raule, N., and Attardi, G. 2004. Discovery of a major D-loop replication origin reveals two modes of human mtDNA synthesis. *Science*, 306(5704), 2098–2101.

Fisher, R.P., Lisowsky, T., Parisi, M.A., and Clayton, D.A. 1992. DNA wrapping and bending by a mitochondrial high mobility group-like transcriptional activator protein. *J Biol Chem*, 267(5), 3358–3367.

Fujie, M., Kuroiwa, H., Kawano, S., and Kuroiwa, T. 1993. Studies on the behavior of organelles and their nucleoids in the root apical meristem of *Arabidopsis thaliana* (L.) Col. *Planta*, 189(3), 443–452.

Fujiwara, M., Nagashima, A., Kanamaru, K., Tanaka, K., and Takahashi, H. 2000. Three new nuclear genes, sigD, sigE and sigF, encoding putative plastid RNA polymerase sigma factors in *Aarabidopsis thaliana*. *FEBS Lett*, 481(1), 47–52.

Gagliardi, D. and Leaver, C.J. 1999. Polyadenylation accelerates the degradation of the mitochondrial mRNA associated with cytoplasmic male sterility in sunflower. *EMBO J*, 18(13), 3757–3766.

Geddy, R., Mahe, L., and Brown, G.G. 2005. Cell-specific regulation of a *Brassica napus* CMS-associated gene by a nuclear restorer with related effects on a floral homeotic gene promoter. *Plant J*, 41(3), 333–345.

Gellert, M., Mizuuchi, K., O'Dea, M.H., and Nash, H.A. 1976. DNA gyrase: An enzyme that introduces superhelical turns into DNA. *Proc Natl Acad Sci USA*, 73(11), 3872–3876.

Giege, P. and Brennicke, A. 1999. RNA editing in *Arabidopsis* mitochondria affects 441 C to U changes in ORFs. *Proc Natl Acad Sci USA*, 96(26), 15324–15329.

Giege, P. and Brennicke, A. 2001. From gene to protein in higher plant mitochondria. *C R Acad Sci III*, 324(3), 209–217.

Giege, P., Hoffmann, M., Binder, S., and Brennicke, A. 2000a. RNA degradation buffers asymmetries of transcription in *Arabidopsis* mitochondria. *EMBO Rep*, 1(2), 164–170.

Giege, P., Hoffmann, M., Binder, S., and Brennicke, A. 2000b. RNA degradation buffers asymmetries of transcription in *Arabidopsis* mitochondria. *EMBO Rep*, 1(2), 164–170.

Giege, P., Sweetlove, L.J., Cognat, V., and Leaver, C.J. 2005. Coordination of nuclear and mitochondrial genome expression during mitochondrial biogenesis in *Arabidopsis*. *Plant Cell*, 17(5), 1497–1512.

Giese, A., Thalheim, C., Brennicke, A., and Binder, S. 1996. Correlation of nonanucleotide motifs with transcript initiation of 18S rRNA genes in mitochondria of pea, potato and *Arabidopsis*. *Mol Gen Genet*, 252(4), 429–436.

Gilmore, R., Blobel, G., and Walter, P. 1982a. Protein translocation across the endoplasmic reticulum. I. Detection in the microsomal membrane of a receptor for the signal recognition particle. *J Cell Biol*, 95(2 Pt 1), 463–469.

Gilmore, R., Walter, P., and Blobel, G. 1982b. Protein translocation across the endoplasmic reticulum. II. Isolation and characterization of the signal recognition particle receptor. *J Cell Biol*, 95(2 Pt 1), 470–477.

Glick, B.S. and Von Heijne, G. 1996. *Saccharomyces cerevisiae* mitochondria lack a bacterial-type sec machinery. *Protein Sci*, 5(12), 2651–2652.

Goremykin, V.V., Salamini, F., Velasco, R., and Viola, R. 2009. Mitochondrial DNA of *Vitis vinifera* and the issue of rampant horizontal gene transfer. *Mol Biol Evol*, 26(1), 99–9110.

Grabau, E.A., Davis, W.H., Phelps, N.D., and Gengenbach, B.G. 1992. Classification of soybean cultivars based on mitochondrial DNA restriction fragment length polymorphisms. *Crop Sci*, 32, 271–274.

Gray, M.W., Burger, G., and Lang, B.F. 1999a. Mitochondrial evolution. *Science*, 283(5407), 1476–1481.

Gray, M.W., Burger, G., and Lang, B.F. 1999b. Mitochondrial evolution. *Science*, 283(5407), 1476–1481.

Gray, M.W. and Spencer, D.F. 1983. Wheat mitochondrial-DNA encodes a eubacteria-like initiator methionine transfer-RNA. *FEBS Lett*, 161(2), 323–327.

Grewe, F., Viehoever, P., Weisshaar, B., and Knoop, V. 2009. A *trans*-splicing group I intron and tRNA-hyperediting in the mitochondrial genome of the lycophyte *Isoetes engelmannii*. *Nucleic Acids Res*, 37(15), 5093–5104.

Grohmann, L., Thieck, O., Herz, U., Schroder, W., and Brennicke, A. 1994. Translation of nad9 mRNAs in mitochondria from *Solanum tuberosum* is restricted to completely edited transcripts. *Nucleic Acids Res*, 22(16), 3304–3311.

Grosskopf, D. and Mulligan, R.M. 1996. Developmental- and tissue-specificity of RNA editing in mitochondria of suspension-cultured maize cells and seedlings. *Curr Genet*, 29(6), 556–563.

Gualberto, J.M., Mileshina, D., Wallet, C., Niazi, A.K., Weber-Lotfi, F., and Dietrich, A. 2013. The plant mitochondrial genome: Dynamics and maintenance. *Biochimie*, 100, 107–120.

Gualberto, J.M., Mileshina, D., Wallet, C., Niazi, A.K., Weber-Lotfi, F., and Dietrich, A. 2014. The plant mitochondrial genome: Dynamics and maintenance. *Biochimie*, 100, 107–120.

Hager, M. and Bock, R. 2000. Enslaved bacteria as new hope for plant biotechnologists. *Appl Microbiol Biotechnol*, 54(3), 302–310.

Hahn, M. and Walbot, V. 1989. Effects of cold-treatment on protein synthesis and mRNA levels in rice leaves. *Plant Physiol*, 91(3), 930–938.

Hanson, M.R. and Bentolila, S. 2004. Interactions of mitochondrial and nuclear genes that affect male gametophyte development. *Plant Cell*, 16 Suppl, S154–S169.

Hecht, J., Grewe, F., and Knoop, V. 2011. Extreme RNA editing in coding islands and abundant microsatellites in repeat sequences of *Selaginella moellendorffii* mitochondria: The root of frequent plant mtDNA recombination in early tracheophytes. *Genome Biol Evol*, 3, 344–358.

Hedtke, B., Borner, T., and Weihe, A. 1997a. Mitochondrial and chloroplast phage-type RNA polymerases in *Arabidopsis*. *Science*, 277(5327), 809–811.

Hedtke, B., Borner, T., and Weihe, A. 1997b. Mitochondrial and chloroplast phage-type RNA polymerases in *Arabidopsis*. *Science*, 277(5327), 809–811.

Hedtke, B., Borner, T., and Weihe, A. 2000. One RNA polymerase serving two genomes. *EMBO Rep*, 1(5), 435–440.

Hedtke, B., Legen, J., Weihe, A., Herrmann, R.G., and Borner, T. 2002. Six active phage-type RNA polymerase genes in *Nicotiana tabacum*. *Plant J*, 30(6), 625–637.

Hedtke, B., Wagner, I., Borner, T., and Hess, W.R. 1999. Inter-organellar crosstalk in higher plants: Impaired chloroplast development affects mitochondrial gene and transcript levels. *Plant J*, 19(6), 635–643.

Hess, W.R. and Borner, T. 1999. Organellar RNA polymerases of higher plants. *Int Rev Cytol*, 190, 1–59.

Hoffmann, M. and Binder, S. 2002. Functional importance of nucleotide identities within the pea atp9 mitochondrial promoter sequence. *J Mol Biol*, 320(5), 943–950.

Holec, S., Lange, H., Canaday, J., and Gagliardi, D. 2008. Coping with cryptic and defective transcripts in plant mitochondria. *Biochim Biophys Acta*, 1779(9), 566–573.

Howad, W. and Kempken, F. 1997. Cell type-specific loss of atp6 RNA editing in cytoplasmic male sterile sorghum bicolor. *Proc Natl Acad Sci USA*, 94(20), 11090–11095.

Ikeda, T.M. and Gray, M.W. 1999. Characterization of a DNA-binding protein implicated in transcription in wheat mitochondria. *Molecular and Cellular Biology*, 19, 8113–8122.

Ibrahim, N.G. and Beattie, D.S. 1976. Formation of yeast mitochondrial-membrane 4. Regulation of mitochondrial protein-synthesis at polyribosomal level. *J Biol Chem*, 251(1), 108–115.

Jia, L., Dienhart, M., Schramp, M., McCauley, M., Hell, K., and Stuart, R.A. 2003. Yeast Oxa1 interacts with mitochondrial ribosomes: The importance of the C-terminal region of Oxa1. *EMBO J*, 22(24), 6438–6447.

Kabeya, Y. and Sato, N. 2005. Unique translation initiation at the second AUG codon determines mitochondrial localization of the phage-type RNA polymerases in the moss *Physcomitrella patens*. *Plant Physiol*, 138(1), 369–382.

Kanazawa, A. and Shimamoto, Y. 1999. Soybean recombination sites are present as dispersed segments in *Arabidopsis* and liverwort mitochondrial DNA. *Plant Mol Biol Report*, 17(1), 19–29.

Kanazawa, A., Tsutsumi, N., and Hirai, A. 1994. Reversible changes in the composition of the population of mtDNAs during dedifferentiation and regeneration in tobacco. *Genetics*, 138(3), 865–870.

Kelly, D.P. and Scarpulla, R.C. 2004. Transcriptional regulatory circuits controlling mitochondrial biogenesis and function. *Genes Dev*, 18(4), 357–368.

Khanam, S.M., Naydenov, N.G., Kadowaki, K., and Nakamura, C. 2007. Mitochondrial biogenesis as revealed by mitochondrial transcript profiles during germination and early seedling growth in wheat. *Genes Genet Syst*, 82(5), 409–420.

Khazi, F.R., Edmondson, A.C., and Nielsen, B.L. 2003. An *Arabidopsis* homologue of bacterial RecA that complements an *E. coli* recA deletion is targeted to plant mitochondria. *Mol Genet Genomics*, 269(4), 454–463.

Khvorostov, I.B., Ivanov, M.K., and Dymshits, G.M. 2002. The mitochondrial genome of higher plants: Gene expression regulation. *Mol Biol (Mosk)*, 36(3), 408–417.

Kmiec, B., Woloszynska, M., and Janska, H. 2006. Heteroplasmy as a common state of mitochondrial genetic information in plants and animals. *Curr Genet*, 50(3), 149–159.

Knoop, V. 2004. The mitochondrial DNA of land plants: Peculiarities in phylogenetic perspective. *Curr Genet*, 46(3), 123–139.

Knoop, V., Schuster, W., Wissinger, B., and Brennicke, A. 1991. *Trans* splicing integrates an exon of 22 nucleotides into the nad5 mRNA in higher plant mitochondria. *EMBO J*, 10(11), 3483–3493.

Kubo, T. and Newton, K.J. 2008. Angiosperm mitochondrial genomes and mutations. *Mitochondrion*, 8(1), 5–14.

Kubo, T., Nishizawa, S., Sugawara, A., Itchoda, N., Estiati, A., and Mikami, T. 2000. The complete nucleotide sequence of the mitochondrial genome of sugar beet (*Beta vulgaris* L.) reveals a novel gene for tRNA(Cys)(GCA). *Nucleic Acids Res*, 28(13), 2571–2576.

Kuhlman, P. and Palmer, J.D. 1995. Isolation, expression, and evolution of the gene encoding mitochondrial elongation factor Tu in *Arabidopsis thaliana*. *Plant Mol Biol*, 29(5), 1057–1070.

Kuhn, J., Tengler, U., and Binder, S. 2001. Transcript lifetime is balanced between stabilizing stem-loop structures and degradation-promoting polyadenylation in plant mitochondria. *Mol Cell Biol*, 21(3), 731–742.

Kuhn, K., Bohne, A.V., Liere, K., Weihe, A., and Borner, T. 2007. *Arabidopsis* phage-type RNA polymerases: Accurate *in vitro* transcription of organellar genes. *Plant Cell*, 19(3), 959–971.

Kuhn, K., Richter, U., Meyer, E.H., Delannoy, E., de Longevialle, A.F., O'Toole, N., Borner, T., Millar, A.H., Small, I.D., and Whelan, J. 2009. Phage-type RNA polymerase RPOTmp performs gene-specific transcription in mitochondria of *Arabidopsis thaliana*. *Plant Cell*, 21(9), 2762–2779.

Kuhn, K., Weihe, A., and Borner, T. 2005. Multiple promoters are a common feature of mitochondrial genes in *Arabidopsis*. *Nucleic Acids Res*, 33(1), 337–346.

Lambowitz, A.M. and Zirnmerly, S. 2004. Mobile group II introns. *Annu Rev Genet*, 38, 1–35.

Lang, B.F., Burger, G., O'Kelly, C.J., Cedergren, R., Golding, G.B., Lemieux, C., Sankoff, D., Turmel, M., and Gray, M.W. 1997. An ancestral mitochondrial DNA resembling a eubacterial genome in miniature. *Nature*, 387(6632), 493–497.

Leaver, C.J., Hack, E., and Forde, B.G. 1983. Protein-synthesis by isolated plant-mitochondria. *Methods Enzymol*, 97, 476–484.

Lehmann, K. and Schmidt, U. 2003. Group II introns: Structure and catalytic versatility of large natural ribozymes. *Crit Rev Biochem Mol Biol*, 38(3), 249–303.

Leino, M., Landgren, M., and Glimelius, K. 2005. Alloplasmic effects on mitochondrial transcriptional activity and RNA turnover result in accumulated transcripts of *Arabidopsis* orfs in cytoplasmic male-sterile *Brassica napus*. *Plant J*, 42(4), 469–480.

Lelandais, C., Gutierres, S., Mathieu, C., Vedel, F., Remacle, C., Marechal-Drouard, L., Brennicke, A., Binder, S., and Chetrit, P. 1996. A promoter element active in run-off transcription controls the expression of two cistrons of nad and rps genes in *Nicotiana sylvestris* mitochondria. *Nucleic Acids Res*, 24(23), 4798–4804.

Li, X.Q., Zhang, M.D., and Brown, G.G. 1996. Cell-specific expression of mitochondrial transcripts in maize seedlings. *Plant Cell*, 8(11), 1961–1975.

Liere, K., Weihe, A., and Borner, T. 2011. The transcription machineries of plant mitochondria and chloroplasts: Composition, function, and regulation. *J Plant Physiol*, 168(12), 1345–1360.

Lilly, J.W., Havey, M.J., Jackson, S.A., and Jiang, J. 2001. Cytogenomic analyses reveal the structural plasticity of the chloroplast genome in higher plants. *Plant Cell*, 13(2), 245–254.

Lind, C., Hallden, C., and Moller, I.M. 1991. Protein synthesis in mitochondria purified from roots, leaves and flowers of sugar beet. *Physiol Plant*, 83, 7–16.

Lippok, B., Brennicke, A., and Unseld, M. 1996. The rps4-gene is encoded upstream of the nad2-gene in *Arabidopsis* mitochondria. *Biol Chem Hoppe Seyler*, 377(4), 251–257.

Lippok, B., Brennicke, A., and Wissinger, B. 1994. Differential RNA editing in closely related introns in *Oenothera* mitochondria. *Mol Gen Genet*, 243(1), 39–46.

Liu, H., Cui, P., Zhan, K., Lin, Q., Zhuo, G., Guo, X., Ding, F. et al. 2011. Comparative analysis of mitochondrial genomes between a wheat K-type cytoplasmic male sterility (CMS) line and its maintainer line. *BMC Genomics*, 12, 163.

Liu, M. and Spremulli, L. 2000. Interaction of mammalian mitochondrial ribosomes with the inner membrane. *J Biol Chem*, 275(38), 29400–29406.

Lizama, L., Holuigue, L., and Jordana, X. 1994. Transcription initiation sites for the potato mitochondrial gene coding for subunit 9 of ATP synthase (atp9). *FEBS Lett*, 349(2), 243–248.

Logan, D.C. 2010. The dynamic plant chondriome. *Semin Cell Dev Biol*, 21(6), 550–557.

Lonsdale, D.M., Hodge, T.P., and Fauron, C.M. 1984a. The physical map and organisation of the mitochondrial genome from the fertile cytoplasm of maize. *Nucleic Acids Res*, 12(24), 9249–9261.

Lonsdale, D.M., Hodge, T.P., and Fauron, C.M. 1984b. The physical map and organisation of the mitochondrial genome from the fertile cytoplasm of maize. *Nucleic Acids Res*, 12(24), 9249–9261.

Lu, B. and Hanson, M.R. 1994. A single homogeneous form of ATP6 protein accumulates in petunia mitochondria despite the presence of differentially edited atp6 transcripts. *Plant Cell*, 6(12), 1955–1968.

Lu, B. and Hanson, M.R. 1996. Fully edited and partially edited nad9 transcripts differ in size and both are associated with polysomes in potato mitochondria. *Nucleic Acids Res*, 24(7), 1369–1374.

Lupold, D.S., Caoile, A.G., and Stern, D.B. 1999a. Genomic context influences the activity of maize mitochondrial cox2 promoters. *Proc Natl Acad Sci USA*, 96(20), 11670–11675.

Lupold, D.S., Caoile, A.G., and Stern, D.B. 1999b. The maize mitochondrial cox2 gene has five promoters in two genomic regions, including a complex promoter consisting of seven overlapping units. *J Biol Chem*, 274(6), 3897–3903.

Lupold, D.S., Caoile, A.G., and Stern, D.B. 1999c. Polyadenylation occurs at multiple sites in maize mitochondrial cox2 mRNA and is independent of editing status. *Plant Cell*, 11(8), 1565–1578.

Maier, R.M., Zeltz, P., Kossel, H., Bonnard, G., Gualberto, J.M., and Grienenberger, J.M. 1996. RNA editing in plant mitochondria and chloroplasts. *Plant Mol Biol*, 32(1–2), 343–365.

Makaroff, C.A., Apel, I.J., and Palmer, J.D. 1989. The atp6 coding region has been disrupted and a novel reading frame generated in the mitochondrial genome of cytoplasmic male-sterile radish. *J Biol Chem*, 264(20), 11706–11713.

Malek, O., Brennicke, A., and Knoop, V. 1997. Evolution of *trans*-splicing plant mitochondrial introns in pre-Permian times. *Proc Natl Acad Sci USA*, 94(2), 553–558.

Maloney, A.P., Traynor, P.L., Levings, C.S., and Walbot, V. 1989. Identification in maize mitochondrial 26s ribosomal-RNA of a short 5′-end sequence possibly involved in transcription initiation and processing. *Curr Genet*, 15(3), 207–212.

Manchekar, M., Scissum-Gunn, K., Song, D., Khazi, F., McLean, S.L., and Nielsen, B.L. 2006a. DNA recombination activity in soybean mitochondria. *J Mol Biol*, 356(2), 288–299.

Manchekar, M., Scissum-Gunn, K., Song, D., Khazi, F., McLean, S.L., and Nielsen, B.L. 2006b. DNA recombination activity in soybean mitochondria. *J Mol Biol*, 356(2), 288–299.

Marchfelder, A., Brennicke, A., and Binder, S. 1996. RNA editing is required for efficient excision of tRNA(Phe) from precursors in plant mitochondria. *J Biol Chem*, 271(4), 1898–1903.

Marechal-Drouard, L., Guillemaut, P., Cosset, A., Arbogast, M., Weber, F., Weil, J.H., and Dietrich, A. 1990. Transfer RNAs of potato (*Solanum tuberosum*) mitochondria have different genetic origins. *Nucleic Acids Res*, 18(13), 3689–3696.

Marechal-Drouard, L., Kumar, R., Remacle, C., and Small, I. 1996. RNA editing of larch mitochondrial tRNA(His) precursors is a prerequisite for processing. *Nucleic Acids Res*, 24(16), 3229–3234.

Martin, W., Rujan, T., Richly, E., Hansen, A., Cornelsen, S., Lins, T., Leister, D., Stoebe, B., Hasegawa, M., and Penny, D. 2002. Evolutionary analysis of *Arabidopsis*, cyanobacterial, and chloroplast genomes reveals plastid phylogeny and thousands of cyanobacterial genes in the nucleus. *Proc Natl Acad Sci USA*, 99(19), 12246–12251.

Matsunaga, M. and Jaehning, J.A. 2004. A mutation in the yeast mitochondrial core RNA polymerase, Rpo41, confers defects in both specificity factor interaction and promoter utilization. *J Biol Chem*, 279(3), 2012–2019.

McKinney, E.A. and Oliveira, M.T. 2013. Replicating animal mitochondrial DNA. *Genet Mol Biol*, 36(3), 308–315.

Menand, B., Marechal-Drouard, L., Sakamoto, W., Dietrich, A., and Wintz, H. 1998. A single gene of chloroplast origin codes for mitochondrial and chloroplastic methionyl-tRNA synthetase in *Arabidopsis thaliana*. *Proc Natl Acad Sci USA*, 95(18), 11014–11019.

Michel, F., Umesono, K., and Ozeki, H. 1989. Comparative and functional-anatomy of group-II catalytic introns—A review. *Gene*, 82(1), 5–30.

Mireau, H., Cosset, A., Marechal-Drouard, L., Fox, T.D., Small, I.D., and Dietrich, A. 2000. Expression of *Arabidopsis thaliana* mitochondrial alanyl-tRNA synthetase is not sufficient to trigger mitochondrial import of tRNAAla in yeast. *J Biol Chem*, 275(18), 13291–13296.

Mireau, H., Lancelin, D., and Small, I.D. 1996. The same *Arabidopsis* gene encodes both cytosolic and mitochondrial alanyl-tRNA synthetases. *Plant Cell*, 8(6), 1027–1039.

Moeykens, C.A., Mackenzie, S.A., and Shoemaker, R.C. 1995. Mitochondrial genome diversity in soybean: Repeats and rearrangements. *Plant Mol Biol*, 29(2), 245–254.

Moller, I.M. and Kristensen, B.K. 2006. Protein oxidation in plant mitochondria detected as oxidized tryptophan. *Free Radic Biol Med*, 40(3), 430–435.

Moriyama, T., Terasawa, K., and Sato, N. 2011. Conservation of POPs, the plant organellar DNA polymerases, in eukaryotes. *Protist*, 162(1), 177–187.

Moschopoulos, A., Derbyshire, P., and Byrne, M.E. 2012. The *Arabidopsis* organelle-localized glycyl-tRNA synthetase encoded by embryo defective development1 is required for organ patterning. *J Exp Bot*, 63(14), 5233–5243.

Mower, J.P., Case, A.L., Floro, E.R., and Willis, J.H. 2012. Evidence against equimolarity of large repeat arrangements and a predominant master circle structure of the mitochondrial genome from a monkeyflower (*Mimulus guttatus*) lineage with cryptic CMS. *Genome Biol Evol*, 4(5), 670–686.

Muise, R.C. and Hauswirth, W.W. 1992. Transcription in maize mitochondria: Effects of tissue and mitochondrial genotype. *Curr Genet*, 22(3), 235–242.

Muise, R.C. and Hauswirth, W.W. 1995. Selective DNA amplification regulates transcript levels in plant mitochondria. *Curr Genet*, 28(2), 113–121.

Mulligan, R.M., Lau, G.T., and Walbot, V. 1988a. Numerous transcription initiation sites exist for the maize mitochondrial genes for subunit 9 of the ATP synthase and subunit 3 of cytochrome oxidase. *Proc Natl Acad Sci USA*, 85(21), 7998–8002.

Mulligan, R.M., Leon, P., Calvin, N., and Walbot, V. 1988b. Transcription of the maize mitochondrial genome. *FASEB J*, 2(5), A1330–A1330.

Mulligan, R.M., Leon, P., and Walbot, V. 1991. Transcriptional and posttranscriptional regulation of maize mitochondrial gene expression. *Mol Cell Biol*, 11(1), 533–543.

Mulligan, R.M., Williams, M.A., and Shanahan, M.T. 1999. RNA editing site recognition in higher plant mitochondria. *J Hered*, 90(3), 338–344.

Nayak, D., Guo, Q., and Sousa, R. 2009. A promoter recognition mechanism common to yeast mitochondrial and phage T7 RNA polymerases. *J Biol Chem*, 284(20), 13641–13647.

Newton, K.J., Gabay-Laughnan, S., and Paepe, R.D. 2004. Mitochondrial mutation in plants. In: Day, D.A., Millar, H.A., and Whelan, J., eds. *Plant Mitochondria: From Genome to Function*. Advances in Photosynthesis and Respiration, Vol. 17. Kluwer Academic Publishers, London, pp. 121–142.

Newton, K.J. and Walbot, V. 1985. Maize mitochondria synthesize organ-specific polypeptides. *Proc Natl Acad Sci USA*, 82(20), 6879–6883.

Newton, K.J., Winberg, B., Yamato, K., Lupold, S., and Stern, D.B. 1995. Evidence for a novel mitochondrial promoter preceding the cox2 gene of perennial teosintes. *EMBO J*, 14(3), 585–593.

Notsu, Y., Masood, S., Nishikawa, T., Kubo, N., Akiduki, G., Nakazono, M., Hirai, A., and Kadowaki, K. 2002. The complete sequence of the rice (*Oryza sativa* L.) mitochondrial genome: Frequent DNA sequence acquisition and loss during the evolution of flowering plants. *Mol Genet Genomics*, 268(4), 434–445.

Nunnari, J., Marshall, W.F., Straight, A., Murray, A., Sedat, J.W., and Walter, P. 1997. Mitochondrial transmission during mating in *Saccharomyces cerevisiae* is determined by mitochondrial fusion and fission and the intramitochondrial segregation of mitochondrial DNA. *Mol Biol Cell*, 8(7), 1233–1242.

Oda, K., Kohchi, T., and Ohyama, K. 1992. Mitochondrial DNA of *Marchantia polymorpha* as a single circular form with no incorporation of foreign DNA. *Biosci Biotechnol Biochem*, 56, 132–135.

Oldenburg, D.J. and Bendich, A.J. 1996. Size and structure of replicating mitochondrial DNA in cultured tobacco cells. *Plant Cell*, 8(3), 447–461.

Oldenburg, D.J. and Bendich, A.J. 2001. Mitochondrial DNA from the liverwort *Marchantia polymorpha*: Circularly permuted linear molecules, head-to-tail concatemers, and a 5′ protein. *J Mol Biol*, 310(3), 549–562.

Oldenburg, D.J., Kumar, R.A., and Bendich, A.J. 2013. The amount and integrity of mtDNA in maize decline with development. *Planta*, 237(2), 603–617.

Ono, Y., Sakai, A., Takechi, K., Takio, S., Takusagawa, M., and Takano, H. 2007. NtPolI-like1 and NtPolI-like2, bacterial DNA polymerase I homologs isolated from BY-2 cultured tobacco cells, encode DNA polymerases engaged in DNA replication in both plastids and mitochondria. *Plant Cell Physiol*, 48(12), 1679–1692.

Ott, M., Prestele, M., Bauerschmitt, H., Funes, S., Bonnefoy, N., and Herrmann, J.M. 2006. Mba1, a membrane-associated ribosome receptor in mitochondria. *EMBO J*, 25(8), 1603–1610.

Palmer, J.D. 1988. Intraspecific variation and multicircularity in Brassica mitochondrial DNAs. *Genetics*, 118(2), 341–351.

Palmer, J.D., Adams, K.L., Cho, Y., Parkinson, C.L., Qiu, Y.L., and Song, K. 2000. Dynamic evolution of plant mitochondrial genomes: Mobile genes and introns and highly variable mutation rates. *Proc Natl Acad Sci USA*, 97(13), 6960–6966.

Palmer, J.D. and Herbon, L.A. 1987a. Unicircular structure of the *Brassica hirta* mitochondrial genome. *Curr Genet*, 11(6–7), 565–570.

Palmer, J.D. and Herbon, L.A. 1987b. Unicircular structure of the *Brassica hirta* mitochondrial genome. *Curr Genet*, 11(6–7), 565–570.

Palmer, J.D. and Shields, C.R. 1984. Tripartite structure of the *Brassica campestris* mitochondrial genome. *Nature*, 307, 437–440.

Parent, J.S., Lepage, E., and Brisson, N. 2011. Divergent roles for the two PolI-like organelle DNA polymerases of *Arabidopsis*. *Plant Physiol*, 156(1), 254–262.

Parisi, M.A., Xu, B., and Clayton, D.A. 1993. A human mitochondrial transcriptional activator can functionally replace a yeast mitochondrial HMG-box protein both *in vivo* and *in vitro*. *Mol Cell Biol*, 13(3), 1951–1961.

Park, A.K., Kim, H., and Jin, H.J. 2009. Comprehensive phylogenetic analysis of evolutionarily conserved rRNA adenine dimethyltransferase suggests diverse bacterial contributions to the nucleus-encoded plastid proteome. *Mol Phylogenet Evol*, 50(2), 282–289.

Peeters, N.M., Chapron, A., Giritch, A., Grandjean, O., Lancelin, D., Lhomme, T., Vivrel, A., and Small, I. 2000. Duplication and quadruplication of *Arabidopsis thaliana* cysteinyl- and asparaginyl-tRNA synthetase genes of organellar origin. *J Mol Evol*, 50(5), 413–423.

Perrotta, G., Grienenberger, J.M., and Gualberto, J.M. 2002. Plant mitochondrial rps2 genes code for proteins with a C-terminal extension that is processed. *Plant Mol Biol*, 50(3), 523–533.

Phreaner, C.G., Williams, M.A., and Mulligan, R.M. 1996. Incomplete editing of rps12 transcripts results in the synthesis of polymorphic polypeptides in plant mitochondria. *Plant Cell*, 8(1), 107–117.

Preuten, T., Cincu, E., Fuchs, J., Zoschke, R., Liere, K., and Borner, T. 2010. Fewer genes than organelles: Extremely low and variable gene copy numbers in mitochondria of somatic plant cells. *Plant J*, 64(6), 948–959.

Pring, D.R., Chen, W., Tang, H.V., Howad, W., and Kempken, F. 1998. Interaction of mitochondrial RNA editing and nucleolytic processing in the restoration of male fertility in sorghum. *Curr Genet*, 33(6), 429–436.

Pring, D.R., Mullen, J.A., and Kempken, F. 1992. Conserved sequence blocks 5′ to start codons of plant mitochondrial genes. *Plant Mol Biol*, 19(2), 313–317.

Rapp, W.D., Lupold, D.S., Mack, S., and Stern, D.B. 1993. Architecture of the maize mitochondrial atp1 promoter as determined by linker-scanning and point mutagenesis. *Mol Cell Biol*, 13(12), 7232–7238.

Reyes, A., Kazak, L., Wood, S.R., Yasukawa, T., Jacobs, H.T., and Holt, I.J. 2013. Mitochondrial DNA replication proceeds via a "bootlace" mechanism involving the incorporation of processed transcripts. *Nucleic Acids Res*, 41(11), 5837–5850.

Richter, U., Kiessling, J., Hedtke, B., Decker, E., Reski, R., Borner, T., and Weihe, A. 2002. Two RpoT genes of *Physcomitrella patens* encode phage-type RNA polymerases with dual targeting to mitochondria and plastids. *Gene*, 290(1–2), 95–105.

Richter, U., Kuhn, K., Okada, S., Brennicke, A., Weihe, A., and Borner, T. 2010. A mitochondrial rRNA dimethyladenosine methyltransferase in *Arabidopsis*. *Plant J*, 61(4), 558–569.

Sage, J.M. and Knight, K.L. 2013. Human Rad51 promotes mitochondrial DNA synthesis under conditions of increased replication stress. *Mitochondrion*, 13(4), 350–356.

Salinas, T., Schaeffer, C., Marechal-Drouard, L., and Duchene, A.M. 2005. Sequence dependence of tRNA(Gly) import into tobacco mitochondria. *Biochimie*, 87(9–10), 863–872.

Sanchez, H., Fester, T., Kloska, S., Schroder, W., and Schuster, W. 1996. Transfer of rps19 to the nucleus involves the gain of an RNP-binding motif which may functionally replace RPS13 in *Arabidopsis* mitochondria. *EMBO J*, 15(9), 2138–2149.

Saraste, M. 1999. Oxidative phosphorylation at the fin de siecle. *Science*, 283(5407), 1488–1493.

Scarpulla, R.C. 2008. Transcriptional paradigms in mammalian mitochondrial biogenesis and function. *Physiol Rev*, 88(2), 611–638.

Schmitz-Linneweber, C. and Small, I. 2008. Pentatricopeptide repeat proteins: A socket set for organelle gene expression. *Trends Plant Sci*, 13(12), 663–670.

Schmitz-Linneweber, C., Williams-Carrier, R., and Barkan, A. 2005. RNA immunoprecipitation and microarray analysis show a chloroplast pentatricopeptide repeat protein to be associated with the 5′ region of mRNAs whose translation it activates. *Plant Cell*, 17(10), 2791–2804.

Schubot, F.D., Chen, C.J., Rose, J.P., Dailey, T.A., Dailey, H.A., and Wang, B.C. 2001. Crystal structure of the transcription factor sc-mtTFB offers insights into mitochondrial transcription. *Protein Sci*, 10(10), 1980–1988.

Schuster, W. and Brennicke, A. 1989. Conserved sequence elements at putative processing sites in plant mitochondria. *Curr Genet*, 15(3), 187–192.

Segui-Simarro, J.M., Coronado, M.J., and Staehelin, L.A. 2008. The mitochondrial cycle of *Arabidopsis* shoot apical meristem and leaf primordium meristematic cells is defined by a perinuclear tentaculate/cage-like mitochondrion. *Plant Physiol*, 148(3), 1380–1393.

Sevarino, K.A. and Poyton, R.O. 1980. Mitochondrial-membrane biogenesis—Identification of a precursor to yeast cytochrome-C oxidase subunit-II, an integral polypeptide. *Proc Natl Acad Sci USA:Biol Sci*, 77(1), 142–146.

Shutt, T.E. and Gray, M.W. 2006. Twinkle, the mitochondrial replicative DNA helicase, is widespread in the eukaryotic radiation and may also be the mitochondrial DNA primase in most eukaryotes. *J Mol Evol*, 62(5), 588–599.

Singh, M. and Brown, G.G. 1991. Suppression of cytoplasmic male-sterility by nuclear genes alters expression of a novel mitochondrial gene region. *Plant Cell*, 3(12), 1349–1362.

Sloan, D.B., Alverson, A.J., Chuckalovcak, J.P., Wu, M., McCauley, D.E., Palmer, J.D., and Taylor, D.R. 2012a. Rapid evolution of enormous, multichromosomal genomes in flowering plant mitochondria with exceptionally high mutation rates. *PLoS Biol*, 10(1), e1001241.

Sloan, D.B., Muller, K., McCauley, D.E., Taylor, D.R., and Storchova, H. 2012b. Intraspecific variation in mitochondrial genome sequence, structure, and gene content in *Silene vulgaris*, an angiosperm with pervasive cytoplasmic male sterility. *New Phytol*, 196(4), 1228–1239.

Sparks, R.B. and Dale, R.M.K. 1980. Characterization of 3H-labeled supercoiled mitochondrial DNA from tobacco suspension culture cells. *Mol Genet Genomics*, 180, 351–355.

Sugiyama, Y., Watase, Y., Nagase, M., Makita, N., Yagura, S., Hirai, A., and Sugiura, M. 2005. The complete nucleotide sequence and multipartite organization of the tobacco mitochondrial genome: Comparative analysis of mitochondrial genomes in higher plants. *Mol Genet Genomics*, 272(6), 603–615.

Sutton, C.A., Conklin, P.L., Pruitt, K.D., and Hanson, M.R. 1991. Editing of pre-mRNAs can occur before *cis*- and *trans*-splicing in petunia mitochondria. *Mol Cell Biol*, 11(8), 4274–4277.

Swiatecka-Hagenbruch, M., Liere, K., and Borner, T. 2007. High diversity of plastidial promoters in *Arabidopsis thaliana*. *Mol Genet Genomics*, 277(6), 725–734.

Szyrach, G., Ott, M., Bonnefoy, N., Neupert, W., and Herrmann, J.M. 2003. Ribosome binding to the Oxa1 complex facilitates co-translational protein insertion in mitochondria. *EMBO J*, 22(24), 6448–6457.

Tada, S.F.S. and Souza, A.P. 2006. A recombination point is conserved in the mitochondrial genome of higher plant species and located downstream from the cox2 pseudogene in *Solanum tuberosum* L. *Genet Mol Biol*, 29, 1415–4757.

Takanashi, H., Ohnishi, T., Mogi, M., Okamoto, T., Arimura, S., and Tsutsumi, N. 2010. Studies of mitochondrial morphology and DNA amount in the rice egg cell. *Curr Genet*, 56(1), 33–41.

Tan, X.Y., Liu, X.L., Wang, W., Jia, D.J., Chen, L.Q., Zhang, X.Q., and Ye, D. 2010. Mutations in the *Arabidopsis* nuclear-encoded mitochondrial phage-type RNA polymerase gene RPOTm led to defects in pollen tube growth, female gametogenesis and embryogenesis. *Plant Cell Physiol*, 51(4), 635–649.

Topping, J.F. and Leaver, C.J. 1990. Mitochondrial gene expression during wheat leaf development. *Planta*, 182(3), 399–407.

Tracy, R.L. and Stern, D.B. 1995. Mitochondrial transcription initiation: Promoter structures and RNA polymerases. *Curr Genet*, 28(3), 205–216.

Unseld, M., Marienfeld, J.R., Brandt, P., and Brennicke, A. 1997a. The mitochondrial genome of *Arabidopsis thaliana* contains 57 genes in 366,924 nucleotides. *Nat Genet*, 15(1), 57–61.

Unseld, M., Marienfeld, J.R., Brandt, P., and Brennicke, A. 1997b. The mitochondrial genome of *Arabidopsis thaliana* contains 57 genes in 366,924 nucleotides. *Nat Genet*, 15(1), 57–61.

Usadel, B., Nagel, A., Thimm, O., Redestig, H., Blaesing, O.E., Palacios-Rojas, N., Selbig, J. et al. 2005. Extension of the visualization tool MapMan to allow statistical analysis of arrays, display of corresponding genes, and comparison with known responses. *Plant Physiol*, 138(3), 1195–1204.

Vaughn, J.C., Mason, M.T., Sper-Whitis, G.L., Kuhlman, P., and Palmer, J.D. 1995. Fungal origin by horizontal transfer of a plant mitochondrial group I intron in the chimeric CoxI gene of *Peperomia. J Mol Evol*, 41(5), 563–572.

Visacka, K., Gerhold, J.M., Petrovicova, J., Kinsky, S., Joers, P., Nosek, J., Sedman, J., and Tomaska, L. 2009. Novel subfamily of mitochondrial HMG box-containing proteins: Functional analysis of Gcf1p from *Candida albicans. Microbiology*, 155(Pt 4), 1226–1240.

Wall, M.K., Mitchenall, L.A., and Maxwell, A. 2004. *Arabidopsis thaliana* DNA gyrase is targeted to chloroplasts and mitochondria. *Proc Natl Acad Sci USA*, 101(20), 7821–7826.

Wang, W., Wu, Y., and Messing, J. 2012. The mitochondrial genome of an aquatic plant, *Spirodela polyrhiza. PLoS One*, 7(10), e46747. doi:10.1371/journal.pone.0046747.

Wanrooij, S., Fuste, J.M., Farge, G., Shi, Y., Gustafsson, C.M., and Falkenberg, M. 2008. Human mitochondrial RNA polymerase primes lagging-strand DNA synthesis *in vitro. Proc Natl Acad Sci USA*, 105(32), 11122–11127.

Warmke, H.E. and Lee, S.L. 1978. Pollen abortion in T cytoplasmic male-sterile corn (*Zea mays*): A suggested mechanism. *Science*, 200(4341), 561–563.

Watson, K. 1972. Organization of ribosomal granules within mitochondrial structures of aerobic and anaerobic cells of *Saccharomyces cerevisae. J Cell Biol*, 55(3), 721–726.

Westermann, B. 2010. Mitochondrial fusion and fission in cell life and death. *Nat Rev Mol Cell Biol*, 11(12), 872–884.

Wissinger, B., Schuster, W., and Brennicke, A. 1991. *Trans* splicing in *Oenothera* mitochondria: nad1 mRNAs are edited in exon and *trans*-splicing group II intron sequences. *Cell*, 65(3), 473–482.

Woloszynska, M. 2010a. Heteroplasmy and stoichiometric complexity of plant mitochondrial genomes—Though this be madness, yet there's method in't. *J Exp Bot*, 61(3), 657–671.

Woloszynska, M. 2010b. Heteroplasmy and stoichiometric complexity of plant mitochondrial genomes—Though this be madness, yet there's method in't. *J Exp Bot*, 61(3), 657–671.

Woloszynska, M., Gola, E.M., and Piechota, J. 2012. Changes in accumulation of heteroplasmic mitochondrial DNA and frequency of recombination via short repeats during plant lifetime in *Phaseolus vulgaris. Acta Biochim Pol*, 59(4), 703–709.

Woloszynska, M., Kieleczawa, J., Ornatowska, M., Wozniak, M., and Janska, H. 2001. The origin and maintenance of the small repeat in the bean mitochondrial genome. *Mol Genet Genomics*, 265(5), 865–872.

Woloszynska, M., Kmiec, B., Mackiewicz, P., and Janska, H. 2006. Copy number of bean mitochondrial genes estimated by real-time PCR does not correlate with the number of gene loci and transcript levels. *Plant Mol Biol*, 61(1–2), 1–12.

Woloszynska, M. and Trojanowski, D. 2009. Counting mtDNA molecules in *Phaseolus vulgaris*: Sublimons are constantly produced by recombination via short repeats and undergo rigorous selection during substoichiometric shifting. *Plant Mol Biol*, 70(5), 511–521.

Xu, B. and Clayton, D.A. 1992. Assignment of a yeast protein necessary for mitochondrial transcription initiation. *Nucleic Acids Res*, 20(5), 1053–1059.

Yan, B. and Pring, D.R. 1997. Transcriptional initiation sites in sorghum mitochondrial DNA indicate conserved and variable features. *Curr Genet*, 32(4), 287–295.

Yang, A.J. and Mulligan, R.M. 1991. RNA editing intermediates of cox2 transcripts in maize mitochondria. *Mol Cell Biol*, 11(8), 4278–4281.

Yin, C., Richter, U., Borner, T., and Weihe, A. 2009. Evolution of phage-type RNA polymerases in higher plants: Characterization of the single phage-type RNA polymerase gene from *Selaginella moellendorffii. J Mol Evol*, 68(5), 528–538.

Yin, C., Richter, U., Borner, T., and Weihe, A. 2010. Evolution of plant phage-type RNA polymerases: The genome of the basal angiosperm *Nuphar advena* encodes two mitochondrial and one plastid phage-type RNA polymerases. *BMC Evol Biol*, 10, 379.

Zaegel, V., Guermann, B., Le Ret, M., Andres, C., Meyer, D., Erhardt, M., Canaday, J., Gualberto, J.M., and Imbault, P. 2006. The plant-specific ssDNA binding protein OSB1 is involved in the stoichiometric transmission of mitochondrial DNA in *Arabidopsis. Plant Cell*, 18(12), 3548–3563.

chapter six

Plant functional genomics
Approaches and applications

Mehboob-ur-Rahman, Zainab Rahmat, Maryyam Gul, and Yusuf Zafar

Contents

Abstract

Deployment of next-generation sequencing (NGS) tools has made it possible to sequence the whole genome within a limited time period. At the moment, more than 100 plant genomes have been sequenced, of these 63% are crop species. The next challenge is to define the function to these sequences (genes) and their regulation and interaction with the other genes. Though the efforts toward the identification of genes function were started in early 1990s, however, after sequencing of *Arabidopsis thaliana* (a model plant species), a mega genome project toward the exploration of genes function was initiated. Both the reverse (gene to phenotype) and forward genetics (phenotype to gene) approaches were used. For example, microarray, virus-induced gene silencing, gene knockout, RNA interference, insertional mutagenesis, Targeting Induced Local Lesions IN Genomes, EcoTILLING, next-generation sequencing technologies, etc., have been extensively used for discovering genes and their functions. All these technologies with their own inherent merits and demerits were supplemented by the emerging bioinformatics tools in translating the information generating on model species to a less-studied species—effective tool in predicting function of unknown sequences within the shortest possible time through homology search. All these efforts set a stage for unraveling the functions of unknown genes involved in conferring various mechanisms including growth, grain development, response to various stresses, etc. Thus, the functional genomic studies would lead to sustain crop productivity on this planet by providing a comprehensive knowledge about the function of genes—a huge genetic resource that can be utilized for improving crop varieties using nonconventional (genetic transformation) and conventional means assisted with DNA markers.

Keywords: Functional Genomics, Expression Profiling, Microarray, Gene Silencing, RNAi, TILLING, DNA Markers.

Introduction

In the twentieth century, an unprecedented progress toward the understanding of genetic code of life was made, which culminated in the demonstration of deoxyribonucleic acid (DNA) model by Watson and Crick in 1953. Later, sequence of the DNA was read using various techniques (Sanger's genome-sequencing technology and next-generation sequencing tools).

In parallel to these discoveries, substantial success has been made in the improvement of genetics of important crop species using various classical breeding and nonconventional genomic tools. Also, various principles for the expression of genes were demonstrated. For example, the "one gene, one enzyme" theory proposed by Beadle and Tatum (1941). Four years later, this hypothesis was modified to "one gene, one polypeptide."

Functional genomic largely deals with the function of various genes and their interactions. For studying the function of genes, sequence or hybridization-based approaches were used extensively. The sequence-based technologies now permit the analysis of whole-genome sequence at once in a single experiment. In multiple investigations, genomic data (nucleotide sequences) is being converted into gene-function data for adding values to the nucleotide sequences. Genes sequence and their location on a chromosome are the preliminary steps for understanding how all parts of a biological system work together.

Functional genomics involves the identification of gene and also various alleles of the same gene conferring different phenotypes. Markers can be developed around such alleles, which may become the basis for initiating marker-assisted selection for improving the genetic makeup of crop varieties—a way of linking functional genomics with the plant breeding.

Currently, genome sequence information of important crop species is available. However, assigning functions to various genes is a big challenge. For example, more than 10 years have been devoted to assign function to the genes of *Arabidopsis thaliana* L. Most crop species share many genes in common, and thus, the information generated on model species can be translated on less-studied species using bioinformatics tools. In future, sequencing different accessions of a species would help in unrevealing the functions of important genes. It would help in sustaining and conserving the biological diversity present in a crop species.

Functional genomics tools and techniques

Understanding the function of every gene product and its interactions with the other genes is important in exploring its role in various biological pathways. Gene function can be explored at different levels such as ribonucleic acid (RNA), protein, and metabolite levels—for exploring possible function of a particular gene and its product, its interaction with the other gene and gene products.

Principally, two basic approaches "gain of function" and "loss of function" have been utilized for defining the gene function. A number of tools or approaches such as Targeting Induced Local Lesions IN Genomes (TILLING) and Eco-TILLING, insertional mutagenesis (T-DNA and transposon mutagenesis), RNA interference (RNAi) and virus-induced gene silencing (VIGS), zinc-finger nuclease (ZFN), Deleteagene, transcription activator-like effector nucleases (TALENs), and clustered regularly interspaced short palindromic repeat (CRISPR) have been used for mutating the genes followed by studying the fluctuations in their expression.

Studying the expression of a genotype

A simple way for assigning function to a gene is to explore the expressed part of the gene (called as expressed sequence tags, ESTs). Gene expresses by making mRNAs (through transcription process), which can be converted into cDNA (relatively a stable molecule). These cDNA molecules can be directly sequenced (using next-generation sequencing tools) and/or by cloning these cDNA molecules followed by sequencing (Bouchez and Hofte 1998). Irrespective of multiple concerns such as poor quality of ESTs and under and overrepresentation of few classes of ESTs (Rudd 2003), these molecules are still useful in assigning function to unknown genes by making comparisons with the characterized genes of the model species. In this case, use of powerful bioinformatics tools may help in assigning the functions to unknown genes. This method is efficient and cost-effective (Bouchez and Hofte 1998).

Currently, millions of ESTs derived from various crop specifies (corn, rice wheat, soybean, etc.) including few model species have been deposited in various databases. Also, the cDNA libraries constructed from different tissues, developmental stages, and/ or exposing the plants with multiple stresses, are the major source of generating EST sequences (Yamamoto and Sasaki 1997). It helps in identifying the tissue or growth stage-specific and stress responsive transcripts. ESTs are shorter in length than that of the corresponding cDNAs. These ESTs sequences can be assembled on a reference genome for drawing some evolutionary consequences. One must expect some misassemblies of sequences because of the occurrence of many paralogous genes especially in polyploids, for example, wheat (Rudd 2003). ESTs were found to be helpful in situations where the whole-genome sequence information was not available. In the present scenario, availability of tools for doing genome sequencing much efficiently, sequencing of ESTs has also been a valid strategy for discovering new genes by comparing the EST sequences derived from genotype(s) grown under normal and stress conditions (Ergen et al. 2009).

Serial analysis of gene expression (SAGE), is another approach that was demonstrated for quantifying the thousands of transcripts instantaneously. Short sequence tags from transcripts are sequenced at once—thus giving the absolute measure of gene expression (Sanchez et al. 2007). These short tags can be used to identify genes and it depends upon the availability of comprehensive EST databases for that particular species (Breyne and Zabeau 2001). Using SAGE approach, new genes and their function were described on rice seedlings (Matsumura et al. 2003). SAGE has also been deployed for investigation of stress responsive genes (Matsumura et al. 2003; Lee and Lee 2003). Later on, massively parallel signature sequencing (MPSS) (another approach) was described—ligation of long-sequence tags to microbeads followed by sequencing in parallel; thus, millions of transcripts can be sequenced simultaneously (Brenner et al. 2000). These advantages empower the researchers for not only identifying the genes with greater specificity and sensitivity but also remained instrumental in capturing the rare transcripts. This approach is extremely helpful for studying the less-studied plant species (not sequenced yet) (Reinartz et al. 2002). In plants, MPSS has also been deployed to study the expression of small RNA (sRNA) molecules (Meyers et al. 2006; Nobuta et al. 2007). MPPS expression data for many important crop species such as soybean, rice, and maize are publically available now in MPPS databases (Nakano et al. 2006). Functional analysis of gene expression can also be carried out by comparing the newly generated MPPS data with the available MPPS data, which can easily be extracted, complied, and compared (Jain et al. 2007).

Use of hybridization-based approaches

In contrast to sequencing of the expressed part of the genes, hybridization-based techniques are useful in determining the expression of the genes. For example, in array-based assays, the target DNA is exposed to the cDNA or oligonucleotide probes attached to a surface to quantify the expression (Schena et al. 1995). These methods require prior knowledge of the transcripts for designing probes (Rensink and Buell 2005). Microarray-based expression studies were reported on rice and *A. thaliana* (Wang et al. 2011; Kumar et al. 2012; Singh et al. 2012). Also, the microarray assays were conducted for studying the expression of stress-responsive genes in barley (Close et al. 2004), wheat (Ergen et al. 2009), corn (Luo et al. 2010; Ranjan et al. 2012), cassava (Utsumi et al. 2012), and tomato (Loukehaich et al. 2012). There are multiple concerns regarding the use of microarray-based studies including cross-hybridization and background noise, which limit their utility in studying the gene expression. Also, the isolation of RNA from whole tissue obscures transcriptional changes occurring in different cell types especially when investigating the stress responsive genes. Thus, such kinds of genes remain unnoticed.

Selection of the tissue and genotypes are important for undertaking such experiments. For example, reproductive tissues should be selected for studying the genes involved in conferring high yields. Similarly, selection of the stress-tolerance genotypes would help in studying the function of genes involved in conferring resistance to the particular stress

(Deyholos 2010). In addition, the response of a genotype to particular stress imposed under controlled conditions (laboratory or in containment) would be different than that of the stress imposed under natural field conditions—there other environmental factors interact together. A microarray study conducted by exposing genotypes with different water regimes revealed that only few genes were commonly regulated (Bray 2004). It has also been shown that the abiotic stresses are complex in nature, thus slight differences in the experimental conditions may change response of the genotype. Also, many transcripts undergo posttranscriptional and posttranslational modifications, thereby further complicating the situation in correlating transcriptomic and proteomic data.

Availability of whole-genome sequence information coupled with the application of bioinformatics tools opened many avenues including the prediction of genes, exons, etc., but still many regions of the genome remain unpredicted due to several limitations pertaining to the use of prediction models. For overcoming such concerns, extension of the microarray assay called as whole-genome tiling arrays has been described. By deploying this approach novel transcriptional units on chromosomes and alternative splice sites can be identified. Consequently, these units (transcripts) and methylation sites can be mapped (Yazaki et al. 2007; Mochida and Shinozaki 2010). Tiling arrays have been applied on *A. thaliana* as well as on other dicots and few grasses for studying the expression in response to the abiotic stresses (Hirayama and Shinozaki 2010; Lopes et al. 2011). In contrast to the response exhibited by the *Arabidopsis* (Skirycz et al. 2010) and maize (Tardieu et al. 2000) under drought stress, the *Brachypodium* leaf meristem remains unaffected and drought-induced growth reduction is caused by the reduced cell expansion. Also the genes involved in responding to various levels of drought stresses at different developmental leaf zones responded differently (Verelst et al. 2013).

Next-generation sequencing

Since the invention of the next-generation sequencing tools, progress for sequencing the whole genome has been accelerated. Sequencing by NGS can be undertaken within much less time period and at very a low cost than that of the conventional technologies. Next-generation sequencers can also yield information about the methylated part of the genome (methylome)—thus enabling us to determine the epigenetic control of various genes and their regulatory regions (Mittler and Shulaev 2013).

Whole-genome shotgun strategy that produces relatively long read lengths, can help in improving the assembly than that of the sequencing methodologies based on bacterial artificial chromosome (BAC) by BAC (Sanger sequencing). Different factors influence the quality of the whole-genome sequences such as coverage of a genome, combinations of sequencing methods employed, complexity, and heterozygosity in addition to the annotation inputs and methodology (Gapper et al. 2014). Also, the availability of high-density genetic maps expedites the process of assigning contigs to the corresponding chromosomes and bridging gaps in the sequenced genome. The genetic map can also be used to pinpoint regions (genes) involved in shaping traits of interest. Thus, identification and/or function of these genes can be defined (Paux et al. 2012). NGS tools are also used for genotyping purposes, which always help in the identification of the genomic regions for which the genotype understudy is different. Using this information, markers can be designed for initiating marker-assisted breeding (Cabezas et al. 2011).

Rice genome sequence information has been extensively used for exploring major quantitative trait loci (QTLs) involved in conferring grain production (Ashikari et al. 2005). The expression of cytokinin oxidase 2 (osCKX2) gene is regulated by a zinc finger transcrition factor drough and salt tolerance (DST) (Li et al. 2013). This gene encodes a transcription factor that has also been reported to regulate drought and salt tolerance in rice (Huang et al. 2009).

Likewise, sequence of maize genome has also revolutionized the development of powerful haplotype maps and predicting the metabolic pathways. Similarly, QTLs were also found for biomass and bioenergy (Gore et al. 2009; Schnable et al. 2009; Riedelsheimer et al. 2012) demonstrating the impact of genome sequence on data integration.

In different tomato species, identification of esterase responsible for the fluctuations in volatile esterase content was made possible by the availability of tomato genome. A gene underlying the uniformly ripening locus in tomato was identified that turned out to be a *Golden 2*-like transcription factor, which determines the chlorophyll distribution in unripe fruits (The Tomato Genome Consortium 2012; Goulet et al. 2012, Powell et al. 2012). Furthermore, the tomato genome along with draft genome of its wild relative *Solanum pennellii* has also illuminated the terpene biosynthesis evolution.

For adaptation to different photoperiod regimes and geographic latitude, control of flowering and maturation time is of recurring interest. The sequencing information of soybean genome was used to identify the maturity locus E_1 that has strong impact on flowering time. Similarly, potato genome sequence information was investigated to find transcription factors that are involved in regulating life cycle and plant maturity, whereas, sugar beet genome has helped in determining the biology of its flowering time control (Xia et al. 2012; Kloosterman et al. 2013; Matsuba et al. 2013). Finally, in rice genome using genome-wide association study, QTL was found for a trait controlling the flowering time (Huang et al. 2012; Bolger et al. 2014).

Due to accessibility of genotyping and sequencing by high-throughput equipments, it is expected that advancement in breeding will flourish in three ways. First, it will be easy to select the parents for new crosses and utilize exotic germplasm for introgression of new alleles by high-resolution fingerprinting, which can be achieved by genotyping of single-nucleotide polymorphism (SNP) or short coverage sequencing; second mapping using bi- or multiparental population and genome-wide association strategy will assist in precise mapping of loci showing linkage with particular concerned trait that will enhance identification of marker associated with the trait of interest; third genotyping or sequencing using high-throughput approach will facilitate genome-assisted breeding in orphan crops also. All these advances will entail improved precision capability of phenotyping in target background (Varshney et al. 2012).

Role of forward and reverse genetics approaches in defining gene function

Forward genetics approach (phenotype to genotype or trait to gene) has been extensively used to study various biochemical processes including signaling process in *A. thaliana* (Meyerowitz and Somerville 1994).

Forward genetics proved to be a powerful approach for elucidating the genetic components of most signaling or biochemical process in *A. thaliana*. In this context, mutants expressing the nuclear chlorophyll a/b binding protein (CAB) gene in the absence of functional chloroplasts were found to be useful in understanding the nucleus–chloroplast communication (Susek et al. 1993). In another study, mutations affecting the specific and marked morphological changes induced in dark-grown seedlings by the ethylene, were studied to explore the ethylene-related loci involved in either biosynthesis, perception, signaling, or response to this gaseous plant hormone (Meyerowitz and Somerville 1994).

A number of genetic resources, for example, Lehle Seeds, Arabidopsis Biological Resource Center Stocks, and the Nottingham Arabidopsis Stock Center are available, which impart different types of "ready-to-screen" mutagenized seed including mutagenized lines developed by exposing to ethylmethanesulfonate (EMS), fast neutrons, T-DNA, and transposons. All these lines are produced using a limited number of genetic backgrounds (such as Columbia, Wassilewskija, and Landsberg erecta), which render them unsuitable for highly specialized screens in which the use of a particular molecular reporter or mutant background is needed. Therefore, research groups often have to develop their own mutagenized genetic resources. Major advantage of using the forward genetic approaches is to discover genes without the preconceived idea about the nature of the gene under study (Alonso and Ecker 2006).

A huge quantum of genome sequencing information of multiple plant species including model plants has been made available after the completion of many whole-genome sequencing projects, which set a platform in defining gene function using reverse genetic approaches. This process was initiated some 15 years ago, and a number of model organisms including *Saccharomyces cerevisiae*, *Drosophila melanogaster*, *Caenorhabditis elegans*, *Mus musculus*, and *A. thaliana* were exposed to this approach. In plants, two features such as the availability of a relatively small- and high-quality genome sequence makes the *A. thaliana* an attractive model for embarking functional genomic studies. A number of gene families showing functional redundancy (difficult to study with the forward genetic approaches) were described using the reverse genetic approaches.

In *A. thaliana*, knocking out of entire gene families uncovered the overlapping and specific functions of their members (Okushima 2005; Prigge 2005). In some cases, size of the gene family is too large that makes it difficult to knockout all members of this family. Then a combination of sequence comparisons and the expression profiles are the best choices to study function of members of such gene families. For example, out of >100 related genes, a small subfamily comprising four transcription factors, involved in the early response to ethylene, were identified (Alonso et al. 2003). Alternatively, gene-silencing approach can be exploited for studying such large gene families through altering expression of several members of such gene families but its utility for handling large gene families are yet to be demonstrated.

In reverse genetic procedure, the genes can be knocked out by deploying RNAi and TILLING approaches. Each technique has its own merits and demerits. TILLING is easy to exercise than the RNAi-based approaches—involving the use of genetic transformation procedures. The success of these procedures also depends upon the availability of highly efficient screening methods for detecting mutations in the genomic DNA. In TILLING experiments, the number of mutants tend to be very high, which can be screened by adopting pooling strategy (pooling of mutants genomic DNA before starting mutant screening).

Gene silencing

In recent years, sRNA molecules (20–24 bp long) have been identified, which influence the gene expression. These molecules regulate the expression of genes at transcriptional and posttranscriptional levels (Khraiwesh et al. 2012; Nakaminami et al. 2012). Many biological and metabolic pathways involved in plant developments are regulated by the micro-RNAs (miRNAs) (Sunkar et al. 2012). A number of stress responsive genes are also regulated by miRNAs (Sunkar et al. 2012; Zhou et al. 2010).

A number of studies were designed for the overexpression and knockdown of miRNAs for studying their role in regulating the target genes using transgenic approaches

(Zhou et al. 2010; Ni et al. 2013). For example, in tomato plant, high tolerance to drought stress is conferred by the overexpression of miR169c. In another study, overexpression of miR319 confers tolerance to cold, salt, and drought in rice (Zhang et al. 2011; Yang et al. 2013; Zhou et al. 2013). Further studies are needed to explore the miRNA gene regulation complex, which would help in avoiding and reducing the off-targets and pleiotropic effect of certain genes (Sunkar et al. 2012; Guan et al. 2013; Yang et al. 2013; Cabello et al. 2014).

Posttranscriptional gene silencing (PTGS)

It regulates the gene by synthesizing sRNA. Posttranscriptional gene silencing (PTGS) occurs due to either RNAi or VIGS. Both are regulated by various types of sRNA including miRNAs 19–25 nt and short interfering (siRNAs 20–40 nt). The siRNAs were first noticed upon studying the transgene silencing and VIGS in plants (Jackson and Linsley 2004). Various types of siRNAs including the endogenous siRNAs (endo-siRNAs) and foreign DNA-induced siRNAs have been reported in *A. thaliana*. These siRNAs include the natural siRNAs (nat-siRNAs), endo-siRNAs, and trans-acting siRNAs (ta-siRNAs). Other types are produced in response to foreign nucleotides (viral RNA or transgenes) in plants. Plant RNA viruses trigger the mechanism of RNA silencing in host through the virus-induced siRNAs, which formed the basis of VIGS.

PTGS has been widely explored for defining the function of various genes conferring resistance to biotic and abiotic stresses. A number of methods have been developed for PTGS based on the study of miRNAs, hairpin RNAs (hpRNAs), and ta-siRNA. The hpRNAs are used for downregulating the genes in plants using hairpin construct, which directly target the gene of interest to be silenced. Vectors such as pHANNIBAL, intron spliced hpRNAi, pRNAi-LIC, and golden gate vectors (pRNAi-GG) have been used in gene silencing (Miki and Shimamoto 2004; Himmelbach et al. 2007).

Virus-induced gene silencing

VIGS is one of the plant defense mechanisms against the invading viruses using the RNAi pathways (Baulcombe 1999; Robertson 2004). The two most commonly used VIGS vectors in plants are tobacco mosaic virus (TMV) and tobacco rattle virus (TRV). TRV-VIGS works well in *Nicotiana benthamiana*. VIGS method is suitable for undertaking reverse genetics studies for defining the function of genes. This method can also perform the large-scale forward genetics screens in plants (Senthil-Kumar and Mysore 2011, 2014). VIGS has many advantages over the other gene-silencing methods because of its sequence homology-dependent process that can silence a large number of genes (e.g., cDNA libraries).

MicroRNA-induced gene silencing

In this method, ta-siRNA is produced by the activation of miR173 gene that is nonconserved. It was reported in *A. thaliana* and its close relatives that without its expression target gene will not be silenced. MicroRNA-induced gene silencing (MIGS) vectors used gateway cloning technology through *Agrobacterium*-mediated plant transformation or biolistic approach. Various MIGS vector are used to silence endogenous gene such as *MIGS2-4*, *MIGS2-1*, and *MIGS1-5*. Various studies showed a successful result in *N. benthamiana* by silencing green fluorescent protein (GFP) gene (de Felippes et al. 2012), and phosphoglycerate dehydrogenase1 (*PGDH1*) gene in *A. thaliana*, showed a strong growth inhibition (Benstein et al. 2013). In *Medicago truncatula*, C-terminally encoding protein1 (*CEP1*) gene was successfully silenced (Imin et al. 2013).

MIR-VIGS is a recently developed plant virus-based miRNA-mediated silencing method (Tang et al. 2010, 2013). Artificial mi-RNAs are used with viral vectors. It is more efficient than the VIGS. It has been exploited efficiently in *N. benthamiana* to silence genes such as phytoene desaturase (PDS), sulfur (SU), and salicylic acid glucosyltransferase (SGT). Most commonly used vectors are pCPCbLCVA.007 and pCPCbLCVB.002 (Tang et al. 2013) derived from the genome sequence of cabbage leaf curl virus component A and B. Simple agro-inoculation approach can be deployed rather than the transformation assay.

Recently, a syringe infiltration approach was demonstrated in *A. thaliana* for delivering the dsRNA to study function of genes of whom downregulation lead to plant lethality. Function of two genes chalcone synthase (CHS) and yellow fluorescent protein (YFP) were defined by silencing these in *A. thaliana* (Numata et al. 2014).

Transcriptional gene silencing

The transcriptional gene silencing (TGS) occurs largely by the methylation of DNA and chromatin remodeling (Waterhouse et al. 2001). TGS in plants also takes place by the RdDM (RNA-directed DNA methylation) pathway, which is induced by viral vectors such as cucumber mosaic virus (CMV), potato virus X (PVX), and TRV. Conventionally, the sRNAs are coined with the posttranscriptional gene silencing but these sRNAs can also trigger the TGS by the expression of inverted repeats present in the promoter sequences and deployment of viral vectors. These viral vectors (e.g., PVX and TRV) induced the RNA-mediated TGS against genes including GFP and β-glucuronidase genes (Jones et al. 1999; 2001).

Several examples of TGS have been found in plants. The promoter of granule-bound starch synthase I gene (*GBSSI*) in potato was induced by the inverted repeats constructs harboring various regions of the respective gene promoter (Heilersig et al. 2006). Similar type of TGS was found for the male sterile 45 (*Ms45*) gene (Cigan et al. 2005).

Among the aforementioned methods exploited for TGS and PTGS, RNAi-based gene silencing and VIGS are the most popular among the research community. VIGS has been extensively used in few plant species for achieving rapid and transient gene silencing as this procedure does not require the stable plant transformation while the RNAi-mediated gene silencing is used for plant species, which can be transformed easily (Pandey et al. 2015).

Targeted genome editing

Major technological advances in genomic science have made possible to edit particular regions of a genome. Targeted genome editing means inducing targeted deletions, insertions, and changing the sequence precisely using customized nucleases (Sander and Joung 2014). Designer nuclease technologies have been utilized for doing site-specific gene targeting and editing. In total, four platforms are available, for example, mega nucleases, ZFNs, TALENs, and a recently developed CRISPR/Cas9 system (Rinaldo and Ayliffe 2015).

Meganucleases are natural restriction endonucleases that produce the double-stranded breaks (DSBs) in the DNA sequences ranging from 12 to 40 bp in size (Pâques and Duchateau 2007). Multiple plant species including *Arabidopsis*, tobacco, and maize have been exposed to these nucleases (D'Halluin et al. 2008; Yang et al. 2009) but limited success was reported in knocking down the genes.

In plants, ZFNs was used for mutating specific loci in *Arabidopsis* (Lloyd et al. 2005; de Pater et al. 2009, 2013; Qi et al. 2013), tobacco (Bibikova et al. 2003; Petolino et al. 2010), soybean (Curtin et al. 2011), petunia (Marton et al. 2010), and maize (Shukla et al. 2009; Ainley et al. 2013). Use of ZFNs in *Arabidopsis* and tobacco produced 3%–7% targeted and heritable mutation frequency in transgene and endogenous gene (Osakabe et al. 2010; Zhang et al. 2010).

TALENs, much like ZFNs, disrupt the gene function by recognizing its specific sequence through a combined catalytic domain of FokI nuclease and transcription activator-like effectors (TALEs) (Cermak et al. 2011). These are the group of proteins (comprising of 30–35 amino acids) discovered in *Xanthomonas oryzae* (Bogdanove et al. 2010; Schornack et al. 2013). These proteins recognize specific DNA sequence located in close vicinity of endogenous plant gene for activating its transcription for stimulating the pathogenicity of the bacterium (Bogdanove et al. 2010). TALENs have been used in a variety of eukaryotic organisms including *Arabidopsis* (Cermak et al. 2011), tobacco (Mahfouz et al. 2011; Zhang et al. 2013), rice (Li et al. 2012), wheat (Shan et al. 2013; Wang et al. 2014), soybean (Haun et al. 2014), maize (Liang et al. 2014), and barley (Wendt et al. 2013; Gurushidze et al. 2014).

Another tool is called as clustered, regularly interspaced, short palindromic repeat (CRISPR), which produces double stranded breaks (DSBs) in a genome. These repeats are 21–47 bp long. The CRISPR-associated protein 9 (Cas9) is used to alter the genome efficiently in the targeted-oriented manner (Gersbach 2014). This system has been evolved in many bacterial species to protect against invading viral and plasmid DNAs3 (Sander and Joung 2014). The Cas9 induces double-stranded breaks (DSBs) in specific region of the genome by specifying a short single-guide RNA (sgRNA, 20 bp) or guided sequences to complement the target DNA. These breaks in the sequences yield knockout genes and alter the gene expression by changing the promoter sequences (Jinek et al. 2012).

Recently, the CRISPR/cas-mediated mutagenesis and gene targeting have been identified in wheat and rice (Shan et al. 2014). Two of the three sgRNAs (sgRNAs 1 through 3) are designed for targeting *PDS* and *DEP1* genes in rice, while the third one was designed for targeting the *LOX2* gene in wheat. In soybean, this system was used to knockout the GFP transgene and also modified the nine endogenous loci (Jacobs et al. 2015). It is concluded that the genome editing tools offer opportunities for defining genes function (either by knocking or gaining of function), modified for conditional control, or labeled for monitoring the protein product (Gersbach 2014).

Targeting-induced local lesions in genomes

Genome-wide expression profiles are useful in identifying the candidate genes conferring various traits. New strategies for inactivating or overexpressing the candidate genes were described for defining their functions. One of these is the use of TILLING method. In TILLING, chemicals induce point mutations in the genome, which are detected using polymerase chain reaction (PCR)-supported selection (McCallum et al. 2000). Genetic stocks, genotypes, improved germplasm, etc., can be used as a starting material to expose with chemicals or radiation treatment, which is found to be very effective than those approaches that involve the use of genetic transformation methods (RNAi, T-DNA, insertion of transposons, etc.) (Ronald 2014). In spite of the complex ploidy level and size of genome, TILLING procedure is very effective in all type of plant species in inducing mutations (Alonso et al. 2003; Hsing et al. 2007; Parry et al. 2009).

TILLING method takes advantage of genome-sequencing information and can investigate the function of specific genes (Slade et al. 2004). Precise phenotyping, especially of complex traits that are difficult to estimate accurately is the prerequisite in this process. Accuracy in phenotyping can be accomplished by conducting experiments in contained environments (lab scale), however, measuring complex traits in the contained environment is not the true indicator of the phenotyping done under natural conditions (Araus et al. 2014).

In TILLING procedure, the gene or part of the gene is amplified using gene-specific primers (can be labeled if a high-throughput system is used). These amplicons are denatured. A cleavage endonuclease CELI (extracted from celery) is used to digest the amplicons at mismatch sites. These amplicons are run on LI-COR gel analyzer system together with gel electrophoresis (McCallum et al. 2000; Colbert et al. 2001). The digested PCR products can also be resolved on 4% metaphor agarose gels (Gul 2014). If mutations are detected, then these fragments can be sequenced to know about the type of mutation. The celery juice can also be used in crude form, making this procedure cost-effective (Till et al. 2004). Many other endonuclease (ENDO I and SURVEYOR/CEL II) can digest the mismatches in the genome (Triques et al. 2008; Okabe et al. 2013). This is a conventional procedure that is limited to eight samples per pool, thus limiting the efficiency of mutation detection.

Some methods do not rely on mismatch cleavage done by endonucleases, such as high-resolution melting analysis (HRM) and next-generation sequencing (NGS) platforms (Parry et al. 2009). For example, a 454-FLX platform was used to discover mutants in tomato *eIF4E* gene by screening more than 3000 families in a single sequencing run (Rigola et al. 2009). Later on, resequencing three amplicons of a subset of 92 lines exhibited the six haplotypes of the *eIF4E* gene (Chen et al. 2014).

In another study, the Illumina GA sequencing platform was used to discover mutations in rice and wheat populations (Tsai et al. 2011). The NGS-TILLING (can use up to 30 PCR amplicons for detecting mutations) is an efficient procedure in detecting mutations accurately in less time period compared to the conventional TILLING approach (Pan et al. 2015). TILLING is applicable to all kinds of genes where mutations can be induced and has been successfully deployed on multiple crops species. For example, identification of MtPT4 (mtpt4-1) allele and its role in symbiotic Pi transport was confirmed in a mutagenized (EMS) population of *M. truncatula* (Javot et al. 2007). Also, the function of *E1* gene in conferring late flowering in soybean was authenticated using TILLING approach (Xia et al. 2012). In another study, two mutant alleles (SBEIIa-A and SBEIIa-B) responsible for a high relative content of amylose and a high level of resistant starch have been bred in durum wheat (Sestili et al. 2015).

In complex polyploids such as wheat detection of mutations through TILLING is always difficult because of the reason that homeolog from other genome in the nucleus can disguise the mutant trait. Mutation induction is essential in all the subgenomes (A, B, and D) in all the corresponding genes—nearly impossible to achieve. In this regard, the mutation induced in each subgenome can be combined in one genotype by selfing/ crossing for three or more generations (Chen et al. 2014).

Another strategy called as EcoTILLING has been demonstrated to identify natural polymorphisms in multiple genomes of a plant species (analogous to TILLING). Polymorphisms that occur in the germplasms are valuable tools for genetic mapping. EcoTILLING can be applied to all types of crop species including polyploids, where it can differentiate between alleles of the homologous and paralogous genes (Comai et al. 2004). In another study, EcoTILLING has not only unraveled the number of alleles of different genes conferring resistance to salt stress but also helped in understanding the complex nature of genetic mechanisms involved in conferring resistance and/or tolerance to salt stress (Negrao et al. 2013). A number of transcription factors have been targeted via

EcoTILLING to investigate the natural variants in rice after exposing it to drought stress (Yu et al. 2012). The availability of comprehensive information about ESTs is vital for the success of the abovementioned approaches to identify genes precisely.

Genome sequencing of plants

Lessons learnt from genome sequences of model plants

Genome sequence of over hundred plant species has been deciphered, and generated useful information about the occurrence of whole-genome duplications (WGDs) followed by the massive rearrangements—thus broken down the collinearity and syntenic relationship among the genomes. At the onset of twenty-first century, genome sequence information about only few plant species including *A. thaliana* followed by rice was elucidated. *A. thaliana*, the most studied plant species, has been extensively used for exploring multiple genetic/biochemical mechanisms shaping different traits because of its small genome size, short life span, ample production of seed, short height, and availability of ample genetic information (high-resolution genetic and physical maps). Above all, it is much more amicable to genetic transformation procedures (Bevan and Murphy 1999). Many important genes have been identified and characterized of this model species (Koornneef et al. 2004). For example, *Arabidopsis* genome harbors ~30 terpene synthase genes (Aubourg et al. 2002).

It has been demonstrated that genome sequencing for the sake of sequencing has no meaning until and unless this information is linked with genetic diversity available in the germplasm of a species, which would set a stage for identifying genetic mechanisms of multiple traits. Thus, additional genetic resources for each plant species should be explored, and upon these resources new mega genome sequencing projects can be initiated for capturing and describing the available genetic diversity (genes and or their allelic variants). For example, 3000 rice genome project has been initiated for defining genes involved in conferring a wide range of traits in the rice germplasm. A total of 18.9 million SNPs were discovered in rice (The 3,000 Rice Genomes Project 2014). This concept is being translated to other crop species such as cotton, maize, etc.

The genome sequencing of many crop species assisted in defining ancestral karyotypes of many plant families. For example, the ancestral karyotype of cereals (evolved all its members from a set of five or seven chromosomes) has been described (Salse et al. 2008). Similarly, members of the Brassicaceae family were evolved from the ancestral karyotype containing eight ancestral chromosomes (Koieng and Weigel 2015).

Syntenic and collinearity relationship among different sequenced genomes have shown that the genomes did not experience the WGDs or experienced one or two could be aligned. However, the genomes that have gone through multiple round of WGD are difficult to compare with the other member of the same family—thus retarding the genetic progress. Large genomes that incurred WGD contain many repetitive genomic elements, which further complicate the situation in assembling the genome (Tang et al. 2008; Paterson et al. 2009).

Although the *A. thaliana* occupied a privileged position as a model plant, useful genetic information was generated that resulted in improving our knowledge about the evolution, defining gene functions, etc. After the discovery of NGS tools, many important members of various families were sequenced. For example, sequencing of *Arabidopsis lyrata* and *A. rubella* made possible to use these as model species because the genomes of these two did not experience recent WGD as incurred in *A. thaliana*. The WGD in *A. thaliana* resulted in reshuffling of the genome and loss of chromosome number—thus not considered a suitable model for undertaking comparative genomic, evolutionary studies, etc. Also, this organism is difficult to cross with the other closely related species, restricting its utility in exploring the genetic basis of interspecies differences (Koieng and Weigel 2015).

In the present age of genomic, the access to sequence the whole genome made it possible to define functions of those genes not present in the model species. Thus, efforts toward defining new model species within the same families have been initiated. For example, for studying the drought tolerant genes in Poaceae, sorghum would be the best choice (Paterson et al. 2009). Otherwise, for genes that confer grain development mechanism in monocot, rice genome can be used as a model.

Overview of the recently sequenced genomes

The genome of *A. thaliana* was sequenced in 2000 (Arabidopsis Genome 2000) followed by the sequencing of rice genome in 2002 (Goff et al. 2002). Till the invention of next-generation sequencing technologies coupled with the parallel evolution in bioinformatics made it possible to sequence the large genomes followed by making their assemblies within limited time period at cheaper cost. All these efforts resulted in sequencing of more than hundred genomes of different plant species including crops. A number of NGS platforms are available for sequencing different organisms. These are Roche 454, Illumina/Solexa Genome Analyzer, and the Applied Biosystems SOLiDTM System.

Roche 454 technology has been exploited for sequencing the genome of *Theobroma cacao* (Scheffler et al. 2009). This technology was also utilized to sequence the whole-grape genome (Velasco et al. 2007)—a combination of 6.5x Sanger paired read sequences and 4.2x unpaired Roche 454 reads were assembled into 209 metacontigs, which together represents approximately 94.6% of the genome.

Roche 454 sequencing has been used to survey the genome of Miscanthus (Swaminathan et al. 2009), while Sanger, Illumina/Solexa and Roche 454 sequencing were used to investigate banana genome. Illumina GAII sequencing platform was used to generate more than 50x coverage of the *Brassica rapa* genome (Wang et al. 2011).

Next-generation sequencing methods alone or in combination with the Sanger sequencing approach were used for a number of crop species including apple genome (Velasco et al. 2010), cocoa genome (Argout et al. 2011), etc. The Sanger sequencing was used just to sequence the BAC ends for acquiring the long-distance structural information of the genome. Later on, the genome of woodland strawberry was sequenced using NGS alone by combing the Roche 454, Illumina, and SOLID platforms. Currently, Illumina sequencing that emerged as a dominant technology over the other NGS has been exploited for sequencing the genomes of many plant species including Chinese cabbage (Wang et al. 2011), potato (Xu et al. 2011), banana (D'Hont 2012), chickpea (Varshney et al. 2013), orange (Xu et al. 2013), watermelon (Guo et al. 2013), etc.

Genomes of the cultivated tetraploid cotton species and its progenitors have also been sequenced using NGS technologies. Information regarding the putative number of genes, karyotype evolution, mechanisms involved in conferring multiple stages of fiber development, and drawing the evolutionary consequence have been described (Li et al. 2012, 2015; Paterson et al. 2012; Wang et al. 2012; Zhang et al. 2015).

Wheat (*Triticum aestivum* L.), hexaploid bread wheat, contains ~17-Gb genome size. All the chromosomes were sequenced in isolation. A total of 124,201 gene loci were annotated, which are distributed evenly on all chromosomes. Unlike many other plant species, limited gene loss retained structural integrity in the hexaploid wheat genome when compared with its diploids and tetraploid relatives. These findings set a stage for gene isolation, development of genetic markers, etc. (The International Wheat Genome Sequencing Consortium 2014). In Table 6.1, genome information of a number of crop species is given.

Limitations of genome sequencing in plants

There is huge variation in the size of plant genomes ranging from 61 (*Genlisea tuberosa*) to over 150,000 Mb (*Paris japonica*) (Michael 2014). Genome size is highly correlated with the repeat contents in genome, which offers a major problem in assembling the plant genome. Through NGS platforms, huge data of the large-sized genomes can be produced but their assembly remained a major challenge for computational procedures. So far, the largest genome of loblolly pine genome (22 Gb) was assembled using a preprocessed condensed set of super reads to lessen computation resources needed for assembly (Zimin et al. 2014).

As indicated earlier, the repetitive elements also present the problem in assembling the short reads, which generated 100–200 bp in length using NGS technologies. These short reads further complicates the process of genome assembly especially when resolving the assembly of complex genomes. All these limitations lead to produce gaps in the assembly of the genomes. For example, gaps (total gap length is 185,644) still exist in the well-assembled *Arabidopsis* genome (Lamesch et al. 2012).

For developing a good assembly of a genome, information about the size of the repetitive element is required. In plant genomes, the average read length of the repeats exceeds 10–20 kb. The most prevalent ones are long terminal repeats (LTRs) whose proliferation leads to fluctuation in genome size. Within plant genomes, cultivars harbor structural variations (SVs) between them due to LTRs movement, and resequencing projects based on the reference genome often miss or inaccurately predict the SVs. *De novo* assembly of three divergent rice strains uncovered several megabases (Mbs) of novel sequences in each strain, with many contigs containing expressed genes (Schatz et al. 2014). This assembly is further complicated by the occurrence of paralogous regions in the genome that result in highly fragmented incomplete assemblies. Polyploidy, a very common phenomenon experienced by multiple plant species, further complicates the situation by increasing the number of duplicated regions.

Occurrence of high level of heterozygosity within genomes including eucalyptus, outcrossing species such as grape, clonally propagated crops such as apple, etc., is another factor that hinders the process of genome assembly by creating bubbles. These concerns can be addressed by deploying their progenitors (diploids) species information in making assemblies, for example, robusta coffee (Denoeud et al. 2014), wheat (Ling et al. 2013), haploid/monoploid lines in case of citrus (Xu et al. 2013), banana (D'Hont et al. 2012), etc.

Another issue is the contamination of organelle DNA. For example, in plant cells ~10,000 of plastid DNA copies per cell are present (Shaver et al. 2006), consequently the organelle derived reads can make 5%–20% of the total sequences in the whole-genome sequences. Such contamination can be minimized considerably using modified DNA extraction protocols—optimized only for nuclei isolation. Another method like qPCR can be undertaken to estimate the organelle contamination (Lutz et al. 2011). A number of organelle-derived regions have been found in the nuclear genomes, which are nearly similar to the corresponding organelle genomes. This issue can be addressed by choosing the appropriate read length of these regions, which can span the insertion junction sites.

Table 6.1 Genome sequencing information of important crop species

Sr. no.	Species name	Family name	Ploidy level	Genome size (Mb)	Chromosome # (1n)	# of genes	Year of sequencing	Approach	References
1	A. thaliana	Brassicaceae	2x	125	5	23,647	2000	Sanger sequencing	Arabidopsis Genome Initiative (2000)
2	Medicago truncatula	Fabaceae	2x	454	8	33,000	2011	Sanger, Roche-454, and Illumina	Young et al. (2011)
3	Zea mays	Poaceae	2x	2300	10	32,540	2009	Sanger sequencing	Schnable et al. (2009)
4	Glycine max	Fabaceae	2x	1115	20	48,164	2010	Sanger sequencing	Schmutz et al. (2010)
5	Triticum urartu (AA)	Poaceae	2x	4940	7	34,879	2013	Illumina	Ling et al. (2013)
6	Aegilops tauschii (DD)	Poaceae	2x	4360	7	43,150	2013	Roche-454, Illumina	Jia et al. (2013)
7	Sorghum bicolor	Poaceae	2x	818	10	34,496	2010	Sanger sequencing	Paterson et al. (2009)
9	Hordeum vulgare	Poaceae	2x	5100	7	30,400	2012	Illumina	International Barley Genome Sequencing Consortium (2012)
10	Oryza sativa	Poaceae	2x	430	12	59,855	2002	Sanger sequencing	Yu et al. (2002)
11	Lotus japonicus	Fabaceae	2x	472	6	30,799	2008	Sanger sequencing	Sato et al. (2008)
12	Solanum tuberosum	solanaceae	4x	844	12	37,368	2011	Sanger sequencing, Roche-454, and Illumina	International Potato Genome Sequencing Consortium (2011)

(Continued)

Table 6.1 (Continued) Genome sequencing information of important crop species

Sr. no.	Species name	Family name	Ploidy level	Genome size (Mb)	Chromosome # (1n)	# of genes	Year of sequencing	Approach	References
13	*Gossypium arboreum (AA)*	Malvaceae	2x	1750	13	41,330	2014	Roche-454, Illumina, and Sanger	Li et al. (2014)
14	*Gossypium raimondii (DD)*	Malvaceae	2x	NG[a]	13	NG	20412	NG	Paterson et al. (2012)
15	*Gossypium raimondii (DD)*	Malvaceae	2x	880	13	40,976	2012	Roche-454, Illumina, and Sanger	Wang et al. (2012)
16	*Gossypium hirsutum (AD)*	Malvaceae	4x	2500	13		2015	Illumina	Li et al. (2015)
17	*Cicer arietinum*	Fabaceae	2x	738	8	28,269	2013	Illumina	Varshney et al. (2013)
18	*T. aestivum*	Poaceae	6x	17,000	7	94,000	2014	Illumina	Mayer et al. (2014)
19	*B. rapa*	Brassicaceae	2x	485	10	41,174	2011	Illumina	Wang et al. (2011)
20	*Brassica napus*	Brassicaceae	4x	NG	19	101,040	2014	454 GS-FLX+ titanium (Roche, Illumina, and Sanger sequence)	Chalhoub et al. (2014)
21	*Brassica oleracea*	Brassicaceae	2x	630	9	NG	2014	Illumina, Roche 454, and Sanger sequence	Liu et al. (2014)

[a] NG = Not given.

The complexities offered by the large genomes can be mitigated by generating longer read lengths, which can be achieved by deploying the third-generation sequencing (TGS) technologies that have the potential to overcome these concerns. It is different from NGS assays that use PCR to grow clusters of the template DNA followed by sequencing them by synthesis. The TGS technologies exploit the single DNA molecules—thus avoiding the biases in sequencing caused by PCR (Schadt et al. 2010). Currently, a number of TGS platforms are available, for example, Helicos true single-molecule sequencing (tsMSTM) (Harris et al. 2008), Pacific BioSciences single-molecule real-time (SMRTTM) sequencing (Eid et al. 2009), and Oxford Nanopores (Clarke et al. 2009). Both Pacific BioSciences and Oxford Nanopores machines can produce multiple kb sequence reads. Thus, the longest repeats (such as telomeres and centromeres) can be resolved, which would pave the way for generating platinum standard genome assemblies (Tilgner et al. 2014). Moreover, SMRTTM sequencing helps in identification of methylated genomic regions, which would allow to study the epigenetic changes in the genome of interest (Flusberg et al. 2010).

Role of functional genomics in crop improvement

Whole-genome sequencing of plant followed by defining function of these sequences laid down the foundation of functional genomic studies and yielded a high level of understanding of different mechanisms involved in shaping phenotypes (Mittler and Shulaeve 2013). A number of tools and methodologies have been described in all the aforementioned sections related to defining function of gene(s), which paved the way for not only acquiring the useful genetic resource but also helped in solving the puzzles pertaining to the evolutionary and developmental biology of various plant species (Ostergaard and Yanofsky 2004). The genomic information has also been investigated to identify DNA markers (EST-SSRs and SNPs) associated with various traits of interest. These markers are being used as diagnostic markers in selecting the correct type of genotype, which could thus lead to develop new cultivars within less time period.

Gene silencing: A tool for improving resistance to biotic and abiotic stresses in crop plants

By understanding the function of genes, various strategies for engineering genome were devised for developing resistance against the pathogens (viruses, bacteria, fungi, nematodes, etc.) and insect pests. For example, RNAi-mediated strategy was used to engineer papaya plant against RNA viruses (Vanitharani et al. 2005). Resistance to root-knot nematode in corn was developed by knocking out the *LOX* gene (Gao et al. 2008). Similarly, these strategies can be used for improving resistance to viruses in tomato (Piron et al. 2010), etc.

In another study, simultaneous editing of all three wheat *Mlo* homoealleles (using TALENs) resulted in broad spectrum resistance to powdery mildew disease (Wang et al. 2014). Similarly, TALENs assays were used in rice to improve resistance to bacterial pathogen by inducing mutations in the endogenous sucrose transporter gene (*Os Sweet14*) (Li et al. 2012). Thus by exploiting the functional genomic approaches, resistance to multiple biotic stresses can be improved in the elite cultivars (Rinaldo and Ayliffe 2015).

Also, these genes have been used for development crop varieties with excellent genetics for mitigating abiotic stress. A number of genes that are differentially expressed in response to drought stress were identified. For example, the expression of *ZmALDH9*, *ZmALDH13*, and *ZmALDH17* genes were identified in maize (Zhou et al. 2012). Similarly, the function of genes responsible for aluminum tolerance was also described (Chen et al. 2012).

Improvement in quality traits

Value addition (quality enhancement) is one of the main focuses of geneticists/molecular biologists, worldwide. The soybean genome was exposed to TALENS, which induced mutations in two genes *FAD2-1A and FAD2-1B*, which resulted in improved oil quality by increasing the oleic acid and simultaneous reduction in the linoleic acid (Haun et al. 2014). In corn, ZFN-mediated cleavage in *IPK1* gene—catalyzing the phytate production—demonstrated resistance to herbicide by depressing the phytate level (Shukla et al. 2009). In another study, the quality of *Brassica napus* oil was improved by reducing the level of saturated fat in transgenic lines by the enhanced expression of *B-ketoacyl-ACP synthase II* (*KASII*) gene (Gupta et al. 2012).

Improvement in agronomic important traits

Genes of plant height were discovered by knocking out the gibberellin 2-oxidase (*GA2ox*) gene (Hsing et al. 2007). Also, the genes encoding yield and kernel size in corn were identified by knocking out *Gln1-3* and *Gln1-4* genes (Martin et al. 2006). Genes that are differentially expressed in F_1s were identified, which assisted in exploring the mechanism of hybrid vigor in rice (Song et al. 2010). Use of these technologies would help in engineering high-yielding cultivars, which can help in sustaining crops productivities on this planet.

Identification of DNA markers

The NGS assays have been used in re-sequencing of different accessions/representative genotypes of multiple plant species including *Arabidopsis*, rice, soybean, etc., which paved the way for identifying novel alleles. DNA markers (SNPs) were developed for tagging QTLs/complex traits (Lam et al. 2010; He et al. 2011; Huang et al. 2012). For example, SNP-associated ebi-1 phenotype was discovered in *Arabidopsis* (Ashelford et al. 2011). Similarly, 2.58 million SNPs were identified by sequencing 916 foxtail millet varieties (Jia et al. 2013). In rice, MutMap was constructed by re-sequencing of whole genome (Abe et al. 2012), which was used for the quick mapping of QTLs (Takagi et al. 2013).

SSR markers have also been designed using the sequence information of expressed part of the gene/genome. These SSRs are designated as EST-SSRs. In multiple investigations, ESTs have been identified in plant species after exposing them to different stresses. These ESTs were not only used to identify genes but also exploited to design DNA markers. These markers can be used in marker-assisted breeding programs (Rahman et al. 2011, 2012, 2014; Tabbasam et al. 2014). A number of markers associated with various genes have been designed. For example, an SNP linked to *TaMYB2*—responsible for dehydration tolerance in bread wheat was identified (Garg et al. 2012). Markers Xucw108 and Xuhw89 associated with the high protein content and stripe rust resistance, respectively, were reported (Distelfeld and Fahima 2007; Brevis and Dubcovsky 2008; Liu et al. 2012). EST-SSRs were identified in lentil (Kaur et al. 2011), fava bean (Kaur et al. 2012), chickpea (Agarwal et al. 2012), radish (Wang et al. 2012), etc. A major handicap with the EST-based markers is that their number is limited than that of structural genomic markers.

Translation of information from model to less-studied species

We have witnessed an unprecedented growth in the generation of genome-sequencing information of many plant species since the invention of NGS tools, and this science was complemented by the bioinformatics procedures, which made us understand various mysteries behind the complex genetic mechanisms responsible for determining a phenotype. *A. thaliana* was explored at length using various genomic assays for describing the functions of its genes. Like many other organisms, this plant species have also been evolved from a common ancestor, which left lot of structural and functional conservativeness among all the plant genomes. Thus the information generated on the model species can be translated to the others using a range of gene prediction bioinformatics tools for predicting genes and gene families (Rahman and Paterson 2010). For this genetic information in the form of a sequenced genome, genetic/physical maps are needed for making comparisons with the model species (Sato et al. 2008).

Rice, the most studied genome in monocots, can serve as model for all the members of plant species of cereals that paved the way for annotating genes and gene prediction involved in grain development in many crop species, including wheat. Similarly, lot of genetic information (ESTs, maps, genome sequence, etc.) has been generated on tomato genome, which can serve as a model for isolating genes involved in fruit development and ripening traits. Thus the science of genomic, together with the computational procedures, can lessen the burden on spending time for studying the mechanisms described on the model species.

Discovery of small/regulatory RNAs

Functional genomic studies have played a key role in understanding the role of epigenetics in conferring a phenotype in response to the fluctuating environment (Piferrer 2013). The NGS platforms have been used in regulation of methylation and its mapping throughout the genome (Lister and Ecker 2009; Simon and Meyers 2011). Also, molecules such as miRNA, siRNA, tRNA, and rRNA, which played an important role in PTGS, were discovered (Xie et al. 2004; Morozova and Marra 2008). For example, miR156, miR159, miR172, miR167, miR158 and miR166, and miRNAs were identified as Brassica sp. (Huang et al. 2013). Similarly, siRNAs responsible for pollen development in rice (Wei et al. 2011) and miRNA responsible for phosphorus requirement in barley were identified (Hackenberg et al. 2013).

Development of new genetic resources

TILLING approach has demonstrated success in creating novel variants, which can generate a number of alleles of a given gene—a novel way to increase the extent of genetic diversity in a germplasm collection. This procedure is user friendly than that of bringing new alleles from wild sources into the cultivated varieties by attempting many backcrosses. The germplasm developed through TILLING is of nontransgenic nature and the mutant phenotype can be genetically transferred by conventional breeding (Slade et al. 2012). This procedure is extremely useful in countries where crop varieties developed through nontransgenic approaches are preferred for cultivation (Varshney et al. 2012).

Conclusions

Many crop genomes have been sequenced using NGS tools (also TGS platforms), which has made possible to generate long read that did not only help in sequencing the large genomes but also facilitated in making useful assembly of the genomes. All this information has direct implications either in the form of development of dense genetic maps or in the annotation of genes. In spite of the fact that the science of genomic has advanced significantly and generated huge amount of sequencing information, a lot of regions/genes of many important plant species have not been assigned functions or explored. Only most genes of the *Arabidopsis* and rice were assigned functions. It is important to capture maximum genetic diversity (germplasm) of the plant/crop species followed by exposing it to TGS assays, which would help in functional characterization of the genes that are responsible for the genetic/phenotypic variations prevalent in that particular species. Also, the alternative approaches can be used for creating genetic variations in the genome. For example, TILLING or any other approaches (RNAi, TALENS, etc.) can be used to induce mutations for studying the function of genes. Choice of the method depends upon the nature of the plant species and objectives of the investigator. For example, TILLING is very helpful on those plant species where transformation approaches do not respond or the genome is complex. Each method has its own advantages and disadvantages. Thus, many approaches together can be used for defining the function of genes.

For taking another step toward analyzing the transcriptomic, proteomic, epigenomic, and metabolomics data, it is imperative to develop platinum standard reference genomes of different crop species. In this regard, multinational genome-sequencing projects should be initiated that would help in complementing the functional characterizations of the genes present in different genetic backgrounds across the world. Development of new computational tools in annotating genes, making perfect genome assemblies, etc., would complement the progress made toward the functional characterization of genes. Once the function of most genes of different crop species would be known, it would be possible to breed cultivars/ideotypes that can mitigate the negative impact of the changing climate, which thus would sustain yield on this planet.

Acknowledgments

This chapter is partially supported by the International Atomic Energy Agency (IAEA) through a national TC project titled "Developing Germplasm through TILLING in Crop Plants Using Mutation and Genomic Approaches (PAK/5/047)." Any opinion, conclusion, or recommendation discussed in this chapter are those of the author(s) and do not necessarily reflect the views of the IAEA.

References

Abe, A., Kosugi, S., Yoshida, K. et al. 2012. Genome sequencing reveals agronomically-important loci in rice from mutant populations. *Nature Biotechnology* 30:174–178.

Agarwal, G., Jhanwar, S., Priya, P., Singh, V.K., Saxena, M.S., Parida, S.K., Garg, R., Tyagi, A.K., and Jain, M. 2012. Comparative analysis of kabuli chickpea transcriptome with desi and wild chickpea provides a rich resource for development of functional markers. *PLoS One* 7:52443.

Ainley, W.M., Sastry-Dent, L., Welter, M.E., Murray, M.G., Zeitler, B., Amora, R., Corbin, D.R., Miles, R.R., Arnold, N.L., and Strange, T.L. 2013. Trait stacking via targeted genome editing. *Plant Biotechnology Journal* 11(9):1126–1134.

Alonso, J.M., and Ecker, J.R. 2006. Moving forward in reverse: Genetic technologies to enable genome-wide phenomic screens in *Arabidopsis*. *Nature Reviews Genetics* 7:524–536.

Alonso, J.M., Stepanova, A.N., Leisse, T.J., Kim, C.J., Chen, H., Shinn, P., Stevenson, D.K., Zimmerman, J., Barajas, P., and Cheuk, R. 2003. Genome-wide insertional mutagenesis of *Arabidopsis thaliana*. *Science* 301:653–657.

Arabidopsis Genome Initiative. 2000. Analysis of the genome sequence of the flowering plant *Arabidopsis thaliana*. *Nature* 408:796–815.

Araus, J.L., Li, J., Parry, M.A., and Wang, J. 2014. Phenotyping and other breeding approaches for a New Green Revolution. *Journal of Integrative Plant Biology* 56(5):422–424.

Argout, X., Salse, J., Aury, J.M., Guiltinan, M.J., Droc, G., Gouzy, J., and Brunel, D. 2011. The genome of *Theobroma cacao*. *Nature Genetics* 43(2):101–108.

Ashelford, K., Eriksson, M.E., Allen, C.M., Amore, R.D., Johansson, M., Gould, P., Kay, S., Miller, A.J., Hall, N., and Hall, A. 2011. Full genome resequencing reveals a novel circadian clock mutation in *Arabidopsis*. *Genome Biology* 12:R28.

Ashikari, M., Sakakibara, H., Lin, S. et al. 2005. Cytokinin oxidase regulates rice grain production. *Science* 309:741–745.

Aubourg, S., Lecharny, A., and Bohlmann, J. 2002. Genomic analysis of the terpenoid synthase (AtTPS) gene family of *Arabidopsis thaliana*. *Molecular Genetics and Genomic* 267:730–745. doi: 10.1007/s00438-002-0709-y.

Baulcombe, D.C. 1999. Fast forward genetics based on virus-induced gene silencing. *Current Opinion in Plant Biology* 2(2):109–113.

Beadle, G.W., and Tatum, E.L. 1941. Genetic control of biochemical reactions in *Neurospora*. *Proceedings of National Academy of Sciences USA* 27:499–506.

Benstein, R.M., Ludewig, K., Wulfert, S., Wittek, S., Gigolashvili, T., Frerigmann, H., Gierth, M., Flügge, U.-I., and Krueger, S. 2013. *Arabidopsis* phosphoglycerate dehydrogenase1 of the phosphoserine pathway is essential for development and required for ammonium assimilation and tryptophan biosynthesis. *The Plant Cell Online* 25(12):5011–5029.

Bevan, M., and Murphy, G. 1999. The small, the large and the wild: The value of comparison in plant genomics. *Trends in Genetics* 15:211–214. doi:10.1016/s0168-9525(99)01744-8.

Bibikova, M., Beumer, K., Trautman, J.K., and Carroll, D. 2003. Enhancing gene targeting with designed zinc finger nucleases. *Science* 300(5620):764–764.

Bogdanove, A.J., Schornack, S., and Lahaye, T. 2010. TAL effectors: Finding plant genes for disease and defense. *Current Opinion in Plant Biology* 13(4):394–401.

Bolger, M.E., Weisshaar, B., Scholz, U., Stein, N., Usadel, B., and Mayer, K.F.X. 2014. Plant genome sequencing-applications for crop improvement. *Current Opinion in Biotechnology* 26:31–37.

Bouchez, D., and Hofte, H.H. 1998. Functional genomics in plants. *Plant Physiology* 118(3):725–732.

Bray, E.A. 2004. Genes commonly regulated by water-deficit stress in *Arabidopsis thaliana*. *Journal of Experimental Botany* 55:2331–2341.

Brenner, S., Johnson, M., and Bridgham, J. 2000. Gene expression analysis by massively parallel signature sequencing (MPSS) on microbead arrays. *Nature Biotechnology* 18:630–634.

Brevis, J.C., and Dubcovsky, J. 2008. Effect of the Gpc-B1 region from *Triticum turgidum* ssp. dicoccoides on grain yield, thousand grain weight and protein yield. Department of Plant Sciences, University of California, Davis.

Breyne, P., and Zabeau, M. 2001. Genome-wide expression analysis of plant cell cycle modulated genes. *Current Opinion in Plant Biology* 4:136–142.

Cabello, J.V., Lodeyro, A.F., and Zurbriggen, M.D. 2014. Novel perspectives for the engineering of abiotic stress tolerance in plants. *Current Opinion in Biotechnology* 26:62–70.

Cabezas, J.A., Ibanez, J., Lijavetzky, D., Velez, D., Bravo, G., Rodriguez, V., Carreno, I., Jermakow, A.M., Carreno, J., and Ruiz-Garcia, L. 2011. A 48 SNP set for grapevine cultivar identification. *BMC Plant Biology* 11(1):153.

Cermak, T., Doyle, E.L., Christian, M., Wang, L., Zhang, Y., Schmidt, C., Baller, J.A., Somia, N.V., Bogdanove, A.J., and Voytas, D.F. 2011. Efficient design and assembly of custom TALEN and other TAL effector-based constructs for DNA targeting. *Nucleic Acids Research* 39(12):e82. doi: 10.1093/nar/gkr218.

Chalhoub, B., Denoeud, F., Liu, S., Parkin, I.A., Tang, H., Wang, X., Chiquet, J., Belcram, H., Tong, C., and Samans, B. 2014. Early allopolyploid evolution in the post-Neolithic *Brassica napus* oilseed genome. *Science* 345(6199):950–953.

Chen, Z.C., Yamaji, N., Motoyama, R., Nagamura, Y., and Ma, J.F. 2012. Up-regulation of a magnesium transporter gene OsMGT1 is required for conferring aluminum tolerance in rice. *Plant Physiology* 159:1624–1633.

Chen, L., Huang, L., Min, D., Phillips, A., Wang, S., Madgwick, P.J., Parry, M.A., and Hu, Y.G. 2012. Development and characterization of a new TILLING population of common bread wheat (*Triticum aestivum* L.). *PLoS One* 7(7). doi: 10.1371/journal.pone.0041570.

Chen, X., Cui, Z., Fan, M. et al. 2014. Producing more grain with lower environmental costs. *Nature* 514:486–489.

Cigan, M.A., Unger-Wallace, E., and Haug-Collet, K. 2005. Transcriptional gene silencing as a tool for uncovering gene function in maize. *The Plant Journal* 43(6):929–940.

Clarke, J., Wu, H.-C., Jayasinghe, L., Patel, A., Reid, S., and Bayley, H. 2009. Continuous base identification for single-molecule nanopore DNA sequencing. *Nature Nanotechnology* 4(4):265–270.

Close, T.J., Wanamaker, S.I., Caldo, R.A., Turner, S.M., Ashlock, D.A., Dickerson, J.A., Wing, R.A., Muehlbauer, G.J., Kleinhofs, A., and Wise, R.P. 2004. A new resource for cereal genomics: 22K barley GeneChip comes of age. *Plant Physiology* 134:960–968.

Colbert, T., Till, B.J., Tompa, R., Reynolds, S., Steine, M.N., Yeung, A.T., McCallum, C.M., Comai, L., and Henikoff, S. 2001. High-throughput screening for induced point mutations. *Plant Physiology* 126(2):480–484.

Comai, L., Young, K., Till, B.J., Reynolds, S.H., Greene, E.A., Codomo, C.A., Enns, L.C., Johnson, J.E., Burtner, C., Odden, A.R., and Henikoff, S. 2004. Efficient discovery of DNA polymorphisms in natural populations by Ecotilling. *Plant Journal* 37(5):78–786.

Curtin, S.J., Zhang, F., Sander, J.D., Haun, W.J., Starker, C., Baltes, N.J., Reyon, D., Dahlborg, E.J., Goodwin, M.J., and Coffman, A.P. 2011. Targeted mutagenesis of duplicated genes in soybean with zinc-finger nucleases. *Plant Physiology* 156(2):466–473.

D'Halluin, K., Vanderstraeten, C., Stals, E., Cornelissen, M., and Ruiter, R. 2008. Homologous recombination: A basis for targeted genome optimization in crop species such as maize. *Plant Biotechnology Journal* 6(1):93–102.

D'Hont, A., Denoeud, F., Aury, J.M. et al. 2012. The banana (*Musa acuminata*) genome and the evolution of monocotyledonous plants. *Nature* 488:213–217.

de Felippes, F.F., Wang, J.W., and Weigel, D. 2012. MIGS: MiRNA-induced gene silencing. *The Plant Journal* 70(3):541–547.

De Pater, S., Neuteboom, L.W., Pinas, J.E., Hooykaas, P.J., and Van Der Zaal, B.J. 2009. ZFN-induced mutagenesis and gene-targeting in *Arabidopsis* through *Agrobacterium*-mediated floral dip transformation. *Plant Biotechnology Journal* 7(8):821–835.

Denoeud, F., Carretero, P.L., Dereeper, A., Droc, G., Guyot, R., Pietrella, M., Zheng, C., Alberti, A., Anthony, F., and Aprea, G. 2014. The coffee genome provides insight into the convergent evolution of caffeine biosynthesis. *Science* 345:1181–1184.

Deyholos, M.K. 2010. Making the most of drought and salinity transcriptomics. *Plant Cell and Environment* 33(4):648–654.

Distelfeld, A., and Fahima, T. 2007. Wild emmer wheat as a source for high-grain-protein genes: Map-based cloning of Gpc-B1. *Israel Journal of Plant Sciences* 55:297–306.

Eid, J., Fehr, A., Gray, J. et al. 2009. Real-time DNA sequencing from single polymerase molecules. *Science* 3235910:133–138.

Ergen, N., Thimmapuram, J., Bohnert, H.J., and Budak, H. 2009. Transcriptome pathways unique to dehydration tolerant relatives of modern wheat. *Functional and Integrative Genomics* 9(3):377–396.

Flusberg, B.A., Webster, D.R., Lee, J.H., Travers, K.J., Olivares, E.C., Clark, T.A., Korlach, J., and Turner, S.W. 2010. Direct detection of DNA methylation during single-molecule, real-time sequencing. *Nature Methods* 7(6):461–465.

Gao, X., Starr, J., Göbel, C., Engelberth, J., Feussner, I., Tumlinson, J., and Kolomiets, M. 2008. Maize 9- lipoxygenase ZmLOX3 controls development, root-specific expression of defense genes, and resistance to root-knot nematodes. *Molecular Plant Microbe Interactions* 21:98–109.

Gapper, N.E., Giovannoni, J.J., and Watkins, C.B. 2014. Understanding development and ripening of fruit crops in an "omics" era. *Horticulture Research* 11403410.1038/hortres.

Garg, B., Lata, C., and Prasad, M. 2012. A study of the role of gene *TaMYB2* and an associated SNP in dehydration tolerance in common wheat. *Molecular Biology Reports* 39(12):10865–10871.

Gersbach, C.A. 2014. Genome engineering: The next genomic revolution. *Nature Methods* 11(10):1009–1011.

Goff, S.A., Ricke, D., Lan, T.H. et al. 2002. A draft sequence of the rice genome (*Oryza sativa* L. spp. Japonica). *Science* 296:92–100.

Gore, M.A., Chia, J.M., Elshire, R.J., Sun, Q., Ersoz, E.S., Hurwitz, B.L., Peiffer, J.A., McMullen, M.D., Grills, G.S., and Ross, J. 2009. A first-generation haplotype map of maize. *Science* 326(5956):1115–1117.

Goulet, C., Mageroy, M.H., Lam, N.B., Floystad, A., Tieman, D.M., and Klee, H.J. 2012. Role of an esterase in flavor volatile variation within the tomato clade. *Proceedings of National Academy of Sciences* 109(46):19009–19014.

Guan, Q., Lu, X., Zeng, H., Zhang, Y., and Zhu, J. 2013. Heat stress induction of miR398 triggers a regulatory loop that is critical for thermo tolerance in *Arabidopsis*. *The Plant Journal* 74(5):840–851.

Gul, M. 2014. Validating candidate genes in *Triticum aestivum using* TILLING populations. M.Phil Thesis. Pakistan Institute of Engineering and Applied Sciences. Islamabad Pakistan.

Guo, S., Zhang, J., Sun, H., Salse, J., Lucas, W.J., Zhang, H., Zheng, Y., Mao, L., Ren, Y., and Wang, Z. 2013. The draft genome of watermelon (*Citrullus lanatus*) and re-sequencing of 20 diverse accessions. *Nature Genetics* 45:51–58.

Gupta, M., DeKelver, R.C., Palta, A., Clifford, C., Gopalan, S., Miller, J.C., Novak, S., Desloover, D., Gachotte, D., and Connell, J. 2012. Transcriptional activation of *Brassica napus* β-ketoacyl-ACP synthase II with an engineered zinc finger protein transcription factor. *Plant Biotechnology Journal* 10(7):783–791.

Gurushidze, M., Hensel, G., Hiekel, S., Schedel, S., Valkov, V., and Kumlehn, J. 2014. True-breeding targeted gene knock-out in barley using designer TALE-nuclease in haploid cells. *PloS One* 9(3):e92046.

Hackenberg, M., Huang, P.J., Huang, C.Y., Shi, B.J., Gustafson, P., and Langridge, P. 2013. A comprehensive expression profile of MicroRNAs and other classes of non-coding small RNAs in Barley under phosphorus-deficient and sufficient conditions. *DNA Research* 20:109–125.

Harris, T.D., Buzby, P.R., Babcock, H., Beer, E., Bowers, J., Braslavsky, I., Causey, M., Colonell, J., DiMeo, J., and Efcavitch, J.W. 2008. Single-molecule DNA sequencing of a viral genome. *Science* 320(5872):106–109.

Haun, W., Coffman, A., Clasen, B.M., Demorest, Z.L., Lowy, A., Ray, E., Retterath, A., Stoddard, T., Juillerat, A., and Cedrone, F. 2014. Improved soybean oil quality by targeted mutagenesis of the fatty acid desaturase 2 gene family. *Plant Biotechnology Journal* 12(7):934–940.

He, Z., Zhai, W., Wen, H., Tang, T., Wang, Y., Lu, X., Greenberg, A.J., Hudson, R.R., Wu, C.I., and Shi, S. 2011. Two evolutionary histories in the genome of rice: The roles of domestication genes. *PLoS Genetics* 7(6). doi: 10.1371/journal.pgen.1002100.

Heilersig, B.H., Loonen, A.E., Janssen, E.M., Wolters, A.-M.A., and Visser, R.G. 2006. Efficiency of transcriptional gene silencing of GBSSI in potato depends on the promoter region that is used in an inverted repeat. *Molecular Genetics and Genomics* 275(5):437–449.

Himmelbach, A., Zierold, U., Hensel, G., Riechen, J., Douchkov, D., Schweizer, P., and Kumlehn, J. 2007. A set of modular binary vectors for transformation of cereals. *Plant Physiology* 145(4):1192–1200.

Hirayama, T., and Shinozaki, K. 2010. Research on plant abiotic stress responses in the post-genome era: Past, present and future. *Plant Journal* 61:1041–1052.

Hsing, Y.I., Chern, C.G., Fan, M.J. et al. 2007. A rice gene activation/knockout mutant resource for high throughput functional genomics. *Plant Molecular Biology* 63:351–364.

Huang, X., Kurata, N., Wei, X. et al. 2012. A map of rice genome variation reveals the origin of cultivated rice. *Nature* 490:497.

Huang, X.Y., Chao, D.Y., Gao, J.P., Zhu, M.Z., Shi, M., and Lin, H.X. 2009. A previously unknown zinc finger protein, dst, regulates drought and salt tolerance in rice via stomatal aperture control. *Genes and Development* 23:1805–1817.

Imin, N., Mohd-Radzman, N.A., Ogilvie, H.A., and Djordjevic, M.A. 2013. The peptide-encoding CEP1 gene modulates lateral root and nodule numbers in *Medicago truncatula*. *Journal of Experimental Botany* 64(17):5395–5409.

International Barley Genome Sequencing Consortium. 2012. A physical, genetic and functional sequence assembly of the barley genome. *Nature* 491(7426):711–716.

International Potato Genome Sequencing Consortium. 2011. Genome sequence and analysis of the tuber crop potato. *Nature* 475(7355):189–195.

International Wheat Genome Sequencing Consortium. 2014. A chromosome-based draft sequence of the hexaploid bread wheat (*Triticum aestivum*) genome. *Science* 345(6194):1251788.

Jackson, A.L., and Linsley, P.S. 2004. Noise amidst the silence: off-target effects of siRNAs. *Trends in Genetics* 20:521–524.

Jacobs, T.B., LaFayette, P.R., Schmitz, R.J., and Parrott, W.A. 2015. Targeted genome modifications in soybean with CRISPR/Cas9. *BMC Biotechnology* 15(1):16.

Jain, M., Nijhawan, A., Arora, R. et al. 2007. F-Box proteins in rice. Genome-wide analysis, classification, temporal and spatial gene expression during panicle and seed development, and regulation by light and abiotic stress. *Plant Physiology* 143(4):1467–1483.

Javot, H., Penmetsa, R.V., Terzaghi, N., Cook, D.R., and Harrison, M.J. 2007. A *Medicago truncatula* phosphate transporter indispensable for the arbuscular mycorrhizal symbiosis. *Proceedings of National Academy of Science, United States* 104:1720–1725.

Jia, G., Huang, X., Zhi, H. et al. 2013. A haplotype map of genomic variations and genome-wide association studies of agronomic traits in foxtail millet (*Setaria italica*). *Nature Genetics* 45:957–961.

Jinek, M., Chylinski, K., Fonfara, I., Hauer, M., Doudna, J.A., and Charpentier, E. 2012. A programmable dual-RNA–guided DNA endonuclease in adaptive bacterial immunity. *Science* 337(6096):816–821.

Jones, L., Hamilton, A.J., Voinnet, O., Thomas, C.L., Maule, A.J., and Baulcombe, D.C. 1999. RNA–DNA interactions and DNA methylation in post-transcriptional gene silencing. *The Plant Cell Online* 11(12):2291–2301.

Jones, L., Ratcliff, F., and Baulcombe, D.C. 2001. RNA-directed transcriptional gene silencing in plants can be inherited independently of the RNA trigger and requires Met1 for maintenance. *Current Biology* 11(10):747–757.

Kaur, S., Cogan, N.O., Pembleton, L.W., Shinozuka, M., Savin, K.W., Materne, M., and Forster, J.W. 2011. Transcriptome sequencing of lentil based on second-generation technology permits large-scale unigene assembly and SSR marker discovery. *BMC Genomics* 12:265.

Kaur, S., Pembleton, L.W., Cogan, N.O., Savin, K.W., Leonforte, T., Paull, J., Materne, M., and Forster, J.W. 2012. Transcriptome sequencing of field pea and faba bean for discovery and validation of SSR genetic markers. *BMC Genomics* 13:104.

Khraiwesh, B., Zhu, J.K., and Zhu, J. 2012. Role of miRNAs and siRNAs in biotic and abiotic stress responses of plants. *Biochimica et Biophysica Acta (BBA)—Gene Regulatory Mechanisms* 1819(2):137–148.

Kloosterman, B., Abelenda, J.A., Gomez, M.C., Oortwijn, M., Boer, J.M., Kowitwanich, Horvath, B.M., van Eck, H.J., Smacmzniak, C., and Prat, S. 2013. Naturally occurring allele diversity allows K potato cultivation in northern latitudes. *Nature* 495(7440):246–250.

Koieng, D., and Weigel, F. 2015. Beyond the thale: Comparative genomics and genetics of *Arabidopsis* relatives. *Nature Reviews Genetics.* 16(5):285–298.

Koornneef, M., Alonso-Blanco, C., and Vreugdenhil, D. 2004. Naturally occurring genetic variation in *Arabidopsis thaliana*. *Annual Review of Plant Biology* 55:141–172. doi: 10.1146/annurev.arplant.55.031903.141605.

Kumar, R., Mustafiz, A., and Sahoo K.K. 2012. Functional screening of cDNA library from a salt tolerant rice genotype Pokkali identifies mannose-1-phosphate guanyltransferase gene (OsMPG1) as a key member of salinity stress response. *Plant Molecular Biology* 79(6):555–568.

Lam, H.M., Xu, X., Liu, X. et al. 2010. Resequencing of 31 wild and cultivated soybean genomes identifies patterns of genetic diversity and selection. *Nature Genetics* 42:1053–1059.

Lamesch, P., Berardini, T.Z., Li, D., Swarbreck, D., Wilks, C., Sasidharan, R., Muller, R., Dreher, K., Alexander, D.L., and Hernandez, G.M. 2012. The *Arabidopsis* information resource (TAIR): Improved gene annotation and new tools. *Nucleic Acids Research* 40:D1202–D1210.

Lee, J.Y., and Lee, D.H. 2003. Use of serial analysis of gene expression technology to reveal changes in gene expression in *Arabidopsis* pollen undergoing cold stress. *Plant Physiology* 132(2):517–529.

Li, F., Fan, G., Wang, K. et al. 2014. Genome sequence of the cultivated cotton *Gossypium arboreum*. *Nature Genetics* 46:567–572.

Li, F., Fan, G., Lu, C. et al. 2015. Genome sequence of cultivated Upland cotton (*Gossypium hirsutum* TM-1) provides insights into genome evolution. *Nature Biotechnology* 33(5):524–530.

Li, F., Fan, G., Wang, K., Sun, F., Yuan, Y., Song, G., Li, Q., Ma, Z., Lu, C., and Zou, C. 2014. Genome sequence of the cultivated cotton *Gossypium arboreum*. *Nature Genetics* 46(6):567–572.

Li, S., Zhao, B., Yuan, D. et al. 2013. Rice zinc finger protein DST enhances grain production through controlling Gn1a/OsCKX2 expression. *Proceedings of National Academy of Sciences United States* 110(8):3167–3172.

Li, T., Liu, B., Spalding, M.H., Weeks, D.P., and Yang, B. 2012. High-efficiency TALEN-based gene editing produces disease-resistant rice. *Nature Biotechnology* 30(5):390–392.

Liang, Z., Zhang, K., Chen, K., and Gao, C. 2014. Targeted mutagenesis in *Zea mays* using TALENs and the CRISPR/Cas system. *Journal of Genetics and Genomics* 41(2):63–68.

Ling, H.-Q., Zhao, S., Liu, D., Wang, J., Sun, H., Zhang, C., Fan, H., Li, D., Dong, L., and Tao, Y. 2013. Draft genome of the wheat A-genome progenitor *Triticum urartu*. *Nature* 496(7443):87–90.

Lister, R., and Ecker, J.R. 2009. Finding the fifth base: Genome wide sequencing of cytosine methylation. *Genome Research* 19:959–966.

Liu, L., Li, Y., Li, S., Hu, N., He, Y., Pong, R., Lin, D., Lu, L., and Law, M. 2012. Comparison of next-generation sequencing systems. *Journal of Biomedicine and Biotechnology* 1–12:11.

Liu, S., Liu, Y., Yang, X., Tong, C., Edwards, D., Parkin, I.A., Zhao, M., Ma, J., Yu, J., and Huang, S. 2014. The *Brassica oleracea* genome reveals the asymmetrical evolution of polyploid genomes. *Nature Communications* 5. doi: 10.1038/ncomms4930.

Lloyd, A., Plaisier, C.L., Carroll, D., and Drews, G.N. 2005. Targeted mutagenesis using zinc-finger nucleases in *Arabidopsis*. *Proceedings of the National Academy of Sciences of the United States of America* 102(6):2232–2237.

Lopes, M.S., Araus, J.L., van Heerden, P.D.R., and Foyer, C.H. 2011. Enhancing drought tolerance in C4 crops. *Experimental Journal of Botany* 62:3135–3153.

Loukehaich, R., Wang, T., and Ouyang, B. 2012. SpUSP, an annexin interacting universal stress protein, enhances drought tolerance in tomato. *Journal of Experimental Botany* 63(15):5593–5606.

Luo, M., Liu, J., Lee, R.D., Scully, B.T., and Guo, B. 2010. Monitoring the expression of maize genes in developing Kernels under drought stress using oligo-microarray. *Journal of Integrative Plant Biology* 52(12):1059–1074.

Lutz, K.A., Wang, W., Zdepski, A., and Michael, T.P. 2011. Isolation and analysis of high quality nuclear DNA with reduced organellar DNA for plant genome sequencing and resequencing. *BMC Biotechnology* 11:54.

Mahfouz, M.M., Li, L., Shamimuzzaman, M., Wibowo, A., Fang, X., and Zhu, J.K. 2011. *De novo*-engineered transcription activator-like effector (TALE) hybrid nuclease with novel DNA binding specificity creates double-strand breaks. *Proceedings of the National Academy of Sciences* 108(6):2623–2628.

Martin, A., Lee, J., Kichey, T. et al. 2006. Two cytosolic glutamine synthetase isoforms of maize are specifically involved in the control of grain production. *Plant Cell* 18:3252–3274.

Marton, I., Zuker, A., Shklarman, E. et al. 2010. Nontransgenic genome modification in plant cells. *Plant Physiology* 154(3):1079–1087.

Matsuba, Y., Nguyen, T.T.H., Wiegert, K., Falara, V., Gonzales, V.E., Leong, B., Schafer, P., Kudrna, D., Wing, R.A., and Bolger, A.M. 2013. Evolution of a complex locus for terpene biosynthesis in *Solanum*. *The Plant Cell* 25(6):2022–2036.

Matsumura, H., Reich, S., Ito, A. et al. 2003. Gene expression analysis of plant host-pathogen interactions by SuperSAGE. *Proceedings of the National Academy of Sciences of the United States of America* 100(26):15718–15723.

Mayer, K.F., Rogers, J., Doležel, J., Pozniak, C., Eversole, K., Feuillet, C., Gill, B., Friebe, B., Lukaszewski, A.J., and Sourdille, P. 2014. A chromosome-based draft sequence of the hexaploid bread wheat (*Triticum aestivum*) genome. *Science* 345(6194):1251788.

McCallum, C.M., Comai, L., Greene, E.A., and Henikof, S. 2000. Targeting induced local lesions IN genomes (TILLING) for plant functional genomics. *Plant Physiology* 123(2):439–442.

Meyerowitz, E.M., and Somerville, C.R. 1994. *Arabidopsis*. Cold Spring Harbor Laboratory Press, Cold Spring Harbor.

Meyers, B.C., Souret, F.F., Lu, C., and Green, P.J. 2006. Sweating the small stuff: MicroRNA discovery in plants. *Current Opinion in Biotechnology* 17:139–146.

Michael, T.P. 2014. Plant genome size variation: Bloating and purging DNA. *Briefings in Functional Genomics* 13(4):308–317.

Miki, D., and Shimamoto, K. 2004. Simple RNAi vectors for stable and transient suppression of gene function in rice. *Plant and Cell Physiology* 45(4):490–495.

Mittler, R., and Shulaev, V. 2013. Functional genomics, challenges and perspectives for the future. *Physiologia Plantarum* 148(3):317–321.

Mochida, K., and Shinozaki, K. 2010. Genomics and bioinformatics resources for crop improvement. *Plant and Cell Physiology* 51(4):497–523.

Morozova, O., Marra, M.A. 2008. Applications of next-generation sequencing technologies in functional genomics. *Genomics* 92:255–264.

Nakano, M., Nobuta, K., Vemaraju, K., Tej S.S., Skogen, J.W., and Meyers, B.C. 2006. Plant MPSS databases: Signature-based transcriptional resources for analyses of mRNA and small RNA. *Nucleic Acids Research* 34:D731–D735.

Nakaminami, K., Matsui, A., Shinozaki, K., and Seki, M. 2012. RNA regulation in plant abiotic stress responses. *Biochimica et Biophysica Acta (BBA) - Gene Regulatory Mechanisms* 1819:149–153.

Negrao, S., Almadanim, M.C., and Pires, I.S. 2013. New allelic variants found in key rice salt-tolerance genes: An association study. *Plant Biotechnology Journal* 11:87–100.

Ni, Z., Hu, Z., Jiang, Q., and Zhang, H. 2013. GmNFYA3, a target gene of miR169, is a positive regulator of plant tolerance to drought stress. *Plant Molecular Biology* 82(1–2):113–129.

Nobuta, K., Venu, R.C., Lu, C. et al. 2007. An expression atlas of rice mRNAs and small RNAs. *Nature Biotechnology* 25:473–477.

Numata, K., Ohtani, M., Yoshizumi, T., Demura, T., and Kodama, Y. 2014. Local gene silencing in plants via synthetic dsRNA and carrier peptide. *Plant Biotechnology Journal* 12(8):1027–1034.

Okabe, Y., Ariizumi, T., and Ezura, H. 2013. Updating the Micro-Tom TILLING platform. *Breeding Science* 63: 42–48.

Okushima, Y. 2005. Functional genomic analysis of the AUXIN RESPONSE FACTOR gene family members in *Arabidopsis thaliana*: Unique and overlapping functions of ARF7 and ARF19. *Plant Cell* 17:444–463.

Osakabe, K., Osakabe, Y., and Toki, S. 2010. Site-directed mutagenesis in *Arabidopsis* using custom-designed zinc finger nucleases. *Proceedings of the National Academy of Sciences* 107(26):12034–12039.

Ostergaard, L., and Yanofsky, M.F. 2004. Establishing gene function by mutagenesis in *Arabidopsis thaliana*. *The Plant Journal* 39(5):682–696.

Pan, L., Shah, A.N., Phelps, I.G., Doherty, D., Johnson, E.A., and Moens, C.B. 2015. Rapid identification and recovery of ENU-induced mutations with next-generation sequencing and paired-end low-error analysis. *BMC Genomics* 16(1):83.

Pandey, P., Senthil-Kumar, M., and Mysore, K.S. 2015. Advances in plant gene silencing methods. *Plant Gene Silencing: Methods and Protocols* 1287:3–23.

Pâques, F., and Duchateau, P. 2007. Meganucleases and DNA double-strand break-induced recombination: Perspectives for gene therapy. *Current Gene Therapy* 7(1):49–66.

Parry, M.A., Madgwick, P.J., Bayon, C., Tearall, K., Hernandez-Lopez, A., Baudo, M., Rakszegi, M., Hamada, W., Al-Yassin, A., and Ouabbou, H. 2009. Mutation discovery for crop improvement. *Journal of Experimental Botany* 60(10):2817–2825.

Pater, S., Pinas, J.E., Hooykaas, P.J., and Zaal, B.J. 2013. ZFN-mediated gene targeting of the *Arabidopsis* protoporphyrinogen oxidase gene through *Agrobacterium*-mediated floral dip transformation. *Plant Biotechnology Journal* 11(4):510–515.

Paterson, A.H., Bowers, J.E., Bruggmann, R. et al. 2009. The Sorghum bicolor genome and the diversification of grasses. *Nature* 457(7229): 551–556.

Paterson, A.H., Wendel, J.F., Gundlach, H. et al. 2012. Repeated polyploidization of *Gossypium* genomes and the evolution of spinnable cotton fibres. *Nature* 492(7429):423–427.

Paux, E., Sourdille, P., Mackay, I., and Feuillet, C. 2012. Sequence-based marker development in wheat: Advances and applications to breeding. *Biotechnology Advances* 30(5):1071–1088.

Petolino, J.F., Worden, A., Curlee, K., Connell, J., Moynahan, T.L.S., Larsen, C., and Russell, S. 2010. Zinc finger nuclease-mediated transgene deletion. *Plant Molecular Biology* 73(6):617–628.

Piferrer, F. 2013. Epigenetics of sex determination and gonadogenesis. *Developmental Dynamics* 242:360–370.

Piron, F., Nicolaï, M., Minoïa, S., Piednoir, E., Moretti, A., Salgues, A., Zamir, D., Caranta, C., and Bendahmane, A. 2010. An induced mutation in tomato eIF4E leads to immunity to two potyviruses. *PloS One* 5(6):11313.

Powell, A.L.T., Nguyen, C.V., Hill, T. et al. 2012. Uniform ripening encodes a Golden 2-like transcription factor regulating tomato fruit chloroplast development. *Science* 336(6089):1711–1715.

Prigge, M.J. 2005. Class III homeodomain-leucine zipper gene family members have overlapping, antagonistic, and distinct roles in *Arabidopsis* development. *Plant Cell* 17:61–76.

Qi, Y., Li, X., Zhang, Y., Starker, C.G., Baltes, N.J., Zhang, F., Sander, J.D., Reyon, D., Joung, J.K., and Voytas, D.F. 2013. Targeted deletion and inversion of tandemly arrayed genes in *Arabidopsis thaliana* using zinc finger nucleases. *G3: Genes, Genomes, and Genetics* 3(10):1707–1715.

Rahman, M., Asif, M., Shaheen, T., Tabbasam, N., Zafar, Y., and Paterson, A.H. 2011. Marker-assisted breeding in higher plants. In: L. Eric (ed.). *Sustainable Agriculture Reviews 6; Alternative Farming Systems, Biotechnology, Drought Stress and Ecological Fertilization*. Springer, Berlin, pp. 39–76.

Rahman, M., Shaheen, T., Tabbasam, N., Iqbal, M.A., Ashraf, M., Zafar, Y., and Paterson, A.H. 2012. Cotton genetic resources. A review. *Agronomy for Sustainable Development* 32:419–432.

Rahman, M., and Paterson, A.H. 2010. Comparative genomics in crop plants. *Molecular Techniques in Crop Improvement* In: S.M. Jains, and D.S. Brar (eds.). London: Springer, pp. 23–60.

Rahman, M., Rahmat, Z., Mahmood, A., Abdullah, K., and Zafar, Y. 2014. Cotton germplasm of Pakistan. In: I. Abdurakhmonov (ed.). *World Cotton Germplasm Resources*, ISBN: 978-953-51-1622-6, InTech. doi:10.5772/58620.

Ranjan, A., Pandey N., Lakhwani, D., Dubey, N.K., Pathre, U.V., and Sawant, S.V. 2012. Comparative transcriptomic analysis of roots of contrasting *Gossypium herbaceum* genotypes revealing adaptation to drought. *BMC Genomics* 13:680.

Reinartz, J., Bruyns, E., Lin, J.Z., Burcham, T., Brenner, S., Bowen, B., Kramer, M., and Woychik, R. 2002. Massively parallel signature sequencing (MPSS) as a tool for in-depth quantitative gene expression profiling in all organisms. *Briefings in Functional Genomic and Proteomic* 1:95–104.

Rensink, W.A., and Buell, C.R. 2005. Microarray expression profiling resources for plant genomics. *Trends in Plant Science* 10(12):603–609.

Riedelsheimer, C., Czedik, E.A, Grieder, C., Lisec, J., Technow, F., Sulpice, R., Altmann, T., Stitt, M., Willmitzer, L., and Melchinger, A.E. 2012. Genomic and metabolic prediction of complex heterotic traits in hybrid maize. *Nature Genetics* 44(2):217–220.

Rigola, D., Oeveren, J.V., Janssen, A. et al. 2009. High-throughput detection of induced mutations and natural variation using KeyPoint™ technology. *PloS one* 4(3). doi : 10.1371/journal.pone.0004761.

Rinaldo, A.R., and Ayliffe, M. 2015. Gene targeting and editing in crop plants: A new era of precision opportunities. *Molecular Breeding* 35(1):1–15.

Robertson, D. 2004. VIGS vectors for gene silencing: Many targets, many tools. *Annual Review of Plant Biology* 55: 495–519.

Ronald, P.C. 2014. Lab to farm: Applying research on plant genet, and genomics to crop improvement. *PLoS Biology* 12(6). doi: 10.1371/journal.pbio.1001878.

Rudd, S. 2003. Expressed sequence tags: Alternative or complement to whole genome sequences. *Trends in Plant Science* 8(7):321–329.

Salse, J., Bolot, S., Throude, M., Jouffe, V., Piegu, B., Quraishi, U.M., Calcagno, T., Cooke, R., Delseny, M., and Feuillet, C. 2008. Identification and characterization of shared duplications between rice and wheat provide new insight into grass genome evolution. *Plant Cell* 20:11–24.

Sanchez, M.V.M., Gowd, M., and Wang, G.L. 2007. Tag-based approaches for deep transcriptome analysis in plants. *Plant Science* 173(4):371–380.

Sander, J.D., and Joung, J.K. 2014. CRISPR-Cas systems for editing, regulating and targeting genomes. *Nature Biotechnology* 32(4):347–355.

Sato, S., Nakamura, Y., Kaneko, T., Asamizu, E., Kato, T., Nakao, M., Sasamoto, S., Watanabe, A., Ono, A., and Kawashima, K. 2008. Genome structure of the legume, *Lotus japonicus*. *DNA Research* 15(4):227–239.

Schadt, E.E., Turner, S., and Kasarskis, A. 2010. A window into third-generation sequencing. *Human molecular genetics* 19(R2):R227–R240.

Schatz, M.C., Maron, L.G., Stein, J.C., Wences, A.H., Gurtowski, J., Biggers, E., Lee, H., Kramer, M., Antoniou, E., and Ghiban, E. 2014. Whole genome *de novo* assemblies of three divergent strains of rice, *Oryza sativa*, document novel gene space of aus and indica. *Genome Biology* 15:506.

Scheffler, B.E., Kuhn, D.N., Motamayor, J.C. et al. 2009. Efforts towards sequencing the Cacao genome (Theobroma cacao). *Plant and Animal Genomes* XVII, San Diego, USA.

Schena, M., Shalon, D., Davis, R.W., and Brown, P.O. 1995. Quantitative monitoring of gene expression patterns with a complementary DNA microarray. *Science* 270:467–470.

Schnable, P.S., Ware, D., Fulton, R.S., Stein, J.C., Wei, F., Pasternak, S., Liang, C., Zhang, J., Fulton, L., and Graves, T.A. 2009. The B73 maize genome: Complexity, diversity, and dynamics. *Science* 326(5956):1112–1115.

Schornack, S., Moscou, M.J., Ward, E.R., and Horvath, D.M. 2013. Engineering plant disease resistance based on TAL effectors. *Annual Review of Phytopathology* 51:383–406.

Senthil-Kumar, M., and Mysore, K.S. 2011. New dimensions for VIGS in plant functional genomics. *Trends in Plant Science* 16(12):656–665.

Senthil-Kumar, M., and Mysore, K.S. 2014. Tobacco rattle virus–based virus-induced gene silencing in *Nicotiana benthamiana*. *Nature Protocols* 9(7):1549–1562.

Sestili, F., Palombieri, S., Botticella, E., Mantovani, P., Bovina, R., and Lafiandra, D. 2015. TILLING mutants of durum wheat result in a high amylose phenotype and provide information on alternative splicing mechanisms. *Plant Science* 233:127–133.

Shan, Q., Wang, Y., Li, J., and Gao, C. 2014. Genome editing in rice and wheat using the CRISPR/Cas system. *Nature Protocols* 9(10):2395–2410.

Shan, Q., Wang, Y., Li, J., Zhang, Y., Chen, K., Liang, Z., Zhang, K., Liu, J., Xi, J.J., and Qiu, J.-L. 2013. Targeted genome modification of crop plants using a CRISPR-Cas system. *Nature Biotechnology* 31(8):686–688.

Shaver, J.M., Oldenburg, D.J., and Bendich, A.J. 2006. Changes in chloroplast DNA during development in tobacco, *Medicago truncatula*, pea, and maize. *Planta* 224:72–82.

Shukla, V.K., Doyon, Y., Miller, J.C., DeKelver, R.C., Moehle, E.A., Worden, S.E., Mitchell, J.C., Arnold, N.L., Gopalan, S., and Meng, X. 2009. Precise genome modification in the crop species *Zea mays* using zinc-finger nucleases. *Nature* 459(7245):437–441.

Simon, S.A., and Meyers, B.C. 2011. Small RNA-mediated epigenetic modifications in plants. *Current Opinion in Plant Biology* 14:148–155.

Singh, A., Pandey, A., Baranwal, V., Kapoor, S., and Pandey, G.K. 2012. Comprehensive expression analysis of rice phospholipase D gene family during abiotic stresses and development. *Plant Signaling and Behavior* 7:847–855.

Skirycz, A., Bodt, D., Obata, S., Clercq, I., Claeys, H., De Rycke, R., Andriankaja, M., Van Aken, O., Van Breusegem, F., and Fernie, A.R. 2010. Developmental stage specificity and the role of mitochondrial metabolism in the response of *Arabidopsis* leaves to prolonged mild osmotic stress. *Plant Physiology* 152:226–244.

Slade, A.J., Fuerstenberg, S.I., Loeffler, D., Steine, M.N., and Facciotti, D. 2004. A reverse genetic, nontransgenic approach to wheat crop improvement by TILLING. *Nature Biotechnology* 23(1):75–81.

Slade, A.J., McGuire, C., Loeffler, D., Mullenberg, J., Skinner, W., Fazio, G., Holm, A., Brandt, K.M., Steine, M.N., and Goodstal, J.F. 2012. Development of high amylose wheat through TILLING. *BMC Plant Biology* 12(1):69.

Song, G.S., Zhai, H.L., Peng, Y.G. et al. 2010. Comparative transcriptional profiling and preliminary study on heterosis mechanism of super-hybrid rice. *Molecular Plant* 3:1012–1025.

Sunkar, R., Li, Y.F., and Jagadeeswaran, G. 2012. Functions of microRNAs in plant stress responses. *Trends in Plant Sciences* 17(4):196–203.

Susek, R.E., Ausubel, F.M., and Chory, J. 1993. Signal transduction mutants of *Arabidopsis* uncouple nuclear *CAB* and *RBCS* gene expression from chloroplast development. *Cell* 74:787–799.

Swaminathan, K., Varala, K., Moose, S.P., Rokhsar, D., Ming, R., and Hudson, M.E. 2009. A genome survey of *Miscanthus Giganteus*. In: *Plant and Animal Genomes* XVII. San Diego, CA.

Tabbasam, N., Zafar, Y., and Rahman, M. 2014. Pros and cons of using genomic SSRs and ESTSSRs for resolving phylogeny of the genus *Gossypium*. *Plant Systematics and Evolution* 3:559–575.

Takagi, H., Abe, A., Yoshida, K. et al. 2013. QTL-seq: Rapid mapping of quantitative trait loci in rice by whole genome resequencing of DNA from two bulked populations. *Plant Journal* 74:174–183.

Tang, Y., Lai, Y., and Liu, Y. 2013. *Virus-induced gene silencing* using artificial miRNAs in Nicotiana benthamiana. In: *Virus-Induced Gene Silencing*. New York:Humana Press, pp. 99–107.

Tang, Y., Wang, F., Zhao, J., Xie, K., Hong, Y., and Liu, Y. 2010. Virus-based microRNA expression for gene functional analysis in plants. *Plant Physiology* 153(2):632–641.

Tang, B., Bowers, J.E., Wang, X. et al. 2008. Synteny and colinearity in plant genomes. *Science* 320(5875):486–488.

Tardieu, F., Reymond, M., Hamard, P., Granier, C., and Muller, B. 2000. Spatial distributions of expansion rate, cell division rate and cell size in maize leaves: A synthesis of the effects of soil water status, evaporative demand and temperature. *Journal of Experimental Botany* 51(530):1505–1514.

The 3,000 Rice Genomes Project: New opportunities and challenges for future rice research. 2014. *Giga Science* 3:7.

The tomato genome consortium. 2012. The tomato genome sequence provides insights into fleshy fruit evolution. *Nature* 485: 635–641.

Tilgner, H., Grubert, F., Sharon, D., and Snyder, M.P. 2014. Defining a personal, allele-specific, and single-molecule long-read transcriptome. *Proceedings of the National Academy of Sciences, United States* 111:9869–9874.

Till, B.J., Reynolds, S.H., Weil, C., Springer, N., Burtner, C., Young, K., Bowers, E., Codomo, C.A., Enns, L.C., and Odden, A.R. 2004. Discovery of induced point mutations in maize genes by TILLING. *BMC Plant Biology* 4:12.

Triques, K., Piednoir, E., Dalmais, M., Schmidt, J. et al. 2008. Mutation detection using ENDO 1:application to disease diagnostics in humans and TILLING and EcoTILLING in plants. *BMC Molecular Biology* 9:42. doi:10.1186/1471-2199-9-42.

Tsai, H., Howell, T., Nitcher, R., Missirian, V., Watson, B., Ngo, K.J., Lieberman, M., Fass, J., Uauy, C., and Tran, R.K. 2011. Discovery of rare mutations in populations: TILLING by sequencing. *Plant Physiology* 156(3):1257–1268.

Utsumi, Y., Tanaka, M., and Morosawa, T. 2012. Transcriptome analysis using a high-density oligo microarray under drought stress in various genotypes of cassava: An important tropical crop. *DNA Research* 19:335–345.

Vanitharani, R., Chellappan, P., and Fauquet, C.M. 2005. Geminiviruses and RNA silencing. *Trends in Plant Sciences* 10:144–151.

Varshney, R.K., Ribaut, J.M., Buckler E.S., Tuberosa, R., Rafalski, J.A., and Langridge, P. 2012. Can genomics boost productivity of orphan crops? *Nature Biotechnology* 30(12):1172–1176.

Varshney, R.K., Song, C., Saxena, R.K., Azam, S., Yu, S., Sharpe, A.G., Cannon, S., Baek, J., Rosen, B. D., and Tar'an, B. 2013. Draft genome sequence of chickpea (*Cicer arietinum*) provides a resource for trait improvement. *Nature Biotechnology* 31(3):240–246.

Verelst, W., Bertolini, E., De Bodt, S. et al. 2013. Molecular and physiological analysis of growth limiting drought stress in *Brachypodium distachyon* leaves. *Molecular Plant* 6:311–322.

Velasco, R., Zharkikh, A., Affourtit, J., Dhingra, A., Cestaro, A., Kalyanaraman, A., and Chu, V.T. 2010. The genome of the domesticated apple (*Malus* [times] *domestica* Borkh.). *Nature Genetics* 42(10):833–839.

Velasco, R., Zharkikh, A., Troggio, M. et al. 2007. A high quality draft consensus sequence of the genome of a heterozygous grapevine variety. *PLoS One* 2:1326.

Wang, K., Wang, Z., Li, F. et al. 2012. The draft genome of a diploid cotton *Gossypium raimondii*. *Nature Genetics* 44:1098–1103.

Wang, X., Wang, H., Wang, J. et al. 2011.The genome of the mesopolyploid crop species *Brassica rapa*. *Nature Genetics* 43:1035–1039.

Wang, K., Wang, Z., Li, F., Ye, W., Wang, J., Song, G., Yue, Z., Cong, L., Shang, H., and Zhu, S. 2012. The draft genome of a diploid cotton *Gossypium raimondii*. *Nature Genetics* 44(10):1098–1103.

Wang, X., Wang, H., Wang, J., Sun, R., Wu, J., Liu, S., Bai, Y., Mun, J.-H., Bancroft, I., and Cheng, F. 2011. The genome of the mesopolyploid crop species *Brassica rapa*. *Nature Genetics* 43(10):1035–1039.

Wang, Y., Cheng, X., Shan, Q., Zhang, Y., Liu, J., Gao, C., and Qiu, J.L. 2014. Simultaneous editing of three homoeoalleles in hexaploid bread wheat confers heritable resistance to powdery mildew. *Nature Biotechnology* 32(9):947–951.

Waterhouse, P.M., Wang, M.B., and Lough, T. 2001. Gene silencing as an adaptive defence against viruses. *Nature* 4116839:834–842.

Watson, J.D., and Crick, F.H.C. 1953. Molecular structure of nucleic acids. *Nature* 171:737–738.

Wei, L.Q., Yan, L.F., and Wang, T. 2011. Deep sequencing on genome-wide scale reveals the unique composition and expression patterns of microRNAs in developing pollen of *Oryza sativa*. *Genome Biology* 12(6):53.

Weinthal, D., Tovkach, A., Zeevi, V., and Tzfira, T. 2010. Genome editing in plant cells by zinc finger nucleases. *Trends in Plant Sciences* 15(6): 308–321.

Wendt, T., Holm, P.B., Starker, C.G., Christian, M., Voytas, D.F., Brinch-Pedersen, H., and Holme, I.B. 2013. TAL effector nucleases induce mutations at a pre-selected location in the genome of primary barley transformants. *Plant Molecular Biology* 83(3):279–285.

Xia, Z., Watanabe, S., Yamada, T. et al. 2012. Positional cloning and characterization reveal the molecular basis for soybean maturity locus E1 that regulates photoperiodic flowering. *Proceedings of the National Academy of Sciences, United States* 109:2155–2164.

Xie, Z., Lisa, K., Johansen, L.K. et al. 2004. Genetic and functional diversification of small RNA pathways in plants. *PLoS Biology* 2:104.

Xu, Q., Chen L.L, Ruan, X., Chen, D., Zhu, A., Chen, C., Bertrand, D., Jiao, W.B., Hao, B.H., and Lyon, M.P. 2013. The draft genome of sweet orange (*Citrus sinensis*). *Nature Genetics* 45:59–66.

Xu, X., Pan, S., Cheng, S., Zhang, B., Mu, D. et al. 2011. Genome sequence and analysis of the tuber crop potato. *Nature* 475:189–195.

Yamamoto, K., and Sasaki, T. 1997. Large-scale EST sequencing in rice. *Plant Molecular Biology* 35:135–144.

Yang, C., Li, D., Mao, D., Liu, X., Ji, C., Li, X., Zhao, X., Cheng, Z., Chen, C., and Zhu, L. 2013. Overexpression of microRNA319 impacts leaf morphogenesis and leads to enhanced cold tolerance in rice (*Oryza sativa* L.). *Plant Cell and Environment* 36(12):2207–2218.

Yang, M., Djukanovic, V., Stagg, J., Lenderts, B., Bidney, D., Falco, S.C., and Lyznik, L.A. 2009. Targeted mutagenesis in the progeny of maize transgenic plants. *Plant Molecular Biology* 70(6):669–679.

Yazaki, J., Gregory, B.D, and Ecker, J.R. 2007. Mapping the genome landscape using tiling array technology. *Current Opinion in Plant Biology* 10 (5):534–542.

Young, N.D., Debellé, F., Oldroyd, G.E., Geurts, R., Cannon, S.B., Udvardi, M.K., Benedito, V.A., Mayer, K.F., Gouzy, J., and Schoof, H. 2011. The *Medicago* genome provides insight into the evolution of rhizobial symbioses. *Nature* 480(7378):520–524.

Yu, J., Hu, S., Wang, J., Wong, G.K.S., Li, S., Liu, B., Deng, Y., Dai, L., Zhou, Y., and Zhang, X. 2002. A draft sequence of the rice genome (*Oryza sativa* L. ssp. indica). *Science* 296(5565):79–92.

Yu, S., Liao, F., and Wang, F. 2012. Identification of rice transcription factors associated with drought tolerance using the Ecotilling method. *PLoS One* 7(2): Article ID e30765.

Zhang, X., Zou, Z., Gong, P., Zhang, J., Ziaf, K., Li, H., Xiao, F., and Ye, Z. 2011. Over-expression of microRNA169 confers enhanced drought tolerance to tomato. *Biotechnology Letters* 33(2):403–409.

Zhang, F., Maeder, M.L., Unger-Wallace, E., Hoshaw, J.P., Reyon, D., Christian, M., Li, X., Pierick, C.J., Dobbs, D., and Peterson, T. 2010. High frequency targeted mutagenesis in *Arabidopsis thaliana* using zinc finger nucleases. *Proceedings of the National Academy of Sciences* 107(26):12028–12033.

Zhang, T., Hu, Y., Jiang, W., Fang, L., Guan, X. et al. 2015. Sequencing of allotetraploid cotton (*Gossypium hirsutum* L. acc. TM-1) provides a resource for fiber improvement. *Nature Biotechnology* 33:531–537.

Zhang, Y., Zhang, F., Li, X., Baller, J.A., Qi, Y., Starker, C.G., Bogdanove, A.J., and Voytas, D.F. 2013. Transcription activator-like effector nucleases enable efficient plant genome engineering. *Plant Physiology* 161(1):20–27.

Zhou, L., Liu, Y., Liu, Z., Kong, D., Duan, M., and Luo, L. 2010. Genome-wide identification and analysis of drought-responsive microRNAs in *Oryza sativa*. *Journal of Experimental Botany* 61(15):4157–4168.

Zhou, M.L., Zhang, Q., Zhou, M. et al. 2012. Aldehyde dehydrogenase protein superfamily in maize. *Functional and Integrative Genomics* 12:683–691.

Zimin, A., Stevens, K.A., Crepeau, M.W. et al. 2014. Sequencing and assembly of the 22-Gb loblolly pine genome. *Genetics* 196:875–890.

chapter seven

Whole-genome resequencing
Current status and future prospects in genomics-assisted crop improvement

**Uday Chand Jha, Debmalya Barh, Swarup K. Parida,
Rintu Jha, and Narendra Pratap Singh**

Contents

Abstract

Tremendous progress in high-throughput sequencing technology has culminated in completion of reference genome sequences of both model and nonmodel plant species dealing with agriculture. Notably, current state-of-the-art of sequencing coupled with reduction of cost of nucleotide sequencing has given a great opportunity to introduce whole-genome resequencing (WGRS) to be performed both in cultivated and wild accession of crop, thereby uncovering the variations at sequence level, those missing in the available reference genome sequence. In this chapter, we discuss the various downstream applications of WGRS providing valuable insights ranging from crop origin, evolution, domestication, and to genetic diversity of crop plants, coupled with identification of novel genes/quantitative trait loci, thus opening up new avenues for the crop improvement program.

Keywords: Whole-genome resequencing, Next-generation sequencing, Genome, Quantitative trait loci.

Introduction

The paradigm shift in next-generation sequencing (NGS)-driven sequencing technology has resulted in remarkable progress in plant genome sequencing, leading to decode more than two dozens of plant genome sequences till date (Deschamps and Campbell 2010, Jackson et al. 2011, Edwards et al. 2012, Michael and Jackson 2013). Importantly, this technology has benefited plant scientists thereby deciphering the structural genome variations either by enabling in *de novo* sequencing or by resequencing of plant genome species with the available "reference genome" sequence (Bhat 2011, Jackson et al. 2011). Subsequently, dramatic evolvement of "second-generation sequencing technologies," the two most commercially used NGS platforms Illumina Genome Analyzer (Illumina-GA) and Roche Genome Sequencer FLX (GS-FLX) have gained much popularity, propelling genome-sequencing research at a greater height in biological science in post-Sanger sequencing era (Mardis 2008, Shendure and Ji 2008, Imelfort and Edwards 2009, Metzke 2010). Interestingly, dropping down in sequencing cost followed by capability in generating longer and high-quality reads, coupled with advancement in short-read assembler tools and algorithms by introducing third-generation sequencing platforms, can further increase the quality of genome assembly at a higher magnitude (Imelfort and Edwards 2009, Schatz et al. 2010, 2012, Thudi et al. 2012). Till date, genome sequences of more than 50 plants belonging to different species have been cracked (Michael and Jackson 2013), thus allowing identification and characterization of genes of interest mostly used in genomics-assisted crop improvement program. Despite the availability of the complete "reference genome" sequence of a specific crop genotype, this is not sufficient to give the complete landscape of all the complex traits and existing structural genome variations causing specific phenotypic variants in crop plants (Kim et al. 2010, Schatz et al. 2014). Evidences of existing genomic variations both between and within the cultivars of crops have been decoded (Springer et al. 2009, Lai et al. 2010, Swanson-Wagner et al. 2010, Haun et al. 2011, Zheng et al. 2011, Lin et al. 2012, McHale et al. 2012). Therefore, understanding of complex traits, genomic diversity, much in-depth study of geographic origin of crop plants in connection with its domestication-related processes and speciation, resequencing of wild species, or crop accessions is necessary (Kim et al. 2010, Sakai et al. 2014).

Why resequencing of genome?

Since the last decade, rapid advancement of NGS techniques has allowed completion of the reference genome sequencing of crop plants having much economic importance (Beavan and Uauy 2013, Michael and Jackson 2013). To this end, whole-genome resequencing (WGRS) has been introduced for offering an in-depth analysis of structure variations, diversity analysis at the genomic level, and comparative genomics analysis in crop plants (Huang et al. 2010, Imelfort and Edwards 2009, Zheng et al. 2011, Morrell et al. 2012, Saxena et al. 2014, Sakai et al. 2014). Furthermore, comparison of sequence reads derived from resequencing of diverse crop accessions with their "reference genome" sequence may provide evolution and domestication-oriented valuable insights, coupled with geographical origin and distribution of crop (Huang et al. 2012, 2013a). Moreover, it can shed light on genomic regions bearing the footprint of human-mediated selection or "signature of selection" occurred during domestication in crop plant (Huang et al. 2012, Li et al. 2013, Mace et al. 2013, Zhou et al. 2014). Importantly, resequencing provides the unprecedented opportunity for investigation of global polymorphism, population structure, and ultimately, bridging the gap of mapping all genetic variants to each corresponding causative phenotypic variations in crop plants (Lister et al. 2009, Hawkins et al. 2010). Equally important, rese-quencing can enable in the detection of gene copy number as well as capturing genetic variation in "homoeologous gene copies" (Beaven and Uauy 2013). Additionally, the WGRS also facilitated in the identification of high-throughput informative single-nucleotide poly-morphism (SNP) markers, insertion–deletions (InDels), copy number variations (CNVs) (Huang et al. 2010, Kim et al. 2010, Zheng et al. 2011, Żmieńko et al. 2013, Guo et al. 2014, Saxena et al. 2014), and gene gain and loss events (Chung et al. 2013, Sakai et al. 2014), coupled with presence/absence variations (PAVs) (Zheng et al. 2011, Saxena et al. 2014) across the genome. Likewise, WGRS can be employed for "Pan genome" analysis (signify-ing "core and dispensable genome," Tettelin et al. 2005) (Hirsch et al. 2014, Saxena et al. 2014, Schatz et al. 2014). WGRS can be applied to detect the "recombination breakpoints" in F_2 or recombinant inbred line (RIL) mapping populations (Huang et al. 2009) and for scan-ning the existing pattern of linkage disequilibrium (LD) throughout the genome (Huang et al. 2010, Chung et al. 2013). Ultimately, the aim of WGRS is to discover "none reference" and/or "cultivar specific sequence" leading to identification of novel genes/quantitative trait loci (QTLs) (Chung et al. 2013, Guo et al. 2014, Sakai et al. 2014) of agronomic impor-tance for crop improvement.

Resolving evolution and domestication trajectories in crop plant

The history of crop evolution, origin, and domestication-oriented events has always remained a debated issue since a long time and are deeply concerned with the modern crop-breeding program in the postdomestication era (Doebley et al. 2006, Molina et al. 2011, Meyer and Purugganan 2013). Thanks to NGS, myriads of population genomic data have been employed for disclosing crop domestication and breeding study (Shi and Lai 2015). Taking note of this, resequencing of different crop strain and wild species can unveil novel insights into understanding the genomic region(s) that underwent human-mediated selection, together with genomic variations during crop domestication coupled with origin of genome structure, phylogenetic study, and demographic history (Lam et al. 2010, Molina et al. 2011, Jiao et al. 2012, Morrell et al. 2012, Li et al. 2013, Kim et al. 2014a, Sakai et al. 2014, Schmutz et al. 2014, Zhou et al. 2014). Resequencing of 1083 accessions belonging to *Oryza sativa*, 446 accessions of *O. rufipogon,* and 15 accessions out of group species provided valu-able insight regarding the origin and domestication process of cultivated rice (Huang et al. 2012). From this study, the authors also advocated an early domestication of cultivated *O. sativa japonica* after their derivation from its wild relative *O. rufipogon* population and subsequently, the domestication of *O. sativa indica* took place. Likewise, resequencing of 66 accessions covering *O. sativa indica, O. sativa japonica,* and *O. rufipogon* species gave a valuable clue regarding the domestication process of rice (He et al. 2011). Additionally, recent *de novo* sequencing of *O. brachyantha* shed important light on evolution of *Oryza* genus (Chen et al. 2013). Subsequently, comparative genomics study suggested only 70% of the coding genes of *O. brachyantha* are in syntenic with the rice genome. While, SNPs derived from resequencing of "Tongil" (IR667-98-1-2), a leading Korean rice cultivar at 47× genome coverage and its parental lines *indica* and *japonica* types at 30× genome coverage, gave an important clue regarding the genome composition of Tongil cultivar (Kim et al. 2014b). The result also suggested that the Tongil harbors more of *indica* genome rather than *japonica* indicating its origin from *indica* type. While in case of maize, resequencing of 75 accessions representing wild, landraces, and improved cultivars offered a valuable clue to domestication and evolution of modern cultivated maize (Hufford et al. 2012). Considering soybean, sequencing of *Glycine soja* var. IT182932 produced similarity of 97.65% sequence with the reference *G. max* genome sequence. However, *G. soja* contained 2.5 Mb substituted bases and 406 kb of small InDels, which were different from *G. max* (Kim et al. 2010). Novel insight of introgression of the total 43.1-Mb genomic regions in C01, C12, and C19 modern cultivars from the wild type of soybean have been reported, indicating the occurrence of a human-driven artificial selection (Lam et al. 2010). Further, this investigation also gave an important clue regarding the divergence of *G. soja/G. max* complex and *G. soja* from a common ancestor. Similarly, to advocate no independent domestication event of the semi-wild soybean, 10 semiwild soybeans including "Maliaodou" and one wild line "Lanxi 1" were sequenced (Qiu et al. 2014). The result supported the fact that the semiwild soybean rose due to hybridization events of a domesticated and wild soybean. Further, the authors uncovered seed size-related selective regions by selective sweep analysis. Importantly, to unfold the evolutionary history of cultivated soybean, 16 accessions including 10 cultivated and six wild species of soybean genome sequenced at >17x coverage. This study gave a significant impetus for genomic regions associated with artificial selection during domestication in soybean (Chung et al. 2013). Interestingly, resequencing of 302 accessions of soybean enabled the identification of 121 "domestication-selective sweeps" and 109 "improvement-selective sweeps" (Zhou et al. 2014). Moreover, evidences of duplicated genes or genomic regions involved in evolution of soybean have been discussed (Kim et al. 2010, Lam et al. 2010). Furthermore, applications of deoxyribonucleic acid (DNA) sequencing unveil novel insights related to evolution and domestication of legume crops has been critically discussed by O'Rourke et al. (2014). Considering common bean, recent WGRS of 160 accessions including wild species and landraces witnessed the occurrence of two independent domestication events (Schmutz et al. 2014). In the recent past, sequencing of *Gossypium hirsutum* TM-1 accession (Li et al. 2015, Zhang et al. 2015) offered important insights of a cotton genome evolution while comparing to the available genome sequence of *G. arboretum* (Li et al. 2014a) and *G. raimondii* (Wang et al. 2012). Therefore, WGRS technique has a tremendous potential to address the evolutionary and domestication-related queries of crop plants, thus offering a scope for improving/accelerating the crop-breeding program during the current scenario of climate change.

Unfolding the hidden genetic diversity at genome level in crop plant

Genetic diversity is the mainstay of any crop improvement program. The multiple causative factors for changes in the genome are evolution, genetic recombination, mutation, and selection, giving rise to genetic diversity in crops (Cao et al. 2011). Indeed, the prospects of whole-genome shotgun-sequencing approach for discovering the underlying diversity in crop plants has been well mentioned (Jackson et al. 2011). Likewise, WGRS can be potentially used to unlock this hidden diversity at the sequence level, both in cultivated and wild species of crops, thus giving ample scope for exploiting them in the crop improvement program. To unfold the genotypic diversity in *Arabidopsis* across the globe, 80 strains representing eight different regions were sequenced (Cao et al. 2011). The authors obtained 4,902,039 SNPs and 810,467 InDels across the given 80 strains, offering the opportunity to study the global pattern of polymorphism and causative variants for adaptation across a wide range of environment in *Arabidopsis* (Table 7.1). In maize, first "HapMap" derived from genotyping 27 inbred lines, witnessed numerous selective sweeps harboring loci for geographic adaptation (Gore et al. 2009). On the basis of a resequencing study of 29 accessions of chickpea including both *kabuli* and *desi* type, Varshney et al. (2013) found evidences of a higher genetic diversity in the *desi* type than in *kabuli* type. The result also advocated the occurrence of a recent domestication bottleneck event in the *kabuli*-type chickpea. While

Table 7.1 Applications of whole-genome resequencing in crop improvement

Crop	Genotype	Sequencing depth/ genome coverage	Number of reads generated	NGS platform used	SNPs/InDels/ CNVs/PAVs identified	Locus/candidate gene/QTL	References
Arabidopsis	Col-0, Bur-0, and Tsu-1	15× to 25×	–	Illumina sequencing by synthesis (SBS)	823325 SNPs, 79,961 InDels	–	Ossowski et al. (2008)
Arabidopsis	Five individuals developed from "Col-0" strain	23× to 31×	–	Illumina Genome Analyzer platform	99 base substitutions 17 small and large InDels	–	Ossowski et al. (2010)
Arabidopsis	Wassilewskija	–	–	–	–	*AtNFXL-2* causing ebi-1 phenotype	Ashelford et al. (2011)
Arabidopsis	80 accessions	–	–	Illumina Genome Analyzer platform	4,902,039 SNPs 810467 InDels	Introduced altered codon, premature codons and extended reading frame in 6197 genes	Cao et al. (2011)
Arabidopsis	Landsberg erecta (Ler) meotic product of Columbia(Col)/Ler hybrid	~18.73	61.6 million reads	Illumina sequencing	349171 SNPs, 58,085 small and 2315 large (InDels)	443 genes showing nonsynonymous substitutions in the coding region and 316 genes affected by large InDels	Lu et al. (2012)
Barley	MHOR474 and Barke	–	82 million and 70 million 2×100 bp	Illumina HiSeq2000	–	Many-noded dwarf (mnd) locus	Mascher et al. (2014)
Brassica rapa	L144 and Z16	–	–	–	108558, 26,795 and 26,693 InDels	–	Liu et al. (2013)

(Continued)

Table 7.1 (Continued) Applications of whole-genome resequencing in crop improvement

Crop	Genotype	Sequencing depth/ genome coverage	Number of reads generated	NGS platform used	SNPs/InDels/ CNVs/PAVs identified	Locus/candidate gene/QTL	References
B. oleracea	C1184 and C1234	17.8× and 17.7×	114 million and 113 million reads		1.20 million and 1.24 million SNPs	One major and three minor QTL for black rot resistance	Lee et al. (2015)
B. napus	Ten elite *B. napus* accessions	5.3× to 37.5×	1600-million paired-end reads of 75 bp	Illumina Golden Gate Assay	900000 SNPs	Identified SNPs will help in genotype screening	Huang et al. (2013)
B. rapa	150 RIL of Bre and Wut cross	0.2×	141 million and 107 million	Illumina Hiseq-2000	>1 million SNPs	18 QTLs for six head traits	Yu et al. (2013)
Commonbean	60 wild and 100 land races	~4×		Illumina sequencing	10,158,326 SNPs from wild sp., 9,661,807 and 3,154,648 SNPs from landraces	Set of genes associated with leaf size and seed size	Schmutz et al. (2014)
Cucumber	115 lines	18.3×	7.275-billion paired-end reads	Illumina-GA IIx	3,305,010 SNPs 336081 InDels, 594 PAVs	b-Carotene hydroxylase gene Five fruit-length QTLs and three leaf-size QTLs	Qi et al. (2013)
Chickpea	90 accessions	9.5×	204.52 Gb + 34.77 Gb	Illumina HiSeq 2000	4.4 million variants (SNPs and InDels)	Candidate gene for disease resistance and other agronomic traits	Varshney et al. (2013)

(Continued)

Table 7.1 (Continued) Applications of whole-genome resequencing in crop improvement

Crop	Genotype	Sequencing depth/genome coverage	Number of reads generated	NGS platform used	SNPs/InDels/CNVs/PAVs identified	Locus/candidate gene/QTL	References
Chickpea	140 cultivars	5X–12X	–	Illumina HiSeq 2000 platform	>1.3 million SNPs	–	Chitikineni et al. (2015)
Maize	27 inbreds	38%	1 billion SBS reads	Sequencing by synthesis	3.3 million SNPs	Genomic regions associated with domestication	Gore et al. (2009)
Maize	Six inbreds	–	–		1,000,000 SNPs 30,000 InDels	–	Lai et al. (2010)
Maize	75 wild, landraces and cultivated	>5×			21141953 SNPs	Domestication candidate genes	Hufford et al. (2012)
Maize	103 lines	~4.2× and ~8×	13 billion reads	Illumina-GA IIx	55 million SNPs	–	Chia et al. (2012)
Maize	15 inbreds	–	4.6 billion	Illumina Hiseq 2000 platform	46556 SNPs	Drought-tolerance candidate gene	Xu et al. (2014)
Maize	–	1×	8,955,215 to 25,196,240 reads	Illumina HiSeq 2000	–	Maize gravitropism gene *lazy plant1*	Howard et al. (2014)
Maize	Chang 7-2 and 787	0.04×	551,114,523		1155158 SNPs	10 QTLs associated with tassel and ear architecture	Chen et al. (2014)
Medicago truncatula	"R108"	11×	4.28 Gb	Illumina Hiseq 2000 sequencer	764154 SNPs, 135,045 InDels	Genomic region accounting for aluminum, sodium toxicity, and iron deficiency response	Wang et al. (2014)

(Continued)

Table 7.1 (Continued) Applications of whole-genome resequencing in crop improvement

Crop	Genotype	Sequencing depth/genome coverage	Number of reads generated	NGS platform used	SNPs/InDels/CNVs/PAVs identified	Locus/candidate gene/QTL	References
Medicago truncatula	26 accessions	~15×	82-million 90-base paired-end reads	GAIIx Illumina sequencing	3,063,923 SNPs	Three gene families associated with resistance against pathogen	Branca et al. (2011)
Rice	150 RILs of Nipponbare/93-11	0.02×		Illumina GA	1226791 SNPs	QTL having a large effect on plant height and "green revolution" gene	Huang et al. (2009)
Rice	128 CSSLs derived from 9311× Nipponbare	0.13×		Illumina-GA IIx	7.68 million	Nine QTLs for culm length and large-effect QTL harboring for "green revolution gene"	Xu et al. (2010)
Rice	59 accessions including cultivated and wild species	–	–	–	2,800 SNPs from indica 2,070 SNPs from *japonica*, 7,274 SNPs from *O. rufipogon*	20 selective sweeps detected in cultivated rice	Molina et al. (2011)
Rice	40 Cultivated rice, 10 wild (*Oryza rufipogon* and *Oryza nivara*)	15×	–	–	6.5 million SNPs	–	Xu et al. (2012)

(Continued)

Table 7.1 (Continued) Applications of whole-genome resequencing in crop improvement

Crop	Genotype	Sequencing depth/genome coverage	Number of reads generated	NGS platform used	SNPs/InDels/CNVs/PAVs identified	Locus/candidate gene/QTL	References
Rice	1083 cultivated and 446 wild rice	>1 ~ 50×		Illumina-GA IIx or HiSeq2000	7970359 SNPs	15 domestication-related traits covered by 58 QTLs	Huang et al. (2012a)
Rice	950 cultivated rice	>1×			4109366 SNPs	32 loci associated with flowering time and 10 grain-related traits	Huang et al. (2012b)
Rice	517 rice landraces	>1×	2.7-billion 73-bp paired-end reads	Illumina-GA II	3625200 SNPs	14 agronomic traits	Huang et al. (2010)
Rice	5 cultivated rice	>58×			1154063 SNPs		Jeong et al. (2013)
Rice	Koshihikari	15.7×			67051 SNPs		Yamamoto et al. (2010)
Rice	A restorer line 7302R and 4 cultivated lines	>13×			307627 SNPs		Li et al. (2012)
Rice	Tongil	47×	17.3 billion reads	Illumina Hiseq	2149991 SNPs	Yield-related gene analysis	Kim et al. (2014a)
Rice	Yukara	34×	12.4 billion	Illumina Hiseq	—		
Rice	IR8	32×	11.7 billion	Illumina Hiseq	—		
Rice	TN1	30×	11.2 billion	Illumina Hiseq	—		

(Continued)

Table 7.1 (Continued) Applications of whole-genome resequencing in crop improvement

Crop	Genotype	Sequencing depth/genome coverage	Number of reads generated	NGS platform used	SNPs/InDels/CNVs/PAVs identified	Locus/candidate gene/QTL	References
Rice	aus rice Kasalath	>6× and >148×	1.73 Gb 57.47 Gb	Roche 454 Illumina GAIIx and HiSeq2000	2 787 250 and 2 216 251 SNPs, 7 393 and 3780 InDels	–	Sakai et al. (2014)
Rice	Hit1917-pl1 and Hit0813-pl2	>12×	70 million and 133 million sequence reads	Illumina GAIIx sequencer	100819 SNPs	Mutations causing pale green leaves semidwarfism trait	Abe et al. (2012)
Rice	RILs of Liang–You–Pei–Jiu and parents	~4× for RILs, ~48× and ~36× for parents	–	Illumina HiSeq 2000	171847 SNPs and 280,055 InDels	43 yield-associated QTLs	Gao et al. (2013)
Rice	F2 population derived from R1128× Nipponbare	>16×	6.17 G	Illumina Hiseq 2000 platform	74329 SNPs	Six novel QTLs	Duan et al. (2013)
Rice	Hitomebore	27.5×	12.25 Gb sequence read	Illumina GAIIx sequencing	124 968 SNPs	Blast-resistant gene *Pii*	Takagi et al. (2013a)
Rice	Hitomebore and Nortai	>6×	57.9 and 62.4 million sequence reads	Illumina GAIIx sequencer	161 563 SNPs	QTLs contributing in partial resistance for rice blast	Takagi et al. (2013b)
Rice	3000 accessions	14×	–	HiSeq2000 platform	18.9 million	–	Li et al. (2014c)

(Continued)

Table 7.1 (Continued) Applications of whole-genome resequencing in crop improvement

Crop	Genotype	Sequencing depth/ genome coverage	Number of reads generated	NGS platform used	SNPs/InDels/ CNVs/PAVs identified	Locus/candidate gene/QTL	References
Rice	Hitomebore	–	7.34 Gb of 75 bp reads	Illumina GAIIx sequencer	1005 SNPs	*hst1*	Takagi et al. (2015)
Rice	PB1 and Dular	24×	–	Illumina HiSeq 2000 NGS	2442979 SNPs and InDels	Phosphate Starvation Responsive (PSR) gene and genes associated with root	Mehra et al. (2015)
Setaria italica	A2	~10×	~5 G raw reads	Illumina GA II	542,322 SNPs 33587 InDels	Sethoxydim (herbicide) resistance	Zhang et al. (2012)
Setaria italica	Shi-Li-Xiang	~11×	74,792,844	Illumina GAIIx sequencer	762,082 SNPs 26,802 InDels	Waxy locus	Bai et al. (2013)
Setaria italica	916 accessions	0.7×	~0.3 Tb of raw reads	Illumina-GA IIx Illumina HiSeq2000	2.58 million SNPs	47 agronomic traits	Jia et al. (2013)
Solanum lycopersicum	8 lines	6.7× to 16.6×	970 million reads	Illumina-GA IIx	>4 million SNPs 128,000 InDels 1,686 putative CNVs	Candidate polymorphism causing phenotypic variation	Causse et al. (2013)
Solanum lycopersicum	84 accessions	–	–	–	>10 million SNPs	–	Aflitos et al. (2014)
Sorghum	5 *S. bicolor* genotypes	6× to 12×	–	–	1 million SNPs	–	Bekele et al. (2013)
Sorghum	654 and LTR108	~25- and 28-fold		Illumina-GA IIx	1 243 151 SNPs	QTLs for eight agronomic traits	Zhou et al. (2012)
Sorghum	44 sorghum accessions	16× to 45×	7.97-billion 90-bp paired-end reads	HiSeq 2,000 platform	8 million SNPs and 1.9 million InDels	725 candidate genes for domestication/ improvement	Mace et al. (2013)

(Continued)

Table 7.1 (Continued) Applications of whole-genome resequencing in crop improvement

Crop	Genotype	Sequencing depth/ genome coverage	Number of reads generated	NGS platform used	SNPs/InDels/ CNVs/PAVs identified	Locus/candidate gene/QTL	References
Sorghum	3 lines	×36.51	620.72-million 44-bp paired-end reads	Illumina Genome Analyzer	1,057,018 99,948 InDels, 17,111 CNVs	1442 genes differentiating sweet and grain sorghum	Zheng et al. (2011)
Soybean	*G. soja* var. IT182932	57% genome coverage	10.9-million GS-FLX reads	Illumina-GA Roche GS FLX	2.5 million SNPs	–	Kim et al. (2010)
Soybean	31 wild and cultivated accessions	×5 depth and >90% coverage	180 Gb of sequence	Illumina-GA II	6,318,109 SNPs 186,177 PAVs	–	Lam et al. (2010)
Soybean	15 accessions including cultivated and wild	>17×	305 Gb of 101-bp paired-end short reads	Illumina HiSeq2000 platform	3.87 million SNPs	Candidate genomic region for domestication	Chung et al. (2013)
Soybean	25 new accessions and 30*		93.55 Gb	Illumina base-calling pipeline	5102244 SNPs	Genomic region contributing in artificial selection	Li et al. (2013)
Soybean	Maliaodou and Lanxi 1	~41×	–	Illumina Hiseq2000	–	144 loci contributing in genetic improvement	Qiu et al. (2014)
Soybean	W05	~1×	73.5 Gb reads	IlluminaGAII platform	1798504 SNPs	GmCHX1, 15 major QTLs	Qi et al. (2014)

(Continued)

Table 7.1 (Continued) Applications of whole-genome resequencing in crop improvement

Crop	Genotype	Sequencing depth/ genome coverage	Number of reads generated	NGS platform used	SNPs/InDels/ CNVs/PAVs identified	Locus/candidate gene/QTL	References
Soybean	302 accessions	>11×	33-billion paired-end reads	Illumina HiSeq 2000	9790744 SNPs, 876,799 InDels 1614 CNVs	Nine domestication traits, 13 loci associated with oil content, plant height, and pubescence	Zhou et al. (2014)
Vigna angularis	49 accessions	5.3× to 27.34×	132.28 Gb sequence	Illumina HiSeq 2000	5,539,411 SNPs	Starch and fatty acid metabolism genes	Yang et al. (2015)
Watermelon	20 accessions including cultivated and wild	5× and 16×	–	Illumina sequencing	6784860 SNPs 965006 InDels	741 candidate genes involved in carbohydrate and nitrogen compound metabolism	Guo et al. (2013)

Note: CNVs, copy number variations; PAVs, presence/absence variations.

in the context of sorghum, resequencing results of its 44 accessions, suggested that wild sorghum is more diverse than improved cultivars, and also provided evidence for the event of a genetic bottleneck during domestication (Mace et al. 2013). Analysis of sequence variants of both wild *Setaria viridis* and improved cultivars of foxtail millet with reference to the genome sequence, supported the existence of both interspecies and intraspecies variants (Jia et al. 2013). Considering soybean, Kim et al. (2010) depicted the presence of 0.3% genomic difference between the *G. soja* var. IT182932 and *G. max* var. William 82 at SNP level. Additionally, 3.8 million SNPs were recovered by resequencing 16 accessions of soybean including wild species and cultivated species (Chung et al. 2013). This study revealed that wild soybean species harbor more diversity than the cultivated species; this has been supported by comparing these SNPs obtained in cultivated and wild accessions. This result was also supported by Lam et al. (2010) suggesting evidences of genetic bottleneck in soybean owing to domestication. Likewise, WGRS has been applied for unveiling the pattern of diversity in cultivated and wild soybean (Kim et al. 2010, Lam et al. 2010, Li et al. 2013) and maize (Lai et al. 2010, Jiao et al. 2012). Results of resequencing of 14 accessions of *Prunus* sp. showed significant nucleotide diversity (π) (Tajima 1983) between the eastern and western varieties (Verde et al. 2013). Importantly, based on population structure and phylogenetic analysis of resequenced 115 cucumber lines resulted in four distinct geographic groups (Qi et al. 2013). The Indian group contained higher nucleotide diversity π than the rest of the other three groups. Given the resequencing of two Perennial and Dempsey cultivars of pepper in combination with *de novo* sequencing of *Capsicum chinense* and whole-genome sequencing of CM334 Mexican landrace, witnessed existence of enough genome variation among the genomes (Kim et al. 2014b). Notably, the presence of higher genetic variation in wild *Citrullus lanatus* subsp. *lanatus* lines than the cultivated *C. lanatus* subsp. *vulgaris* lines was depicted via resequencing 20 lines of watermelon (Guo et al. 2013). Most of the studies based on WGRS given earlier, indicate that a genetic diversity is mostly affected by a human-mediated domestication and allied processes resulting in the loss of haplotype diversity and ultimately narrowing down the genetic diversity in crop plants (Li et al. 2013). Therefore, WGRS can be employed for recovering rare genetic variants existing across the genome at global level in various crops thereby supplanting the crop improvement program.

Genome-wide discovery of markers to expedite genomics-assisted breeding

Comparison of WGRS-derived sequence reads with the available reference sequence of crop plants can facilitate in the discovery of massive SNPs (Jeong et al. 2013). Therefore, WGRS can be effectively applied for mining informative SNP and InDel markers, which can be used for developing high-density genetic map and detection/mapping of candidate genes/QTLs of valuable traits mostly used in the crop-breeding program (Huang et al. 2009, 2010, Schmutz et al. 2014, Zhou et al. 2014, Lee et al. 2015). Taking this into account, in case of rice, WGRS of 150 RILs derived from *indica* cv. 93-11 x *japonica* cv. Nipponbare discovered 1,226,791 SNPs and additionally, a total of 5074 "recombination break points" were elucidated (Huang et al. 2009). In another report, sequencing of 517 landraces mined approximately 3.6 million SNPs that enabled in developing high-density haplotype map in rice genome (Huang et al. 2010). Subjecting the whole-genome sequencing of five Korean cultivars including three anther culture lines, and their progenitors in conjunction with one *japonica* cultivar at over 58x coverage, resulted in the identification of 29,269 nonsynonymous polymorphic SNPs (Jeong et al. 2013). These SNPs may contribute toward identification of genes/QTLs of agronomic importance, whereas, resequencing of elite six inbred lines led to the discovery of 1,000,000 SNP and 30,000 InDel markers, thus giving new horizons for molecular breeding in maize (Lai et al. 2010). In case of sorghum, 1,057,018 SNPs, 99,948 InDels, and 17,111 CNVs were identified by resequencing of two sweet and one grain sorghum genome (Zheng et al. 2011). Most of the SNP obtained from this study was found to be associated with leucine-rich repeats, and disease-resistance genes. Similarly, resequencing of five *Sorghum bicolor* genomes ranging from 6× to 12× coverage provided 1 million quality SNPs for dissecting the genetic basis of marker-trait associations in future (Bekele et al. 2013). To uncover the novel diversity in sorghum, 44 accessions including landraces, improved cultivars, and wild species were resequenced at 16× to 45× coverage, respectively, which offered a total of 8 million high-quality SNPs and 1.9 million InDels (Mace et al. 2013). Additionally, the authors also shed light on complex domestication-related events underwent in sorghum. Resequencing the genomes of 654 and LTR108 parents and the RILs derived from the given parents, discovered 1.2-million high-quality SNPs that are used for the development of a high-density genetic linkage map spanning 1591.4 cM in sorghum (Zhou et al. 2012). Alignment of resequencing reads of "Shi-Li-Xiang" foxtail millet landraces with the two reference genomes "Yugu1 and Zhang gu" resulted in identification of numerous SNPs and InDel polymorphisms, which can be exploited in the detection of agronomic traits in the foxtail millet breeding program (Bai et al. 2013). In soybean, 5,102,244 SNPs were discovered by resequencing of 25 new accessions and 30 accessions (Lam et al. 2010) of published genome sequence. This uncovered 25.5% new SNPs coupled with a small artificial selection causative genomic region, representing 2.99% of the whole genome (Li et al. 2013). Moreover, the study also shed light on understanding the origin of cultivated *Glycine max* from *G. soja* owing to the human-driven artificial selection and domestication process. Importantly, 9,790,744 SNPs and 876,799 InDels in conjunction with 1614 CNVs were recovered by resequencing of 302 soybean accessions (Zhou et al. 2014). These SNPs may allow conducting genomewide association study (GWAS) for unraveling the underlying trait-associated genomic regions (genes) in soybean marker-assisted breeding program. By performing the whole-genome sequencing of L144 and Z16 accessions of *Brassica rapa* identified a useful InDel polymorphism, by comparing the sequences of the given accessions with the reference genome sequence of Chiifu-401-42 (Liu et al. 2013). Considering *Brassica napus*, resequencing of the genome has facilitated in the discovery of 892,536 SNPs distributed throughout its genome, and those may be used for marker-trait association in the crop improvement program (Huang et al. 2013a,b). Considering Chinese cabbage, more than 1 million SNPs were recorded by resequencing the Bre and Wut genome of *B. rapa* (Yu et al. 2013). Efforts of resequencing eight accessions of the tomato genome at a depth ranging from 6.7× to 16.6× resulted in 16,000 unique nonsynonymous SNPs, 128,000 InDels, and 1686 putative CNVs by comparing the sequences of these accessions with the reference genome (Causse et al. 2013). This study also suggested, introgression of target genomic regions from wild species into the cultivated species took place in the course of domestication. Therefore, WGRS can provide a plethora of SNPs, which can be used for strengthening marker-assisted breeding in the coming years.

Uncovering gene/QTLs of agronomic importance

WGRS is of immense importance from the identification of novel candidate genes/QTLs point of view. It can be employed for elucidating the genomic regions/candidate genes responsible for domestication, yield-related QTLs, and complex adaptive traits in crop plants (Chung et al. 2013, Guo et al. 2013, Li et al. 2013). Importantly, given the amenability of high-throughput SNPs, has permitted the discovery of the small-haplotype blocks associated with a valuable loci across the genome (Brachi et al. 2011). Sequencing of *Arabidopsis lyrata* gave valuable insights of loci contributing toward adaptation of this species in serpentine soil-harboring heavy metals (Turner et al. 2010). Whereas, the causal SNP in the *AtNFXL-2* gene was disclosed by resequencing *Arabidopsis* clock mutant early bird (ebi-1) (Ashelford et al. 2011). Analysis of sequence information of 80 strains of *Arabidopsis* suggested that 12,468 SNPs caused a change in start codons, created premature codons, and extended the open-reading frame in 6197 genes (Cao et al. 2011). By sequencing 517 landraces of rice in association with GWAS enabled in understanding the genetic basis of 14 complex agronomic traits controlled by 80 loci in rice (Huang et al. 2010). Significantly associated markers with agronomic traits of interest were disclosed by resequencing 50 accessions of both cultivated and wild rice species (Xu et al. 2012). Additionally, Gao et al. (2013) resequenced 132 RILs of Liang–You–Pei–Jiu as well as their parents 93–11 and Peiai 64s, resulted in identification of 43 yield-related QTLs in rice (Table 7.1). Further, this study permitted to fine map-two important *qSN8* and *qSPB1* QTLs. While, capturing the structural genome variations via NGS-driven *de novo* assembly of three diverse rice genomes IR64, DJ123, and Nipponbare belonging to three different subpopulations, namely, *indica*, *aus*, and temperate *japonica*, respectively, was compared with the available "reference genome" sequences (Schatz et al. 2014). The author obtained valuable insights for the loci "*S5 hybrid sterility, Sub1, LRK* gene cluster, and *Pup1*" from this study. Additionally, they also provided "*Pan genome*" analysis for the given assembles of three strains of rice to identify the "conserved sequence" existing in the genome of all the three strains and "strain specific genome sequence" (Schatz et al. 2014). To this end, the applications of WGRS for elucidating various yield-contributing traits are critically reviewed in rice (Guo et al. 2014). Whereas in maize, resequencing of 75 accessions resulted in shedding light on the candidate genes, namely, *GRMZM2G448355, GRMZM2G036464,* and *GRMZM2G428027* mostly related with domestication and evolution of maize (Hufford et al. 2012). To elucidate the drought-tolerance candidate gene, 15 inbreds of maize were resequenced. Common variants and cluster analysis assisted in the detection of 524 nonsynonymous SNPs concerned with 271 candidate genes from the reference genome sequence of B73 and the given 15 inbreds genome sequences (Xu et al. 2014). In case of sorghum, resequencing of 654 and LTR108 parents and the derived RILs enabled the identification of 57 QTLs governing eight traits of agronomic importance (Zhou et al. 2012). However, resequencing of 16 accessions including the cultivated and wild soybean genome led to the identification of candidate genomic regions harboring the artificially selected genes responsible for soybean domestication (Chung et al. 2013). Likewise, analysis of resequencing of 25 soybean genomes and 30 published genome sequences resulted in shedding light on candidate genes responsible for domestication of soybean and important traits of agronomic interest, namely, seed coat color and flowering time (Li et al. 2013). Importantly, resequencing of 302 accessions of soybean led to identification of genomic regions/QTLs associated with oil content, plant height, pubescence form, and fatty acid biosynthesis (Zhou et al. 2014). By comparing *de novo* sequence data of "W05," a wild soybean accession with a reference genome sequence of "William 82" aided in identification of ion transporter gene *GmCHX1*, contributing in salt tolerance (Qi et al. 2014). The given locus in "W05" genotype lacked retrotransposon insertion otherwise present in cultivars of sensitive "C08" and "William 82" genome in the same locus. In addition, resequencing genome of "R108" strain of *Medicago truncatula* showed genome variations contributing changes in aluminum, sodium toxicity, and iron deficiency responsive genes while comparing the genome sequence of the given strain with "JA17" genome (Wang et al. 2014). Considering Chinese cabbage, genome resequencing of 150 RIL populations developed from Bre × Wat population, enabled in detection of 18 QTLs for six head traits (Yu et al. 2013). Additionally, three candidate genes *BrpGL1*, *BrpESR1*, and *BrpSAW1* contributing in leafy head formation were unfolded. A list of novel genes/QTLs identified in different crops by WGRS has been given in Table 7.1. Henceforth, the applications of resequencing can further aid in identification of genes/QTLs underlying complex traits of agronomic importance in crop improvement program so far, remained unresolved.

Exploring linkage disequilibrium pattern for GWAS

WGRS can help in providing an opportunity to inspect the existing LD pattern throughout the genome, thereby offering a scope of performing GWAS to resolve complex traits of agronomic importance in various crop plants (Huang et al. 2010, Lam et al. 2010, Branca et al. 2011, Jia et al. 2013, Zhou et al. 2014). Recent practices of GWAS have played a significant role in unraveling the genetic basis of phenotypes of underlying complex traits of agronomic importance in diverse crop plants, namely, maize (Buckler et al. 2009, McMullen et al. 2009, Kump et al. 2011, Poland et al. 2011, Tian et al. 2011), barley (Pasam et al. 2012), *Arabidopsis* (Atwell et al. 2010, Bergelson and Roux 2010, Brachi et al. 2010, Li et al. 2010), *B. napus* (Li et al. 2014b), wheat (Cavanagh et al. 2013), rice (Zhao et al. 2011), sorghum (Morris et al. 2013), and in other crops discussed by Ogura and Busch (2015). By performing WGRS in rice, Huang et al. (2010) constructed a high-density haplotype map. Following GWAS of 14 agronomic traits in rice, aided in uncovering the underlying loci for the traits, namely, heading date, grain size, and starch quality. They also analyzed LD pattern in *japonica* and *indica* rice suggesting, LD decay rate was at ~123 kb in *japonica* group and at ~167 kb in *indica* group. Considering sorghum, it remains within ~150 kb (Mace et al. 2013) and ~100 kb in foxtail millet (Jia et al. 2013), whereas in maize LD decays within 2000 bp (Gore et al. 2009). Additionally, in the context of soybean, 10-fold longer LD was identified in wild soybean in comparison to wild maize and rice (Chung et al. 2013). Similarly, Lam et al. (2010) reported this value to be ~150 kb and ~75 kb for cultivated and wild soybean, respectively. In another study, Zhou et al. (2014) estimated LD to be at 83 kb in landraces and 133 kb in improved cultivars by resequencing 302 accessions of soybean (Zhou et al. 2014). Importantly, in this study, GWAS facilitated in unfolding the loci underlying many important agronomic traits, namely, plant height, oil content, and pubescence form. Likewise in maize, deploying WGRS approach helped in developing "Maize HapMap2" (Chia et al. 2012) coupled with laying the foundation for GWAS for five traits (Kump et al. 2011, Poland et al. 2011, Tian et al. 2011) in association with the SNPs developed from this study and SNP developed from "HapMap1" (Gore et al. 2009). Moreover, Jia et al. (2013) developed haplotype map in foxtail millet in conjunction with 0.8-million SNPs derived from low-coverage WGRS of 916 accessions. Simultaneously, they tracked down 512 loci underlie for 47 agronomic traits by conducting GWAS. Therefore, WGRS can be beneficial for elucidating the complete landscape of LD pattern existing in the genome, thus giving an opportunity to conduct association study to uncover the complex traits.

Rare mutation mapping via whole-genome sequencing

Recently, NGS-driven whole-genome sequencing approach is replacing the traditional forward genetic techniques for mapping of mutations (Schneeberger and Weigel 2011). Currently, WGRS has been invested for capturing rare mutations at "single-nucleotide resolution" (Schneeberger 2014) via "SHOREmap (http://1001genomes.org/downloads/shore.html)" (Schneeberger et al. 2009), "next generation mapping" (NGM) (Smith et al. 2008, Zuryn et al. 2010, Austin et al. 2011), "MutMap" (Abe et al. 2012), "Mut Map+" (Fekih et al. 2013), and "MutMap-Gap" (Takagi et al. 2013a) collectively known as "mapping by sequencing" approach (Schneeberger et al. 2009, Galvao et al. 2012, Linder et al. 2012, James et al. 2013, Mascher et al. 2014). NGS-based sequencing approach has been successfully employed to map the candidate causative mutation in pooled F_2 population without requiring mapping information known as "NGM" (Austin et al. 2011). Using this innovative approach, the authors disclosed three genes associated with cell wall biology in *Arabidopsis*. Likewise, an innovative gene isolation technique based on WGRS called "Mut Map" (Abe et al. 2012) has been devoted to capture mutation. This method involves (i) crossing of the recessive mutant with a wild type followed by selfing of F_1, (ii) crossing of the mutant with a wild type to obtain a phenotypic difference at F_2 level, (iii) whole-genome sequencing of pooled DNA of F_2 population containing the mutant phenotype, and (iv) construction of SNP index, which ensures linkage of SNP to the causal mutant phenotype (for details, see Abe et al. 2012). Similarly, "Mut Map+" extension of "Mut Map" enables in the detection of causative mutation, thereby distinguishing the SNP frequency of bulk DNA of mutants and wild-type progeny of M_3 generation (Fekih et al. 2013). In addition, "MutMap-Gap" (Takagi et al. 2013a) has been introduced to facilitate in identification of the nucleotide change in the genomic region of the mutant phenotype, missing from the reference genome, via "MutMap" in conjunction with *de novo assembly* of genomic gap regions. This is successfully applied for isolating blast-resistant *Pii* gene in rice (Takagi et al. 2013a). Likewise, "MutMap" has assisted in the identification of *hst1* mutant in "Hitomebore" cultivar contributing in salt tolerance in rice (Takagi et al. 2015). Interestingly, "modified MutMap" has been applied to isolate "*osms55*" gene causing male sterility in rice (Chen et al. 2014). In addition, this whole-genome sequencing-based approach is used for detecting causative SNPs in ethyl methane sulfonate (EMS)-induced mutation in the nonreference accession of *Arabidopsis* (Laitinen et al. 2010, Uchida et al. 2011, 2014). Moreover, recently, more refinement of "mapping by sequencing" has been introduced by "needle in the *k*-stack (NIKS)" where mutation is detected based on distinguishing whole-genome sequencing data of wild-type individuals and the mutant type based on *k*-mers requiring no reference sequence (Nordström et al. 2013). Therefore, the above-given WGRS-based techniques can expedite in discovering the gene missing in the available "reference genome" sequence in crops.

Major QTL delineation via whole-genome sequencing

To reduce the time and cumbersome process involved in the identification of QTLs, WGRS has been introduced to develop a new scheme called "QTL seq" (Takagi et al. 2013b). In this approach, WGRS is performed in the DNA bulks of two populations showing an "extreme opposite trait value" for the given contrasting traits in F_2 generation, followed by alignment of sequence read to the reference sequence of parental cultivar for SNP index calculation (for details see Takagi et al. 2013b). This approach is used for identification of QTLs for partial resistance of rice blast disease (Takagi et al. 2013b) given in Table 7.1. Recently, by performing "QTL seq" the candidate genes underlying the early flowering *Ef1.1* and seed weight QTLs have been unfolded in cucumber (Lu et al. 2014) and chickpea (Das et al. 2015), respectively. In case of tomato, this approach has facilitated in the identification of three fruit weight-related *QTLs fw11.2, fw1.1, and fw3.3* (Illa-Berenguer et al. 2015).

Conclusion and future prospects

Rapid progress of genome sequencing in various crops has provided us the complete landscape of genomic variations existing in both the cultivated and wild species, thus providing an exciting opportunity to exploit them in trait improvement program (Huang and Han 2014). Concurrently, WGRS approaches are successfully employed for addressing the evolutionary and domestication-related events causing genetic variants in different crops. Notably this approach can be exploited for capturing the hidden genetic diversity for broadening the genetic base of crop improvement. Practices of GWAS are becoming easier for discovering the candidate gene/QTLs for complex traits via a massive number of novel SNPs derived from WGRS. The next grand challenge is to integrate these massive sequence data to the corresponding phenotype, thereby closing the gap between the genotype and phenotype (Schatz et al. 2012) and augmenting genomic selection for crop improvement in the postgenomic era (Morrell et al. 2012).

Acknowledgment

The authors declare no conflict of interest.

References

Abe A, Kosugi S, Yoshida K, Natsume S, Takagi H, Kanzaki H, Matsumura H. et al. 2012. Genome sequencing reveals agronomically important loci in rice using MutMap. *Nature Biotechnology*, 30: 174–178.

Aflitos S, Schijlen E, deJong H, de Ridder D, Finkers R, Wang J, Zhang G. et al. 2014. Exploring genetic variation in the tomato (Solanum section Lycopersicon) clade by whole-genome sequencing. *Plant Biotechnology Journal*, 80: 136–148.

Ashelford K, Eriksson ME, Allen CM, D'Amore R, Johansson M, Gould P, Kay S, Millar AJ, Hall N, and Hall A. 2011. Full genome re-sequencing reveals a novel circadian clock mutation in *Arabidopsis*. *Genome Biology*, 12: R28.

Atwell S, Huang YS, Vilhjálmsson BJ, Willems G, Horton M, Li Y, Meng D. et al. 2010. Genome-wide association study of 107 phenotypes in a common set of *Arabidopsis thaliana* inbred lines. *Nature*, 465: 627–631.

Austin RS, Vidaurre D, Stamatiou G, Breit R, Provart NJ, Bonetta D, Zhang J. et al. 2011. Next-generation mapping of *Arabidopsis* genes. *The Plant Journal*, 67: 715–725.

Bai H, Cao Y, Quan J, Dong L, Li Z, Zhu Y, Zhu L, Dong Z, and Li D. 2013. Identifying the genome-wide sequence variations and developing new molecular markers for genetics research by re-sequencing a Landrace cultivar of foxtail millet. *PLoS One*, 10: e73514.

Beavan MW, and Uauy C. 2013. Genomics reveals new landscapes for crop improvement. *Genome Biology*, 14: 206.

Bekele WA, Wieckhorst S, Friedt W, and Snowdon RJ. 2013. High-throughput genomics in sorghum: From whole-genome resequencing to a SNP screening array. *Plant Biotechnology Journal*, 11: 1112–1125.

Bergelson J, and Roux F. 2010. Towards identifying genes underlying ecologically relevant traits in *Arabidopsis thaliana*. *Nature Genetics Review*, 11: 867–879.

Bhat SR. 2011. Rationalizing investment and effort in whole genome sequencing for harvesting applied benefits. *Current Science*, 100: 1633–1637.

Brachi B, Faure N, Horton M, Flahauw E, Vazquez A, Nordborg M, Bergelson J, Cuguen J, Roux F. 2010. Linkage and association mapping of *Arabidopsis thaliana* flowering time in nature. *PLoS Genetics*, 6: e1000940.

Branca A, Paape TD, Zhou P, Briskine R, Farmer AD, Mudge J, Bharti AK. et al. 2011. Whole-genome nucleotide diversity, recombination, and linkage disequilibrium in the model legume Medicago truncatula. *Proceedings of the National Academy of Sciences United States*, 108: E864–E870.

Buckler ES, Holland JB, Bradbury PJ, Acharya CB, Brown PJ, Browne C, Ersoz E. et al. 2009. The genetic architecture of maize flowering time. *Science*, 325: 714–718.

Cao J, Schneeberger K, Ossowski S, Gunther T, Bender S, Fitz J, Koenig D. et al. 2011. Whole-genome sequencing of multiple *Arabidopsis thaliana* populations. *Nature Genetics*, 43:956–963.

Causse M, Desplat N, Pascual L, Le Paslier MC, Sauvage C, Bauchet G, Bérard A. et al. 2013. Whole genome resequencing in tomato reveals variation associated with introgression and breeding events. *BMC Genomics*, 14:791.

Cavanagh CR, Shiaoman S, Wang S, Huang BE, Stephen S, Kiani S, Forrest K. et al. 2013. Genome-wide comparative diversity uncovers multiple targets of selection for improvement in hexa-ploid wheat landraces and cultivars. *Proceedings of the National Academy of Science United States*, 110: 8057–8062.

Chen J, Huang Q, Gao D, Wang J, Lang Y, Liu T, Li B. et al. 2013. Whole-genome sequencing of *Oryza brachyantha* reveals mechanisms underlying Oryza genome evolution. *Nature Communication*, 4:1595. doi: 10.1038/ncomms2596.

Chen Z, Wang B, Dong X, Liu H, Ren L, Chen J, Hauck A, Song W, and Lai J. 2014. An ultra-high density bin-map for rapid QTL mapping for tassel and ear architecture in a large F2 maize population. *BMC Genomics*, 15:433.

Chia JM, Song C, Bradbury PJ, Costich D, de Leon N, Doebley J, Elshire RJ. et al. 2012. Maize HapMap2 identifies extant variation from a genome in flux. *Nature Genetics*, 44:803–807. doi:10.1038/ng.2313.

Chitikineni A, Gaur PM, Jian J, Doddamani D, Roorkiwal M, Rathore A. et al. 2015. A first genera-tion HapMap of chickpea for mining superior alleles for crop improvement. Plant and Animal Genome XXIII Conference, 2015. pag.confex.com.

Chung WH, Jeong N, Kim J, Lee WK, Lee YG, Lee SH, Yoon W. et al. 2013. Population structure and domestication revealed by high-depth resequencing of Korean cultivated and wild soybean genomes. *DNA Research*, 21:153–167.

Das S, Upadhyaya HD, Bajaj D, Kujur A, Badoni S, Kumar LV, Tripathi S. et al. 2015. Deploying QTL-seq for rapid delineation of a potential candidate gene underlying major trait-associated QTL in chickpea. *DNA Research*, doi: 10.1093/dnares/dsv004.

Deschamps S, and Campbell MA. 2010. Utilization of next-generation sequencing platforms in plant genomics and genetic variant discovery. *Molecular Breeding*, 25: 553–570.

Doebley JF, Gaut BS, Smith BD. 2006. The molecular genetics of crop domestication. *Cell* 127: 1309–1321.

Duan MJ, Sun ZZ, Shu LP, Tan YN, Yu D, Sun XW, Liu RF, Li YJ, Gong SY, and Yuan DY. 2013. Genetic analysis of an elite super-hybrid rice parent using high-density SNP markers. *Rice*, 6: 21.

Edwards D, Henry RJ, and Edwards KJ. 2012. Preface: Advances in DNA sequencing accelerating plant biotechnology. *Plant Biotechnology Journal*, 10: 621–622.

Fekih R, Takagi H, Tamiru M, Abe A, Natsume S, Yaegashi H, Sharma S. et al. 2013. MutMap+: Genetic mapping and mutant identification without crossing in rice. *PLoS ONE*, 8: e68529.

Galvaõ VC, Nordström KJV, Lanz C, Sulz P, Mathieu J, Posé D, Schmid M, Weigel D, and Schneeberger K. 2012. Synteny-based mapping-by-sequencing enabled by targeted enrichment. *The Plant Journal*, 71: 517–526.

Gao ZY, Zhao SC, He WM, Guo LB, Peng YL, Wang JJ. et al. 2013. Dissecting yield-associated loci in super hybrid rice by resequencing recombinant inbred lines and improving parental genome sequences. *Proceedings of the National Academy of Sciences United States*, 110(35): 14492–14497.

Gore MA, Chia JM, Elshire RJ, Sun Q, Ersoz ES, Hurwitz BL, Peiffer JA. et al. 2009. A first-generation haplotype map of maize. *Science*, 326: 1115–1117.

Guo S, Zhang J, Sun H, Salse J, Lucas WJ, Zhang H, Zheng Y. et al. 2013. The draft genome of water-melon (*Citrullus lanatus*) and resequencing of 20 diverse accessions. *Nature Genetics*, 45: 51–58.

Guo L, Gao Z, and Qia Q. 2014. Application of resequencing to rice genomics, functional genomics and evolutionary analysis. *Rice*, 7: 4.

Haun WJ, Hyten DL, Xu WW, Gerhardt DJ, Albert TJ, Richmond T. et al. 2011. The composition and origins of genomic variation among individuals of the soybean reference cultivar Williams 82. *Plant Physiology*, 155: 645–655.

Hawkins RD, Hon GC, and Ren B. 2010. Next-generation genomics: An integrative approach. *Nature Review Genetics*, 11: 476–486.

He Z, Zhai W, Wen H, Tang T, Wang Y, Lu X. et al. 2011. Two evolutionary histories in the genome of rice: The roles of domestication genes. *PLoS Genetics,* 7: 9.

Hirsch CN, Foerster JM, Johnson JM, Sekhon RS, Muttoni G, Vaillancourt, Peñagaricano F. et al. 2014. Insights into the maize pan-genome and pan-transcriptome. *Plant Cell,* 26: 121–135.

Howard TP III, Hayward AP, Tordillos A, Fragoso C, Moreno MA, Tohme J, Kausch AP. et al. 2014. Identification of the maize gravitropism gene lazy plant1 by a transposon-tagging genome resequencing strategy. *PLoS ONE,* 9: e87053.

Huang S, Deng L, Guan M, Li J, Lu K, Wang H, Fu D, Mason AS, Liu S, and Hua W. 2013b. Identification of genome-wide single nucleotide polymorphisms in allopolyploid crop *Brassica napus. BMC Genomics,* 14: 717.

Huang X and Han B. 2014. Natural variations and genome-wide association studies in crop plants. *Annual Review Plant Biology,* 65: 531–551.

Huang X, Kurata N, Wei X, Wang ZX, Wang A, Zhao Q, Zhao Y. et al. 2012. A map of rice genome variation reveals the origin of cultivated rice. *Nature,* 490: 497–501.

Huang X, Zhao Y, Wei X, Li C, Wang A, Zhao Q, Li W. et al. 2012b. Genome-wide association study of flowering time and grain yield traits in a worldwide collection of rice germplasm. *Nature Genetics,* 44: 32–41.

Huang XH, Feng Q, Qian Q, Zhao Q, Wang L, Wang A, Guan JP. et al. 2009. High-throughput genotyping by whole-genome resequencing. *Genome Research,* 19: 1068–1076.

Huang XH, Kurata N, Wei XH, Wang ZX, Wang A, Zhao Q, Zhao Y. et al. 2012a. A map of rice genome variation reveals the origin of cultivated rice. *Nature,* 490: 497–503.

Huang XH, Lu TT, and Han B. 2013a. Resequencing rice genomes: An emerging new era of rice genomics. *Trends Genetics,* 29: 225–230.

Huang XH, Wei XH, Sang T, Zhao Q, Feng Q, Zhao Y, Li CY. et al. 2010. Genome-wide association studies of 14 agronomic traits in rice landraces. *Nature Genetics,* 42: 961–967.

Hufford MB, Xu X, van Heerwaarden J, Pyhäjärvi T, Chia JM, Cartwright RA, Elshire RJ. et al. 2012. Comparative population genomics of maize domestication and improvement. *Nature Genetics,* 44: 808–811.

Illa-Berenguer E, Houten JV, Huang Z, and van der Knaap E. 2015. Rapid and reliable identification of tomato fruit weight and locule number loci by QTL-seq. *Theoritical and Applied Genetics,* 128:1329–1342.

Imelfort M and Edwards D. 2009. De novo sequencing of plant genomes using second-generation technologies. *Briefing in Bioinformatics,* 10: 609–618.

Jackson SA, Iwata A, Lee SH, Schmutz J, and Shoemaker R. 2011. Sequencing crop genomes: Approaches and applications. *New Phytologist,* 191: 915–925.

James GV, Patel V, Nordstrom KJ, Klasen JR, Salome PA, Weigel D, and Schneeberger K. 2013. User guide for mapping-by-sequencing in *Arabidopsis. Genome Biology,* 14: R61.

Jeong IS, Yoon UH, Lee GS, Ji HS, Lee HJ, Han CD, Hahn JH, An GH, and Kim YH. 2013. SNP-based analysis of genetic diversity in anther-derived rice by whole genome sequencing. *Rice,* 6: 6.

Jia G, Huang X, Zhi H, Zhao Y, Zhao Q, Li W, Chai Y. et al. 2013. A haplotype map of genomic variations and genome-wide association studies of agronomic traits in foxtail millet (*Setaria italica*). *Nature Genetics,* 45:957–961.

Jiao Y, Zhao H, Ren R, Song W, Zeng B, Guo J, Wang B. et al. 2012. Genome wide genetic changes during modern breeding of maize. *Nature Genetics,* 44: 812–815.

Kim B, Kim DG, Lee G, Seo J, Choi IY, Yang TJ. et al. 2014a. Defining the genome structure of "Tongil" rice, an important cultivar in the Korean "Green Revolution". *Rice,*7: 22.

Kim S, Park M, Yeom S, Kim Y, Lee JM, Lee H, Seo E. et al. 2014b. Genome sequence of the hot pepper provides insights into the evolution of pungency in *Capsicum* species. *Nature Genetics,* 46: 270–278.

Kim MY, Lee S, Van K, Kim TH, Jeong SC, Choi IY, Kim DS. et al. 2010. Whole-genome sequencing and intensive analysis of the undomesticated soybean (*Glycine soja* Sieb. and Zucc.) genome. *Proceedings of the National Academy of Science United States,* 107: 22032–22037.

Kump KL, Bradbury PJ, Wisser RJ, Buckler ES, Belcher AR, Oropeza-Rosas MA, Zwonitzer JC. et al. 2011. Genome-wide association study of quantitative resistance to southern leaf blight in the maize nested association mapping population. *Nature Genetics,* 43: 163–168.

Lai J, Li R, Xu X, Jin W, Xu M, Zhao H, Xiang Z. et al. 2010. Genome-wide patterns of genetic variation among elite maize inbred lines. *Nature Genetics*, 42: 1027–1030.

Laitinen RAE, Schneeberger K, Jelly NS, Ossowski S, and Weigel D. 2010. Identification of a spontaneous frame shift mutation in a nonreference *Arabidopsis* accession using whole genome sequencing. *Plant Physiology*, 153: 652–654.

Lam HM, Xu X, Liu X, Chen W, Yang G, Wong FL, Li MW. et al. 2010. Re sequencing of 31 wild and cultivated soybean genomes identifies patterns of genetic diversity and selection. *Nature Genetics*, 42: 1053–1059.

Lee J, Izzah NK, Jayakodi M, Perumal S, Joh HJ, Lee HJ, Lee SC. et al. 2015. Genome-wide SNP identification and QTL mapping for black rot resistance in cabbage. *BMC Plant Biology*, 15: 32.

Li F, Fan G, Wang K, Sun F, Yuan Y, Song G, Li Q. et al. 2014a. Genome sequence of the cultivated cotton *Gossypium arboreum*. *Nature Genetics*, 46: 567–572.

Li F, Chen B, Xu K, Wu J, Song W, Bancroft I, Harper AL. et al. 2014b. Genome wide association study dissects the genetic architecture of seed weight and seed quality in rapeseed (*Brassica napus* L). *DNA Research*, 21: 355–367.

Li Z, Fu B-Y, Zhang G, and McNally KL. 2014c. The 3,000 rice genomes project. *Giga Science*, 3: 1–6.

Li SC, Xie KL, LiWB ZT, RenY WSQ, Deng QM, Zheng AP, Zhu J. et al. 2012. Re-sequencing and genetic variation identification of a rice line with ideal plant architecture. *Rice*, 5: 18.

Li Y, Huang Y, Bergelson J, Nordborg M, and Borevitz JO. 2010. Association mapping of local climate-sensitive quantitative trait loci in *Arabidopsis thaliana*. *Proceedings of the National Academy of Science United States*, 107: 21199–21204.

Li YH, Zhao SC, Ma JX, Li D, Yan L, Li J, Qi XT. et al. 2013. Molecular footprints of domestication and improvement in soybean revealed by whole genome resequencing. *BMC Genomics*, 14: 579.

Li F, Fan G, Lu C, Xiao G, Zou C, Kohel RJ. et al. 2015. Genome sequence of cultivated upland cotton (*Gossypium hirsutum* TM-1) provides insights into genome evolution. *Nature Biotechnology*, 33: 524–530.

Lin H, Xia P, Wing AR, Zhang Q, and Luo M. 2012. Dynamic intra-japonica subspecies variation and resource application, *Molecular Plant*, 5: 218–230.

Lindner H, Raissig MT, Sailer C, Shimosato-Asano H, Bruggmann R, and Grossniklaus U. 2012. SNP-ratio mapping (SRM) identifying lethal alleles and mutations in complex genetic backgrounds by next-generation sequencing. *Genetics*, 191: 1381–1386.

Lister R, Gregory BD, and Ecker JR. 2009. Next is now: New technologies for sequencing of genomes, transcriptomes and beyond. *Current Opinion in Plant Biology*, 12: 107–118.

Liu B, Wang Y, Zhai W, Deng J, Wang H, Cui Y, Cheng F, Wang X, and Wu J. 2013. Development of InDel markers for *Brassica rapa* based on whole-genome re-sequencing. *Theoretical and Applied Genetics*, 126: 231–239.

Lu H, Lin T, Klein J, Wang S, Qi J, Zhou Q, Sun J, Zhang Z, Weng Y, Huang S. 2014. QTL-seq identifies an early flowering QTL located near flowering locus T in cucumber. *Theoretical and Applied Genetics* 127, 1491–1499.

Lu P, Han X, Qi J, Yang J, Wijeratne AJ, Li T, and Ma H. 2012. Analysis of *Arabidopsis* genome-wide variations before and after meiosis and meiotic recombination by resequencing Landsberg erecta and all four products of a single meiosis. *Genome Research*, 22: 508–518.

Mace ES, Tai S, Gilding EK, Li Y, Prentis PJ, Bian L, Campbell BC. et al. 2013. Whole-genome sequencing reveals untapped genetic potential in Africa's indigenous cereal crop sorghum. *Nature Communication*, 4: 2320.

Mardis ER. 2008. Next-generation DNA sequencing methods. *Annual Review Genomics Human Genetics*, 9: 387–402.

Mascher M, Jost M, Kuon JE, Himmelbach A, Außfalg A, Beier S, Scholz U, Graner A, and Stein N. 2014. Mapping-by-sequencing accelerates forward genetics in barley. *Genome Biology*, 15: R78.

McHale LK, Haun WJ, Xu WW, Bhaskar PB, Anderson JE, Hyten DL, Gerhardt DJ. et al. 2012. Structural variants in the soybean genome localize to clusters of biotic stress-response genes. *Plant Physiology*, 159: 1295–1308.

McMullen MD, Kresovich S, Villeda HS, Bradbury P, Li H, Sun Q, Flint-Garcia S. et al. 2009. Genetic properties of the maize nested association mapping population. *Science*, 325: 737–740.

Mehra P, Pandey BK, Giri J. 2015. Genome-wide DNA polymorphisms in low phosphate tolerant and sensitive rice genotypes. *Science Reports,* 5: 13090.

Metzke ML. 2010. Sequencing technologies—The next generation. *Nature Review Genetics,* 11: 31–46.

Meyer RS, and Purugganan MD. 2013. Evolution of crop species: Genetics of domestication and diversification. *Nature Genetics Review,* 14: 840–852.

Michael TP, and Jackson S. 2013. The first 50 plant genomes. *The Plant Genome,* 6: 1–7.

Molina J, Martin Sikora M, Nandita Garud N, Flowers JM, Rubinstein S, Reynolds A, Huang P. et al. 2011. Molecular evidence for a single evolutionary origin of domesticated rice. *Proceedings of the National Academy of Science United States,* 108: 8351–8356.

Morrell PL, Buckler ES, and Ross-Ibarra J. 2012. Crop genomics: Advances and applications. *Nature Review Genetics,* 13: 85–96.

Morris GP, Ramu P, Deshpande SP, Hash CT, Shah T, Upadhyaya HD, Riera-Lizarazu O. et al. 2013. Population genomic and genome-wide association studies of agroclimatic traits in sorghum. *Proceedings of the National Academy of Science United States,* 110: 453–458.

Nordström KJV, Albani MC, James GV, Gutjahr C, Hartwig B, Turck F, Paszkowski U, Coupland G, and Schneeberger K. 2013. Mutation identification by direct comparison of whole-genome sequencing data from mutant and wild-type individuals using *k*-mers. *Nature Biotechnology,* 31: 325–330.

O'Rourke JA, Bolon YT, Bucciarelli B, and Vance CP. 2014. Legume genomics: Understanding biology through DNA and RNA sequencing. *Annals of Botany,* 113: 1107–1120.

Ogura T, and Busch W. 2015. From phenotypes to causal sequences: Using genome wide association studies to dissect the sequence basis for variation of plant development. *Current Opinion in Plant Biology,* 23: 98–108.

Ossowski S, Schneeberger K, Clark RM, Lanz C, Warthmann N, and Weigel D. 2008. Sequencing of natural strains of *Arabidopsis thaliana* with short reads. *Genome Research,* 18: 2024–2033.

Ossowski S, Schneeberger K, Lucas-Lledo JI, Warthmann N, Clark RM, Shaw RG, Weigel D, and Lynch M. 2010. The rate and molecular spectrum of spontaneous mutations in *Arabidopsis thaliana. Science,* 327: 92–94.

Pasam RK, Sharma R, Malosetti M, van Eeuwijk FA, Haseneyer G, Kilian B, and Graner A. 2012. Genome-wide association studies for agronomical traits in a worldwide spring barley collection. *BMC Plant Biology,* 12: 16.

Poland JA, Bradbury PJ, Buckler ES, and Nelson RJ. 2011. Genome-wide nested association mapping of quantitative resistance to northern leaf blight in maize. *Proceedings of the National Academy of Science United States,* 108: 6893–6898.

Qi J, Liu X, Shen D, Miao H, Xie B, Li X, Zeng P. et al. 2013. A genomic variation map provides insights into the genetic basis of cucumber domestication and diversity. *Nature Genetics,* 45: 1510–1515.

Qi X, Li MW, Xie M, Liu X, Ni M, Shao G, Song C. et al. 2014. Identification of a novel salt tolerance gene in wild soybean by whole-genome sequencing. *Nature Communication,* 5:4340. doi: 10.1038/ncomms5340.

Qiu J, Wang Y, Wu S, Wang Y-Y, Ye C-Y, Bai X, Li Z. et al. 2014. Genome re-sequencing of semi-wild soybean reveals a complex Soja population structure and deep introgression. *PLoS ONE,* 9: e108479.

Sakai H, Kanamori H, Arai-Kichise Y, Shibata-Hatta M, Ebana K, Oono Y, Kurita K. et al. 2014. Construction of pseudomolecule sequences of the aus rice cultivar Kasalath for comparative genomics of Asian cultivated rice. *DNA Research,* 21: 397–405.

Saxena RK, Edwards D, and Varshney RK. 2014. Structural variations in plant genomes. *Briefings in Functional Genomics,* 13: 296–307.

Schatz MC, Delcher AL, and Salzberg SL. 2010. Assembly of large genomes using second-generation sequencing. *Genome Research,* 20: 1165–1173.

Schatz MC, Maron LG, Stein JC, Wences AH, Gurtowski J, Biggers E, Lee H. et al. 2014. Whole genome *de novo* assemblies of three divergent strains of rice, *Oryza sativa,* document novel gene space of aus and indica. *Genome Biology,* 15: 506.

Schatz MC, Witkowski J, and McCombie WR. 2012. Current challenges in *de novo* plant genome sequencing and assembly. *Genome Biology,* 13: 243.

Schmutz J, McClean PE, Mamidi S, Wu G, Cannon SB, Grimwood J, Jenkins J. et al. 2014. A reference genome for common bean and genome-wide analysis of dual domestications. *Nature Genetics Review,* 46: 707–712.

Schneeberger K. 2014. Using next-generation sequencing to isolate mutant genes from forward genetic screens. *Nature Reviews Genetics,* 15: 662–676.

Schneeberger K, Ossowski S, Lanz C, Juul T, Petersen AH, Nielsen KL, Jorgensen JE, Weigel D, and Andersen SU. 2009. SHOREmap: Simultaneous mapping and mutation identification by deep sequencing. *Nature Methods,* 6: 550–551.

Schneeberger K, and Weigel D. 2011. Fast-forward genetics enabled by new sequencing technologies. *Trends in Plant Science,* 16: 1–7.

Shendure J, and Ji H. 2008. Next-generation DNA sequencing. *Nature Biotechnology,* 26: 1135–1145.

Shi J, and Lai J. 2015. Patterns of genomic changes with crop domestication and breeding. *Current Opinion in Plant Biology,* 24: 47–53.

Smith DR, Quinlan AR, Peckham HE, Makowsky K, Tao W, Woolf B, Shen L. et al. 2008. Rapid whole-genome mutational profiling using next-generation sequencing technologies. *Genome Research,* 18: 1638–1642.

Springer NM, Ying K, Fu Y, Ji T, Yeh CT, Jia Y, Wu W. et al. 2009. Maize inbreds exhibit high levels of copy number variation (CNV) and presence/absence variation (PAV) in genome content. *PLoS Genetics,* 5: e1000734.

Swanson-Wagner RA, Eichten SR, Kumari S, Tiffin P, Stein JC, Ware D, and Springer NM. 2010. Pervasive gene content variation and copy number variation in maize and its undomesticated progenitor. *Genome Research,* 20: 1689–1699.

Tajima, F. 1983. Evolutionary relationship of DNA sequences in finite populations. *Genetics,* 105: 437–460.

Takagi H, Abe A, Yoshida K, Kosugi S, Natsume S, Mitsuoka C, Uemura A. et al. 2013b. QTL-seq: Rapid mapping of quantitative trait loci in rice by whole genome resequencing of DNA from two bulked populations. *The Plant Journal,* 74: 174–183.

Takagi H, Uemura A, Yaegashi H, Tamiru M, Abe A, Mitsuoka C, Utsushi H. et al. 2013a. MutMap-Gap: Whole-genome resequencing of mutant F2 progeny bulk combined with *de novo* assembly of gap regions identifies the rice blast resistance gene Pii. *New Phytologist,* 200: 276–283.

Takagi H, Tamiru M, Abe A, Yoshida K, Uemura A, Yaegashi H, Obara T. et al. 2015. MutMap accelerates breeding of a salt-tolerant rice cultivar. *Nature Biotechnology,* 33: 445–449.

Tettelin H, Masignani V, Cieslewicz MJ, Donati C, Medini D, Ward NL, Angiuoli SV. et al. 2005. Genome analysis of multiple pathogenic isolates of *Streptococcus agalactiae*: Implications for the microbial "pan-genome." *Proceedings of the National Academy of Science United States,* 102: 13950–13955.

Thudi M, Li Y, Jackson SA, May GD, and Varshney RK. 2012. Current state-of-art of sequencing technologies for plant genomics research. *Briefings in Functional Genomics,* 11: 3–11.

Tian F, Bradbury PJ, Brown PJ, Hung H, Sun Q, Flint-Garcia S, Rocheford TR, McMullen MD, Holland JB, and Buckler ES. 2011. Genome-wide association study of leaf architecture in the maize nested association mapping population. *Nature Genetics,* 43: 159–162.

Turner TL, Bourne EC, Wettberg EJV, Hu TT, and Nuzhdin SV. 2010. Population resequencing reveals local adaptation of *Arabidopsis lyrata* to serpentine soils. *Nature Genetics,* 42: 260–263.

Uchida N, Sakamoto T, Kurata T, and Tasaka M. 2011. Identification of EMS-induced causal mutations in a non-reference *Arabidopsis thaliana* accession by whole genome sequencing. *Plant Cell Physiology,* 52: 716–722.

Uchida N, Sakamoto T, Tasaka M, and Kurata T. 2014. Identification of EMS-induced causal mutations in *Arabidopsis thaliana* by next-generation sequencing. *Methods Molecular Biology,* 1062: 259–270.

Varshney RK, Song C, Saxena RK, Azam S, Yu S, Sharpe AG, Cannon S. et al. 2013. Draft genome sequence of chickpea (*Cicer arietinum*) provides a resource for trait improvement. *Nature Biotechnology,* 31: 240–246.

Verde I, Abbott AG, Scalabrin S, Jung S, Shu S, Marroni F, Zhebentyayeva T. et al. 2013. The high-quality draft genome of peach (*Prunus persica*) identifies unique patterns of genetic diversity, domestication and genome evolution. *Nature Genetics,* 45: 487–494.

Wang K, Wang Z, Li F, Ye W, Wang J, Song G, Yue Z. et al. 2012. The draft genome of a diploid cotton *Gossypium raimondii. Nature Genetics,* 44: 1098–1103.

Wang TZ, Tian QY, Wang BL, Zhao MG, and Zhang WH. 2014. Genome variations account for different response to three mineral elements between Medicago truncatula ecotypes Jemalong A17 and R108. *BMC Plant Biology,* 14: 122.

Xu J, Yuan Y, Xu Y, Zhang G, Guo X, Wu F, Wang Q. et al. 2014. Identification of candidate genes for drought tolerance by whole-genome resequencing in maize. *BMC Plant Biology,* 14: 83.

Xu JJ, Zhao Q, Du PN, Xu CW, Wang BH, Feng Q, Liu QQ. et al. 2010. Developing high throughput genotyped chromosome segment substitution lines based on population whole-genome re-sequencing in rice. *BMC Genomics,* 11: 656–669.

Xu X, Liu X, Ge S, Jensen JD, Hu F, Li X, Dong Y. et al. 2012. Resequencing 50 accessions of cultivated and wild rice yields markers for identifying agronomically important genes. *Nature Biotechnology,* 30: 105–111.

Yamamoto T, Nagasaki H, Yonemaru J, Ebana K, Nakajima M, Shibaya T, and Yano M. 2010. Fine definition of the pedigree haplotypes of closely related rice cultivars by means of genome-wide discovery of single-nucleotide polymorphisms. *BMC Genomics,* 11: 267.

Yang K, Tian Z, Chen C, Luo L, Zhao B, Wang Z, Yu L. et al. 2015. Genome sequencing of adzuki bean (*Vigna angularis*) provides insight into high starch and low fat accumulation and domestication. *Proceedings of National Academy of Science USA.* pnas.org/cgi/doi/10.1073/pnas.1420949112.

Yu X, Wang H, Zhong W, Bai J, Liu P, He Y. 2013. QTL mapping of leafy heads by genome resequencing in the RIL population of *Brassica rapa. PLoS ONE,* 8: e76059.

Zhang G, Liu X, Quan Z, Cheng S, Xu X, Pan S, Xie M. et al. 2012. Genome sequence of foxtail millet (*Setaria italica*) provides insights into grass evolution and biofuel potential. *Nature Biotechnology,* 30: 549–554.

Zhang T, Hu Y, Jiang W, Fang L, Guan X, Chen J, Zhang J. et al. 2015. Sequencing of allotetraploid cotton (*Gossypium hirsutum* L. acc. TM-1) provides a resource for fiber improvement. *Nature Biotechnology,* 531: 537.

Zhao K, Tung CW, Eizenga GC, Wright MH, Ali ML, Price AH, Norton GJ. et al. 2011. Genome wide association mapping reveals a rich genetic architexture of complex traits in *Oryza sativa. Nature Communication,* 2011: 2.

Zheng LY, Guo XS, He B, Sun LJ, Peng Y, Dong SS, Liu TF. et al. 2011 Genome-wide patterns of genetic variation in sweet and grain sorghum (*Sorghum bicolor*). *Genome Biology,* 12: R114.

Zhou Z, Jiang Y, Wang Z, Gou Z, Lyu J, Li W, Yu Y. et al. 2014. Re sequencing 302 wild and cultivated accessions identifies genes related to domestication and improvement in soybean. *Nature Biotechnology,* 33: 408–414.

Zou G, Zhai G, Feng Q. et al. 2012. Identification of QTLs for eight agronomically important traits using an ultra-high-density map based on SNPs generated from high-throughput sequencing in sorghum under contrasting photoperiods. *Journal of Experimental Botany,* 63: 5451–5462.

Żmieńko A, Samelak A, Kozłowski P, and Figlerowicz M. 2013. Copy number polymorphism in plant genomes. *Theoretical and Applied Genetics,* doi: 10.1007/s00122-013-2177-7.

Zuryn S, Le Gras S, Jamet K, and Jarriault S. 2010. A strategy for direct mapping and identification of mutations by whole-genome sequencing. *Genetics,* 186: 427–430.

chapter eight

Molecular biotechnology of plant–microbe–insect interactions

Jam Nazeer Ahmad, Samina Jam Nazeer Ahmad, and Sandrine Eveillard

Contents

Abstract

Microorganisms such as bacteria, fungi, nematodes, and viruses may be called plant microbes as they cause many diseases in plant species all over the world. Some fungi are saprophytes that live off dead decaying organic matter while more than 10,000 types of fungi are plant parasites. They spread rust and smut diseases in plants by transmitting spores. Fungus-like organisms such as oomycetes or water molds cause havoc and diseases such as potato blight. Of the 1600 known bacterial species, 100 cause plant diseases. One such bacteria from the class mollicutes is Phytoplasma, which is responsible for many plant diseases with symptoms of virisence, phyllody, yellowing, reddening, stunted growth, and abnormal proliferation. Similarly, certain protozoa transmit plant viruses and colonize in the roots of plant species thereby reducing their growth and development. Plant viruses mostly possess single-strand ribonucleic acid genomes encapsulated in simple protein coats comprising multiple copies of one or a very few coat protein subunits. These particles may be isometric or rod shaped. Like Phytoplasma, they also spread disease from plant to plant by insect vectors such as aphids, whiteflies, leaf, and other hoppers. With the advent of biotechnological tools such as deoxyribonucleic acid (DNA) fingerprinting technology, it has become possible to detect and characterize a wide range of pathogens associated with plants and insects, which were identified through polymerase chain reaction by developing the genomic sequence-specific oligonucleotides. Pathogen attack leads to major alterations in plant gene activities and metabolic processes, gene expression, DNA methylation, and host cell metabolism. Plants tend to defend themselves through a variety of defense strategies such as infestation alert signals, posttranscriptional gene silencing, systemic acquired resistance, pathogenesis-related proteins, and production of signaling phytohormones such as ethylene, jasmonic acid, and salicyclic acid. Although plants have these various resistance mechanisms, which they use against the pathogens, the pathogens have also evolved a number of strategies to defeat the plant defense systems through certain effector molecules that help them to colonize and multiply in host. In addition to these mechanisms, both plants and pathogens possess counterdefense mechanisms by which they can make their survival and growth possible. Thus, the plant–microbe interactions consist of a wide array of mysterious complex connections, most of which are yet to be discovered and unraveled by science. Only then can we make it possible to uplift our agriculture and protect plants from these life-threatening microorganisms.

Key words: Plant microbes, Insect vectors, Genes expression regulation, Biotechnology, Effectors, Defense response.

Introduction to plant microbes

Plant microbes are microorganisms like fungi, oomycetes, protozoa, bacteria, phytoplasma and spiroplasma, viruses, nematodes, etc., that are associated with plants and are causal agents of several diseases worldwide. Among the more than 74,000 known species of fungi that have been described, the majority are saprophytes, living off dead and decaying organic matter. A few are harmful to humans, animals, fish, and insects but more than 10,000 are parasites that can cause varying degrees of damage to living plants by spreading through transmission of spores such as powdery mildews and rust diseases. Mostly, fungi utilize toxins or cell-wall-degrading enzymes to decompose plant cells and then metabolize the nutrients that are released. True fungi are organisms that lack chloroplasts and instead have thread-like mycelia, which are made up of glucans and chitin. Recently, molecular evidence has indicated that some microorganisms because of their similar growth style and infection strategies like Chromista (brown algae and diatoms) should be included into different kingdom. These organisms, the Oomycota, contain the water molds, white rusts, downy mildews, as well as the devastating and notorious plant pathogens such as *Phytophthora infestans,* which causes the potato late blight. A small number of plant diseases are caused by protozoa; for example, *Plasmodiophora brassicae* causes a serious clubroot disease on cabbages and cauliflowers and the *Polymyxa* species are important for their ability to transmit plant viruses and colonize the roots of plants such as wheat and barley, causing detrimental effects on the development of plants. In fungi, more species are parasitic on plants than on animals, however, the opposite is true for bacteria. Of the 1600 bacterial species known, which include many serious and life-threatening disease-causing agents of humans, only about 100 species are known to cause diseases on plants. Most plant-pathogenic bacteria are rod shaped (an exception being the filamentous *Streptomyces scabies* that causes potato scab disease. They have developed some pathogenic and disease-inducing characteristics such as cell-wall-decomposing enzymes or plant hormones to obtain nutrients for their growth and reproduction and interfere in plant signaling and defense response. Among them, Mollicutes are a class of bacteria from Gram-positive ancestors having low guanine and cytosine contents by the so-called regressive evolution (Weisburg et al., 1989). The word "Mollicutes" comes from the Latin word *mollis* (which means soft), and *cutis* (which means skin). They have a relatively small genome size (530–1350 kb), making them particularly fastidious microbes. The smallest known genome (530 kb) among living organisms is of the white leaf phytoplasma occurring on Bermuda grass. Leaf, plant, and treehoppers are its vectors, in which the bacterium is able to multiply and infect internal organs including the salivary glands. The diseases caused by such organisms—aster yellows in many ornamentals and vegetables (lettuce, carrots, celery, etc.), coconut lethal and lethal yellowing diseases, rice yellow dwarf, X-disease of peach and cherry, corn stunt, and grape yellows—are particularly found in tropical/semitropical areas of the world. The means by which these organisms produce their often-devastating symptoms, such as tissue proliferation, phyllody (change of floral parts into leafy structures), abnormal elongation of internodes, stunting, yellowing, and death may suggest the production of toxins and/or modifications to plant hormone levels. Indeed, the Mollicutes do not have cell walls; therefore, the common antibiotics such as penicillin or other β-lactams that target cell walls have no adverse effect on them. Most plant-pathogenic bacteria reside in the apoplastic regions of plant tissues, but the phytopathogenic mollicutes (spiroplasma and phytoplasma) are limited to the phloem sieve tubes where they are passed on from plant to plant through phloem sap-sucking insects (Weintraub and Beanland, 2006).

Divided into two families, the spiroplasmas have a helical structure and the phytoplasmas are pleiomorphic, ranging from 200 μm to 800 μm in diameter. These organisms have not been well characterized at the molecular level because they are generally (apart from some) impossible to grow in an axenic culture.

Many species of plant viruses have been identified and characterized at the molecular level. Plant viruses have been organized into 73 genera, 49 of these genera being classified among 15 families, with the remaining 24 unassigned. The classification of plant viruses is done based on their structural morphology, type of host, and method of transmission. About 50% of the known plant viruses are shaped like rods and are rigid, whereas the remainders are isometric. In addition, there are three genera of *Geminiviridae* that possess geminate (twinned) particles, and a small number of cylindrical bacillus-like rods. Most plant viruses are spread through vectors (in particular insects, nematodes, and fungi), or by seed and pollen. Plant viruses induce various symptoms in different plant hosts, due to which the same virus may be classified under more than one name. The majority of plant viruses contain a single strand of ribonucleic acid (RNA), encapsulated in simple protein coats comprising multiple copies of one or a very few coat protein subunits. These particles may be isometric or rod shaped, and there are only few examples of more complex structures, for example, containing lipids and/or bacilliform structures. There are also some plant viruses with double-stranded RNA (dsRNA), single-stranded DNA (ssDNA), and double-stranded DNA (dsDNA) genomes. Transmission of viruses between plants is mainly by sucking insect vectors such as aphids, whiteflies, leaf, and other hoppers. Some viruses are transmitted by biting or chewing insects such as grasshoppers, by nematodes, by fungi, mechanically by larger animals, vegetative plants, or in seed and pollen. Once inside the plant, the virus replicates in individual cells by modifying and utilizing the plant's replication machinery, and spreads between cells progressively before entering the vascular systems for long-distance systemic spread. Unlike animal viruses, plant viruses can progress in plants without crossing cell membranes, by using cell-to-cell plasmodesmata connections. A further group of plant pathogens, the viroids, consist of single-strand circular RNA molecules with no protein coats and with no capacity to encode proteins.

Plant–microbes-associated insect vectors and diseases

There are many different types of insect vectors that transmit infectious diseases in plants. For example, aphids spread over 150 different types of plant viruses such as cabbage black ring spot, tomato-spotted wilt, turnip yellow mosaic, etc. Diseases such as aster yellows, dwarf disease of rice, and Pierce's disease of grapes are spread by leafhoppers and plant hoppers, viruses, phytoplasma, and spiroplasmas (Bertaccini et al., 2007; Bertaccini and Duduk, 2009). Whiteflies are accountable for transmitting yellow mosaic diseases in at least 20 plant species including cowpeas, roses, soybeans, and tomatoes. Whiteflies are also responsible for leaf curl viruses in cotton, potato, tomato, tobacco, and other plants. Pear decline and greening disease of citrus is caused by Psyllids that are the vectors of mycoplasma-like organisms. Flower thrips are known to transmit bacterial, fungal, and viral pathogens, for example, tomato-spotted wilt virus is spread by onion thrips *Thrips tabaci* and tobacco thrips *Frankliniella fusca*. A total of 35 plant viruses including broad bean mottle, turnip yellow mosaic, southern bean mosaic, and rice yellow mottle are caused by the leaf beetle. Potato flea beetles *Epitrix cucumeris* spread the pathogen of potato scab *Actinomyces scabies* when the larvae enter a tuber. Bark beetles are vectors of fungal pathogens in trees. The elm bark beetle *Scolytus multistriatus* infects elm trees with *Ceratocystis ulmi*, the pathogen of Dutch elm disease. Similarly, these are the pine engravers (*Ips pini*) that spread the blue stain fungus *Ceratocystis ips* in pine trees. The chestnut blight disease *Endothia parasitica* is also caused by the Scolytidae. Apple maggots, *Rhagoletis pomonella*, serve as vectors for the spread of *Pseudomonas melophthora*, the causal agent of bacterial rot in apples. Mealybugs are known vectors of several plant viruses including cocoa swollen shoot virus and cocoa mottle leaf virus. Ash-gray leaf bugs in the genus *Piesma* are vectors of the beet leaf curl virus, the sugar beet savoy virus, and cause beet latent rosette disease.

The type and nature of insect vectors of plant viruses depends on the type of association between the virus and the vector. Stylet-borne viruses that are picked and carried on the mouthparts are termed as nonpersistent because they are lost from the insect once it feeds. Foregut-borne viruses are known as semipersistent, whereas those that pass through the gut into the hemolymph and then travel to the salivary glands are called persistent, because they can often be transmitted for as long as the insect lives. Persistent viruses in turn are subdivided into the propagative viruses, which are able to replicate in both plants and the insect vector, and circulative, which can only replicate in plants. Mutagenic analysis on plant viruses has shown that they encode polypeptides with domains that are essential for insect transmission. In most noncirculative viruses, these domains are found in the coat protein and in a second protein known as the helper component (HC).

Phytopathogenic bacteria such as Phytoplasmas can also be spread or transmitted from plants to plants through sap-sucking insect vectors belonging to the different families of insect vectors such as *Cicadellidea* (leafhoppers), *Fulgoridea* (plant hoppers), and *Psyllidea* (jumping plant lice) (Tsai, 1979; Ploaie, 1981) (Figure 8.1).

In addition to insects, Phytoplasmas can be spread from plant to plant by vegetative propagation through cuttings, storage tubers, rhizomes, or bulbs (Lee and Davis, 1992). Phytoplasmas can also be spread via cuscuta dodder (Carraro et al., 2008) and through grafts but unlike viruses they cannot be transmitted mechanically by inoculation with phytoplasma-containing sap from infected plants.

Insects attain phytoplasmas while feeding on phloem of infected plant. Phytoplasmas contain a major antigenic protein that makes up the majority of their cell surface proteins.

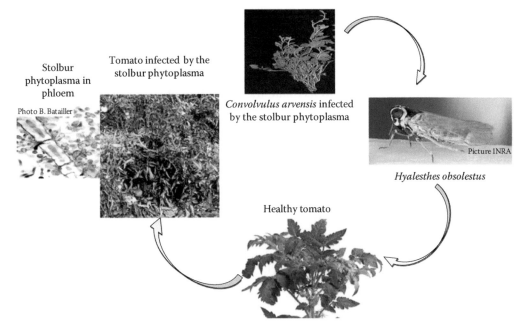

Stolbur phytoplasma in phloem

Photo B. Batailler

Tomato infected by the stolbur phytoplasma

Convolvulus arvensis infected by the stolbur phytoplasma

Picture INRA

Hyalesthes obsolestus

Healthy tomato

Figure 8.1 Transmission of plant microbes through the insect vector (phytoplasma stolbur transmission cycle).

The antigenic membrane protein of *Ca. P. asteris*, protein AMP, was found to interact with the insect cell microfilament and play a role in insect vector specificity (Suzuki et al., 2006). Phytoplasmas may overwinter in insect vectors or perennial plants.

Acquisition feeding is a process in which sap-sucking insects such as plant hoppers and leafhoppers become infected by phytoplasma. Once phytoplasma enters the insect vector from an infected plant, it moves into the intestinal lumen of the host insect, attacks, and replicates at the hemolymph site (Hogenhout et al., 2008) also infecting other organs and tissues including salivary glands. Phytoplasma accumulate into the large vacuoles of salivary glands from where they get access to sieve cells in a process called as inoculation feeding (Nault, 1997).

Phloem-feeding insect vectors belonging to specific groups such as leafhoppers, plant hoppers, or psyllids within the order Hemiptera transmit phytoplasma but other phloem feeders such as aphids are unable to transmit showing the specificity in acquisition and transmission (Nault, 1990).

So, phytoplasma have two distinct kingdoms hosts: Animalia and Plantae. They replicate in different organs and tissues of insects; however, in plants, they only reside in phloem sieve tubes. The capacity of these pathogens to attack and colonize in two different host environments and replicate intracellularly is noteworthy and implies the evolution of the mechanism that enables bacteria to adjust cellular processes in their hosts.

Genomic-based detection, identification, and classification

In the past decades, due to the failure to get pure cultures of any phytoplasma or nonculturable pathogens, their detection and identification was not possible. The presence of specific symptoms in diseased plants and their bodies were the main and important criteria used to diagnose diseases of possible phytoplasmal origin (Shiomi and Sugiura, 1984).

However, in the last 20 years, with the advent of DNA-fingerprinting technology, polymerase chain reaction (PCR) is now widely used for the adequate detection and characterization of particular pathogens. A wide range of pathogens associated with plants and insects were identified through PCR by developing the genomic sequence-specific oligonucleotides using generic or broad-spectrum primers designs based on 16S rDNA (Lee et al., 1992; Namba et al., 1993). To facilitate the separation of such closely related strains and for finer differentiation, several less-conserved genetic markers such as ribosomal protein (rp) genes, secY, tuf, groEL, and 16S-23S rRNA intergenic spacer region sequences have been used as additional tools (Lee et al., 1994, 2000, 2006, 2010; Mitrović et al., 2011). Single-nucleotide polymorphisms (SNPs) have also been exploited as molecular markers separating phytoplasma lineage (Jomantiene et al., 2011).

Plant microbe effectors and associated diseases

Plants have developed powerful strategies to defend themselves against insect and pathogen attack. However, herbivores and plant pathogens try to surmount these physical and biochemical defenses through counterattack. Specific molecules or proteins, called effectors, are produced or secreted by plant pathogens for that purpose. Different insects, nematodes, and pathogens use effectors to facilitate their colonization and multiplication in host. Effectors can be included as elicitors, cell-wall-degrading enzymes, toxins, phytohormone analogs, and other mollicutes that alter host plants (Hogenhout et al., 2009). G proteins are known to regulate many fundamental events in fungi such as mating, pathogenicity, and virulence. G protein subunits have been isolated from many plant pathogens. In the case of the chestnut blight fungus *Cryphonectria parasitica*, targeted gene disruption in one of these has been shown to result in a reduction in virulence.

In genomes of phytoplasmas, Sec-A-secreted proteins are the main effectors that are activated. Recently, 56 candidate effector proteins have been identified and named AY-WB proteins (SAPs) (Bai et al., 2008). Of these, seven out of 56 were encoded on four plasmids and 49 were found on the chromosome (Bai et al., 2006). By a similar approach, 45 in AY, 41 in Australian grapevine yellows (AUGSYs), 13 in AP, and 25 in maize bushy-stunted phytoplasma (MBSP) candidate effectors have been identified. MBSP, OY, and AY-WB belong to group 16SrI and show that even closely related species have different effector contents. Thus, phytoplasma with restricted plant host range may have fewer effectors.

Effector proteins such as SAP11 were detected in cell nuclei (Bai et al., 2008) but TENGU was found at the tip of the stem and was not detected in nuclei (Hoshi et al., 2009). Both were found out of phloem. SAP11 was found to induce witches broom symptom in *Arabidopsis* (Sugio et al., 2010). These plants have curled leaves and an enhanced number of axillary stems that resemble the witches broom symptoms. It has been shown that SAP11 destabilizes the TCP (TEOSINTE BRANCHED1, CYCLOIDEA, PROLIFERATING CELL FACTOR1 and 2) transcription factors that are involved in jasmonic acid (JA) synthesis control of plant development (Martin-Trillo and Cubas, 2010). The undermining of Class11 TCPs causes inhibition of the phytohormone JA, which plays a major role in the defense response against the leafhopper *Macrosteles quadrilineatu*, the AY-WB insect vector (Sugio et al., 2010). It has been shown that survival, reproduction, and fecundity of AY-WB leafhopper was increased when reared upon AY-WB-infected phytoplasma and interestingly more progenies were produced by feeding on SAP11-expressing *Arabidopsis* (Sugio et al., 2010).

Transgenic *Arabidopsis* lines expressing TENGU show certain changes in morphology such as witches broom, dwarfism (short internodes), abnormal leaf arrangement, and production of sterile flowers. It has been shown by microarray that TENGU expressing

Arabidopsis lines revealed a downregulation of several auxin responsive genes and auxin eflux carrier genes. It is thought that TENGU changes plant morphology by manipulating auxin biosynthesis and other signaling pathways. Phytoplasma effector SAP11 targets plant nuclei and destabilizes plant CIN-TCPs and causes morphological changes in the plant suppressing the JA-mediated defense. Some other pathogens also secrete effectors that target cell nuclei but mode of action is distinct or unknown from that of phytoplasma. An effector, PopP2 is secreted by *Ralstonia solanacearum*, which may have virulence function (Deslandes et al., 2003) but in *Ralstonia solanacearum* resistant plants, PopP2 is an avirulence determinant. It colocalizes with corresponding resistance gene (R) in nuclei and prevents degradation of RRS1 (Tasset et al., 2010). *Xanthomonas* species produce TTSS transcription activator-like (TAL) effectors that target plant genes promoter for gene expression. For example, *Xanthomonas oryzae* tal effector, PthXo1, and AvrXa7 seem to pump plant sugar transporter into apoplast and xylem where the pathogen colonizes (Figure 8.2).

Toxins are an important weapon used by many *P. syringae* pathovars that induce chlorosis in plants and increase the severity of disease. The best-studied bacterial toxins in plant pathology have been coronatine, tabtoxin, phaseolotoxin, syringomycin, and syringopeptin produced by a number of *P. syringae* pathovars, which infect soybean, crucifers, ryegrass, *Prunus* spp. and tomato, respectively, that cause necrotic symptoms on plants. In *A. tumefaciens*, the equivalent genes (*tms1* and *tms2*) are present on the tumor inducing (Ti) plasmid. These are used for the expression of genes in plants. Plasmids are extrachromosomal DNA elements that are usually stably inherited within bacterial cell lines and can be transferred between strains, species, or genera and have been found to have a significant role in pathogenicity of many bacteria that infect plants, harboring a number of avirulence genes, a range of toxin and hormone biosynthetic genes, pathogenicity genes, and resistance to chemical control strategies.

In addition to the pathogens, some insects and nematodes also secrete effectors, for example, *Agrobacterium tumefaciens* secrete effector that causes gall making by producing plant growth hormones auxin and cytokinins (Zupan et al., 2000). Whitefly, an important insect, secretes BC1 of the geminivirus and tomato yellow leaf curl China virus, which causes upward curling of leaves. BC1 interferes with myeloblastosis (MYB) transcription factor and suppresses the induction of some JA-responsive genes and promotes the fitness of the whitefly (Yang et al., 2008).

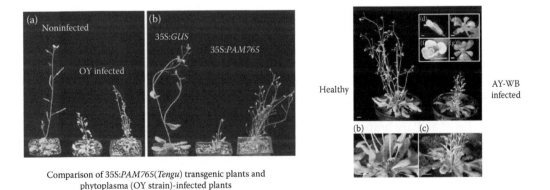

Comparison of 35S:*PAM765*(*Tengu*) transgenic plants and
phytoplasma (OY strain)-infected plants

Figure 8.2 Unique virulence factor for proliferation and dwarfism in plants E factor from phytoplasma (SAP11) in plants. (Adapted from Hoshi A, Oshima K, Kakizawa S, Ishii Y, Ozeki J et al., 2009. *Proceedings of the National Academy of Sciences United States* 106, 6416–6421.)

Interference of plant microbes with plant developmental process

Viral infections of plants modulate their cell metabolism and gene expression thereby activating several defense responses in the plant. The levels of starch, soluble sugars, amino acids and organic acids, and rate of respiration increases in plant tissues whereas the photosynthetic rate of infected plants decreases. As the virus infection front spreads and moves on, the tissues in which these changes are occurring alter. At the same time, the older-infected tissues change to a state in which they are able to survive and perform normal metabolic functions while the virus persists as a nonreplicative entity. The virus must therefore be able to induce a permissive state for its own replication in some cells and at the same time avoid or suppress defense responses. In turn, and as with bacteria and fungi, plants have developed surveillance mechanisms to allow them to detect the presence of viral components and elicit specific defense responses.

The basic infection process involves the virus entering the plant cell through wounds made mechanically or by the vector. In the case of some of the RNA viruses, it has been shown that the capsid is then removed in a process known as cotranslational disassembly, in which the RNA is surrounded by ribosomes to replace the coat protein and protect the RNA from nuclease digestion. The genes encoded by the nucleic acid are then expressed, resulting in genome replication and production of the viral proteins required for pathogenicity. The resultant genomes are then packaged or form ribonucleoprotein complexes, and spread through the plant.

Virus infection results in the diversion of resources away from normal cellular processes as they are recruited for virus replication, for example, the genes encoding the heat shock chaperone protein HSP70 and an NADP +-dependent malic acid enzyme are upregulated during *Cucumber mosaic virus* infection, and during *Pea seed-borne mosaic virus* (PSbMV) infection, mRNAs for HSP70, polyubiquitin, and glutathione reductase 2 were shown to accumulate when other messenger RNAs declined; however, the gene expression for actin and β-tubulin remained unchanged. In *Cucumber mosaic virus*-infected cucurbit, there is an initial localized increase in photosynthesis, NADP +-dependent malic acid enzyme activity, and oxidative pentose phosphate pathway activity in cells, in which the virus is replicating. In healthy plants, sucrose is transported into the phloem. In infected plants, this transport is blocked, and the apoplastic sucrose is converted to hexoses. Having established the role of viral-encoded proteins in replication and assembly of plant viruses, we need to consider the function of other viral genes and proteins in infection. These roles include facilitating the spread of the virus from plant to plant, spread of viruses within plants, and regulation of processes to overcome defense responses within the plant and to ensure a suitable environment for replication and spread.

Phytoplasma-infected plants exhibit a wide range of symptoms on vegetative and nonvegetative parts. The common symptoms caused by phytoplasma disease include phyllody (change of floral parts into leafy structures), virescence (production of green flowers and loss of normal petal pigments), witches broom (proliferation of stem, branches and leaves), proliferation of axillary branches, dwarfism (small-sized flowers and leaves and short internodes), curled leaves, bunchy tips of stems and color changes such as yellowing, pinkish, or reddening of the leaves (Bertaccini, 2007; Ahmad et al., 2015a; Ahmad et al., 2015c).

Phytoplasmas affect the phloem physiology and disturb the transport of carbohydrates (Lepka et al., 1999), thereby causing a yellowish discoloration of foliage leaves (Maust et al., 2003). Phytoplasmas also deteriorate photosynthesis (Lepka et al., 1999); for example, photosystem II is inhibited in many phytoplasma-infected plants (Bertamini et al., 2004). The *Arabidopsis* plant is infected by AY-WB phytoplasma and produces flowers with green

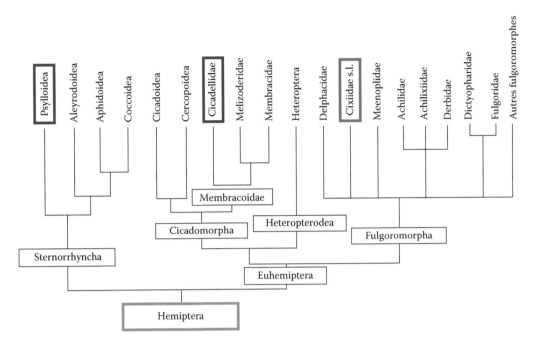

Figure 8.3 Insect vectors of phloem-limited pathogens.

petals. Flower development involves four major stages. These stages have been shown to be altered in the phytoplasma-infected plants. Pracros et al. (2006) and Ahmad et al. (2013a) studied the alteration in flower development by stolbur phytoplasma for the first time. The stolbur phytoplasma-infected tomato showed a sepal hypertrophy, virescence, phyllody, and big bud-like symptoms (Figure 8.3). Often negative for plant health, some symptoms were found positive like the production of more axillary shoots in poinsettia infected with OY phytoplasma, which enables the development of poinsettia plants with more than one flower (Lee et al., 1997) (Figure 8.4).

The mechanism of pathogenesis of phytoplasma is different from those of other Gram-negative phytopathogenic bacteria such as *Pseudomonas, Ralstonia,* and *Xanthomonas,* etc. Such bacteria when it enters the apoplast of infected plants, are inhibited and have to develop a type III secretion system in order to secrete virulence factors. In contrast, phytoplasma live within the sieve cells and their virulence factors or effector proteins are simply secreted through secA-dependent protein translocation system (Kakizawa et al., 2004).

Different phytoplasma effectors have been identified in recent years that alter the cellular processes in plant development playing a role in plant defense (Hogenhout and Loria, 2008; Bai et al., 2008). Whole genome sequence information of various phytoplasma genomes and the use of *Arabidopsis thaliana* in symptom development and effector function analysis has thoroughly helped in the detection and identification of the effectors of phytoplasma (Hogenhout and Music, 2010).

Through semiquantitative RT-PCR, it was demonstrated that tomato homologs WUSCHEL (WUS), CLAVATA1 (CLAV1), APETALA3 (AP3), and AGAMUS (AG) were downregulated, whereas transcription factor LEAFY (LFY) was unchanged or slightly upregulated (Pracros et al., 2006, Ahmad et al., 2013a). According to this research, phytoplasma infection causes decrease in DNA methylation, which in turn downregulates gene

Figure 8.4 Two isolates of bacterial pathogen (Stolbur C and PO) induce different symptoms in infected tomato. (Adapted from Ahmad J N, Pracros P, Garcion C, Teyssier E, Renaudin J, Hernould M, Gallusci P, Eveillard S, 2013a. *Plant Pathology* 62(1), 205–216.)

expression (Pracros et al., 2006, Ahmad et al., 2013a). It was also suggested that homolog of WUS and some class B genes that control floral organ identity are suppressed in OY phytoplasma-infected petunia flowers (Himeno et al., 2010). Further, Cettul and Firrao found that Italian clover phyllody phytoplasma-infected *Arabidopsis* showed changes in flower development due to the downregulation of SEPALATA3 (SEP3) gene (Cettul and Firrao, 2010). Recently, SAP54, an effector of AY-WB, was shown to alter flower development when overexpressed in *A. thaliana* (MacLean et al., 2011).

The different kinds of developmental symptoms observed in phytoplasma-infected plants may be due to the factor that effectors interfere with organ identity. It is suggested that SAP54 may target MADS domain transcription factors. The level of phytoplasmas interference with flower development is yet to be discovered, because many of the flower development genes control each other's expression through a feedback loop mechanism. The effectors may induce pleiotropic effects. Alterations in the levels of plant hormones are particularly important for those bacteria that cause uncontrolled proliferation of plant tissue, resulting in galls and knots, for example, *P. syringae* pv. *savastanoi* (olive and oleander knot), *A. tumefaciens* (crown gall), *Pantoea herbicola* pv. *gypsophilae* (galls on table beet and *Gypsophila paniculata*, a perennial flower), and the Gram-positive nocardiform bacterium *Rhodococcus fascians* (leafy galls on many plants). Among these, *Pseudomonas*, *Agrobacterium*, and *Pantoea* species produce the auxin, indole-3-acetic acid (IAA), which is important for pathogenesis.

Interferences of plant microbes with plant defense response

Plants are continuously attacked by a wide variety of destructive predators, pathogens, and pests such as viruses, bacteria, fungi, and other microorganisms. Each of these potent attackers uses specific strategies to establish a parasitic relationship with its host plant. Plant pathogens secrete diverse virulence factors, which cause disease and severely infect the host plants (Glazebrook, 2005; Göhre and Robatzek, 2008). Phloem feeder insects such as silver whitefly, aphid, and chewing caterpillars have an easy access to food in the form of amino acids and carbohydrates through the phloem tissue. They cause mechanical damage by secreting chemicals from their herbivore saliva and midgut fluids (Walling, 2009; Wu and Baldwin, 2009). Apart from pathogens, plants are commonly attacked by more than one herbivore species at the same time (Vos et al., 2001). To protect themselves from all these different types of pathogens and herbivores, plants have evolved different kinds of structural barriers and antimicrobial metabolites to avert the attackers (De Vos et al., 2005). Plants possess a number of additional defenses against viruses. A line of defense that has become apparent through studies on transgenic plants, is that of posttranscriptional gene silencing (PTGS). PTGS, which incorporates the phenomena termed cosuppression and virus-induced gene silencing (VIGS) is considered to be the plant equivalent of RNA interference (RNAi) (Takahashi et al., 2006). It is believed that the dicer complexes are involved in directing DNA methylation of the transcribed region of the silenced gene, and control transcription of genes through this process, although the role of such control in PTGS resistance to viruses remains unclear.

Plants can distinguish the duration, quality, and quantity of these signals and use "infestation-alert" strategies to resist these external cues by directly controlling insect growth or indirectly through synthesis and release of volatile compounds to attract natural enemies (Dicke and Baldwin, 2010). These strategies protect the plant from attackers. In addition, vegetative growth and reproduction of the plant is maintained (Koornneef and Pieterse, 2008; Mooney et al., 2010). These complex biochemical and physiological responses often lead to a local or systemic resistance (SR) (De Vos and Jander, 2009). Plants can defend themselves with the help of systemic-acquired resistance (SAR) associated with hypersensitive response (HR). SAR is an enhanced level of basal resistance induced in the infected and noninfected parts of the whole plant. During HR, sudden cell death occurs around the infection sites, which prevents the spread of disease from the infection site.

Plants synthesize phytohormones and use them as signaling molecules in defense response. They are very important for proper plant growth and development. These hormones such as salicylic acid (SA), ethylene (ET), and JA play a major role in the plant's counterdefense against pests and pathogens (Glazebrook, 2005; Walling, 2009; Ahmad and Eveillard, 2011; Ahmad et al., 2013b, 2014, 2015b). SA has a role in the defense response against biotrophic pathogens, whereas JA is produced against necrotrophic pathogens and insects (Beckers and Spoel, 2006). Induction of pathogenesis-related proteins (PRs) is one of the best characters associated with defense pathways.

PRs are composed of 17 groups, most often associated with plant defense response (Van loon, 1999). The PR proteins show a wide range of functions but are generally produced when plants face different types of pathogens such as viroids, viruses, bacteria, and fungi or environmental stresses (Van Loon et al., 2006; Ahmad and Eveillard, 2011; Ahmad et al., 2013a, 2014). The SAP11 transgenic *Arabidopsis* lines show a reduced expression of Lox2 and JA accumulation on phytoplasma infection (Sugio et al., 2010).

Two types of immunity systems have been described in plants against different pathogens. Pathogen-associated molecular patterns (PAMPS) or MAPS are recognized by the pattern recognition receptors (PRRs) of a broad host range, which lead to PAMP-triggered immunity (PTI). Strong pathogens are able to secrete virulence effectors that suppress PTI, which results in compatible (susceptible) interaction. When these virulence factors or effectors are recognized by plants, they result in incompatible (resistant) interaction. There are many pathogens that have evolved effectors to suppress PTI or ETI (Dodds and Rathjen, 2010). For instance, *P. syringae* strain DC3000 secretes effectors that suppress PTI or ETI, or both indicating that the suppression of both types of reactions helps to colonize the bacterial pathogens.

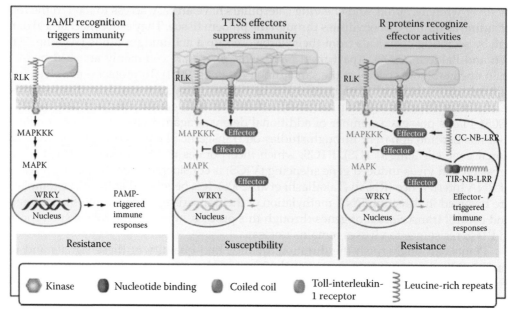

Evolution of bacterial resistance in plants (Chisholm et al., 2006)

Figure 8.5 Plant defense system against plant bacteria.

PAMPS include lipopolysaccharides, peptidoglycans, and conserved domains of fla-gellin but phytoplasma have no such effectors because of the absence of the outer cell wall. Phytoplasma have genes encoding cold shock proteins (CSPs) and the elongation factor Tu (EF-Tu) like PAMPS, which can induce PTI in the host. However, it is not clear how plants respond to phytoplasma by their defense pathways (Figure 8.5).

Cross talk among defense pathways

Plants enhance their defense mechanisms through cross talk between different hormonal-signaling defense pathways, which in addition to growth and development are related to biotic and abiotic stress adaptation (Kazan and Manners, 2008). In nature, the primary defense response induced by the host plant is influenced by concurrent or successive attack of multiple aggressors (Poelman et al., 2008). The cross talk between phytohormonal-sig-naling defense pathways enables the plant to greatly enhance its defense response poten-tial against the attackers (Bostock, 2005; Pieterse and Dicke, 2007; Dick and Baldwin, 2010).

These SA-, JA-, and ET-mediated defense pathways are interconnected and may act antagonistically or synergistically (Spoel et al., 2003; Mur et al., 2006; Ahmad and Eveillard, 2011; Ahmad et al., 2014). It is known that coranatine (COR) produced during JA signaling induces the expression of JA-mediated genes but suppresses the SA-dependent defense genes. According to an earlier report by Doherty et al. (1988), SA and its acetylated deriva-tive aspirin are antagonistic to JA-signaling pathway in tomato and *Arabidopsis.* On the contrary, the JA- and ET-signaling pathways can act synergistically (Penninckx et al., 1998; Ahmad and Eveillard, 2011; Ahmad et al., 2013b). Advancement in the field of genomics has helped scientists investigate how the SA-, JA-, and ET-signaling defense pathways are interconnected. A combined SA-, JA-, or ET-signaling response was observed in response to infection by the hemibiotrophic bacterial pathogen *Pseudomonas syringae* on the whole genome expression profiling of a large set of *Arabidopsis* mutants (Glazebrook et al., 2003), which confirmed that there is a great deal of cross talk between the SA-, JA-, and ET-response pathways, which act as a network in the plant's immune response.

Defense-related signal cross talk can also be observed in antagonistic interaction between the SA- and JA-response pathways. There are many reports on trade-off between SA-dependent resistance against the biotrophic pathogens and JA-dependent defense resistance against necrotrophic pathogens (Pieterse and Dicke, 2007; Poelman et al., 2008). For example, attack of the biotrophic oomycete pathogen *Hyaloperonospora arabidopsidis* on *Arabidopsis* induced the SA pathway which in turn suppressed JA-mediated defenses induced when caterpillars of the small cabbage white *Pieris rapae* fed on it (Koornneef et al., 2008). Likewise, activation of the SA pathway by *P. syringae* downregulated the JA signaling pathway in turn leaving the infected leaves more prone to the necrotrophic fungus *Alternaria brassicicola* (Spoel et al., 2007).

ET is also an important phytohormone involved in plant's defense response to pathogen and insect attack (Van Loon et al., 2006). ET was shown to enhance the response of *Arabidopsis* to SA signaling that results in a potentiated expression of the SA-responsive marker gene *PR-1* (Lawton et al., 1994; De Vos et al., 2006). This synergistic effect of the ET on SA-induced *PR-1* expression was blocked in the ET-insensitive mutant *ein2* (De Vos et al., 2006) indicating that the modulation of the SA pathway by ET is EIN2 dependent and thus functions through the ET-signaling defense pathway.

Network topology study of Glazebrook (2005) highlights the global expression profiles of *P. syringae*-infected *Arabidopsis* wild-type and signaling-defective mutant plants. This study demonstrated the extensive cross talk between the SA and ET-signaling pathways, as evidenced by the fact that the expression of many SA-responsive genes was significantly affected in the *ein2* mutant background. Additionally, these defense-signaling pathways also communicate with other hormones such as abscisic acid (ABA), auxins, gibberellic acid, and brassinosteroid-signaling pathways to enhance the magnitude immune response against attackers (Spoel and Dong, 2008). There are new evidences for additional defense-signaling pathways that are induced by herbivory; however, more detailed studies of these pathways are still required (Glazebrook, 2005; De Vos and Jander, 2009; Bhattarai et al., 2010).

Counterdefense strategies of pathogens

To maintain growth and survival, plants and their respective pathogens have evolved a system of defense and counterdefense strategies (Walling, 2008), for example, some plants can kill cell walls of the pathogens by using the PR proteins b-1,3—triggering the pathogen to release oligosaccharide elicitors that activate plant defense responses making the pathogen more vulnerable to other plant defense responses (Rose et al., 2002). On the contrary, the oomycete pathogen *Phytophthora sojae* is able to slow down the activity of b-1, 3-endoglucanases through its counterdefense response. *P. sojae* secretes glucanase inhibitor proteins (GIPs) that suppress the activity of soybean (*Glycine max*) β-1,3-endoglucanase (Rose et al., 2002). Another plant enzyme–pathogen inhibitor system appears to have coevolved, which includes plant proteases and pathogen protease inhibitors (Rooney et al., 2005; Tian et al., 2005). A diverse family of Kazal-like extracellular Ser protease inhibitors with at least 35 members have been reported in five plant-pathogenic oomycetes: *P. infestans*, *P. sojae*, *Phytophthora ramorum*, *Phytophthora brassicae*, and the downy mildew *Plasmopara halstedii* (Tian et al., 2004). In tomato infected with *P. infestans*, the two Kazal-like Ser protease inhibitors, EPI1 and EPI10, were found to bind and inhibit the PR P69B subtilisin-like Ser protease of the tomato host plant (Tian et al., 2004, 2005). This inhibition of P69B by two structurally different protease inhibitors of *P. infestans* elicits that EPI1 and EPI10 are involved in the counterdefense mechanisms of pathogens (Tian et al., 2004, 2005).

Conclusion

Microbes such as bacteria, fungi, nematodes, protozoans, and viruses are serious and life-threatening disease-causing agents. They are responsible for severe worldwide damage to plants. Most of the plant pathogens are transmitted by biological vectors such as insects, nematodes, and fungi, or through seed and pollen. Transmission of viruses and bacteria between plants is mainly carried out by insect vectors, particularly arthropods such as aphids, whiteflies, beetles, leaf, and other hoppers. Owing to their immobile nature, plants are continuously attacked by a wide variety of destructive pathogens. Specific molecules or proteins, called effectors, are produced or secreted by plant pathogens to facilitate their colonization and multiplication in host. Pathogen attack impairs the morphology and physiology of plants causing developmental abnormalities. Plants activate several complex defense responses to protect themselves from pathogens and insect pests. Major altera-tions in host gene activities and metabolic processes take place due to the toxins/effectors, which are particularly important weapons used by many pathogens (*P. syringae*) to induce chlorosis in plants and increase the severity of disease. Phytohormones such as SA, ET, and JA are the primary signals in the regulation of the plant's immune response against the herbivore insects and pathogens. They are essential for the regulation of plant growth, development, reproduction, and survival. The SA-, JA-, and ET-mediated defense path-ways are interconnected and can act antagonistically or synergistically. Plants also use "infestation-alert" triggers to induce resistance by inhibiting insect growth and releasing volatiles to counterattack the pathogens and defend themselves with the help of SAR asso-ciated with HR through induction of PR proteins. PAMPS or MAPS are recognized by the PRRs of a broad host range, which lead to PTI. In addition to plant responses, pathogenic microbes and insects have also evolved certain counterdefense mechanisms to protect themselves and enable their survival. With the advent of biotechnology, it is now possible to identify and study plantpathogenic microbes and subsequent plant defense response on a molecular level. Although a considerable amount of work has been conducted over the years, we yet have to reveal certain aspects of the complex plant microbe interactions and come up with a sustainable strategy to protect our natural plants.

References

Ahmad J N and Eveillard S, 2011. Study of the expression of defense related protein genes in stolbur C and stolbur PO phytoplasma-infected tomato. *Bulletin of Insectology* 64, S159–S160.

Ahmad J N, Pracros P, Garcion C, Teyssier E, Renaudin J, Hernould M, Gallusci P and Eveillard S, 2013a. Effects of stolbur phytoplasma infection on DNA methylation processes in tomato plants. *Plant Pathology* 62(1), 205–216.

Ahmad J N, Renaudin J and Eveillard S, 2014. Expression of defense genes in stolbur phytoplasma infected tomatoes, and effect of defense stimulators on disease development. *European Journal of Plant Pathology* 139(1), 39–51.

Ahmad J N, Maneta Peyret L, Moreau P and Ahmad R, 2013b. Study of lipid protein interaction in the secretory pathway of plant cell by raising and using antilipid antibodies against par-ticular lipids and proteins in *Arabidopsis* and tobacco plant. *Pakistan Journal of Botany* 459(SI), 509–514.

Ahmad J N, Ahmad S J N, Arif M J and Irfan M, 2015a. First Report of Oil Seed Rape (Brassica napus) associated phytoplasma diseases and their insect vector in Pakistan. *Phytopathogenic Molecule* 5, S89–S90.

Ahmad J N, Renaudin J and Eveillard Sandrine, 2015b. The molecular study of the effect of exogenous Phytohormones application in stolbur phytoplasma infected tomatoes on disease development. *Phytopathogenic Molecule* 5, S121–122.

Ahmad S J N, Ahmad J N, Irfan M, Ahmad M and Aslam M, 2015c. New Reports of phytoplasma occurrence in Pakistan. *Phytopathogenic Molecule* 5, S71–S72.

Bai X, Correa V R, Toruño T Y, Ammar E-D, Kamoun S and Hogenhout S A, 2008. AY-WB phytoplasma secretes a protein that targets plant cell nuclei. *Molecular Plant–Microbe Interactions* 22, 18–30.

Bai X, Zhang J, Ewing A, Miller S, Radek A et al., 2006. Living with genome instability: The adaptation of phytoplasmas to diverse environments of their insect and plant hosts. *Journal of Bacteriology* 188, 3682–3696.

Bertaccini A, 2007. Phytoplasmas: Diversity, taxonomy, and epidemiology. *Frontiers of Bioscience* 12, 673–689.

Bertaccini A and Duduk B, 2009. Phytoplasma and phytoplasma diseases: A review of recent research. *Phytopathologia Mediterranea* 48, 355–378.

Bertaccini A, Botti S, Fiore N, Gayardo A and Montealegre J, 2004. Identification of new phytoplasma(s) infections in grapewine with yellows in Chile. *15th Congress of International Organization of Mycoplasmology*, Athens Georgia, USA, 37, pp. 63–64.

Bertaccini A, Calari A and Felker P, 2007. Developing a method for phytoplasma identification in cactus pear samples from California. *Bulletin of Insectology* 60, 257–258.

Bertamini M, Muthuchelian K and Nedunchezhian N, 2004. Effect of grapevine leafroll on the photosynthesis of field grown grapevine plants (*Vitis vinifera* L. cv. Lagrein). *Journal of Phytopathology* 152, 145–152.

Bostock R M, 2005. Signal crosstalk and induced resistance: Straddling the line between cost and benefit. *Annual Review of Phytopathology* 43, 545–580.

Carraro L, Ferrini F, Martini M, Ermacora P and Loi N, 2008. A serious epidemic of stolbur on celery. *Journal of Plant Pathology* 90(1), 131–135.

Cettul E and Firrao G, 2010. Effects of phytoplasma infection on *Arabidopsis thaliana* development. *Congress of the International Organization for Mycoplasmology*, 18th, Cianciano Terme, 48, 82.

Deslandes L, Olivier J, Peeters N, Feng D X and Khounlotham M, 2003. Physical interaction between RRS1-R, a protein conferring resistance to bacterial wilt, and PopP2, a type III effector targeted to the plant nucleus. *Proc. Natl. Acad. Sci. USA* 100, 8024–8029.

De Vos M and Jander G, 2009. *Myzus persicae* (green peach aphid) salivary components induce defence responses in *Arabidopsis thaliana*. *Plant, Cell and Environment* 32, 1548–1560.

De Vos M, Van Oosten V, Van Poecke R, Van Pelt J, Pozo M, Mueller M, Buchala A, Metraux J-P, Van Loon L and Dicke M, 2005. Signal signature and transcriptome changes of *Arabidopsis* during pathogen and insect attack. *Molecular Plant–Microbe Interactions* 18, 923–937.

De Vos M, Van Zaanen W, Koornneef A, Korzelius J P, Dicke M, Van Loon L C and Pieterse C M J, 2006. Herbivore-induced resistance against microbial pathogens in *Arabidopsis*. *Plant Physiology* 142, 352–363.

Dicke M and Baldwin I T, 2010. The evolutionary context for herbivore-induced plant volatiles: Beyond the "cry for help". *Trends in Plant Science* 15, 167–175.

Dodds P N and Rathjen J P, 2010. Plant immunity: Towards an integrated view of plant–pathogen interactions. *Nature Review of Genetics* 11, 539–548.

Doherty H M, Selvendran R R and Bowles D J, 1988. The wound response of tomato plants can be inhibited by aspirin and related hydroxy-benzoic acids. *Physiological and Molecular Plant Pathology* 33, 377–384.

Glazebrook J, 2005. Contrasting mechanisms of defense against biotrophic and necrotrophic pathogens. *Annual Review of Phytopathology* 43, 205–227.

Glazebrook J, Chen W, Estes B, Chang H-S, Nawrath C, Métraux J-P, Zhu T and Katagiri F, 2003. Topology of the network integrating salicylate and jasmonate signal transduction derived from global expression phenotyping. *The Plant Journal* 34, 217–228.

Göhre V and Robatzek S, 2008. Breaking the barriers: Microbial effector molecules subvert plant immunity. *Annual Review of Phytopathology* 46, 189–215.

Hogenhout S A and Loria R, 2008. Virulence mechanisms of Gram-positive plant pathogenic bacteria. *Current Opinion of Plant Biology* 11, 449–456.

Hogenhout S A and Music M, 2010. *Phytoplasma Genomics, from Sequencing to Comparative and Functional Genomics: What Have We Learnt? Phytoplasmas, Genomes, Plant Hosts and Vectors.* CABI, Wallingford, UK, pp. 19–36.

Hogenhout S A, Oshima K, Ammar L D, Kakizawa S K H N and Namb S, 2008. Phytoplasmas: Bacteria that manipulate plants and insects. *Molecular Plant Pathology* 9, 403–423.

Hogenhout S A, Van d R A, Terauchi R and Kamoun S, 2009. Emerging concepts in effector biology of plant-associated organisms. *Molecular Plant–Microbe Interactions* 22, 115–122.

Hoshi A, Oshima K, Kakizawa S, Ishii Y, Ozeki J et al., 2009. A unique virulence factor for proliferation and dwarfism in plants identified from a phytopathogenic bacterium. *Proceedings of the National Academy of Sciences United States* 106, 6416–6421.

Himeno M, Kojima N, Neriya Y, Sugawara K, Ishii Y et al., 2010. Characterization of floral morphogenesis and expression of floral development genes in phytoplasma-infected *Hydrangea* and petunia. *Congress of the International Organization for Mycoplasmology*, 18th, Cianciano Terme, 222, 198.

Jomantiene R R, Zhao Y, Lee I M and Davis R E, 2011. Phytoplasmas infecting sour cherry and lilac represent two distinct lineages having close evolutionary affinities with clover phyllody phytoplasma. *European Journal of Plant Pathology* 130, 97–107.

Kakizawa S, Oshima K, Nishigawa H, Jung H Y, Wei W et al., 2004. Secretion of immunodominant membrane protein from onion yellows phytoplasma through the Sec protein–translocation system in *Escherichia coli*. *Microbiology* 150, 135–142.

Kazan K and Manners J M, 2008. Jasmonate signaling: Toward an integrated view. *Plant Physiology* 146, 1459–1468.

Koornneef A and Pieterse C M J, 2008. Cross talk in defense signaling. *Plant Physiology* 146, 839–844.

Koornneef A, Verhage A, Leon-Reyes A, Snetselaar R, van Loon L and Pieterse C M, 2008. Towards a reporter system to identify regulators of cross-talk between salicylate and jasmonate signaling pathways in *Arabidopsis*. *Plant Signaling and Behavior* 3(8), 543–546.

Lawton K A, Potter S L, Uknes S and Ryals J, 1994. Acquired resistance signal transduction in *Arabidopsis* is ethylene independent. *The Plant Cell Online* 6, 581–588.

Lee I M and Davis R E, 1992. Mycoplasmas which infect plants and insects. In *Mycoplasmas: Molecular Biology and Pathogenesis*. Eds. J Maniloff, R N McElhansey, R L Finch and J B Baseman. pp. 379–390. American Society for Microbiology, Washington, DC.

Lee I M, Davis R E and Gundersen-Rindal D E, 2000. Phytoplasma: Phytopathogenic mollicutes. *Annual Review of Microbiology* 54, 221–255.

Lee I M, Gundersen D E, Hammond R W and Davis R E, 1994. Use of mycoplasma like organism (MLO) group-specific oligonucleotide primers for nested-PCR assays to detect mixed-MLO infections in a single host plant. *Phytopathology* 84, 559–566.

Lee I M, Zhao Y and Bottner K D, 2006. SecY gene sequence analysis for finer differentiation of diverse strains in the aster yellows phytoplasma group. *Molecular and Cellular Probes* 20, 87–91.

Lee I-M, Bottner-Parker K D, Zhao Y, Davis R E and Harrison N A, 2010. Phylogenetic analysis and delineation of phytoplasmas based on secY gene sequences. *International Journal of Systematic and Evolutionary Microbiology* 60, 2887–2897.

Lee J J, von Kessler D P, Parks S and Beachy P A, 1992. Secretion and localized transcription suggest a role in positional signaling for products of the segmentation gene hedgehog. *Cell* 71, 33–50.

Lepka P, Stitt M, Moll E and Seemüller E, 1999. Effect of phytoplasmal infection on concentration and translocation of carbohydrates and amino acids in periwinkle and tobacco. *Physiological and Molecular Plant Pathology* 55, 59–68.

Martin-Trillo M and Cubas P, 2010. TCP genes: A family snapshot ten years later. *Trends in Plant Science* 15, 31–39.

Maust B E, Espedas F, Talavera C, Aguilar M, Santamaria J M and Oropeza C, 2003. Changes in carbohydrate metabolism in coconut palms infected with the lethal yellowing phytoplasma. *Phytopathology* 93, 976–981.

Maust B E, Espedas F, Talavera C, Aguilar M, Santamaria J M and Oropeza C, 2003. Changes in carbohydrate metabolism in coconut palms infected with the lethal yellowing phytoplasma. *Phytopathology* 93, 976–981.

MacLean A M, Sugio A, Kingdom H N, Grieve V M and Hogenhout S A, 2011. *Arabidopsis thaliana* as a model plant for understanding phytoplasmas interactions with plant and insect hosts. *Bulletin of Insectology*. 64, S173–S174.

Mitrović J, Kakizawa S, Duduk B, Oshima K, Namba S and Bertaccini A, 2011. The groEL gene as an additional marker for finer differentiation of 'Candidatus Phytoplasma asteris'-related strains. *Annals of Applied Biology* 159, 41–48.

Mooney K A, Halitschke R, Kessler A and Agrawal A A, 2010. Evolutionary trade-offs in plants mediate the strength of trophic cascades. *Science* 327, 1642–1644.

Mur L A J, Kenton P, Atzorn R, Miersch O and Wasternack C, 2006. The outcomes of concentration-specific interactions between salicylate and jasmonate signaling include synergy, antagonism, and oxidative stress leading to cell death. *Plant Physiology* 140, 249–262.

Namba S, Kato S, Iwanamic S, Oyaizu H, Shioawa H and Tsuchizaki T, 1993. Detection and differentiation of plant-pathogenic mycoplasma like organisms using polymerse chain reaction. *Phytopathology* 83, 786–791.

Nault L R, 1990. Evolution of an insect pest: Maize and the corn leafhopper, a case study. *Maydica* 35, 165–175.

Nault L R, 1997. Arthropod transmission of plant viruses: A new synthesis. *Annals of the Entomological Society of America* 90, 521–541.

Penninckx I A M A, Thomma B P H J, Buchala A, Métraux J-P and Broekaert W F, 1998. Concomitant activation of jasmonate and ethylene response pathways is required for induction of a plant defensin gene in *Arabidopsis*. *The Plant Cell Online* 10, 2103–2114.

Pieterse C M J and Dicke M, 2007. Plant interactions with microbes and insects: From molecular mechanisms to ecology. *Trends in Plant Science* 12, 564–569.

Ploaie P G, 1981. Mycoplasma-like organisms and plant diseases in Europe. In *Plant Diseases and Vectors: Ecology and Epidemiology*. Eds. K Maramorosch and K F Harris. pp. 61–104. Academic Press, New York.

Poelman E H, van Loon J J A and Dicke M, 2008. Consequences of variation in plant defense for biodiversity at higher trophic levels. *Trends in Plant Science* 13, 534–541.

Pracros P, Hernould M, Teyssier E, Eveillard S and Renaudin J, 2007. Stolbur phytoplasma-infected tomato showed alteration of SlDEF methylation status and deregulation of methyltransferase genes expression. *Bulletin of Insectology* 60, 221–222.

Pracros P, Renaudin J, Eveillard S, Mouras A and Hernould M, 2006. Tomato flower abnormalities induced by stolbur phytoplasma infection are associated with changes of expression of floral development genes. *Molecular Plant–Microbe Interactions* 19, 62–68.

Rooney HC, Van't Klooster JW, van der Hoorn RA, Joosten MH, Jones JD, de Wit PJ, 2005. Cladosporium Avr2 inhibits tomato Rcr3 protease required for Cf-2-dependent disease resistance. *Science* 308, 1783–1786.

Rose J K C, Ham K-S, Darvill A G and Albersheim P, 2002. Molecular cloning and characterization of glucanase inhibitor proteins: Coevolution of a counterdefense mechanism by plant pathogens. *The Plant Cell Online* 14, 1329–1345.

Shiomi T and Sugiora M, 1984. Grouping of mycoplasma like organisms transmitted by the leafhopper vector *Macrosteles orientalis* Virvast, based on host range. *Annals of Phytopathological Society of Japan* 50, 149–157.

Spoel S, Koorneef A, Claessens S, Korzelius J, Van Pelt J et al., 2003. NPR1 modulates cross-talk between salicylate and jasmonate-dependent defense pathways through a novel function in the cytosol. *Plant Cell* 15, 760–770.

Spoel S H, Johnson J S and Dong X, 2007. Regulation of tradeoffs between plant defenses against pathogens with different lifestyles. *Proceedings of the National Academy of Sciences* 104, 18842–18847.

Spoel S H and Dong X, 2008. Making sense of hormone crosstalk during plant immune responses. *Cell Host Microbe* 3, 348–351.

Sugio A, Kingdom H N, Nicholls V M and Hogenhout S A, 2010. The phytoplasma effector protein SAP11 improves vector fitness. *Congress of the International Organization for Mycoplasmology* 18th, Cianciano Terme, 47, 82.

Suzuki S, Oshima K, Kakizawa S, Arashida R, Jung H-Y, Yamaji Y, Nishigawa H, Ugaki M and Namba S, 2006. Interaction between the membrane protein of a pathogen and insect microfilament complex determines insect–vector specificity. *Proceedings of the National Academy of Sciences of the United States* 103, 4252–4257.

Takahashi S, Komatsu K, Kagiwada S, Ozeki J, Mori T, Hirata H, Yamaji Y, Ugaki M and Namba S, 2006. The efficiency of interference of potato virus X infection depends on the target gene. *Virus Research* 116, 214–217.

Tian M, Benedetti B and Kamoun S, 2005. A second Kazal-like protease inhibitor from *Phytophthora infestans* inhibits and interacts with the apoplastic pathogenesis-related protease P69B of tomato. *Plant Physiology* 138, 1785–1793.

Tian M, Huitema E, da Cunha L, Torto-Alalibo T and Kamoun S, 2004. A Kazal-like extracellular serine protease inhibitor from *Phytophthora infestans* targets the tomato pathogenesis-related protease P69B. *Journal of Biological Chemistry* 279, 26370–26377.

Tsai J H, 1979. Vector transmission of mycoplasmal agents of plant diseases. In *The Mycoplasmas.* Eds. R F Whitcomb and J G Tully. Vol. 3, pp. 265–307. New York, Academic Press.

Van Loon L C, 1999. Occurrence and properties of plant pathogenesis related proteins. In *Pathogenesis-Related Proteins in Plants.* Eds. S K Datta and S Muthukrishnan. pp. 1–20. CRC Press, Boca Raton, FL.

Van Loon L C, Geraats B P J and Linthorst H J M, 2006. Ethylene as a modulator of disease resistance in plants. *Trends in Plant Sciences* 11, 184–191.

Vos M, Berrocal S M, Karamaouna F, Hemerik L and Vet L E M, 2001. Plant-mediated indirect effects and the persistence of parasitoid–herbivore communities. *Ecology Letters* 4, 38–45.

Walling L L, 2008. Avoiding effective defenses: Strategies employed by phloem-feeding insects. *Plant Physiology* 146, 859–866.

Walling L L, 2009. Adaptive defense responses to pathogens and pests. *Advances in Botanical Research* 51, 551–612.

Weintraub P G and Beanland L A, 2006. Insect vectors of phytoplasmas. *Annual Revenue of Entomology* 51, 91–111.

Weisburg W G, Tully J G, Rose D L, Petzel J P, Oyaizu H, Yang D, Mandelco L, Sechrest J, Lawrence T G and Van Etten J, 1989. A phylogenetic analysis of the mycoplasmas: Basis for their classification. *Journal of Bacteriology* 171, 6455–6467.

Wu J and Baldwin I T, 2009. New insights into plant responses to the attack from insect herbivores. *Annual Review of Genetics* 44, 1–24.

Yang J Y, Iwasaki M, Machida C, Machida Y, Zhou X and Chua N H, 2008. BetaC1, the pathogenicity factor of TYLCCNV, interacts with AS1 to alter leaf development and suppress selective jasmonic acid responses. *Genes and Development* 22, 2564–2577.

Zupan J, Muth T R, Draper O and Zambryski P, 2000. The transfer of DNA from *Agrobacterium tumefaciens* into plants: A feast of fundamental insights. *Plant Journal* 23, 11–28.

chapter nine

Biotechnology for improved crop productivity and quality

Cassiana Severiano de Sousa, Maria Andréia Corrêa Mendonça, Syed Shah Hassan, Debmalya Barh, and Vasco Ariston de Carvalho Azevedo

Contents

Abstract

New crop varieties with modified characteristics, such as higher yields, tolerance to biotic and abiotic stresses and improved nutritional characteristics are gaining attention in the last decades and this chapter will review the subject. Biotechnology can assist the process of crop breeding. Moreover, selection assisted by molecular markers has been aiding classical crop breeding. Molecular markers and genetic maps will be discussed throughout this chapter. The manipulation of genes in plants started after the discovery of a tool set called recombinant deoxyribonucleic acid (DNA) technology and using this approach, it was possible conferring characteristics of interest to them. Additionally, "omics" tools enabled the knowledge of genes involved with desirable features. Various types of GM plants had been created and they can be categorized according to their purposes. Finally, this chapter will focus on the transgenic genotypes that have been developed, as well as the issues related to biosafety and management of transgenic cultivars.

Keywords: Improved crops, Biotechnology, Plant breeding, Genetically modified plants, Biosafety and management.

Introduction

Since the world population is increasing over the years, a corresponding increase in food production is necessary to feed the population. We are 7.2 billion people around the world and as reported by Food and Agricultural Organization of the United Nations (FAO), the number of people in the world will reach 9.1 billion by 2050. For this reason, food production will need to rise by 60% from its 2005 to 2007 levels. If we consider the role of agriculture to support biofuels, this number will be even higher (FAO, 2013). The challenge is to increase agricultural production without too much increase in the land used for planting. Thus, it is extremely important that new crop varieties with modified/improved characteristics are developed. Among the features that can be highlighted are higher yields, tolerance to various types of biotic and abiotic stresses, as well as improved nutritional characteristics, which will be cited throughout this chapter.

In this context, biotechnology can be inserted, aiding all this process of plant breeding. The genetic breeding assisted by molecular markers and the new biotechnology traits are technologies that were once incorporated into crop improvement programs, which can increase production rates to higher than the ones observed in the last decade. One of the concerns is the effect of climate changes on agricultural production. So, the development of new adapted varieties to the most diverse weather effects, by exhibiting one or more improved features has attracted the attention of researchers (Edgerton, 2009).

Among the main techniques related to biotechnology, the recombinant deoxyribonucleic acid (DNA) technology, which emerged some years ago, allowed the manipulation of genes from various sources and insertion of these genes into plants in order to confer the characteristics of interest. For example, drought and soil salinity resistance, herbicide toleration, disease and pest resistance, enhanced qualities, increased rate of photosynthesis and production of sugar and starch, in the production of medicines and vaccines in crop plants, among others (Sharma et al., 2002). Genetic information has been farmed and applied to many different technologies such as plant tissue culture, genetic engineering of plants, plant transformation for the purpose of breeding, as well as their application to databases and bioinformatics tools.

Since classical breeding, researchers have sought to somehow control and speed up the whole process of getting, farming, and producing plants to attend to the needs of producers and consumers. The use of molecular markers and genetic maps was important to relate the phenotypic with genotypic information and thus accelerate the process of plant breeding step. Molecular markers and genetic maps will be discussed in more detail throughout this chapter. All the biotechnological advances related to the improvement of productivity and quality of crops are of great importance, since the plant breeding can be blamed for more than half of the increase in crop yields over the last century.

DNA is the genetic material that contains the information necessary for the heredity of information and for the genotypes. The elucidation of the structure of the DNA as a double helix by Watson and Crick, in 1950s, was a milestone for understanding important process in the cell related to the biology of life. Biotechnology uses, largely, the knowledge that came from this discovery to act at the gene level by selecting the characteristics of interest and avoiding the nondesirable ones.

Various types of genetically modified (GM) plants with the most diverse objectives had already been created, and they can be classified into three categories: (1) First-generation GM plants, aiming the productive process, as introducing agronomic characteristics, for example, resistance to diseases, insects, and herbicides; (2) second-generation GM plants, aiming the consumer benefits, as nutritional characteristics and better storage after harvest; and (3) third-generation GM plants, aiming their use as biofabrics to synthesize compounds with application in the pharmaceutical, biosanitary, or industrial area (Qaim, 2009). This chapter will discuss the transgenic genotypes that have already been developed, as well as the issues related to biosafety and management of transgenic cultivars.

Genetic and biotechnology evolution in relation to plant breeding

Agriculture emerged about 12,000 years ago, in the prehistoric period, when people observed that some grains could be buried to produce new plants, identical to the ones that originated them. This discovery initiated the process of planting with the aim to provide food for people. In addition, plants began to be planted close to each other to facilitate cultivation.

Over the years, contributions important for the production of plants have increased with the purpose of obtaining plants with characteristics of interest in a single genotype. Table 9.1 lists some of the historical events that contributed to the development in genetics and biotechnology, which can be related to plant breeding.

Classical/conventional improvement × biotechnology

Plant breeding encompasses the origination and selection up to the maintenance of phenotypes with superior characteristics, in the cultivars development, and suits to the farmers and consumer desires (Moose and Mumm, 2008).

Classical/conventional breeding and the breeding aided by biotechnology, with the consequent achievement of genetically modified organisms (GMOs) are two different methodologies that allow the transfer of genes. Although both approaches involve the genetic modification of an organism with respect to DNA sequence, the number of genetic changes introduced by GMO technology are smaller and more defined. This is because in classical breeding a large number of uncharacterized genes may be involved (Datta, 2013).

The characteristics of the cultures that are sought by genetic engineering or by classical breeding are not totally different. However, genetic engineering permits the gene transfer directly between organisms. In addition, many features that with the classical breeding were difficult or even impossible to be transferred from one organism to another can be transferred with the aid of genetic engineering fairly easily (Qaim, 2009).

Classical breeding and the aid of biotechnology and molecular biology

In classical breeding, breeders realize the crossing between correlated species to produce other varieties with wanted features. The crosses are done in order to create a new genetic background. The genomes of the two parents are mixed and randomly distributed in the

Table 9.1 Historical facts important to evolution in the breeding of plants

Date	Milestone fact
1809	Jean Baptiste Lamarck published his theory in the book "Philosophie Zoologigue" based on two principles: (1) living beings have evolved to an increased level of complexity and perfection, the reason why Lamarck believed that living beings had evolved from simple organisms originated from nonliving material (theory of spontaneous generation), (2) the law of use and disuse explains that underutilized organs undergo atrophy and disappear and, on the other hand, the most used are developed and passed on to future generations. Lamarck's ideas influenced the evolutionary studies of Charles Darwin. The theory of acquired characteristics was invalidated by the studies of genetics and heredity.
1859	Charles Darwin published "The Origin of Species," stating that the evolution of species would occur by natural selection and not by use and disuse. According to Darwin's theory some small variations in organisms arise by chance and, if these variations make them more capable than the others, they could survive and transmit their characteristics to their offspring.
1865	Gregor Johann Mendel conducted studies with peas (*Pisum sativum*) and after discovering the genetic variation and appearance of plants, became recognized as the father of genetics. He presented the laws of heredity, known as Mendel's laws.
1910	Thomas Hunt Morgan conducted genetic studies with fruit flies (*Drosophila melanogaster*), which became one of the main animal models in the area. Morgan was able to prove that genes are located on chromosomes.
1928	Through virulence studies with the bacterium *Streptococcus pneumoniae*, Griffith discovered the Principle of Genetic Transformation, which is a way of recombination, exchange of genetic features among organisms.
1941	George Wells Beadle and Edward Lawrie Tatum, in their studies with fungus, concluded that the gene function was to codify an enzyme and, this way, regulate definite chemical events. The work laid the foundations of biochemical genetics.
1944	Avery et al. studies showed that the chemical nature of the bacterial transforming principle was the DNA.
1953	Watson and Crick, using x-ray diffraction studies, have elucidated the double helix DNA structure.
1969	Herbert Boyer discovered the restriction enzymes.
1973	Herbert Boyer and Stanley Cohen used the restriction enzymes to make for the first time a genetic engineering experiment applied to a microorganism, using the recombinant DNA technology. They transferred exogenous genetic material to *Escherichia coli* and created the first GMO.
1974	Paul Berg et al. published in "Science" a moratorium letter proposing to halt all attempts of genetic engineering until an international conference was held to think about the risks that this new science could cause.
1975	Frederick Sanger developed the important procedure for sequencing DNA by adding modified nucleotides (dideoxyribonucleotides) that abort the elongation of the DNA fragment by DNA polymerase replication.
1983	Emergence of first transgenic plant: Belgian researchers have introduced resistance genes to the antibiotic kanamycin in a variety of tobacco.
1985	Genentech, a biotechnology enterprise, made the insulin of humans, using a genetically modified *E. coli*.
1985	A brand new transgenic plant was produced with resistance against insect.
1987	Emergence of the first plant tolerant to glyphosate herbicide.

(Continued)

Table 9.1 (Continued) Historical facts important to evolution in the breeding of plants

Date	Milestone fact
1987	Mullis and Faloona used the thermostable *Taq* DNA Polymerase *in vitro*, which allowed the automation of the PCR reaction, to synthesize the DNA.
1988	Emergence of the Bt corn, the first transgenic cereal.
1990	The National Center of Biotechnology and Information (NCBI) and the Basic Local Alignment Search Tools (BLAST) were created to do the alignment of sequences.
1994	*Flavr-Savr* tomato: For the first time FDA authorized, in the United States, commercial cultivation of a GM food.
1997	First plant with human gene: The tobacco plant producing the human protein c.
1998	The European Union authorized the cultivation of genetically modified maize in Europe.
2000	The complete genome sequence of *E. coli* was done for the first time.
2003	Two independent research groups in the United States reported the sequence of the human genome.
2004	The number of countries that used transgenic varieties reached 18, reaching about 68 million hectares.
2005	The NGS emerged to make the genome sequencing faster than the other available tools.
2011	Large-scale DNA sequencers of second and third generations were created.
2012	Use of technologies for controlling the spatial and temporal expression of genes used in genetic transformation.
2013	Great use of large-scale sequencing, molecular markers for genome-wide selection, expanding the use of omics and bioinformatics.

Source: Extracted from Borém, A.; Fritsche-Neto, R. *Biotecnologia aplicada ao melhoramento de plantas.* Visconde do Rio Branco: Suprema, 2013.

genome of the offspring (Datta, 2013). In practice, this is a process carried out in three steps: (1) Creating or assembling collections of population or germplasm with genetic variations, (2) identification of individuals with superior genotype, and (3) development of improved cultivars from the selected individuals (Moose and Mumm, 2008).

Undesirable genes can be transferred together with the genes of interest and other genes can be lost in offspring. To correct these problems, breeders need to make successive backcrosses. However, this is a time-consuming activity and it is not always possible to separate strongly linked genes (Datta, 2013).

Classical breeding depends on the chromosome recombination, which generates the genetic diversity. Breeders should rely on *in vitro* steps to assist their work. The main characteristics breeders want to improve or modify in plants are improved quality and yield, tolerance to biotic and abiotic stresses, resistance to different microorganisms and other characteristics of commercial value (Moose and Mumm, 2008).

Traditional breeding techniques can be aided by genetic engineering, given that the genetic manipulation allows one or more genes to be transferred from one organism to another that expresses a novel feature of particular interest. Furthermore, it is possible to control the spatial and temporal expression pattern of the gene of interest (Tang and Galili, 2004). Genetic engineering, associated with traditional breeding techniques, enables genes from any organism to be isolated in the laboratory and transferred to another body, breaking the intraspecific barrier and leading to the development of new varieties (Nodari and Guerra, 2001). However, sometimes unknown or unwanted effects in metabolism may occur because the plant transformation methods do not offer control over the insertion site, number of transferred copies, or integrity of the cassette. Thus, unwanted effects may occur due to interruption of a functional gene by inserting the transgene, rearrangements of the gene carried by the cassette, or coexpression of neighboring genes (Dunwell, 2005).

The possibility of producing transgenic plants promoted the progress of cell biology of plants in various fields, as techniques of plant transformation and plant tissue culture, opening up many possibilities in agriculture. With these technologies, it is possible to change the quality of lipids, starch, and protein stored in seeds, besides conferring to certain vegetables, resistance to fungi, viruses, insects, or tolerance to herbicides, among others (Dunwell, 2000).

Plant genetic transformation depends on factors such as variety, target tissues, condition of the plant tissue, methods of injury, regeneration systems, selectable markers, and reporter genes, which have been studied and appropriated to achieve more efficient results (Mello-Farias and Chaves, 2008; Yamada et al., 2012).

Various methods of DNA transfer to plants have been described, with emphasis on the *Agrobacterium*-mediated transformation for excised plant tissue (Olhoft et al., 2006; Tzfira and Citovsky, 2006), electroporation (Fromm et al., 1985, 1986; Shillito et al., 1985), bombardment of particles (Klein et al., 1987; Sanford, 1988; Christou et al., 1992), microinjection (Crossway et al., 1986), protoplasts transformation (Negrutiu et al., 1987; Datta et al., 1990), and transformation mediated by *Agrobacterium* (Bechtold et al., 1993).

Despite the various described methods of introducing DNA, the transformation mediated by *Agrobacterium* and the bombardment of particles are the most common ones (Mello-Farias and Chaves, 2008). They are based on two different methods of DNA transference (biological vs. physical) and the efficiency of them depends on the species and on the type of tissue used. The particle bombardment appropriate to transfer various copies of the gene, can use a DNA mixture containing the different constructs to be used in processing, without the need of complex cloning strategies (Altpeter et al., 2005). *Agrobacterium*-mediated transformation is appropriate to transfer few copies of the gene (Gelvin, 2003).

Obtaining a transgenic plant by *Agrobacterium tumefaciens* infection involves the transfer, the integration of T-DNA into the cells, and the ability of these transformed cells to differentiate into a plant. The ability of differentiation or totipotency allows plant regeneration through tissue culture techniques (Brasileiro and Lacorte, 2000).

The plant tissue culture is the set of techniques that enable the maintenance or cultivation of seedlings, embryos, organs, tissues, and cells *in vitro*, in appropriate and aseptic culture medium, under controlled conditions of temperature, humidity, photoperiod, and light intensity (Torres et al., 1998). These techniques have been used in different ways in the development and propagation of superior genotypes of plants, particularly in the production of GM plants (Olhoft and Somers, 2001; Olhoft et al., 2003, 2006, 2007; Paz et al., 2004, 2006; Zeng et al., 2004; Xue et al., 2006; Yi and Yu, 2006).

The knowledge advancement of molecular biology was essential both to elucidate in detail the molecular basis of the *Agrobacterium*–host interaction process such as for the construction of transformation vectors based on the Ti plasmid. Thus, the techniques of molecular biology in association with techniques of plant tissue culture are the basis for obtaining a transgenic plant (Brasileiro and Lacorte, 2000).

Genome organization and genetic diversity of plants

The genome of plants is organized in chromosomes and can be of various sizes (from 63 to 149,000 Mb) and may present the number of chromosomes from $n = 2$ to $n = 600$. This organization in the chromosomes provides the structure necessary for the occurrence of linkage groups and replication. The chromosomes structural characteristics are well maintained, for example, chromatin folding, telomeres, and centromeres. The difference between the smaller and larger plant genomes can be that smaller genomes are formed by a coding region, a low copy regulatory region, repetitive telomeric DNA, rDNA highly repetitive, centromeres, and transposable elements, whereas the larger plant genomes is formed by basically the same number of genes but have plenty of sequence motifs repeated in tandem. In these genomes, the transposable elements are more than half of the DNA content (Heslop-Harrison and Schwarzacher, 2011).

Sequences of DNA located in organelles (chloroplasts and mitochondria), in nonnuclear genomes, in the genome of viruses, bacteria, fungi, and mycoplasma can be in intimate association with the plant genome. These nonnuclear genomes can be transmitted from one generation to another and should be taken into account in studies of genomics and evolution, given the fact that studies show that despite this transfer of genes to the nuclear genome of the plant is not frequent, such mechanism can occur and has evolutionary importance (Heslop-Harrison and Schwarzacher, 2011).

The first plant to have genome sequenced was *Arabidopsis thaliana*. The genome of *A. thaliana* comprises 125 Mb extended into regions of centromeres. During *Arabidopsis* evolution, its genome was submitted to duplication. After that, some genes were lost and others duplicated. Finally, the genome received genes from a plastid ancestor. *Arabidopsis* genome is constituted by 25,498 genes, which encode approximately 11,000 families of proteins (The *Arabidopsis* Genome Initiative, 2000). One of the most important models for plant genetic studies, *A. thaliana* is used for the identification and function of genes. Studies with this plant can be used to make comparisons between conserved processes in eukaryotes, identification of specific plant genes, including those of interest for breeding (Bevan and Walsh, 2005).

As mentioned, the genome size and the number of chromosomes can differ from plant to plant and in Table 9.2 are shown some examples of plants with the corresponding characteristic of the genome, such as genome size, number of genes, and number of chromosomes.

The events of fission, fusion, duplication, and insertion in chromosomes allow modifications of their number and size. The association with genetic mapping, analysis of sequence, methods of molecular cytogenetic and comparative analysis in revealing the evolution of chromosome and understanding the evolutionary mechanisms in the genome of plants is fundamental to drop development (Heslop-Harrison and Schwarzacher, 2011).

Table 9.2 Characteristics of the genome of some plants of interest for genetic improvement

Plant	Number of genes	Genome size (bp)	Chromosome number	Reference
A. thaliana	25,498	125 million	5	The Arabidopsis Genome Initiative (2000)
Bread wheat				
Triticum aestivum	94,000 up to 96,000 (identified)	17 billion	6×7	Brenchley et al. (2012)
Maize				
Zea mays	32,000	2.3 billion	10	Schnable et al. (2009)
Soybean				
Glycine max	46,340	1.1 billion	20	Schmutz et al. (2010)
Rice				
Oryza sativa	50,000	372 million	12	Goff et al. (2002); Ouyang et al. (2007)

Table 9.3 Morphological, biochemical, and molecular evaluation of genetic diversity, with its positive and negative points

Type of characterization	Positive points	Negative points
Morphological	• No need of expensive technology • Allows evaluation of diversity in the presence of environmental variation	• Requires evaluation of a huge number of individuals, in large extensions of land
Biochemical	• Quick method, based on protein separation in specific patterns of band • Small amounts of sample are necessary	• The number of enzymes available for evaluation is limited, which limits the analysis of diversity
Molecular	• A large number of molecular markers can be used • A huge number of information can be obtained	• As the cost/benefit ratio is satisfactory, the spendings necessary for the execution of analysis should not be considered high

Genetic diversity is the variation within a population or between populations and this pool of genetic variation is the basis for selection and plant breeding. Understanding genetic diversity is of great importance for their conservation and assists in understanding species' taxonomy, their origin, and evolution. To manage conserved germplasm, it is necessary to understand the genetic diversity of the collections because this understanding allows rationalizing the collections, and developing and assisting in the adoption of better protocols for the regeneration of germplasm seeds. Thus, the available resources can be better utilized (Rao and Hodgkin, 2002).

To assess a population, genetic variation can be used for the percentage of polymorphic genes, the allele numbers at them, and the ratio of heterozygous loci. The genetic diversity, both intra and interpopulation can be evaluated at the molecular level through a number of methodologies, for example, by analyzing the DNA or using allozyme. Genetic diversity can also be measured using biochemical and morphological evaluation. In Table 9.3, are listed the positive and negative factors of these studies (Mondini et al., 2009).

Molecular markers as a tool to reduce the time required for improvement

The genetic improvement of plants is of great importance for agricultural production, hence if efficiency improvement is increased, this will be reflected in the production of improved/modified crops. In this context fits the marker-assisted selection (MAS), which is a process that uses morphological, biochemical, and DNA markers as criteria for selecting features of interest for improving and increasing the efficiency and effectiveness of such selection (Ashraf et al., 2012).

In the case of molecular markers, these are tools that can assist genetic breeding, as they allow better understanding of crops domestication, plant evolution, knowledge of the genetic mechanisms involved in traits of agronomic interest, and in a shorter period of time when compared to classical breeding (Borém and Fritsche-Neto, 2013). Molecular markers are heritable differences (follow a Mendelian inheritance pattern) in the DNA sequence in a corresponding position between homologous chromosomes of two individuals. Molecular markers have caused a great impact in the biological sciences (Kesawat and Das, 2009).

One of the advances in the molecular genetics field was to analyze the DNA polymorphism using molecular markers. To choose the better marker to be used, it is required that the researcher knows the characteristics and principles of each of them. Despite the existence of a large number of markers available, there is not one that can meet all the requirements. However, molecular markers have been modified to increase its use and automation of genome analysis and the discovery of the process of polymerase chain reaction (PCR) were critical to the emergence of a new class of DNA markers (Budak et al., 2004; Kesawat and Das, 2009).

Molecular markers can be based on PCR, hybridization, and sequencing. Table 9.4 lists some of the most widely used molecular markers, as well as the advantages and disadvantages of them.

Genetic mapping

Genetic mapping is the graph of chromosomes organization. A genetic map can be constructed to a great variety of species and its basic technology is the use of molecular markers as a reference throughout the chromosomes (Grant and Shoemaker, 2001; Borém and Caixeta, 2009).

The development of genetic maps of plants provides an image of how genes are arranged on chromosomes. Furthermore, the probability of genetic loci to be inherited together can be predicted based on the localization of them on the chromosome. The genetic markers that make up maps can be genes that control classical or molecular markers (Grant and Shoemaker, 2001).

Genetic maps provide the necessary information for the chromosome walk in the direction of the genes of interest, as well as for chromosome landing until the gene of interest (Borém and Caixeta, 2009). With the aid of genomics, high-density genetic maps can be developed. The construction of such maps involves the location of hundreds or thousands of markers on different linkage groups and their covering is very high and without large gaps. The technologies of new sequencing, the next-generation sequencing (NGS), and the genotyping platforms of high yielding led to the development of these maps with high-density markers (Pérez-de-Castro et al., 2012).

Another type of genetic mapping used by researchers is association mapping. It is a promising tool and differs from the classical genetic mapping, because it explores the functional variation in the germplasm. Furthermore, using association mapping is possible to study the diversity and to cross species in a controlled way in order to obtain the progenies (Zhu et al., 2008).

Considering agriculture, association mapping attracts attention, as it can be used in the identification of genes involved in the variation of features in a quantitative way. Progress in genomic expertise propelled the exploration of species variation. The growing of methods of statistical analysis to be used in plant association mapping made them available to breeding programs. By checking the association of marker-trait QTLs (quantitative trait loci) can be identified using these maps (Zhu et al., 2008).

Another variation of genetic mapping is radiation hybrid mapping. In this technique, radiation is used to cause breaks at the chromosome. Theoretically, the more physically close are the adjacent makers, the less is the possibility to separate them with these induced breaks. The genome is randomly affected by the radiation, independent of chromosomal location. This way, radiation hybrid mapping has the same efficiency to order the markers at any region on the chromosome. Moreover, this kind of map does not require allelic polymorphism, as they consider the deletion (absence) or retention (presence) of a gene marker (Kumar et al., 2012).

Table 9.4 Molecular markers most used and their characteristics

Molecular markers	Advantages	Disadvantages
Isozymes	• Useful for evolutionary studies • Does not require radioactive labeling • Applied across species • No need to know the sequence previously	• Weariful • The polymorphism is limited • High cost • Not easily automated
Microsatellites or single sequence repeats (SSR)	• High covering of the genome • The polymorphism is high • The genomic abundance is high • High reproducibility • Multiple alleles • Does not require radioactive labeling • Easy automation	• Not applied across species • Need to know the sequence previously
Inter simple sequence repeat (ISSR)	• Allows simultaneous amplification of different regions of the genome • Simple and fast • Wide coverage of the genome • High reproducibility • Random primers whose sequence is independent of the species under study • Available for several species • Used in studies of diversity, mapping, and MAS	• Dominant marker
Randomly amplified polymorphic DNA (RAPD)	• The genomic abundance is high • High covering of the genome • No need to know the sequence previously • Perfect for automation • Require little DNA concentration • Does not require radioactive labeling • Fast method	• Dominant marker • No information about the primer • Not reproducible • Cannot be used across species
Amplified fragment length polymorphism (AFLP)	• The sensitivity is high • The polymorphism is high • The genomic abundance is high • Applied across species • No need to know the sequence previously • Useful in preparing contig maps	• Not reproducible • Very good primers are necessary
Restriction fragment length polymorphism (RFLP)	• Codominant marker • Highly reproducible • High genomic abundance • High covering of the genome • No need to know the sequence previously • Applied across species • Established for plants • Required to do cloning based on maps	• High DNA concentration is required • Automation is difficult • Weariful • Requires radioactive labeling • Requires probe cloning and characterization

(Continued)

Table 9.4 (Continued) Molecular markers most used and their characteristics

Molecular markers	Advantages	Disadvantages
Sequence-tagged site (STS)	• Good genome coverage • Highly reproducible • Applied to *contig* maps • Do not require radioactive labeling	• Weariful • Need to know the sequence previously • Mutation detection only at the target sites • Requires probe cloning and characterization
Sequence characterized amplified region (SCAR)	• Allows the conversion of dominant markers in codominant • They are loco-specific • They have higher resolution when compared with other classes of markers	• Laborious development • Requires the analysis of genomic DNA with RAPD or AFLP, cloning, and sequencing
Diversity array technology (DArT)	• Allows simultaneous genotyping of hundreds of great length polymorphisms in the genome • Relatively fast • Low cost • No need of previous information sequences of species to be studied • High performance • High reproducibility	• Dominant marker • Involves several steps for establishment • Demand physical and personnel structure
Single nucleotide polymorphism (SNP)	• Can be used in ultra-high-throughput genotyping • Can be used in the construction of genetic maps, high-resolution mapping of features of interest, in MAS, in genetic studies of population structure analysis, phylogenetic analysis, and association mapping based on linkage disequilibrium	• Less informative than SSRs • Need to be identified and validated for the species under study

"Omics": Tools that enabled the knowledge of genes involved with desirable characteristics

Genomics, transcriptomics, proteomics, and other "omics" are being considered as important tools to make programs of crop breeding increasingly faster and, consequently, to improve the production of crops of the next generation (Eldakak et al., 2013).

The advances related to DNA sequencing and information technology have transformed the ability to understand the genetic variation and, moreover, to correlate it with the performance of the plant. Thus, the plant breeding process can be faster and cheaper. This technology is being deployed on food crops and also on other species of plants. In this context, genomics can be inserted, as it can be applied to breeding programs in various ways, such as molecular diagnostics, testing of transgenic, target confirmation of alleles in a population of genetic testing background, genomic identification of the genetic identity test for protection of plant varieties, varietal purity control, organization of genetic variability in germplasm collections, MAS, and phylogenetic studies, among others (Jannink et al., 2010; Borém and Fritsche-Neto, 2013).

As the emergence of genomics has enabled breeders to a new set of tools and techniques to aid in the study of genome as a whole, genomics is leading to a new revolution in the plant genetic improvement. The high-throughput DNA-sequencing technologies development methods such as NGS, is one of the major bases of genetic improvement. These new techniques can help breeders through the availability of large collections of markers, high-throughput genotyping strategies, high-density genetic maps, and new experimental populations, which can be incorporated into the already-existing methods of improvement. The genomic approaches are especially useful for working with complex characteristics, since these generally have a multigene nature (Pérez-de-Castro et al., 2012).

In relation to the transcriptomics, the analyses of transcripts, it covers a large scope of information of the genetic expression. To study the transcriptional profile, the most used tool is microarrays. It is due to the large number of available platforms, which can differ in the sensitivity, specificity, and coverage of genome. Another good option to transcripts sequencing is the RNA-Seq, a high-throughput technology (Baginsky et al., 2010).

The proteomics, in turn, are dynamic. As the cellular processes are mediated by specific molecules, proteomics can be a potent toll to identify them. The analyses of proteins revealed the occurrence of posttranslational modifications (PTMs), which are important during plant growth and development. Differences at the protein expression level can be detected according to different conditions. Proteomics, in association with transcriptomics and genomics, can give a complete view about the gene expression. Furthermore, the emergence of even better bioinformatic tools is making proteomics even more connected to the other "omics." It makes plant breeding more efficient and effective (Eldakak et al., 2013).

Transgenic genotypes designed at aiming the improvement of crop production and quality

Biotechnological proceedings are considered important tools to improve agriculture across the world. It is because the classical methods can be associated with biotechnology, and consequently, better results are obtained. Classical breeding can be assisted by molecular biology and biotechnology to develop plants with superior features and in a relatively fast way (Sartoretto et al., 2008; Cançado et al., 2013). Thus, biotechnology allowed solving important problems in many crops, as susceptibility to pets, diseases, and environmental stress. Moreover, it was possible to develop species with higher productivity and with better nutritional quality. Genetic engineering permits the direct transfer of genes, even if it is difficult to do with the classical breeding tools (Qaim, 2009).

GM traits can be classified into three generations. The first is related to improvement in agronomic characteristics. For example, soybean resistant to glyphosate, canola resistant to pest, and others. The major modified feature is the introduction of insect and herbicide-tolerant plants, especially glyphosate. They confer better yields and are easy to manage. Some first-generation transgenic plants are being commercially used (Que et al., 2010).

The GM plants of the second generation consist of plants with improved quality features. For example, crops with high content of nutrients. There are many examples such as plant oils modified content, increased vitamin content, reduced quantity of unpalatable substances, and resistance against abiotic stresses.

For instance, plant oil is composed of fatty acids with different saturation degrees. The polyunsaturated ones confer better benefits to health, but they can suffer oxidation and, consequently, suffer rancidity. An alternative was to chemically hydrogenate polyunsaturation, but for the healthy it was not a good alternative. So, researchers produced a GM soybean, which managed to produce a soybean with increased levels of oleic acid (revised by Halford et al., 2014).

As a second example, it created rice with the vitamin A precursor β-carotene, known as Golden Rice (Halford et al., 2014).

Finally, the GM plants of the third generation consist of plants producing heterologous proteins for the pharmacy and industry. For example, bananas used in delivery of vaccines (May et al., 1995).

The first GM crop was generated in 1994: The "Flavr-Savr" tomato. In this case, the tomato was transformed with the antisense gene for polygalacturonase, which inhibited cell wall softening (Taiz, 2013). After that, new varieties of other species such as soybean, corn, canola, and cotton were developed (Duijn et al., 1999).

The GM plants were introduced in the 1990s and, since then, this technology was rapidly adopted. The most produced GM crops are soybean, corn, cotton, and canola. The less-produced ones are papaya and potato (Que et al., 2010). According to James (James, 2013), 27 countries planted transgenic crops in 2013. The major producers and exporters are the United States and Canada, and of the developing countries, a few that can be highlighted are Argentina, Brazil, China, and India.

According to International Service for the Acquisition of Agri-biotech Applications (ISAAA), in 2013, 175.2 million hectares of GM plants were cultivated. The total area of cultivated GM crops increased 100-fold in less than 20 years, what makes them the fastest technology to be accepted, considering GM crops, in the recent years (James, 2013).

Using different focuses, some GM crops with different modified features were designed and are being commercialized, as showed in Table 9.5.

Biosafety and management of transgenic cultivars

Undeniably, biotechnology and genetic engineering of plants provided new possibilities in crop improvement and helped in the solution of new requirements. On the other hand, every new technology is not considered totally safe to the environment or to the healthy. So, it is not enough to obtain transgenic plants that express the desired phenotype. For the transgenic plant to be effectively considered as a technology and to be inserted into the productive system, extensive research is needed to ensure that this plant is not harmful to the environment or health.

Since the genetic modified organisms (GMO) entered the commercial marketplace, scientists, scholars, journalists, and consumers have debated about GMO safety and sustainability, which are determinant conditions for the commercialization of these crops.

The association between classical breeding programs and genetic engineering allows the introduction of desired features to a transgenic plant, which can be commercialized in a viable period of time. The possibilities to genetically manipulate crops are substantial and can be used to the most different purposes (Hansen and Wright, 1999). To minimize possible adverse effects of GM plants, evaluation of the new features in GM plants considering the safety of the environment should be included in the transgenic breeding program (Singh et al., 2006).

Considering the rules of health and environment protection, before introducing the transgenic science into agriculture, some analysis should be done (Kuiper et al., 2001). For example, modifications in important fractions of the genome, transcriptomes, proteome, and metabolome should be monitored on transgenic plants (Filipecki and Malepszy, 2006).

Regarding health and food, it is important to verify the possible allergenicity and the toxicity of GM plants and, consequently, of the food made from them. Considering the environment is important to confirm that no undesirable consequences occur, such as

Table 9.5 Major commercialized GM crops

GM crop	Company	Phenotypic trait
Canola (*Brassica napus* L.)	Bayer CropScience	Restoration of fertility, male sterility, control of pollination, and tolerance to glufosinate
	Monsanto	Glyphosate herbicide tolerance
Cotton (*Gossypium hirsutum* L.)	Bayer CropScience	Resistance to Lepidopteran pests; phosphinothricin (PPT) herbicide tolerance (ammonium glufosinate)
	Dow AgroSciences	Resistance to Lepidopteran pests; glyphosate herbicide tolerance
	Monsanto	Resistance to Lepidopteran pests; glyphosate herbicide tolerance
Maize (*Zea mays* L.)	Pioneer Hi-Bred and Dow AgroSciences	Resistance to Coleopteran pests (Corn rootworm); resistance to Lepidopteran pests; tolerance to glyphosate herbicide
	Monsanto and Dow AgroSciences	Resistance to Coleopteran pests; resistance to Lepidopteran and tolerance to glyphosate herbicide
	Syngenta	Resistance to Coleopteran pests; resistance to Lepidopteran pests; tolerance to glyphosate herbicide
Potato (*Solanum tuberosum* L.)	Monsanto	Resistance to Potato virus Y (PVY); glyphosate herbicide resistance; resistance to Colorado potato beetle (*Leptinotarsa decemlineata*)
	BASF	Ratio amylopectin/amylase increased; composition of starch modified
Rice (*Oryza sativa*)	BASF	Tolerance to the imidazolinone herbicide (imazethapyr) synthase
	Bayer CropScience	Glufosinate ammonium herbicide tolerant
Soybean (*Glycine max* L.)	Bayer CropScience	Glufosinate ammonium herbicide tolerant
	BASF Inc.	Tolerance to imidazolinone herbicides
	DuPont Pioneer	High oleic acid soybean by silencing the gene *FAD2-1* (omega-6 desaturase) Two herbicide tolerance genes: detoxifies glyphosate and tolerant to ALS-inhibiting herbicides
	DuPont Canada Agricultural Products	Soybean with high content of oleic acid encoding resulted from "silencing" of the endogenous host gene
	Monsanto	Glyphosate tolerant soybean; resistance to *Pseudoplusia includens* and *Anticarsia gemmatalis*

Source: Adapted from Que, Q. et al. *GM Crops*, 1(4):220–229, 2010.

effect on other organisms, biodiversity loss, gene flow, and others (Andow and Zwahlen, 2006). Moreover, other aspects to be considered about GM crops are the restricted utilization to the new genetic features, monopoly of the private industries, and the costs for small farmers (recently addressed by Johnson, 2014).

Despite the negative impacts that may be caused by GM organisms, these effects have not been proved. Despite this fact, the population does not have an opinion about whether or not to ingest GM foods over the world. In 2003, the Cartagena Protocol on Biosafety was established to guarantee human protection (CBD, 2013).

Horizontal gene transfer (HGT), the genetic material transfer to an organism and subsequent expression, has been considered in some studies and discussions, as it goes against biosafety (Ho et al., 1999; Dale et al., 2002). Another undesirable point is the markers of selection, especially the ones that confer resistance against antibiotics and is much used to select the GM plants (Tuteja et al., 2012).

In this sense, a major trouble could be transference of genes from GM plants to other organisms. The probability of this to occur is very low, as some events should occur, as cited: (1) Output of the plant genome gene; (2) the gene must remain intact to the action of nucleases and other enzymes; (3) genetic transformation of cells; (4) expression of the inserted gene; and (5) the complete establishment of the inserted gene (FAO/WHO, 2000). However, the risk exists.

A proposed solution is to produce GM plants without markers of selection. So, the development of efficient techniques, for obtaining marker-free transgenic crops as well as to control gene insertion, are the next challenges (for further information, see Tuteja et al., 2012).

Perspectives

Since its inception, agriculture has achieved increasing levels of efficiency, due in large part to the advances in genetics and its applications in the development of genetic manipulation techniques. With modern biotechnology, new techniques have been made available, including obtaining GMOs, whose cultivation has steadily increased despite facing strong opposition from certain sectors. Although biotechnology techniques represent an important contribution to the incorporation of desirable traits to crops, conventional breeding techniques are still widely used because of the complexity for obtaining transgenic cultivars, coupled with the delay in approvals from biosafety committees, culminating in delaying their use in productive sectors. On the other hand, conventional techniques of proven efficiency, simpler implementation, present satisfactory results, and wide acceptability by society over the years and generations, cannot be discarded.

The role of GM as supporting conventional breeding has become incontestable. Thus, the use of biotechnological techniques (incorporating specific genes) does not preclude the use of conventional breeding (e.g., selecting superior genotypes) at the launch of new varieties. Based on this, less-demanding cultivars in agrochemical inputs can be produced, and moreover nutritional quality improvement and other desirable characteristics can be incorporated. Therefore, genetic engineering (GE) is actually another tool available to plant breeding and can be used together with conventional breeding techniques, as genetic manipulation of organisms allows genes of any organism to be isolated in a laboratory and transferred to the another one, breaking the intraspecific barrier and leading to the development of new varieties.

Thus, the great challenge is to establish a balance that considers adequately the available technologies that best meet the perspectives of the genetic improvement.

References

Altpeter, F. et al. Particle bombardment and the genetic enhancement of crops: Myths and realities. *Molecular Breeding*, 15:305–327, 2005.

Andow, D.A.; Zwahlen, C. Assessing environmental risks of transgenic plants. *Ecology Letters*, 9: 196–214, 2006.

Ashraf, M.; Akram, N.A.; Rahman, M.U.; Foolad, M.R. Marker-assisted selection in plant breeding for salinity tolerance. *Plant Salt Tolerance—Methods in Molecular Biology*, 913:305–333, 2012.

Baginsky, S.; Hennig, L.; Zimmermann, P.; Gruissem, W. Gene expression analysis, proteomics, and network discovery. *Plant Physiology*, 152:402–410, 2010.

Bechtold, N.; Ellis, J.; Pelletier, G. In plant *Agrobacterium* mediated gene transfer by infiltration of adult *Arabidopsis thaliana* plants. *Comptes Rendus Academy of Science of Paris, Life Sciences*, 316:1194–1199, 1993.

Bevan, M.; Walsh, S. The *Arabidopsis* genome: A foundation for plant research. *Genome Research*, 15:1632–1642, 2005.

Borém, A.; Caixeta, E.T. *Marcadores Moleculares*, 2 ed. Visconde do Rio Branco: Editora Suprema, 2009.

Borém, A.; Fritsche-Neto, R. *Biotecnologia aplicada ao melhoramento de plantas*. Visconde do Rio Branco: Suprema, 2013.

Brasileiro, A.C.M.; Lacorte, C. *Agrobacterium*: Um sistema natural de transferência de genes para plantas [Agrobacterium: A natural system for gene transferring in plants]. *Biotecnologia Ciência and Desenvolvimento*, 15(3):12–15, 2000.

Brenchley, R. et al. Analysis of the bread wheat genome using whole genome shotgun sequencing. *Nature*, 491:705–710, 2012.

Budak, H.; Bölek, Y.; Dokuyucu, T.; Akkaya, A. Potential uses of molecular markers in crop improvement. *KSU Journal of Science and Engineering*, 7(1):75–79, 2004.

Cançado, G.M.A.; Setotaw, T.A.; Ferreira, J.L. Applications of biotechnology in olive. *African Journal of Biotechnology*, 12:767–779, 2013.

Christou, P.; Ford, T.L.; Kofron, M. Rice genetic engineering: A review. *Trends in Biotechnology*, 10:239–246, 1992.

Convention on Biological Diversity (CBD). Cartagena Protocol on Biosafety Ratification List; 2013. http://www.cbd.int/doc/lists/cpb-ratifications.pdf.

Crossway, A.; Oakes, J.V.; Irvine, J.M.; Ward, B.; Knauf, V.C.; Shewmaker, C.K. Integration of foreign DNA following microinjection of tobacco mesophyll protoplasts. *Molecular and General Genetics*, 202:179–185, 1986.

Dale, P.J.; Clarke, B.; Fontes, E.M.G. Potential for the environmental impact of transgenic crops. *Nature Biotechnology*, 20:567–574, 2002.

Datta, A. Genetic engineering for improving quality and productivity of crops. *Datta Agriculture and Food Security*, 2:15, 2013.

Datta, S.K.; Peterhans, A.; Datta, K.; Potrykus, I. Genetically engineered fertile Indica-rice plants recovered from protoplasts. *Biotechnology*, 8:736–740, 1990.

Duijn, G.; Biert, R.; Bleeker-Marcelis, H.; Peppelman, H.; Hessing, M. Detection methods for genetically modified crops. *Food Control*, 10(6):375–378, 1999.

Dunwell, J.M. Transgenic approaches to crop improvement. *Journal of Experimental Botany*, 51:487–496, 2000.

Dunwell, J.M. Transgenic crops: The current and next generations. *Methods in Molecular Biology*, 286:377–398, 2005.

Edgerton, M.D. Increasing crop productivity to meet global needs for feed, food, and fuel—Update on increasing crop productivity. *Plant Physiology*, 149:7–13, 2009.

Eldakak, M.; Milad, S.I.M.; Nawar, A.; Rohila, J.S. Proteomics: A biotechnology tool for crop improvement. *Frontiers in Plant Science*, 4(35):1–12, 2013.

FAO Statistical Yearbook. 2013. *World Food and Agriculture*. Food and Agriculture Organization of the United Regions. Rome, 2013.

Filipecki, M.; Malepszy, S. Unintended consequences of plant transformation: A molecular insight. *Journal of Applied Genetics*, 47(4):277–286, 2006.

Food and Agriculture Organization/World Health Organization (FAO/WHO). Safety aspects of genetically modified foods of plant origin. Report of a Joint FAO/WHO Consultation on Foods Derived from Biotechnology. World Health Organization, Geneva, 2000.

Fromm, M.; Taylor, L.P.; Walbot, V. Expression of genes transferred into monocot and dicot plant cells by electroporation. *Proceedings of the National Academy of Sciences, USA*, 82:5824–5828, 1985.

Fromm, M.E.; Taylor, L.P.; Walbot, V. Stable transformation of maize after gene-transfer by electroporation. *Nature*, 319:791–793, 1986.

Gelvin, S.B. *Agrobacterium*-mediated plant transformation: The biology behind the "gene-jockeying" tool. *Microbiology and Molecular Biology Reviews*, 1:16–37, 2003.

Goff, S.A. et al. A draft sequence of the rice genome (*Oryza sativa* L. ssp. *japonica*). *Science*, 296:92–100, 2002.

Grant, D.; Shoemaker, R.C. Plant gene mapping techniques. *Encyclopedia of Life Sciences*, 1:1–6, 2001.

Halford, N.G.; Hudson, E.; Gimson, A.; Weightman, R.; Shewry, P.R.; Tompkins, S. Safety assessment of genetically modified plants with deliberately altered composition. *Plant Biotechnology Journal*, 12(6):651–654, 2014.

Hansen, G.; Wright, M.S. Recent advances in the transformation of plants. *Trends in Plant Science*, 4(6):26–231, 1999.

Heslop-Harrison, J.S. (Pat); Schwarzacher, T. The plant genome: An evolutionary view on structure and function—Organisation of the plant genome in chromosomes. *The Plant Journal*, 66:18–33, 2011.

Ho, M.W.; Ryan, A.; Cummins, J. Cauliflower mosaic viral promotor—A recipe for disaster? *Microbial Ecology in Health and Disease*, 11:194–197, 1999.

James, C. *Global Status of Commercialized Biotech/GM Crops: 2013. International Service for the Acquisition of Agri-Biotech Applications (ISAAA) Brief 46-2013*. ISAAA, Ithaca, NY, 2013.

Jannink, J.; Lorenz, A.J.; Iwata, H. Genomic selection in plant breeding: From theory to practice. *Briefings in Functional Genomics*, 9(2):166–177, 2010.

Johnson, S. Genetically modified food: A golden opportunity? *Sustainable Development Law and Policy*, 14(1):34, 69–70, 2014.

Kesawat, M.S.; Das, B.K. Molecular markers: It's application in crop improvement—Review article. *Journal of Crop Science and Biotechnology*, 12(4):169–181, 2009.

Klein, T.M.; Wolf, E.D.; Sanford, J.C. High-velocity microprojectiles for delivering nucleic acids into living cells. *Nature*, 327:70–73, 1987.

Kuiper, H.A.; Kleter, G.A.; Noteborn, H.P.J.M.; Kok, E.J. Assessment of the food safety issues related to genetically modified foods. *Plant Journal*, 27:503–528, 2001.

Kumar, A. et al. Physical mapping resources for large plant genomes: Radiation hybrids for wheat D-genome progenitor *Aegilops tauschii*. *BMC Genomics*, 13:597, 2012.

Ma, J.K.C.; Hiatt, A.; Hein, M.; Vine, N.D.; Wang, F.; Stabila, P.; Vandolleweerd, C.; Mostov, K.; Lehner, T. Generation and assembly of secretory antibodies in plants. *Science*, 268:716–719, 1995.

May, G.D. et al. Generation of transgenic banana (*Musa acuminata*) plants via *Agrobacterium*-mediated transformation. *Biotechnology*, 13:486–492, 1995.

Mello-Farias, P.C.; Chaves, A.L.S. Advances in *Agrobacterium*-mediated plant transformation with emphasis on soybean. *Scientia Agricola*, 65(1):95–106, 2008.

Mondini, L.; Noorani, A.; Pagnotta, M.A. Assessing plant genetic diversity by molecular tools—Review. *Diversity*, 1:19–35, 2009.

Moose, S.P.; Mumm, R.H. Molecular plant breeding as the foundation for 21st century crop improvement—Editor's choice series on the next generation of biotech crops. *Plant Physiology*, 147:969–977, 2008.

Negrutiu, I.; Shillito, R.D.; Potrykus, I.; Biasini, G.; Sala, F. Hybrid genes in the analysis of transformation conditions. I. Setting up a simple method for direct gene transfer in plant protoplasts. *Plant Molecular Biology*, 8:363–373, 1987.

Nodari, R.O.; Guerra, M.P. Avaliação de riscos ambientais de plantas transgênicas [Assessment of environmental risks of genetically modified plants.]. *Caderno de Ciência and Tecnologia*, 18(1):81–116, 2001.

Olhoft, P.M. et al. A novel *Agrobacterium rhizogenes*-mediated transformation method of soybean [*Glycine max* (L.) Merrill] using primary-node explants from seedlings. *In Vitro Cellular and Developmental Biology—Plant*, 43:536–549, 2007.

Olhoft, P.M.; Donovan, C.M.; Somers, S.A. Soybean (*Glycine max*) transformation using mature cotyledonary node explants. In: Wang, K. (ed.) *Agrobacterium Protocol*, 2nd ed. Springer, Dordrecht, pp. 473–498, 2006.

Olhoft, P.M.; Flagel, L.E.; Donovan, C.M.; Somers, D.A. Efficient soybean transformation using hygromycin B selection in the cotyledonary-node method. *Planta*, 216:723–735, 2003.

Olhoft, P.M.; Somers, D.A. L-Cysteine increases *Agrobacterium*-mediated T-DNA delivery into soybean cotyledonary-node cells. *Plant Cell Reports*, 20:706–711, 2001.

Ouyang, S. et al. The TIGR rice genome annotation resource: Improvements and new features. *Nucleic Acids Research*, 35:883–887, 2007.

Paz, M.M.; Martinez, J.C.; Kalvig, A.B.; Fonger, T.M.; Wang, K. Improved cotyledonary node method using an alternative explant derived from mature seed for efficient *Agrobacterium*-mediated soybean transformation. *Plant Cell Reports*, 25:206–213, 2006.

Paz, M.M.; Shou, H.; Guo, Z.; Zhang, Z.; Banerjee, A.K.; Wang, K. Assessment of conditions affecting *Agrobacterium*-mediated soybean transformation using the cotyledonary node explant. *Euphytica*, 136:167–179, 2004.

Pérez-de-Castro, A.M.; Vilanova, S.; Cañizares, J.; Pascual, L.; Blanca, J.M.; Díez, M.J.; Prohens, J.; Picó, P. Application of genomic tools in plant breeding. *Current Genomics*, 13:179–195, 2012.

Qaim, M. The economics of genetically modified crops. *Annual Review of Resource Economics*, 1: 665–693, 2009.

Que, Q. et al. Trait stacking in transgenic crops: Challenges and opportunities. *GM Crops*, 1(4): 220–229, 2010.

Rao, V.R.; Hodgkin, T. Genetic diversity and conservation and utilization of plant genetic resources. *Plant Cell, Tissue and Organ Culture*, 68:1–19, 2002.

Sanford, J.C. The biolistic process. *Trends in Biotechnology*, 6:299–302, 1988.

Sartoretto, M.L.; Saldanha, C.W.; Corder, M.P.M. Genetic transformation: Strategies for forest species breeding. *Ciência Rural*, 38(3):861–871, 2008.

Schmutz, J. et al. Genome sequence of the palaeopolyploid soybean. *Nature*, 463:175–183, 2010.

Schnable, S. et al. The B73 maize genome: Complexity, diversity, and dynamics. *Science*, 326: 1112–1115, 2009.

Sharma, H.C., Crouch, J.H., Sharma, K.K., Seetharama, N., Hash, C.T. Applications of biotechnology for crop improvement: Prospects and constraints. *Plant Science*, 163:381–395, 2002.

Shillito, R.D.; Saul, M.W.; Paszkowski, J.; Muller, M.; Potrykus, I. High efficiency direct gene transfer to plants. *Biotechnology*, 3:1099–1103, 1985.

Singh, O.V.; Ghai, S.; Paul, D.; Jain, R.K. Genetically modified crops: Success, safety assessment, and public concern. *Applied Microbiology Biotechnology*, 71(5):598–607, 2006.

Taiz, L. Agriculture, plant physiology, and human population growth: Past, present, and future. *Theoretical and Experimental Plant Physiology*, 25(3): 167–181, 2013

Tang, G.; Galili, G. Using RNAi tom improve plant nutritional value: From mechanism to application. *Trends in Biotechnology*, 22:09, 2004.

The Arabidopsis Genome Initiative. Analysis of the genome sequence of the flowering plant *Arabidopsis thaliana*. *Nature*, 408:796–815, 2000.

Torres, A.C.; Caldas, L.S.; Buso, J.A. *Cultura de tecidos e transformação genética de plantas*. Embrapa, Brasília, 1998.

Tuteja, N.; Verma, S.; Sahoo, R.K.; Raveendar, S.; Reddy, I.N.B.L. Recent advances in development of marker-free transgenic plants: Regulation and biosafety concern. *Journal of Biosciences*, 37: 167–197, 2012.

Tzfira, T.; Citovsky, V. *Agrobacterium*-mediated genetic transformation of plants: Biology and biotechnology. *Current Opinion in Biotechnology*, 17:147–154, 2006.

Xue, R.; Xie, H.; Zhang, B. A multi-needle-assisted transformation of soybean cotyledonary node cells. *Biotechnology Letters*, 28:1551–1557, 2006.

Yamada, T.; Takagi, K.; Ishimoto, M. Recent advances in soybean transformation and their application to molecular breeding and genomic analysis. *Breeding Science*, 61:480–494, 2012.

Yi, X.; Yu, D. Transformation of multiple soybean cultivars by infecting cotyledonary-node with *Agrobacterium tumefaciens*. *African Journal of Biotechnology*, 5:1989–1993, 2006.

Zeng, P.; Vadnais, D.A.; Zhang, Z.; Polacco, J.C. Refined glufosinate selection in *Agrobacterium*-mediated transformation of soybean [*Glycine max* (L.) Merrill]. *Plant Cell Reports*, 22:478–482, 2004.

Zhu, G.; Gore, M.; Buckler, E.S.; Yu, J. Status and prospects of association mapping in plants. *The Plant Genome*, 1:5–20, 2008.

chapter ten

Overview of methods to unveil the epigenetic code

Sarfraz Shafiq and Abdul Rehman Khan

Contents

Abstract

Epigenetics refers to the heritable changes in gene expression that are not caused by change in deoxyribonucleic acid (DNA) sequence. Histone modifications and DNA methylation are ubiquitous mechanisms to regulate gene expression. A wide range of techniques have been used to analyze locus-specific as well as genome-wide epigenetic processes. An exponential increase in novel techniques has been witnessed in recent years, to elucidate the molecular enigma of epigenetic inheritance and change in gene expression. In this chapter, we summarize the techniques being commonly used to study the locus-specific and genome-wide epigenetic modifications in plants and discuss the future challenges in technology development.

Keywords: Epigenetics, Histone modifications, DNA methylation, Techniques.

Introduction

Epigenetics refers to the change in gene expression that are mitotically and/or meiotically transmissible while deoxyribonucleic acid (DNA) sequence does not change (Holliday 1994). The size of DNA in one cell's nucleus is quite large compared to the place it occupies, for example, a human cell's DNA is about 3 m in length. So, to fit this DNA in the nucleus of a cell, it is packed in a highly organized nucleoprotein complex whose fundamental unit is nucleosome. Nucleosome comprises DNA of ~147 base pairs, wrapped around the histone proteins (a core of eight histone molecules). This packing is rather dynamic in the sense that it can move, assemble/disassemble to specific genomic regions in response to developmental cues and environmental stimuli and regulates the processes, including transcription, DNA replication, DNA repair, recombination, transposition or chromosome segregation through modulating the DNA accessibility and affect the plant development by influencing various processes such as root growth, flowering timing, floral organogenesis, gametophyte or embryo formation, as well as the response to pathogens or environmental changes (Berr et al. 2011). Along with other processes, epigenetic modifications, i.e., DNA methylation/demethylation, posttranscriptional covalent histone modifications (e.g., acetylation, ubiquitination, methylation, phosphorylation, sumoylation) and expression of noncoding ribonucleic acids (RNAs), are involved in the regulation of gene expression (Feng et al. 2012). Different techniques are being used to study these epigenetic modifications depending upon the conditions (Figure 10.1), therefore, in this chapter we discuss the techniques being used to study the DNA methylation and posttranscriptional covalent histone modifications.

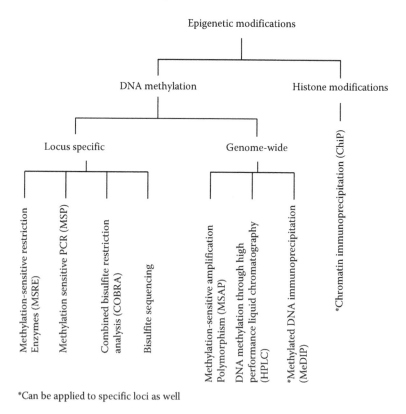

Figure 10.1 Overview of methods to study the epigenetic modifications.

DNA methylation

DNA methylation is the addition/attachment of methyl group (–CH₃) to the specific nucleotide base, either cytosine (C) or adenine (A) making it 5-methylcytosine (5-mC), and N6-methyladenine (6 mA) respectively, or a more recently discovered minor base N4-methylcytosine (4-mC) (Ratel et al. 2006). However, the most studied and reported is 5-mC. Because DNA methylation provides an additional layer of information for gene regulation, and plays an important role in an organism's normal development (Laird 2010), it is very important to profile the DNA methylation. Many techniques along with their various modifications have been developed and used by scientists in the last few decades. Depending upon the goals and objectives of the study, these techniques can be broadly classified into two categories; (i) locus specific and (ii) genome wide.

Locus-specific DNA methylation techniques

Methylation-sensitive restriction enzymes

Methylation-sensitive restriction enzymes (MSRE) is among the earliest techniques used for the detection of DNA methylation at specific locus. It was first reported in the 1970s (Bird and Southern 1978; Cedar et al. 1979). Although restriction enzymes have the ability to cleave the DNA by recognizing the specific nucleotide motifs, some restriction enzymes have been found to be sensitive to cytosine methylation while their isoschizomers do not have this capability. Using this property, a technique to detect DNA methylation at a specific region/site in the genome was developed.

Principle 5-mC shows resistance to cleavage by certain restriction enzymes while some isoschizomers (methylation nonsensitive restriction enzymes [MNSREs]) do not have the ability to differentiate the nonmethylated cytosines and 5-mC, and will cut the DNA irrespective of its methylation state (Figure 10.2). By using this principle, a pair of such isoshizomers (like HpaII/MspI or EcoRII/BstNI etc.) is used to detect DNA methylation by comparing the pattern of digested DNA given by either of the restriction enzymes. This pattern can be identified by different techniques such as southern blotting or polymerase chain reaction (PCR) amplification.

Procedure As a prerequisite, extraction of good quality DNA with sufficient quantity and selection of a proper pair of restriction enzymes (depending upon the experimental requirements) is essential. A variety of these restriction enzymes are commercially available and various suppliers recommend different protocols based on their optimization results. Prepare the mix (reaction buffers, water, etc.) as recommended by the manufacturer (one for each isoschizomer such as *Msp*I and *Hpa*II). Add DNA in the mix with equal final concentration. At the end, add the restriction enzymes required to completely digest the added DNA. Incubate the tubes at the recommended temperature for a recommended duration. Initially digested DNA was detected by radiolabeling and two-dimensional thin layer chromatography (TLC). This method was then replaced by southern blotting technique (Southern 1975). But currently, PCR-based method (Singer-Sam et al. 1990) for the detection of digested fragment and their quantification is being used. The digested DNA from both the MSRE and MNSRE is amplified with primers flanking the restriction site(s). In addition, nondigested DNA (treated same but without any restriction enzyme) is also amplified with the same primers. Since the MNSRE will cleave the fragment irrespective of the methylation state, there will never be a band in the PCR reaction. The absence of the

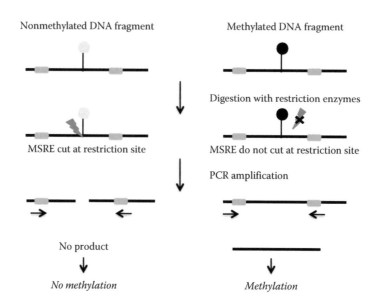

Nonmethylated DNA fragment   Methylated DNA fragment

Digestion with restriction enzymes

MSRE cut at restriction site   MSRE do not cut at restriction site

PCR amplification

No product

No methylation   Methylation

Figure 10.2 Schematic view of methylation-sensitive restriction enzyme (MSRE) followed by PCR method. MSR enzymes cannot digest the methylated DNA which result in PCR amplification. On the other hand, nonmethylated DNA can be digested by MSRE which lead to no PCR amplification.

band with PCR product of MSRE indicates that the site is hypomethylated, because MSRE can cleave the fragment. Whereas, the presence of the band represents the hypermethylation at the specific site(s). In case of partial methylation, the quantification is done by comparing the intensity of band between samples and control. This can be done using the softwares such as ImageJ (Rasband 1997). However, increase intensity of lower or upper bands after digestion represent the hypomethylation or hypermethylation state of DNA, respectively.

Advantages and disadvantages   It is one of the earliest techniques to study the locus-specific DNA methylation and is still in use. It is relatively quantitative in nature with reduced cost. It can be used in experiments or analysis where the methylation state of a specific site is already known and there is a need for an easy and economical technique for epigenotyping of samples. It can also be used to confirm the data acquired by some other techniques (bisulfite sequencing), which are suspected to be biased. However, there are some major drawbacks in this technique as well. It provides the limited amount of information on the cytosine site(s) and depends upon the existence of the restriction site in the fragment of interest. The digestion efficiency and incubation time of the restriction enzyme is also an issue of concern with this technique.

Bisulfite conversion/treatment

Epigenetic marks like DNA methylation are removed during the processes of amplification, i.e., PCR. So, to fix the epigenetic marks into the DNA sequence before the amplification techniques pushed the scientists toward the development of bisulfite treatment. It is one of the major advances in the field of DNA methylation analysis, which was based on discovery made in 1970 (Hayatsu et al. 1970; Shapiro et al. 1970) and was first reported to detect 5-mC by Frommer et al. (1992). This serves as one of the basic techniques with many modifications.

Principle   Bisulfite treatment is based on the deamination reactions of cytosine and 5 mC after the treatment of sodium bisulfite (Frommer et al. 1992; Tollefsbol 2011), which allow us to discriminate between methylated (5 mC) or unmethylated cytosine (Figure 10.3). In single-stranded DNA, the cytosine is converted to uracil in the presence of sodium bisulfite, while leaving 5 mC intact. This deamination effectively converts the epigenetic modification at cytosine into genetic information which can then be studied through regular molecular techniques in use for genetic analysis (Esteller 2004). After this modification, the specific region of interest (region of a gene) in the genome can be amplified through PCR, where uracil residues are converted to thymine. There are numerous modifications of this technique to study DNA methylation status, i.e., usage of selective or nonselective amplification and then direct PCR sequencing or cloning followed by sequencing. These modifications will be discussed in detail later in this chapter. All these modifications, allow the differentiation of methylated and nonmethylated DNA at the single-nucleotide base level.

Basic procedure   After the extraction, the genomic DNA is converted into single-stranded bisulfite-treated DNA. Fragmentation of the genomic DNA is sometimes recommended to facilitate and improve the quality of conversion of genomic DNA into bisulfite-treated DNA. Various DNA methylation kits by different companies are available in the market with varying specifications and optimized protocols but all of them involve the following steps; the first step is denaturation; it is critical to have a single-stranded DNA because this deamination reaction specifically works on single-stranded DNA. This

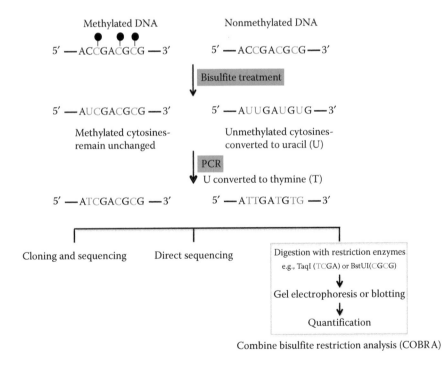

Figure 10.3 Schematic view of bisulfite treatment and related methods. Bisulfite treatment can only convert nonmethylated cytosine into uracil while it does not affect methylated cytosine. After PCR, uracil is converted to thymine. The PCR product can be then analyzed by various methods.

can be done by chemical denaturation with sodium hydroxide or through temperature denaturation. The time and temperature (for temperature denaturation) varies according to the recommendations of DNA methylation kit in use. The second step is sulfonation; sodium bisulfite treatment on denatured DNA deaminates the cytosine–bisulfite derivative to uracil–bisulfite derivatives. However, in some more advanced kits the denaturation and deamination steps are combined. The third step is desulfonation; to get uracil from the uracil–bisulfite derivatives, the sulfonate group is removed by increasing the pH by the subsequent alkali treatment.

Advantages and disadvantages   Treatment of genomic DNA with bisulfite allows distinction between methylated (5-mC) and unmethylated cytosines (C), thus it provides information about the single nucleotide polymorphism after bisulfite treatment. It is, however, challenging to achieve complete and highly selective bisulfite conversion of nonmethylated Cs without significant side reactions or extensive cleavage of genomic DNA. The rate of this chemical conversion can fluctuate depending upon the quality of DNA used, incomplete denaturation, or rapid renaturation process. Since the presence of proteins can also affect the rate of conversion, it is recommended to use protein-free DNA. In case of incomplete or less efficient conversion, larger number of cytosines will appear in the resulting data, which could be misleading and represent higher level of hypermethylation than actual condition. Therefore, keeping a check on the proper conversion should also be considered while doing this procedure. If used properly, following the recommended protocol, the DNA methylation kits in the market give very high conversion rate upto 99.9%.

Detection methods after bisulfite treatment   Since only one of the alleles of the gene is methylated in this method, changes in alleles after denaturation can be studied by methylation-specific PCR (that selectively amplifies the methylated or unmethylated allele) or non-methylation-specific PCR procedures (which amplify both the methylated and unmethylated alleles with equal efficiency).

Methylation-specific PCR procedures

Methylation-sensitive PCR   In methylation-sensitive PCR (MSP), two sets of primers are used for PCR and the primer pair which is designed to specifically amplify methylated allele can be termed as methylation-specific primers (MSp). The other primer pair that specifically amplifies the unmethylated allele can be termed as unmethylated-DNA-specific primers (USps). In MSp, all the CGs in the sequence remain as CGs, which favors the amplification of methylated allele, whereas in USp all the CGs in the sequence are converted into TGs to favor amplification of unmethylated/hypomethylated allele, thus, ability of primer amplification shows the status of DNA methylation.

This method is suitable to know the methylation level of a specific locus. Moreover, it is a quick and relatively cheap technique without going into the trouble of sequencing and analyzing the sequenced data. However, some modifications like MethyLight (Eads et al. 2000), QAMA (Zeschnigk 2004), and melting curve analysis have been reported in this technique.

Nonmethylation-specific PCR procedures

Combined bisulfite restriction analysis   Another modification/derivation of quantitative approach to avoid the sequencing is combined bisulfite restriction analysis (COBRA) (Xiong and Laird 1997). It combines bisulfite conversion-based PCR (Figure 10.3). Originally developed to reliably handle minute amounts of genomic DNA from microdissected paraffin-embedded tissue samples, the technique has since seen the widespread usage in cancer research and epigenetics studies.

This technique works on the principle that the sequence conversion through bisulfite treatment could cause the methylation-dependent creation of a new restriction enzyme site or it could lead to the methylation-dependent retention of a preexisting site such as *Bst*UI (CGCG). The differential pattern of bands on the gel will lead toward the identification of DNA methylation pattern of the fragment under investigation.

In this technique, the PCR of bisulfite-treated DNA is done with nonselective primers (NSps) that amplify bisulfite-treated DNA without distinguishing between hypomethylation and hypermethylation states. Then, instead of cloning and sequencing or direct bisulfite sequencing (DBS), the PCR amplicon is digested by the restriction enzyme (like *Bst*UI or TaqI, etc.), which can only cleave the sites if the cytosine is intact, i.e., the site is methylated and bisulfite treatment has not converted this cytosine into uracil. After this RE digestion, the amplicon is run on the gel. The pattern of band(s) of this RE-digested PCR amplicon on the gel is compared with the nondigested PCR amplicon.

If the results indicate a partial methylation pattern, then the quantification of the bands is necessary to find out the level of methylation at the given CG site. By electroblotting from polyacrylamide gel, oligo hybridization and phosphor-imager quantification could be a good approach to achieve the desired results. The flexibility of choosing the probe position in relation to the restriction sites is an advantage of this hybridization strategy. Results obtained with different probes can be used to corroborate DNA methylation values obtained for a particular CG site.

Bisulfite sequencing   This approach involves amplification by the use of primers, which are nonsensitive to either methylation state (i.e., hypermethylation or hypomethylation) of the allele. Classically, the amplified PCR product is cloned and these clones are sequenced ranging from 20 to 100 clones depending upon the complexity of the fragment and quality of the amplicon. The sequencing data from these clones is then analyzed to quantify the level of DNA methylation at each of cytosine in the desired fragment. Nowadays, different softwares such as KISMETH (Gruntman et al. 2008), R package MethVisual (Zackay and Steinhoff 2010), etc., are available to facilitate this kind of data analysis.

Direct bisulfite sequencing

Cloning and sequencing is a costly, laborious, and time-consuming process. Therefore, in search of a solution to these problems, bioinformatic tools, i.e., various softwares such as ESME (Lewin et al. 2004), Mquant (Leakey et al. 2008), Mutation Surveyor (SOFTGENETICS®), etc., have been developed. These softwares compare the relative peak heights of both cytosine and thiamine to quantify the methylation level at a given cytosine. Increasing number of publications using this software-based DBS approach shows the increasing level of confidence of scientific community on this adaptation (Cortese et al. 2011; Steinfelder et al. 2011; Khan et al. 2013). In mammalian DNA, the DNA methylation is restricted to CG, so it is easier to quantify and normally ESME software is used. However in plants, the occurrence of DNA methylation at all three contexts, that is, CG, CHG, and CHH (where H stands for any nucleotide but G) make the quantification more complicated, so a software like Mutation Surveyor that has the capability to quantify all three contexts is used.

Bisulfite and next-generation sequencing   The advancements in the sequencing techniques (next-generation sequencing [NGS]) has allowed scientists to get a large amount of sequencing data in a very short time and the cost of it is decreasing with every new technology in this field. These NGS techniques have also been put into use in DNA methylation analysis as well. Most of the time these techniques are used on the PCR product of bisulfite-treated DNA. Depending upon the NGS technique, there could be various modifications.

Bisulfite pyrosequencing   Bisulfite pyrosequencing is an advancement/modification in bisulfite sequencing techniques (sometimes also acknowledged as modification in the COBRA technique) for the quantification of DNA methylation after bisulfite treatment and PCR amplification where the restriction analysis is substituted with the use of high-throughput sequencing (pyrosquencing) for the quantification of DNA methylation at cytosine sites (mainly CGs) (Colella et al. 2003; Tost and Gut 2007).

The process starts with the two steps having done bisulfite treatment (as described above). The desired fragment (of up to 350 bp) is then amplified through PCR by using the nonspecific primers (NSps), which amplify bisulfite-converted DNA (both methylated and nonmethylated fragments). One of these two primers is labeled with biotin at its 5′ terminal. This facilitates the rendering of PCR product to become single stranded by immobilization of this (biotin labeled) primer on streptavidin-coated bead during the purification process. This immobilization is subsequently followed by alkaline treatment. Usage of the vacuum preparation tool (Biotage) is recommended to capture and hold the bead during the differential purification step while allowing the solution to pass through the filter. Now the sequencing primer is hybridized to the complementary DNA template. It has to be kept in mind to design these primers so that it can hybridize several bases (30–120 bases) 5′ to the regent/site of interest due to the limitation in the read length of pyrosequencing. Then the pyrosequencing reaction takes place. After that, data analysis completes the process by providing the quantitative information about the methylation state of the sites of interest.

Genome-wide DNA methylation techniques

Methyl-sensitive amplification polymorphism

Methyl-sensitive amplification polymorphism (MSAP) is a modification of AFLP technique to study genome-wide DNA methylation. Reyna-López et al. in 1997 used this technique in fungi for the first time followed by its modified version to be used in plant species (Xiong et al. 1999). It is a very useful technique for developmental biology studies, population genetics studies, and even in plant breeding (Long et al. 2011; Meng et al. 2011; Morán and Pérez-Figueroa 2011; Schulz et al. 2013).

Principle   It is a modification of AFLP in which rare cutter and frequent cutter are used for the digestion of genomic DNA followed by ligation with adopters and then PCR amplified (Figure 10.4). In both AFLP and MSAP, the rare cutter is same, i.e., *Eco*RI whereas for frequent cutter instead of *Mse*I, MSAP uses *Msp*I and *Hpa*II in parallel. As both these enzymes recognize the same restriction site 5′–CCGG–3′, their sensitivity toward the methylation state of cytosine is different. This difference is used to identify the DNA methylation pattern. The comparison of the resulting *Eco*RI/*Hpa*II and *Eco*RI/*Msp*I fragment profiles allows the detection of particular methylation status of the restriction sites and comparison of DNA methylation profile of different individuals of a population or different population of a species.

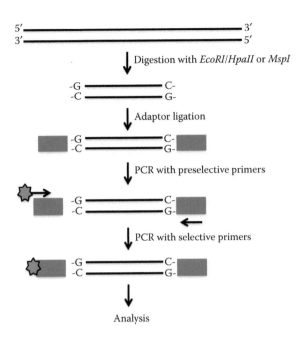

Figure 10.4 Schematic view of methyl-sensitive amplification polymorphism (MSAP). DNA is subjected to adaptor ligation after digestion with restriction enzymes. Then the fragment is amplified by using preselective primers, followed by selective primers and analysis.

Procedure   The main steps are as follows:
DNA extraction and digestion: The genomic DNA is first digested with the rare cutting restriction enzyme, i.e., *EcoRI*. This digestion should be performed following the protocol provided by the manufacturer. This digested DNA (by *EcoRI*) is then divided into two equal parts and each half is further digested with either *HpaII* or *MspI* following the provider's recommendations.

Adaptor ligation and amplification: Then, fragments are ligated with adapters complementary to the *EcoRI* or *MspI/HpaII* cohesive ends by following the recommendations of manufacturers. Then the preamplification PCR is performed by using the primers complementary to the ligated adapters for *EcoRI* and *MspI/HpaII* followed by selective amplification.

Analysis: The resulting PCR product is resolved on gels to differentiate the different patterns of methylation. MSAP fragments are carefully excised and purified from gel. The purified fragments are subjected to either cloning followed by sequencing or directly sequencing.

Advantages and disadvantages   MSAP is a very effective, economical and fast technique because it is easy to perform, can provide information about the DNA methylation variation for the large number of samples in a limited time. It can serve as the starting point toward isolation and characterization of regions, which show a shift in DNA methylation pattern after a certain treatment or stress. Another advantage of this technique is that it can work with a low quantity of DNA.

Apart from these advantages, this technique also has some shortcomings; it provides a limited amount of information since the restriction enzymes are very specific toward their recognition sites. In addition, different combinations, i.e., methylation stats like double-stranded methylated $^mC^mCGG$, double-stranded methylated ^mCCGG cannot be differentiated. This technique cannot provide site-per-site analysis of DNA methylation shift and percentage of DNA methylation.

DNA methylation analysis through high-performance liquid chromatography

One of the earliest techniques reported for the quantitative determination of global DNA methylation is through the use of high performance liquid chromatography (HPLC) (Singer et al. 1977). The decade of 1980s witnessed various improvements and modifications in this technique (Kuo et al. 1980; Catania et al. 1987).

Principle   HPLC is a highly sensitive chromatographic technique, which is used for the separation of compounds based on their affinity and polarity toward mobile phases and stationary phases, then these separated compounds are detected by using different mode and types of detectors. Using this basic principle of HPLC, the global content of DNA methylation can be measured.

Procedure   The main steps are as follows:

DNA extraction and hydrolysis: The procedure starts with the extraction of high-quality RNA-free DNA from the samples (the sample could be animal or plant tissues) followed by denaturation by heating and chemical hydrolysis by using enzymes such as nuclease PI, alkaline phosphatase.

Nucleoside separation: Nucleoside separation is carried out by using hydrophobic-binding interaction between the sample's solute molecule in the mobile phase and the stationary phase in the column, which are basically immobilized hydrophobic ligands. The eluted molecules from the column are then introduced in the electrospray system of a mass spectrometer (ESI/MS). This should give the quantitative analysis of DNA methylation by using the HPLC peaks by ESI/MS spectra.

Analysis: The quantification can be done by using the following equation: 5mdC peak area * 100/(dC peak area + 5mdC peak area); where 5mdC is 5-methyl-2'-deoxycytidine and dC is 2'-deoxycytidine.

Advantages and disadvantages   The advantages of this technique are that it involves biological and biochemical testing, structure elucidation, and characterization of side products from production, metabolites from the biological matrix, and natural products. The disadvantages include the difficulty to detect coelution with HPLC, which may lend to inaccurate compound categorization and high-cost equipment and requirement of a trained technician to operate.

Methylated DNA immunoprecipitation

Methylated DNA immunoprecipitation (MeDIp) can be used for locus-specific detection of DNA methylation as well as genome-wide determination of DNA methylation.

Principle   The MeDIP method uses the basic principle of the affinity purification in which the methylated DNA is precipitated by an antibody directed against 5-mC (Figure 10.5). The enriched DNA fragments can be analyzed by PCR for locus-specific studies or by microarrays (MeDIP-chip) and massively parallel sequencing (MeDIP-seq) for whole-genome studies.

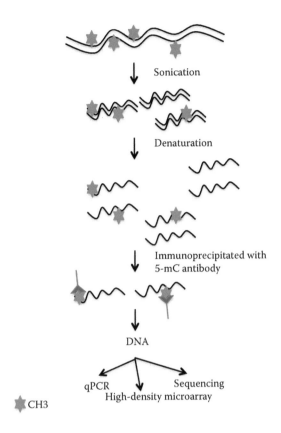

Figure 10.5 Schematic view of methylated DNA immunoprecipitation. DNA is sonicated to get smaller fragments. After denaturation, DNA is precipitated by using 5-mC antibody. Finally, DNA is recovered to analyze by using various methods.

Procedure   The main steps used in this technique are as follows:

Extraction of genomic DNA: DNA from a variety of desired tissues or cell lines can be used for MeDIP. High-quality DNA should be extracted and it should be free from associated proteins such as histones and RNAs as much as possible since these can interfere with both DNA quantitation and antibody binding.

Shearing of genomic DNA: Random shearing genomic DNA can be done by sonication or by the usage of restriction enzymes to produce fragments between 300 and 1000 bp. But restriction enzymes are not generally recommended in cases where unbiased microarray is under consideration. Various factors such as quality and quantity of DNA, sonicator settings, and size and quality of the sonication tip may cause fluctuations in the sonication efficiency. Therefore, proper monitoring through gel electrophoresis and uniform sonication settings are recommended.

Immunoprecipitation of methylated DNA: The sonicated DNA is then immunoprecipitated by using a monoclonal antibody against 5-methylcytidine (5-mC). Various commercial companies provide these monoclonal antibodies. The inclusion of untreated DNA is recommended, which will serve the purpose of input control. Immunoprecipitation provides the enriched DNA methylated fragments.

Analysis of MeDIP fragments: Depending upon the requirements of the experiment, the DNA methylation data can be acquired both for specific loci and genome-wide level. For locus-specific analysis, PCR or real-time PCR could be done and relative enrichment of MeDIP fraction can be calculated in comparison with unmethylated control. For genome-wide analyses, two dyes, i.e., Cy3 and Cy5 are used to differentially label the input and MeDIP fractions. These are cohybridized to microarrays, which serve as a two-color experiment to measure the methylation level of DNA, which depends on the intensity ratio of immunoprecipitated to input DNA.

Advantages and disadvantages This technique has advantages—it is relatively fast; can be both locus specific and genome wide; compatible with array-based analysis; and can be applicable for high-throughput sequencing. The drawbacks or the potential problems of this technique include the antibody specificity, the fragments with low levels of DNA methylation can hardly be detected, and availability of skilled persons.

Histone modifications

Chromatin immunoprecipitation

Chromatin immunoprecipitation (ChIP) is an *in vivo* method used to unveil the link between the genome and proteome. It is used to detect the association of individual protein with the specific genomic regions and histone modifications in epigenetics (Figure 10.6).

Procedure The main steps of the procedure are

Cross-linking: Because many DNA–protein interactions are important for different biological functions, the ChIP requires the cross-linking of DNA–protein complexes with

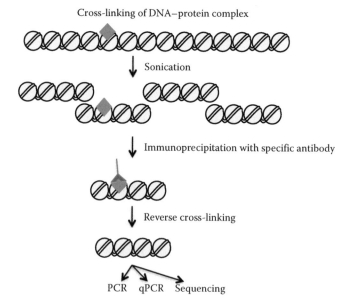

Figure 10.6 Schematic view of chromatin immunoprecipitation (ChIP). DNA–protein complexes are cross-linked with cross-linkers. The cross-linked DNA is subjected to sonication which result in smaller fragments. Then the DNA is precipitated by using specific antibodies. Finally, DNA is recovered after reverse cross-linking which is then used for analysis by using various methods.

cross-linkers (formaldehyde, EGS, DGS, etc.) that permeate directly into the cells to covalently stabilize DNA–protein interactions.

Cell lysis: The DNA–protein complexes are extracted by lysing the cell membrane either with detergents or mechanical method. To increase the sensitivity and remove background in later steps, nuclei extraction and removal of other cellular comparments is recommended.

Chromatin shearing/digestion: After following the extraction step, genomic DNA is sheared into small pieces (200 –> 1000 bp) either with micrococcal nuclease (MNase) or by using the sonicator. Both MNase and sonicator have their limitations such as sonication time; MNase concentration and time; etc.

Immunoprecipitation with specific antibody: Then the DNA–protein complex is precipitated with specific antibodies against histone modifications or transcription factors. Moreover, antibodies against fusion proteins such as HA, Myc, or GST can also be used.

De-cross-linking and DNA recovery: Because the precipitated DNA contains DNA–protein complex, therefore, DNA–protein interaction is reversed by heat incubations along with protein digestion with proteinase K. To cleanup protein and get DNA, the phenol–chloroform followed by standard ethanol precipitation is performed.

Analysis: The ChIP data can be acquired for both locus specific and genome wide. PCR or real-time PCR could be done and relative enrichment can be calculated in comparison with control (InPut or Mock) for locus specific. For genome-wide analysis, library preparation followed by amplification with different barcodes that can be specifically recognized during sequencing is performed.

Advantages and disadvantages Coupling with qPCR makes ChIP a quantitative tool to capture a snapshot of specific DNA–protein interactions in living cells. ChIP can provide the locus specific as well as genome-wide information. However, it needs a specialized person to perform the experiments as well as the specific antibody.

Future challenges

Epigenetics regulators modulate chromatin packaging and nuclear organization to control different cellular processes by regulating gene expression. The recent advances in technology enabled us to survey genome-wide epigenetic modifications; thereby, locus-specific as well as genome-wide epigenetic modifications may lead us to an alternative way of crop improvement by manipulating epigenetic regulators. Unlike genetic information which is static, epigenetic information is dynamic that requires an adaptation process; this shows variability among different cells, even between same cell type at different developmental stages, and/ or in response to different environmental stimuli. The relationship between epigenetic modifications and phenotype is stronger than DNA mutation and phenotype in many instances, therefore, studying this phenotypic plasticity encoded by epigenetic modifications is very challenging in terms of technology and resources. Thus, we need to establish new robust methods to study the genome-wide epigenetic modification in a single cell. Moreover, we also need to upgrade the existing methods that can detect minute quantity of signals for a certain epigenetic modification to enhance their sensitivity, specificity, and application.

References

Berr A, Shafiq S, Shen W-H. 2011. Histone modifications in transcriptional activation during plant development. *Biochimica et Biophysica Acta* 1809: 567–576.

Bird AP, Southern EM. 1978. Use of restriction enzymes to study eukaryotic DNA methylation: I. The methylation pattern in ribosomal DNA from *Xenopus laevis*. *Journal of Molecular Biology* 118: 27–47.

Catania J, Keenan BC, Margison GP, Fairweather DS. 1987. Determination of 5-methylcytosine by acid hydrolysis of DNA with hydrofluoric acid. *Analytical Biochemistry* 167: 347–351.

Cedar H, Solage A, Glaser G, Razin A. 1979. Direct detection of methylated cytosine in DNA by use of the restriction enzyme MspI. *Nucleic Acids Research* 6: 2125.

Colella S, Shen L, Baggerly KA, Issa JP, Krahe R. 2003. Sensitive and quantitative universal Pyrosequencing methylation analysis of CpG sites. *Biotechniques* 35: 146–150.

Cortese R, Lewin J, Bäckdahl L, Krispin M, Wasserkort R, Eckhardt F, Beck S. 2011. Genome-wide screen for differential DNA methylation associated with neural cell differentiation in mouse. *PLoS ONE* 6: e26002.

Eads CA, Danenberg KD, Kawakami K, Saltz LB, Blake C, Shibata D, Danenberg PV, Laird PW. 2000. MethyLight: A high-throughput assay to measure DNA methylation. *Nucleic Acids Research* 28: e32.

Esteller M. 2004. *DNA Methylation: Approaches, Methods, and Applications*. CRC Press, USA, ISBN 9780849320507.

Feng W, Dong Z, He B, Wang K. 2012. Analysis method of epigenetic DNA methylation to dynamically investigate the functional activity of transcription factors in gene expression. *BMC Genomics* 13: 532.

Frommer M, McDonald LE, Millar DS, Collis CM, Watt F, Grigg GW, Molloy PL, Paul CL. 1992. A genomic sequencing protocol that yields a positive display of 5-methylcytosine residues in individual DNA strands. *Proceedings of the National Academy of Sciences* 89: 1827–1831.

Gruntman E, Qi Y, Slotkin RK, Roeder T, Martienssen RA, Sachidanandam R. 2008. Kismeth: Analyzer of plant methylation states through bisulfite sequencing. *BMC Bioinformatics* 9: 371.

Hayatsu H, Wataya Y, Kazushige K. 1970. The addition of sodium bisulfite to uracil and to cytosine. *Journal of the American Chemical Society* 92: 724–726.

Holliday R. 1994. Epigenetics: An overview. *Developmental Genetics* 15: 453–457.

Khan AR, Enjalbert J, Marsollier A-C, Rousselet A, Goldringer I, Vitte C. 2013. Vernalization treatment induces site-specific DNA hypermethylation at the VERNALIZATION-A1 (VRN-A1) locus in hexaploid winter wheat. *BMC Plant Biology* 13: 209.

Kuo KC, McCune RA, Gehrke CW, Midgett R, Ehrlich M. 1980. Quantitative reversed-phase high performance liquid chromatographic determination of major and modified deoxyribonucleosides in DNA. *Nucleic Acids Research* 8: 4763–4776.

Laird PW. 2010. Principles and challenges of genome-wide DNA methylation analysis. *Nature Reviews Genetics* 11: 191–203.

Leakey TI, Zielinski J, Siegfried RN, Siegel ER, Fan C-Y, Cooney CA. 2008. A simple algorithm for quantifying DNA methylation levels on multiple independent CpG sites in bisulfite genomic sequencing electropherograms. *Nucleic Acids Research* 36: e64.

Lewin J, Schmitt AO, Adorján P, Hildmann T, Piepenbrock C. 2004. Quantitative DNA methylation analysis based on four-dye trace data from direct sequencing of PCR amplificates. *Bioinformatics* 20: 3005–3012.

Long Y, Xia W, Li R, Wang J, Shao M, Feng J, King GJ, Meng J. 2011. Epigenetic QTL mapping in *Brassica napus*. *Genetics* 189: 1093–1102.

Meng F-R, Li Y-C, Yin J, Liu H, Chen X-J, Ni Z-F, Sun Q-X. 2011. Analysis of DNA methylation during the germination of wheat seeds. *Biologia Plantarum* 56: 269–275.

Morán P, Pérez-Figueroa A. 2011. Methylation changes associated with early maturation stages in the Atlantic salmon. *BMC Genetics* 12: 86.

Rasband WS. 1997. *ImageJ*. U.S. National Institutes of Health, Bethesda, Maryland.

Ratel D, Ravanat J-L, Berger F, Wion D. 2006. N6-methyladenine: The other methylated base of DNA. *Bioessays: News and Reviews in Molecular, Cellular and Developmental Biology* 28: 309–315.

Reyna-López GE, Simpson J, Ruiz-Herrera J. 1997. Differences in DNA methylation patterns are detectable during the dimorphic transition of fungi by amplification of restriction polymorphisms. *Molecular and General Genetics* 253: 703–710.

Schulz B, Eckstein RL, Durka W. 2013. Scoring and analysis of methylation-sensitive amplification polymorphisms for epigenetic population studies. *Molecular Ecology Resources* 13: 642–653.

Shapiro R, Servis RE, Welcher M. 1970. Reactions of uracil and cytosine derivatives with sodium bisulfite. *Journal of the American Chemical Society* 92: 422–424.

Singer J, Stellwagen RH, Roberts-Ems J, Riggs AD. 1977. 5-Methylcytosine content of rat hepatoma DNA substituted with bromodeoxyuridine. *The Journal of Biological Chemistry* 252: 5509–5513.

Singer-Sam J, Grant M, LeBon JM, Okuyama K, Chapman V, Monk M, Riggs AD. 1990. Use of a HpaII-polymerase chain reaction assay to study DNA methylation in the Pgk-1 CpG island of mouse embryos at the time of X-chromosome inactivation. *Molecular and Cellular Biology* 10: 4987–4989.

Southern EM. 1975. Detection of specific sequences among DNA fragments separated by gel electrophoresis. *Journal of Molecular Biology* 98: 503–517.

Steinfelder S, Floess S, Engelbert D, Haeringer B, Baron U, Rivino L, Steckel B et al. 2011. Epigenetic modification of the human CCR6 gene is associated with stable CCR6 expression in T cells. *Blood* 117: 2839–2846.

Tollefsbol TO (ed.). 2011. *Epigenetics Protocols.* Humana Press, Totowa, NJ.

Tost J, Gut IG. 2007. DNA methylation analysis by pyrosequencing. *Nature Protocols* 2: 2265–2275.

Xiong LZ, Xu CG, Saghai Maroof MA, Zhang Q. 1999. Patterns of cytosine methylation in an elite rice hybrid and its parental lines, detected by a methylation-sensitive amplification polymorphism technique. *Molecular and General Genetics* 261: 439–446.

Xiong Z, Laird PW. 1997. COBRA: A sensitive and quantitative DNA methylation assay. *Nucleic Acids Research* 25: 2532–2534.

Zackay A, Steinhoff C. 2010. MethVisual - visualization and exploratory statistical analysis of DNA methylation profiles from bisulfite sequencing. *BMC Research Notes* 3: e337–e337.

Zeschnigk M. 2004. A novel real-time PCR assay for quantitative analysis of methylated alleles (QAMA): Analysis of the retinoblastoma locus. *Nucleic Acids Research* 32: e125–e125.

Animal biotechnology

chapter eleven

Animal models and biomedical research

Adeena Shafique, Azka Khan, Kinza Waqar,
Aimen Niaz, and Alvina Gul

Contents

Abstract

Genetically engineered animals or transgenic animals represent such organisms in which specific genetic modifications are introduced or cloning is done to generate animal models with desired genotypes and phenotypes. The major consideration involved in the art of creating transgenic models is that they are bona fide biological models that best report the biological function of the gene that they represent. This innovative technology has opened up new horizons in all fields of biomedical research as transgenic and knockout animals have provided novel insight and greatly increased the understanding and knowledge of several human diseases. This technology has provided invaluable biological tools for the production of animal models of human diseases that allow the development and validation of new medical drugs for the treatment of human diseases. Transgenic mouse models have been the most common models for revealing the gene function. Over the past couple of decades, researchers have advanced toward creating complex transgenic animal models of human diseases, including genetically modified rabbits, pigs, and even cows. Moreover, they also proved to be instrumental for target discovery, authentication, and production of therapeutic proteins. The technology entailing engineered animal models promises significant benefits for human health via its application in biomedical research. This chapter provides an in-depth knowledge about genetically engineered animals and illustrates their significance for biomedical research highlighting their multiple uses for humans.

Keywords: Engineered animal models, Transgenic animal models, Engineered cell lines, Biomedical research.

Introduction

Scientists around the world have been working to treat the pain and suffering of humans for about two centuries. Extensive research in biomedicine has considerably offered a range of novel potentials to cure fatal diseases and to improve public health. One of the major concerns about biomedical research these days is to convert sequence information of genes and analytical data into understandable knowledge to comprehend their functions. However, with the advent of biotechnological breakthroughs, complex processes of many severe genetic and other diseases have been clearly understood to some extent. Usage of animal models to explore the basic pathophysiological mechanisms has been a cornerstone of biomedical research [1].

Recent developments in molecular and cellular biology have helped in production of genetically engineered animal models, presenting most invaluable tools for biomedical research. Genetic engineering has led to the production of transgenic animal models by manipulating particular genes of their genome. These manipulations result into the alterations of genetic function by either introducing some exogenous genes or removing specific endogenous genes [2]. This magnificent approach has provided a deep understanding of normal physiological and pathological phenomena of human beings. These engineered animal models are being used to indicate the most important biological processes to reveal the functioning of genes in the milieu of susceptibility and prognosis of many diseases and response to new therapeutic treatments.

The production of a transgenic mouse that overexpressed growth hormone in the early 1980s revolutionized the era of molecular biology and molecular physiology [3]. After a few years, the introduction of gene targeting and knockout animals took another step toward progress of biomedical research [4,5].

Animal models possibly more close to humans are being used as an efficient tool for vaccine development and to evaluate their efficacy as well as safety prior to clinical studies and commercialization. This approach has been used to generate many exogenous or endogenous proteins, including many important hormones, which are easily isolated and purified [6].

Tissue engineering covers a variety of applications in terms of bone and cartilage regeneration and repair or replacement of a portion or a whole tissue, for example, skin, muscles, and blood vessels. At present, over 95% of transgenic animals utilized as a part of biomedical research are rodents. By virtue of comprehensive analysis of the whole genome sequence, analogy to the human genome, accessibility of sturdy, latest methodologies that enable manipulations at the genetic level in cells and embryos, potential to perform physiological and behavioral tests that can be extrapolated straight to human diseases and short life cycle, mouse is one of the best animal models of choice [7]. In last few years, hundreds of genetically engineered mouse models have been developed that carry the genetic predisposition to many fatal diseases like Alzheimer disease as well as hereditary diseases like cystic fibrosis or obesity, etc. These mice models develop the same clinical symptoms as humans and have been extensively utilized to study other characteristics pertaining to human diseases.

Transgenic rodent models have provided insights to study arthritis presenting pathological features which are more specific than already-existing animal models. These models are crucial for human immunodeficiency virus and acquired immune deficiency syndrome (HIV/AIDS) research because they have been used to study the functioning of the immune system and role of many important genes. To study carcinogenicity and neurodegenerative disorders, transgenic mice models represent the most efficient models as less time is consumed for experiments [8].

In the last two decades, genetically engineered pig models have been developed for many biomedical applications. Recent advancements in pig genome research and the generation of genetically modified primary pig cells and pig cloning have greatly facilitated the engineered pig models for xenotransplantation and many other human diseases. Owing to the similarities of pigs to humans with respect to genetics, physiology, and anatomy, pigs are highly appropriate animal models for biomedical research [9]. As compared to rodents, pigs have been shown to share many important resemblances with human beings in context to the gastrointestinal, respiratory, hepatic, cardiovascular, renal, reproductive, central nervous, endocrine, immune, dermal, optical, skeletal, and muscular systems. They have also become important large animal models for understanding inherited metabolic disorders and preclinical testing of preventive drugs as well as other novel therapeutic interventions.

Studies on other genetically engineered animal models such as zebra fish, sheep, goats, dogs, cattle, and large primates such as monkeys have been done. Currently, these animal models pose many challenging situations that need more efficient techniques and methods to use them as ideal animal models for specific human diseases. Nevertheless, these genetically modified animal models have positively influenced other biological fields, including human medicine, oncology, pathophysiology, vaccinology, and pharmacology [10].

Cell lines (human, mouse, rat) as model for biomedical research

Human cell lines

Development of cell culture led to the development of human cell lines, which are important for biomedical research. The first human cell line is the HeLa cell line, which was developed in 1951 from a cervical cancer patient [11]. These cell lines were proved to be very successful *in vitro* and it was a novel step to achieve future benefits in biomedical research. Many other cell lines were also derived from humans and they have been extensively used to study various fatal diseases to date. In order to obtain the same disease conditions *in vitro*, these cell lines are genetically engineered to express the desired genes, which show maximum similarity to the original disease conditions. Genetically engineered cell lines have proved to be a promising tool for medical research. Some of the important human cell lines used in cancer research are discussed below.

Tumor cell lines

Human epithelial carcinoma cell line A549

Human epithelial carcinoma cell line A549 is a model cell line for cancer research [12]. These cell lines have been used to study many types of tumors. As epidermal growth factor receptor (EGFR) is overexpressed in many solid tumors, particularly in non-small-cell lung cancer, overexpression of EGFR in tumor cells was studied for therapeutic purposes since it is an excellent target for tumor therapy. An anti-EGFR single-chain variable fragment (scFv) with high antigen specificity and affinity has also been generated. A study has been conducted to investigate the antitumor potential of the anti-EGFR scFv and to elucidate whether chimeric antigen receptor (CAR)-modified T lymphocytes have potential application in the treatment of EGFR-positive tumors.

CAR-transfected lymphocytes and A549 tumor cells were inoculated in mice models to develop tumors. CAR-transfected T lymphocytes showed a remarkable antitumor efficacy both *in vitro* and *in vivo* [13].

Human bladder carcinoma cell lines and human fibroblast IMR90 cell lines

Human bladder cell line T24 and human fibroblast IMR90 cell line have been shown to be effective tools for gene therapy to treat tumors [14]. These cell lines have been used to determine the cell-specific activity of a fragment of diphtheria toxin DT-A chain gene driven by the H19 promoter region. These two cell lines were co-transfected by a Luc-4 control vector along with the SV40 promoter and enhancer and DTAPBH19 vector.

A dose-dependent decline in the expression of the luciferase gene was observed in IMR-90 cells and T24P cells. There was a minimal inhibition at 0.05 g of toxin gene-carrying construct in IMR-90 cells (H19 regulatory sequences used were not expressed) cotransfected with DTA-PBH19 and almost complete inhibition was observed with 1 g. As compared to IMR-90 cells the expression of luciferase was fivefold higher in the T24P cells expressing H19 regulatory sequences. The antitumor activity of DTA protein has shown great therapeutic potential for the treatment of bladder cancer in preclinical studies [15].

SH-SY5Y human neuroblastoma cells

SH-SY5Y is a human-derived cell line used in biomedical research [16]. This cell line was subcloned from another cell line named SK-N-SH isolated from human bone marrow biopsy of a 4-year-old female patient of neuroblastoma. These cell lines are important *in vitro* model for investigation of neuronal differentiation and function. Besides being adrenergic in phenotype, these cell lines also express dopaminergic markers that are used extensively to study Parkinson disease (PD) [17]. Human embryonic kidney 293 cell lines, also known as HEK239 cell, are the particular cell lines derived from embryonic kidney cells of humans that are grown on tissue culture. These cell lines are commonly used worldwide for cell biology research due to their easy-to-grow and fast transfection properties [18].

The two aforementioned cells lines have been used to study one of the genetic causes of PD. Research has shown that the leucine-rich repeat kinase 2 (*LRRK2*) gene plays an important role in the progression of PD. Mutations in *LRRK2* can result into late-onset PD. To investigate the role of this gene in PD etiology, plasmid containing full-length *LRRK2* with a C-terminal myc-His tag was transfected into HEK-293T cell lines and SH-SY5Y cells lines.

The expression of a single protein that corresponds to the *LRRK2* gene was observed upon transfection in both cell lines. Results have shown a gain-of-function mechanism for *LRRK2*-associated disease along with a vital role for kinase activity in the progression of PD. In future, the development of such agents that can revoke *LRRK2* kinase activity may present a therapeutic prospective for the treatment of PD [19]. Table 11.1 shows some of the cell lines and their use in different diseases.

Table 11.1 Human cell lines used in different diseases

Name of cell line	Method of production	Disease	Use	References
Human epithelial carcinoma cell line A549, T lymphocytes	Derived from human epithelial cell carcinoma	Lung carcinoma	Antitumor potential of the anti-EGFR	[12,13]
Human bladder carcinoma cell lines and human fibroblast IMR90 cell lines	Derived from human bladder carcinoma	Bladder cancer	Antitumor activity of DTA protein	[14,15]
SH-SY5Y human neuroblastoma cells and HEK239 cell lines	Derived from human neuroblastoma cells and human embryonic kidney cells	Parkinson disease	Mutations in *LRRK2* gene	[16–19]

Mouse cell lines

Murine 3T3-L1 cell line

3T3-L1 cell line is a mouse cell line that is derived from mouse 3T3 cells. These cells have a morphology similar to the fibroblast cells. These cells have a tendency to differentiate into adipocytes under suitable conditions. Owing to this feature, this cell line is used for understanding many metabolic diseases.

PRKAR2B, an important subunit of protein kinase A (PKA), regulates the complex process of adipogenesis. 3T3-L1 cells and primary human preadipocytes were used to investigate the role of PKA regulatory subunits in differentiation along with their expression in human adipose tissue during fetal growth. siRNA were used against PRKAR2B and cell lines were allowed to transfect with siRNA. Cell culture was induced to differentiate. In *in vitro* differentiation, the expression of PRKAR2B has to be gradually increased. The silencing of PRKAR2B led to the reduction of certain enzymes vital to lipid metabolism resulting in intracellular accumulation of triglycerides with impaired differentiation of adipocytes in both murine as well as human cell lines [20].

Leukotrienes (LTs) play an important role in adipocyte differentiation and maturation. The role of leukotriene B_4 (LTB_4)–leukotriene receptor (BLT) signaling has been studied in mouse 3T3-L1 cell lines. To block the LTB_4–BLT signaling pathway, the mouse 3T3-L1 preadipocytes were mixed with small interfering RNA (to knockdown the BLT signaling) along with lipoxygenase (LOX) inhibitors and the BLT antagonist. It was observed that siRNA dramatically suppressed the differentiation of preadipocytes into mature adipocytes [21].

Other mouse cell lines

Genetically engineered mouse models have been developed to study pancreatic cancers. These engineered mouse models produce pancreatic adenocarcinoma (PDAC). Though these models provide detailed understanding of the pathology of disease, there is still a need to establish syngeneic cell lines from such models. Three mouse cell lines are successfully established from two (PDAC) mouse models. The UN-KC-6141, UN-KPC-960, and UN-KPC-961 cell lines were derived from the pancreatic tumors of two transgenic mice. These cell lines showed cancer mutations that were present in parent mice. These cell lines expressed a number of markers and they were also resistant to chemotherapeutic drugs. The ability of these cell lines to mimic the genetic mutation occurring in human PDAC make them a potential biological tool for therapeutic research [22]. Table 11.2 shows some of the mouse cell lines and their uses in diseases.

Rat cell lines

PC12 cell line

In 1996, PC12 cell lines were cloned from a pheochromocytoma of the rat adrenal medulla. On treatment with certain growth factors, these cell lines cease proliferation. These cell lines are one of the important model cell lines used for neuronal differentiation. Such cell lines have been extensively used for many life-threatening neurological disorders and help in providing insights to develop new strategies to cure such diseases [23].

Von Hippel–Lindau (VHL) disease is a genetic disorder that occurs due to mutations in the VHL gene. The manifestations of such mutations are the formation of tumors, that is, pheochromocytoma. The exact mechanisms underlying this disease condition are not yet well understood. Rat pheochromocytoma PC12 cell lines were used to understand

Table 11.2 Mouse cell lines used in different diseases

Name of cell line	Method of production	Disease	Use	References
Murine 3T3-L1 cell line	Derived from mouse embryonic fibroblasts	Adipogenesis	Investigating the PRKAR2B subunit activity, blockage of LTB_4-BLT signaling pathway	[20,21]
UN-KC-6141, UN-KPC-960, and UN-KPC-961	Derived from pancreatic tumor of two transgenic mice	Pancreatic adenocarcinoma	Therapeutic model cell line for PADC	[22]

these mechanisms that lead to VHL-linked pheochromocytoma. To reduce levels of VHL in pheochromocytoma, VHL protein levels were allowed to decrease in rat PC12 cell lines by using RNA interference. High levels of substrate of VHL protein were observed in cell lines. The morphology of PC12 cell line with reduced levels of VHL protein seemed to be large in size and less differentiated as compared to control cell lines. Results have suggested that decreased level of VHL protein in VHL disease lead to a loss of differentiation. Loss of differentiation in chromaffin precursor cells could be one of the factors that results in cancer phenotype [24].

Rat C6 glioma cell lines
Rat C6 glioma cell line is an important experimental cell line that has been used for biological research to elucidate the mechanisms of tumor formation [25].

Rat C6 glioma has been used to investigate the functional association of netrin receptors in glioma cell migration and the formation of tumors. Rat C6 glioma cells were engineered to express netrin-1, netrin-3, and the netrin receptor unc5B. It was evident from this research that netrin and unc5B may enhance cell migration by reducing the interaction of tumor cells. In rat glioma cell lines, loss of deleted in colorectal carcinoma (DCC) may increases the growth and invasion of the tumor [26].

Other rat cell lines
During the last three decades, research has been conducted to find the cure for diabetes. Many efforts have been put to create cell lines that secrete insulin, that is, beta cell lines with normal secretion of insulin but only some research has been successful. Most important cell lines derived from transgenic mouse used for insulin secretion are RIN, HIT, MIN, INS-1, and βTC cells. These cell lines have the ability to produce insulin along with a low amount of glycogen and somatostatin. Even though many problems have been related to the culturing of beta cell lines, these cell lines have made available very important information about many physiological phenomena associated with diabetes. Moreover, there is an utter need to develop some beta cell lines derived from humans and pigs that provide more suitable conditions to secrete insulin-like normal cells [27]. Table 11.3 shows rat cell lines and their use in diseases.

Transgenic monkey model
A foreign gene is inserted in the viral system and then into monkey embryos that are cultivated in sucrose syrup. A gene extracted from jelly fish is used as a marker. The animal

Table 11.3 Rat cell lines and their use in diseases

Name of cell line	Method of production	Disease	Use	References
PC12 cell line	Derived from pheochromocytoma of the rat adrenal medulla	Von Hippel–Lindau disease	Mutations in VHL gene	[23,24]
Rat C6 glioma cell lines	Derived from rat glioma	Various tumors	Functional association of netrin receptors in glioma cell migration and formation of tumor	[25,26]

Table 11.4 Different monkey species for different models

Diseases	Animal model	Method of creating model	System used	References
Huntington disease	*Macaca fuscata*, vervet monkey (*Chlorocebus aethiops*)	DNA recombinant technology	Lentivirus system	[28]
Autism	Marmoset	Gene-editing strategy, DNA-swapping, CRISPR	DNA-swapping, CRISPR, zinc finger nucleases	[29]
Obesity and cardiovascular disease	Hominoidea *Macaca fuscata*	–	Electrocardiography and echocardiography	[30,31]
Schizophrenia	Pigtail (*Macaca nemestrina*)	EEG method (ketamine-administration)	Oddball paradigm	[29]

is exposed to UV light to verify that the gene is integrated in its position. The embryos are then injected in monkeys. It is a breakthrough in the medical field as most of the neurodegenerative diseases do not show symptoms in rodent models. Table 11.4 shows different monkey species used for modeling different diseases.

Monkey model and Huntington disease

The similarity in the genetic makeup of monkeys and humans as compared to rodents makes them a better model to monitor disease development and efficacy for trial of drugs. A genetically engineered monkey with a transformed Huntingtin gene shows similarity with Huntington disease in humans and signify the physical characteristics and mental symptoms closer to humans than any other model. To assess the monkey several cognitive and behavioral tests are performed. Of these tests, the Huntington disease primate model scale is an important test whose scale ranges from 0 to 80, and 80 shows the worse symptoms and is used to describe the involuntary movements in the engineered monkey. This test indicates that the monkey will show clear symptoms of chorea and dystonia.

The functional magnetic resonance imaging test, in addition to physical and cognitive tests, is performed to observe the neurodegeneration and intranuclear Huntingtin inclusions. Due to the large brain size of monkeys as compared to rodents, it is easy to monitor the neural changes in the brain. The development of a germline of macaque pluripotent stem cells with a mutant Huntingtin gene led to new hope for Huntington disease patients [28].

Method of transgenesis

Many models have been designed to study Huntington disease but the monkey model is the only transgenic model so far. To create the transgenic model, the Huntingtin gene that has 84 GAC repeats is injected into monkey egg cells by using a viral vector. The most commonly used viral vector is lentivirus. Lentivirus is classified into the Retroviridae family (these are enveloped double-stranded RNA viruses). These viruses have reverse transcriptase that helps them to transcribe their RNA into DNA. The problem to infect nondividing cells is overcome by the possible use of lentivirus, especially in patients with neurodegenerative disorders. Due to the incapability of neurons to proliferate or divide at a very low rate, the lentivirus system is the most efficient system for delivery of genes in other systems. The transfer of the gene to the monkey by using lipofectamine or electroporation is difficult as compared to mouse embryonic stem cells. This may be due to inefficient standard culture conditions. Though the success rate for homologous recombination is low, the transfection of the monkey is nevertheless reported. The main efficiency in transgenic technology is due to the lentivirus system and the green fluorescence protein gene that can transfer the gene into non-mitotic cells. The only disadvantage of the lentiviral vector (LV) system is the random integration of genes, which leads to overexpression of other genes. This problem is overcome by using RNA interference technology.

Production of transgenic HD monkeys by lentivirus

The engineered monkey model for Huntington disease was created by using the lentivirus system. It is then injected into the perivitelline space of oocytes. After that sperm is injected into the embryo in *in vitro* culture till it is converted to four to eight cells. In the lentivirus system, two important genes are inserted: The first is human Huntingtin with 84 repeats of CAG. The second is the green fluorescence protein gene with its regulator human polyubiquitin C promoter. It is confirmed by Southern blotting and polymerase chain reaction (PCR) [32].

Generation of HD monkey ES-like cells by induced pluripotency

A skin fibroblast culture is formed from the male monkey model at the fourth month of gestation. Seventy-two CAG repeats are present in the mutant HTT gene and its expression is confirmed by immunohistochemistry and Western blot with monoclonal antibodies named mEM48, which reacts with human HTT that is boosted by its regulator. Huntington disease monkey cells are also infected with a retrovirus with an expression of Oct4, Sox2, and Klf4 [33]. Embryonic stem cells are distinguished on the basis of their nuclear to cytoplasmic ratio. Four pluripotent cell lines are formed from which RiPS-3 are picked out randomly. To verify the expression of stem cell markers and antigens, immunochemistry and an alkaline phosphatase test is performed.

Monkey model and schizophrenia

Schizophrenia is a neural disorder in which a person has hallucinations and illusions of extreme fear. Schizophrenics have jumbled speech and series of thoughts [29,34]. There is a lack of coordination in their posture and imbalance in the body. This is due to blockage of multiple signals in the cerebral cortex or the loss of the glutamate neurotransmitter. The formation of the monkey model for schizophrenia is due to the homology between human and monkey. The noninvasive electroencephalography (EEG) system is used to study the neurophysiological changes of sensory and mental functions in both the human and animal model. Cognitive and sensory impairment is due to two signals: mismatch negativity (MMN) and P3a. The reduction of signals is due to the ketamine antagonist of the glutamate receptor (NMDA—*N*-methyl-D-aspartate receptor). To verify it, an oddball test was performed on five humans and two monkeys to check the EEG response. A series of tones with the interruption of oddball louder were stimulated for monkeys and humans [30]. This caused the high-voltage signal relative to the standard one. The MMN signals were produced in accordance with the nonstandard stimulus that indicated the brain activity toward the deviant stimulus. It was observed that the oddball impetus gave rise to P3a signals that suggest the reorienting response of the brain. So these are used to observe the symptoms and their cause [31].

Monkey model and autism

Autism is a neural disorder with reduced social life and impaired speech and typecast behavior. It is a cognitive disorder mostly diagnosed at the age of 3 [35]. To study autism, the transgenic monkey model is used in which an enzyme is injected in embryonic cells that will cause the autism-related syndrome named the tuberous sclerosis complex. Marmosets are the species used for autism study [36]. Mutations in TSC1 and TSC2 genes will lead to convulsions, autism, and tumors in brain and body [37]. Researchers are trying to extract these cells and incorporate them in a mice model to see whether they cause tumors in them or not. The marmoset model was created by injecting the fluorescence gene in 91 embryos, out of which the gene was integrated into 80 embryos. By creating the model, the aim was to generate the knockout effect in marmosets [38].

Monkey model and heart disease

Induced coronary atherosclerosis was established in a rhesus monkey so as to study the symptoms of the disease. The symptoms are tested electrocardiographically and echocardiographically. The following symptoms are considered: arterial blood pressures, lower high-density lipoprotein cholesterol concentrations, and apolipoprotein concentrations. Diseased monkeys have high cholesterol; with higher LDL than normal. Similar symptoms were observed in the diseased model [39].

Transgenic rat model

Rats are more identical to humans physiologically as well as behaviorally. Rats have more advantage over mice because of their large size. It is more feasible to understand the diseases and examine the pharmacological effects in rats as compared to mice (Tables 11.5 and 11.6).

Rat model and obesity

Obesity is the major culprit for many cardiovascular and brain-related diseases [53]. It also plays its role in hypertension in people. To study the relationship between obesity

Table 11.5 Method of creating rat model

Cells used	Method of induction	References
Embryonic stem cells	Lentivirus	[40]
Embryonic stem cells	Microinjection	[40]
Embryonic stem cells	Plasmid	[40]
Fibroblast growth factor	Plasmid	[40]
Embryonic stem cells	Mutagenesis (zinc fingers)	[41]
Embryonic stem cells	Mutagenesis (transposons)	[41]

and hypertension, diet-induced rats, that is, the Sprague–Dawley (SD) rat and chow-fed rat models are used because they have more similar symptoms to humans [54]. These rats have the ability to show some of the hypertension hormones and enzymes, that is, increased plasma norepinephrine to intravenous glucose, increased plasma leptin concentration, and decreased growth hormone (GH) secretion [55]. While performing experiments, three responses are observed: short-term responses, long-term responses, systolic blood pressure, lipid peroxidase, and lipid profile. Due to the activation of dyslipidemia,

Table 11.6 Different rat models used for different diseases

Rat models	Diseases	References
Goto–Kakizaki (GK) rat	Type 2 diabetes, hyperinsulinemia, insulin resistance, mild hyperglycemia, neuropathy, osteopathy, retinopathy, nephropathy	[42,43]
Zucker diabetic fatty (ZDF) rat	Obesity, insulin resistance, hyperinsulinemia, type 2 diabetes, hyperglycemia, hypertriglyceridemia, hypercholesterolemia, nephropathy, impaired wound healing, mild hypertension, neuropathy	[44,45]
Zucker rat	Insulin resistance, hyperinsulinemia, hypertriglyceridemia, hypercholesterolemia, metabolic syndrome	[46]
Obese-prone (OP-CD) rat	Obesity, metabolic syndrome, hypertension, hypertriglyceridemia, hyperinsulinemia, insulin resistance	[43]
Obese-resistant (OR-CD) rat	Obesity non-responder, hyperinsulinemic, insulin resistant	[47]
ZSF1 rat	Nephropathy, congestive heart failure, hypertension, obesity, type 2 diabetes, insulin resistance, hyperinsulinemia, hypertriglyceridemia, hypercholesterolemia	[48,49]
Dahl/salt-sensitive (Dahl/SS) rat	Hypertension, insulin resistance, hyperinsulinemia, hypertriglyceridemia, hypercholesterolemia, nephropathy, cardiac hypertrophy, heart failure	[50]
SS-13BN rat	Insulin resistant, hyperinsulinemia, hypertriglyceridemia, normotensive control	[51]
Spontaneous hypertensive stroke prone (SHRSP) rat	Hypertension, nephropathy, insulin resistance, hyperinsulinemia, hypertriglyceridemia, hypercholesterolemia	[52]

increased oxidative stress associated vascular and renal pathology shows the relationship between hypertension and obesity. It also shows the relationship of hypercholesterolemia to cardiovascular disease [56].

Rat model and hypoxic brain damage

To study the disease 120 SD rats were used. To examine the disease, the rats were grouped into three groups. One was the control group, and the second was the hypoxic group in which brain injuries were induced and they were incubated at room temperature, that is, 37°C with proper aeration [57,58]. The other group contained both the hypoxic injuries and ischemia with ligation of the bilateral cephalic artery. To justify the experiment, two behavior tests were performed. In the first test, the right reflex, the rats were placed on their backs and the time taken to flip over onto their paws was calculated [59]. In the other geotactic test, the rats were placed on a wooden slope with their heads facing downward, and the time taken to turn on their faces was calculated. In a vertical test, the clinging of rats to a screen was observed [60,61]. In a grid test, the rats were placed on a wire mesh screen and the time that their paws were stuck on the screen was calculated. The short-term reflexes showed that primary reflexes in rats were suppressed. The long-term reflexes showed impaired motor coordination in hypoxic brain diseases [62].

Transgenic rat model and antiepileptic drug production

Transgenic animals are used to find causes, symptoms, and treatment of disease so as to serve mankind. The drawback in using animal models is that sometimes they are not affected by the disease or do not present symptoms as in humans. To overcome this problem, scientists are trying to produce better designed animals. The most common technique used in transgenesis is knockout. In knockout animals, some of the genes are deleted from the DNA so as to observe its effect and why the gene fails to work. To create knockout rats, zinc finger technology is used.

Antiepileptic drugs are used to treat most common neural disorder epilepsy as well as nonepileptic convulsive disorders. Drugs are designed to inhibit the abnormal potential discharge instead of its cause. They work on three principles, that is, (1) enhancement of gamma-aminobutyric acid (GABA) action; (2) inhibition of sodium channel function; and (3) inhibition of calcium channel function [63].

Enhancement of GABA action

Several antiepileptic drugs (e.g., phenobarbital and benzodiazepines) enhance the activation of GABA receptors, thus facilitating the GABA-mediated opening of chloride channels.

Inhibition of sodium channel function

A large number of antiepileptic drugs affect membrane excitability by an action on voltage-dependent sodium channels, which carry the inward membrane current necessary for the generation of an action potential. Their blocking action shows the property of use-dependence; in other words, they preferentially block the excitation of cells that are firing repetitively, and the higher the frequency of firing, the greater the block produced, for example, carbamazepine and phenytoin.

Inhibition of calcium channels

Drugs that are effective against seizures (ethosuximide and valproate) all appear to share the ability to block T-type low-voltage-activated calcium channels. T-type channel activity is important in determining the rhythmic discharge of thalamic neurons associated with the absence of seizures.

Other mechanisms

Many of the newer antiepileptic drugs were developed based on activity in animal models. Their mechanism of action at the cellular level is not fully understood. Levetiracetam appears to act in a manner different from all other antiepileptic drugs, its target being a synaptic vesicle protein involved in neurotransmitter release.

Engineered rat model of antiepileptic drugs

To study the epileptic drugs female rats were used that weighed between 210 and 230 g. Before performing the experiment, the rats were kept in controlled conditions, that is, temperature 24–25°C, humidity 50%–60%, and 12/12 h light/dark cycle for 1 week. Rats were provided with special food and tap water. First of all, the pharmacokinetics of levetiracetam was tested in normal female rats and then in kindle rats.

The kindling process

Kindling is the phenomena in which an erratic low-threshold electric or chemical stimulus is given and the electrophysiological changes in the brain are checked. The stimulus is given until the brain responses decrease toward threshold. Kindling of subcortical areas at low intensity and discrete interims produce seizures with an irregular threshold. It is the best way to demonstrate that convulsions can be a permanent behavioral sign. The changes that occur during the process can be demonstrated as electrographic and neurochemical responses [64].

Properties of kindling

It is the model method for epilepsy study and the best way to demonstrate that convulsions can be a permanent behavioral sign. The changes that occur during the process can be demonstrated as electrographic and neurochemical responses.

- After discharge is drawn through low threshold with stimulating convulsions in the kindling process.
- Synaptic potential in the limbic system is long lasting in the kindling process.
- Impulsive seizures are produced in the kindling process.

An experiment was performed in which bipolar electrodes were inserted in the amygdala of the rat and stimulated the rat with different tones to check the rate and specificity of epileptogenesis. Rats were divided into three categories of tone stimulus [65]. There was no considerable difference in the three groups showing that delayed response appeared in the kindling process at an early stage. Seizures were produced during amygdalostriatal transition in the kindling process. Inhibitory effects in epilepsy play an important role because if the limbic system has a proper function, then the anticonvulsant agents will block the intermittent excitation of the neural circuit [66]. The limbic system has great importance in the epileptic condition because its structure can be kindled easily, and has an important role in behavioral manifestations such as excitement and memory. Symptoms produced at this site were easily compared with other neurological disorders. The arrangement of the limbic system helped in the easy demonstration of different effects, such as that more emotions were seen in the right lobe in the condition of epilepsy as compared to the left lobe [67].

Levetiracetam was administered in plasma by gas chromatography. Electrodes were implanted as discussed in the kindling process earlier. They were fixed by dental fixture material. Kindling was performed in eight rats. There were two important drugs, first was antiepileptogenic that denotes the inhibition of pathways involved in epilepsy and the other one was anticonvulsant that inhibits the fits type condition [68]. It was observed in the experiment that the one that blocks the kindling process was antiepileptogenic and the one that blocks the convulsions was anticonvulsant. The administration of levetiracetam in rats showed effective antiepileptogenic effects without showing tranquilizing and narcotic effects. The rats with drug administration showed fewer convulsions as compared to normal ones when kindled after termination of the medication. So levetiracetam is the new drug with both properties. After termination of the drug for a long time, no seizures were produced in the kindling process. The antiepileptic effect was shown by previous drugs but in controlled parameters. The mechanism of the drug is still unknown but it might also act on GABAergic inhibition. *In vivo* studies show that it prevents CA3 neuronal excitability effect in the rat's hippocampus through a pathway of non-GABAergic effect [69].

The presence of stereoselective binding sites for levetiracetam in the rat's hippocampus is confirmed by the binding assay.

The affinity for the binding site of levetiracetam is enhanced by its homolog. Levetiracetam is inactive in more than 30 receptors in the brain and that is confirmed through the binding assay. It also does not bind to ion channels like T-type calcium channels that show that the levetiracetam binding site is different from that of other antiepileptic drugs. This shows its unique character with the least side effects [70].

Porcine xenotransplantation

Porcine xenotransplantation is a suitable option because of the easy availability and handling of pigs and the comparable size and physiological structure of the transplantable organs [71]. Studies indicated that if the immune response and risk of infection due to the transplanted porcine organs can be decreased, pigs can be used as organ donors to cater to organ failure in humans. [72]. Researchers suggest that clinical acceptance to xenotransplantation can only be considered after controlled solid organ transplant from pigs to humans to evaluate the risk of cross-species spread of infection and the safety and efficacy of transplanted organs [73].

Organs for xenotransplantation

Liver transplant   A report published in 2009 indicated that 30,000 patients died waiting on liver transplants during the last 13 years. [73]. In 1995, a porcine liver transplant to a human being was attempted by Makowa et al. [74], which resulted in the failure of the transplant and eventually in death. In 2000, livers from transgenic pigs were transplanted into immunosuppressed baboons. Later in 2010, research conducted by Ekser et al. [73] indicated that livers from transgenic pigs were transplanted to baboons, which led to the normal performance of many liver functions, including coagulation. Although there was the production of pig proteins that included coagulation factors, they functioned normally in the host; the researcher suggested the evaluation of cross-species compatibility. A study conducted in 2007 suggested the possibility of transplant of porcine hepatocytes [75].

Cardiac transplant   Bauer et al. [76] reported a heart transplant from pig to baboon in 2010. Heart transplant survival as long as 6 months has been reported [46]. According to a study conducted in 2009, complement dysregulation has been reported as a result of a heart transplant from a pig to a human. In the pig-to-primate model, the left ventricular pressure is the most valuable tool for the assessment of the grafted heart model [77].

Renal transplant   The potential agents for the failure of pig to primate renal transplant are reported to be alpha Gal and the non-alpha Gal T-antigens that are present in the pig kidney [78]. In an attempt to avoid hyperacute rejection, the use of Gal T knockout kidneys has been tried [79]. In a study conducted in 2009, it was observed that with the help of complete immune suppression the survival of the transplanted porcine kidney extended up to 3 days [80].

Pancreatic islet xenotransplantation

The survival of neonatal and adult porcine islets xenotransplantation to nonhuman primates under immunosuppressive treatment has been reported to survive more than 6 months. Due to the long survival of porcine pancreas over other organs, they may be considered as a model for the treatment of type I and type II diabetes.

Risk of infection in porcine xenotransplantation

Porcine xenotransplantation is associated with a massive threat of infection with endogenous porcine retroviruses. Among other viruses, these retroviruses, which are normal part of the pig genome, pose a greater risk of infection. One way to reduce the risk of infection is the careful selection of pig models and their genetic engineering to control the risk of infection and pathogenicity. Endogenous porcine retroviruses type A and type B can infect humans while porcine endogenous retroviruses type C infection is limited only to porcine cells [81]. In one study, researchers utilized porcine endogenous retroviruses for *in vitro* infection of HEK-293 cells which indicated that this infection caused no change in host cells or in the porcine endogenous retroviruses long terminal repeat (LTR) region [82].

An important strategy to reduce the risk of hyperacute rejection is the transgenesis of the pig genome that would express human complement regulatory proteins to reduce the deposition of the complement and by the knocking out of the genes that encode for Gal [83]. Porcine xenotransplantation with Gal T knockout grafts having increased levels of circulating porcine endogenous retroviruses has been reported in immunosuppressed baboons and Gal T knockout swine [84].

Strategies for controlling the risk of infection

In addition to the use of knockout strains, which encode for antigens that elicit massive and detrimental immune responses, other strategies to control the risk of infection by porcine xenograft in the host include [83]

1. The use of antibody treatment
2. Vaccinations
3. By increasing expression of the antiviral restriction factor

4. Antiretroviral therapy in the recipient of the graft
5. *In vitro* reduction of replication of viruses

Other than porcine endogenous retrovirus, viruses to be considered as a threat of infection in the host are

Hepatitis E virus (HEV)
Cytomegalovirus (CMV)
Herpes virus

Porcine xenograft rejection

Rejection of the transplanted xenograft can take place by four mechanisms:

1. Hyperacute rejection
2. Acute vascular rejection
3. Cell-mediated rejection
4. Chronic rejection

The first barrier toward successful transplant eventually leading to the survival of the transplanted xenograft is hyperacute rejection, which takes place in minutes and results in the rejection of the transplant. When pig organs are transplanted in human and nonhuman primates, the carbohydrates Gala1 and Gal 3 are primarily responsible for rejection. To decrease the incidence of hyperacute rejection, genetically engineered α1,3-galactosyltransferase gene knockout pigs have been used that lack the ability to make the Gal oligosaccharide responsible for the rejection of the xenograft [85]. Acute vascular rejection is initiated within 24 h and results in destruction of the graft [86]. Acute vascular rejection is elicited due to the formation of xeno-reactive antibodies, activation of the xenograft endothelium, complement, and coagulation system. Pig antigens are exposed to the primate T cells via direct and indirect xenorecognition [87]. Studies indicate that rejection of porcine xenografts is mediated by NK cells in human hosts [88].

Genetically engineered porcine models for xenotransplantation

There are a number of techniques that have been used for the genetic modification of the porcine xenograft models.

Some examples are given in Table 11.7.

Different strains of pigs are available with different genetic modifications. A few examples are listed in Table 11.8.

Diseases treated with porcine xenograft

Porcine xenotransplantation in Huntington disease   Huntington disease is characterized by the expansion of the trinucleotide repeat of CAG, which leads to neurodegeneration.

Table 11.7 Techniques for genetic modification of porcine xenograft model

Techniques for genetic modifications	Vector system used	References
Sperm-mediated gene transfer	Nonviral episomal vector	[89]
Germline gene transfer	Retroviral vector	[90]
Cre site-specific DNA recombinase system		[91]

Table 11.8 Genetically engineered porcine models

Genetic modifications of pig model	Indications/reason for modification	References
Human H-transferase8	Decrease in Gal antigen expression	[92]
Human CD55, CD59, CD46	Complement regulation	[93,94]
CIITA-DN20	T-cell activation suppression	[95]
vWF-deficient14	Platelet activation inhibition	[96]
Human hemeoxygenase 123	Antiapoptotic; anti-inflammatory	[97]

It is an autosomal dominant condition that causes degeneration of neurons in the cerebral cortex, caudate nucleus, and putamen in the brain. Experiments conducted to transplant neuronal ganglionic tissues in the animal model of Huntington disease showed improvement in motor activity. Neuroblasts transplantation from the porcine model to human was conducted with immunosuppressive therapy via cyclosporine or monoclonal antibodies provided safe results but did not improve the condition [98]. Over a 1 year span from the time of transplantation, there were no signs of complications by the xenograft. However, the xenograft did not improve the condition in Huntington disease patients and these patients progressed with the disease at the same pace as those that were not given the xenograft [98]. The clinical applicability in terms of gene therapy of the xenograft model or dose has to be evaluated.

Porcine xenotransplantation in PD PD is a neurodegenerative disease that causes the death of the dopaminergic neuron and thus very low levels of dopamine. It is a neurodegenerative disorder. In a study conducted on 12 patients who were given transplants of neuronal cells from the porcine model to rat model of PD showed that the grafted cells survived for 15–20 weeks with immunosuppressive therapy. In another study, which was a clinical trial, 12 patients were involved that were divided into two groups, both groups were given immunosuppressive therapy and showed improvements in their clinical scores [99].

Conclusion

The discovery of engineered animal models has played a significant role in biomedical research. In almost every field of biomedical research, genetically modified or engineered animal models have remained fundamental to study the biology of many human diseases. These animal models have massively assisted in understanding the molecular causes of many genetic diseases. Most animal models have been to study physiological and molecular mechanisms of underlying pathological conditions. Development and testing of novel drugs and responses to these therapeutic interventions has also been done in many genetically modified animal models. Genetic disorders are one of the most challenging scenarios to deal with in the field of biomedicine. Genetic manipulations of laboratory animals have made it easier to integrate molecular biology with whole-animal physiology, developing a resource to understand the molecular regulation of composite physiology and its behavior. Since animals and humans share a huge percentage of genes, disease models based on genetic engineering of animals have proved to be a promising source to provide links between the genetic makeup and pathogenesis of diseases. Moreover, it will lead to better understanding of molecular genetics and will help in the treatment of disorders in humans.

References

1. Racay P. 2002. Genetically modified animals and human medicine. *Bratislava Medical Journal* 103(3): 121–126.
2. Brown SD, Balling R. 2001. Systematic approaches to mouse mutagenesis. *Current Opinion in Genetics & Development* 11, 268–273.
3. Palmiter RD, Brinster RL, Hammer RE, Trumbauer ME, Rosenfeld MG, Birnberg NC, Evans RM. 1982. Dramatic growth of mice that develop from eggs microinjected with metallothionein-growth hormone fusion genes. *Nature* 300: 611–615.
4. Bernstein A, Breitman M. 1989. Genetic ablation in transgenic mice. *Journal of Molecular Medicine* 6: 523–530.
5. Thomas KR, Capecchi MR. 1987. Site-directed mutagenesis by gene targeting in mouse embryo-derived stem cells. *Cell* 51: 503–512.
6. Bockamp E, Maringer M, Spangenberg C, Fees S, Fraser S, Eshkind L, Oesch F, Zabel B. 2002. Of mice and models: Improved animal models for biomedical research. *Physiological Genomics* 11: 115–132.
7. Francia G, Kerbel RS. 2010. Raising the bar for cancer therapy models. *Nature Biotechnology* 28(6): 561–562.
8. Adam SJ, Rund LA, Kuzmuk KN, Zachary JF, Schook LB, Counter CM. 2007. Genetic induction of tumorigenesis in swine. *Oncogene* 26: 1038–1045.
9. Palumbo A, Anderson K. 2011. Multiple myeloma. *New England Journal of Medicine* 364(11): 1046–1060.
10. Luo Y, Lin L, Bolund L, Jensen TG, Sørensen CB. 2012. Genetically modified pigs for biomedical research. *Journal of Inherited Metabolic Disease* 35: 695–713.
11. Scherer WF, Syverton JT, Gey GO. 1953. Studies on the propagation *in vitro* of poliomyelitis viruses. IV. Viral multiplication in a stable strain of human malignant epithelial cells (strain HeLa) derived from an epidermoid carcinoma of the cervix. *Journal of Experimental Medicine* 97(5): 695–710.
12. Rutishauser BR, Schurch S, Gehr P. 2007. Interaction of particles with membranes. In: *Particle Toxicology*, Donaldson K, Borm P (eds), Vol. 7, CRC Press, United States, pp. 150–151.
13. Zhou X, Li J, Wang Z, Chen Z, Qiu J, Zhang Y, Wang W et al. 2013. Cellular immunotherapy for carcinoma using genetically modified EGFR-specific T lymphocytes. *Neoplasia* 15: 544–553.
14. Bai Y, Mao Q-Q, Qin J, Zheng X-Y, Wang Y-B, Yang K, Shen H-F, Xie L-P. 2010. Resveratrol induces apoptosis and cell cycle arrest of human T24 bladder cancer cells *in vitro* and inhibits tumor growth *in vivo*. *Cancer Science* 101(2): 488–493.
15. Ohana P, Bibi O, Matouk I, Levy C, Birman T, Ariel I, Schneider T et al. 2002. Use of H19 regulatory sequences for targeted gene therapy in cancer. *International Journal of Cancer* 98(5): 645–665.
16. Russo M, Cocco S, Secondo A, Adornetto A, Bassi A, Nunziata A, Polichetti G et al. 2011. Cigarette smoke condensate causes a decrease of the gene expression of Cu–Zn superoxide dismutase, Mn superoxide dismutase, glutathione peroxidase, catalase, and free radical-induced cell injury in SH-SY5Y human neuroblastoma cells. *Neurotoxicity Research* 19(1): 49–54.
17. Grassi D, Bellini MJ, Acaz-Fonseca E, Panzica G, Garcia-Segura LM. 2013. Estradiol and testosterone regulate arginine-vasopressin expression in SH-SY5Y human female neuroblastoma cells through estrogen receptors-α and -β. *Endocrinology* 154(6): 2092–2100.
18. Geiser V, Cahir-McFarland E, Kieff E. 2011. Latent membrane protein 1 is dispensable for Epstein–Barr virus replication in human embryonic kidney 293 cells. *PLoS One* 6(8): e22929.
19. West AB, Moore DJ, Biskup S, Bugayenko A, Smith WW, Ross CA, Dawson VL, Dawson TM. 2005. Parkinson's disease-associated mutations in leucine-rich repeat kinase 2 augment kinase activity. *PNAS* 102(46): 16842–16847.
20. Peverelli E, Ermetici F, Corbetta S, Gozzini E, Avagliano LMA, Bulfamante G, Beck-Peccoz1 PA, Mantovani G. 2013. PKA regulatory subunit R2B is required for murine and human adipocyte differentiation. *Endocrine Connections* 2(4): 196–207.
21. Hirata K, Wada K, Murata Y, Nakajima A, Yamashiro T, Kamisaki Y. 2013. Critical role of leukotriene B4 receptor signaling in mouse 3T3-L1 preadipocyte differentiation. *Lipids in Health and Disease* 12: 122, http://www.lipidworld.com/content/12/1/122.

22. Torres MP, Rachagani S, Souchek JJ, Mallya K, Johansson SL, Batra SK. 2013. Novel pancreatic cancer cell lines derived from genetically engineered mouse models of spontaneous pancreatic adenocarcinoma: Applications in diagnosis and therapy. *PLoS One* 8(11): e80580.
23. Grau CM, Greene LA. 2012. Use of PC12 cells and rat superior cervical ganglion sympathetic neurons as models for neuroprotective assays relevant to Parkinson's disease. *Methods in Molecular Biology* 846: 201–211.
24. Deshpande P. 2011. The effects of lowering VHL protein in rat pheochromocytoma PC12 cell line. Theses. Adelphi University, 61pp, 1493199.
25. Grobben B, De Deyn PP, Slegers H. 2002. Rat C6 glioma as experimental model system for the study of glioblastoma growth and invasion. *Cell and Tissue Research* 310(3): 257–270.
26. Durko M, Koty Z, Zhu L, Marçal N, Kennedy TE, Nalbantoglu J. 2013. Rat C6 glioma cell motility and glioma growth are regulated by netrin and netrin receptors unc5B and DCC. *Journal of Cancer Therapeutics & Research* 2: 18.
27. Skelin M, Rupnik M, Cencic A. 2010. Pancreatic beta cell lines and their applications in diabetes mellitus research. *ALTEX* 27(2): 105–113.
28. Claire-Anne G, Li S, Yi H, Mulroy JS, Kuemmerle S, Jones R, Rye D, Ferrante RJ, Hersch SM, Li XJ. 1999. Nuclear and neuropil aggregates in Huntington's disease: Relationship to neuropathology. *Journal of Neuroscience* 19(7): 2522–2534.
29. Umbricht D. 2000. Ketamine-induced deficits in auditory and visual context dependent processing in healthy volunteers: Implications for models of cognitive deficits in schizophrenia. *Archives of General Psychiatry* 57(12): 1139–1147.
30. Luby ED, Gottlieb JS, Cohen BD, Rosenbaum G, Domino EF. 1962. Model psychoses and schizophrenia. *American Journal of Psychiatry* 119: 61–67.
31. Swerdlow NR. 2011. Are we studying and treating schizophrenia correctly? *Schizophrenia Research* 130(1–3): 1–10.
32. Cha JH. 2007. Transcriptional signatures in Huntington's disease. *Progress in Neurobiology* 83(4): 228–248.
33. Sapp E, Schwarz C, Chase K, Bhide PG, Young AB, Penney J, Vonsattel JP, Aronin N, DiFiglia M 1997. Huntingtin localization in brains of normal and Huntington's disease patients. *Annals of Neurology* 42(4): 604–612.
34. Rissling AJ, Light GA. 2010 Neurophysiological measures of sensory registration, stimulus discrimination, and selection in schizophrenia patients. *Current Topics in Behavioral Neurosciences* 4: 283–309.
35. Crawley JN. 2004. Designing mouse behavioral tasks relevant to autistic-like behaviors. *Mental Retardation and Developmental Disabilities Research Reviews* 10: 248–258.
36. Nestler EJ, Hyman SE. 2010. Animal models of neuropsychiatric disorders. *Nature Neuroscience* 13: 1161–1169.
37. Martin LA, Ashwood P, Braunschweig D, Cabanlit M, Van de Water J, Amaral DG. 2008. Stereotypies and hyperactivity in rhesus monkeys exposed to IgG from mothers of children with autism. *Brain Behavior, and Immunity* 22: 806–816.
38. Karley WK, Platt ML. 2012. Of mice and monkeys: Using non-human primate models to bridge mouse- and human-based investigations of autism spectrum disorders. *Journal of Neurodevelopmental Disorders* 4: 21.
39. Brown BG, Zhao X-Q, Sacco DE, Albers JJ. 1993. Lipid lowering and plaque regression: New insights into prevention of plaque disruption and clinical events in coronary disease. *Circulation* 87: 1781–1791.
40. Donaldson HH. 2012. The history and zoological position of the albino rat. *Journal of the Academy of Natural Sciences* 15: 365–369.
41. Pinto M. 1983. The relative roles of MHC and non-MHC antigens in bone marrow transplantation in rats: Graft acceptance and antigenic expression on donor red blood cells. *Transplantation* 35(6): 607–611.
42. Goto Y, Kakizaki M, Masaki N. 1975. Spontaneous diabetes produced by selective breeding of normal Wistar rats. *Proceedings of the Japan Academy* 51: 80–85.
43. Ostenson CG. 2007. *The Goto-Kakizaki Rat. Animal Models of Diabetes Frontiers in Research*. CRC Press, United States, pp. 119–137.

44. Belin de Chantemèle EJ. 2009. Type 2 diabetes severely impairs structural and functional adaptation of rat resistance arteries to chronic changes in blood flow. *Cardiovascular Research* 81(4): 788–796.

45. Leonard BL. 2005. Insulin resistance in the Zucker diabetic fatty rat: A metabolic characterisation of obese and lean phenotypes. *Acta Diabetologica* 42(4): 162–170.

46. Frisbee JC. 2005. Hypertension-independent microvascular rarefaction in the obese Zucker rat model of the metabolic syndrome. *Microcirculation* 12(5): 383–392.

47. Flegal KM, Carroll MD, Kuczmarski RJ, Johnson CL. 1998. Overweight and obesity in the Unites States: Prevalence and trends, 1960–1994. *International Journal of Obesity* 22: 39–47.

48. Dominguez JH. 2006. Renal injury: Similarities and differences in male and female rats with the metabolic syndrome. *Kidney International* 69: 1969–1976.

49. Tofovic SP, Salah EM, Jackson EK, Melhem M. 2007. Early renal injury induced by caffeine consumption in obese, diabetic ZDF1 rats. *Renal Failure* 29: 891–902.

50. Chen PY. 1993. Hypertensive nephrosclerosis in the Dahl/Rapp rat. Initial sites of injury and effect of dietary L arginine supplementation. *Laboratory Investigation* 68(2): 174–184.

51. Liang M. 2002. Renal medullary genes in salt-sensitive hypertension: A chromosomal substitution and cDNA microarray study. *Physiological Genomics* 28(8):139–149.

52. Sepehrdad R, Chander PN, Singh G, Stier CT, Jr. 2004. Sodium transport antagonism reduces thrombotic microangiopathy in stroke-prone spontaneously hypertensive rats. *American Journal of Physiology Renal Physiology* 286(6): 1185–1192.

53. Hall JE, Brands MW, Henegar JR, Shek EW. 1998. Abnormal kidney function as a cause and a consequence of obesity hypertension. *Clinical and Experimental Pharmacology and Physiology* 25: 58–64.

54. Tuck ML. 1992. Obesity, the sympathetic nervous system, and essential hypertension. *Hypertension* 19(1): 67–77.

55. Russo C, Olivieri O, Girelli D, Faccini G, Zenari ML, Lombardi S, Corrocher R. 1998. Antioxidant status and lipid peroxidation in patients with essential hypertension. *Journal of Hypertension* 16: 1267–1271.

56. Lauterio TJ, Davies MJ, DeAngelo M, Peyser M, Lee J. 1999. Neuropeptide Y expression and endogenous leptin concentrations in a dietary model of obesity. *Obesity Research* 7: 498–505.

57. Douglas-Escobar M, Weiss MD. 2012. Biomarkers of hypoxic-ischemic encephalopathy in newborns. *Frontiers in Neurology* 3: 144.

58. Wallin A, Ohrfelt A, Bjerke M. 2012. Characteristic clinical presentation and CSF biomarker pattern in cerebral small vessel disease. *Journal of the Neurological Science* 322: 192–196.

59. Van Laerhoven H, de Haan TR, Offringa M, Post B, van der Lee JH. 2013. Prognostic tests in term neonates with hypoxic-ischemic encephalopathy: A systematic review. *Pediatrics* 131: 88–98.

60. Kim JJ, Buchbinder N, Ammanuel S, Kim R, Moore E. 2013. Cost-effective therapeutic hypothermia treatment device for hypoxic ischemic encephalopathy. *Medical Devices* 6: 1–10.

61. Peliowski-Davidovich A. 2012. Hypothermia for newborns with hypoxic ischemic encephalopathy. *Paediatrics and Child Health* 17: 41–46.

62. Robertson NJ, Faulkner S, Fleiss B, Bainbridge A, Andorka C. 2013. Melatonin augments hypothermic neuroprotection in a perinatal asphyxia model. *Brain* 136: 90–105.

63. Mack RS. 2001. Overview of the current animal models for human seizure and epileptic disorders. *Epilepsy Behavior* 2: 201–216.

64. Emilio P, French J, Bialer M. 2007. Development of new antiepileptic drugs: Challenges, incentives, and recent advances. *Lancet Neurology* 6: 793–804.

65. Jack PS, Bertram E, Dudek FE, Holmes G, Mathern G, Pitkänen A. 2003. Therapy discovery for pharmacoresistant epilepsy and for disease-modifying therapeutics: Summary of the NIH/NINDS/AES models II workshop. *Epilepsia* 44: 1472–1478.

66. Coulter DA, McIntyre DC, Löscher W. 2002. Animal models of limbic epilepsies: What can they tell us? *Brain Pathology* 12: 240–256.

67. Dalby NO, Mody I. 2009. The process of epileptogenesis: A pathophysiological approach. *Current Opinion in Neurology* 14: 187–192.

68. Dudek FE, Staley K, Sutula TP. 2002. The search for animal models of epileptogenesis and pharmacoresistance: Are there biologic barriers to simple validation strategies? *Epilepsia* 43: 1275–1277.

69. Maru E, Kanada M, Ashida H. 2002. Functional and morphological changes in the hippocampal neuronal circuits associated with epileptic seizures. *Epilepsia* 43: 44–49.
70. Postma T, Krupp E, Li XL, Post RM, Weiss SR. 2000. Lamotrigine treatment during amygdalakindled seizure development fails to inhibit seizures and diminishes subsequent anticonvulsant efficacy. *Epilepsia* 41: 1514–1521.
71. Vodicka P, Smetana K, Jr., Dvoránková B, Emerick YZ, Xu, J, Ourednik, V. Motlík J. 2005. The miniature pig as an animal model in biomedical research. *Annals of the New York Academy of Sciences* 1049: 161–171.
72. Puga Y, Schneider GMK, Seebach JD. 2009. Immune responses to alpha1, 3 galactosyltransferase knockout pigs. *Current Opinion in Organ Transplantation* 14: 154–160.
73. Ekser B, Gridelli B, Tector AJ, Cooper DK. 2009. Pig liver xenotransplantation as a bridge to allotransplantation: Which patients might benefit? *Transplantation* 88: 1041–1049.
74. Makowa L, Cramer DV, Hoffman A, Breda M, Sher L, Eiras-Hreha M, Tuso PJ, Yasunaga C, Cosenza CA, Wu GD. 1995. The use of pig liver xenograft for temporary support of a patient with fulminant hepatic failure. *Transplantation* 59: 1654–1659.
75. Gewartowska M, Olszewski WL. 2007. Hepatocyte transplantation-biology and application. *Annals of Transplantation* 12: 27–36.
76. Bauer AJ, Postrach, M, Thormann S, Blanck C, Faber B, Wintersperger S, Michel S et al. 2010. First experience with heterotopic thoracic pig-to-baboon cardiac xenotransplantation. *Xenotransplantation* 17: 243–249.
77. Horvath KA, Corcoran AK, Singh RF, Hoyt C, Carrier ML, Thomas ML, Mohiuddin MM. 2010. Left ventricular pressure measurement by telemetry is an effective means to evaluate transplanted heart function in experimental heterotopic cardiac xenotransplantation. *Transplantation Proceedings* 42: 2152–2155.
78. Kirkeby S, Mikkelsen HB. 2008. Distribution of the alphaGal- and the non-alphaGal T-antigens in the pig kidney: Potential targets for rejection in pig-to-man xenotransplantation. *Immunology Cell Biology* 86: 363–371.
79. Diswall M, Angström J, Karlsson H, Phelps CJ, Ayares D, Teneberg S, Breimer ME. 2010. Structural characterization of alpha1,3-galactosyltransferase knockout pig heart and kidney glycolipids and their reactivity with human and baboon antibodies. *Xenotransplantation* 17: 48–60.
80. Ezzelarab MB, Garcia A, Azimzadeh H, Sun CC, Lin H, Hara S, Kelishadi T et al. 2009. The innate immune response and activation of coagulation in alpha1,3-galactosyltransferase gene-knockout xenograft recipients. *Transplantation* 87: 805–812.
81. Denner J, Schuurman HJ, Patience C. 2009. The International Xenotransplantation Association consensus statement on conditions for undertaking clinical trials of porcine islet products in type 1 diabetes-chapter 5: Strategies to prevent transmission of porcine endogenous retroviruses. *Xenotransplantation* 16: 239–248.
82. Yu P, Zhang P, Zhang L, Li SF, Cheng JQ, Lu YR, Li, YP, Bu H. 2009. Studies on long-term infection of human cells with porcine endogenous retrovirus. *Acta Virologica* 53: 169–174.
83. Meije Y, Tönjes RR, Fishman JA. 2010. Retroviral restriction factors and infectious risk in xenotransplantation. *American Journal of Transplantation* 10: 1511–1516.
84. Issa NC, Wilkinson RA, Griesemer A, Cooper DK, Yamada K, Sachs DH, Fishman JA. 2008. Absence of replication of porcine endogenous retrovirus and porcine lymphotropic herpes virus type 1 with prolonged pig cell microchimerism after pig-to-baboon xenotransplantation. *Journal of Virology* 82: 12441–12448.
85. Hisashi Y, Yamada K, Kuwaki K, Tseng YL, Dor FJ, Houser SL, Robson SC et al. 2008. Rejection of cardiac xenografts transplanted from alpha1,3-galactosyltransferase gene-knockout (GalT-KO) pigs to baboons. *American Journal of Transplantation* 8: 2516–2526.
86. Platt JL. 1998. New directions for organ transplantation. *Nature* 392: 11–17.
87. Pierson RN. 2009. Antibody-mediated xenograft injury: Mechanisms and protective strategies. *Transplantation Immunology* 21: 65–69.
88. Khalfoun B, Barrat D, Watier H, Machet MC, Arbeille-Brassart B, Riess JG, Salmon H, Gruel Y, Bardos P, Lebranchu Y. 2000. Development of an *ex vivo* model of pig kidney perfused with human lymphocytes. Analysis of xenogeneic cellular reactions. *Surgery* 128: 447–457.

89. Manzini S, Vargiolu A, Stehle IM, Bacci ML, Cerrito MG, Giovannoni R, Zannoni A et al. 2006. Genetically modified pigs produced with a nonviral episomal vector. *Proceedings of the National Academy of Sciences* 103: 17672–17677.

90. Cabot RA, Kuhholzer B, Chan AW, Lai L, Park KW, Chong KY, Schatten G et al. 2001. Transgenic pigs produced using *in vitro* matured oocytes infected with a retroviral vector. *Animal Biotechnology* 12: 205–214.

91. Prather RS, Hawley RJ, Carter DB, Lai L, Greenstein JL. 2003. Transgenic swine for biomedicine and agriculture. *Theriogenology* 59: 115–123.

92. Costa C, Zhao L, Burton WV. 1999. Expression of the human alpha1, 2–fucosyltransferase in transgenic pigs modifies the cell surface carbohydrate phenotype and confers resistance to human serum-mediated cytolysis. *FASEB Journal* 13: 1762–1773.

93. Cozzi E, White DJG. 1995. The generation of transgenic pigs as potential organ donors for humans. *Nature Medicine* 1: 964–969.

94. Fodor WL, Williams BL, Matis LA. 1994. Expression of a functional human complement inhibitor in a transgenic pig as a model for the prevention of xenogeneic hyperacute organ rejection. *Proceedings of the National Academy of Sciences* 91: 11153–11157.

95. Hara H, Koike N, Long C. 2010. *In vitro* investigation of the human immune response to corneal cells from genetically-modified pigs. *American Journal of Transplantation* 10: 298.

96. Cantu E, Balsara KR, Li B. 2007. Prolonged function of macrophage, von Willibrand factor-deficient porcine pulmonary xenografts. *American Journal of Transplantation* 7: 66–75.

97. Casu A, Bottino R, Balamurugan AN. 2008. Metabolic aspects of pig-to-monkey (*Macaca fascicularis*) islet transplantation: Implications for translation into clinical practice. *Diabetologia* 51: 120–129.

98. Fink JS. 2000. Porcine xenografts in Parkinson's disease and Huntington's disease patients: Preliminary results. *Cell Transplant* 9: 273–278.

99. Schumacher JM. 2000. Transplantation of embryonic porcine mesencephalic tissue in patients with PD. *Neurology* 54: 1042–1050.

chapter twelve

Variations in our genome
From disease to individualized cure

Bishwanath Chatterjee and Cecilia W. Lo

Contents

Abstract

Each of us has a unique, signature genome. The decreasing cost of whole-genome sequencing via next-generation sequencing (NGS) platforms not only allows for the use of NGS technologies to elucidate variation among healthy individuals, but also allows comparisons between individuals with disease or individuals with susceptibilities to disease. In this new era in genomics, huge quantities of data are being generated from various diseased individuals, disease-resistant groups of individuals, and ethnic groups around the world in order to compare and understand the complexity of what constitutes a "normal" genome. Variations in healthy individuals' genomes are particularly useful, as these variations may be responsible for a susceptibility to disease, although on the flip side some particular variations in a particular geographical area may really be protective against certain diseases. These variations may also predict if a particular therapy or treatment will be useful for a patient, and these variations may even give insight into the prognosis of the disease. The number of known variation within individuals and ethnic groups has been on the rise since the discovery of single-nucleotide polymorphisms. These vast inventories of data may help in determining the treatment strategy for an individual in the future. Currently, scientists are also facing huge challenges—generating faster data, storing and analyzing huge data sets, creating meaningful inferences from the data—all in time for a patient to receive treatment tailored for him or her.

Keywords: Genome structural variation, Single nucleotide variation, Copy number variation, Disease susceptibility, Drug efficacy, Adverse drug reaction.

Introduction

Every individual has a unique signature genome. The signatures of bases can determine whether a person has a genetic disorder that is linked to a disease or whether they have a genetic variation that causes them to be nonresponsive to certain drugs. Alternatively, an individual can have signature of bases that assures that they do not have a particular genetic disorder or that they will respond positively to a given treatment or drug. The sequencing and mapping of human genome, which took place more than a decade ago,[1–3] revolutionized the field of genetics. Since then, the introduction of modern technologies[4] has not only decreased the cost of sequencing, but also shortened the time with which the human genome can be sequenced. The post-human genome-sequencing era has revealed several challenges for biomedical researchers. The central question is, do we know enough if someone is susceptible to a disease or will respond to a medication through the information stored in their genome? Does some kind of "perfect" signature genome that is free from all kinds of genetic disorder exist? Do particular variations that are protective toward certain diseases exist?[5] To answer this question, we need to extract useful information from the genomic data of several individuals from each race and ethnicity. By getting useful data sets, we can increase our ability to determine whether health disparities (e.g., cardiovascular disease [CVD]) are due to biological differences between the individuals, so that in the future these data sets can help us to predict disease susceptibility and design treatments that combine genetics with other general factors (such as metabolic activity). In this chapter, we will discuss what has been elucidated so far from data obtained from genomic studies, challenges associated with resources, biobanking, and storage and analysis of data.

Genetic variation in human population

The first human genome sequencing was completed in 2001.[1–3] Within a couple of years, genetic variation was reported between individuals.[6,7] Gross chromosomal variations in the form of additions, deletions, or rearrangements were known for some time prior to human genome sequencing, but the discovery of single-nucleotide polymorphisms (SNPs) led the way to estimate genetic variation among populations and individuals within a population.[8,9] As more and more data are being accumulated on SNPs through an increasing number of studies, we understand more about the human genome and its dynamic structure. It is proposed that 10 million SNPs may be present within the human genome, according to one estimation.[10] The International HapMap project[11] (http://www.hapmap.org/) is storing database of variations in human genome. Initially, this project had collected data from few continents such as Africa, Asia, and Europe to enable identification of most common haplotypes that are present in populations worldwide. Such data allowed researchers to find genetic variation within or near the genes that may affect their disease susceptibility and their responses to a particular medication, their health, and gene–environment interactions. For many years, a number of studies have been focused on SNP and phenotypes, and consequently, several databases such as dbSNP,[12] HGVbase,[13] F-SNP,[14] and MedRefSNP[15] were created to link these variations to various data dealing with function or disease. A second class of genetic variation, called copy number variation (CNV), has evolved consequently. An increasing number of variations due to copy number (insertion or deletion) are now being recorded in normal and diseased populations.[16–19] These are the regions of DNA generally larger than 1000 bases long, but not long enough to be visualized by karyotyping. To record and document these variations, scientists from a number of countries came together to establish databases to keep up with the fast-accumulating data on SNPs and copy number variations (CNVs). Consequently, a unique global project called The Human Variome Project[20] (http://www.humanvariomeproject.org/) was launched to identify and curate all human genetic variation that can affect human health directly or indirectly. The curated data of this project will be used to improve health outcomes by providing human genetic variation data set and its possible impact on health and disease. The Human Variome Project is pursuing this mission by creating system and standards for the storage and transmission of genetic variation and its translation to global health care. Other useful database for such variation can be found, such as Database of Genomic Variants[21] (http://projects.tcag.ca/variation/) or Leidenopen Variation Database[22] (http://www.lovd. nl/2.0/). An ambitious 1000 Genome Project (www.1000genomes.org/),[23,24] an international collaborative project, was initiated in the year 2000 so that an extensive public catalog of human genetic variations could be created, including SNPs, structural variants, and their haplotype contexts.[22] This resource also supports genome-wide association studies (GWAS) as well as other medical research studies. The data from the 1092 individual genomes from this initiative serves as a reference point for analysis of disease-causing variations for large clinical studies involving whole-genome or whole-exome sequencing (refer our studies). Currently, 2500 individuals are included in this project to accommodate 25 different ethnic groups. Data from both the 1000 Genome Project and the HapMap project will be very crucial for future clinical applications such as disease association, disease susceptibility, and drug response.[25] Even though such databases are useful for researchers, right now there is a great need to optimize the data output so that databases can be compared. A recent study has compared the cataloged variations between HapMap and the 1000 Genome Project and found that not all variants cataloged in HapMap are cataloged in the 1000 Genome Project, leading to the ambiguity in decision making about which resources to use for future studies.[26] The National Institute of Standard Technology (NIST) launched a project called Genome in a Bottle,[27] a public–private consortium to develop reference material for human genome sequencing. Data coordination from different centers and studies will require the standardization of data analysis with a standardized protocol for output so that this protocol can be used for straightforward comparison.[28] According to an estimate,[29] two out of three people will be affected by a gene variation at some point in their life. These projects will have an impact upon health care and create a global resource for clinicians, researchers, and those patients affected by a genetic disorder.

Genome variation and susceptibility to disease

Single base change was postulated to be the cause of the bulk of phenotypic variability in the human population. As a result, there is a tremendous growth in high-throughput technology, including instrumentation and statistical analysis, which can easily detect these changes. The introduction of number of modern genetic tools, including high-density SNP and CNV arrays, has given researchers the necessary tools to explore the role of these variations in disease. Industries have also kept up with growth in this field—the newest arrays from Affymetrix and Illumina can genotype more than one million SNPs. The use of arrays is now on the decrease, given the reduction in the cost of whole-genome sequencing, but arrays are still very useful for larger studies, for which whole-genome sequencing is not an affordable option. With the tremendous increase in data, a number of databases were developed to store the genotype information, so that these can be available to researchers. These advances were translated quickly in several gene-mapping studies that were done using SNPs. SNP-based susceptibility variants were detected from such population-based studies on common, complex diseases—for example, diabetes,[30,31] heart disease,[32,33] and Alzheimer disease.[34] With the discovery of the role of the structural variations in gene expression level and disease phenotype,[35] new CNV-detecting platforms have been introduced by several companies, leading to an increasing number of studies that link disease to structural genetic variations (mostly CNVs). Many SNP microarray chips also contain CNV modules. Recently, several GWAS were also extended to CNVs.[36,37] The National Institutes of Health (NIH) has started a project called ClinSeq (http://www.genome.gov/20519355)[38] to investigate the use of whole-genome-sequencing data set as a tool for future clinical research. This project will extract large-scale medical sequencing (LSMS) data in a clinical research setting. These targeted sequenced regions of a person's genome will be used by researchers to investigate the technical, medical, and genetic counseling issues that accompany the implementation of LSMS in the clinical setting. Disease-specific targeted sequence have been used recently to look into specific mutations associated with diseases.[39] For such studies, disease-specific databases[40] are essential, along with locus-specific database[41] for real scientific use of the large data sets available to the scientific community. Such efforts are being made and are available to researchers, worldwide. For example, various tumor and normal tissue data are compiled together in TCGA Data Portal (https://tcga-data.nci.nih.gov/tcga/tcgaHome2.jsp).[40] Interestingly, these databases not only have whole-genome-sequencing data, but also gene expression data either through transcriptome or microarray analysis. Some other samples may also have whole-genome methylation data sets.

Genome variation and response to drugs: The era of personalized medicine

Traditionally, understanding and treating disease was to treat the patient just based on symptoms and investigative results obtained through biochemical or imaging tests. The success rate of such a traditional approach to treatment varies widely from patient to patient. Drug-related adverse events affect approximately 2 million patients in the United States, resulting in about 100,000 deaths per year[42] with similar numbers from Europe.[43,44] The patient's genetic variation profile can be used to select drug(s) or treatment protocols that may minimize harmful side effects and possibly provide a better outcome. Personalized medicine is therefore a going-forward extension of traditional treatment approaches. Recent studies have recognized the effect of genetic variation on the efficacy of drugs as well as medical procedures such as the response to anesthesia[45] or chemotherapy.[46] Drug metabolism genes are the natural targets to understand how a particular drug could be metabolized in a patient with a particular genotype. Examples include Tamoxifen (for the treatment of breast cancer)[47,48] and Clopidogrel (used to reduce risk of heart attack or stroke),[49] where variations in these genes contribute toward the efficacy of the drug. A handful of drugs in the United States have labels that recommend genetic testing before the drug is actually administrated to the patient. This may well affect the individual's treatment plan.[50] Currently, more than 200 drug labels in the United States have pharmacogenetic test information. There are three categories of these test information on the label for making therapeutic decision. Broadly they are test required, test recommended, and for information only.[51] Even though there is a long list of genotypes that are predictive of the efficacy of drug use, the risk predictions are not always good. The NIH started a pharmacogenomics knowledge base (PharmGKB) in the year 2000, which is one of the first post-genomic databases.[52] This repository of primary data provides a tool for tracking associations between different genes and drugs and provides a catalog of the location of genetic variation, its frequency in population, and its impact of drug response. With more and more data available for genes in question, this site provides more complex information on genes, variants, drugs, diseases, and pathways. The new challenges are now to provide the tools to plan and interpret genome-wide pharmacogenomics studies to predict gene–drug interactions based on shared mechanism of action.[53]

It may be noted that most of the studies available were done in population of European ancestry, and the predictions from these studies may not transfer from one population to other. It is expected that the gene variations are involved in metabolism, transport, and action and are different for various populations due to unique variation that may be present in one ethnic population.[54] So to effectively assay the clinical validity of these genetic tests, one should look into more prospective studies with populations from other non-European groups in order to find out whether these associated markers are also replicated in other ethnic groups, especially in countries where considerable diversity exists.[55]

Genome-wide association studies

In the past few years, human geneticists concentrated around the concept that compares the genomes of normal individuals and in individuals with disease using thousands of SNPs. This was based on the assumption that DNA variants can be associated with the risk of developing disease; it signals that something in the area of genome may be responsible for the disease and these hits will give a better understanding of disease, diagnosis, and treatment. These studies found one or more hits on variations associated with complex diseases. A lot of these data were always not reproducible, and new regions have been found along with or without known genomic regions in other prospective studies for the same disease. For example, the first GWAS studies in asthma found the variant in ORMDL3 contribute to childhood onset of asthma,[56] and this variation has been confirmed in some subsequent studies in different populations. Since then, the association of several other genes has been reported and a reproducibility study has found asthma associations for a minority of candidate genes.[57] Currently, more than 100 genes are known to be associated with asthma.[58] It is also postulated that so far, only a small number of variants have been identified for some complex diseases.[59] The National Human Genome Research Institute has made a catalog of all published genome-wide studies, and this database can be searched for genes that are linked to disease[60] (available at: http://www.genome.gov/gwastudies/). This resource, together with bioinformatics predictions, will give researchers access to complex disease etiology due to common/specific variants present in the population/individual. Although GWAS was key in several studies to find the mutation-causing diseases, this process is now increasingly being replaced by whole-exome or targeted resequencing.

Transcriptional profiling

Even with profiling the variations in the genomes of healthy and diseased individuals, a full understanding of the effect of the variation is not always clear. The question remains as to whether a genomic variation can lead to phenotypic differences, encompassing the pathological variation that predisposes to or causes disease and the polymorphic variation that may act as passenger variation. For example, a genomic amplification in the alveolar rhabdomyosarcoma is always not associated with the altered expression of miRNA in the amplified area of the diseased genome.[61] One way to tackle this issue is to get genome-wide whole expression pattern of the genes. A high-throughput quantitation of the transcriptome was possible with the development of expression microarray.[62] Recent advances in the next-generation sequencing (NGS) technologies enabled researchers to get the massively parallel whole-exome sequencing by exome capture.[63] The whole-genome cDNA sequencing or RNA-seq not only allowed quantitation of transcriptomes, but also characterized the transcripts for alternative splicing and mutations in exomes if there were any.[64] The combination of DNA and gene expression alteration is now becoming a golden standard along with epigenetic changes (which is not covered in this review) for studying how genetic variations can influence diseases.

Regulating RNAs and enhancers

The genetic variation that we see in individuals' genomes guides us to the elements in the genome which directly or indirectly regulate a number of basic events by regulating gene expression. These elements are usually present several mega bases away from the site of action or even sometimes present on other chromosomes altogether. The ENCODE (Encyclopedia of DNA Elements) Project funded by the National Human Genome Research Institute with international collaborations aims to document these elements. The goal of this project was to build a comprehensive list of functional elements in the human genome. Such data are useful in understanding DNA variations that were previously thought to have no functional significance.[65] The project delivered cell type-specific sites for biochemical activity that regulate gene expression and interpretation of its association with disease.[66]

Data storage and analysis challenges

Over the last few years, the decreasing cost of NGS technologies has spurred the generation of data, and such data are growing at the rate of 100+% per year. A typical NGS whole-genome data are about 80–200 GB depending upon the read length and coverage. The challenge is then how to store and analyze such large data sets in a time when both computing power (Moore's law) and Internet bandwidth (Nielsen's law) are only growing at a rate of 45%–50% per year. This is a big challenge for the laboratories that are generating data and for the bioinformaticians who will be analyzing such huge pools of data. The problem is partially solved by cloud computing. All the top computing companies such as Microsoft, IBM, etc., and others such as Amazon have been generating cloud computing facilities that the researcher can use to not only save their data, but to also run their data analyses on different servers available for their disposal. Galaxy[67] (http://galaxyproject.org/) is one such free public server from The Center for Comparative Genomics and Bioinformatics, Penn State University, which can be accessed to submit data using cloud services. Another solution that is currently being explored is the development of powerful algorithms.[68]

Omics data sharing

The availability of a patient's omics data for a particular disease is an important step in the future of genomic medicine, and has a great implication for the advancement of genomic science.[69] A very good coordination is required between the phenotype and the geno-type correlation. The patient's privacy becomes a big obstacle in the way of such an effort. Recently, 69 medical and clinical institutions around the world declared an alliance for sharing of data sets.[70] Increasing the number of samples can also provide information about the timeline of a mutation in diseases such as cancer, and whether any of many mutations that are found in such a disease could be an early event, so that such a mutation spectrum can be used as a diagnostic tool and treatment. It may be noted that there is a growing concern about the mosaicism in the human body in regard to genetic variations. Cancer is one example, but recent studies have exposed the mosaicism in the human body with reference to other diseases.[71] With each cell division, there is a very small probability of acquiring a mutation, but during development, while there are millions of cell division taking place, the chances of acquiring mutation in a cell is proportionately high, which can lead to heterogeneity within a normal tissue.[72] Frequently, such tissue heterogeneity can lead to disease.[73] Single-cell genomics[74] is an important step in analyzing the relation of such mutation to disease, though the technology and the cost of such analysis is still very high.

Challenges of personalized medicine in developed and developing countries

Traditional approaches to understanding and treating disease have been practiced in all the third-world countries. Personalized medicine is the extension of these approaches by considering individual variation that may be found in each individual or group.[75] The traditional way of diagnosis is to consider observable evidence and prescribe a treatment, which is not made for each individual. Introducing personalized medicine, physicians have to not only consider the visually obvious evidence, such as a tumor, pathological appearance, or biochemical results, but also consider the molecular makeup of the individual patient. Patient's genetic variation profile will be taken care of to provide a guideline for the selection of drugs or treatment which will fit the individual with minimum known side effects. Genetic testing may also provide information about predisposition to disease, so that future testing and monitoring plan could be generated in consultation with the physician. However, these are hard to come by for third-world countries where there is no ability to profile the structural variation in DNA sequence, gene and protein expression, and metabolite profile. These informations are redefining the classification of diseases and their treatment. This enables physicians to make the most effective clinical decisions for each patient. Since there is substantial benefit of personalized medicine, an infrastructure is required to practice personalized medicine that includes law, policy, and education. Medical institutions in the United States are committed to implement personalized medicine into practice. However, in reality, the typical training that a physician receives in American medical institutes reveals that this training is not suited for promoting and practicing personalized medicine.[76]

Data from the year 2007 revealed that very small number of U.S. and Canadian school have practical training in medical genetics as part of their curricula (11%).[77] Another survey of families with genetic conditions in United States revealed that most of the physicians are poorly prepared to deal with issues related to genetics and genomics.[78] A 2011 survey consisting of 800 physicians revealed that only 10%–30% of the specialty physicians are considering the use of personalized medicine,[79] although most of them agree that personalized medicine will play a crucial role in treatment decisions in the years to come. Another survey reveals that physicians have a strong interest in using genetic information for making treatment decision if they know how to get hold of them.[80] The waiting time for test results and cost of such genetic testing to the patients is generally a problem for physicians who want to incorporate genetic medicine into their practice. It may be noted that during emergency medical situations there may not be enough time to get the genetic testing done for a patient, and it might lead to under/overdosing to the patient, which could be life threatening.

The scenario of medical treatment in third-world countries is still considered primitive, as physicians rely mostly on visual evidence for treatment. Though there is substantial improvement in the urban areas, the rural areas need a thorough overhaul. Given this, there is an immediate need to modify the medical education curriculum to make sure that future physicians have proper knowledge of the genomic and genetic implications of the medical practice. Medical education in these countries should also aim to organize workshops that educate current physicians so that they can work with the future challenges. In the United States, Department of Health and Human Services, the President's Council of Advisors in Science and Technology, and the Personalized Medicine Coalition have recommended wide-ranging policy for implementation of personalized medicine.[81] For example, in the United States, a genetic privacy law has been established to tackle the ethical issues associated with genetic testing.[82] Even with growing knowledge of genomics/genetics in general, medical education has been lagging behind in this fast-growing area.

Patient cohorts and biobanking

Personalized medicine calls for necessary infrastructure that is required to go forward. The primary goal of such medical advancement is to treat the patients in terms of their own genome. To do this, the number one objective would be to create patient cohorts for particular diseases with strict monitoring and evaluation of the physical, biochemical, and pathological symptoms and history of disease documented, so that subjects may be grouped according to the need of the study. Applications in medicine and biomedical research require identification of large patient cohorts with specific clinical characteristics to generate power of statistical significance of results. Besides enrolling the diseased individual, there is a need to register the so-called "normal" subject. The normal individual in a cohort serves as a case control, and would not have the symptoms related to the disease cohorts. In developed countries, creating cohorts and pooling cohorts for future studies has become a part of the genomic era. The importance of such cohorts is very evident in one of the oldest cohort in heart study, called Framingham Heart Study (http://www.framinghamheartstudy.org/about/history.html). This cohort was started in 1948 with the objective of identifying the common factors or characteristics that contribute to CVD. This goal was achieved by following development of CVD over a long period of time for individuals who had not yet developed overt symptoms of CVD or suffered a heart attack or stroke. For this study, 5209 men and women between the ages of 30 and 62 were recruited from the town of Framingham, Massachusetts. These people were subjected to extensive physical examinations and lifestyle interviews so that their common patterns related to CVD development could be analyzed. All subjects have been asked to return to the study every 2 years for physical examination and laboratory tests. Later on in 1971, the second generation of the original participants—5124 adult children and their spouses—were recruited for similar examinations. In 1994, a new study reflecting a more diverse community of Framingham was recruited for Omni cohort of the Framingham Heart Study. In April 2002, the study enrolled a third generation of participants, the grandchildren of the Original Cohort, followed by recruitment of second group of Omni participant. Researchers from all over the world are using such a great collection of diseased and normal individuals for heart studies. In the year 2009 alone, 902 peer-reviewed articles were published using this great cohort.

The second aspect of the infrastructure for genomics studies is biobanking. This is such an important aspect of personalized medicine that *Time* magazine has put biobanking among the top 10 ideas changing the world right now (March 23, 2009 edition of *Time*). Biospecimen procurement, processing, banking, and distribution is an essential component of the genomic medicine research infrastructure and provides an invaluable resource when specimens are associated with clinical information from patients treated or untreated.[83] A very large infrastructure for storage of samples under ideal condition and tracking of sample inventory and their medical records is essential for such studies. Though these efforts are considered as the golden standard, ethical use of such biobanks needs to be deliberated and considered.[84]

Conclusion

A lot more has to be done to understand and take advantage of personalized medicine throughout the world. The identification of diseased loci, drug development, and treatment due to complications in genetic variation delineate the need for multidirectional genomic analyses of each individual (Figure 12.1). Given the differences in makeup of genomes in term of structural variations in ethnic groups, a more defined research setup has to be established worldwide to address these issues. With decreasing cost and faster rate of data collection for genomic sequencing, achieving the targets will not be difficult. Over the 10 years of human genome sequencing, the cost of sequencing has decreased about 10,000-fold and will catch up to $1000 per genome in the years to come. On the other hand, there are new obstacles in terms of storage, tracking, and analysis of the data, which pose a real challenge to the scientific world. At present time, handling of data is not keeping up with the generation of data. A big IT infrastructure will be required to tackle this problem that is coming up very fast. Even though we have whole-genome-sequencing data

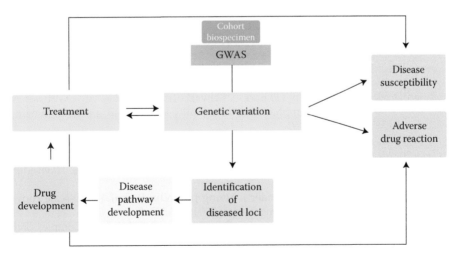

Figure 12.1 Forming disease cohort, collection of biospecimen, and GWAS/transcriptome analysis can lead to identification of gene(s) in the pathway that can lead to drug development and treatment. The identification of diseased loci may be complicated by the genetic variation that may exist in the population of patients. Normally, genetic variation can lead to susceptibility to disease/adverse drug reaction and can interact with new drug development and treatment of complex diseases.

available for several thousand individuals, we are far from the basic understanding for signature genome of an individual for use in personalized medicine. It may be noted that the signature genome of an individual is not the only factor in elucidating the diseased/wellness state of an individual. Multiple genetic factors are known to interact with environmental factors such as what you eat, where you live, and so on, and these factors have been known to play a greater role in combination with genetic factor for your day-to-day life, and these must also be considered in the future of personalized medicine.

References

1. Lander E, Linton L, Birren B, Nusbaum C, Zody M, Baldwin J et al. Initial sequencing and analysis of the human genome. *Nature* 2001; 409:860–921.
2. Venter J, Adams M, Myers E, Li P, Mural R, Sutton G et al. The sequence of the human genome. *Science* 2001; 291:1304–51.
3. Consortium IHGS. Finishing the euchromatic sequence of the human genome. *Nature* 2004; 431:931–45.
4. Metzker ML. Sequencing technologies—The next generation. *Nat. Rev. Genet.* 2010; 11:31–46.
5. Cockburn IA, Mackinnon MJ, O'Donnell A, Allen SJ, Moulds JM, Baisor M, Bockarie M, Reeder JC, Rowe JA. A human complement receptor 1 polymorphism that reduces *Plasmodium falciparum* rosetting confers protection against severe malaria. *Proc. Natl. Acad. Sci. U. S. A.* 2004; 101:272–7.
6. Disotell T. Human genomic variation. *Genome Biol.* 2000; 1.
7. Shriver M, Mei R, Parra E, Sonpar V, Halder I, Tishkoff S et al. Large-scale SNP analysis reveals clustered and continuous patterns of human genetic variation. *Hum. Genomics* 2005; 2:81–9.
8. Altshuler D, Pollara V, Cowles C, Van Etten W, Baldwin J, Linton L et al. An SNP map of the human genome generated by reduced representation shotgun sequencing. *Nature* 2000; 407:513–6.
9. Sachidanandam R, Weissman D, Schmidt S, Kakol J, Stein L, Marth G et al. A map of human genome sequence variation containing 1.42 million single nucleotide polymorphisms. *Nature* 2001; 409:928–33.
10. Rosenfeld JA, Mason CE, Smith TM. Limitations of the human reference genome for personalized genomics. *PLoS ONE* 2012; 7:e40294.
11. Thorisson G, Smith A, Krishnan L, Stein L. The international HapMap project website. *Genome Res* 2005; 15:1592–3.
12. Sherry S, Ward M, Kholodov M, Baker J, Phan L, Smigielski E et al. dbSNP: The NCBI database of genetic variation. *Nucleic Acids Res.* 2001; 29:308–11.
13. Fredman D, Siegfried M, Yuan YP, Bork P, Lehvaslaiho H, Brookes AJ. HGVbase: A human sequence variation database emphasizing data quality and a broad spectrum of data sources. *Nucleic Acids Res.* 2002; 30:387–91.
14. Lee PH, Shatkay H. F-SNP: Computationally predicted functional SNPs for disease association studies. *Nucleic Acids Res.* 2008; 36(Database issue):D820–4.
15. Rhee H, Lee J. MedRefSNP: A database of medically investigated SNPs. *Hum. Mutat.* 2009; 30:E460–6.
16. Sebat J, Lakshmi B, Troge J, Alexander J, Young J, Lundin P et al. Large-scale copy number polymorphism in the human genome. *Science* 2004; 305:525–8.
17. Fredman D, White SJ, Potter S, Eichler EE, Den Dunnen JT, Brookes AJ. Complex SNP-related sequence variation in segmental genome duplications. *Nat. Genet.* 2004; 36:861–6.
18. Hollox E, Armour J, Barber J. Extensive normal copy number variation of a beta-defensin antimicrobial-gene cluster. *Am. J. Hum. Genet.* 2003; 73:591–600.
19. Iafrate AJ, Feuk L, Rivera MN, Listewnik ML, Donahoe PK, Qi Y, Scherer SW, Lee C. Detection of large-scale variation in the human genome. *Nat. Genet.* 2004; 36:949–51.
20. Ring HZ, Kwok PY, Cotton RG. Human Variome Project: An international collaboration to catalogue human genetic variation. *Pharmacogenomics* 2006; 7:969–72.

21. Zhang J, Feuk L, Duggan GE, Khaja R, Scherer SW. Development of bioinformatics resources for display and analysis of copy number and other structural variants in the human genome. *Cytogenet. Genome Res.* 2006; 115:205–14.

22. Fokkema IF, Taschner PE, Schaafsma GC, Celli J, Laros JF, den Dunnen JT. LOVD v.2.0: The next generation in gene variant databases. *Hum. Mutat.* 2011; 32:557–63.

23. The 1000 Genomes Project Consortium. An integrated map of the genetic variation from 1,092 human genomes. *Nature* 2012; 491:56–65.

24. Via M, Gignoux C, Burchard EG. The 1000 genomes project: New opportunities for research and social challenges. *Genome Med.* 2010; 2:3.

25. Zhang W, Dolan ME. Impact of the 1000 genomes project on the next wave of pharmacogenomic discovery. *Pharmacogenomics* 2010; 11:249–56.

26. Buchanan CC, Torstenson ES, Bush WS, Ritchie MD. A comparison of cataloged variation between international HapMap consortium and 1000 genomes project data. *J. Am. Med. Inf. Assoc.* 2012; 19:289–94.

27. Zook JM, Chapman B, Wang J, Mittelman D, Hofmann O, Hide W, Salit M. Integrating human sequence data sets provides a resource of benchmark SNP and indel genotype calls. *Nat. Biotechnol.* 2014; 32:246–251.

28. Magnant G. Opinion: Standard needed. *The Scientist*, 2013; Sep 18. http://www.the-scientist.com/?articles.view/articleNo/37347/title/Opinion--Standards-Needed/

29. Baird PA, Anderson TW, Newcombe HB, Lowry RB. Genetic disorders in children and young adults: A population study. *Am. J. Hum. Genet.* 1988; 42:677–93.

30. Hakonarson H, Grant SF, Bradfield JP, Marchand L, Kim CE, Glessner JT et al. A genome-wide association study identifies KIAA0350 as a type 1 diabetes gene. *Nature* 2007; 448:591–4.

31. Sladek R, Rocheleau G, Rung J, Dina C, Shen L, Serre D et al. A genome-wide association study identifies novel risk loci for type 2 diabetes. *Nature* 2007; 445:881–5.

32. Wellcome Trust Case Control Consortium. Genome-wide association study of 14,000 cases of seven common diseases and 3,000 shared controls. *Nature* 2007; 447:661–78.

33. Samani NJ, Burton P, Mangino M, Ball SG, Balmforth AJ, Barrett J et al. A genomewide linkage study of 1,933 families affected by premature coronary artery disease: The British Heart Foundation (BHF) Family Heart Study. *Am. J. Hum. Genet.* 2005; 77:1011–20.

34. Waring SC, Rosenberg RN. Genome-wide association studies in Alzheimer disease. *Arch. Neurol.* 2008; 65:329–34.

35. Stranger BE, Forrest MS, Dunning M, Ingle CE, Beazley C, Thorne N et al. Relative impact of nucleotide and copy number variation on gene expression phenotypes. *Science* 2007; 315:848–53.

36. McCarroll, SA. Extending genome-wide association studies to copy-number variation. *Hum. Mol. Genet.* 2008; 17:R135–42.

37. Ionita-Laza I, Rogers AJ, Lange C, Raby BA, Lee C. Genetic association analysis of copy-number variation (CNV) in human disease pathogenesis. *Genomics* 2009; 93: 22–6.

38. Biesecker LG, Mullikin JC, Facio FM, Turner C, Cherukuri PF, Blakesley RW et al. The ClinSeq project: Piloting large-scale genome sequencing for research in genomic medicine. *Genome Res.* 2009; 19:1665–74.

39. Nakhleh N, Francis R, Giese RA, Tian X, Li Y, Zariwala MA et al. High prevalence of respiratory ciliary dysfunction in congenital heart disease patients with heterotaxy. *Circulation* 2012; 125: 2232–42.

40. Howard HJ, Beaudet A, Gil-da-Silva Lopes V, Lyne M, Suthers G, Van den Akker P et al. Disease-specific databases: Why we need them and some recommendations from the Human Variome Project Meeting, May 28, 2011. *Am. J. Med. Genet.* 158A: 2763–6.

41. Collins FS, Barker AD. Mapping the cancer genome. Pinpointing the genes involved in cancer will help chart a new course across the complex landscape of human malignancies. *Sci. Am.* 2007; 296:50–7.

42. Lazarou J, Pomeranz BH, Corey PN. Incidence of adverse drug reactions in hospitalized patients: A meta-analysis of prospective studies. *JAMA* 1998; 279:1200–5.

43. Scheiber J, Chen B, Milik M, Sukuru SC, Bender A, Mikhailov D et al. Gaining insight into off-target mediated effects of drug candidates with a comprehensive systems chemical biology analysis. *J. Chem. Inf. Model.* 2009; 49:308–17.

44. Pirmohamed M, James S, Meakin S, Green C, Scott AK, Walley TJ et al. Adverse drug reactions as cause of admission to hospital: Prospective analysis of 18 820 patients. *BMJ* 2004; 329:15–9.

44. Searle R, Hopkins PM. Pharmacogenomic variability and anaesthesia. *Br. J. Anaesth.* 2009; 103:14–25.

46. Ansari M, Krajinovic M. Pharmacogenomics in cancer treatment defining genetic bases for inter-individual differences in responses to chemotherapy. *Curr. Opin. Pediatr.* 2007; 19:15–22.

47. Sangkuhl K, Berlin DS, Altman RB, Klein TE. PharmGKB: Understanding the effects of individual genetic variants. *Drug Metab. Rev.* 2008; 40:539–51.

48. Hoskins JM, Carey LA, McLeod HL. CYP2D6 and tamoxifen: DNA matters in breast cancer. *Nat. Rev. Cancer.* 2009; 9:576–86.

49. Shuldiner AR, O'Connell JR, Kevin P, Gandhi A, Ryan K, Horenstein RB et al. Association of cytochrome P450 2C19 genotype with the antiplatelet effect and clinical efficacy of clopidogrel therapy. *JAMA* 2009; 302:849–57.

50. http://www.pathway.com/more_info/drug_responses.

51. Ikediobi ON, Shin J, Nussbaum RL, Phillips KA. Addressing the challenges of the clinical application of pharmacogenetic testing. *Clin. Pharmacol. Ther.* 2009; 86:28–31.

52. Klein TE, Chang JT, Cho MK, Easton KL, Fergerson R, Hewett M et al. Integrating genotype and phenotype information: An overview of the PharmGKB project. Pharmacogenetics Research Network and Knowledge Base. *Pharmacogenomics J.* 2001; 1:167–70.

53. Thorn CF, Klein TE, Altman RB. Pharmacogenomics and bioinformatics: PharmGKB. *Pharmacogenomics* 2010; 114:501–5.

54. Eichelbaum M, Ingelman-Sundberg M, Evans WE. Pharmacogenomics and individualized drug therapy. *Annu. Rev. Med.* 2006; 57:119–37.

55. Hardy BJ, Seguin B, Singer PA, Mukerji M, Brahmachari SK, Daar AS. From diversity to delivery: The case of Indian genome variation initiative. *Nat. Rev. Genet.* 2008; Suppl 1:S9–14.

56. Moffatt MF, Kabesch M, Liang L, Dixon AL, Strachan D, Heath S et al. Genetic variants regulating ORMDL3 expression contribute to the risk of childhood asthma. *Nature* 2007; 448:470–3.

57. Rogers AJ, Raby BA, Lasky-Su JA, Murphy A, Lazarus R, Klanderman BJ et al. Assessing the reproducibility of asthma candidate gene associations, using genome-wide data. *Am. J. Respir. Crit. Care. Med.* 2009; 179:1084–90.

58. Wjst M, Sargurupremraj M, Arnold M. Genome-wide association studies in asthma: What they really told us about pathogenesis. *Curr. Opin. Allergy Clin. Immunol.* 2013; 13(1):112–8.

59. Manolio TA, Collins FS, Cox NJ, Goldstein DB, Hindorff LA, Hunter DJ et al. Finding the missing heritability of complex diseases. *Nature* 2009; 461:747–53.

60. Hindorff LA, Sethupathy P, Junkins HA, Ramos EM, Mehta JP, Collins FS, Manolio TA. Potential etiologic and functional implications of genome-wide association loci for human diseases and traits. *Proc. Natl. Acad. Sci. U. S. A.* 2009; 106:9362–7.

61. Reichek JL, Duan F, Smith LM, Gustafson DM, O'Connor RS, Zhang C, Dunlevy MJ, Gastier-Foster JM, Barr FG. Genomic and clinical analysis of amplification of the 13q31 chromosomal region in alveolar rhabdomyosarcoma: A report from the children's oncology group. *Clin. Cancer Res.* 2011; 17(6):1463–73.

62. Schena M, Shalon D, Davis RW, Brown PO. Quantitative monitoring of gene expression patterns with a complementary DNA microarray. *Science* 1995; 270:467–70.

63. Ng SB, Turner EH, Robertson PD, Flygare SD, Bigham AW, Lee C et al. Targeted capture and massively parallel sequencing of 12 human exomes. *Nature* 2009; 461:272–6.

64. Marioni JC, Mason CE, Mane SM, Stephens M, Gilad Y. RNA-seq: An assessment of technical reproducibility and comparison with gene expression arrays. *Genome Res.* 2008; 18:1509–17.

65. The Encode Project Consortium. The ENCODE (ENCyclopedia of DNA elements) Project 2004. *Science* 306:636.

66. Kellis M, Wold B, Snyder MP, Bernstein BE, Kundaje A, Marinov GK et al. 2014. Defining functional DNA elements in the human genome. *Proc. Natl. Acad. Sci.* 111:6131–8.

67. Goecks J, Nekrutenko A, Taylor J. Galaxy team. Galaxy: A comprehensive approach for supporting accessible, reproducible, and transparent computational research in the life sciences. *Genome Biol.* 2010; 11(8):R86.

68. Berger B, Peng J, Singh M. Computational solution for Omics data. *Nat. Rev. Genet.* 2013; 14:333–46.
69. Callier S, Husain R, Simpson R. Genomic data sharing: What will be our legacy? *Frontier Genet.* 2014; 5:34.
70. Hayden EC. Geneticists push for global data-sharing. *Nature* 2013; 498:16–7.
71. Biesecker LG, Spinner NB. A genomic view of mosaicism and human disease. *Nat. Rev. Genet.* 2013; 14:307–20.
72. De S. Somatic mosaicism in healthy human tissue. *Trends Genet.* 2011; 27:217–23.
73. Podury A, Evrony GD, Cai X, Walsh CA. Somatic mutation, genomic variation and neurologic disease. *Science* 2013; 34:1237758.
74. Macaulay IC, Voet T. Single cell genomics: Advances and future prospectives. *PLoS Genet.* 2014; 10:e1004126.
75. The case of personalized medicine. Personalized medicine coalition report, May 2009, available at http://www.ageofpersonalizedmedicine.org.
76. Salari K. The dawning era of personalized medicine exposes a gap in medical education. *PLoS Med.* 2009; 6:e1000138.
77. Thurston VC, Wales PS, Bell MA, Torbeck L, Brokaw JJ. The current status of medical genetics instruction in US and Canadian medical schools. *Acad. Med.* 2007; 82:441–5.
78. Harvey EK, Fogel CE, Peyrot M, Christensen KD, Terry SF, McInerney JD. Providers' knowledge of genetics: A survey of 5915 individuals and families with genetic conditions. *Genet. Med.* 2007; 9:259–67.
79. Ray T. Genome web Oct 2012 Reality Check: Educating Physicians on Genomic Medicine.
80. Najafzadeh M, Davis JC, Joshi P, Marra C. Barrier for integrating personalized medicine into clinical practice: A quantitative analysis. *Am. J. Med. Genet.* 2013; 161:758–63.
81. Carroll JC, Rideout AL, Wilson BJ, Allanson JM, Blaine SM, Esplen MJ et al. Genetic education for primary care providers: Improving attitudes, knowledge, and confidence. *Can. Fam. Physician* 2009; 55:e92–9.
82. President's Council of Advisors on Science and Technology. *Priorities for Personalized Medicine.* September 2008 (available online at: http://www.ostp.gov/galleries/PCAST/pcast_report_v2.pdf).
83. Wadman M. Genetics bill cruises through the senate. *Nat. News* 2008; 453:9.
84. Hewitt HE. Biobanking: The foundation of personalized medicine. *Curr. Opin. Oncol.* 2011; 23:112–9.
85. Haldeman KM, Cadigan RJ, Davis A, Goldenberg A, Henderson GE, Lassiter D, Reavely E. Community engagement in US biobanking: Multiplicity of meaning and method. *Public Health Genomics* 2014; 17(2):84–94.

chapter thirteen

Molecular biotechnology for diagnostics

Shailendra Dwivedi, Saurabh Samdariya, Gaurav Chikara, Apul Goel, Rajeev Kumar Pandey, Puneet Pareek, Sanjay Khattri, Praveen Sharma, Sanjeev Misra, and Kamlesh Kumar Pant

Contents

Abstract

The current advent of molecular techniques with the multidisciplinary relationship of numerous fields led to the development and progress of genomics, which focuses on the finding of events at the genome level. Structural and functional genomic tactics have now pinpointed the obstacles in the investigation of disease-related genes and their structural and functional variations. Various new technologies and diagnostic applications of structural genomics are currently preparing a large database of disease genes and genetic alterations by mutation scanning, sequencing, and deoxyribonucleic acid chip technology. Functional genomics is exploring gene expression by traditional methods (hybridization-, polymerase chain reaction [PCR]-, and sequence-based technologies), two-hybrid technology, along with advanced next-generation sequencing. Developments in microarray chip technology have taken place as microarrays have permitted the concurrent analysis of gene expression patterns of several genes in real time. Further advances of genetic engineering have also revolutionized immunoassay biotechnology via the engineering of genes-encoded antibodies and phage-screening technology. Biotechnology plays an important role in the development of diagnostic biomarkers in response to an outburst or against a dangerous disease in times of need. A well-known example is the field validation of the single-nucleotide polymorphism (SNP) assay for equine herpesvirus-1 when there was an outbreak on 3 California race in 2006/2007, and the recent improvement and validation of a real-time PCR assay for the recognition of a virulent variant of the infectious bursal disease virus in 2009.

However, there is a dire need to identify the many obstacles related to the commercialization and global dispersal of genetic knowledge generated from the exploitation of the biotechnology industry as well as the development and marketing of diagnostic services. Application of stringent genetic criteria for patient selection and sensible screening of the risks and benefits of treatment raise a major challenge to the pharmaceutical industry. Thus, this field is being revolutionized in the current era and may further open new vistas in the field of disease management.

Keywords: SNPs, Microarray, Real-time PCR, Genetic diseases, Disorders.

Introduction

Diagnostics is the most essential and critical part of the present-day healthcare system, as its results guide physicians in making various decisions for the treatment of patients. From the genetic tests that advise personalized treatment to the microbial culture that identifies the correct antibiotic to fight against an infection, diagnostics deliver critical insights at all stage of medical care prevention, diagnosis, and successful management of diseases. The comprehensive categories of diagnostics are medical chemistry, immunology, hematology, microbiology, and molecular diagnostics. Molecular diagnostics has kept a specific direction in recent years because of its deep insights and the increasing clinical practice of these tests. The recognition and fine description of the genetic basis of the disease in any disease condition is essential for the accurate diagnosis of many diseases. Gene discovery provides priceless insights into the mechanisms of disease and gene-based markers that permit clinicians not only to screen the disease predisposition but also to design and utilize improved diagnostic methods. The latter is of great significance, as the surplus and variety of molecular defects requires the use of several rather than a single mutation-detection equipment. Molecular diagnostics is an incredible reality and its foundation is based on the study of gene expression and function. The techniques are utilized to diagnose and screen disease states along with the prediction of risk and to choose personalized medicine for individual patients (Poste 2001, Carl et al. 2012). By studying the characteristics of the patient and their disease, molecular diagnostics makes the platform for personalized medicine (Hamburg and Collins 2010). Molecular diagnostic tests are precious in a range of clinical specialties including communicable disease, oncology, hematology, immunology, and pharmacogenomics I. For example, genetic forecast of which drugs will be suitable for treatment (Grody et al. 2010).

Molecular diagnostics is a vibrant and transformative technique of diagnostics, which is now evolving after relentless research and is employed in the treatment of several diseases. It explores the genetic material or proteins related with the health condition or disease, provides the path to explore the pinpointed mechanisms of disease, and permits clinicians to plan care at an individual level, thus enabling the practice of personalized medicine. Relentless innovation and development in tools and technology in this field is now accelerating the progress in molecular diagnostics and very soon a day will come when the whole genome sequencing will be in routine practice. Increasing use of computerized software and automation technology is enabling sophisticated molecular tests to be achieved in the full scope of healthcare settings, extending state of the art diagnostics to all nations of the world. This chapter will provide an overview of the current scenario of molecular diagnostics, with the glimpse of the main technologies that are driving the molecular revolution with its use in some specific disorders. At the end, emphasis has been given to describe the novel progress in the therapeutics approach of pharmacogenomics and nutrigenomics that have come in a big way to influence the final management of diseases.

History of molecular diagnostics: Advancements and discovery

The past five decades have seen remarkable progress in the field of molecular biology. The whole journey of its progress and inventions with its clinical implications are summarized in Table 13.1. Kan et al. in 1980 suggested a prenatal genetic test for thalassemia, which relied on restriction endonuclease digestion that cut DNA at targeted specific short sequences, generating various bands of DNA strand depending on the presence of allele (genetic variation). In the 1980s (Kan et al. 1980), the phrase was tagged as a brand name of companies such as Molecular Diagnostics Incorporated (Cohn et al. 1984) and Bethseda Research Laboratories Molecular Diagnostics (Kaplan and Gaudernack 1982, Persselin and Stevens 1985). Further, newly explored genes and DNA sequencing techniques led to the launch of a distinctive field of molecular and genomic laboratory medicine in the 1990s. Furthermore in 1995, the association for molecular pathology (AMP) was founded to provide organization to the field and later in 1999, the AMP launched The *Journal of Medical Diagnostics* (Fausto and Kaul 1999) and *Informed Health Care* founded expert reviews in medical diagnostics in 2001 (Poste 2001). Moreover with the beginning of the twenty-first century, the Hap Map Project presented information on the one-letter genetic differences that persist in the human population, as single-nucleotide polymorphisms (SNPs) also showed their association with disease (Carl et al. 2012). In 2012, the blotting and hybridization test were evolved and thus genetic hybridization tests were employed to characterize the specific SNP for thalassemia.

In 1949, Pauling and coworkers introduced the term "molecular disease" in the medical database, and on the basis of their findings, a change in a single amino acid at the β-globin chain that causes sickle cell anemia was reported. Consequently, their innovations have set the frame of molecular diagnostics, although the revolutionary changes occurred many years later. The foundation of molecular diagnostics was led in the beginning period of recombinant DNA technology, with numerous scientists from varied specialties working in concert. For exploring the various sequences of genes, cDNA cloning and sequencing emerged as precious tools at that time. Then the blotting device came into reality that provided a number of DNA probes, permitting the investigation via Southern blotting of genomic regions that was based on the principle and concept of restriction fragment length polymorphism (RFLP) to trail a mutant allele from heterozygous parents to a high-risk pregnancy. In 1976, Kan et al. performed first time prenatal diagnosis of α-thalassemia by utilizing hybridization of DNA isolated from fetal fibroblasts. Moreover, Kan et al. (1980), applied RFLP technique to pinpoint sickle cell alleles of African descent in 1978. This innovation provided the fuel for the establishment of similar diagnostic methods for the recognition of other genetic diseases, such as phenylketonuria (Woo et al. 1983), cystic fibrosis (CF) (Farrall et al. 1986), etc. The first requirement to identify any mutation causing disease was the construction of the genomic DNA library from the affected individuals that is only possible after the cloning of the mutated allele after which sequence could be deciphered. Moreover, this approach was utilized in exploring the human globin gene mutations for the first time (Treisman et al. 1983). Further, Orkin et al. (1983) reported a number of sequence alternations that were linked to specific β-globin gene mutations. Mutations causing

Table 13.1 Historical significance-development and progression
of molecular biology/techniques

Year/decades	Discovery/event	Discoverer/company	Remarks
1869	Deoxyribonucleic acid or DNA	Johann Friedrich Miescher	–
1944	Transforming material is DNA	Oswald Avery, McCarty, and Colin MacLeod	DNA seems to be genetic material
1928	Transformation	Franklin Griffith	Genetic material is a heat-stable chemical
1949	DNA composition was species specific	Erwin Chargaff	A = T; G = C
1949	Characterization of sickle cell anemia as a molecular disease	Linus Pauling	Discovery that a single amino acid change at β-globin chain leads to sickle cell anemia
1953	Double-helical model of DNA	Watson–Crick	Led the foundation of molecular biology
1958	Isolation of DNA polymerases	Arthur Kornberg	Important milestone for DNA replication
1960	First-hybridization techniques	Roy Britten	–
1969	*In situ* hybridization	Gall and Pardue	
1970	Isolated the first restriction enzyme	Hamilton Smith	An enzyme that cuts DNA at a very specific nucleotide sequence
1972	Assembled the first DNA molecule	Paul Berg	Crucial steps in subsequent development of recombinant genetic engineering
1961	First "triplet"—a sequence of three bases of DNA	Marshall Nirenberg	Triplet codes for one of the 20 amino acids
1961	Theory of genetic regulatory mechanisms	François Jacob and Jacques Monod	Showed on a molecular level, how certain genes are activated and suppressed
1973	Efforts to create the construction of functional organisms	Stanley Cohen and Herbert Boyer	Experiments try to demonstrate the potential impact of DNA recombinant engineering
1977	Developed new techniques for rapid DNA sequencing	Walter Gilbert (with graduate student Allan M. Maxam) and Frederick Sanger	Made it possible to read the nucleotide sequence for entire genes
1970s	Nucleic acid hybridization methods and DNA probes	–	Highly specific for detecting targets

(Continued)

Table 13.1 (Continued) Historical significance-development and progression
of molecular biology/techniques

Year/decades	Discovery/event	Discoverer/company	Remarks
1983	PCR	Kary Mullis	For rapidly multiplying fragments of DNA
1985	New method to detect patient's β-globin gene for diagnosis of sickle cell anemia	Saiki et al.	–
1987	Identified human immunodeficiency virus (HIV) by using PCR method	Kwok and colleagues	The first report of the application of PCR in clinical diagnosis infectious disease
1992	Conception of real-time PCR	Higuchi et al.	Amplification in real time
1996	First application of DNA microarrays	Derisi et al.	DNA arrays to be made on glass substrates
2001	First-draft version of human genome sequence	International Human Genome Sequencing Consortium	–

disease were primarily discovered by applying RFLPs screening techniques and termed as haplotypes (both intergenic and intragenic). Thus, a data of mutations were developed but it was not possible to establish the exact relationship to diseases with mutations as they varied on the basis of ethnicity of the population (Hardison et al. 2002, Patrinos et al. 2004). In the meantime, various other methods to identify the sequence and mutations were tried and developed as mismatch mutations and were screened by the well-known DNA/DNA or RNA/DNA heteroduplexes method (Myers et al. 1985a,b), or distinction mismatched DNA heteroduplexes were also performed by using gel electrophoresis, according to their melting profile (Myers et al. 1987). Thus, a number of mutations or SNPs data base was generated which further paved the route to design synthetic oligonucleotides which can differentiate and recognize allele-specific probes onto genomic Southern blots which were utilized to detect β-thalassemia mutations (Orkin et al. 1983, Pirastu et al. 1983). It was only after a few decades that molecular diagnosis entered its golden era with the discovery of the most powerful molecular biology tool since cloning and sequencing, the polymerase chain reaction (PCR).

Molecular diagnostic techniques in identification and characterization

Specificity and sensitivity are two important tools in diagnosis as identification of microbes in microbial diseases, particular genetic sequences in genetic diseases, and protein levels are essential for the management of these patients. Classical molecular techniques such as normal PCR and blotting played a huge role in diagnosis in the past, but presently there are new techniques like the gene and peptide sequencer, real-time PCR, and microarrays that act more precisely and specifically without consuming much time.

Polymerase chain reactions

The first discovery of polymerization (Mullis and Foolna 1987) and its rapid optimization by a thermo stable Taq DNA polymerase which is extracted from *Thermus aquaticus* (Saiki et al. 1988) and PCR has transfigured molecular diagnostics. The distinguishing attributes of PCR is that a small amount of DNA can be amplified into a large amount of DNA by using an appropriate primer which is possible mainly due to its exponential amplification. Thus, this technique permits the characterization of a known mutation within a few hours. This quality of PCR actually reformed molecular diagnosis and allowed it to cross the threshold of the clinical laboratory for the provision of genetic services, and be employed in the screening of mutations, carriers, prenatal diagnosis, etc. PCRs working is based on an enzymatic process, which involves various repeated cycles and each cycle consists of various phases of DNA denaturation, primer annealing, and extension of the prime DNA sequence. Each cycle exponentially replicates the targeted DNA sequences thus accumulation of several copy numbers is possible in a very short time. The polymerization reaction involves the two primers (forward and reverse), Taq polymerase and this *in vitro* amplification process is temperature dependent, so the whole amplification reaction can be categorized into:

1. Denaturation: In this phase, the temperature is raised up to 95°C and the separation of the targeted DNA double strand into two single strands takes place.
2. Annealing: In this phase, the temperature reduces to ~55°C and two specific oligonucleotide primers attach to the DNA template complementarily.
3. Extension: In this phase, the temperature rises to 72°C, the DNA polymerase proceeds to extend the primers at the 3' terminus and synthesizes the complementary strands along 5' to 3' terminus of each template DNA using deoxynucleotides and magnesium chloride that is present in the reaction mixture.

Thus, after extension the whole process is repeated several times and ultimately each strand of the DNA produces a new DNA template as mentioned in the reaction above.

Multiplex PCR

In this method, two or more DNA templates can be amplified by utilizing two or more primers in a single tube simultaneously. This technique is useful in the detection of a few diverse bacteria in one PCR reaction. The important thing for this technique is that the primer pairs should be specific to the target DNA and PCR products should be in different sizes. Thus, this method saves time and reagents and amplifies the two different gene sequences by implying only two corresponding sets of DNA.

Reverse transcription-polymerase chain reaction

In this technique, the synthesis of complementary DNA from RNA is done by utilizing a reverse transcription (RT) first, and then the corresponding cDNA is amplified by using a targeted primer in PCR. Currently, this technique is a more precise, accurate, and sensitive form of RNA detection and quantitation. It is most useful in the detection of viruses and the viability of microbial cells by evaluation of microbial mRNA. The invention of PCR also has set the platform for the design and development of many mutation-detection approaches, based on amplified DNA. Commonly, PCR is applied for the amplification of the DNA fragments to be assessed, which is a part of the characterization method. In the beginning, the first effort was to use the restriction enzymes (Saiki et al. 1985) or oligonucleotide probes, immobilized onto membranes, or in solution (Saiki et al. 1986) in order to characterize the presence of genetic variation, in particular the sickle cell disease-causing mutation.

Real-time PCR

PCR machines and amplification processes evolved continuously in the last decade of the twentieth century and in 1993, Higuchi et al. (1992) first demonstrated real-time PCR, which is a simple and quantitative assay for any amplifiable DNA sequence. The precision and accuracy of this method is greater than earlier methods as this method utilizes fluorescent labeled probes for detection, confirmation, and quantification the PCR products as they are being produced in real time. The real-time PCR method is a more rapid process than traditional PCR assays and the hybridization of specific DNA probes takes place as the amplification reaction proceeds. A fluorescent dye is coupled to the probe and fluoresces only when the hybridization takes place. The amplified product is monitored automatically by fluorescence in real time, so the post PCR processing is not required for quantification and characterization. Recently, commercial automated real-time PCR systems have been available (LightCycler and TaqMan) and these machines utilized SYBR Green fluorescence chemistry which binds nonspecifically to double-stranded DNA produced during the PCR amplification. Others, such as the TaqMan probe, use florescent probes that bind specifically to amplification target sequences.

These techniques can be separated into three categories, depending on the basis for discriminating the allelic variants:

1. *Enzymatic-based methods.* RFLP study was historically the most extensively utilized method, and explored the alterations in restriction enzyme sites, leading to the gain or loss of restriction events (Saiki et al. 1985). Further, a number of enzymatic processes for mutation detection have been employed, based on the dependence of a secondary structure on the primary DNA sequence. These methods employed the activity of resolvase enzymes T4 endonuclease VII, and recently, T7 endonuclease I to digest heteroduplex DNA formed by annealing wild type and mutant DNA (Mashal et al. 1995). Digested fragments point out the existence and the position of any mutations. Further, chemical reagents were utilized to screen out the variation (Saleeba et al. 1992). One more enzymatic method had shown potency to detect mutations by oligonucleotide ligation assay (Landegren et al. 1988). In this assay, two oligonucleotides are hybridized to complementary DNA stretches at sites of possible mutations. The oligonucleotide primers are designed in such a way that the 3" end of the first primer is immediately near the 5" end of the second primer. Therefore, if the first primer matches completely with the target DNA, then the primers can be ligated by DNA ligase. Moreover, if a mismatch occurs at the 3" end of the first primer, then no ligation products will be obtained.

2. *Electrophoretic-based techniques.* This subcategory is illustrated by a number of different approaches intended for the screening of known or unknown mutations, based on the mobility under the denaturing condition, the screening of mutant allele can be obtained. Single-strand conformation polymorphism (SSCP) and heteroduplex (HDA) methods (Orita et al. 1989) were also used to trace the spot molecular of defects in genomic loci. In combination with capillary electrophoresis, SSCP, and HDA analysis, it now offers the good opportunity of having an excellent, accurate, sensitive, simple, and fast mutation-detection platform with low processing costs and most intriguingly, the potential of easily being automated, thus allowing for high-throughput analysis of a patient's DNA.

3. *Solid phase-based techniques/hybridization or blotting techniques.* These techniques are currently the key of modern mutation-detection technologies, since they have the extra advantage of being automated and hence are highly recommended for high-throughput mutation detection or screening.

An appropriate, rapid, and precise method for the detection of known mutations is reverse dot-blot, originally developed by Saiki et al. (1989), and employed for the detection of β-thalassemia mutations. The method includes the utilization of oligonucleotides, bound to a membrane, as hybridization targets for amplified DNA. Relentless progress has given rise to allele-specific hybridization of amplified DNA (PCR-ASO), on filters and newly extended on DNA oligonucleotide microarrays for high-throughput mutation analysis (Gemignani et al. 2002, Chan et al. 2004).

Microarrays

A microarray is an assembly of improved characteristics of microscopic technique, wherein DNA is hybridized with a target molecule for quantitative (gene expression) or qualitative (diagnostic) of large numbers of genes concurrently or to genotype multiple regions of a genome. Each DNA spot covers approximately picomoles (10^{-12} moles) of a definite DNA sequence, known as probes (or reporters). Improvements in fabrication, robotics, and bioinformatics microarray technology is constantly refining its efficiency, discriminatory power, reproducibility, sensitivity, and specificity. The enhancements have permitted the evolution of microarrays from the solid research bench site to the bedside in clinical diagnostic applications. Microarrays can be differentiated on behalf of the features such as the nature of the probe, the solid-surface support used, and the specific method used for probe addressing and/or target detection (Miller 2009). The fundamental principle behind microarrays is the hybridization between two DNA strands, one unknown and another known probe, which are the characteristics of complementary nucleic acid sequences to exclusively pair with each other by making hydrogen between nucleotide base pairs. A larger number of complementary base pairs in a nucleotide sequence means stronger and tighter noncovalent association between the two strands. It can be understood as probes are synthesized and immobilized as distinct features or spots and each feature includes millions of similar probes. The target is fluorescently labeled and then hybridized to the probe microarray. A fruitful hybridization phenomenon between the labeled target and the immobilized probe will report an increase of fluorescence intensity over a background level, which can be evaluated by using a fluorescent scanner (Miller 2009). The total strength of the signal, from a spot (feature), depends on the amount of target sample binding to the probes present on that spot. *In situ*-synthesized arrays are extremely high-density microarrays that utilized oligonucleotide probes, of which Gene Chips (Affymetrix, Santa Clara, CA) are generally used. The oligonucleotide probes are manufactured directly on the surface of the microarray, which is typically a 1.2-cm^2 quartz wafer. Because *in situ*-synthesized probes are generally short (20–25 bp), multiple probes per target are included to improve sensitivity, specificity, and statistical accuracy. Similar to the earlier mentioned printed and *in situ*-hybridized microarrays, Bead Arrays (Illumina, San Diego, CA) provide a patterned substrate for the high-density detection of target nucleic acids. However, instead of glass slides or silicon wafers as direct substrates, Bead Arrays rely on 3-μm silica beads that randomly self-assemble onto one of the two available substrates: the sentrix array matrix (SAM) or the sentrix bead chip (Oliphant et al. 2002, Fan et al. 2006). Unlike the other array, the exclusive feature of Bead Arrays is that they rely on passive transport for the hybridization of nucleic acids. Another type of array, electronic microarrays utilize active hybridization via electric fields to control nucleic acid transport. Microelectronic cartridges (NanoChip 400; Nanogen, San Diego, CA) use complementary metal oxide semiconductor technology for the electronic addressing of nucleic acids (Sosnowski et al. 1997). Each NanoChip cartridge has 12 connectors that control 400 individual test sites.

The innovation of DNA sequencing has drastically accelerated biological research and discovery at the molecular levels. The fast speeds of sequencing have generated the complete DNA sequences of many genomes. In 1977, two different methods for sequencing DNA were reported, that is, the chain-termination method and the chemical degradation method. In 1976–1977, A. Maxam and W. Gilbert presented a DNA sequencing method that relied on chemical modification of DNA and subsequent cleavage at specific bases. Maxam–Gilbert sequencing quickly became more established, since purified DNA could be used directly, while the initial Sanger method required that each read start be cloned for production of single-stranded DNA. However, with the improvement of the chain-termination method, Maxam–Gilbert sequencing became outdated due to its practical complexity, extensive use of harmful chemicals, and difficulties with scale-up, barring its utilization in standard molecular biology kits. Chemical treatment generates breaks at a small proportion of one or two of the four nucleotide bases in each of four reactions (G, A+G, C, C+T). Thus, a series of labeled fragments is produced, from the radio labeled end to the first "cut" site in each molecule. The fragments in the four reactions are electrophoreses parallel in acrylamide gel and separated on the basis of their size. Since, the chain-terminator method or Frederick Sanger method is more competent and uses fewer toxic chemicals and lower amounts of radioactivity than the Maxam–Gilbert method of, it quickly became the method of choice. The key principle of the Sanger method was the use of dideoxynucleotide triphosphates (ddNTPs) as DNA chain terminators.

The traditional chain-termination technique requires a single-stranded DNA template, a DNA primer, a DNA polymerase, normal deoxynucleotide triphosphates (dNTPs), and modified nucleotides (dideoxyNTPs) that terminate DNA strand elongation. These ddNTPs will also be radioactively or fluorescently labeled for detection in automated sequencing machines.

Both methods were uniformly popular in the beginning, but the chain-termination method soon became more accepted and this method is more normally used. This method is based on the principle that single-stranded DNA molecules that differ in length by just a single nucleotide can be separated from one another using polyacrylamide gel electrophoresis.

Automated DNA sequencing is keenly automated by a variation of Sanger's sequencing method in which dideoxynucleotides used for each reaction are labeled with a different colored fluorescent tag. Sequencing reactions by thermo cycling, cleanup, and resuspension in a buffer solution are performed separately before loading onto the sequencer.

Molecular diagnostics in the postgenomic era

Molecular biology has entered a new era with unique opportunities and challenges since the declaration of the first-draft sequence of the human genome in February 2001 (International Human Genome Sequencing Consortium 2001, Venter et al. 2001). These fabulous developments have inspired the intensification of research efforts in the accessible methods for mutation detection, to make available data sets with genomic variation and to examine these sets using specialized software, to standardize and commercialize genetic tests for regular diagnosis, and to improve the available technology in order to provide state-of-the-art automated devices for high-throughput genetic analysis. After the publication of the human genome draft sequence, the major thrust was to advance the existing mutation-detection tools and technologies to attain a robust, cost effective, rapid, sensitive, and high-throughput analysis of genomic variation. Since then, technology has enhanced swiftly and novel mutation-detection techniques have become available, whereas old tactics have evolved and become compatible as per the increasing demand for automated and high-throughput screening. The chromatographic screening of polymorphic changes of disease-causing mutations using denaturing high-performance liquid chromatography (DHPLC) is emerging as a new technology on demand. DHPLC works by the differential retention of homo- and heteroduplex DNA on reversed-phase chromatography in partial denaturation condition. Single-base substitutions, deletions, and insertions can be analyzed effectively by UV or fluorescence monitoring within 2–3 mins in unpurified PCR products as large as 1.5-kilo bases. These characteristics, together with its economical cost, make DHPLC one of the most influential tools for mutational analysis. Also, pyrosequencing, a nongel-based genotyping technology, offers a very trustworthy method and an attractive alternative to DHPLC. Pyrosequencing detects *de novo* incorporation of nucleotides based on the specific template. The incorporation process releases a pyrophosphate that gets converted into ATP and followed by luciferase stimulation. The light produced and detected by a charge couple device camera is translated to a pyrogram, from which the nucleotide sequence can be deducted (Ronaghi et al. 1998).

Current scenario of molecular diagnosis in various diseases

Molecular tools and techniques such as plasmid profiling, methods for genotyping RFLP, PCR, and microarrays are quickly making their way into clinical settings. PCR-based systems to characterize the biological causative agents of disease directly from clinical samples, without the need for culture, have been helpful in the quick discovery of fastidious microorganisms. Moreover, sequence assessment of amplified microbial DNA permits recognition and better characterization of the pathogen. Subspecies variation, identified by various techniques has been shown to be significant in the prognosis and diagnosis of certain diseases. Further characterization of various classes of mRNA like noncoding (lnc and snc), microRNA, and genetic variants in various cancers and other diseases like Alzheimer, Parkinson, etc., may be fruitful in management of these patients. The relentless progress in molecular diagnostic techniques has opened new vistas in the diagnosis of not only microbial diseases but also in innumerable diseases like cancer, neurodegenerative disorders, and genetic diseases.

Application of various molecular techniques in microbial diseases

Role of hybridization technique in microbial disease
Currently, this technique is used for screening DNA and RNA for microbial identification. After few modifications in classical probe hybridization, the fluorescent *in situ* hybridization (FISH) technique evolved, which is a highly valuable tool for the specific rapid detection of pathogenic bacteria in clinical samples without culture (DeLong et al. 1989, Amann et al. 1995, Trebesius et al. 1998). Another study demonstrated that in patients with *Haemophilus influenzae*, *Staphylococcus aureus*, and *Pseudomonas aeruginosa*, FISH was used for the rapid detection of microorganisms that cause acute pulmonary infections with 100% sensitivity (Michael et al. 2000).

DNA probe methods that detect *Chlamydia trachomatis* or *Neisseria gonorrhoeae* are good examples of low-cost molecular tests. For hybridization assays such as the INNO-LiPA®

Table 13.2 Molecular techniques utilized for characterization of various microbes

Organism	Techniques	Infections
CMV	Qualitative PCR Real-time PCR	CNs infection Congenital infection
Influenza and para influenza viruses	RT-PCR Real-time PCR Multiples PCR	Flu Bronchiolitis Croup
HIV	Real-time quantitative PCR (viral load detection)	HIV/AIDS
Hepatitis	Real-time PCR Hybridization	Hepatitis (chronic)
Middle-East respiratory syndrome Coronavirus (MERS-CoV)	RT-PCR Real-time PCR	MERS-CoV pneumonia Vaginitis
Group-B Streptococcus (GBS)	16s rDNA PCR	Meningitis
Neisseria meningitides *Helicobacter* species	16s rDNA PCR	Osteomyelitis
Plasmodium falciparum Methicillin-resistant *Staphylococcus aureus (MRSA)* Multidrug resistant *Mycobacterium* tuberculosis	Nested PCR Multiples PCR Real-time PCR (Gene Xpert)	Malaria Healthcare associated Infections Tuberculosis MDR-TB
Brucella spp., *Brucella abortus*, or *Brucella melitensis*	Real-time PCR	Brucellosis
Stenotrophomonas maltophila	23S rRNA based Specific oligonucleotide probes Real-time PCR	CF

Rif.TB (Innogenetics) and GenoType® MTBDR(*plus*) (Hain Life Science GmbH) line-probe assays represent a pooled good sensitivity and a specificity of 0.99 for detecting rifampin resistance in isolates or directly from clinical specimens. Amplification of the detection signal after probe hybridization improves sensitivity to as low as 500 gene copies per microliter and provides quantitative capabilities. This method is widely utilized for quantitative assays of viral load (hepatitis B virus [HBV] and hepatitis C virus [HCV]). The molecular techniques currently in use for characterization of various microbes are shown in Table 13.2.

The commercial probe generally available utilizes solution-phase hybridization and chemiluminescence for direct recognition of microbial agents in sample material like PACE2 products of Gen-Probe and the hybrid capture assay systems of Digene and Murex. These systems are easy to handle and work, have a long shelf life, and are adaptable to small or large numbers of specimens. The PACE2 products help in screening of both *N. gonorrhoeae* and *C. trachomatis* in a single specimen (one specimen, two separate probes). The hybrid capture systems identify human papillomavirus (HPV) in cervical scrapings, herpes simplex virus (HSV) in vesicle material, and cytomegalovirus (CMV) in body fluids. These methods are less sensitive in comparison to target amplification-based methods for the detection of viruses, but the quantitative results have proven useful for determining viral load, prognosis, and for monitoring response to therapy (Nolte 1999). Probe hybridization is valuable for the screening of slow-growing organisms. Identification of mycobacteria and other slow-growing dimorphic fungi (*Histoplasma capsulatum*, *Coccidioides immitis*, and *Blastomyces dermatitidis*) has been facilitated by commercially available probes.

All commercial probes for screening of organisms are produced by Gen-Probe and use acridinium ester-labeled probes directed at species-specific rRNA sequences. Gen-Probe products are existing for the culture screening of *Mycobacterium tuberculosis*, *Mycobacterium avium-intracellulare* complex, *Mycobacterium gordonae*, *Mycobacterium kansasii*, *Cryptococcus neoformans*, the dimorphic fungi, *N. gonorrhoeae*, *S. aureus*, *Streptococcus pneumoniae*, *Escherichia coli*, *H. influenzae*, *Enterococcus* spp., *Streptococcus agalactiae*, and *Listeria monocytogenes*. The sensitivity and specificity of these probes are outstanding, and they give species identification within 4–6 h. The mycobacterial probes, on the other hand, are accepted as a common method for the identification of *M. tuberculosis* and related species (Woods 2001, Dwivedi et al. 2012).

Role of microarray technique in microbial disease

PCR amplification in combination with oligonucleotide microarray, was employed to recognize *Bacillus anthracis* based on the rRNA ITS region (Nubel et al. 2004). Studies have demonstrated the application of microarrays to recognize pathogenic yeasts and molds by targeting the ITS regions in fungal rRNA genes (Hsiao et al. 2005, Leinberger et al. 2005, Huang et al. 2006). In recent times, a DNA microarray has been launched which can identify 14 commonly encountered fungal pathogens in clinical specimens collected from neutropenic patients (Spiess et al. 2007).

A microarray technique for the screening and recognition of enteropathogenic bacteria was developed, covering pathogenic *E. coli*, *Vibrio cholerae*, *Vibrio parahaemolyticus*, *Salmonella enterica*, *Campylobacter jejuni*, *Shigella* spp., *Yersinia enterocolitica*, and *L. monocytogenes* (You et al. 2008). A microarray-based multiplexed assay was also established to screen foot-and-mouth disease virus with rule-out assays for two other foreign animal diseases and four domestic animal diseases that cause vesicular or ulcerative lesions that are identical to those of foot-and-mouth disease virus infection in cattle, sheep, and swine (Lenhoff et al. 2008).

Role of PCRs (multiplex, nested/seminested, broad range, reverse transcription, and real time) in microbial disease

The marvelous capacity of PCR to amplify minute amounts (less than three copies) of specific microbial DNA sequences in a background mixture of host DNA makes it an exclusive diagnostic tool. Several microorganisms, such as *Mycobacterium tuberculosis*, pneumococci and *Burkholderia cenocepacia*, can be directly recognized by their specific PCR. Acute viral encephalitis in humans can be caused by more than a hundred viruses. Viruses which infect the central nervous system (CNS) can exclusively occupy the spinal cord (myelitis), brain stem (e.g., rhombencephalitis), cerebellum (cerebellitis), or cerebrum (encephalitis). In all these conditions, some degree of meningeal as well as parenchymal inflammation occurs. Rarely, such as in the West Nile virus (WNV) or CMV infections, polymorphonuclear cells rather than lymphocytes may be the predominant cell type, and this may give diagnostic guidelines. However, despite this variation, routine cerebrospinal fluid (CSF) studies are rarely used in diagnostics for a specific pathogen. Diagnosis of viral infections of the CNS has been reformed by the inventions of novel molecular diagnostic techniques such as the PCR to amplify viral nucleic acid from CSF (DeBiasi and Tyler 1999, Zunt and Marra 1999, Thomson and Bertram 2001). The technique is fast and reasonably priced and has become an integral component of diagnostic medical practice in the United States and other developed countries. PCR is also fruitful in research to distinguish active versus postinfectious immune-mediated disease, recognize determinants of drug resistance, and examine the etiology of neurologic disease of an uncertain cause.

Further, real-time PCR is a more sensitive and precise method than conventional simple PCR, as it require significantly less time and reduces contamination risks. *Neisseria meningitidis* (Nm) is the etiologic agent of epidemic bacterial meningitis and quickly fatal sepsis throughout the world. For its recognition from clinical and carriage specimens, a very rapid, sensitive, and specific real-time PCR assay was developed and validated. This novel diagnostic tool will be very fruitful in confirming isolate species identification and characterizing meningococci and Nm from carriage and clinical specimens, regardless of the organism's encapsulation status (Dolan et al. 2011). The real-time multiplex PCR assay allows the confirmation of bacterial isolates as *Brucella* spp., *B. abortus*, or *B. melitensis* within 2–3 h and the addition of a genus specific primers-probe set assists in the identification of rare and atypical *Brucella* strains. Conservative methods for *Brucella* isolation may take days to weeks and need preparation of heavy suspensions of these highly infectious pathogens (William et al. 2004). Studies have shown that utilization of real-time PCR quantitation has proven valuable when studying the role of viral reactivation or persistence in the progression of disease as viral disease severity and viral load are linked (Lallemand et al. 2000, Nitsche et al. 2000, Tanaka et al. 2000, Najioullah et al. 2001).

Real-time PCR assays have also noticeably contributed in diagnosing invasive infections of *Aspergillus fumigatus* and *Aspergillus flavus* (Brandt et al. 1998, Costa et al. 2002). Real-time PCR can do a rapid quantitation and differentiation of some more exotic pathogenic bacteria, such as the tick-borne spirochete *Borrelia burgdorferi* (Pietila et al. 2000, Rauter et al. 2002), and the methanotropic bioremediating *Methylocystis* spp. (Kikuchi et al. 2002).

Diagnosing efflux-mediated resistance in *P. aeruginosa* has also been possible using a molecular diagnostic approach. Mex efflux pumps are the underlying cause for multidrug resistance in *P. aeruginosa*, which may result in an incompetent antibiotic treatment. A study group has shown genotypic detection by semiquantitative reverse transcription PCR (RT-PCR) for mexC and mexE, and by quantitative competitive RT-PCR and real-time PCR for mexA and mexX (correlation between both methods: >88%; over expression levels ranging between 4.8 and 8.1) (Mesaros et al. 2007).

HIV qualitative nucleic acid assays

The qualitative detection of HIV nucleic acids is used in the following areas: identification of an acute infection, assurance of blood safety, and in early diagnosis of HIV in infants. An amended testing algorithm employing HIV qualitative nucleic acid testing was recently proposed to address the shortcomings of Western blot analysis (Branson 2010). The APTIMA® HIV-1 RNA qualitative assay and the Procleix HIV-1/HCV assay (Gen-Probe, San Diego, CA, USA) are both Food and Drug Administration (FDA) approved for blood-donor screening for HIV infection (Fiebig et al. 2003). They employ transcription-mediated amplification technology (Giachetti et al. 2002). HIV infected infants have a high morbidity and mortality in the first 2 years of life; thus, it is important to establish the infection status of the exposed infant in order to employ appropriate antiretroviral therapy (ART) early in the course (Resino et al. 2006). The persistence of maternal antibodies directed against HIV in exposed infants up to 18 months of age prevents the use of antibody-based assays for the early diagnosis of HIV infection. Qualitative nucleic acid assays for the detection of HIV proviral DNA, (Sherman et al. 2005), viral RNA, and total nucleic acids (Rouet et al. 2001) have become the investigations of choice for diagnosis in infants born to HIV-1-infected mothers (Okonji et al. 2012).

HIV-1 infection results in lifelong viral persistence. In chronically HIV infected patients, HIV RNA viral load in the plasma and CD4 T-cell numbers are two important markers used to guide ART initiation, monitor treatment effectiveness, determine clinical progression (Ford et al. 2009), and treatment regimens (Kitchen et al. 2001). An HIV RNA level below the detection limit is a sign of ART efficacy and excellent drug intake adherence by patients (Ford et al. 2009). HIV-1 viral load estimation is performed with HIV RNA amplification by RT-PCR, nucleic acid sequence-based amplification (NASBA) or branched chain DNA tests. Typically, viral load falls by at least 10-fold in 1 month after the initiation of ART and below the detection limit by 4–6 months, usually to less than 50–75 copies/mL (Kitchen et al. 2001). New microarray techniques such as from Affymetrix (Santa Clara, CA, USA), have allowed host transcriptome analyses in individuals infected with HIV-1 (Wang et al. 2004). A comprehensive review of the 34 studies involving HIV-1 and microarrays from 2000 to 2006 concluded that important data obtained in these studies on HIV-1-mediated gene expression effects provided new insights into the intricate interactions occurring during infection (Giri et al. 2006). Multiple studies have demonstrated progress in expanding the pool of target genes and understanding the functional correlates of gene modulation to HIV-1 pathogenesis *in vivo* (Rotger et al. 2011). Such host transcriptome profiles will be utilized in future for the evaluation of disease progression and prognosis. The precision of transcriptome analyses will be significantly improved through the added resolution of the RNA-Seq approach using deep-sequencing technologies for transcriptome profiling (Wang et al. 2009). Currently Gen-Probe's (APTIMA) HIV-1 RNA qualitative assay (Aptima 2009), is the only molecular assay that is FDA approved as a confirmatory test for diagnosing HIV-1 in samples that test reactive for HIV-1 antibodies (Stevens et al. 2009).

Role of sequencing in microbial disease

By sequencing broad range PCR products, it is possible to spot DNA from almost any bacterial species. Only after exploring the nucleotide sequences by any device when comparison is done with known sequences in GenBank or other databases, is identification of unknown bacteria possible. Since the 1990s, 16S rDNA sequencing has become a significant tool in analysis, which is used more commonly in microbial recognition, especially for nonculturable, fastidious, and slow-growing pathogens, or after antibiotics have been administered to the patient. This technique is becoming more fruitful for detecting microbes, thus it is providing a helping hand to fight infectious diseases. Additionally in such a condition, a consistent screening of bacteria in specimens should be critical in diagnosis and planning therapy. So, PCR is the utmost reliable assay for screening microbes in clinical specimens. Latest studies of microbial species have demonstrated considerable genetic diversity within microbial species with a very few or one nucleotide change in the whole genome. The main obstacles in the fast screening, recognition, and characterization of microbial species lies in the accurate recognition of traits, species, subspecies and genus, or a combination of traits, which is exclusive to a specific microbial strain (Cowan et al. 2003, Lipkin 2008). Traditional laboratory methods commonly utilize different types of phenotypic assays, as serotyping based assay for identification and characterization, although this approach is only applicable to the organisms that can be cultured in a laboratory. DNA-based assays that identify known genomic signatures based on 16s rRNA are fast and trustworthy for screening of microbial pathogens in an accurate manner (Van Ert et al. 2007, Zwick et al. 2008). Recent upgradation in gene sequencing techniques (Mardis 2008) makes more nucleotide sequence consideration at a much lower cost per base, it does not require any prior knowledge of strain-specific variants. New DNA sequencing machines are enabling novel approaches to discover and characterize microbial genomes (Srivatsan et al. 2008), and further, they simultaneously enhance our knowledge of the natural genetic diversity exist in microbial populations (Chen et al. 2010). Simple orthodox diagnostic tests based on *B. anthracis* markers have the capability to eliminate these organisms as nonanthracis isolates, while the sequencing can reveal the presence of virulence factors similar to those present in *B. anthracis* (Steinberg et al. 2008, Morozova and Marra 2008).

Role of ribosomal RNA in microbial diseases

In a few cases, when we may not have any information regarding a bacterial strain, that is, it is a stranger in the sense of an accessible organism's database, then amplification of DNA encoding ribosomal RNA genes in conjunction with DNA sequencing of the amplicon is very important (Kolbert and Persing 1999). In bacteria, the 16S ribosomal RNA (Rrna) out of three genes which make up the rRNA functionality, that is, 5S, 16S, and 23S rRNA has been most commonly applied for screening and recognition processes, since these regions are highly conserved (Woese 1987). Highly variable portions of the 16S rRNA sequence offer an important phylogenetic relationship with other bacteria. Since 16S rRNA molecules have critical structural constraints, certain conserved regions of sequence are found in all known bacteria. "Broad-range" PCR primers are focused to highly conserved regions of the 16S and 23S rRNA genes and these may either be universal or specific, targeting a specific genus, for example, *Bartonella* spp. (Houpikian and Raoult 2001), *Chlamydia* spp. (Madico et al. 2000), *Tropheryma whipplei* (Geissdorfer et al. 2001), *Mycobacterium* spp. (Roth et al. 2000), and *Salmonella* spp. (Bakshi et al. 2002), and if still any interference is seen, then sequencing analysis may be utilized. Both the techniques are able to recognize and characterize these conserved bacterial 16S rRNA gene sequences and used to amplify intervening, variable, or diagnostic regions, without the need to know any prior sequence or phylogenetic information about the unknown bacterial isolate.

Genetic diseases and disorders and molecular diagnosis

The interface between basic science and clinical practice has a close linking which has become even more obvious in the past decades with the remarkable rate of development of molecular genetics.

Fragile X syndrome

The trinucleotide repeat nucleotide sequences positioned within the transcribed region of a gene can enlarge, by a process of vigorous mutation, by "strand slippage" mechanism during DNA replication and eventually compromise the function of the gene dynamic mutations of CGG triplets that give rise to folate-sensitive fragile sites of human chromosomes. The first of these fragile sites to be cloned and characterized, which is known for the most general type of inherited mental retardation, was the fragile X syndrome. It is mainly defined by variable level of mental retardation, modest facial dimorphisms, and macroorchidism in adult males. It has been reported several years back as an X-linked Mendelian condition presenting abnormal and confusing characteristics of inheritance, penetrance, cytogenetic expression, and clinical variability. The major populations of the males who have fragile X mutation (FRAXA) are influenced by mental retardation and represent a cytogenetically inducible fragile site, but 20% of the obligate male carrier, the so called "normal transmitting males," are both clinically and cytogenetically negative. These characteristics features are now known as FRAXA dynamic mutation and the inactivation of the FMR1 (fragile X mental retardation 1) gene, coding for the RNA-binding protein (FMRP), have been reported in human brain and testes. At current 59 untranslated region of the FMR1 gene harbors, a series of CGG triplet repeats which is highly polymor-phic in normal individuals; repetition of these triplets have shown variations from 4 to about 54 (Brown 1996). FRAXA mutation can be demarcated as a "permutation" when the repeats cross the threshold of normal limits to a size of about 200 triplets that may be car-rier females and normal transmitting males, despite the fact that structural modification, gene expression, and protein functions remain intact. The altered allele transmitted by a permutated mother to her offspring rarely showed any variation in number of repeats, that is, they remain intact (Brown 1996).

Molecular diagnosis of FRAXA mutations

Direct molecular diagnosis of fragile X syndrome can be performed by screening and exami-nation of abnormal expansion of the CGG repeats and characterization on the basis of degree of the FMR1 promoter, a reflection of the residual function of the gene. Generally, there is global consensus on the point that DNA-based testing for the FRAXA mutation should be performed and confirmed by use of two techniques: PCR amplification and Southern blot/hybridization. This is a general way to screen and manage the molecular diagnosis of FRAXA mutations that is regularly performed by diagnostic service providers for the screen-ing of various types of mental retardation of unknown cause. An alternative way is Southern blotting of genomic DNA digested with a combination of restriction enzymes that includes methylation-sensitive rare cutters, followed by hybridization to a suitable labeled probe. This is more a laborious and time consuming procedure, and it is not so accurate and precise in distinguishing small permutations from large normal alleles, though it can provide an over-view of the diagnostic conditions. The initial technique, PCR amplification of the repeats, has the chief advantage of requiring less time, labor, and can be used on small amounts of DNA. For more accuracy, sensitivity, and specificity real-time PCR could be utilized.

Familial cancer syndromes: The von Hippel-Lindau disease

The molecular genetics of cancer has been elaborated due to relentless work in this field with the discovery of genes involved in these processes directly or indirectly from onco-genes to tumor suppressor genes (Strachan and Reed 1996). Oncogenes are transformed version of proto-oncogenes, which are generally needed for growth and development of cells and tissues. Tumor suppressor genes or anti-oncogenes act as inhibitors of cell prolif-eration. Tumor suppressor genes (TSG) are recessive in nature and for their expression the mutations in the both alleles are required, whereas in proto-oncogene the mutations in a single allele is sufficient to express (Knudson 1971). Tumor suppressor gene mutations are transferred as constitutional and are known to be accountable for many familial cancer syndromes. The diverse jobs of known TSG, that include DNA-binding transcriptional factors (p53, WT1), genes which modify transcriptional regulation (RB, APC, BRCA1), genes that are responsible for the functioning of signaling pathways (NF1), or genes that have role in mRNA processing (VHL), intensely support that they may actually be holdups in several cellular pathways. Mutations of a gene mapped to the short arm of chromosome 3 (3p25-26), have been now known to be a cause for the von Hippel–Lindau disease (Latif et al. 1993), a highly penetrant, dominantly inherited syndrome described by susceptibility to a variety of benign and malignant tumors. Recently it is recognized by retinal and CNS hemangioblastomas, pheochromocytomas, and renal and pancreatic cancer (Neumann et al. 1995). In von Hippel–Lindau disease both sexes are involved, with a 90% penetrance at the age of 65 years. The VHL gene is 1810 bp, having an open reading frame (ORF) of 852 bp and two in-frame starting codons; it is composed of three exons, with exon 2 alternatively spliced. The various alternative transcripts of the genes have been shown in different tissue and it also depends upon a developmentally selective manner (Latif 1993). The study of the normal function of the VHL gene provides an extensive knowledge mainly in the field of gene transcription: VHL acts as a competitor of Elongin (SIII), one of the transcription elongation factors (Kibel et al. 1995), and can modulate vascular endothelial growth factor and neo-angiogenesis, upon which tumor growth is dependent (Wizigmann-Voos et al. 1995). The evaluation is done by using PCR amplification of partially overlapping genomic DNA fragments covering the entire coding sequence of the VHL gene and including exon-flanking intronic regions and then this product is run on polyacrylamide gel to find out any mutation on the basis of the change in molecular weight. The SSCP method precisely screens out single-base substitutions, which is the most commonly employed mutation scanning technique (Cotton 1996). It is basically oriented to the fact that for single-stranded DNA molecules to assume a three-dimensional conformation which is dependent on the primary sequence is very precise and can be easily detected on a native polyacrylamide gel. DNA fragments that show size alterations on PAGE analysis or aberrant SSCP bands are subsequently subjected to direct automated sequencing to confirm and recognize the mutation. This approach has permitted the detection of diverse mutations, in unrelated VHL patients, and to screen a case of somatic mosaicism in an asymptomatic subject (Murgia et al. 2000).

Cystic fibrosis

CF is one of the most well-known and common autosomal recessive inherited diseases in Caucasians having an incidence of approximately 1 in 2500 individuals. It is a complex multifactorial disorder influencing the pulmonary, pancreatic, gastrointestinal, and reproductive organ systems. The major cause of the disease is mutations in the gene encoding cystic fibrosis transmembrane conductance regulator (CFTR), which is a membrane chloride channel present in the apical membrane of secretory epithelia. The CFTR protein is a cyclic-adenosine monophosphate (c-AMP)-dependent channel, which is required for the activation of protein kinase A and its activation occurs only when there is an increase in the levels of c-AMP inside a secretory epithelial cell; eventually it binds to the phosphorylation site on the (regulatory) R-domain of the CFTR protein thus opening the channel (Collins 1992). The CFTR chloride channel fundamentally works as an electrostatic attractant by directing intracellular and extracellular anions toward the positively charged transmembrane domains inside the channel. The CFTR protein includes 12 transmembrane (TM) domains. Two of these (TM1 and TM6) domain attract and bind chloride (and/or bicarbonate) ions. After binding chloride ions to these sites in the pore, the mutual repulsion accelerates expulsion of the ions from the cell (Linsdell 2006).

Activation of CFTR gene is accompanied by the secretion of chloride ions out of the cell. But, in addition to chloride ion secretion, inhibition of the epithelial sodium channel (ENaC) is also regulated by CFTR (Konig et al. 2001), thus a minimal amount of sodium is absorbed into the cell, upholding a higher combined ionic gradient to permit water to depart the cell by osmosis providing fluid for epithelial tissue secretions. In CF, mucus secretions become hyperviscous and that accounts for the main features of CF. Till date, more than 1600 individual CFTR mutations have been demonstrated to cause CF (http://www.genet.sickkids. on.ca/cftr/app and http://www.hgmd.cf.ac.uk/ac/index.php). These are inactivating (loss of function) or change in expression of gene have been reported by deletions, insertions, splice site mutations, nonsense mutations as well as more than 650 missense mutations. The degree of severity and presence of the disease may depend on the type of mutation.

Testing of CF

Molecular diagnosis is only a way to screen the disease. First, the identification of two mutant allele allows a molecular confirmation for the diagnosis of CF. Moreover, the parental presence of these mutations confer a risk and should be ruled out during pregnancy and will simultaneously help in exploring the carriers of these alleles in family and close relatives. Further, there is no standard procedure for routine testing but initial screening is basically done by commercial kits available in the market that roughly allows evaluation of approximately 30 sequences variants, which are responsible for more than 90% of CF disease-causing mutations. The mutations tested should be capable of identifying at least 80% of mutations for example, at least p.Phe508del (F508del), p.Gly551Asp, (G551D), p.Gly542X (G542X), and c.489+1G>T (621+1G>T). Modern CFTR screening methods can be divided into two groups: (1) hitting on established mutations and (2) scanning methods. The current database has been extended and now it includes larger unknown CFTR rearrangements, large deletions, insertions, and duplications by semiquantitative PCR experiments, that is, multiplex ligation-dependent probe. Amplification CFTR mutations may be missed by scanning techniques, especially when homozygous, and even direct sequencing cannot identify 100% of mutations. Hidden CFTR mutations may penetrate deep within introns or regulatory regions which are not commonly evaluated. For example 3849+10kbC>T (c.3718-2477C>T) and 1811+1.6kbA>G (c.1679+1.6kbA>G), the detection of which needs careful methodologies. Mainly heterogeneity should be addressed, as it is more evident in the traditional form of CF, including a positive sweat test, but this possibly concerns less than 1% of cases. Moreover, mutations in the sodium channel nonneuronal 1 (SCNN1) genes and ENaC subunits have been shown in nonclassic CF cases where no CFTR mutations have been screened out by extensive mutation scanning (Langfelder-Schwind et al. 2014). Recently, CFTR gene analysis revealed 13 different CFTR gene mutations and 1 intronic variant that led to aberrant splicing. p.Phe508del (n = 16) and p.Arg117His (n = 4) were among the most common severe forms of CFTR mutations identified in the Indian population (Sharma et al. 2014).

Eye diseases and molecular diagnostics

Current progress in molecular diagnosis has unraveled several monogenetic and multifactorial relationships in eye diseases. In this section, we will focus on common genetic eye diseases and disorders that have known causative genes and their available measures for prevention or treatment.

Glaucoma

Presently, the prevalence of glaucoma is in more than 2% of the global population over 4 years of age. This disease is known for its heterogeneity and the fundamental molecular mechanism for glaucoma is still under exploration. Currently, 15 genetic loci have been plotted for POAG and two loci for PCG, of which mainly GLC1A (myocilin, MYOC) (Stone et al. 1997), optineurin (OPTN, GLC1E) (Rezaie et al. 2002), GLC1G (WD repeat domain 36, WDR36) (Monemi et al. 2005), and GLC3A (cytochrome P4501B1, CYP1B1) (Stoilov et al. 1997) have been reported. Mutations in MYOC are accountable for about 2%–4% of POAG cases according to European and American studies (Fingert et al. 1999, Faucher et al. 2002). In Chinese populations, prevalence of MYOC mutations is 1.1%–1.8% and OPTN mutations is 16.7% in POAG patients (Lam et al. 2000). Mutations in CYP1B1 were seen in 48% of French PCG patients, and about 20% of Japanese patients (Kakiuchi-Matsumoto et al. 2001).

Further, CYP1B1 mutations were also recognized in early-onset POAG, whereas MYOC mutations were associated with PCG. These data suggested that CYP1B1 may act as a modifier of MYOC expression and these two genes may interact through a common pathway (Vincent et al. 2002).

Age-related macular degeneration

Age-related macular degeneration (AMD) is a progressive destruction of the macula that is a major cause of central vision loss. It affects 1.5% of the overall population in Western Europe (Klein et al. 1999). Stargardt macular dystrophy (STGD) is the most general hereditary form (Klein et al. 1992). STGD is an autosomal recessive or less often dominant disorder of the retina and is responsible for early-onset macular degeneration. Mainly three genes have been associated with AMD; complement factor H (CFH), ATP-binding cassette transporter (ABCA4), and Apo lipoprotein E (APOE) in which CFH is known to be a main gene for AMD. One mutation in CFH, Y402H, was found to be linked for up to 50% of the attributable risk of AMD (Edwards et al. 2005, Klein et al. 2005, Zareparsi et al. 2005). The APOE allele ε2 also presented an increased risk for AMD while ε4 which conferred a protective role against AMD and ABCA4 mutation, T1428M, was reported and characterized in 8% of Japanese STGD patients (Klaver et al. 1998).

Retinitis pigmentosa

It is a heterogeneous group of degenerating diseases of the retina affecting the rod photoreceptors. Retinitis pigmentosa (RP) affects night blindness, loss of peripheral vision, and ultimately leads to a loss of central vision. The prevalence of RP is 1 in 3500 globally (Rivolta et al. 2002). Currently, 40 genetic loci have been linked to nonsyndromic RP, from which 32 genes have been explored [http://www.sph.uth.tmc.edu/Retnet/home.htm]. Rhodopsin (RHO) mutations are known for more than 25% of RP cases (Sohocki et al. 2001). Mutations in RP1 account for 6%–8% of adRP cases and Retinitis pigmentosaGTPase regulator (RPGR) is the most significant gene for X-linked retinitis pigmentosa (XLRP).

Retinoblastoma

The main intraocular malignancy in children is retinoblastoma (RB) which may be familial or sporadic, with an incidence of 1 in 15,000–20,000 live births in almost all populations [http://www.genetests.org/query?dz=retinoblastoma]. Generally familial, bilateral, or unilateral multifocal RB is regarded as a carrier of a RB1 germ-line mutation. RB is transferred by autosomal dominant manner with 80%–90% penetrance (Naumova and Sapienza 1994). Secondary genetic and epigenetic alterations in another gene(s) are also required for the formation and development of tumors (Lohmann and Gallie 2004). Among the Chinese population, the germ-line mutation rate in RB is 19% (8/42), with 11% (3/28) among unilateral cases. The straight genetic screening can be performed by a sequencer, after exploring the nucleotide sequence. However, optional screening methods are SSCP or confirmation-sensitive gel electrophoresis (CSGE), and DHPLC. Other techniques can also be utilized to explore the complete gene mutations, such as FISH or methylation-specific PCR. Recently, the Taqman probe based on real time has also been utilized to explore the nucleotide sequence at a particular locus. The 27 known RB1 coding exons, splice boundaries, and the promoter can be amplified by PCR followed by direct sequencing (Fan et al. 2006).

Molecular diagnosis in neurological disorders and diseases

Transmissible spongiform encephalopathies

Central nervous system (CNS) diseases transmissible spongiform encephalopathies (TSEs) or prion are caused by unconventional infectious agents which do not have any nucleic acid and also do not induce any immunological reaction in the host. As proper nucleic acid is not present in these disorders, therefore PCR is not suitable for diagnosis. Common prion diseases are: (1) CJD (Creutzfeldt–Jakob disease), (2) GSS (Gerstmann–Sträussler–Scheinker syndrome), (3) FFI (fatal familial insomnia), (4) vCJD (variant Creutzfeldt–Jakob disease), (5) scrapie, (6) CWD (chronic wasting disease), and (7) BSE (bovine spongiform encephalopathy).

Conformational change in the host protein is the main pathological event as cellular prion protein (PrPc), encoded by the prion gene PRNP, into a pathological isoform. This conformer, called PrPSc (after its first identification in experimentally scrapie-infected rodents) aggregates into amyloid fibrils and stores into neural and, often, lymph reticular cells (Wadsworth et al. 2001). Further, blood taken from vCJD patients and inoculated into susceptible mice failed to cause disease every time (Bruce et al. 2001). Numerous animal studies have shown that the pathogenesis of TSE after experimental inoculation of rodents/mice with TSE agents, PrP^{Sc} can be typically demonstrated in the CNS weeks before the expression of disease and its level enhances until the animal dies. As the rise of PrP^{Sc} corresponds to infectivity, PrP^{Sc} is commonly used as a surrogate marker for assessing the amount of infectivity in biological samples. In recent times, few approaches at preclinical levels have enough potency to be proved as good biomarkers as they can evaluate very low levels of these proteins, like chemicophysical precipitation-based protocols, affinity chromatography, or affinity precipitation techniques. The improved isolation method for PrP^{Sc} with sodium phosphotungstate (Wadsworth et al. 2001), novel molecules plasminogen (Fischer et al. 2000), and protocadherin-2 binding with high affinity to PrP^{Sc} might open new diagnosis approaches of TSEs. An exclusive technique to detect a very detectable level of PrP^{Sc} has been described by Saborio et al. (2001). Among immunological methods of PrPSc evaluation, Western blotting is the most preferable method. It provides the benefit of recognizing diverse forms of PrPSc through the analysis of the molecular mass and the relative abundance of di-, mono-, and nonglycosylated bands. These factors characterize the so-called PrPglycotype, a kind of "PrP signature," which shows a variation in diverse form of TSEs. PrPScglycotyping has been employed in differentiating TSEs (e.g., scrapie from BSE (Kuczius and Groschup 1999), sporadic from variant CJD, and for improving the classification of human TSEs (Collinge et al. 1996)). For example in sporadic CJD, the combination of the two most recurrent PrPScglycotypes (I and II) with the three probable genotypes of PrP at the polymorphic codon 129 (methionine homozygous, valine homozygous, or heterozygous) allows the subclassification of this form into six distinct groups, each of which presents distinct clinical and pathological features (Parchi et al. 1999). The drawback of this method is that it is time consuming and only a few samples can be evaluated at a time, so ELISA can surmount these problems. Moreover, progress and upgradation in immunoassay in the form of dissociation-enhanced lanthanide fluorescence immunoassay/conformation-dependent immunoassay (DELFIA/CDI) is the modern immunoassay with an ELISA format in which the evaluation system is sensitive time-resolved lanthanide fluorescence instead of chemiluminescence (Völkel et al. 2001). It can assess picograms (10–12 g) of PrPSc per milliliter and thus represents one of the most sensitive techniques for the detection of PrPSc. Another procedure named MUFS (multispectral ultraviolet fluorescence spectroscopy) illustrates proteins by their specific fluorescent pattern of emission after excitation by ultraviolet radiation and bypasses the requirement for pretreatment steps to eliminate PrPc or for antibody binding. It showed the capability to distinguish cellular from pathological prion protein, and various forms of PrPSc from different strains. Moreover, a most sensitive and specific technique which uses a fluorescent antibody with confocal microscopy in FCS (fluorescence correlation spectroscopy) can identify single fluorescent molecules in solution as they pass between the exciting laser beam and the objective of a confocal microscope, equipped with a single-photon counter. It is performed quickly and requires only small amounts of samples. The assay solution is mixed with anti-PrP antibodies tagged with fluorophores that bind strongly to PrPSc aggregates, which become highly fluorescent and easily visible against the background of monomeric PrPc (Giese et al. 2000). This technique is ~20-fold more sensitive than the Western blot. In addition, 14-3-3 proteins were reported in the CSF patients with genetic CJD having PRNP gene mutation in the codon 200 or 210, this occurrence is reported in 50% of CSF samples (Zerr et al. 2000, Green et al. 2001). Further, tau proteins, neuron-specific enolase (NSE), and S-100 also get aggravated in CJD patients. Genetic risk is very common in 70% of affected persons who were homozygous for methionine at the genetic variants at 129 site of PrP both sporadic and iatrogenic CJD (The EUROCJD Group 2001) whereas in vCJD, 100% of patients were seen as methionine homozygous (Will et al. 2000). Further, discovery of a PrP-like gene (PRNDgene) in mammalian species has shown a probability to resolve the pathogenesis of TSE; thus the people with risk could be identified (Mastrangelo and Westaway 2001). Similarly, a reduced frequency of the human leukocyte antigen (HLA) class-II-type DQ7 of HLA typing of vCJD patients in comparison with sporadic CJD cases and unaffected Caucasian British controls (Jackson et al. 2001).

Alzheimer disease and molecular diagnostics

Alzheimer disease (AD) is a very common aging-associated neurodegenerative syndrome described by irreversible loss of intellectual functions and memory. The deposition of extracellular amyloid plaques, cerebrovascular amyloidosis, and intracellular NFTs (neurofibrillary tangles) are generally seen in the pathogenesis of AD. Hyperphosphorylation of the microtubule-associated protein tau (MAPT) are responsible for the formation of NFTs, while amyloid-beta precursor protein's (APP) proteolytic cleavage generates the neurotoxic Ab peptide, which has an established role in the formation of neuritic amyloid plaques. Salvadores et al. reported that the detection of misfolded $A\beta$ oligomers could be used for the screening of AD. Finally, his group also differentiated AD from other similar neurodegenerative disorders or nondegenerative neurological diseases with a higher degree of specificity and sensitivity (Salvadores et al. 2014).

Recently, a study has shown a reduced expression of novel gene P9TLDR, which has a significant role in neuronal migration and is known for coding microtubule-associated protein, when extracted nucleic acid and its expression were compared by using PCR-select cDNA suppression subtractive hybridization (PCR-cDNA-SSH) analysis from an intracerebral brain-site-specific (AD temporal lobe versus AD occipital lobe). Further, similar results were also seen when these findings were validated by an *in vitro* AD-related cell model, amyloid-beta peptide (Ab)-treated neurons that decreased P9TLDR expression, and also associated with aggravated tau protein phosphorylation (Yokota et al. 2012).

Late-onset Alzheimer's disease

The molecular interaction and organization of cellular pathways and its components is generally deranged in this type of complex disease. Zhang et al. have explored and reported that in late-onset Alzheimer's disease (LOAD), the reconfiguration of molecular interaction structure is seen in some specific portion of the brain while studying 1647 postmortem brain tissues from LOAD patients and nondemented subjects. The immune- and microglia-specific module focused results also claimed that the gene associated with phagocytosis contains an aggravated expression of TYROBP in LOAD, which is a key regulator of phagocytosis (Zhang et al. 2013).

Molecular diagnosis of genetically transmitted cardiovascular diseases

Recent progress in the field of molecular biology and its technique have explored fundamental mechanisms involved in the pathogenesis of many cardiovascular disease conditions and their phenotypic expression, consequently, several new molecular biomarkers have been framed and formulated.

Hypertrophic cardiomyopathy

Hypertrophic cardiomyopathy (HCM) is a well-known familial cardiac disease recognized by multifactorial pathophysiology in its morphological, functional aspect, and clinical progression. This wide diversity of this disease is known by the fact that HCM may exist in all stages of life, from the infant to the old. The clinical progression is quite inconstant, as few individuals remain asymptomatic throughout life while some others may undergo severe symptoms of heart failure or die abruptly without prior symptoms or undergo progressive heart attacks. It follow Mendelian autosomal dominant trait rule for its inheritance or its expression, and its appearance is marked by a mutation in any 1 of 5 genes that encode proteins of the cardiac sarcomere: b-myosin heavy chain, (Geisterfer-Lowrance et al. 1990, Marian et al. 1995) cardiac troponin T, troponin I, a-tropomyosin (Watkins et al. 1995, Forissier et al. 1996), and cardiac myosin-binding protein C located on chromosome 14; 1; 19; 15; and 11, respectively (Bonne et al. 1995, Niimura et al. 1998). Moreover, mutations in two genes expression that have an importance in the regulation of myosin light chains have been demonstrated in a few extreme forms of HCM. This intragenic database is now elaborated with more than 100 disease-causing mutations in HCM. Mutations in the b-myosin heavy chain gene that code for contractile protein of thick myofibril filaments seems to affect about 35% of familial HCM, generally all mutations in myosin genes reported were of missense type of mutations. Similarly, cardiac troponin T mutations have been reported for about 10%–20% of familial HCM, as this protein along with tropomyosin has a key role in controlling cardiac contraction and relaxation (Forissier et al. 1996).

Long-QT syndrome

The long-QT syndrome (LQTS; Romano–Ward) is an uncommon familial disease which shows its autosomal dominant inheritance (Ward 1964), generating a risk of sudden cardiac arrest or cardiac death often seen during stress and vigorous muscular activity. Sudden cardiac arrest is mediated via ventricular tachyarrhythmias such as polymorphic ventricular tachycardia and ventricular fibrillation (Moss et al. 1991). The established fact for diagnosis of LQTS is irregular perpetuation of ventricular depolarization in which QT intervals get elongated on the 12-lead ECG.

Marfan syndrome

Marfan syndrome (MFS) first demonstrated by Antoine Marfan in 1896, is a systemic connective tissue disorder which follows an autosomal dominant inheritance (Marfan 1896). Involvement of the cardiovascular system and continuation in aortic root dilations, or valvular regurgitations further reduces life span or expectancy (Glesby and Pyeritz 1989, Roman et al. 1993). The first defect is noticed in the FBN1 gene which codes for fibrillin-1 a well-known connective tissue protein which has a significant role in the independent formation of elastic fibers. Studies have shown the contributory role of fibrillin mutations with Marfan syndrome and no locus heterogeneity is noted (Dietz et al. 1991, Aoyama et al. 1993, Milewicz et al. 1995, Nijbroek et al. 1995).

Molecular diagnosis and various cancers

Advanced molecular diagnostics has helped to solve the puzzle of carcinogenesis into significant hallmarks such as sustaining proliferative signaling, escaping growth suppressors, preventing cell death, allowing replicative immortality, persuading angiogenesis and triggering invasion, and metastasis. Genomic instability generates diversity that expedites above hallmarks of carcinogenesis. Cancerous cells occasionally have mutations in oncogenes, such as KRAS and CTNNB1 (β-catenin) (Minamoto et al. 2002). After evaluating the DNA mutations and mRNA expression of cancerous cells, certain molecular signatures can be characterized that would help clinicians to better management of these patients. Further, many antibodies array against specific protein markers have been developed; these multiplex assays could examine many markers at a time (Brennan et al. 2010). Similarly, few microRNA molecules have been characterized that can differentiate cancerous cells from normal healthy ones (Ferracin et al. 2010). This section will highlight the current progress in the molecular characterization of most frequent cancers and its mechanism.

Prostate carcinoma

Prostate carcinoma (PCa) is one of the main cancers in men influencing one third of men globally. The pathological testing (biopsy) is an invasive method for detecting carcinoma of prostate as others, at present the serum prostate-specific antigen (PSA) is used as one of the finest available cancer markers for identifying PCa at early stage. However, there are few drawbacks of PSA as it is prostate-specific and not cancer specific. In addition, its serum level also upsurges in several circumstances like benign prostate hyperplasia and prostatitis. The sensitivity and specificity of PSA is also debatable as it differs from 0.78 to 1.00 and 0.06 to 0.66, respectively (Philip et al. 2009).

Genomics/epigenetic and SNPs
Epigenetic gene regulations are exclusively noncoded heritable variations in gene expression, for example DNA methylation, histone modifications, and noncoding RNA-induced transcriptional changes. Two histone modifications HAT p300 and HDM EZH2 are potential biomarkers which have been found to be over expressed in PCa and its expression levels were also linked with diverse disease stages (Isharwal et al. 2008).

Hypermethylation and gene silencing have a well-established role in cell cycle regulation like anaphase promoting complex (APC) and Ras association domain-containing protein 1 (RASSF1a), and detoxification enzymes glutathione S-transferase Pi 1 (GSTP1). Mixed assays of GSTP1 and APC hypermethylation have shown its significance for discriminating PCa from normal to healthy with 100% sensitivity (Yegnasubramanian et al. 2011). The risk of PCa could be estimated after exploration of genetic variance of alleles in different regions of chromosomes (EHBP1, THADA, ITGA6, EEFSEC, PDLIM5, FU20032, SLC22A3, JAF1, LMTK2, NKX3, CMYC, MSMB, CTBP2, HNF1B, KLK2-3, TNRC6B, BIK, NUDT10-11) which have the capability to characterize the disease and its stage progression by changing expressions of mRNA and protein (Willard and Koochekpour 2012).

Transcriptomics

The noncoding RNA (ncRNA) is a novel field of research, the term ncRNA includes the well-studied RNAs like rRNA and tRNA, with microRNA (miRNA) including long ncRNA (lncRNA) and small interfering RNA (siRNA). Three explored lncRNAs have a capability to detect, screen, and monitor PCa (Gibb et al. 2011) with a greater degree of specificity and sensitivity: (1) PCa noncoding RNA-1 (PRNCR1), (2) prostate-specific gene 1 (PSGEM1), and (3) PCa antigen 3 (PCA3); which is also known as display 3 (DD3). Evaluation of TMPRSS2-ERG fusion (TEF) approaches have been explored by immunohistochemistry, FISH, and RT PCR and the results obtained have shown their importance in the diagnosis of PCa. Moreover, TE fusion in combination with PCA3 mRNA have also shown to be fruitful in diagnosis (Tomlins et al. 2011).

Circulating miRNA has newly emerged as a good biomarker for noninvasive screening of various tumors. Research on oligonucleotide array hybridization miRNA profiling has recognized 51 miRNAs which have the capability to discriminate between benign and malignant prostate cancers, a further 37 have shown their lower expression and 14 have higher expression in PCa. These variations in miRNAs expression lead to alternation in expression and activity of their targets in PCa (Porkka et al. 2007). Further, Mitchell et al.'s work on a mouse xenograft model have shown that miRNAs existing in PCa xenografts finally come into the circulation, results also showed that miR-141 was over expressed in malignancy of PCa patients which can screen PCa patients from healthy controls with high sensitivity and more accuracy (Mitchell et al. 2008).

Proteomics

Proteomics also has plethora of newly characterized biomolecules which showed a good power in diagnosis and prognosis. For example CGRP, VEGF, endoglin (CD105), chromogranin-A, neuron-specific enolase, interleukin-6, transforming growth factor-b, other methylated genes including RASSF1a, APC, RARB2 and CDH1, prostate-specific cell antigen, testosterone, estrogen, sex hormone binding globulin, caveolin-1, E-cadherin, b-catenin, MMP-9, tissue inhibitor of MMPs (TIMP 1, 2), progastrin-releasing peptide (ProGRP [31–98]) and PSP94, ZAG, prostasome (auto-antibodies), huntingtin interacting protein 1 (auto-antibodies), TSP-1, leptin, ILGF-1, -2, human kallikrein 2, a-methylacyl-CoA racemase (auto-antibodies), early prostate cell antigen-1, -2, GSTP1 hypermethylation, cytokine macrophage MIF, hK11, and apolipoprotein A-II for diagnosis.

Moreover, urokinase-type plasminogen activator system, prostate membrane-specific antigen, hepatocyte growth factor, MIC-1, EGFR family (c-erbB-1 (EGFR), c-erbB-2 (HER2/neu), c-erbB-3 (HER3), and c-erbB-4 (HER4) (Ramírez et al. 2008), have the capability to predict diagnosis as well as prognosis of the patients suffering from prostate cancer. Recently, Dwivedi et al. have proposed a circulating serum interleukin-18 as a good diagnostic biomarker and interleukin-10 for prognosis. This group further explored the promoter genetic variations of these interleukins and correlated them with corresponding serum levels (Dwivedi et al. 2011, 2012, 2013, 2014a,b, 2015a,b). Metastatic castration resistant PCa (MCRPCa) and metastasis-associated protein-1 (MTA-1) have also been evaluated for their significance in vascularization of an advancing tumor (Kai et al. 2011). The role of WNT5A, EZH2, MAPK signaling cascade members, AR, and various androgen metabolism genes expression were reported to be over expressed in malignancy of PCa whereas c-FOS and jun B have shown their decreased expression, therefore they have potency to work as a good biomarker (Kattan et al. 2003).

Breast cancer

Breast cancer is the most common cancer in women worldwide. In 2004, the database has shown 1.15 million new breast carcinoma patients and approximately 500,000 deaths globally (Parkin et al. 2005). The specific method of immunology to screen breast cancer cases is by characterizing hormone receptor by immunohistochemistry (IHC). IHC employed antibodies and enzymes, such as horseradish peroxidase, to stain tissue sections for the targeted identification of specific tumor antigens (Taylor et al. 1994). IHC decides the percentage of positive nuclei along with intensity of staining in individual nuclei. However, the interpretation of IHC differs pathologist to pathologist due to lack of proper protocol and standardization (Layfield et al. 2000). Instead of such discrimination in staining intensity, it is still the most common screening method in breast cancer. Additionally, the human epidermal growth factor receptor 2 (HER2) oncogene of EGF family that is used in evaluation of primary invasive breast cancer can also use in predicting the prognosis of these patients (King et al. 1985). Further, gene amplification of HER2 has been reported in 10%–40% of primary tumors and HER2 protein over expression is seen in about 25% of individuals suffering from breast cancer (Paik et al. 1990).

HERmark™ assay

This is a new upgraded method developed by monogram biosciences to well differentiate breast cancer patients from the normal. This assay evaluate the total HER2 protein (H2T) and functional HER2 homodimer (H2D) levels present on the cell surface of formalin-fixed paraffin-embedded (FFPE) breast cancer tissue. This assay has a dual antibody system, one fluorescent tagged antibody for providing an amplified signal which is activated by the second antibody having a photo activated molecule. Capillary electrophoresis is basically utilized to quantify the signals generated by these fluorescent tags. It only suggests whether the individual sample is HER2-negative, -positive or -equivocal based on quantified signals of HER2 protein levels provided in some numerical value (HERmark, monogram biosciences, Inc. www.hermarkassay.com).

Transcriptomics-based biomarkers: Theros H/ISM and MGISM

Theros H/ISM is a newly explored test that asses the ratio of HOXB13:IL17BR gene expression and that has the capability to predict the efficacy of treatment in breast cancer patients treated with tamoxifen. An increased two-gene ratio is linked with tumor aggression and resistance to tamoxifen treatment (Jansen et al. 2007). Theros MGISM is also another test that utilizes a panel of five-gene expression index to categories ER+ breast cancer patients into greater or lower risk of recurrence by subgrouping grade 2 (intermediate proliferative) tumors into grade 1-like or grade 3-like outcomes (Ma et al. 2008).

MammaPrint™

Recently another molecular test named MammaPrint test has also been introduced that can screen tumor recurrence in breast cancer patients. MammaPrint includes 70 gene signature panels that can predict the survival of patients with node-negative breast cancer (Buyse et al. 2006). For this test, a fresh tissue sample usually 3 mm in diameter is isolated and preserved in RNA later or RNA-stabilizing solution and sent to the Agendia laboratory in Amsterdam for analysis. Thus, amplified extracted RNA sample is amplified and then hybridized with standard of the MammaPrint microarray to find the 70-gene expression profile. It has shown a greater degree of precision and accuracy for prediction of tumor recurrence (p < 0.0001) (Glas et al. 2006). U.S. FDA approved the molecular testing of MammaPrint for use on freshly frozen tissue in 2007.

Oncotype DX®

Quantitative real-time PCR and microarray-based Oncotype DX also includes a 21-gene expression assay panel that categorized the patients into various groups on the basis of the probability of benefiting from chemotherapy. This test utilizes FFPE tissue blocks that can be shipped from anywhere in the United States and worldwide. Currently, Oncotype DX is the standard breast cancer assay for patients with early-stage (Stage I or II), node-negative, and ER+ invasive breast cancer. The Oncotype DX test actually provides a recurrence score from 0 to 100 and predicts the recurrence probability within 10 years of the original diagnosis. Further it can be grouped into low, medium, or higher risk, so that the better treatment plan of early diagnosis can be tailored specifically to each patient. American Society of Clinical Oncology (ASCO) and the National Comprehensive Cancer Network (NCCN) have permitted inclusion of the test into their protocols and guidelines (Cronin et al. 2007).

MicroRNA

Mi-RNA deregulation was firstly demonstrated in 2005 in breast cancer, (Iorio et al. 2005). After this exploration, many reports have been published on various mi-RNAs and their roles in breast cancer have been proposed. MicroRNA-21 has been explored in several studies and its significant enhanced expression was reported in breast cancer cell lines and human tissue in comparison to normal cells and tissues (Iorio et al. 2005, Zhu et al. 2007, Lu et al. 2008). Moreover, Iorio et al. reported miR-10b was found consistently decreased in breast cancer cells when compared with primary human mammary epithelial cells (HMECs) (Iorio et al. 2005). However, in a later study, miR-10b expression was found aggravated in metastatic cancer cells. Well-designed *in vivo* and *in vitro* studies have demonstrated that miR-10b increased expression endorses cell migration and invasion (Ma et al. 2007). Further, several studies reported a significant association between miR-206 and estrogen receptors expression in breast cancer. Moreover, Iorio et al. (2005) were the first to report that estrogen receptor (ER)-tumors which showed a higher expression of miR-206, similarly miR-125a and miR-125b, were first reported to be significantly decreased in HER2-positive breast cancers. Further, *in silico* studies confirmed their target site at the 3′-UTR regions of HER2 and HER3 from the earlier discussed miRNAs (Mattie et al. 2006). A culture study also showed that increased expression of miR-125a or miR-125b in an ErbB2-dependent cancer cell line (SKBR3) inhibited HER2 and HER3 transcription and translation, which ultimately suppressed cell motility and invasiveness (Scott et al. 2007). Recently, new advancements in liquid biopsy-based biomarkers, especially DTCs and CTCs bearing molecular signature have the capability to behave as potential biomarkers and can discriminate between localized breast cancer and a metastasizing one (Dwivedi et al. 2014a).

Lung cancer

Currently, lung cancer is the top most cause of cancer-associated death globally. In lung cancer like other malignancies, carcinogenesis narrates to the activation of growth promoting proteins (for example, v-Kiras2 Kirsten rat sarcoma viral oncogene homolog [KRAS], ALK, HER2, MET, BRAF, MEK-1, and are rearranged during transfection [RET]) in addition to inactivation of tumor suppressor genes (for example, P53, phosphatase with tensin homology [PTEN], and LKB-1 [Larsen and Minna 2011]). Further, a large scale exome sequencing study of 31 non-small cell lung cancer (NSCLC) recognized 727 newly mutated genes. Moreover, genomic studies have confirmed KRAS, EGFR, and BRAF mutations and other targetable mutations of signaling pathways include genetic variations in JAK2, ERBB4, RET, fibroblast growth factor receptor 1 (FGFR1) (Hammerman et al. 2011, Pao and Girard 2011, Shaw and Solomon 2011, Sriram et al. 2011, Cancer Genome Atlas Research Network 2012) and discoid in the domain receptor 2 (DDR2) (Hammerman et al. 2011). Lung cancer mutations have been recognized in v-Ki-ras2 Kirsten rat sarcoma viral oncogene homolog (KRAS), APC, ACVR1B, AKT1, BAI3, BAP1, MEK, ALK, ROS1 and the parallel phosphatidylinositol 3-kinase (PI3K) pathway, thus these mutations help in providing new therapy targets along with risk predisposition. The importance of tumor suppressor genes is gradually accepted with mutations in TP53, PTEN, RB1, LKB11, and p16/CDKN2A although the prevalence of these mutations varies with ethnicity (Pao et al. 2011). Genetic variation of EFGR varied from approximately 20% in Caucasians to 40% in Asian patients with NSCLC (Sriram et al. 2011). These ethnic differences in frequency of mutations in EFGR along with ALK, KRAS, MET may be a reason for varied results with various therapies and treatment (Pao et al. 2011). The ALK rearrangement exists in up to 7% of patients of lung adenocarcinoma (ADC) among Asians, and it is established that the EML4-ALK rearrangement is more sensitive to a specific tyrosine kinase inhibitor named crizotinib and these variations affect drug sensitivity (Shaw and Solomon 2011, Travis et al. 2013). Thus various categories can be subgrouped on the basis of their ethnicity, mutations, and rearrangement, and ultimately this will help in designing targeted therapy and treatment for better management of these patients (Zakowski et al. 2009). The response rate in lung adenocarcinoma is up to 70% in patients having active EGFR mutations (Mok et al. 2009) and median survival is usually 9–11 months in patients bearing various mutations in tyrosine kinase inhibitors (Zhou et al. 2011, Yang et al. 2012).

Oral cancer

Oral cancer is among the 10 most common cancers worldwide, which generally exist in older males. It is supposed that initial screening of potentially malignant leucoplakia (PML) can decrease death worldwide (Sankaranarayanan 2005a,b, Massano et al. 2006). The most commonly used predictive molecular markers so far are TSG p53 protein expression, DNA ploidy, and loss of heterozygosity; LOH) in chromosomes 3p or 9p in oral squamous cell carcinoma (OSCC) development. Moreover, cell proliferation (Ki-67 antigen) and apoptosis (Bax, Bcl-2) molecular biomarkers are utilized currently in which apoptotic Bcl-2 expression decreased in dysplastic and early invasive lesions and Ki-67 expression enhanced sharply in early stages of OSCC (Derka et al. 2006).

The conventional method of brush and scalpel biopsy were utilized in some studies but were not so appealing although the concept of a saliva test to diagnose OSCC is even more appealing (Li et al. 2004a,b, Hu et al. 2006, 2007, Park et al. 2006, Wong 2006) in which the promoter hypermethylation noticed in tumor suppressor gene p16,

O6-methylguanine-DNA-methyltransferase, and death-associated protein kinase have been screened in the saliva of head and neck cancer patients (Rosas et al. 2001). Forensic science has already noted the presence of a number of mRNA fragments with histatin 3, salivary specific statherin, and the proline-rich proteins PRB1, PRB2, PRB3spermidine and N1 acetyl transferase (SAT), β-actin, and glyceraldehyde-3-phosphate dehydrogenase (GAPDH) (Juusola and Ballantyne 2003). Various mRNAs molecules have been explored in the saliva of 300 OSCC patients and compared with that of normal healthy individuals, it showed a higher expression of a few mRNA in cancer patients and was exclusively able to discriminate with about 85% accuracy. Braakhuis et al. reported four salivary mRNAs OLF/EBF associated zinc finger protein [OAZ], SAT, IL8, and IL1b collectively can differentiate with a power of 91% sensitivity and specificity for OSCC to normal healthy individuals (Braakhuis et al. 2003). Common alternations in host DNA of dysplastic or cancer cells marks with point mutations, deletions, translocations, amplifications and methylations, cyclin D_1, epidermal growth factor receptor (EGFR), microsatellite instability, and HPV existence. Further allelic loss on chromosomes 9p has been screened in OSCC (Nawroz et al. 1994), and few mitochondrial DNA mutations have also been recognized by sequencing of 67% saliva OSCC patients (Fliss et al. 2000). Moreover, Rosas et al. recognized aberrant methylation of three genes p16, MGMT, or DAP-K in OSCC and abnormal promoter hypermethylation was also noticed in the saliva sample of 65% of OSCC patients (Rosas et al. 2001). Expression of several mRNA molecules were found aggravated in the saliva of OSCC patients (Li et al. 2004a) as mRNA expression of: (1) IL8 (interleukin 8) which have established a role in angiogenesis, replication, calcium-mediated signaling pathway, cell adhesion, chemotaxis, cell cycle arrest, and immune response, (2) IL1B (interleukin 1B) that is well characterized in signal transduction, proliferation, inflammation, and apoptosis, (3) DUSP1 (dual specificity phosphatase 1) is known for protein modification, signal transduction, and oxidative stress, (4) H3F3A (H3 histone, family 3A) required for DNA-binding activity, (5) OAZ1 (ornithine decarboxylase antizyme 1) significant contribution in polyamine biosynthesis, (6) S100P (S100 calcium binding protein P) have a role in protein binding and calcium ion binding, and (7) SAT (spermidine/spermine N1-acetyltransferase) which is needed for transferase activity were found significantly enhanced in OSCC patients when compared with healthy controls (Zimmermann et al. 2007).

Future perspectives

Molecular diagnostic tools as DNA or RNA sequences have been explored due to relentless progress in the field of molecular biology and biotechnology, and this knowledge is further going to provide a helping hand in the field of diagnosis of microbial, cancerous, neurological, and genetic diseases. The current era has been blessed with several automated and advanced systems that have been utilized in the amplification and detection of the nucleic acid sequence of microbial agents and they can also be used in the characterization of changes in the expression pattern of the few biomolecules during the course of the disease. These automated systems like PCR and microarray have shown a better assay efficiency and sensitivity in characterization of targeted biomolecules. It is established since the time of Gregor Mendel that various factors are responsible for development of characters and after the draft of human genome, the picture of the genotype to phenotype relationship appears more vivid and diverse. The genotype and phenotype association can be explored and characterized more smoothly than earlier due to availability of modern sophisticated instruments like microarray, sequencing and *in silico* tools, techniques, and approaches. Now, modern molecular techniques have opened new vistas in medical research especially in the field of diagnosis and this will further help in designing treatment plans after integrating with novel therapeutics such as pharmacogenomics and nutrigenomics that may revolutionize the management of untreated disease and disorders.

Personalized medicine: An integration of diagnostics with therapeutics

Management of disease often represents such situations where a medicine or drug that is fruitful for many patients often fails to work in other patients, or in cases where it does work it may then produce several adverse effects and may even cause death in some cases. Variability in drug efficacy has been documented since the existence of medicinal treatment. However, it has received more attention after the human genome draft has been completed. The diversity in the genome is such that each and every individual has some difference in their gene content that is expressed differentially so it ultimately affects its behavior in terms of functioning. Each individual's response in terms of drug efficacy, drug side effects, or both may vary person to person (Lu 1998, Meyer 2000, Evans and McLeod 2003, Weinshilboum 2003, Evans and Relling 2004). The speedy increase of molecular knowledge and the database on genome–disease and genome–drug interfaces has also impelled the transformation of lab science into clinical settings, simultaneously it provided a justification for the hope that the individualized medicine concept could be accomplished in the future.

The utilization of extraordinary throughput sequencing and genotyping techniques for the characterization of SNPs ultimately may help in determining an exclusive molecule in very little time and thus able to predict the risk of diseases. For example, two mutations have been characterized, A to T transversion at 1762 locus and G to A transition at 1764 locus, and are mostly screened in individuals having hepatocellular carcinoma, chronic hepatitis, and hepatitis (Kramvis and Kew 1999). Thus, each drug response can be predicted from known genetic variances and their adverse effect can be assembled so as to be helpful for better treatment. Further, this will permit clinicians to adapt a selective drug treatment to the patient. Few pharmaceutical companies and research development agencies are developing a precise haplotyping/genotyping scheme to identify individuals/patients who will benefit from a particular type of drug in a particular disease and disorder. Further, nutrigenomics is also evolving and promising the better management of patients. Numerous bioactive food components, including both essential and nonessential nutrients have a profound effect in the regulation of gene expression patterns. Thus, nutrigenomics is providing the effects of consumed nutrients and other food components on gene expression and gene regulation, that is, diet–gene interaction will help in understanding how health is affected. Nutritional genomics (nutrigenomics), the junction between health, diet, and genomics is influenced via the epigenetic, transcriptomic, and proteomics processes of biology. One study has reported that sulforaphane, butyrate, and allyl sulfur are potent inhibitors of histone deacetylase (HDAC), these HDAC inhibitors have been linked with enhanced acetylation and acetylation in promoter regions of the P21 and BAX genes consequently increase the expression of p21Cip1/Waf1 and BAX proteins (Myzak et al. 2006). Notably, sulforaphane demonstrated its activity in inhibiting HDAC activity in humans (Ross 2003). Further, a few studies demonstrated the role of lycopene, an important component of tomato which is well known for preventing breast cancer by up regulating the expression of apoptosis-related genes such as PIK3C3 and Akt1 that occur on lycopene intake. PIK3C3 is a member of the phosphoinositide (PI)3-kinase family and has an established role in the signaling cascade and is known for the regulation of cell growth, proliferation, survival, differentiation, and cytoskeletal changes

(Dwivedi et al. 2014a). Thus, molecular diagnostics is going to shift the whole paradigm of screening and treatment. Currently, nucleic acid extracted from several components such as blood, hair, semen, and tissue samples are explored and recognized globally in laboratories and the genetic tests and risk predicting database are continuously growing each year. So, the population prone to any particular diseases and disorders will not only be diagnosed earlier but treatment too can be planned at the prenatal stage.

Conclusions

In the future, molecular diagnostics will show its critical importance in public health worldwide. Molecular diagnostics will generate necessary critical information from large samples of diseased individuals and ultimately it will help not only in diagnosis but also in selecting appropriate therapies and monitoring of disease progression. However, still there are several obstacles that need to be overcome so that this diagnosis can be implemented in laboratories and hospitals, for example, which type of machine and technology should be preferred and the high cost of these tests so they cannot access them. These tests should also have good sensitivity, specificity, and robustness. Further, the development of integrated silicon chips mounted with biomolecules is now going to change the concept of the traditional wet lab to the "lab-on-a-chip." It will then be possible to analyze thousands of genes/proteins in hours from small amount/single cell samples. Therefore, the coming era will be revolutionary, it will not only change our diagnostic systems but also plans for treatment and therapy.

Acknowledgments

The authors are grateful to Shashi Dwivedi, Dr. Sumbul Samma, Dr. Himanhu Sharma, Jatin Joshi, Pradeep Sharma, and Prakash Jagson for their help in editing this chapter.

References

Amann RI, Ludwig W, Schleifer KH. 1995. Phylogenetic identification and *in situ* detection of individual microbial cells without cultivation. *Microbiol Rev*, 59, 225–229.

Aoyama T, Tynan K, Dietz HC, Francke U, Furthmayr H. 1993. Missense mutations impair intracellular processing of fibrillin and microfibril assembly in Marfan syndrome. *Hum Mol Genet*, 2, 2135–2140.

Bakshi CS, Singh VP, Malik M, Sharma B, Singh RK. 2002. Polymerase chain reaction amplification of 16S-23S spacer region for rapid identification of *Salmonella serovars*. *Acta Vet Hung*, 50, 161–166.

Bonne G, Carrier L, Bercovici J. et al. 1995. Cardiac myosin binding protein-C gene splice acceptor site mutation is associated with familial hypertrophic cardiomyopathy. *Nat Genet*, 11:438–440.

Braakhuis BJ, Tabor MP, Kummer JA, Leemans CR, Brakenhoff RH. 2003. A genetic explanation of Slaughter's concept of field cancerization: Evidence and clinical implications. *Cancer Res*, 63, 1727–1730.

Brandt ME, Padhye AA, Mayer LW, Holloway BP. 1998. Utility of random amplified polymorphic DNA PCR and TaqMan automated detection in molecular identification of *Aspergillus fumigatus*. *J Clin Microbiol*, 36, 2057–2062.

Branson BM. 2010. The future of HIV testing. *J Acquir Immune Defic Syndr*, 55(Suppl 2), S102–S105.

Brennan DJ, O'Connor DP, Rexhepaj E, Ponten F, Gallagher WM. 2010. Antibody-based proteomics: Fast-tracking molecular diagnostics in oncology. *Nat Rev Cancer*, 10(9), 605–617.

Brown TW. 1996. The molecular biology of the fragile X mutation. In: Hagerman RJ, Cronisier A, eds. *Fragile X Syndrome: Diagnosis, Treatment and Research*. Baltimore: The Johns Hopkins University Press.

Bruce M, McConnell I, Will R, Ironside J. 2001. Detection of variant Creutzfeldt-Jakob disease infectivity in extraneural tissues. *The Lancet*, 358, 208–209.

Buyse M, Loi S, van't Veer L. et al. 2006. Validation and clinical utility of a 70-gene prognostic signature for women with node-negative breast cancer. *J Natl Cancer Inst*, 98(17), 1183–1192.

Cancer Genome Atlas Research Network. 2012. Comprehensive genomic characterization of squamous cell lung cancers. *Nature*, 489, 519–25.

Carl AB, Edward RA, David EB. 2012. Tietz textbook of clinical chemistry and molecular diagnostics. *Elsevier*. ISBN 1-4557-5942-2. Retrieved 2013-09-26.

Chan K, Wong MS, Chan TK, Chan V. 2004. A thalassaemia array for Southeast Asia. *Br J Haematol*, 124, 232–239.

Chen PE, Cook C, Stewart AC, Nagarajan N, Sommer DD. et al. 2010. Genomic characterization of the *Yersinia* genus. *Genome Biol*, 11(1), R1.

Cohn DV, Elting JJ, Frick M, Elde R. 1984. Selective localization of the parathyroid secretory protein-I/adrenal medulla chromogranin a protein family in a wide variety of endocrine cells of the rat. *Endocrinology*, 114(6), 1963–1974.

Collinge J, Sidle KC, Meads J, Ironside J, Hill AF. 1996. Molecular analysis of prion strain variation and the aetiology of "new variant" CJD. *Nature*, 383, 685–690.

Collins FS. 1992. Cystic fibrosis: Molecular biology and therapeutic implications. *Science*, 2569(5058), 774–779.

Costa C, Costa JM, Desterke C, Botterel F, Cordonnier C, Bretagne S. 2002. Real-time PCR coupled with automated DNA extraction and detection of galactomannan antigen in serum by enzyme-linked immunosorbent assay for diagnosis of invasive aspergillosis. *J Clin Microbiol*, 40, 2224–2227.

Cotton RGH. 1996. Detection of unknown mutations in DNA: A catch 22. *Am J Hum Genet*, 59, 289–91.

Cowan S, Barrow G, Steel K, Feltham R Cowan and Steel's. 2003. In: Barrow G, Feltham R, eds. *Manual for the Identification of Medical Bacteria*. Cambridge, UK: Cambridge University Press. ISBN-10 0521543282 ISBN-13 9780521543286. Edition 3.

Cronin M, Sangli C, Liu ML. et al. 2007. Analytical validation of the oncotype DX genomic diagnostic test for recurrence prognosis and therapeutic response prediction in node-negative, estrogen receptorpositive breast cancer. *Clin Chem*, 53(6), 1084–1091.

DeBiasi RL, Tyler KL. 1999. Polymerase chain reaction in the diagnosis and management of central nervous system infections. *Arch Neurol*, 56, 1215–1219.

DeLong EF, Wickham GS, Pace NR. 1989. Phylogenetic strains: Ribosomal RNA-based probes for the identification of single cells. *Science*, 243, 1360–1363.

Derka S, Vairaktaris E, Papakosta V et al. 2006. Cell proliferation and apoptosis culminate in early stages of oral oncogenesis. *Oral Oncol*, 42, 540–550.

Dietz HC, Cutting GR, Pyeritz RE. et al. 1991. Marfan syndrome caused by a recurrent *de novo* missense mutation in the fibrillin gene. *Nature*, 352, 337–339.

Dolan TJ, Hatcher CP, Satterfield DA, Theodore MJ, Bach MC. et al. 2011. SodC-based real-time PCR for detection of *Neisseria meningitidis*. *PLoS ONE*. 6(5), e19361.

Dwivedi S, Goel A, Khattri S, Mandhani A, Sharma P, Pant KK. 2014a. Tobacco exposure by various modes may alter pro-inflammatory (IL-12) and anti-inflammatory (IL-10) levels and affects the survival of prostate carcinoma patients: An explorative study in North India. *Biomed Res Int*, 2014, 11. doi: dx.doiorg/10.1155/2014/158530.

Dwivedi S, Singh S, Goel A, Khattri S, Mandhani A, Sharma P, Misra S, Pant KK. 2015a. Pro-(IL-18) and anti-(IL-10) inflammatory promoter genetic variants (intrinsic factors) with tobacco exposure (extrinsic factors) may influence susceptibility and severity of prostate carcinoma: A prospective study. *Asian Pac J Cancer Prev*, 16(8), 3173–3181.

Dwivedi S, Goel A, Khattri S, Mandhani A, Sharma P, Misra S, Pant KK. 2015b. Genetic variability at promoters of IL-18 (pro-) and IL-10 (anti-) inflammatory gene affects susceptibility and their circulating serum levels: An explorative study of prostate cancer patients in North Indian populations. *Cytokine* 74(1):117–122.

Dwivedi S, Goel A, Khattri S, Sharma P, Pant KK. 2015. Aggravation of inflammation by smokeless tobacco in comparison of smoked tobacco. *Indian J Clin Biochem*, 30(1), 117–119.

Dwivedi S, Goel A, Mandhani A, Khattri S, Pant KK. 2012. Tobacco exposure may enhance inflammation in prostate carcinoma patients: An explorative study in north Indian population. *Toxicol Int*, 19(3), 310–8.

Dwivedi S, Goel A, Natu SM, Mandhani A, Khattri S, Pant KK. 2011. Diagnostic and prognostic significance of prostate specific antigen and serum interleukin 18 and 10 in patients with locally advanced prostate cancer: A prospective study. *Asian Pac J Cancer Prev*, 12, 1843–8.

Dwivedi S, Khattri S, Pant KK. 2012. In: Tiwari SP, Sharma R, Singh RK, eds. *Recent Advances in Molecular Diagnostic Approaches for Microbial Technology*, 133–154. Hauppauge, NY: Nova Science Publishers, Inc.

Dwivedi S, Shukla KK, Gupta G, Sharma P. 2013. Non-invasive biomarker in prostate cancer: A novel approach. *Indian J Clin Biochem*, 28(2), 107–9.

Dwivedi S, Goel A, Sadashiv, Verma A, Shukla S, Sharma P, Khattri S. et al. 2014a. Molecular diagnosis of metastasizing breast cancer based upon liquid biopsy. *Debmalya Barh* (ed.), 425–460. Springer India, doi: 10.1007/978-81-322-0843-3_22.

Dwivedi S, Shukla S, Goel A, Sharma P, Khattri S, Pant KK. 2014b. In: Barh D, ed. *Nutrigenomics in Breast Cancer*, 105–126. New Delhi, India: Springer India. doi: 10.1007/978-81-322-0843-3_6.

Dwivedi S, Yadav SS, Singh MK, Shukla S, Khattri S, Pant KK. 2013. Pharmacogenomics of viral diseases. In: Barh D, ed., *Omics for Personalized Medicine*, 637–676. New Delhi, India: Springer India. doi: 10.1007/978-81-322-1184-6_28.

Edwards AO, Ritter III R, Abel KJ, Manning A, Panhuysen C, Farrer LA. 2005. Complement factor H polymorphism and age-related macular degeneration. *Science*, 308, 421–4.

Evans WE and McLeod HL. 2003. Pharmacogenomics—Drug disposition, drug targets, and side effects. *N Engl J Med*, 348, 538–549.

Evans WE and Relling MV. 2004. Moving towards individualized medicine with pharmacogenomics. *Nature*, 429, 464–468.

Fan BJ, Tam PO, Choy KW, Wang DY, Lam DS, Pang CP. 2006. Molecular diagnostics of genetic eye diseases. *Clinical Biochemistry*, 39, 231–239.

Fan JB, Gunderson KLM, Bibikova JM. et al. 2006. Illumina universal bead arrays. *Methods Enzymol*, 410, 57–73.

Farrall M, Rodeck CH, Stanier P. et al. 1986. First-trimester prenatal diagnosis of cystic fibrosis with linked DNA probes. *Lancet*, 327, 1402–1405.

Faucher M, Anctil JL, Rodrigue MA. et al. 2002. Founder TIGR/myocilin mutations for glaucoma in the Quebec population. *Hum Mol Genet*, 11, 2077–2090.

Fausto N, Kaul, KL. 1999. Presenting the journal of molecular diagnostics. *J Mol Diagn*, 1, 1. doi: 10.1016/S1525-1578(10)60601-0.

Ferracin M, Veronese A, Negrini M. 2010. Micromarkers: MiRNAs in cancer diagnosis and prognosis. *Expert Rev Mol Diagn*, 10(3), 297–308.

Fiebig EW, Wright DJ, Rawal BD et al. 2003. Dynamics of HIV viremia and antibody seroconversion in plasma donors: Implications for diagnosis and staging of primary HIV infection. *AIDS*, 17, 1871–1879.

Fingert JH, Heon E, Liebmann JM. et al. 1999. Analysis of myocilin mutations in 1703 glaucoma patients from five different populations. *Hum Mol Genet*, 8, 899–905.

Fischer MB, Roeckl C, Parizek P, Schwarz HP, Aguzzi A. 2000. Binding of disease-associated prion protein to plasminogen. *Nature*, 408, 479–483.

Fliss MS, Usadel H, Caballero OL. et al. 2000. Facile detection of mitochondrial DNA mutations in tumors and bodily fluids. *Science*, 287, 2017–2019.

Ford N, Nachega JB, Engel ME, Mills EJ. 2009. Directly observed antiretroviral therapy: A systematic review and meta-analysis of randomised clinical trials. *Lancet*, 374, 2064–2071.

Forissier J-F, Carrier L, Farza H. et al. 1996. Codon 102 of the cardiac troponin T gene is a putative hot spot for mutations in familial hypertrophic cardiomyopathy. *Circulation*, 94, 3069–3073.

Geissdorfer W, Wittmann I, Rollinghoff M, Schoerner C, Bogdan C. 2001. Detection of a new 16S-23S rRNA spacer sequence variant (type 7) of *Tropheryma whippelii* in a patient with prosthetic aortic valve endocarditis. *Eur J Clin Microbiol Infect Dis*, 20, 762–763.

Geisterfer-Lowrance AA, Kass S, Tanigawa G. et al. 1990. A molecular basis for familial hypertrophic cardiomyopathy: A b-cardiac myosin heavy chain gene missense mutation. *Cell*, 62, 999–1006.

Gemignani F, Perra C, Landi S. et al. 2002. Reliable detection of beta-thalassemia and G6PD mutations by a DNA microarray. *Clin Chem*, 48, 2051–2054.

Gen-Probe Incorporated. 2009. Product insert, Aptima HIV-1 RNA qualitative assay. Gen-Probe Incorporated, San Diego, CA.

Giachetti C, Linnen JM, Kolk DP. et al. 2002. Highly sensitive multiplex assay for detection of human immunodeficiency virus type 1 and hepatitis C virus RNA. *J Clin Microbiol*, 40, 2408–2419.

Gibb EA, Brown CJ, Lam WL. 2011. The functional role of long noncoding RNA in human carcinomas. *Mol Cancer*, 10, 38.

Giese A, Bieschke J, Eigen M, Kretzschmar HA. 2000. Putting prions into focus: Application of single molecule detection to the diagnosis of prion diseases. *Arch Virol*, Suppl 2000(16), 161–171.

Giri MS, Nebozhyn M, Showe L, Montaner LJ. 2006. Microarray data on gene modulation by HIV-1 in immune cells: 2000–2006. *J Leukoc Biol*, 80, 1031–1043.

Glas AM, Floore A, Delahaye LJ. et al. 2006. Converting a breast cancer microarray signature into a high throughput diagnostic test. *BMC Genomics*, 7, 278. [PubMed: 17074082].

Glesby MJ, Pyeritz RE. 1989. Association of mitral valve prolapse and systemic abnormalities of connective tissue: A phenotypic continuum. *JAMA*, 262, 523–528.

Green AJ, Thompson EJ, Stewart GE. et al. 2001. Use of 14-3-3 and other brain-specific proteins in CSF in the diagnosis of variant Creutzfeldt–Jakob disease. *J Neurol Neurosurg Psychiatry*, 70, 744–748.

Grody WW, Nakamura, RM, Strom CM, Kiechle FL. 2010. *Molecular Diagnostics: Techniques and Applications for the Clinical Laboratory*. Boston MA: Academic Press Inc. ISBN 978-0-12-369428-7.

Hamburg MA, Collins FS. 2010 The path to personalized medicine. *N Engl J Med*, 363(4), 301–304. doi: 10.1056/NEJMp1006304.PMID20551152.

Hammerman PS, Sos ML, Ramos AH. et al. 2011. Mutations in the DDR2 kinase gene identify a novel therapeutic target in squamous cell lung cancer. *Cancer Discov*, 1, 78–89.

Hardison RC, Chui DH, Giardine B. et al. 2002. HbVar: A relational database of human hemoglobin variants and thalassemia mutations at the globin gene server. *Hum Mutat*, 19, 225–233.

HERmark, Monogram Biosciences, Inc. www.hermarkassay.com.

Higuchi R, Dollinger G, Walsh PS, Griffith R. 1992. Simultaneous amplification and detection of specific DNA sequences. *Biotechnology, (N Y)*. 10(4), 413–7.

Houpikian P, Raoult D. 2001. 16S/23S rRNA intergenic spacer regions for phylogenetic analysis, identification, and subtyping of *Bartonella* species. *J Clin Microbiol*, 39, 2768–2778.

Hsiao CR, Huang L, Bouchara JP, Barton R, Li HC, Chang TC. 2005. Identification of medically important molds by an oligonucleotide array. *J Clin Microbiol*, 43, 3760–3768.

Hu S, Li Y, Wang J, Xie Y, Tjon K, Wolinsky L, Loo RR, Loo JA, Wong DT. 2006. Human saliva proteome and transcriptome. *J Dent Res*, 85, 1129–1133.

Hu S, Loo JA, Wong DT. 2007. Human saliva proteome analysis. *Ann N Y Acad Sci*, 1098, 323–329.

Huang A, Li J-W, Shen Z-Q, Wang X-W, Jin M. 2006. High-throughput identification of clinical pathogenic fungi by hybridization to an oligonucleotide microarray. *J Clin Microbiol*, 44, 3299–3305.

International Human Genome Sequencing Consortium. 2001. Initial sequencing and analysis of the human genome. *Nature*, 409, 860–921.

Iorio MV, Ferracin M, Liu CG. et al. 2005. MicroRNA gene expression deregulation in human breast cancer. *Cancer Res*, 65(16), 7065–7070. [PubMed: 16103053].

Isharwal S, Miller MC, Marlow C, Makarov DV, Partin AW, Veltri RW. 2008. p300 (histone acetyltransferase) biomarker predicts prostate cancer biochemical recurrence and correlates with changes in epithelia nuclear size and shape. *Prostate*, 68, 1097–104.

Jackson GS, Beck JA, Navarrete C. et al. 2001. HLA-DQ7 antigen and resistance to variant CJD. *Nature*, 414, 269–270.

Jansen MP, Sieuwerts AM, Look MP. et al. 2007. HOXB13-to-IL17BR expression ratio is related with tumor aggressiveness and response to tamoxifen of recurrent breast cancer: A retrospective study. *J Clin Oncol*, 25(6), 662–668. [PubMed: 17308270].

Juusola J, Ballantyne J. 2003. Messenger RNA profiling: A prototype method to supplant conventional methods for body fluid identification. *Forensic Sci Int*, 135, 85–96.

Kai L, Wang J, Ivanovic M. et al. 2011. Targeting prostate cancer angiogenesis through metastasis-associated protein 1 (MTA1). *Prostate*, 71, 268–80.

Kakiuchi-Matsumoto T, Isashiki Y, Ohba N, Kimura K, Sonoda S, Unoki K. 2001. Cytochrome P450 1B1 gene mutations in Japanese patients with primary congenital glaucoma(1). *Am J Ophthalmol*, 131, 345–50.

Kan YW, Lee KY, Furbetta M, Angius A, Cao A. 1980. Polymorphism of DNA Sequence in the β-Globin Gene Region. *N Engl J Med*, 302(4), 185–188.

Kaplan G, Gaudernack G. 1982. *In vitro* differentiation of human monocytes. Differences in monocyte phenotypes induced by cultivation on glass or on collagen. *J Exp Med*, 156(4), 1101–1114.

Kattan MW, Shariat SF, Andrews B. et al. 2003. The addition of interleukin-6 soluble receptor and transforming growth factor beta1 improves a preoperative nomogram for predicting biochemical progression in patients with clinically localized prostate cancer. *J Clin Oncol*, 21, 3573– 3579.

Kibel A, Iliopulos O. et al. 1995. Binding of the von Hippel–Lindau tumor suppressor protein to Elongin B and C. *Science*, 269, 1444–6.

Kikuchi T, Iwasaki K, Nishihara H, Takamura Y, Yagi O. 2002. Quantitative and rapid detection of the trichloroethylene degrading bacterium *Methylocystis* sp. M in groundwater by real-time PCR. *Appl Microbiol Biotechnol*, 59, 731–736.

King CR, Kraus MH, Aaronson SA. 1985. Amplification of a novel v-erbB-related gene in a human mammary carcinoma. *Science*, 229(4717), 974–976.

Kitchen CM, Kitchen SG, Dubin JA, Gottlieb MS. 2001. Initial virological and immunologic response to highly active antiretroviral therapy predicts long-term clinical outcome. *Clin Infect Dis*, 33, 466–472.

Klaver CC, Kliffen M, van Duijn CM. et al. 1998. Genetic association of apolipoprotein E with age-related macular degeneration. *Am J Hum Genet*, 63, 200–206.

Klein R, Klein BE, Cruickshanks KJ. 1999. The prevalence of age-related maculopathy by geographic region and ethnicity. *Prog Retin Eye Res*, 18, 371–389.

Klein R, Klein BE, Linton KL. 1992. Prevalence of age-related maculopathy, The Beaver Dam Eye Study. *Ophthalmology*, 99, 933–943.

Klein RJ, Zeiss C, Chew EY. et al. 2005. Complement factor H polymorphism in age-related macular degeneration. *Science*, 308, 385–389.

Knudson AG. 1971. Mutation and cancer: Statistical study of retinoblastoma. *Proc Natl Acad Sci USA*, 68, 820–823.

Kolbert CP, Persing DH. 1999. Ribosomal DNA sequencing as a tool for identification of bacterial pathogens. *Curr Opin Microbiol*, 2, 299–305.

Konig J, Schreiber R, Voelcker T, Mall M, Kunzelmann K. 2001. The cystic fibrosis transmembrane conductance regulator (CFTR) inhibits ENaC through an increase in the intracellular Cl-concentration. *EMBO Reports*, 2(11), 1047–1051.

Kramvis A, Kew MC. 1999. The core promoter of hepatitis B virus. *J Viral Hepat*, 6, 415–427.

Kuczius T, Groschup MH. 1999. Differences in proteinase K resistance and neuronal deposition of abnormal prion proteins characterize bovine spongiform encephalopathy (BSE) and scrapie strains. *Mol Med*, 5, 406–418.

Lallemand F, Desire N, Rozenbaum W, Nicolas J-C, Marechal V. 2000. Quantitative analysis of human herpesvirus 8 viral load using a real-time PCR assay. *J Clin Microbiol*, 38, 1404–1408.

Lam DS, Leung YF, Chua JK. et al. 2000. Truncations in the TIGR gene in individuals with and without primary open-angle glaucoma. *Invest Ophthalmol Vis Sci*, 41, 1386–1391.

Landegren U, Kaiser R, Sanders J, and Hood L. 1988. A ligase-mediated gene detection technique. *Science*, 241, 1077–1080.

Langfelder-Schwind E, Karczeski B, Strecker MN. et al. 2014. Molecular testing for cystic fibrosis carrier status practice guidelines: Recommendations of the National Society of Genetic Counselors. *J Genet Couns*, 23(1), 5–15. doi: 10.1007/s10897-013-9636-9.

Larsen JE, Minna JD. 2011. Molecular biology of lung cancer: Clinical implications. *Clin Chest Med*, 32, 703–740.

Latif F. et al. 1993. Identification of the von Hippel–Lindau disease tumor suppressor gene. *Science*, 260, 1317–1320.

Layfield LJ, Gupta D, Mooney EE. 2000. Assessment of tissue estrogen and progesterone receptor levels: A survey of current practice, techniques, and quantitation methods. *Breast J*, 6(3), 189–196.

Leinberger DM, Schumacher U, Autenrieth IB, Bachmann TT. 2005. Development of a DNA microarray for detection and identification of fungal pathogens involved in invasive mycoses. *J Clin Microbiol*, 43, 4943–4953.

Lenhoff RJ, Naraghi-Arani P, Thissen JB et al. 2008. Multiplexed molecular assay for rapid exclusion of foot-and-mouth disease. *J Virol Methods*, 153, 61–69.

Li Y, St John MA, Wong DT. 2004a. RNA profiling of cell-free saliva using microarray technology. *J Dent Res*, 83, 199–203.

Li Y, St John MA, Zhou X. et al. 2004b. Salivary transcriptome diagnostics for oral cancer detection. *Clin Cancer Res*, 10, 8442–50.

Linsdell P. 2006. Mechanism of chloride permeation in the cystic fibrosis transmembrane conductance regulator chloride channel. *Exp Physiol*, 91(1), 123–129.

Lipkin WI. 2008. Pathogen discovery. *PLoS Pathog*, 4, e1000002.

Lohmann DR, Gallie BL. 2004. Retinoblastoma: Revisiting the model prototype of inherited cancer. *Am J Med Genet C Semin Med Genet*, 129, 23–28.

Lu AY. 1998. Drug-metabolism research challenges in the new millennium: Individual variability in drug therapy and drug safety. *Drug Metab Dispos*, 26, 1217–1222.

Lu Z, Liu M, Stribinskis V. et al. 2008. MicroRNA-21 promotes cell transformation by targeting the programmed cell death 4 gene. *Oncogene*, 27(31), 4373–4379. [PubMed: 18372920].

Ma L, Teruya-Feldstein J, Weinberg RA. 2007. Tumour invasion and metastasis initiated by microRNA-10b in breast cancer. *Nature*, 449(7163), 682–688. [PubMed: 17898713].

Ma XJ, Salunga R, Dahiya SA. et al. 2008. Five-gene molecular grade index and HOXB13:IL17BR are complementary prognostic factors in early stage breast cancer. *Clin Cancer Res*, 14(9), 2601–2608. [PubMed: 18451222].

Madico G, Quinn TC, Boman J, Gaydos CA. 2000. Touchdown enzyme time release-PCR for detection and identification of *Chlamydia trachomatis, C. pneumoniae*, and *C. psittaci* using the 16S and 16S-23S spacer rRNA genes. *J Clin Microbiol*, 38, 1085–1093.

Mardis ER. 2008. Next-generation DNA sequencing methods. Annu Rev Genomics Hum Genet 9, 387–402. http://www.plosone.org/article/findArticle.action?author=Mardis&title=Next-generation%20DNA%20sequencing%20methods.

Marfan AB. 1896. A case of congenital formation of four more pronounced Membership charactérisée ends by the elongation of the bones with a certain degree of thinning. *Bull Mem Soc Med Hosp Paris*, 13, 220–226.

Marian AJ, Mares A Jr., Kelly DP. et al. 1995. Sudden cardiac death in hypertrophic cardiomyopathy: variability in phenotypic expression of b-myosin heavy chain mutations. *Eur Heart J*, 16, 368–376.

Mashal RD, Koontz J, Sklar J. 1995. Detection of mutations by cleavage of DNA heteroduplexes with bacteriophage resolvases. *Nat Genet*, 9, 177–183.

Massano J, Regateiro FS, Januario G, Ferreira A. 2006. Oral squamous cell carcinoma: Review of prognostic and predictive factors. *Oral Surg Oral Med Oral Pathol*, 102, 67–76.

Mastrangelo P. and Westaway D. 2001. The prion gene complex encoding PrPc and Doppel: Insights from mutational analysis. *Gene*, 275, 1–18.

Mattie MD, Benz CC, Bowers J. et al. 2006. Optimized high-throughput microRNA expression profiling provides novel biomarker assessment of clinical prostate and breast cancer biopsies. *Mol Cancer*, 5, 24. [PubMed: 16784538].

Mesaros N, Glupczynski Y, Avrain L, Caceres NE, Tulkens PM, Van Bambeke F. 2007. A combined phenotypic and genotypic method for the detection of Mex efflux pumps in *Pseudomonas aeruginosa*. *J Antimicrob Chemother*, 59(3), 378–386.

Meyer UA 2000. Pharmacogenetics and adverse drug reactions. *Lancet*, 356, 1667–1671.

Michael H, Karlheinz T, Anna MG et al. 2000. Specific and rapid detection by fluorescent insitu hybridization of bacteria in clinical samples obtained from cystic fibrosis patients. *Clin Microbiol*, 38, 2818–2825.

Milewicz DM, Grossfield J, Cao S-N, Kielty C, Covitz W, Jewett T. 1995. A mutation in FBN1 disrupts profibrillin processing and results in isolated skeletal features of the Marfan syndrome. *J Clin Invest*, 95, 2373–2378.

Miller MB. 2009. Solid and liquid phase array technologies. In D Persing, F Tenover, R Hayden, F Nolte, YW Tang, and A Van Belkum (ed.), *Molecular Microbiology: Diagnostic Principles and Practice*, 2nd ed., in press. ASM Press, Washington, DC.

Minamoto T, Ougolkov AV, Mai M. 2002. Detection of oncogenes in the diagnosis of cancers with active oncogenic signaling. *Expert Rev Mol Diagn*, 2(6), 565–575.

Mitchell PS, Parkin RK, Kroh EM. et al. 2008. Circulating microRNAs as stable blood-based markers for cancer detection. *Proc Natl Acad Sci USA*, 105, 10513–10518.

Mok T, Wu YL, Zhang LA. 2009. Small step towards personalized medicine for non-small cell lung cancer. *Discov Me*, 8, 227–231.

Monemi S, Spaeth G, DaSilva A. et al. 2005. Identification of a novel adultonset primary open-angle glaucoma (POAG) gene on 5q22.1. *Hum Mol Genet*, 14, 725–733.

Morozova O, Marra MA. 2008. Applications of next-generation sequencing technologies in functional genomics. *Genomics*, 92, 255–264. http://www.plosone.org/article/findArticle.action?author=Hoffmaster&title=Identification%20of%20anthrax%20toxin%20genes%20in%20a%20Bacillus%20cereus%20associated%20with%20an%20illness%20resembling%20inhalation%20anthrax.

Moss AJ, Schwartz PJ, Crampton RS. et al. 1991. The long QT syndrome: Prospective longitudinal study of 328 families. *Circulation*, 8, 1136–1144.

Mullis KB, Faloona FA. 1987. Specific synthesis of DNA *in vitro* via a polymerase-catalyzed chain reaction. *Methods Enzymol*, 155, 335–350.

Murgia A, Martella M, Vinanzi C, Polli R, Perilongo G, Opocher G. 2000. Somatic mosaicism in von Hippel–Lindau Disease. *Hum Mutat*, 15(1), 114.

Myers RM, Larin Z, Maniatis T. 1985a. Detection of single base substitutions by ribonuclease cleavage at mismatches in RNA:DNA duplexes. *Science*, 230, 1242–1246.

Myers RM, Lumelsky N, Lerman LS, Maniatis T. 1985b. Detection of single base substitutions in total genomic DNA. *Nature*, 1985, 313, 495–498.

Myers RM, Maniatis T, Lerman LS. 1987. Detection and localization of single base changes by denaturing gradient gel electrophoresis. *Methods Enzymol.*, 155, 501–527.

Myzak MC, Hardin K, Wang R, Dashwood RH, Ho E. 2006. Sulforaphane inhibits histone deacetylaseactivity in BPH-1, LnCaP and PC-3 prostate epithelial cells. *Carcinogenesis*, 27, 811–819.

Najioullah F, Thouvenot D, Lina B. 2001. Development of a real-time PCR procedure including an internal control for the measurement of HCMV viral load. *J Virol Meth*, 92, 55–64.

Naumova A, Sapienza C. 1994. The genetics of retinoblastoma, revisited. *Am J Hum Genet*, 54, 264–73.

Nawroz H, van der Riet P, Hruban RH, Koch W, Ruppert JM, Sidransky D. 1994. Allelotype of head and neck squamous cell carcinoma. *Cancer Res*, 54, 1152–1155.

Neumann HP, Lipps CJ, Ilsia YE, Zbar B. von Hippel. 1995. Lindau syndrome. *Brain Pathol*, 5, 181–93.

Niimura H, Bachinski LL, Sangwatanaroj S. et al. 1998. Mutations in the gene for cardiac myosin binding protein C and late-onset familial hypertrophic cardiomyopathy. *N Engl J Med*, 338, 1248–1257.

Nijbroek G, Sood S, McIntosh I. et al. 1995. Fifteen novel FBN1 mutations causing Marfan syndrome detected by heteroduplex analysis of genomic amplicons. *Am J Hum Genet*, 57, 8–21.

Nitsche A, Steuer N, Schmidt CA. et al. 2000. Detection of human cytomegalovirus DNA by real-time quantitative PCR. *J Clin Microbiol*, 38, 2734–2737.

Nolte FS. Impact of viral load testing on patient care.1999. *Arch Pathol Lab Med*, 123, 1011–14.

Nubel U, Schmidt PM, Reiss E, Bier F, Beyer W, Naumann D. 2004. Oligonucleotide microarray for identification of *Bacillus anthracis* based on intergenic transcribed spacers in ribosomal DNA. *FEMS Microbiol Lett*, 240, 215–223.

Okonji JA, Basavaraju SV, Mwangi J. et al. 2012. Comparison of HIV-1 detection in plasma specimens and dried blood spots using the Roche COBAS Ampliscreen HIV-1 test in Kisumu, Kenya. *J Virol Methods*, 179, 21–25.

Oliphant AD, Barker L, Stuelpnagel JR, Chee MS. 2002. Bead array technology: Enabling an accurate, cost-effective approach to high-throughput genotyping. *Bio Techniques*, 56–58, 60–61.

Orita M, Iwahana H, Kanazawa H, Hayashi K, Sekiya, T. 1989. Detection of polymorphisms of human DNA by gel electrophoresis as single-strand conformation polymorphisms. *Proc Natl Acad Sci USA*, 86, 2766–2770.

Orkin SH, Markham AF, Kazazian HH Jr. 1983. Direct detection of the common Mediterranean beta-thalassemia gene with synthetic DNA probes: An alternative approach for prenatal diagnosis. *J Clin Invest*, 71, 775–779.

Paik S, Hazan R, Fisher ER. et al. 1990. Pathologic findings from the national surgical adjuvant breast and bowel project: Prognostic significance of erbB-2 protein overexpression in primary breast cancer. *J Clin Oncol*, 8(1), 103–112.

Pao W, Girard N. 2011. New driver mutations in non-small-cell lung cancer. *Lancet Oncol*, 12, 175–180.

Parchi P, Giese A, Capellari S, Brown P. et al. 1999. Classification of sporadic Creutzfeldt–Jakob disease based on molecular and phenotypic analysis of 300 subjects. *Ann Neurol*, 46, 224–233.

Park NJ, Li Y, Yu T, Brinkman BM, Wong DT. 2006. Characterization of RNA in saliva. *Clin Chem*, 52, 988–994.

Parkin DM, Bray F, Ferlay J, Pisani P. 2005. Global cancer statistics, 2002. *CA Cancer J Clin*, 55(2), 74–108. [PubMed: 15761078].

Patrinos GP, Giardine B, Riemer C. et al. 2004. Improvements in the HbVar database of human haemoglobin variants and thalassemia mutations for population and sequence variation studies. *Nucleic Acids Res*, 32, D537–D541.

Persselin JE, Stevens RH. 1985. Anti-Fab antibodies in humans. Predominance of minor immunoglobulin G subclasses in rheumatoid arthritis. *J Clin Invest*, 76(2), 723–730.

Philip H, Amman B, Deborah E, Ben C, Aphrodite I, Mary W. 2009. A systematic review of the diagnostic accuracy of prostate specific antigen. *BMC Urol*, 9, 14.

Pietila J, He Q, Oksi J, Viljanen MK. 2000. Rapid differentiation of *Borrelia garinii* from *Borrelia afzelii* and *Borrelia burgdorferi* sensu stricto by LightCycler fluorescence melting curve analysis of a PCR product of the recA gene. *J Clin Microbiol*, 38, 2756–2759.

Pirastu M, Kan YW, Cao A, Conner BJ, Teplitz RL, Wallace RB. 1983. Prenatal diagnosis of beta-thalassemia. Detection of a single nucleotide mutation in DNA. *N Engl J Med*, 309, 284–287.

Porkka KP, Pfeiffer MJ, Waltering KK, Vessella RL, Tammela TLJ, Visakorpi T. 2007. MicroRNA expression profiling in prostate cancer. *Cancer Res*, 67, 6130–6135.

Poste G. 2001. Molecular diagnostics: A powerful new component of the healthcare value chain. *Expert Rev Mol Diagn*, 1(1), 1–5.

Rauter C, Oehme R, Diterich I, Engele M, Hartung T. 2002. Distribution of clinically relevant *Borrelia* genospecies in ticks assessed by a novel, single-run, real-time PCR. *J Clin Microbiol*, 40, 36–43.

Resino S, Resino R, Maria Bellon J. et al. 2006. Clinical outcomes improve with highly active antiretroviral therapy in vertically HIV type-1-infected children. *Clin Infect Dis*, 43, 243–252.

Rezaie T, Child A, Hitchings R. et al. 2002. Adult-onset primary open-angle glaucoma caused by mutations in optineurin. *Science*, 295, 1077–1079.

Rivolta C, Sharon D, DeAngelis MM, Dryja TP. 2002. Retinitis pigmentosa and allied diseases: Numerous diseases, genes, and inheritance patterns. *Hum Mol Genet*, 11, 1219–1227.

Roman MJ, Rosen SE, Kramer-Fox R, Devereux RB. 1993. The prognostic significance of the pattern of aortic root dilatation in the Marfan syndrome. *J Am Coll Cardiol*, 22, 1470–1476.

Ronaghi M, Uhlen M, Nyren P. 1998. A sequencing method based on real-time pyrophosphate. *Science*, 281, 363–365.

Rosas SL, Koch W, Carvalho MGC. et al. 2001. Promoter hypermethylation patterns of p16, O6-methylguanine-DNA-methyltransferase, and death-associated protein kinase in tumors and saliva of head and neck cancer patients. *Cancer Res*, 61, 939–942.

Ross SA. 2003. Diet and DNA methylation interactions in cancer prevention. *Ann N Y Acad Sci*, 983, 197–207.

Rotger M, Dalmau J, Rauch A. et al. 2011. Comparative transcriptomics of extreme phenotypes of human HIV-1 infection and SIV infection in sooty mangabey and rhesus macaque. *J Clin Invest*, 121, 2391–2400.

Roth A, Reischl U, Streubel A. et al. 2000. Novel diagnostic algorithm for identification of mycobacteria using genus-specific amplification of the 16S-23S rRNA gene spacer and restriction endonucleases. *J Clin Microbiol*, 38, 1094–1104.

Rouet F, Montcho C, Rouzioux C et al. 2001. Early diagnosis of paediatric HIV-1 infection among African breast-fed children using a quantitative plasma HIV RNA assay. *AIDS*, 15, 1849–1856.

Saborio GP, Permanne B, Soto C. 2001. Sensitive detection of pathological prion protein by cyclic amplification of protein misfolding. *Nature*, 411, 810–813.

Saiki RK, Bugawan TL, Horn GT, Mullis KB, Erlich HA. 1986. Analysis of enzymatically amplified beta-globin and HLA-DQ alpha DNA with allele-specific oligonucleotide probes. *Nature*, 324, 163–166.

Saiki RK, Gelfand DH, Stoffel S. et al. 1988. Primerdirected enzymatic amplification of DNA with a thermostable DNA polymerase. *Science*, 239, 487–491.

Saiki RK, Scharf S, Faloona F. et al. 1985. Enzymatic amplification of beta-globin genomic sequences and restriction site analysis for diagnosis of sickle cell anemia. *Science*, 230, 1350–1354.

Saiki RK, Walsh PS, Levenson CH, Erlich HA. 1989. Genetic analysis of amplified DNA with immobilized sequence-specific oligonucleotide probes. *Proc Natl Acad Sci USA*, 86, 6230–6234.

Saleeba JA, Ramus SJ, Cotton RG. 1992. Complete mutation detection using unlabeled chemical cleavage. *Hum Mutat*, 1, 63–69.

Salvadores N, Shahnawaz M, Scarpini E, Tagliavini F, Soto C. 2014. Detection of misfolded Aβ oligomers for sensitive biochemical diagnosis of Alzheimer's disease. *Cell Rep*, 10,7(1), 261–8.

Sankaranarayanan R, Ramadas K, Thomas G. et al. 2005b. Oral cancer screening study group. Effect of screening on oral cancer mortality in Kerala, India: A cluster-randomised controlled trial. *Lancet*, 4–10; 365(9475), 1927–1933.

Sankaranarayanan R. 2005a. Screening for cervical and oral cancers in India is feasible and effective. *Natl Med J India*, 18, 281–284.

Scott GK, Goga A, Bhaumik D et al. 2007. Coordinate suppression of ERBB2 and ERBB3 by enforced expression of micro-RNA miR-125a or miR-125b. *J Biol Chem*, 282, 1479–1486.

Sharma H, Mavuduru RS, Singh SK, Prasad R. 2014. Heterogeneous spectrum of mutations in CFTR gene from Indian patients with congenital absence of the vas deferens and their association with cystic fibrosis genetic modifiers. *Mol Hum Reprod*, 20(9):827–35.

Shaw AT, Solomon B. 2011. Targeting anaplastic lymphoma kinase in lung cancer. *Clin Cancer Res*, 17, 2081–6.

Sherman GG, Cooper PA, Coovadia AH. et al. 2005. Polymerase chain reaction for diagnosis of human immunodeficiency virus infection in infancy in low resource settings. *Pediatr Infect Dis J*, 24, 993–997.

Sohocki MM, Daiger SP, Bowne SJ. et al. 2001. Prevalence of mutations causing retinitis pigmentosa and other inherited retinopathies. *Hum Mutat*, 17, 2–51.

Sosnowski RG, Tu E, Butler WF, O'Connell JP, Heller MJ. 1997. Rapid determination of single base mismatch mutations in DNA hybrids by direct electric field control. *Proc Natl Acad Sci USA*, 94, 1119–1123.

Spiess B, Seifarth W, Hummel M. et al. 2007. DNA microarray-based detection and identification of fungal pathogens in clinical samples from neutropenic patients. *J Clin Microbiol*, 45, 3743–3753.

Sriram KB, Larsen JE, Yang IA. et al. 2011. Genomic medicine in non-small cell lung cancer: Paving the path to personalized care. *Respirology*, 16, 257–263.

Srivatsan A, Han Y, Peng J, Tehranchi AK, Gibbs R. et al. 2008. High-precision, whole-genome sequencing of laboratory strains facilitates genetic studies. *PLoS Genet* 4, e1000139.

Steinberg KM, Okou DT, Zwick ME. 2008. Applying rapid genome sequencing technologies to characterize pathogen genomes. *Anal Chem*, 80, 520–528.

Stevens WS, Noble L, Berrie L, Sarang S, Scott LE. 2009. Ultra-high throughput, automated nucleic acid detection of HIV for infant diagnosis using the GEN-PROBE APTIMA HIV-1 screening assay. *J Clin Microbiol JCM*, 00317–09.

Stoilov I, Akarsu AN, Sarfarazi M. 1997. Identification of three different truncating mutations in cytochrome P4501B1 (CYP1B1) as the principal cause of primary congenital glaucoma (buphthalmos) in families linked to the GLC3A locus on chromosome 2p21. *Hum Mol Genet*, 6, 641–647.

Stone EM, Fingert JH, Alward WLM. et al. 1997. Identification of a gene that causes primary open angle glaucoma. *Science*, 27, 668–70.

Strachan T, Reed AP. 1996. Somatic mutations and cancer. In: Strachan T, Read AP, eds., *Human Molecular Genetics*, Oxford, Great Britain: Bios Scientific Publishers, 458–476.

Tanaka N, Kimura H, Iida K. et al. 2000. Quantitative analysis of cytomegalovirus load using a eal-time PCR assay. *J Med Virol*, 60, 455–462.

Taylor CR, Shi SR, Chaiwun B, Young L, Imam SA, Cote RJ. 1994. Strategies for improving the immunohistochemical staining of various intranuclear prognostic markers in formalin-paraffin sections: Androgen receptor, estrogen receptor, progesterone receptor, p53 protein, proliferating cell nuclear antigen, and Ki-67 antigen revealed by antigen retrieval techniques. *Hum Pathol*, 25(3), 263–270.

The EUROCJD Group 2001. Genetic epidemiology of Creutzfeldt–Jakob disease in Europe. *Rev Neurol, (Paris)*, 157, 633–637.

Thomson RB, Bertram H. 2001. Laboratory diagnosis of central nervous system infections. *Infect Dis Clin N Am* 15, 1047–1071.

Tomlins SA, Aubin SM, Siddiqui J. et al. 2011. Urine TMPRSS2: ERG fusion transcript stratifies prostate cancer risk in men with elevated serum PSA. *Sci Transl Med*, 3, 94ra72.

Travis WD, Brambilla E, Riely GJ. 2013. New pathologic classification of lung cancer: Relevance for clinical practice and clinical trials. *J Clin Oncol*, 31, 992–1001.

Trebesius K, Harmsen D, Rakin A, Schmelz J, Heesemann J. 1998. Development of RNA-targeted PCR and *in situ* hybridization with fluorescently labelled oligonucleotides for detection of Yersinia species. *J Clin Microbiol*, 36, 2557–2564.

Treisman R, Orkin SH, Maniatis T. 1983. Specific transcription and RNA splicing defects in five cloned beta-thalassaemia genes. *Nature*, 302, 591–596.

Van Ert MN, Easterday WR, Simonson TS. et al. 2007. Strain-specific single-nucleotide polymorphism assays for the *Bacillus anthracis* Ames strain. *J Clin Microbiol*, 45, 47–53.

Venter JC, Adams MD, Myers EW. et al. 2001. The sequence of the human genome. *Science*, 291, 1304–1351.

Vincent AL, Billingsley G, Buys Y. et al. 2002. Digenic inheritance of earlyonset glaucoma: CYP1B1, a potential modifier gene. *Am J Hum Genet*, 70, 448–460.

Völkel D, Zimmermann K, Zerr I. et al. 2001. Immunochemical determination of cellular prion protein in plasma from healthy subjects and patients with sporadic CJD or other neurologic diseases. *Transfusion*, 41, 441–448.

Wadsworth JD, Joiner S, Hill AF. et al. 2001. Tissue distribution of protease resistant prion protein in variant Creutzfeldt–Jakob disease using a highly sensitive immunoblotting assay. *Lancet* 358, 171–180.

Wang Z, Gerstein M, Snyder M. 2009. RNA-Seq: A revolutionary tool for transcriptomics. *Nat Rev Genet*, 10, 57–63.

Wang Z, Trillo-Pazos G, Kim SY. et al. 2004. Effects of human immunodeficiency virus type 1on astrocyte gene expression and function: Potential role in neuropathogenesis. *J Neurovirol*, 10(Suppl 1), 25–32.

Ward OC. 1964. A new familial cardiac syndrome in children. *J Irish Med Assoc*, 54, 103–106.

Watkins H, Conner D, Thierfelder L. et al. 1995. Mutations in the cardiac myosin binding protein-C gene on chromosome 11 cause familial hypertrophic cardiomyopathy. *Nat Genet*, 11, 434–437.

Weinshilboum R. 2003. Inheritance and drug response. *N Engl J Med*, 348, 529–537.

Will RG, Zeidler M, Stewart GE, Macleod MA, Ironside JW, Cousens SN, Mackenzie J, Estibeiro K, Green AJ, Knight RS. 2000. Diagnosis of new variant Creutzfeldt-Jakob disease. *Ann Neurol*, 47(5), 575–582.

Willard SS, Koochekpour S. 2012. Regulators of gene expression as biomarkers for prostate cancer. *Am J Cancer Res*, 2(6), 620–57.

William S, Probert KN, Schrader NY. et al. 2004. Graves real-time multiplex PCR assay for detection of *Brucella* spp., *B. abortus*, and *B. melitensis*. *J Clin Microbiol*, 42(3), 1290–1293.

WizigmannVoos S, Breier G. et al. 1995. Upregulation of vascular endothelial growth factor and its receptors in von Hippel–lindau disease—Associated and sporadic hemangioblastomas. *Cancer Res*, 55, 1358–1364.

Woese CR. 1987. Bacterial evolution. *Microbiol Rev*, 51, 221–271.

Wong DT. 2006. Salivary diagnostics for oral cancer. *J Calif Dent Assoc*, 34, 303–308.

Woo SL, Lidsky AS, Guttler F, Chandra T, Robson KJ. 1983. Cloned human phenylalanine hydroxylase geneallows prenatal diagnosis and carrier detection of classical phenylketonuria. *Nature*, 306, 151–155.

Woods GL. Molecular techniques in mycobacterial detection. 2001. *Arch Pathol Lab Med*, 125,122–126.

Yang JC, Shih JY, Su WC. et al. 2012. Afatinib for patients with lung adenocarcinoma and epidermal growth factor receptor mutations (LUXLung 2): A phase 2 trial. *Lancet Oncol*, 13, 539–548.

Yegnasubramanian S, Wu Z, Haffner MC, Esopi D, Aryee MJ, Badrinath R. et al. 2011. Chromosome-wide mapping of DNA methylation patterns in normal and malignant prostate cells reveals pervasive methylation of gene-associated and conserved intergenic sequences. *BMC Genomics*, 12, 313.

Yokota T, Akatsu H, Miyauchi T, Heese K. 2012. Characterization of the novel protein P9TLDR (temporal lobe down-regulated) with a brain-site-specific gene expression modality in Alzheimer's disease brain. *FEBS Lett*, 586(24), 4357–4361.

You YC, Fu X, Zeng D. et al. 2008. A novel DNA microarray for rapid diagnosis of enteropathogenic bacteria in stool specimens of patients with diarrhea. *J Microbiol Methods*, 75, 566–571.

Zakowski MF, Hussain S, Pao W. et al. 2009. Morphologic features of adenocarcinoma of the lung predictive of response to the epidermal growth factor receptor kinase inhibitors erlotinib and gefitinib. *Arch Pathol Lab Med*, 133, 470–477.

Zareparsi S, Branham KE, Li M. et al. 2005. Strong association of the Y402H variant in complement factor H at 1q32 with susceptibility to age-related macular degeneration. *Am J Hum Genet*, 77, 149–53.

Zerr I, Pocchiari M, Collins S. et al. 2000. Analysis of EEG and CSF 14-3-3 proteins as aids to the diagnosis of Creutzfeldt–Jakob disease. *Neurology*, 55, 811–815.

Zhang B, Gaiteri C, Bodea LG. et al. 2013. Integrated systems approach identifies genetic nodes and networks in late-onset Alzheimer's disease. *Cell*, 153(3), 707–720.

Zhou C, Wu YL, Chen G. et al. 2011. Erlotinib versus chemotherapy as first-line treatment for patients with advanced EGFR mutation-positive non-small cell lung cancer (OPTIMAL, CTONG-0802): A multicentre, open-label, randomised, phase 3 study. *Lancet Oncol*, 12, 735–742.

Zhu S, Si ML, Wu H, Mo YY. 2007. MicroRNA-21 targets the tumor suppressor gene tropomyosin 1 (TPM1). *J Biol Chem*, 282(19), 14328–14336. [PubMed: 17363372].

Zimmermann BG, Park NJ, Wong DT. 2007. Genomic targets in saliva. *Ann N Y Acad Sci*, 1098, 184–191.

Zunt JR, Marra CM. 1999. Cerebrospinal fluid testing for the diagnosis of central nervous system infection. *Neurol Clin*, 17, 675–689.

Zwick ME, Kiley MP, Stewart AC, Mateczun A, Read TD. 2008. Genotyping of *Bacillus cereus* strains by microarray-based resequencing. *PLoS ONE*, 3, e2513.

chapter fourteen

Techniques for cervical cancer screening and diagnosis

Haq Nawaz, Nosheen Rashid, Hugh J. Byrne, and Fiona M. Lyng

Contents

Abstract

In this chapter, different techniques used for cervical cancer, including Papanicolaou (Pap) smear, colposcopy, human papillomavirus
deoxyribonucleic acid/ribonucleic acid-based techniques, visual
inspection with acetic acid, histological examination, confocal
microscopy, optical coherence tomography, and vibrational spectroscopy, are discussed. These techniques are currently in use for
research or clinical purposes. The advantages and disadvantages
of these techniques, with particular focus on vibrational spectroscopy, are elaborated. Moreover, the use of Fourier transform infrared (FTIR) and Raman microspectroscopy is demonstrated and the
spectral data acquired by using these techniques are compared to
demonstrate that Raman microspectroscopy has great potential to
be used for the early diagnosis of cancer. The results presented and
discussed so far for FTIR and Raman microspectroscopy and their
comparison point toward the greater suitability of the latter technique for the current study.

Keywords: Cervical cancer, Early diagnosis, Different techniques,
Comparison of FTIR and Raman spectroscopy.

Introduction

Cervical cancer is reported to be the third most common cancer among women worldwide. Notably, 85% of cases arise in developing countries (Jemal et al., 2011). However, the
mortality linked with this cancer can be considerably reduced if the disease is detected at
the early stages of its expansion or at the premalignant state, which is called as cervical
intraepithelial neoplasia (CIN) (Parkin et al., 2005).

Current methods for identifying neoplastic cells and differentiating them from
their normal counterparts are often nonspecific, slow, invasive, or a combination thereof
(Chan et al., 2006). Papanicolaou (Pap) smear test is considered to be a primary tool for
the screening of cervical neoplasia, which is a microscopic evaluation of the exfoliated

cells for morphological abnormalities (Papanicolaou and Traut, 1941). Another extensively used tool is colposcopy, which typically is recommended after an abnormal Pap smear. The incidences of the disease and associated deaths have been reducing in developed countries, where the above-mentioned screening tools are commonly practiced. However, these methods have a number of shortcomings, consisting of high rate of false-negative or false-positive results that might be due to subjective interpretations of the cytologist/ pathologist who is diagnosing the disease on the basis of morphological abnormalities. There are difficulties in separation of normal, basal cell hyperplasia, immature squamous metaplasia, and inflammation-associated changes from true koilocytes, which are indicative of low-grade squamous metaplasia (McCluggage et al., 1998). There are also difficulties in differentiating low-grade squamous intraepithelial lesions (LSILs) from immature metaplastic squamous with atypia, mild abnormality, resulting in inter- and intraobserver difference of opinion (McCluggage et al., 1998).

There is an urgent requirement for effective screening tools which approximately can reduce 90% of deaths by early diagnosis (Bazant-Hegemark et al., 2008). Optical diagnostic methods, including confocal microscopy and optical coherence tomography (OCT) are potential techniques that can be used to visualize the neoplastic changes more sensitively. These tools are emerging as nondestructive, real-time imaging modalities for early diagnosis of precancerous lesions of the cervix (Bazant-Hegemark et al., 2008). Potentially offering higher sensitivity and specificity based on biomolecular analyses, spectroscopic methods such as infrared (IR) absorption and Raman scattering are attracting increased attention. The suitability of these techniques, particularly Raman spectroscopy, for rapid noninvasive screening of cervical biopsies will be explored in this chapter.

Cervix

Cervix is the lower fibro-muscular part of the uterus. The upper portion adjoining the uterus is known as the endocervix and the lower end adjoining the vagina is known as the exocervix. The lining of the endocervix is covered by a single layer of tall columnar cells with dark staining nuclei forming the columnar epithelium (Figure 14.1). The exocervix, on the other hand, is followed by the squamous epithelium layer, which has glycogen as a major component. The point where the two epithelial layers meet is known as the

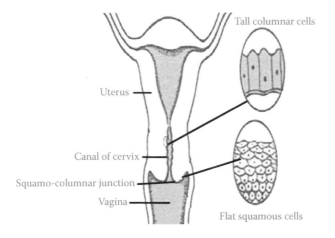

Figure 14.1 Diagrammatical representation of human cervix.

transformation zone (Sellors and Sankaranarayanan, 2003). Initially, the epithelium at the transformation zone comprises a single layer of mucus-secreting columnar epithelium, but the hormonal changes at the onset of puberty make the environment for the exposed epithelium more acidic, causing it to change from a single-layer epithelium to a more protective, stratified squamous epithelium.

Cervical cancer

Cervical cancer is the major cause of gynecologic-related death worldwide, but is absolutely avoidable with regular cytological screening (Shepherd and Bryson, 2008). Cervical cancer impacts the lives of women worldwide. Recent reports on cervical cancer indicate that 275,100 women died of cervical cancer in 2008 and 85% of these deaths occurred in developing countries (Jemal et al., 2011). Cervical cancer is only the 11th most common cancer among women in developed countries, however, due to existing screening practices. In the absence of such established screening procedures, it becomes the second most common cancer (Peto et al., 2004). Most of the precancerous and cancerous lesions arise in the transformation zone, which is the area where metaplasia occurs at the onset of puberty (Cuzick et al., 2008).

Types of cervical cancer

Cervical cancer can be divided into the following types.

Squamous cell carcinoma

Squamous cell carcinoma (SCC) is the most common carcinoma of the cervix (about 90%) and it originates from the squamous epithelium of the exocervix. It is subdivided into three classes:

1. Large cell keratinizing tumors
2. Large cell nonkeratinizing tumors
3. Small cell nonkeratinizing tumors

Adenocarcinoma

Adenocarcinoma is a type of cancer that arises from the glandular epithelium. Although it is rarely found in the cervix, it develops in the endocervix. It accounts for 10% of cervical cancers and primarily affects women over the age of 60 years (Mera, 1997). Different reports suggest that there has been a recent increase of up to 15% in its occurrence, especially in young women who take oral contraceptives (Quinn, 1998).

Mixed adenosquamous tumors are rare (about 5%). Other malignant tumors such as lymphomas, melanoma, and sarcomas can be diagnosed but are rare (Mera, 1997).

Risk factors

The risk factors for cervical cancer according to epidemiological studies are discussed in detail below.

Sexual activity

The risk relates to the number of sexual partners for either the woman or her partner. The greater the number of sexual contacts, the greater can be the risk involved (Mera, 1997).

The age of first sexual intercourse is not a risk factor in itself, other than the likelihood that initiation at an earlier age may result in the accumulation of more sexual partners. Most of the other risk factors are associated with this factor.

Infection with human papillomavirus

Infection with the human papillomavirus (HPV) is the most commonly cited cause of cervical cancer, occurring in 99.7% of cases, especially with types 16 and 18 of the virus (Bosch et al., 1995; Walboomers et al., 1999). Over 100 types of HPV have been identified, of which 40 are known to infect the genital tract. Some of them, such as type 6 and 11, cause only warts and are known as low-risk strains, while others such as type 16 and 18 can lead to cervical precancerous states and are thus classed as high-risk strains. Type 16 followed by 18 are the most frequently detected types of HPV at the time of diagnosis of SCC (Woodman et al., 2007). Other high-risk HPV types include 31, 33, 35, 39, 45, 51, 52, 56, 58, 59, 68, 73, and 82 and the low-risk types include 6, 11, 40, 42, 43, 44, 54, 61, 70, 72, and 81 (Villa and Denny, 2006). The integration rate of the HPV 16 genome with that of the infected basal cells increases with severity of the disease. However, in some women with invasive disease, only episomal forms, genetic material existing as a free autonomously replicating deoxyribonucleic acid (DNA), are identified (Woodman et al., 2007). On the other hand, HPV 18 is found in an integrated form in most of the cases with high-grade CIN or invasive disease (Woodman et al., 2007). Infection with HPV 18 is considered to be the cause of adenocarcinoma, which is only 10% of cervical cancer (Woodman et al., 2007).

   Pathological examination of cervix cone biopsies reveals that, in many cases, more than one grade of cancer can coexist in the same sample and this is considered to be due to the infection of the cervix by different strains of HPV and not due to the diverse stages of evolution in an ongoing process that is caused by a single viral-type infection (Agorastos et al., 2005).

Irregular cervical cytological screening

Those women who do not undergo cervical cytological examination on a regular basis may have four times more risk of developing cervical cancer than those who are regularly screened (Mera, 1997).

Cigarette smoking

Cigarette smokers have higher risk of developing cervical cancer than nonsmokers (Mera, 1997). The metabolites of tobacco smoke are found in the cervical mucus of smokers. The presence of constituents of tobacco smoke in the cervix may favor infection by HPV. The DNA of the cells in the cervix forms adducts with some metabolites of tobacco and can act as a tumor promoter after infection with HPV (Collins et al., 2010).

Use of oral contraceptives

The risk of cervical cancer is increased due to the use of oral contraceptives, whereas use of barrier contraceptive methods result in a decreased risk and the reason for these methods being safe is that they do not allow seminal fluid to come in contact with the cervix (Mera, 1997). Prolonged use of oral contraceptives lead to increased risk of cervical cancer in women who are HPV positive (Smith et al., 2003).

Classification system

Cervical cancer arises after progressing through a long preinvasive period. Microscopically, this is characterized as a sequence of events ranging from cellular atypia to various grades of dysplasia or CIN before the cancer progresses to an invasive stage (Richart, 1969). It has been observed that some cases of dysplasia regress, while some persist, some progressing to carcinoma *in situ* (CIS). The theory of cervical cancer originated back in the late nineteenth century, when noninvasive atypical changes were found in tissue samples adjoining invasive cancer (Sellors and Sankaranarayanan, 2003). The term "carcinoma *in situ*" was developed in 1932 to represent those lesions that had undifferentiated carcinoma cells in the full thickness of the epithelium without disturbing the basement membrane (Sellors and Sankaranarayanan, 2003). A relationship between CIS and invasive cervical cancer was reported (Reagan et al., 1953). In late 1950, the term "dysplasia" was coined to indicate cervical epithelial atypia, a missing step between normal epithelium and CIS (Reagan et al., 1953). Dysplasia was further classified into three groups: mild, moderate, and severe, depending on the levels of atypical cells in the epithelium. As a result, cervical precancerous lesions were described using the categories of dysplasia and CIS for many years. A system that separates the classification of dysplasia from CIS was increasingly accepted on the basis of observations from a number of follow-up cases consisting of women with such lesions (Sellors and Sankaranarayanan, 2003). A direct relationship between progression of disease and histological grade of tissue led to the idea of a single continuous process of the disease (Sellors and Sankaranarayanan, 2003). This continuous process leads the normal epithelium through various grades of dysplasia into invasive cancer. On this basis, the term "CIN" was developed by Richart in 1968 for cellular atypia confined to the epithelium (Richart, 1969). It classified mild dysplasia as CIN I, moderate as CIN II, with CIN III describing severe dysplasia to CIS.

HPV was recognized, in 1980, as an infectious agent responsible for pathological changes such as koilocytic atypia (Richart, 1990), which means cells with irregular cytoplasm or a halo; koilocytes are cells with a luminous circle in the cytoplasm (Sellors and Sankaranarayanan, 2003).

The hypothesis that cervical cancer may arise from infections with HPV was made by Hausen (Hausen et al., 1974a,b; Zurhausen, 1976). The isolation of HPV 16 and later HPV 18 and characterization as infectious agents responsible for cervical cancer (Durst et al., 1983; Boshart et al., 1984) and a detailed study of cervical dysplasia suggesting that its development is not a continuous event, as only 5% of CIN II and less than 12% of CIN III progress to invasive cancer if left untreated (Pinto and Crum, 2000), paved the way for the creation of the Bethesda system (TBS) of classification. TBS of classification was created in 1988 and revised in 1991 in a workshop organized by the US National Cancer Institute (Jastreboff and Cymet, 2002). This system classifies CIN according to the following categories: atypical squamous cells of undetermined significance (ASCUS) refers to squamous atypia, LSIL applies to lesions with condylomatous (a skin tumor near the anus or genital organs) atypia, mild dysplasia or CIN I, and high-grade squamous intraepithelial lesion (HSIL) applies to lesions with moderate to severe or CIN II and CIN III lesions and CIS (Kaufman and Adam 1999; Scott et al., 1999; Sigurdsson, 1999). TBS was reevaluated and revised in a workshop organized in 2001 by the US National Cancer Institute, at which more than 20 countries were represented (Solomon et al., 2002). The revisions of the 2001 TBS included explicit statements about sample adequacy, general categorization, and interpretation and results. In the adequacy category, "satisfactory" and "unsatisfactory" are kept, but "satisfactory but limited by" is removed. The category of "atypical squamous cells" (ASC) is replaced by "atypical squamous cells of undetermined significance" (ASCUS) while dividing it into qualifiers of (a) ASC of "undetermined significance" (ASC-US) and (b) "cannot exclude high-grade squamous intraepithelial lesion (HSIL)" or (ASC-H). The categories of ASCUS, "favor reactive" and "favor neoplasia" were removed. The terms including LSILs and HSILs remained the same as previous. The category of "atypical glandular cells of undetermined significance" (AGUS) is removed to avoid confusion with ASCUS and replaced by the term "atypical glandular cells" (AGC), in an effort to recognize whether the cells were originated from endometrial, endocervical, or unqualified. "Endocervical adeno CIS" and "AGC, favor neoplastic" were included as separate AGC categories (Apgar et al., 2003).

Diagnostic methods

The molecular and biochemical changes that ultimately led to the occurrence of cancer include the disturbance of the basic organization of nucleic acids, proteins, lipids, and carbohydrates (Ooi et al., 2008) and these changes can be used as diagnostic markers of cancer. Historically, many techniques have targeted the above-mentioned biomolecules and subcellular structures, nucleus-to-cytoplasm ratio, or cellular morphology for diagnosis as well as for the prognosis of cancer. However, changes at this level are detectable only after they have already caused considerable gross morphological changes in the tissue. Due to technological advances, more opportunities have arisen to investigate the biomolecular components in tissue samples and to examine malignant abnormalities on this basis. As a result, the mutations/biochemical changes in biomolecules appearing before manifestation of disease on a cellular level can be detected earlier, facilitating timely and precise diagnosis, leading to increased patient survival and quality of life (Ooi et al., 2008).

Techniques currently used

Currently, screening for cervical cancer is preliminarily done by examining the exfoliated cells of the cervix for cytological abnormalities with microscope (Ronco et al., 2006) and the technique is commonly known as the Papanicolaou test or Pap smear test (Papanicolaou and Traut, 1941). Pap smear test showing abnormality is followed by the assessment of the cervix with a low-power light microscope (colposcope) for localized variations in the tissue reflectance, a marker for precancerous lesions (Thekkek and Richards-Kortum, 2008). The agents, including acetic acid and Lugol's iodine and use of a green illumination filter, can enhance the contrast of suspicious regions. Due to the low specificity of such a visual examination, the ultimate diagnosis is recognized by examining a punch biopsy or excised tissue sample by the pathologist (Mitchell et al., 1998). HPV infection is accepted as a root cause of such pathological changes (Richart, 1990). Thus, the identification of abnormalities in exfoliated cells (smears) may be followed by screening using the HPV DNA or ribonucleic acid (RNA) detection test (Cuzick et al., 2008). Currently, HPV testing is being introduced as a primary screening tool and an adjunct to cervical cytology for women over 30 years as recommended by the American Society for Colposcopy and Cervical Pathology (Wright et al., 2007). The recommendations have been updated to address the age-appropriate screening methodologies, including cytology and HPV testing, follow-up of women after screening, including management of screen positives and screening interval for screen negatives, age at which to exit screening, future considerations regarding HPV testing alone as a primary screening approach, and screening strategies for women vaccinated against HPV16/18 infections (Saslow et al., 2012).

Pap smear test

In order to determine cytological abnormalities in the cells of the cervix, the Pap smear is the test that is used for the preliminary screening of cervical cancer. This is done by the microscopic examination of the exfoliated cells of the cervix (Ronco et al., 2006).

The collection of cells from the squamo-columnar junction and outer aperture of the cervix of the uterus is done by the use of a cyto brush. The cells are then stained by the Papanicolaou technique and are examined under a microscope for abnormalities in the nucleus and cytoplasm, such as nuclear-to-cytoplasm-size ratio. Increased ratio is linked with a severe degree of CIN. Symptomatic changes to the nucleus include (a) increase of nuclear size with difference in shape (common feature of all dysplastic cells), (b) intense staining pattern of nucleus, (c) irregular chromatin distribution, and (d) uncommon mitotic figures and visible nucleoli.

The proportion of the thickness of the epithelium with mature and differentiated cells is used for grading CIN. More severe degrees of CIN may have greater proportion of the thickness of epithelium (Sellors and Sankaranarayanan, 2003).

Advantages

1. The sensitivity and specificity of Papanicolaou test are reported to be 11%–99% and 14%–97%, respectively (Boyko, 1996).
2. The technique is relatively noninvasive.

Disadvantages

1. Sampling errors may occur and consistent and reliable sampling cannot be guaranteed.
2. The technique is subjective and may lead to wrong interpretation.

HPV-based screening

A large amount of epidemiology and laboratory-based research indicates that infection with oncogenic HPV, especially with types 16 and 18, is frequently associated with SCC in the cervix (Woodman et al., 2007). This link between infection with high-risk HPV in the lower genital tract and cervical cancer has led to the establishment of a detection system which is based on the presence of the HPV DNA or RNA for cervical cancer screening and of vaccines against high-risk HPV types. There are currently two vaccines available; Gardasil (Merck and Co.) and Cervarix (GlaxoSmithKline). Gardasil provides immunity against the two high-risk HPV types, 16 and 18, which are linked with the development of cervical cancer and two low-risk HPV types, 6 and 11, responsible for the majority of genital warts and recurrent respiratory papillomatosis. Cervarix claims to provide immunity against the two high-risk HPV types, 16 and 18 (Madrid-Marina et al., 2009).

DNA-based techniques

Currently, two DNA-based tests for HPV detection are approved by the US Food and Drug Administration (FDA); Hybrid Capture II (HC II) (Qiagen Gaithersburg, Inc., Maryland), and the Hologic Cervista high-risk HPV DNA test (Hologic Incorporated, Bedford, Maryland) (Clad et al., 2011). Other DNA-based tests available include Roche Amplicor, Roche Linear Array, and Cobas 4800 (Ovestad et al., 2011).

Hybrid capture-II has a sensitivity of 93.1% and 95.5% for detecting CIN II and CIN III, respectively (Dufresne et al., 2011). The detection of a virus in women with transient stages of infection may increase the burden of screening of HPV-positive woman, as detection of HPV DNA prompts immediate treatment of the lesions, although follow up of these lesions suggest that 45.7% of CIN I and 21% of CIN II lesions regress (Insinga et al., 2009). The technique can be used as a follow-up test for women treated for high-grade lesions with localized treatment such as ablation or excision therapy to identify women who have been or have not been cured by their treatment (Cuzick et al., 2008). This technique is able to distinguish between HPV types, including high-risk as well as low-risk one, on the basis of nucleic acid hybridization of the bases of the E-1 gene but does not allow the identification of specific genotype.

Amplicor® is another DNA-based HPV test, which can target and detect DNA from 13 high-risk genotypes. This test allows simultaneous PCR amplification of the target DNA from 13 high-risk HPV genotypes (16, 18, 31, 33 35, 39, 45, 51, 52, 56, 58, 59, and 68) and from β-globin DNA as a cellular control. The test uses amplification of target DNA by the polymerase chain reaction (PCR), using target specific complementary biotinylated primers. After PCR amplification, the oligonucleotide probes are used to hybridize to the amplified products and they are detected by colorimetric determination (Ovestad et al., 2010).

In the Roche Linear Array test, the biotinylated primers are used to define about 450-bp sequence within the L1 region of the HPV genome. A pool of primers is used to amplify the HPV DNA targets from 37 HPV genotypes (6, 11, 16, 18, 26, 31, 33, 35, 39, 40, 42, 45, 51, 52, 53, 54, 55, 56, 58, 59, 61, 62, 64, 66, 67, 68, 69, 70, 71, 72, 73, 81, 82, 83, 84, 89, and IS39 [subtype of 82]). It should be noted that the group of 16 high-risk genotypes (16, 18, 31, 33, 35, 39, 45, 51, 52, 56, 58, 59, 68, 73, 82, and IS39) includes the 13 genotypes targeted by the Amplicor test. Moreover, groups of possible high-risk (26, 53, and 66), unclassified (55, 62, 64, 67, 69, 71, 83, 84, and 89), and low-risk (6, 11, 40, 42, 54, 61, 70, 72, and 81) genotypes can also be detected. The β-globin DNA can be used as a cellular control (Ovestad et al., 2010).

The Cobas 4800 system, another DNA-based test for HPV detection (Roche Molecular Diagnostics, Pleasanton, California, USA) provides automated sample preparation combined with real-time PCR technology, which can amplify and simultaneously detect 14 high-risk HPV genotypes, identification of HPV 16 and 18, and β-globin used as an internal control for sample validity (Ovestad et al., 2011).

The Hologic Cervista high-risk HPV DNA and Cervista HPV 16/18 tests (Third Wave Technologies, Madison, Wisconsin; now owned by Hologic, Bedford, Massachusetts) were approved by the US FDA in March 12, 2009. Cervista HPV HR (high risk), for use in cervical cancer screening, is a DNA test for 14 carcinogenic HPV genotypes. Cervista HPV 16/18 is approved for the DNA detection of HPV-16 and HPV-18, the two most carcinogenic HPV types (Kinney et al., 2010).

RNA-based techniques

The progression of cervical cancer is dependent on the continuous expression of the E6/E7 genes of HPV, essential for the successful transformation and maintenance of the neoplastic phenotypes in cervical carcinoma cells. To detect these genes, on the basis of their mRNA, two techniques are available: Pre Tect HPV Proofer and APTIMA (Clad et al., 2011).

The Pre Tect HPV Proofer test is a real-time multiplex assay for full-length amplification of the mRNA of five high-risk HPV types and is carried out by extraction of the total RNA from the samples.

The APTIMA HPV messenger RNA assay works on the target capture transcription-mediated amplification of E6/E7 mRNA from 14 high-risk HPV types and is able to differentiate between these types and identify which HPV type is present as well as which are more likely to persist.

Advantages
1. HPV DNA detection can be used as a primary screening method.
2. It is an adjunctive test to cytology; can be used as a classification tool for the administration of women having ASCUS.
3. HPV DNA genotyping can be used for the recognition of different types of HPV.
4. Sensitivity and specificity for detection of high-risk HPV are found to be >92% and 99% for the APTIMA HPV Assay and 93% and 82% for the HC II test (Dockter et al., 2009).

Disadvantages
1. HPV testing identifies transient infections that are common in women under 30 years and hence no important information is provided below this age.
2. None of these techniques can predict regression of CIN-II and CIN-III.

Colposcopy

Colposcopy is a technique of visualizing abnormalities in the tissue using light microscopy. Contrasting agents such as 3%–5% dilute acetic acid, resulting in the so-called aceto whitening, are used as contrast agents to help differentiate abnormal areas of the tissue from normal. Aceto whitening is caused by reversible coagulation of intracellular proteins. An increase in intracellular proteins during neoplasia results in a dense white spot (Gaffikin et al., 2003). Recognition of well-defined areas near to the squamo-columnar junction indicates a positive test. The colposcopist extracts biopsies from any abnormal area found and sends them to the pathologist for further examination.

Advantages
1. This is a relatively noninvasive technique.
2. Detailed and careful examination is possible as the technique provides good magnification.
3. It has sensitivity greater than 90% (Mitchell, 1994).

Disadvantages
1. The technique/instrumentation is expensive.
2. Minor signs may be overinterpreted and may lead to wrong decisions.
3. It has poor specificity (greater than 50%) (Mitchell, 1994).
4. It is subjective.

Visual inspection with acetic acid

In addition to more advanced techniques, simple visualization methods have also been investigated to facilitate cervical cancer screening in resource poor settings. For example, the use of visual inspection with acetic acid (VIA) is also under study as a substitute to Pap smear and colposcopy in many developing countries (Sankaranarayanan et al., 1998, 2007; Aggarwal et al., 2010; Cremer et al., 2010; Ekalaksananan et al., 2010; Murillo et al., 2010; Muwonge et al., 2010; Nessa et al., 2010; Ngoma et al., 2010). VIA involves the evaluation of cervix with unaided eye with the help of a light source after application of contrasting agents such as 3%–5% acetic acid for about 1 min time. A latest study of the performance of this technique in 26 experiments reported 80% sensitivity and 92% specificity (Sauvaget et al., 2011).

Advantages
1. VIA requires minimum support infrastructure.
2. The patient can be referred immediately for testing.

Disadvantages
1. It is a subjective technique.
2. There is a need to identify consistent protocol for the doubtful lesions and to train the provider.

Histopathology

The examination of the processed tissue samples by a pathologist for diagnosis is called histopathology. The processing of the tissue samples includes a series of steps, which preserves the anatomy of the tissue samples as close as possible to their structure *in vivo*. After taking a biopsy, the tissue samples are passed through different processes which include fixation, dehydration, clearing, embedding, section cutting, and staining. Tissue processing can be done manually or can be automated. After the final step, staining, the tissue sections on slides are examined by the pathologist. The diagnosis and grading of CIN by a pathologist is done on the basis of histological features, including differentiation, maturation, and stratification of the cells, as well as nuclear abnormalities. The samples with more severe degrees of CIN are expected to have a higher proportion of the thickness of the epithelium composed of undifferentiated cells with only a thin layer of mature, differentiated cells on the surface. Nuclear abnormalities, including enlarged nuclei, increased nucleus-to-cytoplasm ratio, increased intensity of stained nuclei, as well as variation in the size of nuclei are also taken into account (Sellors and Sankaranarayanan, 2003).

Advantages
1. Exact diagnosis of the severity of the cancer is possible.
2. The margin line of diseased area can be determined.

Disadvantages
1. False-negative and false-positive results may be possible, especially in the case of CIN or atypical cells.
2. The technique is subjective and interpretation of the results is in the hands of the pathologist.

New techniques based on optical methods

There has been a tremendous increase in the development of more complex optical methods for the detection and diagnosis of cervical cancer and they are now considered as techniques of choice for early detection, because these are fast, can be automated (Chidananda et al., 2006). The optical techniques currently under development include confocal microscopy, OCT, and fluorescence spectroscopy. In the next sections of this chapter, these techniques are discussed in detail with a particular focus on the use of these techniques in studies related to the detection of cervical cancer along with advantages and disadvantages of each technique.

Confocal microscopy

A confocal microscope works on the principle of optical sectioning and is normally explained as a microscope with light coming only from the focal plane. It allows noninvasive high-resolution imaging at the cellular level at varying depths of the cervical epithelium. The light for detection is created by either fluorescence or reflection. Confocal fluorescence microscopy, by probing for fluorophoric metabolites (NADH, FAD), can differentiate fluorescence from the cytoplasm and peripheral membrane of the cell. It can also be employed to differentiate between different layers of the tissue samples, including basal cells from parabasal, intermediate, and superficial cells (Pavlova et al., 2003).

Confocal microscopy, while working on the basis of the reflectance, can visualize the indicators of cancer onset and progression, such as morphological changes, nuclear-to-cytoplasm ratio, or irregularity of cell spacing (Carlson et al., 2005; Bazant-Hegemark et al., 2008). The confocal technique has been applied *in vivo* to acquire images of the cervical epithelium with subcellular resolution by using a fiber-bundle confocal reflectance microscope (Carlson et al., 2005) in order to differentiate between normal and cervical precancers.

The technique has the following advantages and disadvantages.

Advantages

1. It can provide an axial resolution of ≤ 1 μm.
2. Light required for the detection is produced by reflection as well as fluorescence, providing double selectivity of target volume. Simple chemicals such as acetic acid can be used as contrast agents.
3. The technique has high sensitivity (100%) and specificity (91%) (Bazant-Hegemark et al., 2008).

Disadvantages

1. Contrast agents can make the focusing complicated by causing the tissue to swell.
2. The field of inspection is of the order of a few hundred micrometers.
3. The penetration depth of modern probes is ~250 μm but can be impaired by the presence of scatterers in the overlying tissue (Bazant-Hegemark et al., 2008).

Optical coherence tomography

OCT is an imaging tool that uses the scattering of the incident light by the tissue to obtain quantitative information of the microstructural changes in the tissue sample. The technique is based on low coherence interferometry, which is obtained by using typically a near-IR light source and an interferometer and usually its working principle is explained by analogy to ultrasound. The changes in the refractive index determine the contrast for the structural features (Bazant-Hegemark et al., 2008). The higher resolution of OCT compared to ultrasound is due to the use of light, rather than sound or radio waves. The light beam is focused on the tissue, and a tiny portion of this light which is reflected from subsurface features is collected. It should be noted that most light is not reflected but, rather, scatters and loses its original direction and does not contribute to forming an image but rather contributes to glare. By using the OCT technique, scattered light is filtered out with the help of an interferometer to completely remove the glare. Even the very tiny proportion of reflected light that is not scattered can then be detected and used to form the image.

The diagnostic potential of the technique has been shown by different researchers, assessing normal and abnormal cancerous tissues (Boppart et al., 1998; Clark et al., 2004;

Maitland et al., 2008). Endoscopic OCT was used for the first time for the evaluation of cervical malignancy (Sergeev et al., 1997; Feldchtein et al., 1998). The intactness of the basement membrane has been determined to be a marker for a healthy epithelium, by the evaluation of cervical tissue samples from the genital tract of the female, 21 healthy and 11 CIN III samples, using OCT (Pitris et al., 1999). In another study of samples, including CIN I and CIN II/III, it was reported that the thickness of the squamous epithelium varies for healthy samples and a healthy epithelium can be used as a marker for an intact basement membrane (Escobar et al., 2004, 2005).

The technique has the following advantages and disadvantages.

Advantages
1. The technique can show diagnostic capability at a tissue depth of 2–3 mm because it uses near-IR where tissue is more transparent.
2. The technique is noninvasive, as it makes use of near-IR light sources in the range of 800–1500 nm, at powers that are harmless to skin (Fujimoto et al., 2000).
3. The interferometric detection is not disturbed by ambient light.

Disadvantages
1. Although it has sensitivity of 95%, it has a specificity of only 46% for the differentiation of precancerous lesions from cancerous lesions of cervical epithelium (Gallwas et al., 2010).
2. The choice of the wavelength of light is limited by the strong absorbers in tissue samples, namely, melanin (visible range) and water (IR range).
3. The light sources to be used need to fulfill the stringent criteria with respect to partial coherence, for the interferometric detection.
4. Low contrast of the technique.
5. Poor spatial resolution and subcellular structures or even single cells cannot be resolved.

Spectroscopy

Spectroscopy is defined as the interaction of light with matter in terms of wavelength (λ). The three main types of spectroscopy are absorption, emission, and scattering spectroscopy. Absorption spectroscopy usually entails ultraviolet, visible, and/or IR absorption. It measures the spectrum of light that a substance absorbs to get information about its structure. Emission spectroscopy measures the spectrum of light that a substance emits upon excitation, for example, fluorescence and phosphorescence spectroscopy. Scattering spectroscopy is similar to emission spectroscopy but it detects and measures all the wavelengths that a substance scatters upon excitation, for example, Raman spectroscopy (Latimer, 1967).

Fluorescence spectroscopy

This technique is a form of electromagnetic spectroscopy and utilizes light, usually in the UV region of the spectrum, to excite electrons in molecules of specific compounds, resulting in the emission of light of lower energy than the incident light, called fluorescence, which is measured for the analysis of the samples.

Fluorescence spectroscopy has been employed to demonstrate the potential of the technique for detection of precancers in the cervix (Ramanujam et al., 1996; Georgakoudi et al., 2002; Zlatkov, 2009). The fluorescence spectra of normal and cancerous tissue samples have been studied (Hubmann et al., 1990; Ramanujam et al., 1994; Agrawal et al., 1999; Brookner et al., 1999; Zuluaga et al., 1999; Grossman et al., 2001; Ramanujam et al., 2001) in an effort to establish an optical pathology method for the early detection of cancer.

By acquiring the laser-induced fluorescence spectra from normal and cancerous stage-III B cervical tissue samples and employing principal component analysis (PCA), Chidananda et al. (2006) have concluded that fluorescence spectroscopy can be used as a technique to predict the pathology of samples with high sensitivity and specificity, by performing comparisons with the calibration sets of different stages of cervical cancer.

Advantages

1. The technique can provide very specific information as specific molecules will give fluorescence at a specific excitation wavelength.
2. The technique has great sensitivity to changes in the structural and dynamic properties of biomolecules.
3. The technique has made it possible to identify cellular components with a high degree of specificity.

Disadvantages

1. Consumables like the fluorescent tags may be required for the samples to be studied, which are expensive.
2. As it uses UV or visible light, confocal fluorescence spectroscopy has a penetration depth of only ~0.1 mm.

Vibrational spectroscopy

IR spectroscopy

IR radiation belongs to that part of the electromagnetic spectrum that lies between visible and microwave region. The IR region is divided into three regions (near-, mid-, and far-IR) according to its distance from the visible region in the electromagnetic spectrum, as illustrated in Table 14.1.

Frequency (v) is defined as the number of wave cycles that pass through a point in 1 s. The unit for the frequency is hertz, 1 Hz = 1 cycle s^{-1}. The length of one wave cycle is defined as the wavelength (λ) and is measured in micrometers (μm). Wavelength and frequency are inversely related:

$$v = c/\lambda \qquad (14.1)$$

IR spectroscopy is based on the absorption of IR radiation by the sample under study and the fact that molecules absorb specific frequencies of the incident light, which are characteristic of their structure.

Table 14.1 Regions of the IR spectrum in commonly employed units

Region	Wavelength range (λ, μm)	Wavenumber range (λ, cm^{-1})	Energy (E) (kcal mol^{-1})
Near	0.78–3	12,820–4,000	10–37
Middle	3–30	4,000–400	1–10
Far	30–300	400–33	0.1–1

IR spectroscopy optically probes the molecular vibrations that depend on the composition and structure of the material under study. IR light is used to irradiate the sample, and changes in the vibration induced are recorded. Disease and other pathological conditions lead to changes in structure and chemical composition. These changes produce alterations in the vibrational spectra that can be used as sensitive phenotypic marker of disease (Krafft et al., 2009). Near- or mid-infrared light sources give the best contrast for medical applications.

Fourier transform infrared (FTIR) spectroscopy is a measurement technique for the acquisition of IR spectra. Frequency differentiation is performed with an interferometer. After passing through the sample, the measured signal is the interferogram. The mathematical Fourier transform of this signal produces a spectrum similar to that from conventional (dispersive) IR spectroscopy (Gremlich and Yan, 2001).

The use of FTIR spectroscopy for cervical cancer diagnosis has been demonstrated by studying the exfoliated cells from 156 samples from female patients (Wong et al., 1991) and differentiating them into normal, dysplastic, and malignant. The cancerous progression and normal proliferation has been associated with glycogen reduction, increased hydrogen bonding from phosphodiester groups, and reduced hydrogen bonding in alcoholic groups of the amino acids.

Analysis of FTIR spectra of over 2000 individual cells from 10 normal females, 7 females with dysplasia, and 5 females with SCC has revealed that the spectra of normal-appearing intermediate and superficial cells of the cervix obtained from women with either dysplasia or cancer are different from those of normal women. It is reported that the structural changes underlying the spectroscopic variations may be more representative of the state of the disease than the morphological changes (Cohenford and Rigas, 1998). This information strongly supports the potential of FTIR spectroscopy for the diagnosis of cervical cancer. As cervical cells mature from basal to parabasal, intermediate, and then superficial cells, an increase of glycogen content is reported (Koss, 1992). This normal progression is disrupted in diseased tissue and hence can be used as a diagnostic marker for cervical cancer by FTIR spectroscopy. Moreover, changes in the levels of biomolecules, including RNA and DNA, can also be helpful for diagnostic purposes (Mordechai et al., 2004). The IR spectroscopy has been employed for the analysis of Pap smear cells and an algorithm is developed called the Pap Map to process the imaging data, which is reported to lead toward the automation of the screening process (Schubert et al., 2010).

Advantages

1. Larger sample area can be scanned in less time as compared to Raman spectroscopy.
2. Spatial resolution is ≥ 10 µm.
3. Possibility to measure the secondary structure of protein noninvasively in a native environment.
4. Same samples can be used for further studies.
5. No interference from fluorescence of biological sample during measurements.

Disadvantages

1. The interference from water at 1640 cm^{-1} overlaps IR absorbance of protein. Modern FTIR instrument and software can overcome this problem to a great extent.
2. There is strong interference from glass. There is need for alternative materials that do not interfere during IR spectroscopy to be used as substrates for biological samples.
3. It needs good data analysis techniques to acquire useful and meaningful information from the spectral data.

Raman spectroscopy

The working principle of Raman spectroscopy is based on a scattering phenomenon. Scattering results from the interaction between the photons and materials, such as molecules. As shown schematically in Figure 14.2, most photons colliding with molecules do not change their energy after collision. These collisions are called elastic collisions and the result of these collisions is elastic scattering called *Rayleigh* scattering. Rayleigh scattering is a two-photon process as a result of which there may be a change in direction of light but no net change in frequency.

On the other hand, inelastic collisions, after which the incident photons change their energy, lead to inelastic scattering during which exchange of vibrational energy occurs. The process is defined as Raman scattering. For a molecular vibration to be Raman active, the net molecular polarizability must be changed. The polarizability (α) is the ability of an applied electric field, E, to induce a dipole moment, μ_0, in an atom or molecule.

The Raman shift of the photon energy is a measure of the energy of vibration of the molecule. A complex molecule will have many vibrational modes, and the intensity of the scattered rays versus Raman shift gives a Raman spectrum, which is characteristic for each individual substance, and for this reason, Raman spectral features can be the identification markers of a substance and can be used to analyze its structure (Ekins and Sasic, 2008).

Studies ranging from pure biological molecules in isolation to all aspects of the human body have been a matter of strong interest in the field of spectroscopy of biomolecules. FTIR spectroscopy was used to study human and animal tissues (Blout and Mellors, 1949; Woernely, 1952). Due to the lack of well-established instrumentation at that time, researchers who attempted to record the Raman spectra of such samples were facing problems of overwhelming fluorescence from biological molecules, long integration times, and high power density requirements. Important developments in the field of diode lasers and charged coupled device (CCD) cameras in 1969 (Boyle and Smith, 1970) finally made it possible to record high-quality Raman spectra of biological samples in the 1970s (Lord and Yu, 1970). Vibrational spectroscopy has emerged as a major tool for biochemical applications as it is able to provide direct information regarding the molecular constituents of tissue samples at the cellular level.

There has been extensive use of Raman spectroscopy for various applications, including studies made on biological tissue. These include bone (Rehman et al., 1995), cornea (Kachi et al., 2000), epithelial tissue, including the larynx, tonsil, esophagus, stomach, bladder, and prostate (Stone et al., 2002, 2004; Keller et al., 2008), and lung (Kaminaka et al.,

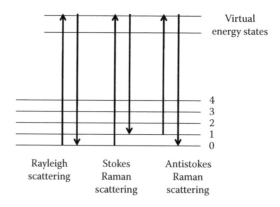

Figure 14.2 Schematic presentation of Rayleigh and Raman scattering of incident light.

2001; Huang et al., 2003a,b; Notingher et al., 2003; Liu et al., 2005; Min et al., 2005; Krafft et al., 2008; Qian et al., 2008; Short et al., 2008; Bonnier et al., 2010). The technique has also been applied to study the composition of lymphocytes (Puppels et al., 1993) and human red blood cells (Deng et al., 2005) as well as to differentiate between different types of cells, including mixed cancer cells (Krishna et al., 2005), human living cells (Kuhnert and Thumser, 2004), and individual cells (Chan et al., 2006). In addition, it has been employed to study different biological/macromolecules, including DNA (Ruiz-Chica et al., 2004), bodily fluids such as saliva (Farquharson et al., 2005a,b), tissue processing (O'Faolain et al., 2005), and raft cultures (Viehoever et al., 2003).

The use of Raman spectroscopy to distinguish cervical cancer from normal tissue was first reported in 1992 (Liu et al., 1992). Three significant peaks at 1262, 1445, and 1657 cm^{-1} were consistently observed in all gynecologic tissues studied using FTIR and Raman spectroscopy and it was observed that the intensity ratio of 1445/1657 cm^{-1} could be used to differentiate between normal and malignant cervical tissue samples. The intensity ratio of 1445 and 1657 cm^{-1} was found to be higher in normal tissue samples (Liu et al., 1992). Resonance Raman spectroscopy, using a 254 nm laser, was used to differentiate between malignant and normal cells of the cervix and the breast (Yazdi et al., 1999). Differentiation was possible due to changes detected in the peak ratios of different nucleic acid contents in normal and tumor cells. Changes in the vibrational structure of nucleic acids associated with the malignant phenotype of the cell were also detected. The peak ratio 1480/1614 cm^{-1} was found to have a higher value in malignant samples and lower for normal samples (Yazdi et al., 1999).

Spectra obtained from human cervical tissue *in vitro* using an NIR Raman spectroscopic probe could differentiate between normal and precancerous tissues with a sensitivity of 82% and specificity of 92% (Mahadevan-Jansen et al., 1998). An increase in ratio of intensities at 1454–1656 cm^{-1} and decrease in ratio of intensities at 1330–1454 cm^{-1} was used to separate the cervical dysplasia from normal tissue types *in vivo* (Utzinger et al., 2001). DNA, phospholipids, and collagen were identified as diagnostically significant (Utzinger et al., 2001). The analysis of distinctive peaks and fingerprint region agrees with the portrait provided by IR spectroscopy, that is, basal cells have characteristic bands of DNA, epithelial cells show high level of glycogen in normal cells (Wong et al., 1991). A statistical method that combines the method of maximum representation and discrimination feature (MRDF) and sparse multinomial logistic regression (SMLR) was developed for multiclass analysis of Raman spectra. Application of this method and incorporation of patient hormonal status increased the sensitivity to 98%, specificity to 96%, and accuracy to 94% (Kanter et al., 2009a,b). Four different cell lines, HPV negative C33A, HPV 18 positive HeLa which has 20–50 integrated copies of HPV per cell, HPV 16 positive SiHa with 1–2 HPV copies per cell, and CaSki with 60–600 integrated copies of HPV per cell, were studied using Raman and FTIR spectroscopy. PCA clearly differentiated the C33A, SiHa group from HeLa and CaSki group (Ostrowska et al., 2010). A recent study of cervical tissue samples from which wax was removed followed by Raman mapping and hierarchical cluster analysis (HCA) was performed. This helped to distinguish between normal squamous epithelium and CIN (Tan et al., 2011).

In the next section, FTIR and Raman mapping of tissue sections from cervical tissue biopsies is described.

Advantages of the technique
1. This is a nondestructive technique; the sample is not damaged during the analysis by this technique and can be used again.
2. The technique is fast; about 1 min per acquisition is required.

3. It is a noninvasive method.
4. There is minimal interference from water and CO_2.
5. No labeling is required during the use of this technique.
6. *In vivo* and real-time measurements from the biological systems are possible.

Disadvantages of the technique
1. High laser intensities are required because Raman scattering is a relatively weak phenomenon and thus the equipment cost can be quite high depending on their applications.
2. The technique, generally, is less sensitive for quantitative analysis as compared to chromatographic techniques such as HPLC.
3. It needs good data analysis techniques to acquire useful and meaningful information from the Raman spectra.

Optimization of vibrational spectroscopy for cervical tissue imaging
One of the several advantages of FTIR and Raman spectroscopy is the ability of the techniques to provide information about the biochemical composition of biological specimens with little or no sample preparation. In the current study, both techniques have been demonstrated for the imaging of cervical tissue section and compared to choose the best one for further analysis.

Comparison of FTIR and Raman spectral mapping of cervical tissue sections
For this study, two parallel tissue sections from an FFPP cervical tissue block were cut, one used for H&E staining and other for FTIR and Raman spectral mapping.

For H&E staining, first, the tissue section was dewaxed using xylene and then stained in Harris' hematoxylin for 5 min, and then washed in running tap water for 3 min. After washing, the sample was differentiated in 1% acid alcohol for 1 or 2 s. The cellular nuclei should be blue and the background should be clear when observed under the light microscope. This was followed by staining in 1% eosin for 1 min and then rinsing well in tap water. The sample was dehydrated by dipping three times in methylated spirit followed by two baths of absolute alcohol for 3 and 2 min, respectively, and then two baths of histoclear for 4 and 5 min, respectively. Finally, the stained tissue section was mounted with glass cover slip using DPX mounting medium. The stained tissue section was used as a morphological standard to identify abnormal and normal epithelium against the unstained tissue sections.

FTIR data acquisition
FTIR images were recorded using a Perkin Elmer Spotlight 400 N FTIR imaging system, incorporating a liquid-nitrogen-cooled mercury cadmium telluride 16×1, 6.25 μm pixel array detector, and were acquired by the spectral image software. The FTIR image from dewaxed cervical tissue section mounted on a CaF_2 slide was recorded over the range 4000–800 cm^{-1} in transmittance mode with a resolution of 4 cm^{-1} and interferometer speed of 1.0 cm^{-1} s^{-1} at continuously varying magnification. The scans per pixel for background were 120 and, for images, 16 per pixel, respectively.

Raman data acquisition
Raman maps were recorded using a HORIBA Jobin Yvon HR 800 Raman microscope (LabSpec V5.58) with a 785-nm laser as source. The 100X dry objective was used to focus on the sample and collect the Raman scattered light in a backscattering geometry. Raman

scattering was collected through a 100-μm confocal hole onto a Synapse air-cooled CCD detector for the range of 400–1800 cm^{-1} using a 300 Gr mm^{-1} diffraction grating. The instrument was calibrated using the 520.7 cm^{-1} peak of silicon. Raman spectral mapping was done using 15 s × 2 acquisition with a step size of 18 μm.

Data preprocessing of FTIR and Raman maps

For the FTIR spectra, the correction to atmospheric CO_2 absorption bands in the spectra was done by a built-in function of Perkin Elmer Spotlight software. All data processing (FTIR and Raman spectral data) was done directly on the spectral image using MATLAB® 7.2 and the protocols that have been established in the Dublin Institute of Technology (DIT) based on the internationally recognized methods (Kniefe, 2010). Data preprocessing included smoothing, baseline correction, and normalization. All spectra, including calibration and substrate backgrounds, were vector normalized and smoothed using a Savitzky Golay smoothing method. A rubber band correction for baseline removal for all the spectra was carried out and the substrate spectra were subtracted from each spectrum. This has been demonstrated by presenting the raw Raman spectral data and the processed data in Figures 14.3 and 14.4, respectively.

All the data sets have been processed by using the same protocols and there is no manipulation of the data done, which could potentially contribute to the changes in the Raman spectra and into the results of the PCA loading and KMCA. For example, the vector normalization of the spectra, in fact removes the intensity variation of different peaks (if it exists) and all the data would reflect only the changes associated with the changes occurring as a result of the disease progression.

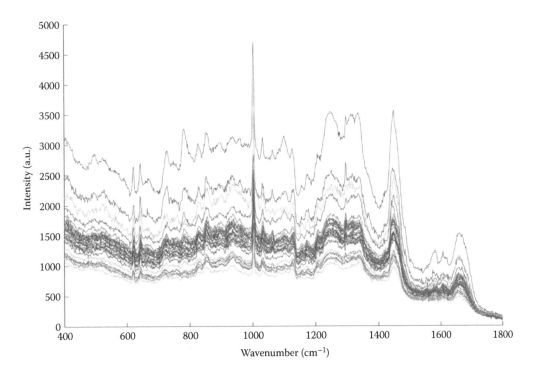

Figure 14.3 Raw Raman spectral data of basal-true normal cervical tissue sample.

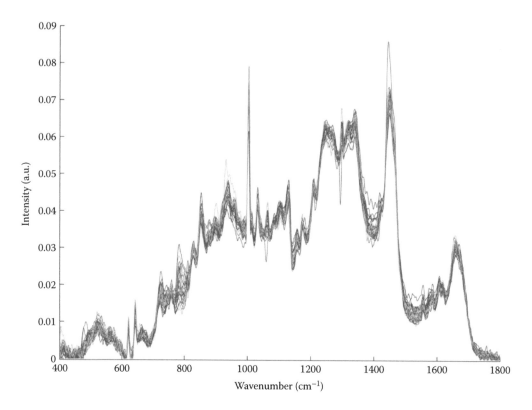

Figure 14.4 Processed Raman spectral data of basal-true normal cervical tissue sample.

Data analysis

K-means cluster analysis (KMCA) was used to analyze the spectral data sets obtained from the FTIR and Raman maps. The results of the KMCA for the initial study are presented in this chapter. KMCA is an unsupervised nonhierarchical method of clustering the cases randomly into user predefined clusters and cycles until a local minimum is found by using the Euclidean sum of squares as a descriptor (Wang and Mizaikoff, 2008). The false color map generated then shows the clusters with similar spectral and hence biological properties.

Results of FTIR and Raman spectral mapping of cervical tissue sections

Figure 14.5a shows an H&E stained cervical tissue section, the red box indicating the area from where Raman and FTIR maps are acquired. The tissue region has been diagnosed as normal by the histopathology department of the Coombe Women and Children's Hospital, Dublin. Figure 14.5b and d shows the optical images of the parallel unstained tissue section as viewed by the FTIR and Raman microspectrometer. KMCA of the FTIR and Raman maps of the same sample are shown in Figure 14.5c and e, respectively. KMCA of the FTIR image divides the map into four parts, indicated by blue and brown and orange clusters representing stroma, basal, and substrate, respectively, while green and mid-blue both associated with the superficial layer.

Figure 14.6 shows the mean spectra for the five K-means clusters in the FTIR map. The specific vibrational modes are labeled with solid lines. The bands at 1028, 1080, and 1152 cm^{-1} are assigned to glycogen and are only observed in the superficial layer (green and mid-blue). These bands are absent in the stroma (blue). This is understandable, as no glycogen

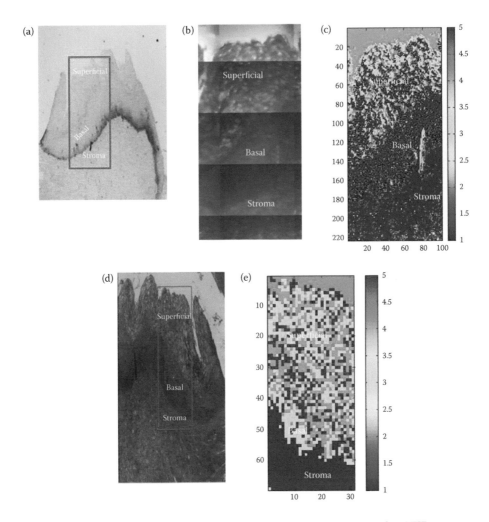

Figure 14.5 (a) H&E-stained cervical epithelium, (b) unstained tissue under FTIR spectroscope, (c) five cluster KMCA map generated from the FTIR map, (d) unstained tissue under Raman spectroscope, and (e) five cluster KMCA map generated from the Raman map.

is expected in this layer (Chowdhury and Chowdhury, 1981). The FTIR feature at 1236 cm^{-1} can be assigned to the asymmetric stretching of the O-P-O of DNA/RNA in the basal layer (brown) and for collagen proteins in the stroma (blue) (Andrus and Strickland, 1998). This FTIR band is present in both these layers. The stromal layer is mainly composed of fibroblasts that are rich in collagen and the basal layer contains more dividing cells, which leads to a greater concentration of DNA and hence this band is more prominent in these layers.

The bands at 1402 and 1454 cm^{-1} can be assigned to the CH deformation modes of carbohydrates (glycogen) (Wood et al., 1998; Dovbeshko et al., 2000). These FTIR bands are dominant in the squamous epithelium (green and mid-blue spectra) as expected. The difference between green and mid-blue is only in the intensity/concentration of glycogen, but otherwise they give the same biochemical information.

The features of the FTIR spectra at 1654 cm^{-1} (amide-I) and 1546 cm^{-1} (amide-II), both assigned to proteins, are present in all three layers, but their intensity seems to decrease

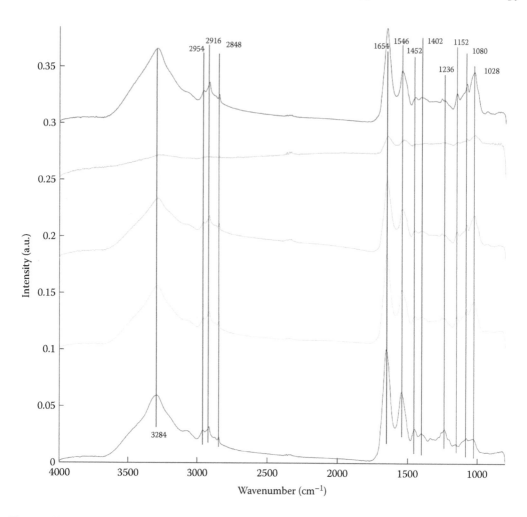

Figure 14.6 K-means cluster spectra for the FTIR map, blue cluster corresponding to stroma, brown to basal layer, green and mid-blue colors to superficial epithelium, and orange representing substrate of region in Figure 14.5c.

from epithelial, to stroma, and to superficial layers, respectively. In the epithelial layer, these FTIR spectral features can be associated to the presence of the proteins, which are also present inside the nucleus of the epithelial cells required for folding of the DNA. Hence, these FTIR features of the proteins are almost always seen in every spectra taken from the cells or tissues. The band at 2300 cm^{-1} cannot be assigned to any biological feature and is probably due to the incomplete atmospheric correction, therefore assumed to be spectral artifacts. The sharp FTIR bands at 2848, 2916, and 2956 cm^{-1} are due to the lipids (Fabian et al., 1995; Huleihel et al., 2002).

The KMCA of the FTIR map shows the separation of cervical epithelium into three distinct layers, but the analysis of the mean representative spectrum of each cluster does not elucidate the biological differences underlying this separation. As the FTIR marker bands for DNA (basal layer) and collagen (stromal layer) at ~1236 cm^{-1} are not resolved, there is no band to differentiate between these two layers.

KMCA of the Raman map is shown in Figure 14.5e. The mid-blue cluster corresponds to the substrate. The KMCA Raman map differentiates the tissue section into stroma (blue), with characteristic Raman bands of collagen including 849, 921, 938, and 1245 cm^{-1}, basal (green) with Raman bands for DNA bases, thymine (755 cm^{-1}), adenine (722 cm^{-1}), and cytosine (782 cm^{-1}) and superficial layers (orange and brown) with Raman bands of glycogen at 480, 849, and 938 cm^{-1}, as shown in Figure 14.7.

The layer differentiating bands for the mean spectra of FTIR map were limited to those at 1028, 1080, and 1152 cm^{-1}, related to glycogen, and 1236 cm^{-1} related to asymmetric stretching of the O-P-O of DNA/RNA and amide-III mainly due to collagen. However, in the case of Raman spectroscopy, there are distinctly resolved bands for collagen, DNA bases, and glycogen. It should be noted that these bands related to DNA were not observed in the FTIR spectra of this sample. Thus, a comparison of Figures 14.6 and 14.7 indicates the rationale for preference of Raman microspectroscopy over FTIR for the current studies. Moreover, the spatial resolution of Raman spectroscopy is less than 1 µm while FTIR is ~5–10 µm, offering the prospect of investigations at subcellular level.

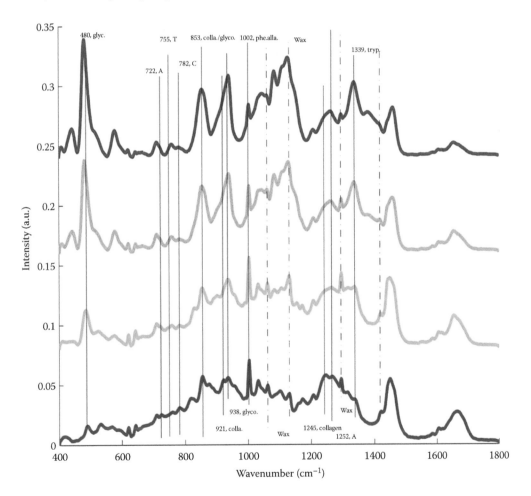

Figure 14.7 Mean Raman spectra from KMCA of Raman map, blue cluster corresponding to stroma, green to basal layer, and brown and orange colors to superficial epithelium from region in Figure 14.5e.

Conclusion

Different techniques currently in use for research or clinical purposes for the screening of cervical cancer have been introduced in this chapter. These techniques, Pap smear, colposcopy, HPV DNA/RNA-based techniques, VIA, histological examination, confocal microscopy, OCT, and vibrational spectroscopy are discussed along with their advantages and disadvantages, with particular focus on vibrational spectroscopy. Moreover, the use of the FTIR and Raman microspectroscopy is demonstrated and the spectral data acquired by using these techniques is compared to demonstrate that Raman microspectroscopy has great potential to be used for the early diagnosis of the cancer.

FTIR and Raman microspectroscopy have been employed to acquire spectral maps from the cervical tissue sections. This is done to compare the results of both the techniques and choose one for the further studies based on the biochemical information attained.

The results presented and discussed so far for FTIR and Raman microspectroscopy and their comparison points toward the greater suitability of the latter technique for the current study. The parallel analysis of the spectral data revealed more biochemical changes in the Raman spectra, particularly more DNA-related bands are detected in the case of the Raman spectral data. The higher content Raman spectral information achieved is due to the higher spatial and spectral resolution of Raman microspectroscopy compared to FTIR spectroscopy.

References

Aggarwal, P., Batra, S., Gandhi, G., and Zutshi, V. 2010. Comparison of Papanicolaou test with visual detection tests in screening for cervical cancer and developing the optimal strategy for low resource settings. *International Journal of Gynecological Cancer*, 20, 862–868.

Agorastos, T., Miliaras, D., Lambropoulos. A. F., Chrisafi, S., Kotsis, A., Manthos, A., and Bontis J. 2005. Detection and typing of human papillomavirus DNA in uterine cervices with coexistent grade I and grade III intraepithelial neoplasia: Biologic progression or independent lesions? *European Journal of Obstetrics & Gynecology and Reproductive Biology*, 121, 99–103.

Agrawal, A., Utzinger, U., Brookner, C., Pitris, C., Mitchell, M. F., and Richards-Kortum, R. 1999. Fluorescence spectroscopy of the cervix: Influence of acetic acid, cervical mucus, and vaginal medications. *Lasers in Surgery and Medicine*, 25, 237–249.

Andrus, P. G. L. and Strickland, R. D. 1998. Cancer grading by Fourier transform infrared spectroscopy. *Biospectroscopy*, 4, 37–46.

Apgar, B. S., Zoschnick, L., and Wright, T. C. 2003. The 2001 Bethesda system terminology. *American Family Physician*, 68, 1992–1998.

Bazant-Hegemark, F., Edey, K., Swingler, G. R., Read, M. D., and Stone, N. 2008. Optical micrometer resolution scanning for non-invasive grading of precancer in the human uterine cervix. *Technology in Cancer Research & Treatment*, 7, 483–496.

Blout, E. R. and Mellors, R. C. 1949. Infrared spectra of tissues. *Science*, 110, 137–138.

Bonnier, F., Knief, P., Lim, B., Meade, A. D., Dorney, J., Bhattacharya, K., Lyng, F. M., and Byrne, H. J. 2010. Imaging live cells grown on a three dimensional collagen matrix using Raman microspectroscopy. *Analyst*, 135, 3169–3177.

Boppart, S. A., Brezinski, M. E., Pitris, C., and Fujimoto, J. G. 1998. Optical coherence tomography for neurosurgical imaging of human intracortical melanoma. *Neurosurgery*, 43, 834–841.

Bosch, F. X., Manos, M. M., Munoz, N., Sherman, M., Jansen, A. M., Peto, J., Schiffman, M. H. et al. 1995. Prevalence of human papillomavirus in cervical cancer: A worldwide perspective. *Journal of the National Cancer Institute*, 87, 796–802.

Boshart, M., Gissmann, L., Ikenberg, H., Kleinheinz, A., Scheurlen, W., and Hausen, H. Z. 1984. A new type of papillomavirus DNA, its presence in genital cancer biopsies and in cell-lines derived from cervical-cancer. *Embo Journal*, 3, 1151–1157.

Boyko, E. J. 1996. Meta-analysis of Pap test accuracy. *American Journal of Epidemiology*, 143, 406–407.

Boyle, W. and Smith, G. 1970. Charge coupled devices. *Bell System Technical Journal*, 49, 587–593.

Brookner, C. K., Utzinger, U., Staerkel, G., Richards-Kortum, R., and Mitchell, M. F. 1999. Cervical fluorescence of normal women. *Lasers in Surgery and Medicine*, 24, 29–37.

Carlson, K., Pavlova, I., Collier, T., Descour, M., Follen, M., and Richards-Kortum, R. 2005. Confocal microscopy: Imaging cervical precancerous lesions. *Gynecologic Oncology*, 99, S84–S88.

Chan, J. W., Taylor, D. S., Zwerdling, T., Lane, S. M., Ihara, K., and Huser, T. 2006. Micro-Raman spectroscopy detects individual neoplastic and normal hematopoietic cells. *Biophysical Journal*, 90, 648–656.

Chidananda, S. M., Satyamoorthy, K., Rai, L., Manjunath, A. P., and Kartha, V. B. 2006. Optical diagnosis of cervical cancer by fluorescence spectroscopy technique. *International Journal of Cancer*, 119, 139–145.

Chowdhury, T. R. and Chowdhury, J. R. 1981. Significance of the occurrence and distribution of glycogen in cervical cells exfoliated under different physiologic and pathologic conditions. *Acta Cytologica*, 25, 557–565.

Clad, A., Reuschenbach, M., Weinschenk, J., Grote, R., Rahmsdorf, J., and Freudenberg, N. 2011. Performance of the aptima high-risk human papillomavirus mRNA assay in a referral population in comparison with hybrid capture 2 and cytology. *Journal of Clinical Microbiology*, 49, 1071–1076.

Clark, A. L., Gillenwater, A., Alizadeh-Naderi, R., El-Naggar, A. K., and Richards-Kortum, R. 2004. Detection and diagnosis of oral neoplasia with an optical coherence microscope. *Journal of Biomedical Optics*, 9, 1271–1280.

Cohenford, M. A. and Rigas, B. 1998. Cytologically normal cells from neoplastic cervical samples display extensive structural abnormalities on IR spectroscopy: Implications for tumor biology. *Proceedings of the National Academy of Sciences of the United States of America*, 95, 15327–15332.

Collins, S., Rollason, T. P., Young, L. S., and Woodman, C. B. J. 2010. Cigarette smoking is an independent risk factor for cervical intraepithelial neoplasia in young women: A longitudinal study. *European Journal of Cancer*, 46, 405–411.

Cremer, M., Bullard, K., Maza, M., Peralta, E., Moore, E., Garcia, L., Masch, R., Lerner, V., Alonzo, T. A., and Felix, J. 2010. Cytology versus visual inspection with acetic acid among women treated previously with cryotherapy in a low-resource setting. *International Journal of Gynecology & Obstetrics*, 111, 249–252.

Cuzick, J., Arbyn, M., Sankaranarayanan, R., Tsu, V., Ronco, G., Mayrand, M. H., Dillner, J., and Meijer, C. 2008. Overview of human papillomavirus-based and other novel options for cervical cancer screening in developed and developing countries. *Vaccine*, 26, K29–K41.

Deng, J. L., Wei, Q., Zhang, M. H., Wang, Y. Z., and Li, Y. Q. 2005. Study of the effect of alcohol on single human red blood cells using near-infrared laser tweezers Raman spectroscopy. *Journal of Raman Spectroscopy*, 36, 257–261.

Dockter, J., Schroder, A., Hill, C., Guzenski, L., Monsonego, J., and Giachetti, C. 2009. Clinical performance of the APTIMA® HPV Assay for the detection of high-risk HPV and high-grade cervical lesions. *Journal of Clinical Virology*, 45, (Supplement 1), S55–S61.

Dovbeshko, G. I., Gridina, N. Y., Kruglova, E. B., and Pashchuk, O. P. 2000. FTIR spectroscopy studies of nucleic acid damage. *Talanta*, 53, 233–246.

Dufresne, S., Sauthier, P., Mayrand, M. H., Petignat, P., Provencher, D., Drouin, P., Gauthier, P. et al. 2011. Human papillomavirus (HPV) DNA triage of women with atypical squamous cells of undetermined significance with amplicor HPV and hybrid capture 2 assays for detection of high-grade lesions of the uterine cervix. *Journal of Clinical Microbiology*, 49, 48–53.

Durst, M., Gissmann, L., Ikenberg, H., and Zurhausen, H. 1983. A papillomavirus DNA from a cervical-carcinoma and its prevalence in cancer biopsy samples from different geographic regions. *Proceedings of the National Academy of Sciences of the United States of America—Biological Sciences*, 80, 3812–3815.

Ekalaksananan, T., Pientong, C., Thinkhamrop, J., Kongyingyoes, B., Evans, M. F., and Chaiwongkot, A. 2010. Cervical cancer screening in north east Thailand using the visual inspection with acetic acid (VIA) test and its relationship to high-risk human papillomavirus (HR-HPV) status. *Journal of Obstetrics and Gynaecology Research*, 36, 1037–1043.

Ekins, S. and Sasic, S. 2008. *Pharmaceutical Applications of Raman Spectroscopy*, John Wiley & Sons, Hoboken, New Jersey.

Escobar, P. F., Belinson, J. L., White, A., Shakhova, N. M., Feldchtein, F. I., Kareta, M. V., and Gladkova, N. D. 2004. Diagnostic efficacy of optical coherence tomography in the management of preinvasive and invasive cancer of uterine cervix and vulva. *International Journal of Gynecological Cancer*, 14, 470–474.

Escobar, P. F., Rojas-Espaillat, L., and Belinson, J. L. 2005. Optical diagnosis of cervical dysplasia. *International Journal of Gynaecology and Obstetrics*, 89, 63–64.

Fabian, H., Jackson, M., Murphy, L., Watson, P. H., Fichtner, I., and Mantsch, H. H. 1995. A comparative infrared spectroscopic study of human breast-tumors and breast-tumor cell xenografts. *Biospectroscopy*, 1, 37–45.

Farquharson, S., Gift, A. D., Shende, C., Maksymiuk, P., Inscore, F. E., and Murran, J. 2005a. Detection of 5-fluorouracil in saliva using surface-enhanced Raman spectroscopy. *Vibrational Spectroscopy*, 38, 79–84.

Farquharson, S., Shende, C., Inscore, F. E., Maksymiuk, P., and Gift, A. 2005b. Analysis of 5-fluorouracil in saliva using surface-enhanced Raman spectroscopy. *Journal of Raman Spectroscopy*, 36, 208–212.

Feldchtein, F., Gelikonov, G., Gelikonov, V., Kuranov, R., Sergeev, A., Gladkova, N., Shakhov, A. et al. 1998. Endoscopic applications of optical coherence tomography. *Optics Express*, 3, 257–270.

Fujimoto, J. G., Pitris, C., Boppart, S. A., and Brezinski, M. E. 2000. Optical coherence tomography: An emerging technology for biomedical imaging and optical biopsy. *Neoplasia*, 2, 9–25.

Gaffikin, L., Blumenthal, P. D., Emerson, M., Limpaphayom, K., Lumbiganon, P., Ringers, P., Srisupundit, S. et al. 2003. Safety, acceptability, and feasibility of a single-visit approach to cervical-cancer prevention in rural Thailand: A demonstration project. *Lancet*, 361, 814–820.

Gallwas, J., Turk, L., Friese, K., and Dannecker, C. 2010. Optical coherence tomography as a non-invasive imaging technique for preinvasive and invasive neoplasia of the uterine cervix. *Ultrasound in Obstetrics & Gynecology*, 36, 624–629.

Georgakoudi, I., Sheets, E. E., Muller, M. G., Backman, V., Crum, C. P., Badizadegan, K., Dasari, R. R., and Feld, M. S. 2002. Trimodal spectroscopy for the detection and characterization of cervical precancers in vivo. *American Journal of Obstetrics and Gynecology*, 186, 374–382.

Gremlich, H. U. and Yan, B. 2001. *Infrared and Raman Spectroscopy of Biological Materials*, Marcel Dekker, New York.

Grossman, N., Ilovitz, E., Chaims, O., Salman, A., Jagannathan, R., Mark, S., Cohen, B., Gopas, J. and Mordechai, S. 2001. Fluorescence spectroscopy for detection of malignancy: H-ras overexpressing fibroblasts as a model. *Journal of Biochemical and Biophysical Methods*, 50, 53–63.

Hausen, H. Z., Meinhof, W., Scheiber, W., and Bornkamm, G. W. 1974a. Attempts to detect virus-specific DNA in human tumors. 1. Nucleic-acid hybridizations with complementary RNA of human wart virus. *International Journal of Cancer*, 13, 650–656.

Hausen, H. Z., Schulteh.H, Wolf, H., Dorries, K., and Egger, H. 1974b. Attempts to detect virus-specific DNA in human tumors. 2. Nucleic-acid hybridizations with complementary RNA of human herpes group viruses. *International Journal of Cancer*, 13, 657–664.

Huang, Z., Mcwilliams, A., Lui, H., Mclean, D. I., Lam, S., and Zeng, H. 2003a. Near-infrared Raman spectroscopy for optical diagnosis of lung cancer. *International Journal of Cancer*, 107, 1047–1052.

Huang, Z. W., Mcwilliams, A., Lam, S., English, J., Mclean, D. I., Lui, H., and Zeng, H. 2003b. Effect of formalin fixation on the near-infrared Raman spectroscopy of normal and cancerous human bronchial tissues. *International Journal of Oncology*, 23, 649–655.

Hubmann, M. R., Leiner, M. J. P., and Schaur, R. J. 1990. Ultraviolet fluorescence of human sera. 1. Sources of characteristic differences in the ultraviolet fluorescence-spectra of sera from normal and cancer-bearing humans. *Clinical Chemistry*, 36, 1880–1883.

Huleihel, M., Salman, A., Erukhimovitch, V., Ramesh, J., Hammody, Z., and Mordechai, S. 2002. Novel spectral method for the study of viral carcinogenesis in vitro. *Journal of Biochemical and Biophysical Methods*, 50, 111–121.

Insinga, R. P., Dasbach, E. J., and Elbasha, E. H. 2009. Epidemiologic natural history and clinical management of human papillomavirus (HPV) disease: A critical and systematic review of the literature in the development of an HPV dynamic transmission model. *BMC Infectious Diseases,* 9, 119.

Jastreboff, A. M. and Cymet, T. 2002. Role of the human papillomavirus in the development of cervical intraepithelial neoplasia and malignancy. *Postgraduate Medical Journal,* 78, 225–228.

Jemal, A., Bray, F., Center, M. M., Ferlay, J., Ward, E., and Forman, D. 2011. Global cancer statistics. *CA—A Cancer Journal for Clinicians,* 61, 69–90.

Kachi, S., Hirano, K., Takesue, Y., and Miura, M. 2000. Unusual corneal deposit after the topical use of cyclosporine as eye drops. *American Journal of Ophthalmology,* 130, 667–669.

Kaminaka, S., Yamazaki, H., Ito, T., Kohda, E., and Hamaguchi, H. O. 2001. Near-infrared Raman spectroscopy of human lung tissues: Possibility of molecular-level cancer diagnosis. *Journal of Raman Spectroscopy,* 32, 139–141.

Kanter, E. M., Majumder, S., Vargis, E., Robichaux-Viehoever, A., Kanter, G. J., Shappell, H., Jones, H. W., and Mahadevan-Jansen, A. 2009a. Multiclass discrimination of cervical precancers using Raman spectroscopy. *Journal of Raman Spectroscopy,* 40, 205–211.

Kanter, E. M., Vargis, E., Majumder, S., Keller, M. D., Woeste, E., Rao, G. G., and Mahadevan-Jansen, A. 2009b. Application of Raman spectroscopy for cervical dysplasia diagnosis. *Journal of Biophotonics,* 2, 81–90.

Kaufman, R. H. and Adam, E. 1999. Is human papillomavirus testing of value in clinical practice? *American Journal of Obstetrics and Gynecology,* 180, 1049–1053.

Keller, M. D., Kanter, E. M., Lieber, C. A., Majumder, S. K., Hutchings, J., Ellis, D. L., Beaven, R. B., Stone, N., and Mahadevan-Jansen, A. 2008. Detecting temporal and spatial effects of epithelial cancers with Raman spectroscopy. *Disease Markers,* 25, 323–337.

Kinney, W., Stoler, M. H., and Castle, P. E. 2010. Special commentary patient safety and the next generation of HPV DNA tests. *American Journal of Clinical Pathology,* 134, 193–199.

Knief, P. 2010. Interactions of carbon nanotubes with human lung epithelial cells in vitro, Assessed by Raman Spectroscopy. BioPhysic, Dublin Institute of Technology, Dublin.

Koss, L. G. 1992. *Diagnostic Cytology and Its Histopathology,* 4th ed. Lipincott, Philadelphia.

Krafft, C., Codrich, D., Pelizzo, G., and Sergo, V. 2008. Raman and FTIR imaging of lung tissue: Methodology for control samples. *Vibrational Spectroscopy,* 46, 141–149.

Krafft, C., Steiner, G., Beleites, C., and Salzer1, R. 2009. Disease recognition by infrared and Raman spectroscopy. *Journal of Biophotonics,* 2(1–2), 13–28.

Krishna, C. M., Sockalingum, G. D., Kegelaer, G., Rubin, S., Kartha, V. B., and Manfait, M. 2005. Micro-Raman spectroscopy of mixed cancer cell populations. *Vibrational Spectroscopy,* 38, 95–100.

Kuhnert, N. and Thumser, A. 2004. An investigation into the use of Raman microscopy for the detection of labelled compounds in living human cells. *Journal of Labelled Compounds & Radiopharmaceuticals,* 47, 493–500.

Latimer, P. 1967. Absolute absorption and scattering spectrophotometry. *Archives of Biochemistry and Biophysics,* 119, 580–581.

Liu, C. H., Das, B. B., Glassman, W. L. S., Tang, G. C., Yoo, K. M., Zhu, H. R., Akins, D. L. et al. 1992. Raman, fluorescence, and time-resolved light-scattering as optical diagnostic-techniques to separate diseased and normal biomedical media. *Journal of Photochemistry and Photobiology B-Biology,* 16, 187–209.

Liu, G., Liu, J. H., Zhang, L., Yu, F., and Sun, S. Z. 2005. Raman spectroscopic study of human tissues. *Spectroscopy and Spectral Analysis,* 25, 723–725.

Lord, R. C. and Yu, N. T. 1970. Laser-excited Raman spectroscopy of biomolecules: I. Native lysozyme and its constituent amino acids. *Journal of Molecular Biology,* 50(2), 509–524.

Madrid-Marina, V., Torres-Poveda, K., López-Toledo, G., and García-Carrancá, A. 2009. Advantages and disadvantages of current prophylactic vaccines against HPV. *Archives of Medical Research,* 40, 471–477.

Mahadevan-Jansen, A., Mitchell, M. F., Ramanujam, N., Malpica, A., Thomsen, S., Utzinger, U., and Richards-Kortum, R. 1998. Near-infrared Raman spectroscopy for *in vitro* detection of cervical precancers. *Photochemistry and Photobiology,* 68, 123–132.

Maitland, K. C., Gillenwater, A. M., Williams, M. D., El-Naggar, A. K., Descour, M. R., and Richards-Kortum, R. R. 2008. *In vivo* imaging of oral neoplasia using a miniaturized fiber optic confocal reflectance microscope. *Oral Oncology,* 44, 1059–1066.

Mccluggage, W. G., Walsh, M. Y., Thornton, C. M., Hamilton, P. W., Date, A., Caughley, L. M., and Bharucha, H. 1998. Inter- and intra-observer variation in the histopathological reporting of cervical squamous intraepithelial lesions using a modified Bethesda grading system. *British Journal of Obstetrics and Gynaecology*, 105, 206–210.

Mera, S. L. 1997. *Pathology and Understanding Disease Prevention*, Stanley Thornes , Cheltenham, Great Britain.

Min, Y. K., Yamamoto, T., Kohda, E., Ito, T., and Hamaguchi, H. 2005. 1064 nm near-infrared multichannel Raman spectroscopy of fresh human lung tissues. *Journal of Raman Spectroscopy*, 36, 73–76.

Mitchell, M. F. 1994. Accuracy of colposcopy. *Clinical Consulting Obstetrics Gynecology*, 6, 70–73.

Mitchell, M. F., Schottenfeld, D., Tortolero-Luna, G., Cantor, S. B., and Richards-Kortum, R. 1998. Colposcopy for the diagnosis of squamous intraepithelial lesions: A meta-analysis. *Obstetrics and Gynecology*, 91, 626–631.

Mordechai, S., Sahu, R. K., Hammody, Z., Mark, S., Kantarovich, K., Guterman, H., Podshyvalov, A., Goldstein, J., and Argov, S. 2004. Possible common biomarkers from FTIR microspectroscopy of cervical cancer and melanoma. *Journal of Microscopy-Oxford*, 215, 86–91.

Murillo, R., Luna, J., Gamboa, O., Osorio, E., Bonilla, J., Cendales, R., and Study, I. N. C. C. C. S. 2010. Cervical cancer screening with naked-eye visual inspection in Colombia. *International Journal of Gynecology & Obstetrics*, 109, 230–234.

Muwonge, R., Manuel, M. D., Filipe, A. P., Dumas, J. B., Frank, M. R., and Sankaranarayanan, R. 2010. Visual screening for early detection of cervical neoplasia in Angola. *International Journal of Gynecology & Obstetrics*, 111, 68–72.

Nessa, A., Hussain, M. A., Rahman, J. N., Rashid, M. H. U., Muwonge, R., and Sankaranarayanan, R. 2010. Screening for cervical neoplasia in Bangladesh using visual inspection with acetic acid. *International Journal of Gynecology & Obstetrics*, 111, 115–118.

Ngoma, T., Muwonge, R., Mwaiselage, J., Kawegere, J., Bukori, P., and Sankaranarayanan, R. 2010. Evaluation of cervical visual inspection screening in Dar es Salaam, Tanzania. *International Journal of Gynecology & Obstetrics*, 109, 100–104.

Notingher, I., Verrier, S., Haque, S., Polak, J. M., and Hench, L. L. 2003. Spectroscopic study of human lung epithelial cells (A549) in culture: Living cells versus dead cells. *Biopolymers*, 72, 230–240.

O'Faolain, E., Hunter, M. B., Byrne, J. M., Kelehan, P., Mcnamara, M., Byrne, H. J., and Lyng, F. M. 2005. A study examining the effects of tissue processing on human tissue sections using vibrational spectroscopy. *Vibrational Spectroscopy*, 38, 121–127.

Ooi, G. J., Fox, J., Siu, K., Lewis, R., Bambery, K. R., Mcnaughton, D., and Wood, B. R. 2008. Fourier transform infrared imaging and small angle x-ray scattering as a combined biomolecular approach to diagnosis of breast cancer. *Medical Physics*, 35, 2151–2161.

Ostrowska, K. M., Malkin, A., Meade, A., O'leary, J., Martin, C., Spillane, C., Byrne, H. J., and Lyng, F. M. 2010. Investigation of the influence of high-risk human papillomavirus on the biochemical composition of cervical cancer cells using vibrational spectroscopy. *Analyst*, 135, 3087–3093.

Ovestad, I. T., Gudlaugsson, E., Skaland, I., Malpica, A., Kruse, A. J., Janssen, E. A. M., and Baak, J. P. A. 2010. Local immune response in the microenvironment of CIN2–3 with and without spontaneous regression. *Modern Pathology*, 23, 1231–1240.

Ovestad, I. T., Vennestrom, U., Andersen, L., Gudlaugsson, E., Munk, A. C., Malpica, A., Feng, W. W., Voorhorst, F., Janssen, E. A. M., and Baak, J. P. A. 2011. Comparison of different commercial methods for HPV detection in follow-up cytology after ASCUS/LSIL, prediction of CIN2–3 in follow up biopsies and spontaneous regression of CIN2–3. *Gynecologic Oncology*, 123, 278–283.

Papanicolaou, G. N. and Traut, H. F. 1941. The diagnostic value of vaginal smears in carcinoma of the uterus. *American Journal of Obstetric and Gynecology*, 42, 193–206.

Parkin, D. M., Bray, F., Ferlay, J., and Pisani, P. 2005. Global cancer statistics, 2002. *CA—A Cancer Journal for Clinicians*, 55, 74–108.

Pavlova, I., Sokolov, K., Drezek, R., Malpica, A., Follen, M., and Richards-Kortum, R. 2003. Microanatomical and biochemical origins of normal and precancerous cervical autofluorescence using laser-scanning fluorescence confocal microscopy. *Photochemistry and Photobiology*, 77, 550–555.

Peto, J., Gilham, C., Fletcher, O., and Matthews, F. E. 2004. The cervical cancer epidemic that screening has prevented in the UK. *Lancet*, 364, 249–256.

Pinto, A. P. and Crum, C. P. 2000. Natural history of cervical neoplasia: Defining progression and its consequence. *Clinical Obstetrics and Gynecology*, 43, 352–362.

Pitris, C., Goodman, A., Boppart, S. A., Libus, J. J., Fujimoto, J. G., and Brezinski, M. E. 1999. High-resolution imaging of gynecologic neoplasms using optical coherence tomography. *Obstetrics and Gynecology*, 93, 135–9.

Puppels, G. J., Garritsen, H. S. P., Kummer, J. A., and Greve, J. 1993. Carotenoids located in human lymphocyte subpopulations and natural-killer-cells by Raman microspectroscopy. *Cytometry*, 14, 251–256.

Qian, X. M., Peng, X. H., Ansari, D. O., Yin-Goen, Q., Chen, G. Z., Shin, D. M., Yang, L., Young, A. N., Wang, M. D., and Nie, S. M. 2008. *In vivo* tumor targeting and spectroscopic detection with surface-enhanced Raman nanoparticle tags. *Nature Biotechnology*, 26, 83–90.

Quinn, M. A. 1998. Adenocarcinoma of the Cervix. *Annual Academy of Medicine Singapore*, 27, 662–665.

Ramanujam, N., Mitchell, M. F., Mahadevan, A., Warren, S., Thomsen, S., Silva, E., and Richards-kortum, R. 1994. In-vivo diagnosis of cervical intraepithelial neoplasia using 337-nm-excited laser-induced fluorescence. *Proceedings of the National Academy of Sciences of the United States of America*, 91, 10193–10197.

Ramanujam, N., Mitchell, M. F., Mahadevan-Jansen, A., Thomsen, S. L., Staerkel, G., Malpica, A., Wright, T., Atkinson, N., and Richards-kortum, R. 1996. Cervical precancer detection using a multivariate statistical algorithm based on laser-induced fluorescence spectra at multiple excitation wavelengths. *Photochemistry and Photobiology*, 64, 720–735.

Ramanujam, N., Richards-Kortum, R., Thomsen, S., Mahadevan-Jansen, A., Follen, M., and Chance, B. 2001. Low temperature fluorescence imaging of freeze-trapped human cervical tissues. *Optics Express*, 8, 335–343.

Reagan, J. W., Seidemann, I. L., and Saracusa, Y. 1953. The cellular morphology of carcinoma in situ and dysplasia or atypical hyperplasia of the uterine cervix. *Cancer*, 6, 224–235.

Rehman, I., Smith, R., Hench, L. L., and Bonfield, W. 1995. Structural evaluation of human and sheep bone and comparison with synthetic hydroxyapatite by FT-Raman spectroscopy. *Journal of Biomedical Materials Research*, 29, 1287–1294.

Richart, R. M. 1967. Natural history of cervical intraepithelial neoplasia. *Clinical Obstetrics and Gynecology*, 10(4), 748–784.

Richart, R. M. 1990. A Modified terminology for cervical intraepithelial neoplasia. *Obstetrics and Gynecology*, 75, 131–133.

Ronco, G., Giorgi-Rossi, P., Carozzi, F., Dalla Palma, P., Del Mistro, A., De Marco, L., De Lillo, M. et al. 2006. Human papillomavirus testing and liquid-based cytology in primary screening of women younger than 35 years: Results at recruitment for a randomised controlled trial. *Lancet Oncology*, 7, 547–555.

Ruiz-Chica, A. J., Medina, M. A., Sanchez-Jimenez, F., and Ramirez, F. J. 2004. Characterization by Raman spectroscopy of conformational changes on guanine-cytosine and adenine-thymine oligonucleotides induced by aminooxy analogues of spermidine. *Journal of Raman Spectroscopy*, 35, 93–100.

Sankaranarayanan, R., Esmy, P. O., Rajkumar, R., Muwonge, R., Swaminathan, R., Shanthakumari, S., Fayette, J. M., and Cherian, J. 2007. Effect of visual screening on cervical cancer incidence and mortality in Tamil Nadu, India: A cluster-randomised trial. *Lancet*, 370, 398–406.

Sankaranarayanan, R., Wesley, R., Somanathan, T., Dhakad, N., Shyamalakumary, B., Amma, N. S., Parkin, D. M., and Nair, M. K. 1998. Visual inspection of the uterine cervix after the application of acetic acid in the detection of cervical carcinoma and its precursors. *Cancer*, 83, 2150–2156.

Saslow, D., Solomon, D., Lawson, H. W., Killackey, M., Kulasingam, S. L., Cain, J. M., Garcia, F. A. R. et al. 2012. American cancer society, American society for colposcopy and cervical pathology, and American society for clinical pathology screening guidelines for the prevention and early detection of cervical cancer. *Journal of Lower Genital Tract Disease*, 16, 175–204.

Sauvaget, C., Fayette, J. M., Muwonge, R., Wesley, R., and Sankaranarayanan, R. 2011. Accuracy of visual inspection with acetic acid for cervical cancer screening. *International Journal of Gynecology & Obstetrics*, 113, 14–24.

Schubert, J. M., Mazur, A. I., Bird, B., Miljkovic, M., and Diem, M. 2010. Single point vs. mapping approach for spectral cytopathology (SCP). *Journal of Biophotonics*, 3, 588–596.

Scott, M., Stites, D. P. et al. 1999. Th1 cytokine patterns in cervical human papillomavirus infection. *Clinical and diagnostic laboratory immunology*, 6(5), 751–755.

Sellors, J. and Sankaranarayanan, R. 2003. *Colposcopy and Treatment of Cervical Intraepithelial Neoplasia.* Geneva: WHO.

Sergeev, A., Gelikonov, V., Gelikonov, G., Feldchtein, F., Kuranov, R., Gladkova, N., Shakhova, N. et al. 1997. *In vivo* endoscopic OCT imaging of precancer and cancer states of human mucosa. *Optics Express*, 1, 432–440.

Shepherd, L. J. and Bryson, S. C. P. 2008. Human papillomavirus—Lessons from history and challenges for the future. *Journal of Obstetrics and Gynaecology Canada*, 30, 1025–1033.

Short, M. A., Lam, S., Mcwilliams, A., Zhao, J. H., Lui, H., and Zeng, H. S. 2008. Development and preliminary results of an endoscopic Raman probe for potential *in vivo* diagnosis of lung cancers. *Optics Letters*, 33, 711–713.

Sigurdsson, K. 1999. Cervical cancer, Pap smear and HPV testing: An update of the role of organized Pap smear screening and HPV testing. *Acta Obstetricia Et Gynecologica Scandinavica*, 78, 467–477.

Smith, J. S., Green, J., De Gonzalez, A. B., Appleby, P., Peto, J., Plummer, M., Franceschi, S., and Beral, V. 2003. Cervical cancer and use of hormonal contraceptives: A systematic review. *The Lancet*, 361, 1159–1167.

Solomon, D., Davey, D., Kurman, R., Moriarty, A., O'connor, D., Prey, M., Raab, S. et al. The 2001 Bethesda system: Terminology for reporting results of cervical cytology. *JAMA*, 287, 2114–2119.

Stone, N., Kendall, C., Shepherd, N., Crow, P., and Barr, H. 2002. Near-infrared Raman spectroscopy for the classification of epithelial pre-cancers and cancers. *Journal of Raman Spectroscopy*, 33, 564–573.

Stone, N., Kendall, C., Smith, J., Crow, P., and Barr, H. 2004. Raman spectroscopy for identification of epithelial cancers. *Faraday Discussions*, 126, 141–157.

Tan, K. M., Herrington, C. S., and Brown, C. T. 2011. Discrimination of normal from pre-malignant cervical tissue by Raman mapping of de-paraffinized histological tissue sections. *Journal of Biophotonics*, 4, 40–8.

Thekkek, N. and Richards-Kortum, R. 2008. Optical imaging for cervical cancer detection: Solutions for a continuing global problem. *Nature Reviews Cancer*, 8, 725–731.

Utzinger, U., Heintzelman, D. L., Mahadevan-Jansen, A., Malpica, A., Follen, M., and Richards-Kortum, R. 2001. Near-infrared Raman spectroscopy for *in vivo* detection of cervical precancers. *Applied Spectroscopy*, 55, 955–959.

Viehoever, A. R., Anderson, D., Jansen, D., and Mahadevan-Jansen, A. 2003. Organotypic raft cultures as an effective *in vitro* tool for understanding Raman spectral analysis of tissue. *Photochemistry and Photobiology*, 78, 517–524.

Villa, L. L. and Denny, L. 2006. Methods for detection of HPV infection and its clinical utility. *International Journal of Gynecology & Obstetrics*, 94, S71–S80.

Walboomers, J. M. M., Jacobs, M. V., Manos, M. M., Bosch, F. X., Kummer, J. A., Shah, K. V., Snijders, P. J. F., Peto, J., Meijer, C., and Munoz, N. 1999. Human papillomavirus is a necessary cause of invasive cervical cancer worldwide. *Journal of Pathology*, 189, 12–19.

Wang, L. Q. and Mizaikoff, B. 2008. Application of multivariate data-analysis techniques to biomedical diagnostics based on mid-infrared spectroscopy. *Analytical and Bioanalytical Chemistry*, 391, 1641–1654.

Wong, P. T. T., Wong, R. K., Caputo, T. A., Godwin, T. A., and Rigas, B. 1991. Infrared-spectroscopy of exfoliated human cervical cells—Evidence of extensive structural-changes during carcinogenesis. *Proceedings of the National Academy of Sciences of the United States of America*, 88, 10988–10992.

Wood, B. R., Quinn, M. A., Tait, B., Ashdown, M., Hislop, T., Romeo, M., and Mcnaughton, D. 1998. FTIR microspectroscopic study of cell types and potential confounding variables in screening for cervical malignancies. *Biospectroscopy*, 4, 75–91.

Woodman, C. B. J., Collins, S. I., and Young, L. S. 2007. The natural history of cervical HPV infection: Unresolved issues. *Nature Reviews Cancer*, 7, 11–22.

Wright, T. C., Massad, S., Dunton, C. J., Spitzer, M., Wilkinson, E. J., and Solomon, D. 2007. 2006 consensus guidelines for the management of women with abnormal cervical cancer screening tests. *American Journal of Obstetrics and Gynecology*, 197, 346–355.

Yazdi, Y., Ramanujam, N., Lotan, R., Mitchell, M. F., Hittelman, W., and Richards-Kortum, R. 1999. Resonance Raman spectroscopy at 257 nm excitation of normal and malignant cultured breast and cervical cells. *Applied Spectroscopy*, 53, 82–85.

Zlatkov, V. 2009. Possibilities of the TruScreen for screening of precancer and cancer of the uterine cervix. *Akush Ginekol (Sofiia)*, 48, 46–50.

Zuluaga, A. F., Utzinger, U., Durkin, A., Fuchs, H., Gillenwater, A., Jacob, R., Kemp, B., Fan, J., and Richards-Kortum, R. 1999. Fluorescence excitation emission matrices of human tissue: A system for *in vivo* measurement and method of data analysis. *Applied Spectroscopy*, 53, 302–311.

Zurhausen, H. 1976. Condylomata acuminata and human genital cancer. *Cancer Research*, 36, 794–794.

chapter fifteen

Type 2 diabetes mellitus, obesity, and adipose tissue biology

Fazli Rabbi Awan and Syeda Sadia Najam

Contents

Abstract

Diabetes mellitus is the most common metabolic disorder characterized by persistently high blood glucose levels in the uncontrolled
condition, and accounts for more than 90% of all diabetics. It is
mainly caused by insulin resistance by the target tissues and in later
stages reduced secretion of insulin owing to dysfunction or damage of pancreatic beta cells. Although it can be managed and treated
by modifications in lifestyle, dietary habits and medications, as well
as insulin therapy, however, in the uncontrolled condition, several
grave complications arise in vital tissues and organs of the body,
which lead to disability or early death. The main tissues and organs
affected by diabetes are kidneys, eyes, heart, brain, peripheral neurons, and lower limbs.

Obesity, which is characterized by excess fat in the body especially in the abdominal region, is one of the precursors for type 2 diabetes (T2D) as more than 80% type 2 diabetics are obese. Body fat or adipose tissue was once considered as a depot of energy, and is now an established endocrine organ and active target for investigating the pathobiology of obesity and T2D. As a number of hormones called adipokines are secreted from adipose tissue and have been providing insight into insulin resistance and T2D pathophysiology, some of those adipokines such as adiponectin, leptin, and resistin are discussed in this chapter. Moreover, a number of whole-body and tissue-specific knockout mice are now available to provide new insights into the mechanisms of disease and testing novel drug targets for obesity and T2D. One such promising drug target from Forkhead box (FoxO) family of transcription factors has also been briefly discussed at the end.

Keywords: Type 2 diabetes, Obesity, Insulin resistance, Adipocyte, Adipokines, FoxO.

Diabetes mellitus

Diabetes mellitus (DM) is a metabolic disorder in which the body fails to regulate blood glucose levels and results in chronically increased levels of glucose also termed as hyperglycemia. Hyperglycemia can be either due to insufficient or even complete absence of insulin (glucose-regulating hormone), or due to blunt response of insulin responsive tissues. In the latter case, insulin is produced in normal amounts by the pancreatic β-cells but the target tissues for insulin action do not respond to it. This state is termed as insulin resistance and it is the hallmark of one of the several types of DM, which is type 2 diabetes mellitus (T2DM or T2D). T2D results from insulin resistance, which is a condition in which the insulin responsive cells fail to respond and use insulin properly and sometimes it can be due to a combined effect of insulin deficiency as well. This type is most prevalent among all diabetes cases and about 90%–95% of diabetic patients have T2D (Rubino 2008).

Prevalence of diabetes

According to a recent estimate by International Diabetes Federation (IDF), globally around 387 million people have diabetes, which is expected to rise to 592 million by 2035. The update also reports increasing number of people with T2D in every country (IDF 2014). In Pakistan, the prevalence of diabetes is also alarming and it was reported to be 22%–25%, which included both diabetics and prediabetics (Basit and Shera 2008). According to World Health Organization (WHO), Pakistan currently ranks at the seventh position in the list of countries with diabetes burden, which is expected to move to fourth position if the number of diabetics increases at the same rate in this country (Qidwai and Ashfaq 2010).

Etiology and diagnosis of T2D

T2D is the most prevalent type of diabetes and is characterized by high glucose levels in the blood. It is also a component of metabolic syndrome along with other contributing factors such as hypertension, hyperlipidemia, central obesity, and insulin resistance (Lee and Sanders 2012). Genetics and environmental factors such as lifestyle, food intake, and energy

Table 15.1 Diagnosis criteria for prediabetes and diabetes according to WHO

Status	Fasting blood glucose level	Oral glucose tolerance test (2-hour plasma glucose)
Normal	<6.1 mmol/L (110 mg/dL)	<7.8 mmol/L (140 mg/dL)
Diabetes	≥7.0 mmol/L (126 mg/dL)	≥11.1 mmol/L (200 mg/dL)

expenditure play an important role in diabetes development (Hu 2011). Obesity is one of the major environmental factors linked to insulin resistance and ultimately has a role in diabetes development (Qatanani and Lazar 2007; Zhuang, Zhao et al. 2009). T2D symptoms include polyuria (increased urination), polyphagia (increased hunger), and polydipsia (increased thirst) (Kumar, Bharti et al. 2014). Poor glycemic control in T2D leads to certain complications of T2D, which are classified as macro- and microvascular complications. Macrovascular complications are coronary artery disease, peripheral arterial disease, and stroke, while microvascular complications include diabetic nephropathy, neuropathy, and retinopathy. Diagnosis of diabetes according to WHO criteria is shown in Table 15.1 (WHO 2006).

Pathophysiology of T2D

Glucose is the primary source of energy in the human body and it is balanced in a very narrow range of 4.4–6.1 mmol/L (79.2–110 mg/dL). Requirements of glucose in the body are met exogenously by diet and endogenously mostly by liver. Glucose from the diet is taken up by cells with insulin's action, and then metabolized to meet energy needs in the body while in case of glucose deficiency liver produces glucose by the breakdown of glycogen stored within it. The homeostasis of glucose is very crucial and is regulated by two hormones, insulin and glucagon. Insulin and glucagon action, interaction, and mechanism are important to have an insight into the pathophysiology of T2D.

An overview of pancreas functional biology

Pancreas is an organ in the body that plays an important role in the metabolism by secreting important hormones such as insulin, glucagon, etc. The endocrine portion of pancreas comprises a cluster of cells collectively called as islet of Langerhans. This cluster has mainly four types of cell: α-cells that release glucagon; β-cells that release insulin; delta cells that release somatostatin; and PP cells that secrete pancreatic polypeptides (Longnecker 2014). Insulin and glucagon hormones play important but antagonistic role in the glucose homeostasis.

Insulin and glucagon

Both insulin and glucagon are released from pancreatic cells in response to blood glucose levels. Insulin is secreted in response to high glucose levels, which are usually after meal intake while glucagon is secreted in response to low glucose levels (Quesada, Tuduri et al. 2008). As T2D is characterized by high levels of glucose, insulin mechanism is of more importance and will be discussed in detail. The major functions of insulin are transport of glucose from blood to muscle, liver, and adipose tissue. In muscles and liver, insulin stimulates glucose metabolism and glucose storage, while in adipose tissue it stimulates dietary fat storage. It also inhibits the breakdown of already stored glucose, proteins, and fats, which happens in glucose-deficient state to overcome the energy requirements of body (Porte 2006). The balance between insulin and glucagon levels is crucial to maintain normal glucose levels in the blood.

Insulin secretion in response to glucose

Insulin is secreted in response to high glucose levels. The mechanism of insulin secretion involves the action of glucose transporter 2 (GLUT2). GLUT2 is a member of family of proteins that facilitates the entry of glucose into β-cells of pancreas. In β-cells, the glycolytic phosphorylation of glucose occurs resulting in increased adenosine triphosphate (ATP):adenosine diphosphate (ADP) ratio. Potassium channels of pancreatic β-cells are ATP sensitive and increased ATP production leads to shutting down of potassium channel. The depolarization of plasma membrane due to shutting down of potassium channel opens up the calcium channel in plasma membrane. This opening up of calcium channel allows the inward flow of calcium ions into the cells and triggers the secretion of insulin by exocystosis (Rorsman, Ramracheya et al. 2014).

Insulin mechanism of action

Insulin after secretion from β-cells of pancreas in response to high blood glucose levels binds to its receptors. The initiation of certain signal transduction cascade on binding of insulin to its receptors promotes the uptake of glucose into target tissues. These tissues include skeletal muscles where glucose is metabolized for energy and adipose tissue where glucose is converted to triglycerides for storage. The major step in glucose metabolism is the increased number and immediate activation of type 4 glucose transporter (GLUT4). GLUT4 protein is insulin-regulated glucose transporter and mainly found in adipose tissue and skeletal muscles.

Insulin activates the GLUT4 via signal transduction mechanism. Insulin binds to its receptor's α-subunit, which is embedded in the cell membrane. This binding triggers the tyrosine kinase activity in the β-subunit of insulin receptor (attached to the α-subunit). This activity leads to the phosphorylation of two enzymes, which are mitogen-activated protein kinase (MAPK) and phosphatidylinositol-3-kinase (PI-3K). These enzymes are responsible for the mitogenic (cell growth and gene expression) and metabolic actions (synthesis of lipids, proteins, and glycogen) of insulin, respectively. The PI-3K pathway is also responsible for isolating the GLUT4 from glucose as the GLUT4 binds to the PI-3K after bringing glucose in the cell. The glucose that is isolated undergoes glycolysis process and excess glucose is stored in the cell as glycogen while GLUT4 is translocated back to cell membrane (Bevan 2001).

Insulin resistance

Insulin's primary metabolic action is to facilitate the postprandial disposition of glucose. Defects in β-cells insulin secretion and insulin action on its target tissues lead to insulin resistance and diabetes (Porte 2006). Insulin resistance is the condition in which the cellular response to insulin diminishes and activation of GLUT4 is ceased, which leads to high levels of glucose in blood. Insulin resistance involves a very complicated interaction of signaling pathways that are regulated by a number of hormones in diabetes (Kahn, Hull et al. 2006). Although the blood glucose levels are high, the uptake by tissues is affected (Rosenblatt-Velin, Lerch et al. 2004). Hence, the cells cannot sense glucose and so liver produces more glucose irregularly and therefore the blood glucose levels rise even higher, which is a hallmark of T2D (Guilherme, Virbasius et al. 2008).

Management and treatment of diabetes

Diabetes management and treatment is very important for a healthy lifestyle and this can be achieved with right diet, exercise, and antidiabetic drugs as per physician's instructions.

Diet management for a diabetic patient is important and it must meet energy needs, weight control, and a proper glucose level regulation. Exercise is also a significant part in diabetes management. It has effects on lowering of blood glucose level and improvement of insulin utilization by increasing glucose uptake by the muscles.

There are several oral antidiabetic drugs available, which along with diet and exercise are used by diabetics to regulate elevated blood glucose levels. Some major antidiabetic drugs are metformin, which decreases the amount of glucose released from liver; sulfonylurea, which stimulates the pancreas to release more insulin; and thiazolidinediones (TZDs), which makes the body more sensitive to the effects of insulin. When the metabolic control cannot be achieved with diet, exercise, or antidiabetic drugs, then insulin therapy is advised.

Obesity

Obesity is a condition characterized by the accumulation of excessive adipose tissue in the body. This excess of adipose tissue in obesity affects both the physical and psychosocial health of individuals and is also considered as a health disaster in both developing and developed countries (Naser, Gruber et al. 2006). Obesity is calculated with body mass index (BMI) scale by the National Institutes of Health (NIH). It is calculated as body weight in kilograms divided by square of height in meters (kg/m²). The values of BMI for different weight categories are normal weight (18.5–24.9), overweight (25–29.9), and obese (30 or above). Environmental and genetic factors play a major role in the development of obesity (Ershow 2009). Disruption in balance between energy expenditure and food intake leads to obesity, which can be a reason for many health issues. Some genetic factors are also responsible of causing obesity (Hebebrand and Hinney 2009; Farooqi 2014).

Obesity, insulin resistance, and T2D

Obesity and T2D are linked to insulin resistance and mechanisms linking obesity and T2D are not fully elucidated as all obese are not hyperglycemic and all diabetics are not obese (Neeland, Turer et al. 2012). Obesity is one of the major risk factors for T2D as obese people are three times more at risk of developing T2D as compared to normal weight people (Abdullah, Peeters et al. 2010). There are some notions that link obesity and diabetes, like release of proinflammatory cytokines such as IL-6, TNF-α from adipose tissue that makes the body cells less sensitive to insulin and this disrupts the action of insulin in the uptake of glucose by the body cells. This leads to hyperglycemia and ultimately T2D (Freemantle, Holmes et al. 2008; Despres 2012). Obesity can also trigger changes in the body's metabolic processes and it can be a cause of release of certain hormones, increased amount of fatty acids, proinflammatory cytokines from adipose tissue, which affect insulin's normal action and leads to insulin resistance. The development of insulin resistance accompanied by the β-cell dysfunction disrupts the control of normal blood glucose levels, ultimately leading to chronic condition of T2D (Kahn, Hull et al. 2006).

Adipose tissue and adipocytes

For many years, the adipose tissue was considered as a fat depot but endocrine function of adipose tissue has been established since the discovery of leptin in 1994 (Zhang, Proenca et al. 1994; Galic, Oakhill et al. 2010; Adamczak and Wiecek 2013). Adipocytes are secretory cells of adipose tissue (Figure 15.1), which release fatty acids to meet the energy balance

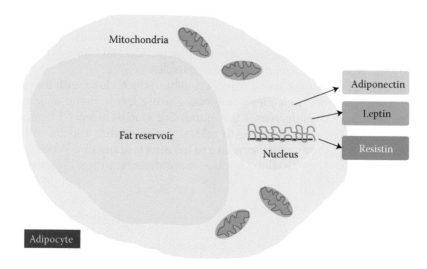

Figure 15.1 Adipocyte structure.

of the body, different lipid molecules, variety of proinflammatory and anti-inflammatory factors, as well as cytokines and chemokines, such as TNF-alpha, IL-6, and several others (Fantuzzi 2005). A number of hormones and proteins are also secreted by adipocytes, which are termed as adipocytokines or adipokines (Trayhurn 2007).

An adipocyte is responsible for the expression of certain hormones that are termed as adipokines such as adiponectin, leptin, and resistin.

Adipokines

A variety of protein-based hormones are released from adipocytes of adipose tissue and termed as adipokines. The major adipokines that are released by the adipose tissue are leptin, resistin, adiponectin, visfatin, etc. Leptin, adiponectin, and resistin are the adipokines, which are extensively studied and have an established link with insulin resistance and T2D. These adipokines circulate and communicate with other organs, including brain, liver, muscle, and adipose tissue itself (Kwon and Pessin 2013). The deregulated production or secretion of these adipokines due to excess adipose tissue can lead to the pathogenesis of various obesity-associated complications such as insulin resistance, T2D, and cardiovascular diseases (Antuna-Puente, Feve et al. 2008; Ouchi, Parker et al. 2011).

Leptin

Leptin is a hormone derived from adipose tissue and is encoded by obese gene (*Ob*) located on chromosome 7, and consists of 167 amino acids (Zhang, Proenca et al. 1994). The expression of leptin in adipocytes and its plasma concentration are both positively correlated with adiposity (Woods and D'Alessio 2008). It is the main regulator of the balance between food intake and energy expenditure through communication with the central nervous system (Farooqi and O'Rahilly 2009). Due to defect in either secretion of leptin or abnormal leptin there is leptin deficiency, which is a hallmark of certain rare cases of obesity. Leptin deficiency or resistance can result in profound obesity and diabetes. Leptin-deficient mice and humans are severely obese and have several metabolic and endocrine alterations, such as hyperglycemia, insulin resistance, hypertriglyceridemia, and central hypothyroidism

(Flier 1998). If the defect is on leptin receptors in brain or in the downstream signaling pathways on hypothalamic neuronal cells, which are the target sites for leptin action, it would not exert its effects and will be unable to regulate the food intake and energy homeostasis. In this situation, brain is irresponsive for the intake of food and energy, and thus elicits a response to the body for more food intake, which can cause weight gain and obesity. Since obesity is a major risk factor for T2D, the increase in adipose tissue may lead to diabetes.

Adiponectin

Adiponectin is another adipose tissue-derived protein that contains 244 amino acids with a molecular weight of 30 kDa (Huerta 2006). The levels of adiponectin are variable with respect to gender as males usually have low adiponectin concentration than females. Moreover, age and pregnancy also affect the adiponectin expression (Haluzik 2005). It is a multifunctional hormone and it has been shown to act in the brain to decrease the body weight (Nawrocki and Scherer 2005). Its direct actions are in the liver, skeletal muscles, and vasculature with prominent roles in improvement of hepatic insulin sensitivity, increased fuel oxidation, and decreased vascular inflammation (Sun, Xun et al. 2009). Adiponectin is an important protein hormone implicated in obesity and T2D. Its levels in the blood show an inverse relationship with body fat composition and it is thought to be the only adipose-derived hormone that is decreased in obesity and T2D (Haluzik, Parizkova et al. 2004; Salmenniemi, Zacharova et al. 2005).

Resistin

Resistin such as adiponectin and leptin is one of the several adipokines and was discovered in a screen for adipocyte gene products (Steppan, Brown et al. 2001). Human resistin gene is located on chromosome 19p13.3 and it is encoded by three exons (Ukkola 2002). The length of human resistin prepeptide is 108 amino acids and its molecular weight is 12.5 kDa (Steppan and Lazar 2004). Resistin also has been identified to be expressed by white blood cells in humans (Nagaev and Smith 2001). The role of resistin in obesity and T2D is extensively studied. In rodent models, resistin has a confirmed link to insulin resistance, as the administration of antiresistin antibody to mice with diet-induced obesity and hyperglycemia partially improved insulin sensitivity and administration of recombinant resistin to normal mice led to impaired glucose tolerance (Steppan and Lazar 2002). However, the role of human resistin in obesity and diabetes is still controversial (Schwartz and Lazar 2011).

Forkhead box transcription factors and their importance in obesity and diabetes

Transcription factors are the proteins that bind to specific DNA sequences and control information that is transcribed by specific genes. Forkhead box transcription factors (Fox) family is a class of transcriptional factors that is characterized by a conserved DNA-binding domain (the Forkhead box) (Benayoun, Caburet et al. 2011). Many key members of Forkhead transcription factors, especially the FoxO, are strongly associated with important biological processes such as cancer, drug resistance, stress response, cell cycle regulation, glucose metabolism, and apoptosis (Carlsson and Mahlapuu 2002). FoxO family of transcription factors is one of the major subfamilies of FOX proteins and has been identified in species ranging from yeast to humans. The FoxO family is regulated by insulin/PI-3k/AKT signaling pathways and therefore has a very important role in glucose metabolism (Carter and Brunet 2007; Nakae, Oki et al. 2008). FoxO1 proteins are members of FoxO family and are encoded by FoxO1 gene. FoxO1 has a very important role in metabolic regulation. It is expressed mainly in the insulin responsive tissue, that is, liver, pancreas, adipose tissue, and muscles (Nakae, Kitamura et al. 2001). FoxO1 is involved in adipocyte differentiation and pancreatic β-cell failure, which makes it a possible cause of obesity and related insulin resistance and diabetes (Kitamura 2013). Study of functional biology of a gene is important, so the mechanism of a particular pathway can be elucidated. Hence, investigation of a gene function in a live biological system gives meaningful information regarding its mechanism. Therefore, knocking out a particular gene from a live biological specimen (cells or tissues or whole animal) can help in understanding the function of that particular gene. Such animal-based studies for genes like *FoxO1* are the current focus of identifying and validating novel drug targets for obesity and T2D. It is hoped that uncovering the molecular biology of such drug targets will provide new opportunities for developing potent drugs for such diseases.

References

Abdullah, A., A. Peeters et al. 2010. The magnitude of association between overweight and obesity and the risk of diabetes: A meta-analysis of prospective cohort studies. *Diabetes Res Clin Pract* **89**(3): 309–319.

Adamczak, M. and A. Wiecek. 2013. The adipose tissue as an endocrine organ. *Semin Nephrol* **33**(1): 2–13.

Antuna-Puente, B., B. Feve et al. 2008. Adipokines: The missing link between insulin resistance and obesity. *Diabetes Metab* **34**(1): 2–11.

Basit, A. and A. S. Shera. 2008. Prevalence of metabolic syndrome in Pakistan. *Metab Syndr Relat Disord* **6**(3): 171–175.

Benayoun, B. A., S. Caburet et al. 2011. Forkhead transcription factors: Key players in health and disease. *Trends Genet* **27**(6): 224–232.

Bevan, P. 2001. Insulin signalling. *J Cell Sci* **114**(Pt 8): 1429–1430.

Carlsson, P. and M. Mahlapuu. 2002. Forkhead transcription factors: Key players in development and metabolism. *Dev Biol* **250**(1): 1–23.

Carter, M. E. and A. Brunet. 2007. FOXO transcription factors. *Curr Biol* **17**(4): R113–R114.

Despres, J. P. 2012. Body fat distribution and risk of cardiovascular disease: An update. *Circulation* **126**(10): 1301–1313.

Ershow, A. G. 2009. Environmental influences on development of type 2 diabetes and obesity: Challenges in personalizing prevention and management. *J Diabetes Sci Technol* **3**(4): 727–734.

Fantuzzi, G. 2005. Adipose tissue, adipokines, and inflammation. *J Allergy Clin Immunol* **115**(5): 911–919; quiz 920.

Farooqi, I. S. 2014. Defining the neural basis of appetite and obesity: From genes to behaviour. *Clin Med* **14**(3): 286–289.

Farooqi, I. S. and S. O'Rahilly. 2009. Leptin: A pivotal regulator of human energy homeostasis. *Am J Clin Nutr* **89**(3): 980S–984S.

Flier, J. S. 1998. Clinical review 94: What's in a name? In search of leptin's physiologic role. *J Clin Endocrinol Metab* **83**(5): 1407–1413.

Freemantle, N., J. Holmes et al. 2008. How strong is the association between abdominal obesity and the incidence of type 2 diabetes? *Int J Clin Pract* **62**(9): 1391–1396.

Galic, S., J. S. Oakhill et al. 2010. Adipose tissue as an endocrine organ. *Mol Cell Endocrinol* **316**(2): 129–139.

Guilherme, A., J. V. Virbasius et al. 2008. Adipocyte dysfunctions linking obesity to insulin resistance and type 2 diabetes. *Nat Rev Mol Cell Biol* **9**(5): 367–377.

Haluzik, M. 2005. Adiponectin and its potential in the treatment of obesity, diabetes and insulin resistance. *Curr Opin Investig Drugs* **6**(10): 988–993.

Haluzik, M., J. Parizkova et al. 2004. Adiponectin and its role in the obesity-induced insulin resistance and related complications. *Physiol Res* **53**(2): 123–129.

Hebebrand, J. and A. Hinney. 2009. Environmental and genetic risk factors in obesity. *Child Adolesc Psychiatr Clin N Am* **18**(1): 83–94.

Hu, F. B. 2011. Globalization of diabetes: The role of diet, lifestyle, and genes. *Diabetes Care* **34**(6): 1249–1257.

Huerta, M. G. 2006. Adiponectin and leptin: Potential tools in the differential diagnosis of pediatric diabetes? *Rev Endocr Metabol Disord* **7**(3): 187–196.

IDF. 2014. Key findings. From http://www.idf.org/diabetesatlas/update-2014.

Kahn, S. E., R. L. Hull et al. 2006. Mechanisms linking obesity to insulin resistance and type 2 diabetes. *Nature* **444**(7121): 840–846.

Kitamura, T. 2013. The role of FOXO1 in beta-cell failure and type 2 diabetes mellitus. *Nat Rev Endocrinol* **9**(10): 615–623.

Kumar, A., S. K. Bharti et al. 2014. Type 2 diabetes mellitus: The concerned complications and target organs. *Apollo Medicine* **11**: 161–166.

Kwon, H. and J. E. Pessin. 2013. Adipokines mediate inflammation and insulin resistance. *Front Endocrinol* (Lausanne) **4**(71): 1–13. doi: 10.3389/fendo.2013.00071.

Lee, L. and R. A. Sanders. 2012. Metabolic syndrome. *Pediatr Rev* **33**(10): 459–466; quiz 467–458.

Longnecker, D. 2014. Anatomy and histology of the pancreas. *Pancreapedia: Exocrine Pancreas Knowledge Base*, doi: 10.3998/panc.2014.3.

Nagaev, I. and U. Smith. 2001. Insulin resistance and type 2 diabetes are not related to resistin expression in human fat cells or skeletal muscle. *Biochem Biophys Res Commun* **285**(2): 561–564.

Nakae, J., T. Kitamura et al. 2001. The forkhead transcription factor Foxo1 (Fkhr) confers insulin sensitivity onto glucose-6-phosphatase expression. *J Clin Invest* **108**(9): 1359–1367.

Nakae, J., M. Oki et al. 2008. The FoxO transcription factors and metabolic regulation. *FEBS Lett* **582**(1): 54–67.

Naser, K. A., A. Gruber et al. 2006. The emerging pandemic of obesity and diabetes: Are we doing enough to prevent a disaster? *Int J Clin Pract* **60**(9): 1093–1097.

Nawrocki, A. R. and P. E. Scherer. 2005. Keynote review: The adipocyte as a drug discovery target. *Drug Discov Today* **10**(18): 1219–1230.

Neeland, I. J., A. T. Turer et al. 2012. Dysfunctional adiposity and the risk of prediabetes and type 2 diabetes in obese adults. *JAMA* **308**(11): 1150–1159.

Ouchi, N., J. L. Parker et al. 2011. Adipokines in inflammation and metabolic disease. *Nat Rev Immunol* **11**(2): 85–97.

Porte, D. 2006. Central regulation of energy homeostatsis: The key role of insulin. *Diabetes* **55**(Suppl. 2): S155–S160.

Qatanani, M. and M. A. Lazar. 2007. Mechanisms of obesity-associated insulin resistance: Many choices on the menu. *Genes Dev* **21**(12): 1443–1455.

Qidwai, W. and T. Ashfaq. 2010. Imminent epidemic of diabetes mellitus in Pakistan: Issues and challenges for health care providers. *JLUMHS* **9**(3): 112–113.

Quesada, I., E. Tuduri et al. 2008. Physiology of the pancreatic alpha-cell and glucagon secretion: Role in glucose homeostasis and diabetes. *J Endocrinol* **199**(1): 5–19.

Rorsman, P., R. Ramracheya et al. 2014. ATP-regulated potassium channels and voltage-gated calcium channels in pancreatic alpha and beta cells: Similar functions but reciprocal effects on secretion. *Diabetologia* **57**(9): 1749–1761.

Rosenblatt-Velin, N., R. Lerch et al. 2004. Insulin resistance in adult cardiomyocytes undergoing dedifferentiation: Role of GLUT4 expression and translocation. *FASEB J* **18**(7): 872–874.

Rubino, F. 2008. Is type 2 diabetes an operable intestinal disease? A provocative yet reasonable hypothesis. *Diabetes Care* **31**(Suppl 2): S290–296.

Salmenniemi, U., J. Zacharova et al. 2005. Association of adiponectin level and variants in the adiponectin gene with glucose metabolism, energy expenditure, and cytokines in offspring of type 2 diabetic patients. *J Clin Endocrinol Metab* **90**(7): 4216–4223.

Schwartz, D. R. and M. A. Lazar. 2011. Human resistin: found in translation from mouse to man. *Trends Endocrinol Metab* **22**(7): 259–265.

Steppan, C. M., E. J. Brown et al. 2001. A family of tissue-specific resistin-like molecules. *Proc Natl Acad Sci U S A* **98**(2): 502–506.

Steppan, C. M. and M. A. Lazar. 2002. Resistin and obesity-associated insulin resistance. *Trends Endocrinol Metab* **13**(1): 18–23.

Steppan, C. M. and M. A. Lazar. 2004. The current biology of resistin. *J Intern Med* **255**(4): 439–447.

Sun, Y., K. Xun et al. 2009. Adiponectin, an unlocking adipocytokine. *Cardiovasc Ther* **27**(1): 59–75.

Trayhurn, P. 2007. Adipocyte biology. *Obes Rev* **8**(Suppl 1): 41–44.

Ukkola, O. 2002. Resistin—A mediator of obesity-associated insulin resistance or an innocent bystander? *Eur J Endocrinol* **147**(5): 571–574.

WHO. 2006. Definition and diagnosis of diabetes mellitus and intermediate hyperglycaemia: Report of a WHO/IDF consultation. Geneva, World Health Organization.

Woods, S. C. and D. A. D'Alessio. 2008. Central control of body weight and appetite. *J Clin Endocrinol Metab* **93**(11 Suppl 1): S37–S50.

Zhang, Y., R. Proenca et al. 1994. Positional cloning of the mouse obese gene and its human homologue. *Nature* **372**(6505): 425–432.

Zhuang, X. F., M. M. Zhao et al. 2009. Adipocytokines: A bridge connecting obesity and insulin resistance. *Med Hypotheses* **73**(6): 981–985.

chapter sixteen

Human tissue banking and its role in biomedical research

Shahid Mian and Ibraheem Ashankyty

Contents

Abstract

Biorepository of human tissues provides baseline information to understand molecular pathways and that how they communicate with each other. Further, the application of human tissue repositories to twenty-first-century medicine is broad and can impact areas such as drug development pipelines, novel diagnostic, and prognostic tools. This will ultimately lead toward medical care of patients based on their specific genomic signatures. Consequently, a tissue repository may require support at a variety of levels to ensure that it has the strongest chance of success of achieving its goals and appropriate buy-in by society and public/private sector organizations will be essential in meeting these objectives.

Keywords: Biobanking, Biorepository, Genomic signatures, Microarrays, Personalized medicine, Pathology.

Human tissue requirements and application to biomedical research

High-throughput screening technologies such as mass spectrometry, deoxyribonucleic acid (DNA) microarrays, multiplex PCR, and the unprecedented advances in DNA sequencing throughput/speed/accuracy/cost have led to a paradigm shift for both research and development (R&D) and clinical translational environments. Given that it is now possible to sequence a whole patient genome in 24 hours for a cost in the range of $1500–$2000 (depending on volume, platform utilized, and whether the laboratory is accredited—2015 pricing), genomic interrogation can be conducted at various levels. These include whole genome [1]; exonic coding that may also extend to noncoding untranslated regions (UTR) and miRNA (exome) [2]; targeted resequencing of gene loci or hotspot mutations; and even epigenetic mapping studies [3] are actively being explored for drug and diagnostic development. The ability to rapidly map biological pathways implicated in disease development, drug sensitivity/toxicity, or even drug dosing levels based upon a specific genomic feature set for a given patient would have significant impact upon both drug development programs and ultimately management of patient healthcare [4].

There has been a concerted effort to improve outcomes for patients through examination of their own specific genomic [5,6], transcriptomic [7], proteomic [8–12], and metabolomic [13] expression patterns and correlating these to clinical phenotypes of interest. For example, in patients some cancerous tumors metastasize and settle in distal tissues is not entirely clear. In colorectal cancer patients with epidermal growth factor receptor (EGFR) receptor expression, a wild-type k-ras gene is required in order to derive benefit from panitumumab/cetuximab inhibition. Sequence analysis to identify patients having wild-type k-ras allows a rapid demarcation between patients who are potentially capable of benefitting from therapy versus those with low response probabilities and thus who should/ should not receive therapy [14–16]. This scenario provides an example of a companion diagnostic (sequencing of k-ras) that is required to facilitate the triage of patients who are likely to have positive outcomes to drug treatment (panitumumab/cetuximab). In order to identify biological pathways that might act in a diagnostic capacity and/or drug target, an R&D pipeline encompassing discovery and validation is required. At the heart of this process is a fundamental requirement for human tissue. This material will provide the necessary resource of biomolecules to allow discovery and/or validation to be undertaken [17].

Human tissues form the basis for biomarker discovery program irrespective of the physical nature of the biomarker (e.g., DNA, RNA, and protein) [18]. The material has a twofold purpose:

1. Providing a source of biomarkers in which high-throughput screening tools (e.g., DNA sequencers for mutation/single nucleotide polymorphism [SNP] discovery) can reveal candidates for downstream validation.
2. Sufficient numbers of tissue samples in a variety of forms will allow statistical power to be attained. Corroborating/refuting a link between the biomarker and the clinical phenotype under investigation with expedite development of diagnostic tests and/ or new drug targets.

The use of patient-matched tissue samples (e.g., control and tumor) will ascertain whether the biomarker is potentially differentially expressed (e.g., quantitative polymerase chain reaction [PCR], DNA microarrays, and ribonucleic acid [RNA] sequencing) or mutated (e.g., confirmed through SNP arrays and genomic DNA sequencing) between the control and disease conditions as examples. These candidates will then be explored further. Biomarkers with the strongest potential for differential occurrence between the two states will then be screened for. This selection is generally performed in conjunction with *in silico* methodologies (e.g., pattern recognition software capable of performing supervised/unsupervised classification and Benjamini and Hochberg multiple testing correction) [19,20]. *In vitro* methodologies will be exploited to further refine and reduce biomarker candidates for expanded (and costly) validation studies. *In vitro* approaches commonly used to confirm differential expression include tissue microarrays (combined with image analysis software for semiquantitation/quantitation) or enzyme-linked immunosorbent assay (ELISA) for antigen quantitation from protein extracts derived from tissue/primary cell lines. Stringent validation using independent tissue samples representative of the same tissue types used for the discovery process will then be required to confirm the general reproducibility of biomarker expression [18].

A prerequisite to the development of experimental workflows predicated upon the use of human tissue are the data fields that accompany these samples. Information will be required to triage and select specific tissue samples for experimental use based upon clearly defined parameters [21]. Each tissue sample must be annotated with

1. Sample-specific data (e.g., date of collection, volume, and barcode)
2. Clinical data (e.g., what is the disease stage for the patient)
3. Metadata about data structures that can be used for communication/searching/audit (e.g., user who created the file/modified a parameter and date of file creation)

The provision of structured data fields that allow samples to be easily categorized and referenced is an essential requirement for basic R&D and clinical translational programs. Structure query language (SQL) search tools can then be applied by clinical scientists to rapidly search tissue repositories for samples that meet specific selection criteria and therefore can be considered for inclusion into R&D programs.

It is important to be cognizant that tissue acquisition programs will be responsible for the collection, processing, and storage of material in a wide variety of forms. These include (but are not limited to) formalin-fixed paraffin-embedded (FFPE) blocks used in routine pathological assessments (e.g., tumor grade), serum, plasma (EDTA/heparin treated), frozen, buccal swabs, blood, cerebral spinal fluid, sputum, semen, and urine. The heterogeneity of tissue types, variable volumes, and potential for serial collections from the same patient provides the research scientist with a plethora of invaluable material in which to identify novel biomarkers with clinical applications (e.g., surrogate markers for disease risk), but these same benefits also provide significant challenges for the biobanking operations, which will now be discussed, including variable-volume storage sites.

Infrastructure requirements for human tissue biobanking

Biomedical research is predicated upon the interrogation of tissue as part of the discovery process (primary R&D), validation (confirmation of initial findings through robust testing of independent tissue cohorts), and verification studies (confirmation of validated findings for clinical testing facilities, e.g., those with Clinical Laboratory Improvement Amendments [CLIA] accreditation). Given the potential impact that novel companion diagnostics may have upon patient therapeutic selection and ultimately drug development pipelines, human tissue repositories will form an essential bedrock for basic and applied biomedical research. In order to achieve this goal, human tissue biobank facilities will require a significant investment covering both physical assets as well as appropriately trained and experienced personnel. The major facets of these areas will now be explored.

Physical infrastructure

Biobank repositories are responsible for the collection, processing, storage, and ultimately retrieval of the correct tissue sample stored within their facilities. Large-scale biobanking projects such as the UK Biobank have collected 500,000 samples over a period of 5 years. The samples were varied and included blood, urine, and saliva. As described above, many human tissue biobanks have other types of macromolecular isolates, including DNA, RNA, protein, or even cells (e.g., lymphocytes). Each sample will vary in volume, number of samples collected per patient, and the particular storage conditions that are likely to be required for housing. For example, with frozen tissue specimens, there are two primary methods that are routinely used for long-term storage:

1. Liquid nitrogen containers (–196°C)
2. Low-temperature storage freezers (–150°C)

Both methodologies are widely used to facilitate the preservation of tissue samples in a state that is believed to be representative of the point indicative of the metabolic, transcriptomic, proteomic, and genomic state at which the tissue was sampled from the patient. This is essential in order to provide research/validation material for biomedical research/ diagnostic testing and will increase the confidence that experimental data derived from the use of this material is valid and can ultimately be extrapolated to the parameter of interest.

Other tissue storage methodologies include FFPE blocks. These samples can be stored at room temperature for many decades and can be a useful sources to examine antigen expression patterns (e.g., immunohistochemistry and *in situ* hybridization). For the collection of material such as blood, serum, or plasma, there may be a requirement for temporary storage at 4°C prior to transportation and deposition in long-term storage containers. For plasma, there are two main mechanisms that are utilized in order to prevent coagulation of the sample and include heparin and EDTA. Each anticoagulant may have compatibility issues for certain biochemical assays and/or impact on long-term storage and thus the type of assay must be carefully aligned with the mechanism of tissue processing/storage procedure. Failure to do so could result in erroneous results and thus impact the biomarker discovery/validation pipelines.

Each type of storage methodology will bring with it its own logistical challenges. For example, with liquid nitrogen containers, there is a requirement to provide physical rooms to house the storage vats. Liquid nitrogen evaporates over a period of time and there will therefore be a requirement to replace lost fluid, which represents a financial cost. With respect to freezers and fridges, these instruments are electrical devices and require a constant power source. As with the replacement of liquid nitrogen, electrical power has a financial implication and over a period of 12 months may lead to a significant overhead cost depending upon the number and types of freezers that are stored in the biobank. The –80°C and –150°C freezers generate considerable heat from their motors and rooms housing these devices will require good ambient cooling (air-conditioning units). Such a physical setup will produce additional costs for power supply as part of the process of maintaining tissue in a frozen state, including rental costs for rooms.

With respect to the storage of FFPE blocks, cassettes can be stored at room temperature for many years. While it may appear an easy storage option for repositories, there are significant challenges that accompany the banking of these items. For example, regulatory requirements may demand that pathology blocks are retained for a period of 30+ years after the point of extraction and fixation. This generates challenges with respect to the buildup and storage of many tens of thousands of potential paraffin blocks. Blocks are stable at room temperature (e.g., 22°C) and remain in a hardened state, whereas those that are housed in warmer environments (e.g., Saudi Arabia with an average summer time temperature 45–50°C) may be subject to higher rates of instability. Macromolecules such as

RNA and protein can be subject to degradation over many years and consequently affect the results of biomedical research projects or validation results using these samples. It is imperative therefore that optimization studies be piloted on any sample (frozen/room temperature) prior to any experimental analyses.

Physical infrastructure requirements also extend to support areas such as

1. Tissue receipt and holding areas prior to sample logging and barcoding
2. Laboratories for tissue processing (e.g., automated liquid handling devices and tissue microarray robotics)
3. Office space for data entry clerks to record relevant sample information (e.g., patient ID, hospital number, referring clinical center, how the sample is to be processed, and what is the biobanking project number [if any])

Physical support infrastructure must also take into account, including computational hardware requirements, for example, whether desktop machines and/or terminal nodes will be the primary user interface to accessing the data repository. Additional hardware factors that will need to be addressed by biobanking facilities include the types of data storage devices (e.g., "hot data" [such as solid-state disks] versus "cold data" [such as tape drives]) and the relative distribution of data between the two types; amount of storage that will be required to record biobank data (e.g., gigabytes, terabytes, and petabytes); how much storage space will be needed initially and does the infrastructure allow for expansion over time as the amount of data grows; will the data be stored locally or will renting facilities over a "cloud"-based infrastructure be more cost effective; and is the data duplicated to enable disaster recovery in the event of a physical malfunction or algorithm corruption (e.g., computer viruses).

Computing infrastructure will be required to record not only information about the sample such as storage location, barcode number, how many samples, and what type of samples have been collected [22] but also clinical/pathological information that may have direct relevance to a biomedical research scientist. Commonly collected annotation variables normally associated with tissue specimens include age, sex, ethnicity, disease diagnosis/staging, and drug treatment as limited but important examples.

While data annotation may sound a relatively simple task, there are several factors that need to be addressed. These include

1. What databases will be used to house the clinical data (e.g., open source programs, proprietary commercial databases, or a combination of both)?
2. What operating systems will the database require?
3. How will the database communicate with other databases (if at all)?
4. How will it potentially be accessed by users, for example, web-based interfaces or intranet? [23]
5. Will there be different users (e.g., research scientists, healthcare workers, and database administrators), and if so, will they have different levels of data access?
6. If the data are to be communicated, will they be encrypted to ensure patient confidentiality?
7. Does the database have audit-level control to track changes to the system? [24]
8. Does the database permit tracking of tissue samples, including their assignment and usage in R&D projects?
9. Does the database facilitate compliance required by regulatory authorities (e.g., the UK Human Tissue Authority, US FDA, and CF21 part 11) if demanded?

All these factors will potentially have a direct impact upon the implementation and expansion of biobank data repositories and thus need to be considered carefully by staff who are operating/about to initiate biobanking facilities [25]. One final area that needs careful consideration is the bandwidth and transfer rate of information that must be installed in order to meet user requirements. For example, Mb/s transfer speeds are likely to be sufficient for simple alphanumeric data but Gb/s rates could be required for high-resolution magnetic resonance imaging (MRI) image files that are linked with tissue samples.

Personnel infrastructure

While physical infrastructure is an essential prerequisite for human tissue repositories, the personnel operating the biobank represent an equally vital resource for success. Given the breadth of activities that are required to collect, process, store, retrieve, annotate, and administer (computationally and business), it will therefore have a requirement for a multidisciplinary team with proportionate skills and experience. Some of the essential staff include

1. Consent nurses: Informed consent is a necessary requirement of all tissue donations made by patients [26–28]. It demands that the patient is made aware of the conditions upon which their tissue will be stored and utilized in biomedical operations.
2. Laboratory technicians: Technical staff will be responsible for the collection, processing, and maintenance of correct storage of human tissue. They will also be required to undertake proficiency/competency training on a yearly basis in order to maintain their registration. Staff training and yearly assessment are standard requirements of United Kingdom Human Tissue Authority (UK HTA)-accredited biobanks.
3. Data entry clerks: Clinical and tissue sample data will be entered into relational databases to allow audit and tracking of samples. SQL can be used to quickly search and retrieve appropriate tissue samples for R&D/validation studies.
4. Database administrators: Database administrators will require staff with appropriate computational skills to maintain, backup, and ensure downtime is minimized, thus maximizing database and biobank operational efficiency.
5. Business staff: Operations managers, clerical staff, accounts clerks, and secretaries represent some of the business personnel that will be needed in order to effect the smooth delivery of biobanking services.

Centralization versus localized tissue repositories

There is a significant investment required to establish and maintain human tissue repositories. As described above, this investment covers capital infrastructure, capital build, and personnel. Regulatory authorities given legal oversight over biobanking activities must ensure the highest levels of transparency, quality, and operational governance for the facilities they regulate. In this manner, the public who make an invaluable contribution to biomedical research, clinical diagnostics, and drug development through tissue and medical data donation can be assured that both their confidentiality remains secured and that their tissue will be utilized in the manner to which they gave consent [29]. Given the social, ethical, and legal requirements that are needed, human tissue repositories are moving away from local (e.g., departmental) storage facilities (with basic data capture) and more toward institutional centralized operations. While local facilities offer significant advantages in the short term (e.g., minimum capital infrastructure and limited technical capability requirement by staff), significant limitations will develop once the tissue repository begins to grow. Departmental biobanks are unlikely to have the necessary resources (physical and personnel) to maintain and guarantee the level of stringency required by the increased demands of accrediting authorities. For example, with respect to freezers, it is generally considered that uninterrupted power supply (UPS) backup combined with external power generators be deployed should there be a major power supply failure. Freeze/thawing of tissue samples would lead to degradation of sample quality and therefore variability of results derived from analyses involving these samples. The downstream impact are potential false positive/negative results with a concomitant loss of time and financial resources in the generation of erroneous data are potentially significant.

Regulatory compliance requires significant adherence to standard operating procedures, proficiency testing, documentation, audit, and reporting. They are therefore likely to place significant burden upon both financial as well as personnel resources for small departments that operate biobanks. A more centralized approach is likely to represent a higher benefit:cost ratio. Amalgamation of resources, adoption of best practice, benchmarking to other high-quality tissue repositories, and sharing of personnel/finances would enable the biobank to

1. Maximize its processing/storage efficiency and storage capacity
2. Reduce lead time in gaining operational performance to national/international standards
3. Minimize duplication of activities and maximize return on investment
4. Enable it to harmonize its activities with those of other national/international biobanks with greater efficiency

Standardization

The premise of any tissue biobank is that the material stored within the repository (including the clinical/metadata associated with each sample) represents a true reflection of the biological state of the sample at the point of removal from the patient. In order to achieve this goal, there will be several factors that must be addressed to minimize the risk of sample/data artifacts being introduced that are specifically attributable to the collection and storage processes. This ideal position is likely to be achieved through the implementation of standard operating procedures, personnel training/annual competency assessments, and ultimately benchmarking against best practices by the national/international biobanking community to ensure reproducibility of results [30,31].

Standardization will need to be addressed at several levels:

1. Tissue collection and processing: As stated, a human biobank repository must be predicated upon one fundamental tenet—the tissue (and therefore the biomolecules derived from that specimen) must be representative of the state at which the tissue was removed from the patient. From the moment a sample is extracted, there are several factors capable of introducing variation in biomolecular composition/concentration that are not related to a sample's actual physiological/pathological state. For example, with respect to tissue, it is a highly complex 3D architecture that comprises supporting connective stromal cells, endothelial cells, smooth muscle for vascular constriction/dilation, infiltrating red blood cells, immune cells (T-, B-, NK, and dendritic), and circulating cells that have sloughed off from other tissue sites. The tissue is highly innervated with blood supplying nutrients, oxygen, and growth

regulatory signals in addition to the usual products associated as metabolic waste. Upon excision of the mass, the tissue will undergo stress and potentially activate pathways linked to apoptosis (e.g., NF-kappa b), resulting in potential differential expression of

a. RNA
b. Proteins
c. Phosphorylation sites
d. Metabolites
e. Chromatin localization

The end result is that differential expression may potentially arise as a result of cellular stress rather than the true biological state of the tissue in question. In order to minimize the risk of artifacts developing, standard operating procedures (SOPs) for collection, transport, and processing are essential.

SOPs can include ensuring factors such as staff collecting the sample on ice; that the tissue is frozen/fixated in a given timeframe window; cryopreservation is optimized [32]; if serum is collected that blood is clotted at a specific temperature and is spun within a standard time to minimize the risk of hemolysis, which may impact downstream analyses; for plasma collection that the appropriate anticoagulant needs to be used to ensure compatibility with downstream biochemical assays; for FFPE samples, variation in genomic DNA-sequencing output and quality is greatly impacted by the methods used to fix the tissue samples (e.g., is the formalin buffered and if so what buffer is used; a constant time used for fixation must be implemented; and the temperature applied for paraffin embedding can impact the stability of biomolecules stored in the block). Optimization studies may have to be implemented in order to ensure that the processing and storage conditions applied by the biobank are appropriate and have been checked for quality by the biobank for the specific experimental procedure required by the research community [33].

2. Personnel training: The heart of any biobanking operation will be the quality of the staff operating the facility. Ensuring adequate capacity and capability to deliver on its obligations will be key factors in the long-term success of the biobank. As part of its requirements therefore, staff will have to be subjected to systems and processes that are capable of training (e.g., staff induction programs, equipment training, and annual proficiency testing) and monitoring performance. If deficiencies are found then to record and action measures to rectify the problem. Effective communication strategies between all aspects of the biobanking team will be required. Attendance of conferences and meetings will be key for the dissemination of best practice.

3. Databases and sample annotation: The information recorded for each donated sample must be consistent with a standard ontology (naming and definition of data types).
   a. If there are alternative disease classification systems available, then adherence to a single system must clearly be identified by the database administrators and used by all staff.
   b. If drug treatments are given to a patient, then how the regimen is deployed and what drugs are utilized (e.g., does the database used brand and/or generic drug name) will need to be defined prior to data entry.
   c. How are ethnic groups defined, including those of mixed race?

4. Data protection: Laws will vary from country to country and the ability to access, interrogate, and share information will be tightly controlled [34,35]. Data sharing with external groups [36] is an extremely important consideration for federated biobanks that wish to create a single "virtual biobank" and thus increased their effective

capacity. Internal accessibility requirements will define the information available to specific personnel within a given tissue repository (e.g., research nurses may be given rights to view all information including patient name and home address whereas a research scientist working in the same center may not). Which users have authority to enter data and modify fields must be carefully controlled through the database software. Many biobank facilities seek accreditation and/or are controlled by regulatory authorities to ensure that patient tissue samples/medical data records are used in the manner for which consent was given [37].

5. Compliance through audit: Regulatory authorities will have a requirement to provide legal oversight that biobanks are fulfilling their obligations. These include ensuring that patient confidentiality is fully protected and tissue is used/not used for specific applications. To deliver on these obligations, a biobank database can also be used as part of the audit/regulatory compliance process. For example, SQL can be used to provide reports regarding how many tissue samples have been deposited within the biobank, within what period of time, and where the samples are stored for all patients. This provides transparency of the biobank's activities and ensures the highest levels of procedural and ethical practice. It is also a mechanism that can be used to monitor performance and provide information rapidly when required as part of audit and benchmarking.

While variation in tissue sample collection, processing, storage, and clinical annotation are inherent in any systematic process, the standardization of procedures, practices, and training programs will serve to minimize the risk of this variation. A constant evaluation of the center's performance (e.g., annual review and statistical interrogation through audit) should lead to constant improvement of key metrics through feedback mechanisms (e.g., how many tissue samples were not frozen in the appropriate time window, why and what corrective measures have to be implemented to reduce this from repeating again). The use of international standards such as ISO 9001, ISO 15189, and CAP (College of American Pathologists) would provide a useful framework for biobanking facilities to measure their internal performance against international standards. Through partnerships with other biobanking facilities, communication at conferences and production of scientific publications, the operation can both benchmark and acquire best practice in order to improve its performance and also assist other tissue repositories improve theirs.

National/international examples of biobanking operations

Human tissue biobanking is now gaining such momentum that consortia of biobanks are being formed. They do this as mentioned previously in order to enable the development and rapid dissemination of best practices as well as to provide a quality framework in which to benchmark their activities [38]. By coordinating activities, it allows harmonization of processes and increases the quality of the tissue stored within the repository. The downstream impact is that scientific/clinical studies utilizing this material have a significantly higher probability of producing higher-quality research findings. Biobanks are now registered and successfully operate as central facilities in hospitals (e.g., the UK Nottingham Health Science Biobank, which has been operating for over 40 years from its first inception; http://nuhrise.org/nottingham-health-science-biobank/) and their tissue results in the production of new information regarding the development of breast cancer [39,40]. One ambitious project was established in 2005 by the Medical Research Council (MRC) to collect specimens from 500,000 patients over a 5-year period (2006–2010) to

explore diseases that are commonly associated with an ageing population (https://www
.ukbiobank.ac.uk/). These diseases encompass cancer, diabetes, dementia, and arthritis
among others and have focused on collecting a breadth of samples (including blood and
urine) from an age group ranging between 40 and 69 years. Volunteers have provided lifestyle and clinical data in conjunction to their samples. In this manner, genetic risk factors
and disease-associated genes can be potentially identified from across the UK population
base. This represents an example where a single biobanking initiative has been implemented at a national level in order to achieve a common goal of tissue and data collection
followed by scientific analyses through satellite collection and storage centers. Scientific
analyses of tissue/data held within the repository is starting to yield new insight into disease development [41–43].

Another model that has been successfully implemented is the formation of a consortium of biobanks that are focused upon a common disease area. The UK National Cancer
Research Institute (NCRI) has brought together over 30 biobanks as part of the "NCRI
Confederation of Cancer Biobanks" (http://ccb.ncri.org.uk/). The rationale is to coordinate
their activities in order to enhance the quality of cancer research programs. Such a strategy
may also provide the mechanism in which to effectively increase the number of samples
that are now available for scientific/clinical research. Through strategic partnerships such
as this, there is a positive impact in two areas. The first results in an effective increase in
the number of cases that are available for analyses potentially resulting in increased statistical power for the study being proposed. A second equally important factor from the harmonization of activities is that variation in results may be the consequence of processing
artifacts between centers, that is, center to center variation. Through the standardization of
procedural elements the risk of findings being the result of nonspecific systematic artifacts
becomes greatly (although not entirely) reduced.

A European initiative has been established that involves biobanking facilities in over
16 countries. The Biobanking and Biomolecular Resources Research Initiative (BBMRI-
ERIC) is focused on providing an international resource of tissue and other biomolecules
of potential interest/value pertinent to human diseases. Once again the harmonization of
activities will enable this multicenter initiative to effectively increase the number of tissue
samples available for biomedical research and ensure consistency of sample acquisition
and storage within each collection facility [44,45].

The National Cancer Institute (NCI) in the United States has established the
Biorepositories and Biospecimen Research Branch (BBRB), which is a nationally funded
program to collect, annotate, and provide research scientists with high-quality tissue samples for biomedical interrogation (http://biospecimens.cancer.gov/default.asp). As with
the initiatives above, there is a strong focus on developing and optimizing processes that
will allow tissue samples to be stored effectively while minimizing the risk of biomolecular degradation/variation due to systematic artifacts. Interestingly, large-scale biobanking
programs have been formed using animal models. The Canine Comparative Oncology
Genomics Consortium is another NCI initiative that was the direct result of veterinary
and medical oncologists wishing to explore the similarity/dissimilarity of human cancers
compared to their canine (dog) counterparts (https://ccrod.cancer.gov/confluence/dis-
play/CCRCOPWeb/Canine+Comparative+Oncolgy+Genomics+Consortium). The rationale behind such an initiative is that canines develop cancers similar to those observed in
humans. These include, for example, lymphoma, osteosarcoma, and melanoma with the
premise being that genetic analyses of canine tumors may provide insight into the development and treatment of human tumors [46]. Sample processing techniques are also being
explored by this consortium, including, for example, different methodologies for freezing

tissues (flash frozen versus OCT-embedded formats). Thus, the CCOGC program intends to develop best practice for its national partners in the acquisition, processing, storage, and analyses of canine tumors with the aim of ultimately translating these findings to improve the healthcare of both canine and human cancer patients.

Challenges for human tissue repositories: Current and future

This final section will explore factors that present challenges as well as opportunities for human tissue repositories and if navigated effectively, are likely to lead to sustainability of the biobank. As with all businesses (whether charitable, government, or private sector), there is no one single solution to the establishment and operation of a biobank. A set of guiding principles can be formulated (e.g., standard operating procedures on how tissue is to be collected; how tissues should be processed; what clinical data should be entered into relational databases; and supporting ontologies that will be used by the biobank) and thus can assist with harmonizing best practice internally and between tissue repositories. There are however parochial challenges (e.g., center-to-center operational procedures, country-specific laws, cultural differences, and logistical challenges) that will have significant impact upon both implementation and ultimately delivery of biobanking services [47]. Thus, it will be the responsibility of the management team to develop a detailed business plan in order to provide the strategic guidance for growth, personnel requirements, type of tissue to be collected, and even the cost associated with operating such a facility.

Some of the issues that will need to be considered include the following:

1. Regulation: Many countries have introduced laws controlling the donation, storage, and use of human tissue [48]. The UK biobanking industry is tightly regulated by the Human Tissue Authority (HTA) and provides oversight regarding all aspects of tissue donation, storage, and dissemination, and even extends to patient consent and intellectual property ownership rights resulting from the study of donated tissue. The HTA website (https://www.hta.gov.uk/) states that

   > The HTA is a regulator set up in 2005 following events in the 1990s that revealed a culture in hospitals of removing and retaining human organs and tissue without consent.

   Once the practice was discovered, it resulted in significant damage to the UK National Health Service (NHS) human tissue donation/pathology service and the biomedical research industry as a whole. After government, public, and industry consultation, the HTA was established to provide regulation of all human tissue donation activities. Through concerted action, extensive dialog, and transparency, the United Kingdom has now become a world leader in the donation and supply of human tissue to biomedical research. A number of stakeholders have been involved in the establishment of this policy, including medical charities, patients, government, industry, regional healthcare providers, and even economic development agencies. The reasoning behind such an extensive consultation process was to ensure that a wide-ranging policy was developed rigorous enough to restore public faith in tissue donation, but at the same time had sufficient flexibility to allow biomedical research to proceed to the highest levels of quality. With the construction of the UK Biobank (see above), it signified a model of where regulatory oversight actually increased public confidence in the tissue donation/exploitation process and led to nationwide support for large-scale biomedical research programs. Other challenges that biobanks may have to deal with is that regulation may change over time and as such business operations in the biobank would have to respond accordingly to these new requirements. The outcome of these changes may impact several areas, including financial (e.g., more expense to deliver and maintain accreditation), operational practice, personnel retraining, and even the acquisition of new personnel skills that currently do not exist in order to fulfil regulatory requirements.

2. Multidisciplinary team requirements: This aspect represents a significant challenge as the collection/storage/annotation of tissue necessitates a multidisciplinary team. For example, patient advocates (volunteers and/or hospital staff) will begin the process of counselling potential donors with respect to the implication of their donation. There must therefore be a strong interplay between the healthcare professional, public volunteers, and the patient. Additional skills input will be required from certified pathologists, technical staff for collecting and processing samples (e.g., fixation/paraffin embedding, cellular isolation, and DNA/RNA extraction), and computer specialists (e.g., those with expertise in systems administration and data backup while others will be necessitated having skills in configuring databases/SQL report writing); personnel will also be required for office duties such as general secretarial work, writing business contracts, accreditation specialists, marketing specialists, and even accounting clerks. The depth and breadth of activities will therefore require significant resources to be put in place if the biobank is to meet the demands of regulators, scientific quality assurance, and public expectations.

3. Targeted biobanking versus "catch all" tissue collection: A significant question that all biobanking operations will need to address relates to the tissue that will be collected. As seen in the previous two sections, the investment required to establish and maintain a quality biobank is not trivial. The time, resources, personnel, and funding need careful planning. There may even be the need to integrate and/or share services in order to accomplish the goals of a tissue repository (which may suggest that centralization of biobanking facilities represents the best long-term solution for achieving all the requirements of a quality-assured facility). Given this commitment, there will need to be a mechanism in place for cost recovery. Such funding may be through a variety of financial instruments, including
    a. Institutional funding as a core facility
    b. Grant income
    c. Charitable support
    d. Commercial contributions

To this end, the tissue that will form the basis of such a strategic investment must be utilized with maximum return on that investment by all stakeholders. It is essential that the types of tissues, its method of storage, volume, and sample types be compatible with end-user requirements. For these reasons, a targeted approach to tissue collection is likely to ensure maximum exploitation and rapid turnover of samples (due to user demand), which would eventually translate into maximum benefit for patients. A "catch all" strategy could result in the collection and long-term storage (with associated costs) of tissue that is neither required nor in high demand by the biomedical/medical community. In this scenario, the material would be stored without guarantee of use and potentially be counter to public expectation that their tissue would benefit future generations of patients with the same disease. Thus, targeted tissue collection programs are likely to offer significantly higher return on investment in both the short and long term.

4. Stakeholder engagement: Given the disparate groups that need to be actively engaged in the process of tissue storage and exploitation in biomedical research, there will be a requirement to consult with key stakeholders, canvass their views, and gain their collective buy-in [49]. Such a strategy would assist in avoiding potential issues that may either slow or even halt biobanking operations. Stakeholders are broad in nature and include, for example, the patient group donating tissue [50–52]; health authorities and hospital ethics panels; nurses/clinicians/surgeons/laboratory staff; pharmaceutical/diagnostic companies that will ultimately have a requirement for tissue as part of drug/diagnostic test development; and political support from the highest levels of government, which may include several departments/ministries, for example, health, law, education, and business. Given the diverse nature of stakeholders, early engagement and continued dialog will be important for all biobanking facilities.

5. Public awareness: A crucial aspect of any biobanking operation is promoting the activities and objectives of the facility to the wider public. Engagement can occur at all levels of society (e.g., not just patients but the broader population who may eventually be impacted by disease). In multiethnic societies, an awareness of religious beliefs and cultural etiquette is essential if a broad spectrum of individuals from different genetic backgrounds is required for scientific study and eventually served from a healthcare perspective. The ability to not only gain support from the public but also have them take ownership/contribution of donation services will lead to a greater chance of success in the biobank achieving its overall targets.

6. Cost: A major impact upon any biobanking project will be the cost. Cost covers areas such as financial, resources, infrastructure, and personnel. The cost will not only be limited to initial start-up costs but also include the period of time that the biobank is intended to operate. How will the facility be funded, what overheads will be charged (e.g., will lighting, heating, office/lab space rent be charged separately to the biobank or will a standard rate [if any] be applied), and will funding be purely from designated sources (e.g., endowments, hospital core unit funding, private investment, and fee for service) will need to be addressed. Funding models vary and will thus need to be discussed, assessed, and a sound fiduciary plan developed based upon the strengths/weakness/opportunities/threats (SWOT) analysis biobanking business plan. Financial planning will be a responsibility of the management team as part of the overall business plan.

Summary

A tissue repository represents a significant investment to society. The benefits for basic scientific research with respect to understanding molecular pathways and how they communicate with each other in order to produce both physiological functioning are profound. Using this information for baseline data can also provide information on how pathologies develop. The application of human tissue repositories to twenty-first century medicine is broad and can impact areas such as drug development pipelines, novel diagnostic, and prognostic tools. The goal of this investment is to ultimately personalize medical care of the patient based upon their specific genomic signature.

It is important to acknowledge that biobanks should also be considered as long-term investments by all stakeholders. Consequently, a tissue repository may require support at a variety of levels to ensure that it has the strongest chance of success of achieving its goals and appropriate buy-in by society and public/private sector organizations will be essential in meeting these objectives.

References

1. Gilissen C, Hehir-Kwa JY, Thung DT, van de Vorst M, van Bon BW, Willemsen MH, Kwint M, Janssen IM, Hoischen A, Schenck A et al.: Genome sequencing identifies major causes of severe intellectual disability. *Nature* 2014, 511(7509):344–347.

2. Moltke I, Grarup N, Jorgensen ME, Bjerregaard P, Treebak JT, Fumagalli M, Korneliussen TS, Andersen MA, Nielsen TS, Krarup NT et al.: A common Greenlandic TBC1D4 variant confers muscle insulin resistance and type 2 diabetes. *Nature* 2014, 512(7513):190–193.

3. Kundaje A, Meuleman W, Ernst J, Bilenky M, Yen A, Heravi-Moussavi A, Kheradpour P, Zhang Z, Wang J, Ziller MJ et al.: Integrative analysis of 111 reference human epigenomes. *Nature* 2015, 518(7539):317–330.

4. Juric D, Castel P, Griffith M, Griffith OL, Won HH, Ellis H, Ebbesen SH, Ainscough BJ, Ramu A, Iyer G et al.: Convergent loss of PTEN leads to clinical resistance to a PI(3)K alpha inhibitor. *Nature* 2015, 518(7538):240–244.

5. Inaki K, Menghi F, Woo XY, Wagner JP, Jacques PE, Lee YF, Shreckengast PT, Soon WW, Malhotra A, Teo AS et al.: Systems consequences of amplicon formation in human breast cancer. *Genome Research* 2014, 24(10):1559–1571.

6. Pinto EM, Chen X, Easton J, Finkelstein D, Liu Z, Pounds S, Rodriguez-Galindo C, Lund TC, Mardis ER, Wilson RK et al.: Genomic landscape of paediatric adrenocortical tumours. *Nature Communications* 2015, 6:6302.

7. Ongen H, Andersen CL, Bramsen JB, Oster B, Rasmussen MH, Ferreira PG, Sandoval J, Vidal E, Whiffin N, Planchon A et al.: Putative cis-regulatory drivers in colorectal cancer. *Nature* 2014, 512(7512):87–90.

8. Gholami B, Norton I, Eberlin LS, and Agar NY: A statistical modeling approach for tumor-type identification in surgical neuropathology using tissue mass spectrometry imaging. *IEEE Journal of Biomedical and Health Informatics* 2013, 17(3):734–744.

9. Zhang B, Wang J, Wang X, Zhu J, Liu Q, Shi Z, Chambers MC, Zimmerman LJ, Shaddox KF, Kim S et al.: Proteogenomic characterization of human colon and rectal cancer. *Nature* 2014, 513(7518):382–387.

10. Wilhelm M, Schlegl J, Hahne H, Moghaddas Gholami A, Lieberenz M, Savitski MM, Ziegler E, Butzmann L, Gessulat S, Marx H et al.: Mass-spectrometry-based draft of the human proteome. *Nature* 2014, 509(7502):582–587.

11. Chang YH, Gregorich ZR, Chen AJ, Hwang L, Guner H, Yu D, Zhang J, and Ge Y: New mass-spectrometry-compatible degradable surfactant for tissue proteomics. *Journal of Proteome Research* 2015, 14(3):1587–1599.

12. Keshishian H, Burgess MW, Gillette MA, Mertins P, Clauser KR, Mani DR, Kuhn EW, Farrell LA, Gerszten RE, and Carr SA: Multiplexed, quantitative workflow for sensitive biomarker discovery in plasma yields novel candidates for early myocardial injury. *Molecular & Cellular Proteomics: MCP* 2015, 14:2375–2393.

13. Winnike JH, Wei X, Knagge KJ, Colman SD, Gregory SG, and Zhang X: Comparison of GC–MS and GCxGC–MS in the analysis of human serum samples for biomarker discovery. *Journal of Proteome Research* 2015, 14:1810–1817.

14. Misale S, Yaeger R, Hobor S, Scala E, Janakiraman M, Liska D, Valtorta E, Schiavo R, Buscarino M, Siravegna G et al.: Emergence of KRAS mutations and acquired resistance to anti-EGFR therapy in colorectal cancer. *Nature* 2012, 486(7404):532–536.

15. Wormald S, Milla L, and O'Connor L: Association of candidate single nucleotide polymorphisms with somatic mutation of the epidermal growth factor receptor pathway. *BMC Medical Genomics* 2013, 6:43.

16. Li X, Pezeshkpour G, and Phan RT: KRAS mutation status impacts diagnosis and treatment decision in a patient with two colon tumours: A case report. *Journal of Clinical Pathology* 2015, 68(1):83–85.

17. Ding XF, Yin DQ, Chen Q, Zhang HY, Zhou J, and Chen G: Validation of p27KIP1 expression levels as a candidate predictive biomarker of response to rapalogs in patient-derived breast tumor xenografts. *Tumour Biology: The Journal of the International Society for Oncodevelopmental Biology and Medicine* 2015, 36:1463–1469.

18. Hsu CY, Ballard S, Batlle D, Bonventre JV, Bottinger EP, Feldman HI, Klein JB, Coresh J, Eckfeldt JH, Inker LA et al.: Cross-disciplinary biomarkers research: Lessons learned by the CKD biomarkers Consortium. *Clinical Journal of the American Society of Nephrology: CJASN* 2015, 10:894–902.

19. Mian S, Ugurel S, Parkinson E, Schlenzka I, Dryden I, Lancashire L, Ball G, Creaser C, Rees R, and Schadendorf D: Serum proteomic fingerprinting discriminates between clinical stages and predicts disease progression in melanoma patients. *Journal of Clinical Oncology: Official Journal of the American Society of Clinical Oncology* 2005, 23(22):5088–5093.

20. Kauffmann A and Huber W: Microarray data quality control improves the detection of differentially expressed genes. *Genomics* 2010, 95(3):138–142.

21. Cui W, Zheng P, Yang J, Zhao R, Gao J, and Yu G: Integrating clinical and biological information in a shanghai biobank: An introduction to the sample repository and information sharing platform project. *Biopreservation and Biobanking* 2015, 13(1):37–42.

22. Nussbeck SY, Skrowny D, O'Donoghue S, Schulze TG, and Helbing K: How to design biospecimen identifiers and integrate relevant functionalities into your biospecimen management system. *Biopreservation and Biobanking* 2014, 12(3):199–205.

23. Eder J, Gottweis H, and Zatloukal K: IT solutions for privacy protection in biobanking. *Public Health Genomics* 2012, 15(5):254–262.

24. Norling M, Kihara A, and Kemp S: Web-based biobank system infrastructure monitoring using Python, Perl, and PHP. *Biopreservation and Biobanking* 2013, 11(6):355–358.

25. Tukacs E, Korotij A, Maros-Szabo Z, Molnar AM, Hajdu A, and Torok Z: Model requirements for Biobank Software Systems. *Bioinformation* 2012, 8(6):290–292.

26. Stewart C, Fleming J, and Kerridge I: The law of gifts, conditional donation and biobanking. *Journal of Law and Medicine* 2013, 21(2):351–356.

27. Reichel J, Lind AS, Hansson MG, and Litton JE: ERIC: A new governance tool for biobanking. *European Journal of Human Genetics: EJHG* 2014, 22(9):1055–1057.

28. Beskow LM, Dombeck CB, Thompson CP, Watson-Ormond JK, and Weinfurt KP: Informed consent for biobanking: Consensus-based guidelines for adequate comprehension. *Genetics in Medicine: Official Journal of the American College of Medical Genetics* 2015, 17(3):226–233.

29. Muller TH and Thasler R: Separation of personal data in a biobank information system. *Studies in Health Technology and Informatics* 2014, 205:388–392.

30. Salvaterra E, Giorda R, Bassi MT, Borgatti R, Knudsen LE, Martinuzzi A, Nobile M, Pozzoli U, Ramelli GP, Reni GL et al.: Pediatric biobanking: A pilot qualitative survey of practices, rules, and researcher opinions in ten European countries. *Biopreservation and Biobanking* 2012, 10(1):29–36.

31. Mee B, Gaffney E, Glynn SA, Donatello S, Carroll P, Connolly E, Garrigle SM, Boyle T, Flannery D, Sullivan FJ et al.: Development and progress of Ireland's biobank network: Ethical, legal, and social implications (ELSI), standardized documentation, sample and data release, and international perspective. *Biopreservation and Biobanking* 2013, 11(1):3–11.

32. Bissoyi A and Pramanik K: Role of the apoptosis pathway in cryopreservation-induced cell death in mesenchymal stem cells derived from umbilical cord blood. *Biopreservation and Biobanking* 2014, 12(4):246–254.

33. Xie R, Chung JY, Ylaya K, Williams RL, Guerrero N, Nakatsuka N, Badie C, and Hewitt SM: Factors influencing the degradation of archival formalin-fixed paraffin-embedded tissue sections. *The Journal of Histochemistry and Cytochemistry: Official Journal of the Histochemistry Society* 2011, 59(4):356–365.

34. Williams BA and Wolf LE: Biobanking, consent, and certificates of confidentiality: Does the ANPRM muddy the water? *The Journal of Law, Medicine & Ethics: A Journal of the American Society of Law, Medicine & Ethics* 2013, 41(2):440–453.

35. Loukides G and Gkoulalas-Divanis A: Utility-aware anonymization of diagnosis codes. *IEEE Journal of Biomedical and Health Informatics* 2013, 17(1):60–70.

36. Izzo M, Mortola F, Arnulfo G, Fato MM, and Varesio L: A digital repository with an extensible data model for biobanking and genomic analysis management. *BMC Genomics* 2014, 15 (Suppl 3):S3.

37. Hirschberg I, Kahrass H, and Strech D: International requirements for consent in biobank research: Qualitative review of research guidelines. *Journal of Medical Genetics* 2014, 51(12):773–781.

38. Lablans M, Bartholomaus S, and Uckert F: Providing trust and interoperability to federate distributed biobanks. *Studies in Health Technology and Informatics* 2011, 169:644–648.
39. Curtis C, Shah SP, Chin S-F, Turashvili G, Rueda OM, Dunning MJ, Speed D, Lynch AG, Samarajiwa S, Yuan Y et al.: The genomic and transcriptomic architecture of 2000 breast tumours reveals novel subgroups. *Nature* 2012, 486(7403):346–352.
40. Ross-Innes CS, Stark R, Teschendorff AE, Holmes KA, Ali HR, Dunning MJ, Brown GD, Gojis O, Ellis IO, Green AR et al.: Differential oestrogen receptor binding is associated with clinical outcome in breast cancer. *Nature* 2012, 481(7381):389–393.
41. Tyrrell JS, Taylor MS, Whinney D, and Osborne NJ: Associations of leg length, trunk length, and total adult height with Meniere's: Cross-sectional analysis in the UK biobank. *Ear and Hearing* 2015, 36: e122–e128.
42. Ntuk UE, Gill JM, Mackay DF, Sattar N, and Pell JP: Ethnic-specific obesity cutoffs for diabetes risk: Cross-sectional study of 490,288 UK biobank participants. *Diabetes Care* 2014, 37(9):2500–2507.
43. Moore DR, Edmondson-Jones M, Dawes P, Fortnum H, McCormack A, Pierzycki RH, and Munro KJ: Relation between speech-in-noise threshold, hearing loss and cognition from 40–69 years of age. *PloS One* 2014, 9(9):e107720.
44. Zika E, Paci D, Braun A, Rijkers-Defrasne S, Deschenes M, Fortier I, Laage-Hellman J, Scerri CA, and Ibarreta D: A European survey on biobanks: Trends and issues. *Public Health Genomics* 2011, 14(2):96–103.
45. van Ommen GJ, Tornwall O, Brechot C, Dagher G, Galli J, Hveem K, Landegren U, Luchinat C, Metspalu A, Nilsson C et al.: BBMRI-ERIC as a resource for pharmaceutical and life science industries: The development of biobank-based expert centres. *European Journal of Human Genetics: EJHG* 2015, 23:893–900.
46. Paoloni M, Webb C, Mazcko C, Cherba D, Hendricks W, Lana S, Ehrhart EJ, Charles B, Fehling H, Kumar L et al.: Prospective molecular profiling of canine cancers provides a clinically relevant comparative model for evaluating personalized medicine (PMed) trials. *PloS One* 2014, 9(3):e90028.
47. Calzolari A, Napolitano M, and Bravo E: Review of the Italian current legislation on research biobanking activities on the eve of the participation of national biobanks' network in the legal consortium BBMRI-ERIC. *Biopreservation and Biobanking* 2013, 11(2):124–128.
48. Soini S: Finland on a road towards a modern legal biobanking infrastructure. *European Journal of Health Law* 2013, 20(3):289–294.
49. Gaskell G, Gottweis H, Starkbaum J, Gerber MM, Broerse J, Gottweis U, Hobbs A, Helen I, Paschou M, Snell K et al.: Publics and biobanks: Pan-European diversity and the challenge of responsible innovation. *European Journal of Human Genetics: EJHG* 2013, 21(1):14–20.
50. Ahram M, Othman A, Shahrouri M, and Mustafa E: Factors influencing public participation in biobanking. *European Journal of Human Genetics: EJHG* 2014, 22(4):445–451.
51. Caenazzo L, Tozzo P, and Pegoraro R: Biobanking research on oncological residual material: A framework between the rights of the individual and the interest of society. *BMC Medical Ethics* 2013, 14:17.
52. Husedzinovic A, Ose D, Schickhardt C, Frohling S, and Winkler EC: Stakeholders' perspectives on biobank-based genomic research: Systematic review of the literature. *European Journal of Human Genetics: EJHG* 2015, 23:1607–1614.

section three

Industrial and environmental biotechnology

chapter seventeen

Microbial biotechnology

Margarita Aguilera and Jesús Manuel Aguilera-Gómez

Contents

Abstract

The reader through this chapter will acquire the basics of essential knowledge of microorganisms and their impact on biotechnology. The uses of microorganisms such as bacteria, archaea, fungi, and viruses in different domains of life have been contributing continuously to the progress of biotechnology. They have been involved since the first classical biotechnological processes such as the manufacture of fermented foods until today with complex designed genetic therapy for curing diseases as modern biotechnology advances. The easy use of microorganisms since the discovery of deoxyribonucleic acid and genetic engineering applications has greatly allowed the development of this science. This chapter will explain the key points regarding the main types of bacteria, viruses, and fungi used in biotechnology. It will also review the importance of microorganisms as factories, tools, and producers of biotechnological products themselves and of microbial products of interest as biopolymers. The important details of the bioprocesses, bioreactors, and high-scale production using microorganisms have also been reviewed.

Keywords: Biotechnology, Bacteria, Fungi, Virus, Industrial bioprocess, Biopolymers.

Microorganisms and biotechnology

Introduction

Biotechnology makes use of different types of microorganisms depending on the application that needs to be achieved. In most cases, if we use molecular biology for obtaining a product, the preferable microorganism used is *Escherichia coli*. Its origin, the taxonomic variability of this species, and its history of use are well known. This variability is given by the differential genetic characteristics of variable strains, which are described by their genotype. Thus, biotechnologists may use a specific genotype to select the strain of *E. coli* required for a specific use. Moreover, these genotypic characteristics will control and lead the culture conditions. The culture collection and the strain biobanks are entities that act as safe depositors of the strains used in biotechnology. Moreover, a long list of microorganisms has been isolated, identified, and strategically used in biotechnology for the production of interesting biotechnological products (Glazer and Nikaido 2007; Iáñez-Pareja 2009), for example, citric acid, which is normally made by *Aspergillus niger*; antibiotic substances produced by the group of actinomycetes, specifically genus *Streptomyces*, or the production of xanthan typically from *Xanthomonas campestris*.

Main bacteria of biotechnological interest

Escherichia coli: Model organism in biotechnology

Escherichia coli is the microorganism most studied in recent times and is regarded as the model microorganism in bacteriology, allowing its use as one of the tools widely applied in modern biotechnology and forming the basis of work for the first sequencing of the genome. Thus, the first genome sequencing was performed on this microorganism (Blattner et al. 1997).

Over time, a number of variants of different strains of *E. coli* have developed by natural evolution in the environment or through a series of mutations that occurred in different laboratories. Thus, there are different lines of *E. coli* carrying different mutations, many of them unknown until now, which may or may not affect research work. The genomes of different strains of *E. coli* of interest were recently published, not only for their virulence or pathogenicity, but also for their application in the field of biotechnology depending on their specific genetic background (http://openwetware.org/wiki/E._coli_genotypes; http://cgsc.biology.yale.edu/).

Genotypes of Escherichia coli

The gene pool is determined by a specific genotype, and the genotype describes the entire genetic information from a particular organism contained in the deoxyribonucleic acid (DNA). This genotype explains the phenotypic properties of a strain. However, in practical terminology, the genotypes of microorganisms are explained and thus e.g., *E. coli* is showing a list of genes in which the cell is deficient, conversely if a gene is not listed, then the organism is a wild type for that gene (the gene is perfectly functional). These lists of genes that constitute a strain genotype or line generally indicate the genes that carry a mutation that prevents the proper functioning of the protein for encoding. Sometimes there are multiple genes involved in a function so in such cases each of these genes is usually distinguished by an end with capital letter and in italics before the including gene. Thus, for example, *rec*A, *rec*B, and *rec*C genes are the three genes coding for proteins involved in the recombination process. Moreover, if the gene is preceded by the letter delta (Δ), then it indicates that the gene has been deleted. Δ, if the letter preceding the name of genes such as brackets Δ (lac-pro), then it indicates that all the genes included between these brackets that are named, have been deleted. Whereas, if a particular gene is not named, it is understood that it is the wild-type strain for that gene.

The strains most frequently used in molecular biology are *E. coli* K12 derivatives (*E. coli* K12), *E. coli* DH5-alpha, and *E. coli* XL1Blue.

Mutations in common strains of Escherichia coli for improvement of the strain in biotechnology

To insert foreign DNA into the *E. coli* strain, it should be considered whether the gene pool contains restrictases, in addition to this process of cloning, if the use of X-gal as a substrate for the selection of white or blue cloning colonies, and therefore the presence of intact *lacZ* gene in the genotype of *E. coli* and whether by this genotype can be complemented for a correct use of the targeting strategy. We must pay equal attention to the presence of recombinases if the insertion of genes into the genome of the microorganism needs to be achieved, or for its absence if prevention of gene insertion into the chromosome is required. Protein production requires decreasing the cytoplasmic proteases during production and secretion to the extracellular medium, according to the destination of the synthesized protein. Finally, the presence of genes resistant to certain antibiotics in the genome of the bacteria being used as the genetic background can prevent the selection process of transfer of certain vectors from being crippled selection markers of these vectors.

The presence of specific enzymes with important roles in recombinant DNA products is important as methylases, restrictases, recombinases, and proteases. Moreover, the presence or absence of genes involved in the lac operon, specifically the vector-based detection system in the presence of the intact lacZ gene with a multiple cloning site within the same (e.g., pUC19, pGEMT) is used. Thus, if the gene inserted in the site lacZ encoding galactosidase β, it will not be expressed correctly, no catalysis of X-gal (5-bromo-4-chloro-3-indolyl-β-d-galactopyranoside) and no blue product (5-bromo-4-chloro-3-hidroxindol), substance responsible for blue staining colonies. Thus, a successful ligation indicated by the appearance of white colonies (where the insert disrupts the lacZ gene) with respect to the blue indicates the absence of the insert and therefore error in cloning (since the lacZ gene is intact). Moreover, it is sometimes important to repress the P-lac promoter with the repressors lacI, or improved versions as lacIq gene.

Antibiotic resistance genes (ARG) are important in selecting a strain in a recombinant process or other transfer vector markers. When it is necessary to follow the strain it is interesting to use antibiotics that allow us to keep it free of contamination by other microorganisms, but on the other hand, the presence of certain genes of resistance may prevent selection of different exogenous input DNA markers such as those present in plasmids, phages, cosmids, or transposons. Some of the more common antibiotics found in different genetic backgrounds can be streptomycin (Str), kanamycin (Kan), chloramphenicol (Cam), nalidixic acid (Nal), or tetracycline (Tet). We must remember that current legislation prohibits the intentional release of microorganisms with antibiotic resistance markers (genes) to the environment, in order to avoid the dispersion of these genes between pathogenic microorganisms with the consequent loss of the repertoire of tools and possible treatments to combat these microorganisms.

Genetic backgrounds are taken into account also for the experimental design processes for the expression of certain recombinant proteins. In specific cases, it will be needed to seek the cheapest means of cultivation and strains must be able to grow on these media.

Other Gram-positive bacteria such as *Bacillus subtilis* and *Bacillus licheniformis* have been used as models for use in biotechnology bioprocesses, mainly when is not possible to use *E. coli*.

Culture collections of strains

Microbial culture collections and cell biobanks prepare strains for being distributed against specific request. Strains not protected commercially can be purchased from type culture collections or through trading companies in the area of molecular biology. Regarding the type culture collections we can highlight the American Type Culture Collection (ATCC) (www.atcc.org); the German Collection of Microorganisms and Cell Cultures (www.dsmz .de); the National Institute of Genetics, Japan (http://www.shigen.nig.ac.jp/ecoli/strain/top/top.jsp); or the Spanish Type Culture Collection (http://www.cect.org). There is a wide collection of microorganisms from different companies that sell cells ready for distribution and use. There are also culture collection of yeasts and fungi, including the British

National Collection of Yeast Cultures (http://www.ncyc.co.uk). There are also specific culture collections for filamentous fungi: "The Centraal bureau voor Schimmelcultures" (CBS) Fungal Biodiversity Centre (http://www.cbs.knaw.nl).

Main virus biotechnological interest

Introduction

Certain viruses are important in the field of biotechnology, including those that fight against infectious agents, those involved in the production of vaccines, knowing viral vectors and their construction, or those that are in the production of variable proteins.

Beyond the importance of bacteriophages in the maintenance of the existing natural ecological balance in microbiomes, phages were used directly as therapeutic agents for the control of bacteria converted to infectious agents since early in the last century. From a detailed study of some such phages as the T7 or λT4 phages new drugs are being designed based on the mechanism of action of some of their proteins to control these bacteria. However, the use of viruses overshoot and can be used as a protein expression system, noting in this respect the use of other viruses such as baculovirus to obtain modified proteins during and after the translation.

Viruses are also useful tools for the selection of variants of interest based on techniques such as phage display, with examples in pest control and the fight against cancer. Viruses can be modified to make them more effective in the treatment of cancer, with special attention to increasing the selectivity for cancer cells, reduction of virulence in finding appropriate cellular targets, in the search for transcriptional targets, and strategies to protect virus immune system attack so that more effective systems can be achieved. Recently, researchers have designed recombinant viruses for being applied in specific gene therapy.

Phage as therapeutic agents

The main advantages in the use of phages as a control system of infectious microorganisms lies in the specificity of "phage" action (as only infecting a given species or genus, not affecting the rest of the microbiota), direct action on the focus of infection, rapid mode of action, and above all the mechanism is independent of the existence of antibiotic resistance.

Several recognized pharmaceutical and biotechnological companies marketed phage preparations as therapeutic agents against infectious diseases with various preparations. Examples of biotechnological use of phages as drugs for several treatments of infectious diseases (Mathur et al. 2003) are Phage BioDerm products, consisting of biodegradable polymer plasters impregnated with various substances, including bacteriophages against *Pseudomonas aeruginosa*, *Staphylococcus aureus*, *E. coli*, *Streptococcus* species, and *Proteus* species. The use of bacteriophages to treat *Clostridium difficile* is associated with very promising results being extremely effective against diseases. Currently, genetically engineered bacteriophages are proposed to be used as adjuvant along with antibiotic therapy. Phages encoded with lysosomal enzymes are also effectual in the treatment of infectious diseases (Qadir 2015).

Mechanisms of action of the phage to search for new antibiotics

The mechanisms of action of bacteriophages allow the design and search for drugs aimed specifically at targeting viruses used to attack microbial cells, for example, expression of the protein kinase or proteins or molecules that mimic the mechanism of action that can be considered as antimicrobials. Some examples, as the factor σ70, which is controlling and dealing with the expression of housekeeping or essential genes, the expression of the AsiA, phage protein λP or B phage protein P2. In both cases, these proteins are associated with DnaB helicase that are essential for replication. The system is distinguished from phagotherapy (phage supply directly controlling the microorganism) in which these tools are effective in removing bacteria, since the virus does not pass through lysogen states that are not effective for the process control or death of the bacteria.

Virus as protein expression systems

T7 Virus polymerase and promoters

T7 RNA transcribed only from T7 promoters, which are used in combination in expression vectors. Furthermore, T7 polymerase transcribing is especially efficient, has a very low error rate in transcription, and ribonucleic acids (RNAs) allow the production of polymerases longer than others, which is particularly useful for the expression of large proteins, in high concentration. Furthermore, this polymerase, unlike bacterial is not inhibited by the addition of rifampicin.

Baculovirus

Baculoviruses are rod-shaped viruses containing covalently closed circular double-stranded DNA that normally infect insect cells. They are used when expression of proteins require modifications as glycosylations, phosphorylations, processing of signal peptides, proteolytic processing, acetylations, carboxyl methylations, or disulfide processing. These eukaryotic changes are the key to proper functioning of the protein, which allows an additional level of specificity for protein interactions. Vectors based on the use of other viruses, known as baculoviruses, are used for inserting the gene for the protein to be expressed in insect cells.

Phage display

The viruses can also be used for the selection of protein variants of interest. This is the case of the virus M13, a filamentous phage and covalently closed circular single-stranded DNA which is coated by several proteins. These proteins have a minority or PIII protein called P3, which binds to the receptor of *E. coli* found in the tip of the pili F. DNA library fragments are cloned after coding the gene coding for protein III of the M13 phage, so that a collection of phage M13 in which each differs from the other variant of the single chain containing the specific fragment obtained will be exhibited. Then proceed to the selection of the variant of interest by exposing this collection to antigens that are to be recognized.

Oncolytic viruses

Viruses are also being used in the fight against cancer more directly, rather than through processes as described in the process of phage display.

Wild oncolytic virus

In 1991, one of the viruses tested cured the Ehrlich ascites carcinoma: Newcastle disease virus (NDV).

Modified oncolytic viruses

Modified oncolytic viruses as therapeutic resources must be based on viruses that replicate exclusively in cancer cells. One way in which the virus can be modified is by inserting tumor-specific promoters for the expression of essential genes in the virus, or by altering the surface receptors for virus-selective ligands that are specifically expressed in cancer cells or tumor environments.

The advantages of oncolytic virus-based therapies are that they are not inconsistent with the traditional methods of combating cancer, such as chemotherapy or radiotherapy. Moreover, the virus from replicating in the tumor cell increases the number of virus particles that are specifically localized in the tumor, and also many viruses can produce a strong immune response within the tumor, overcoming immune suppression and creating a vaccine by exposing the tumor-associated antigens that can respond to the patient's own immune system.

Viruses used as vectors in gene therapy

DNA can be used to replace defective genes within the cells of the patient, thereby curing the disease caused by those genes. Thus, there are two different types of DNA that can be introduced: first, the corrected DNA replacing the mutated DNA of the host, thereby synthesizing the protein encoded correctly, or if the disease is complex and affects several genes it is unknown precisely how to replace these genes, and can include a DNA encoding for the synthesis of a drug to counteract the effect.

Some researchers prefer another type of gene therapy known as silencing or antisense therapy, whereby a nucleic acid (DNA or RNA) is inserted into the patient and thereby prevent the expression of the gene that is causing the disease. Inserting the nucleic acid into the cell can be done by the use of two strategies, called *in vivo* or *ex vivo*. On the one hand, insert the nucleic acid directly into the cells, which can be done by using different techniques such as electroporation, cannons DNA (gene gun), or using multiple other methods possible today. However, although these methods solve problems such as immune rejection, they present great disadvantage as they are not very effective in the transfection or expression thereof. Therefore, the aim is to use viruses as vectors most suitable in gene therapy.

Today, clinical trials are being conducted on patients with Leber congenital diseases (consisting of a type of congenital blindness), severe combined immunodeficiency or SCID (also known as "bubble kids" disease, which often leads to death of the child before the age of 2 years), adrenoleukodystrophy (known as Lorenzo's disease), chronic myeloid leukemia, and Parkinson disease.

Eukaryotic microorganisms with biotechnology interest

Fungi and yeast have been essential in classic biotechnology due to their fundamental role in fermentation processes.

Fungi are classified into groups known as eukaryotes and heterotrophs. These organisms present a wide biodiversity; they can be reproduced by spores by sexual or asexual reproduction. They can use substrates of large size, and produce extracellular enzymes of great biotechnological interest. One of the main roles approached in biotechnology is the production of recombinant enzymes.

Myxomycota fungi are also named "mucilaginous fungi" and are a special cluster. Eumycota fungi contain the "true" fungi that will be approached in this part of the chapter.

Classification and biotechnological applications of fungi

More described species of fungi are classified based on polyphasic studies, based on molecular analyses, morphological, reproductive structures, and the nature of their cell walls, six divisions have been made: Chytridiomycota, Glomeromycota, Zygomycota, Ascomycota, Basidiomycota, and Deuteromycota.

- Chytridiomycota is morphologically diverse with about 1000 known species. *Rhizophlyctis rosea* which is one of most common fungi found in the cellulose decomposition in soil has been extracted and patented as an endoglucanase for the treatment of fibers, paper pulp, etc. Another example of this division is *Phytophthora infestans* in the fungus, the pathogen that infects potatoes.
- Glomeromycota consists of 150 described species grouped into 10 genera.
- Zygomycota constitutes 1% of fungal species described and are those that reproduce by asexual spores, also called zygospores with the application of biotechnology, entomopathogenic strains are being used in the biological control of insects.
- Ascomycota contains the highest number of species, about 15,000 described species. Highlights of this group include some yeasts such as *Saccharomyces* or *Pichia*, which will be discussed later in more detail. Another species of *Fusarium venenatum*, which is extracted mycoprotein, has a fibrous net similar to those present in animal muscle bundles.
- Basidiomycota forms sexual spores (basidiospores) with a special microscopic structure called basidia. As representatives of this division are mushrooms of the genus *Agaricus* and *Lentinus edodes* formed or phytopathogenic *Ustilago maydis* (which attacks maize) or *Puccinia graminis* (which attacks cereals).
- Deuteromycota is an artificial group formed to include all fungi known only by its asexual phase. Their vegetative structures, as in the case of both ascomycetes and basidiomycetes are unicellular (yeast), as septated mycelia. The cell wall polysaccharides are composed of glucans and chitin. Examples of great economic importance include species of the genera *Aspergillus* and *Penicillium*, including *A. niger* (in the production of citric and gluconic acid). *Aspergillus oryzae* is used in the food industry in the fermentation of soybeans and rice. There are also species of the genus *Aspergillus* that are plant pathogens such as *Aspergillus flavus* infecting some nuts and as a result aflatoxin B1, a mycotoxin which can induce liver cancer.

However, one of the most popular genus is *Penicillium*, through which Alexander Fleming discovered penicillin by observing how *Penicillium* spores inhibited the growth of staphylococci when they polluted their plates, which can be considered the beginning of antibiotic therapy. Other species of the genus *Penicillium* are also of great importance in the food industry such as *Penicillium roqueforti* and *Penicillium camemberti* for the production of cheeses that bear his name.

Yeasts are the group of fungi widely used for biotechnological purposes

Yeasts are not a taxonomic group by itself, but belong to Ascomycota, Basidiomycota, and Deuteromycota divisions, which present a common growth and contain about 1500 described species. In the field of modern biotechnology, yeast species have occupied an important role in the industrial production of recombinant proteins because they are easy to manipulate genetically, they rapidly multiply, and have the capacity to produce typical posttranslational modifications of eukaryotes, at a lower cost, achieving a higher cell density and thus generating greater concentration of protein (Çelik and Çalık 2012).

The design of an optimum system for the production of recombinant proteins in yeast (Delic et al. 2013) requires the following steps:

1. Selection of the host strain to facilitate proper protein folding and posttranslational modifications
2. Selection of an appropriate vector (integrative or episomal) with a suitable promoter (constitutive, inducible, or repressible) and with a selection marker (antibiotic or metals such as copper or tellurite, etc.)
3. Codon optimization of the gene to be expressed
4. Fusion of the gene, either for affinity purification or detection of the recombinant protein
5. Selection of an appropriate signal peptide if you want to direct the protein to the extracellular medium
6. Protection system proteolytic attack
7. Designing a culture medium (carbon and nitrogen source) and induction systems
8. Optimization of the production system (temperature, pH, oxygen transfer, etc.)

Morphological, structural, and physiological features of main yeasts

Saccharomyces cerevisiae

Yeast is traditionally used in biotechnology, which was also the first sequenced eukaryotic genome, and therefore was proposed as the broader expression system for recombinant proteins in the last three decades. This strain is used as a model organism to study human genetic diseases or disorders related to degenerative diseases, and has recently been used for the delivery of oral therapeutic proteins (probiotic).

Moreover, the first effective vaccine against hepatitis B was produced intracellularly by this strain. The Food and Drug Administration (FDA) and the European Medicines Agency (EMA) approved recombinant drugs if they were produced by *S. cerevisiae*. Given the numerous molecular studies of this strain, genomic studies including metabolomics, transcriptomics, proteomics, and others have allowed this strain to be considered as generally accepted as safe or GRAS strain (generally regarded as safe). Many biotechnology drugs on the market come from *S. cerevisiae* including insulin, the surface antigen of hepatitis B, urate oxidase, glucagon, granulocyte-macrophage colony-stimulating factor (GM-CSF), and platelet-derived growth factor, among others.

Pichia pastoris

This methylotrophic yeast can use methanol as the sole carbon source. When *P. pastoris* is grown on a medium containing methanol, a high induction of the genes encoding alcohol oxidase 1 (AOX1) and the dihydroxyacetone synthase occurs. After induction of these genes, encoding proteins form 30% of *P. pastoris* biomass. This effect has been used to substitute these genes by the gene encoding the protein to be obtained. The AOX1 promoter gene is one of the strongest and best regulated promoters, which has led to use as a promoter in expression systems for many proteins. This strain has a respiratory rather than a fermentative growth metabolism, which prevents the accumulation of fermentation products such as ethanol or acetic acid, allowing the achievement of high cell densities. Moreover, *P. pastoris* is able to secrete recombinant proteins, including high-molecular weight, unlike *S. cerevisiae* typically held in the periplasm, which combined with their

ability to grow in a medium with low protein content, greatly facilitates the protein purification processes.

Expression systems of *P. pastoris* are patented by Research Corporation Technologies
and kits are available for research use by Invitrogen. The first drug expressed in *P. pastoris*
was Ecallantide, a polypeptide of 60 amino acids used to treat attacks of hereditary disorder of the immune system, which has been approved by the FDA in 2009. One advantage
of this strain with respect to *S. cerevisiae* is the absence of the glycosylation type the latter
produces.

Hansenula polymorpha

H. polymorpha is another strain able to metabolize methanol as carbon and energy source
as *P. pastoris*; moreover, it has proven its value as a protein expression system to be in
the market as recombinant proteins of hepatitis B vaccine, interferon alfa-2a, anticoagulant hirudin, or insulin. It is also used to produce food supplements such as oxidases and
lipases and hexoses.

The most common system used for expression in *H. polymorpha* is also based on methanol utilization genes, and genes specifically in methanol oxidase (MOX) and formate
dehydrogenase (FMD). The promoters of these genes can be regulated as required.

Moreover, *H. polymorpha* can also achieve high cell densities, efficiently secrete high-
molecular weight proteins (up to 150 kDa), and has one of the highest productivity values
expressed protein from yeast (13.5 g/L phytase production). Additionally, the vectors used
for this strain are integrated into the chromosome for more than 800 generations, allowing
expression systems for certain sufficiently stable protein production processes.

Other advantages in using *H. polymorpha* is that the strain is heat resistant (up to 49°C)
which allows purification of thermostable proteins that are ready for further crystallization studies. And as in the case of *P. pastoris*, *H. polymorpha* has a highly present perixo-
somal membrane, allowing expression of heterologous membrane proteins when they are
merged as a signal peptide Pex3p.

Kluyveromyces lactis

K. lactis has been used in the food industry for the production of β-galactosidase since
1950. In addition, it has also been used for the heterologous expression of bovine chymosin
(rennin), used for curdling in the cheese industry. The factors that have made *K. lactis* good
for being background for the expression of recombinant proteins include the fact that it
contains a strong and inducible promoter LAC4 repressed in the presence of low concentrations of glucose, plus it can use economic substrates as lactose present in the whey, as
well as gained approval as a GRAS strain, its ability to secrete high-molecular weight proteins, the availability of the genome sequence, and the existence of protein expression kits.

Schizosaccharomyces pombe

S. pombe was initially isolated from fermented millet beer in East Africa, to which it owes
its name (beer in Swahili is pombe). This is a fission yeast that grows preferentially as haploid strains. It was the sixth eukaryotic organism whose genome was sequenced. It was
also the third organism with proteome, after *S. cerevisiae* and *Homo sapiens*. The intense
characterization and several important properties are shared with higher eukaryotes such
as the regulation of cell cycle, transcription initiation, organization of chromosomes, their
RNA processing, and RNA interference (RNAi) pathways. Besides the Golgi apparatus
being morphologically well defined, the process of protein folding of is more like in mammals than in *S. cerevisiae*. For these reasons, *S. pombe* is considered today one of the most

attractive for mammalian protein expression systems. However, there is still much work to be done to bring *S. pombe* research to the industrial level.

Arxula adeninivorans

A. adeninivorans is a dimorphic yeast, as it is both able to multiply by budding (when incubated under 42°C) as by mycelium (when grown above 42°C). Besides presenting temperature-dependent dimorphism it is also thermoresistant (tolerates up to 48°C) in addition to being halotolerant. This yeast is haploid, nonpathogenic, and can grow in a wide variety of carbon and energy sources including adenine (hence its name) and other purine, *n*-alkanes, and starch. Moreover, *A. adeninivorans* makes and secretes extracellular enzymes into the culture medium during growth, including RNAses, proteases, and glycosidases different from trehalose and cellobiose. The presence of stable plasmids that multiply, together with the above requirements indicate the great biotechnological potential of *A. adeninivorans*.

Yarrowia lipolytica

Yarrowia has only one species described to date, *Y. lipolytica*, with two varieties. This yeast is being increasingly used as popular system for the expression of heterologous proteins because of its inherent ability to secrete high-molecular weight proteins in large quantities, because proteins can be secreted by cotranslational translocation associated with, and analogous to, that which occurs in higher eukaryotes and in contrast to posttranslational translocation that occurs predominantly in *S. cerevisiae*. Another reason why it is becoming a popular expression system protein is because it was not able to ferment sugars; its genome is fully sequenced; the existence of commercial kits; the fact that it can reach high cell densities; and it has been approved as a GRAS strain by the FDA for various industrial processes. Furthermore, it has already taken the first step in the humanization of the glycosylation pathway by OCH1 gene deletion to prevent hypermannosylation proteins.

Biotechnological products from microorganisms

Many products with commercial interest are produced by microorganism such as polymers, proteins, vaccines, and metabolites (primary or secondary). One example will be discussed in this chapter.

Production of microbial polymers (polysaccharides and poly-beta-hydroxyalkanoates) for use as excipients in medication

Polymers of microbiological origin offer additional advantages over those of chemical origin in certain fields of industry. Biopolymers are being extensively used in many different areas; these polymers can be either natural polysaccharides or polyester products. Knowledge of their functions in the cell allows us to explore their biosynthesis, gene organization, and regulation of the expression of these genes. The building blocks (monomers) that join them together often determine the physical properties that conditioned their industrial applicability. These properties can be achieved by engineering techniques (thanks to finding appropriate culture conditions that allow the production of a particular polymer), by seeking alternative producing microorganisms (with biosynthetic capabilities not well described or advantageous to cultivation), or by genetic engineering (allowing expression of a particular enzymatic machinery and their control systems) in a suitable background.

Polysaccharides

Polymers of carbohydrates or polysaccharides of microbial origin are often cellular components or secreted for different purposes by different microorganisms (Woodward et al. 2010). However, secretion of polysaccharides to the medium is also produced by certain microorganisms without these polysaccharides which constitute part of the wall microorganism. These compounds in general are often referred to as extracellular polysaccharides (EPS). With respect to the capsule (König et al. 2010), it has traditionally been studied in microbiology as a tool developed by microorganisms for forming biofilms, protection against desiccation, resistance systems innate to host defense (the complement system), immune system acquired resistance, the intracellular survival, and for phage attack.

To avoid immune recognition, microorganisms have developed specificity due to the different chemical compositions of these polysaccharides which constitute their covers.

EPS with industrial interest must achieve lower production costs (obtaining strains with increased production capacity and less cost) as well as postprocessing (Table 17.1 and Figure 17.1) (Freitas and Reis 2011). Another alternative is to search for products with an interest in high value markets such as the cosmetic, pharmaceutical, or biomedical fields (Nwodo et al. 2012).

An innovative application proposed for microbial polysaccharide is as a biomaterial to be used in a mixture of materials for fine arts in order to increase the plasticity of materials and other physical properties such as ductility, hardness, elasticity, and modability (Figure 17.2a).

Other microbial polymers: Polyesters

The other polymers of biological origin with great relevance in biotechnology are polyesters. These have special interest in the field of biomedicine given the importance of developing materials with plastic and elastomeric properties, that are consistent with body molding properties, but are mostly biodegradable so they can be reabsorbed by the body. In this way, it can be developed films or solid plates that are both waterproof and porous patches, no tissue made of fibers, screws, pipes for guiding channels and ducts, no filamentous structures and microspheres which can produce materials bioabsorbable thread-type suture, and screws for fixation of cartilage and bone (avoiding having to open by surgery to remove metal parts when they are not absorbable), biodegradable membranes for their use in tissue engineering,

Table 17.1 Polysaccharides produced by microorganisms

Polysaccharide	Microorganism	Molecular weight
Xanthan	*Xanthomonas campestris*	2×10^6–5×10^7
Dextran	*Leuconostoc mesenteroides* and *Acetobacter* sp.	1×10^5–2×10^7
Gellan	*Pseudomonas elodea*	
Pullulan	*Aureobasidium pullulans*	1×10^4–1×10^5
Alginate	*Azotobacter vinelandii* and *Pseudomonas* sp.	5×10^5
Scleroglucan	*Sclerotium* sp.	1.9×10^4–2.5×10^4
Curdlan	*Alcaligenes faecalis* and *Agrobacterium* sp.	1.9×10^4–2.5×10^4
Gellan	*Sphingomonas elodea*	–
Levan	*Aerobacter levanicum*, *Zymomonas mobilis*, and *Bacillus licheniformis*	–
Succinoglycan	*Rhizobium, Alcaligenes, Agrobacterium*, and *Pseudomonas* sp.	–
GalactoPol	*Pseudomonas oleovorans*	–
FucoPol	*Enterobacter* A47	–

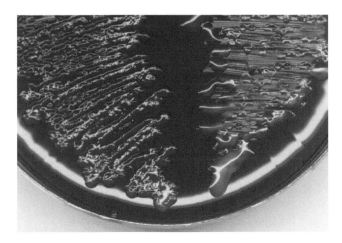

Figure 17.1 Soluble and insoluble polysaccharides produced by bacteria belonging to *Paenibacillus* genus.

periodontal treatments, surgical meshes for cover injury, or to repair intestine, bone, or the pericardium, forming guide channels for nerve systems controlled release of drugs, etc. In nature there is a material produced by bacteria that meets these properties and which is developing under intense research. These compounds are called microbial polyesters.

The bioplastic produced by microorganisms are the polyhydroxyalkanoates (PHA) and polyhydroxybutyrates (PHB). The chemical nature of these polymers of acetyl-CoA makes it possible to be polymerized for the different enzyme activities: beta-ketothiolase reductase, and polymerase or synthetase: *Azotobacter* sp. FA8, *Ralstonia eutropha (Alcaligenes eutrophus), Alcaligenes faecalis, Pseudomonas oleovorans, Bacillus cereus,* and *Bacillus megaterium.*

Industrial fermentations and bioprocess

The most common fermentation processes can be clustered into five main groups (Stanbury et al. 2003; Waites et al. 2009):

1. Production of microbial cells, also known by the term biomass product
2. Production of microbial enzymes
3. Production of metabolites (primary, secondary, or polymeric)
4. Production of recombinant products (such as proteins)
5. Biotransformation or modified transformation processes

Components and design of the fermentation process

Basic components in a bioprocess

The bioprocess contains six basic components:

1. Design an appropriate culture medium to cultivate the microorganism, both the seed culture and for the production process
2. Sterilization of the medium, the bioreactor, and complementary equipment
3. Production of active and pure cultures on sufficient concentration to inoculate into bioreactors (preinoculum)

(a)

(b)

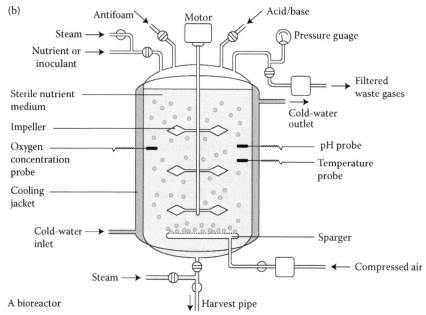

Figure 17.2 (a) Modeled materials containing a mixture of exopolysaccharides: xanthan, gellan, and pullulan. (b) Bioreactor: stirred tank reactor (STR).

4. Culture of the organism in the production bioreactor under optimal conditions for product generation
5. Product extraction and purification
6. Elimination of effluents generated during the process

Bioreactors

The design of a bioreactor involves a series of steps that must take into account multiple factors, always keeping in mind the ultimate purpose for which the microorganisms fulfil their function with maximum efficiency and with minimum cost under optimal conditions. It should be also protected the external environment, controlling the appearence of other microorganisms, through specific physicochemical conditions.

To achieve this, they must closely follow and adapt conditions such as gas flow (such as the introduction of air, oxygen, nitrogen, carbon dioxide, or any gas which is involved in the production process), temperature, pH, and the stirring speed or movement. However, most of the initial processes make use of the same type of bioreactor from which adaptations will lead to an optimization of the production system. The bioreactor most commonly used is stirred tank reactor or STR.

Stirred tank reactor   Tools and parameters that should be controlled:

- Temperature control.
- Gas control: Aerobic cultures require oxygen transfer to the culture medium.
- Agitation with deflectors, are for the agitation and include various strategies that increase the engine power to increase the stirring rate.
- Air diffusers or aerators.
- In relation to the auxiliary systems to oxygen diffusion, agitators or impellers ensure that the motor shaft is coupled out and would enhance the dispersion of gas (oxygen) in the liquid (culture medium), and also favors dissolving solids in liquids (including culture cells), liquid in liquid (the mixture of the various medium components and additives, e.g., to keep the pH at a particular range) and enables heat transfer.
- Agitators: The three main types of agitators used in industry are the propeller, pallet, and turbine.
- Rotors.
- Sampling and probes: In a standard bioreactor it is normal to find probes for control of temperature, pH, and dissolved oxygen.
- Gas analyzer: Measuring and controlling gas flow to the bioreactor is important in the control of bioreactors.
- Pressure. The pressure measurement is especially important for safety reasons, since there is risk of an explosion if it exceeds certain limits.
- Measurement systems foam and foam control.
  Sensors *in situ* and *ex situ* are used in the bioreactors.
- Central control system. Automatic systems are usually of four types:
  - Two-position control systems (ON/OFF).
  - Proportional control systems.
  - Integrative drivers.
  - Derivative controllers (Figure 17.2b).

Other alternative bioreactors to STR   There are certain types of bioreactors whose operation is more suitable for other types of processes. Depending on the distribution of gas, classification is done as follows: (1) bioreactors by stirring (STR); (2) bioreactors that use pumps; (3) bioreactors by air pressure; (4) gas phase continuous bioreactors; (5) Waldhof fermenter type; (6) Acetators and cavitators; (7) tower bioreactors; (8) cylindrical vessels; (9) air lift fermenter; (10) fermenter deep jet; (11) column cyclone; (12) packaged tower; (13) rotating disk, etc.

This type of classification has been organized into three types of bioreactors considering the most widely used methods for the generation of biotech products: growing by lot or batch, fed-batch, and continuous or chemostat culture bioreactors.

Culture media   The types of culture medium may be classified into solid, semi-solid, or liquid, being the most widely used we will focus first on the liquid medium.

The media can also be classified in terms of their composition as synthetic, semisynthetic, and complex.

Once the bioreactor and the operating mode is selected, we define the medium used for culturing the microorganism and production of the byproduct of interest. Cultivating microorganisms have in common certain features depending on the cell requirements which typically include water, carbon sources (in crops that are nonphotosynthetic), nitrogen (non-crops nitrogen fixers), energy (if not autotrophic), minerals, and vitamins, and oxygen (depending on whether it is aerobic metabolism) or carbon dioxide (if phototrophs).

The selection of the substrate should be done according to (1) economical as possible; (2) generate maximum cell biomass production per gram of substrate used; (3) generate high product concentration (or biomass if the ultimate goal is to produce it) at the maximum rate of formation of this product; (4) generate the minimum rate of formation of undesired by-products; and (5) suited to the production process and sterilization, as well as elements of aeration, agitation, extraction, purification, and waste treatment.

Some of the substrates that meet these criteria are used as carbon sources in the culture media include cane molasses, beet, cereals, starch, glucose, sucrose, and lactose. As the nitrogen source, ammonium salts, urea, nitrates, corn steep liquor, soybean meal, slaughterhouse waste, and fermentation residues are used.

Sterilization   Regarding the sterilization of the medium, one can resort to methods such as filtration, radiation, ultrasound, chemical treatment, or heat treatment. However, the most commonly used method is the autoclave, together with the filter for those components that are heat sensitive.

Sterilization processes can be divided into two main groups. One of them is the sterilization process in lot (where the sterilization process is performed within the same vessel bioreactor (*in situ*) or by inserting the bioreactor into a pressure cooker (*ex situ*), and the other process involves continuous sterilization (in which the medium is sterilized and will be used in the bioreactor) by a device known as spiral heat exchanger.

References

Blattner FR et al. The complete genome sequence of *Escherichia coli* K-12. *Science*. 1997;277(5331):1453–1462. PubMed PMID: 9278503.

Çelik E and Çalık P. Production of recombinant proteins by yeast cells. *Biotechnology Advances*. 2012;30:1108–1118.

Delic M, Valli M, Graf AB, Pfeffer M, Mattanovich D, and Gasser B. The secretory pathway: Exploring yeast diversity. *FEMS Microbiology Reviews*. 2013;37(6):872–914. doi: 10.1111/1574-6976.12020. Epub April 12, 2013. PubMed PMID: 23480475.

Freitas F, Alves VA, and Reis MA. Advances in bacterial exopolysaccharides: From production to biotechnological applications. *Trends in Biotechnology*. 2011;29:388–398.

Glazer AN and Nikaido H. *Microbial Biotechnology. Fundamentals of Applied Microbiology*. 2nd ed. Cambridge University Press, NY, US, 2007.

Iáñez-Pareja E. *Biotechnology of Microorganisms*. University of Granada, Granada, Spain, 2009.

König et al. Prokaryotic cell wall compounds. *Structure and Biochemistry*. Springer: Heidelberg, Germany. 2010.

Mathur MD, Vidhani S, and Mehndiratta PL. Bacteriophage therapy: An alternative to conventional antibiotics. *Journal of the Association of Physicians of India*. 2003;51:593–596. Review. PubMed PMID: 15266928.

Nwodo U, Green E, and Okoh AI. Bacterial exopolysaccharides: Functionality and prospects. *International Journal of Molecular Sciences*. 2012;13:14002–14015. doi:10.3390/ijms131114002.

Qadir MI. Review: Phage therapy: A modern tool to control bacterial infections. *Pakistan Journal of Pharmaceutical Sciences*. 2015;28(1):265–270. PubMed PMID: 25553704.

Stanbury PF, Whitaker A, and Hall SJ. *Principles of Fermentation Technology*. 2nd ed. Butterworth Heinemann, Elsevier Science: Oxford, UK, 2003.

Waites MJ, Morgan NL, Rockey JS, and Higton G. *Industrial Microbiology. An Introduction*. Blackwell Science: Oxford, UK, 2009.

Woodward R et al. *In vitro* bacterial polysaccharide biosynthesis: Defining the functions of Wzy and Wzz. *Nature Chemical Biology*. 2010;6:418–423.

chapter eighteen

Molecular biology of viruses
Disease perspective

Muhammad Mubin, Sehrish Ijaz, Sara Shakir,
and Muhammad Shah Nawaz-ul-Rehman

Contents

Abstract

Viruses are found wherever there is life and have probably existed since living cells first evolved. During the course of evolution, viruses have plagued almost all kingdoms of life, but most importantly from a human perspective, viruses infecting plants and animals are relatively well characterized. Viruses can be broadly divided into three categories on the basis of genome nature: deoxyribonucleic acid viruses, ribonucleic acid viruses, and retroviruses. In the recent past, there have been deadly global outbreaks of several viruses, such as human immunodeficiency virus, dengue, hepatitis, and severe acute respiratory syndrome (SARS) in the animal kingdom, and among plants, cassava mosaic disease, tomato yellow leaf curl disease, and cotton leaf curl disease-related viruses have created havocs in food shortage. Several studies revealed that plants and animal viruses bring major changes in the cell cycle by reprogramming host's machinery. Viruses have fewer genes in limited-size genome, but are still able to replicate and complete infection cycle in different types of differentiated cells. Viruses face a common challenge of suppressing the host defense mechanism by targeting important pathways inside the cell. Therefore, the emerging picture for viral biology depicts viruses as playing tricks to modify the cellular pathways instead of targeting a particular individual gene. Despite significant efforts to understand plants and animals viruses, our current knowledge about virus–host interaction, viral diversity, and developing resistance strategies is incomplete. One of the important frontiers in research is to understand the major common circuits in the cellular machinery, which can be better tailored to avoid damage caused by the viruses to their hosts. Recent data of epidemics show that viruses are always out there in the environment and appear as epidemic when conditions are favorable. Mutations, recombination, and changing environment are the driving forces behind evolution of viruses. In this chapter, we describe the most important viral pathogens that caused serious losses to human beings.

Keywords: Begomoviruses, CLCuD, Retroviruses, Molecular biology, Brome mosaic virus, Herpes simplex virus, Host factors.

Introduction

Viruses were discovered in the nineteenth century and were soon recognized as severe pathogenic agents of almost all forms of life. In fact, history records show that viral diseases such as small pox, influenza, and acquired immunodeficiency syndrome (AIDS) have drastic effects on human population. Similarly, CLCuD, maize streak virus (MSV), mungbean yellow mosaic virus (MBYVV), wheat dwarf virus (WDV), and several other diseases in cultivated, noncultivated, and ornamental plants have adversely affected the human population through diseases and famines. Such disease epidemics invited molecular biologists to study extensively these viral pathogens in order to find ways to eradicate them. As a result, the new areas of molecular biology in respect of interaction between virus and host cells have been developed. Viruses are the simplest organisms; therefore, they are very important to biologists for several reasons. Viruses are the simplest form of life. They exist on the borderline between living and nonliving worlds. Viruses do not contain any organelles; therefore, they are not cellular forms of life. Due to the fact that viruses can reveal complex biological phenomenon, they are used as model organisms for cellular studies. According to evolutionary biologists, all living organisms on this planet are complex machines whose only purpose is to produce copies of their genomes. Viruses are also not an exception; however, they require host machinery for their replication. They are mostly nucleic acids, which are wrapped in a coating (termed a capsid). The capsid of viruses has a dual function, first of all it protects the viral nucleic acids from degradation and second, it allows viral nucleic acids to gain entrance to appropriate host cells. As viruses are obligate cellular parasites, we can accept that they only evolved later than cells, either as mobile genetic elements or as defective cellular genes that attained the capacity to manipulate the replication machinery of the host cells. Viral genomes evolve more rapidly than the genomes of cellular organisms. With the passage of time, the high mutation and recombination rate in viruses have erased any relation that may have existed between various types of viruses and their ancestors.

Mature virus particles remain dormant; they become alive and gain the ability to reproduce once they enter a cell. Viruses are a kind of obligate parasite that cannot be cultivated using any growth media. None of the viruses has the ability to synthesize protein or energy-producing apparatuses. Viruses remain immobile outside the infected host and rely on a variety of other organisms such as arthropods, nematodes, fungi, or the environmental factors for their dissemination. Virions of some viruses remain stable outside their hosts, while some are degraded after a short period of time. For example, tobacco mosaic virus (TMV) can remain stable for years, whereas tomato spotted wilt virus (TSWV) cannot survive for more than few hours. Despite the fact that several viral proteins are part of the virion, biological function of viruses is strictly the property of their nucleic acids. The role of coat protein is to provide the protective sheathing for viral nucleic acids and interaction with the vectors for further transmission. The coat protein itself has no role in infectivity, but it is important for the infectivity cycle of the virus. The coat protein of virus is also called as structural protein, while viral polymerases, helicases, or transcription factors are called as nonstructural proteins. Vectors play an important biological role in transmission of viruses. The mode of transmission determines the success of a plant virus. Viruses use different vectors for their successful transmission. For example, the poty-viruses (Potyviridae) spread through aphids, while those in the Geminiviridae like the white fly transmits begomoviruses.

Viruses cannot be controlled through application of pesticides or chemicals in the farmer's fields. They expose their presence by manifestation of specific symptoms. Due to the fact that they cause heavy yield losses, viruses remain challenging pathogens and attract the attention of scientists to investigate their various functions.

Recent global outbreaks of viruses

The global warming, intercontinental trade, quick transfer of fresh agricultural products to the neighboring countries, and resistance against pesticides in the insect vectors have invited more viral outbreaks in the last two decades. The spread of the viral diseases across the globe are mainly related to the nature of the food crops grown in the region and the agricultural practices used. In Asian countries, wheat, rice, maize, cotton, sugarcane, tobacco, and vegetables are the common crops, while in Africa, cassava is an important agricultural commodity, which is most vulnerable to different viral diseases.

Global outbreaks of plant viruses

Cassava mosaic viruses and cassava brown streak virus in Africa

Cassava mosaic viruses are single-stranded begomoviruses of African origin. They are bipartite in nature and are transmitted through whitefly. Cassava brown streak disease (CBSD) is caused by two different species of ipomoeaviruses, which are also transmitted by whiteflies. Other than whiteflies, they also spread through infected cuttings. CBSD was first reported in coastal East Africa in the year 2003. Since then the disease is reportedly spreading in the Great Lakes region. CBSD has no effect on the apparent phenotype and growth of the plants; however, the roots (the edible part of cassava) is destroyed in such a way that they become inedible.

Presently, the CBSD has been reported in many countries of the African continent, such as Angola, Central African Republic, Gabon, Burundi and democratic republic of Congo, Kenya, Zambia, Sudan, and Uganda. Several outbreaks of CBSV have occurred in East Africa around the Great Lakes region and have destroyed up to 80% of the cassava plantation, and mosaic disease of the cassava has caused 90% losses in cassava production.

Tomato yellow leaf curl disease around the globe

In Middle East (Israel), Southwest Europe, tropical Africa, Southeast Asia, and the Caribbean Islands, tomato yellow leaf curl virus (TYLCV) epidemic has destroyed the tomato crop many times. The viruses causing tomato yellow leaf curl disease can poten-tially destroy 100% of the tomato fruit in Middle East. The virus is transmitted through silver leaf or B biotype whitefly. The symptoms include flower abscission, leaf yellowing, and stunted growth of plants.

CLCuD in Asia and Africa

In the early 1990s and 2000, the CLCuD became epidemic in Pakistan, causing severe losses to cotton crop and ultimately to the agriculture and economy of Pakistan. At the beginning, the infecting virus was identified as cotton leaf curl Multan virus (CLCuMuV) and the resulting strategies were at that time to engineer cotton crop to bring resistance against the CLCuD. But later on, in spite of a number of resistant varieties in the field, there was an epidemic again in the central and the southern Punjab, especially in Vehari and Burewala districts during 2000. After isolation and further investigations, it was proved that the new epidemic was not due to CLCuMuV but rather due to another recombinant virus that resulted in the possible breakage of resistance against the CLCuD.

Global outbreaks of animal or human viruses

Dengue virus epidemic

Dengue virus (DENV) is a mosquito-borne (*Aedes aegypti*) disease and exclusively includes four closely but antigenically different serotypes, designated as DEN-1, DEN-2, DEN-3, and DEN-4. DENV belongs to the positive-strand ribonucleic acid (RNA) viruses and is categorized under the family Flaviviridae. The vaccine for DENV is difficult to make, as any of the four serotypes can cause infection. At present, several efforts are in progress to make a vaccine against all DENV. However, escape from mosquitoes and early diagnosis procedures are used to minimize the risk of heavy infections. Once the mosquito acquires the DENV through blood feeding, the virus can stay in the mosquito for rest of its life. Studies also show that infected female mosquitoes can potentially transmit the virus by transovarial (via egg) transmission.

As a matter of fact, most urbanized cities in the world's tropics or semitropics (such as Brazil, Pakistan, China, and Malaysia) are more affected by DENV infection. Due to the absence of a proper model to study the dengue disease, very little is known about the viruses that cause severe damage to human life.

For other flaviviruses and dengue, the first step inside the cell is to produce viral proteins, including an RNA-dependent RNA polymerase to switch the viral RNA from translation to RNA replication. DENV recognizes dendritic cells of the body underlying the skin. The replication of viral genome happens on the endoplasmic reticulum of the cell. The mature virus particles bud out of the cell through exocytosis and enter into the blood stream (monocytes and macrophages).

Middle East respiratory syndrome coronavirus

Middle East respiratory syndrome (MERS) is a relatively new viral disease in humans, first reported from the Middle East. It was first reported from Saudi Arabia in 2012 and since then it has been reported from several countries, including the United States. The patients infected with MERS virus develop severe respiratory illness, with high fever, coughing, and shortness of breath. Since April 2012, out of the 212 people detected with MERS virus infection, 88 died. Although the source of MERS virus has not been identified, it has been speculated that camels are its alternative hosts. Indeed, several MERS viral genomes have been sequenced from camels in Egypt and Saudi Arabia. The phylogenetic analysis reveals that MERS virus originated from bat-associated clade-2C betacoronaviruses. The genome size of MERS virus is ~30,000 nt. The genome contains 10 predicted open reading frames (ORFs). Nine ORFs are predicted to be nested genes and are expressed from seven subgenomic RNA molecules.

Molecular biology of single-stranded DNA viruses

Classification of the viruses is based on the viral genomes and proteins encoded by them. Viruses having deoxyribonucleic acid (DNA) as a genome fall into group I and II of Baltimore classification. DNA viruses use host DNA polymerase for their replication. The replication of single-stranded DNA (ssDNA) viruses generally happens in the nucleus through rolling circle amplification mechanism. The ssDNA viruses also produce double-stranded DNA (dsDNA) as their intermediate for replication. Initially, it was thought that DNA viruses only infect bacteria (bacteriophages). But recently, a large number of DNA viruses infecting animals, plants, and murine viruses have been identified.

Epidemiology and taxonomy of ssDNA viruses

Single-stranded viruses have been classified on the basis of the nature of the genome (circular or linear) and the host range. ssDNA viruses are classified into 10 different families:

- Family Anelloviridae
- Family Bacillariodnaviridae
- Family Bidnaviridae
- Family Circoviridae
- Family Geminiviridae
- Family Inoviridae
- Family Microviridae
- Family Nanoviridae
- Family Parvoviridae
- Family Spiraviridae

Weeds as alternative hosts for begomoviruses

Due to their diverse environmental adaptability, weeds are found in all parts of the world. Weeds are primary sources for inoculums of begomoviruses and can potentially act as alternative hosts for viral spread. Native weed plants are considered as important reservoirs of begomoviruses and they gave the platform to the virus for recombination due to multiple infections. Every crop has a particular season, that is, Rabi or Kharif for its life cycle. A crop that is grown in Rabi season is absent in Kharif season from the field. Due to their perennial nature, different weeds are grown within or around agricultural crops through the year. The previous observations suggest that weeds often show vein yellowing or yellow mosaic symptoms. Historically, cotton, cassava, and tomato are introduced crops and are heavily infected by different viruses. It is presumed that the viruses originated from weeds were transferred to new susceptible hosts. Some common weeds are *Ageratum conyzoides, Alternanthera philoxeroides, Sida acuta, Amaranthus spinosus, Amaranthus viridis, Ageratum conyzoides, Bidens pilosa*, and *Mimosa invisa*.

Global warming and spread of begomoviruses

Climate change refers to global average temperature increases, which results in changes in insect vector populations. Whitefly, which is the insect vector for begomoviruses, can survive and reproduce at high temperatures. However, the prolonged winter season can adversely affect the whitefly population. The geographical distribution of whitefly depends largely on climatic conditions, which favor its rate of reproduction. Whitefly populations are often greatest in areas where there is high temperature and low rainfall. As the temperature rises in the tropical, subtropical, and arid areas, the chances of begomoviruses also increase. Global warming progresses over the areas in middle latitudes and hence chances of begomoviruses emergence are more there.

Role of human in spread of begomoviruses

Humans play a pivotal role in the spread of begomoviruses, by introducing exotic germplasm from different environments, by transferring plants from their centers of diversity or origin to different climatic regions, by disseminating viruses through clonally propagated infected plants, or by spreading of viruliferous insect vectors. The begomovirus epidemics on introduced crops such as tomato, cotton, cassava, or chilies are associated with human activities. The spread of Begomovirus across the globe has a direct correlation with the increased population of the B biotype of *Bemisia tabaci* (Gennadius).

In the United States, a huge number of crops such as cotton, tomato, beans, and cucurbits have been infected by geminiviruses and the reason was the spread of sweet potato whitefly. The global introduction of tomato crop represents the best example of human impact on the spread of begomoviruses. During the early 1990s, the spread of TYLCV in the New World (NW) happened due to the export of tomato seedlings from Israel to the Caribbean. It took about 10 years to spread the TYLCV from the Middle East to the rest of the world. The available evidences suggest that the crops that have been badly infected by viruses were introduced crops, for example, cassava plant originated and exported to Africa from South America. There is no cassava mosaic disease in South America, but it is widely present in Africa. As these viruses are not native to America, therefore it is assumed that the cassava mosaic disease moved from indigenous infected plants to cassava.

Similarly, American cotton was introduced from Mexico to the Indian subcontinent in 1818. The incidence of CLCuD was not noticed in the Indian subcontinent till 1967. Due to industrialization and high yield of NW cotton, the area under cotton cultivation was increased. This increase in area and introduction of foreign varieties resulted in the first record of CLCuD in Pakistan during the 1960s. Later in 1985, the disease spread to central Pakistan and within 5 years, the disease appeared in an epidemic form and by 2000 it was also reported in neighboring countries such as India, China, and Philippines. Taxonomically, there are several known species of geminiviruses, which infect cotton plant in Pakistan.

During the last century, a huge adaptation of geminiviruses to new hosts and new ecological zones has been observed. Theoretically, there are at least eight major centers of crop diversifications or origins in the world. In every center of origin, there is a huge diversity of particular plant families and comparatively less diversity in the periphery. In an analogy to crops centers of diversifications, we have identified centers of diversification for geminiviruses in the world.

Eight different geographical regions have been identified as centers for diversification of "geminiviruses." These centers are

1. Australia
2. Japan
3. South China
4. Indian subcontinent
5. Sub-Saharan Africa

6. Mediterranean–European region
7. South America
8. Central America

Evolution of CLCuD

Origin of CLCuD

Cotton has long been a major crop providing food, feed, and fiber to humanity. The history of using cotton fibers to make fabrics leads back to at least 7000 years. There are 51 different species in the genus *Gossypium*. Out of 51, only *Gossypium hirsutum*, *Gossypium barbadense* (allotetraploid), *Gossypium arboreum*, and *Gossypium herbaceum* (dipoid) are cultivated species. The tetraploid species originated in Central America and northern Peru, respectively. The diploid species are native to the Old World (OW) but with low fiber quality.

CLCuD severely infects different plant species belonging to the family Malvaceae; especially cotton (genus: *Gossypium* L.). Affected cotton plants show very unusual symptoms such as leaf enation (formation of leaf-like structure underside the leaf), upward and downward cupping of leaf, vein swelling, and stunted growth. Dark green-colored leaves of cotton as compared to healthy ones indicate CLCuD infection because chloroplast containing tissues start proliferating toward leaves during infection. However, symptoms are highly dependent on cotton variety as well as the age of the plant. Infection soon after seed germination is represented by severe leaf curling and stunted growth resulting in drastic yield loss while late season infections appeared to be less dangerous for yield in association with mild symptoms.

In 1992, identification of whitefly *B. tabaci* as a vector of CLCuD suggested that the causing agent of the disease is a geminivirus (family Geminiviridae, genus *Begomovirus*). Begomoviruses causing CLCuD are associated with pathogenicity-determinant betasatellites and an evolutionarily distinct group of alphasatellites. In 1993, it was reported that CLCuD is not seed-transmissible.

Over the past 25 years, the world has encountered two epidemics due to CLCuD. The first one was in 1990, but conventional breeding for viral resistance saved the farmer from heavy loss, but a more recent one involving a virus–satellite complex, which is resistance breaking, completely destroyed the crop. This epidemic left the farmer with poor host plant tolerance to encounter the unholy disease effects. For years, CLCuD remained a continuous threat to Africa and Asia, that is, Pakistan and north-western India. There has always been a fear that this devastating disease can spread from limited geographical regions to other agricultural areas of the world, where this disease is still absent. Unfortunately, recent published reports have been showing this fear coming true, with CLCuD appearing in China.

Components of the CLCuD begomovirus complex

CLCuD-associated begomoviruses

Viruses belonging to family Geminiviridae consist of one or two single-stranded circular DNA genomes ranging from 2.5 to 5.6 kb in size. The circular genomes are encapsidated in geminate capsid with a size of ~18 × 30 nm. So far, more than 200 species of geminiviruses have been reported, which are further divided into seven genera, namely, *Begomovirus*, *Mastrevirus*, *Curtovirus*, *Topocovirus*, *Turncurtovirus*, *Eragrovirus*, and *Becurtovirus*; based on host range, diversity in genome organization, and mode of transmission through insect vectors (Figure 18.1). Each genera is named after the type member and their properties such as Mastrevirus derives its name from its species type *maize streak virus* (MSV), Curtovirus—*Beet curly top virus* (BCTV), Begomovirus—*Bean golden mosaic virus* (BGMV), Topocuvirus—*Tomato pseudo-curly top virus* (TPCTV), and Becurtovirus—*Beet curly top Iran virus*; other species include *spinach severe curly top virus*, Eragrovirus—*Eragrostis curvula streak virus*, and Turncurtovirus—*Turnip curly top virus*.

Begomoviruses can be divided into two different categories, namely, monopartite and bipartite. The monopartite begomoviruses are single-component viruses, which induce disease in plants. The example of monopartite viruses include TYLCV. In some instances, the monopartite begomoviruses are associated with pathogenicity-determinant molecules, known as betasatellites and alphasatellites (they have no known role yet).

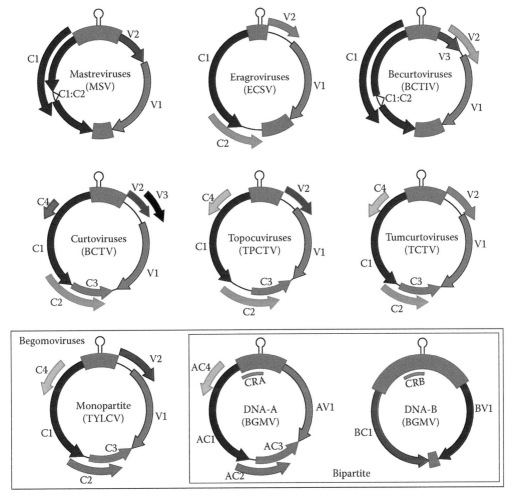

Figure 18.1 The genome organization of different geminiviruses species. There are seven different genera of geminiviruses. Their names are mentioned in the middle of each circle. The genes of the virion strand are mentioned as C1 or C2, while the genes of the complementary strand are mentioned. All the genera except begomoviruses have single circular molecules. The genus *Begomovirus* comprises either monopartite (single circular molecule) or bipartite (two circular molecules).

In NW, bipartite begomoviruses infect malvaceous plant species, including cotton. DNA-A encodes replication-associated protein and replication-enhancer protein, required for replication as well as the coat protein, which is required for insect transmission. The second component encodes two proteins, that is, movement protein and nuclear shuttle protein, which enable viral movement within and between cells in host tissues. Bipartite begomovirus, cotton leaf crumple virus (CLCrV), was isolated from cotton in the southern United States.

In OW, two distinct CLCuD complexes (the African and Asian complexes) are present. In 1912, CLCuD was first reported from Africa but the causative pathogen was not identified for a long time. From Africa, only cotton leaf curl Gezira virus (CLCuGeV) has been isolated from cotton. CLCuGeV has a wide host range infecting a number of plant species, including cotton, *Sida* spp., hollyhock, and okra.

During the 1990s, when CLCuD was in epidemic form in Pakistan and India, six different viruses were isolated from cotton. These species were named as cotton leaf curl Multan virus (CLCuMuV), cotton leaf curl Alabad virus (CLCuAlV), cotton leaf curl Rajasthan virus (CLCuRaV), cotton leaf curl Kokhran virus (CLCuKoV), tomato leaf curl Bangalore virus (ToLCBaV), and papaya leaf curl virus (PaLCuV). For infection of these begomoviruses to cotton, the presence of cotton leaf curl Multan betasatellite (CLCuMuB) is necessary.

Cotton leaf curl Burewala virus (CLCuBuV) was responsible for resistance breakdown in 2001 in Punjab, Pakistan. However, Sindh province of Pakistan was free of CLCuD during the Buewala epidemic in Punjab. CLCuD appeared as a major problem during 2004 in Sindh province. Recent diversity analysis revealed that relatively more diverse begomoviruses are present in Sindh as compared to Punjab. CLCuKoV, CLCuGeV, and a new species, cotton leaf curl Shahdadpur virus (CLCuShV), were identified from Sindh. The recombination of CLCuMuB and CLCuKoV resulted in the generation of CLCuBuV and CLCuShV.

During CLCuD epidemics in Pakistan, due to CLCuMuV and CLCuBuV, the disease perpetuated into neighboring countries such as India and China. Many virus species (CLCuMuV, CLCuKoV, and CLCuBuV) identified in Pakistan were then isolated from cotton in India. However, one species CLCuRaV was identified extensively in cotton in India that did not appear first on cotton in Pakistan. This virus was only identified from wild cotton species in Multan and tomato plants in Faisalabad. This might be the case where a cotton-infecting geminivirus moved from India to Pakistan. Similarly, there was a report of only one-time infection of cotton with ToLCBaV in India, which may suggest that this virus is not a serious pathogen. CLCuBaV may be an accidental infection on cotton, which was maintained by other cotton-infecting viruses. Likewise, there is a single report of the bipartite begomovirus tomato leaf curl New Delhi virus (ToLCNDV; GenBank accession no. EF063145) infecting cotton in India. The most recent diversity study of begomovirus infecting cotton in India reveals the presence of CLCuBuV and CLCuRaV, while in Pakistan only CLCuBuV is the dominant begomovirus on cotton. The possible reason for this smaller difference could be the high-scale development of CLCuD-resistant lines in Pakistan.

Betasatellites

The betasatellites are circular ssDNA molecules, first identified as recently as 1999–2000. The size of a betasatellite is approximately ~1350 nt with no sequence homology to the helper component except the region containing the nonanucleotide sequence TAATATTAC in the potential stem loop structure. The stem loop structure is present in the satellite-conserved region (SCR), which is highly conserved among all betasatellites. Additionally, betasatellites contain an adenine-rich sequence (A-rich) and a single ORF (the βC1 gene) in the complementary sense.

A helper virus is needed for replication and trans-encapsidation of betasatellite. βC1 protein encoded by betasatellite is the only protein that performs all the functions ascribed to betasatellite. For CLCuMuB, the βC1 protein, pathogenicity determinant, performs several functions. This protein determines the symptoms of the infection, act as a suppressor of PTGS (overcoming host plant defense), and is also involved in movement of virus within plants, modulating the levels of developmental micro-RNAs and upregulating viral DNA levels in plants. βC1 has also been shown to bind DNA/RNA and interact with a variety of host factors such as coat protein (CP) of the helper virus. The evolutionary origin of DNA betasatellite remains unclear.

Alphasatellites

Alphasatellites are capable of self-directed replication; that is why they are not true satellites. Therefore, they are characterized as satellite-like molecules. Alphasatellites are about half the size (~1400 nt) of the genomes of begomoviruses, slightly bigger than betasatellites. They contain an A-rich, a highly conserved gene in the virion-sense strand, encoding Rep protein. They also possess a predicted hairpin loop structure with nonanucleotide sequence (TAGTATTAC), which is just like nanoviruses (another class of ssDNA viruses). The initial sequencing results of alphasatellites associated with CLCuD-affecting cotton suggested that these molecules are genetically related. However, a recent diversity study conducted on alphasatellites associated with CLCuD-affecting cotton from Pakistan showed that these molecules are much diverse than previously assumed.

Although alphasatellites were originally identified from the complex of OW monopartite begomoviruses, there are recent reports indicating alphasatellite presence in combination with the NW bipartite viruses. The advantage of the presence of alphasatellite is still unknown for the pathogenicity of begomovirus–betasatellite complex.

Alphasatellites reduce the virus titer and allow the host to survive longer so that it can act as a source for onward transmission for a longer period. This suggestion came from initial studies where viral (begomovirus–betasatellite) DNA level was reduced due to the presence of alphasatellite.

Satellites are an unholy alliance with begomoviruses

Some monopartite begomoviruses such as TYLCV do not require any satellite molecule. Such viruses perhaps encode all the required proteins for successful infection. However, in case of many other viruses, such as CLCuMuV and Kenaf leaf curl virus or ageratum yellow vein virus, the presence of betasatellite is necessary for the development of symptoms. Although the presence of a DNA-B component could be a logical explanation for this after experimentation, no such entity was identified.

The first bonafide satellite-DNA identified was associated with ToLCV infection and its nucleotide sequence was completely different from ToLCV except the origin of replication or stem loop structure in the nonanucleotides. This satellite was ~1/4 (682 nt) of genome size of virus. The satellite associated with ToLCV was not involved in any symptoms development. It was proven that this particular satellite-DNA is likely to represent a parasitic element, as it has no role in disease development.

The second type of satellite molecule was isolated and characterized from cotton plants. Apparently, they have no role in the development of symptoms. This satellite molecule was found to encode its own Rep-like protein, similar to the nanoviruses. Initially, such molecules were named as DNA-1, but nowadays the term "alphasatellite" is used.

Alphasatellites are autonomously replicating molecules, but for insect transmission and movement, they depend on helper component.

Center of diversity for CLCuD

Eight major centers of plant origin have been established so far. The diversity of viruses in the centers of origin is always huge and decreases while moving toward periphery. If we talk about geminiviruses, there are eight different geographical locations that have been recognized as centers of diversification for geminiviruses. These centers are Australia, Indian subcontinent, Japan, sub-Saharan Africa, South China, South America, Central America, and Mediterranean–European region. Out of the total, 98% of alphasatellites, 94% of betasatellite, and 46% of geminiviruses are present in the Chinese and Indian centers only. Based on the presence of huge diversity for geminiviruses and associated satellites Indo-China has been declared as a center of origin for plants infecting ssDNA viruses.

The center of diversity for CLCuD can be determined by phylogenetic analysis of all the viral sequences available in the data bank. The center of origin can be described where a maximum number of CLCuD-associated geminiviruses have been identified. Similar studies can be designed for CLCuD-associated satellites.

Surprisingly, Indo-China as a center of diversity for cotton leaf curl-associated viruses is not the center of diversity for cotton. Cultivated cotton is native to Central America but there it is infected by an NW begomovirus, namely, CLCrV. However, at least seven different begomoviruses have been reported to infect cotton plant in the Indian subcontinent.

Molecular biology of herpes simplex virus

Herpes viruses are classified among order Herpesvirales, which is further classified into three families. These are among the most complex viruses containing several viral and cellular proteins that assemble into tegument, envelop, and nucleocapsid. Their capsid formation and encapsidation of newly replicated viral genome occurs through autocatalytic activity, which resembles tailed dsDNA bacteriophages. The host specific occurrence of herpes viruses indicates that they have evolved with their hosts for a long time and are well adapted to them. The most important herpes viruses are herpes simplex virus 1 (HSV1), which causes cold sores, and HSV2, which causes genital herpes. Both these diseases are contagious and can spread easily through physical contact, for example, by saliva while sharing drinks. Herpes viruses are among the largest and most complex of viruses with the virion of 200–250 nm in diameter, and their dsDNA genome packaged within an icosahedral capsid approximately 125 nm in diameter.

Types of herpes simplex virus

Herpes simplex virus (HSV) has been categorized into two types. HSV1 causes cold sore around the mouth and lips. Genital herpes can be caused by both HSV1 and HSV2. In case of HSV2, the affected people develop sores around the genitals or rectum. Although different body parts may develop sore, they are usually found below the waist.

Taxonomy and genome organization of HSV

HSV belongs to herpes virus family, Herpesviridae, that infect human and are also known as human herpes virus (HHV). There are three subfamilies of the Herpesviridae family:

Alphaherpesvirinae, Betaherpesvirinae, and Gammaherpesvirinae. These three subfamilies belong to mammalian, avian, and reptilian origin, respectively. Extensive sequencing data provided new basis of classification and a new order was proposed, "Herpesvirales," which includes three families, Herpesviridae, Alloherpesviridae, and Malacoherpesviridae. Amphibian and piscine herpes viruses are classified in Alloherpesviridae while the only herpes virus reported from invertebrates is classified among Malachoherpesviridae. Gene rearrangements and genome sequencing data revealed that at least 40 conserved genes are present among all families of herpes viruses. Most of these genes encode for proteins essential for replication of respective virus. While most of the genome of these three families does not show much homology among nucleotides and amino acid sequences, the basic mechanisms involved in viral replication, encapsidation, and movement remain the same, which provides common morphology to these viruses.

The genome of herpes virus comprises large dsDNA molecule surrounded by an icosahedral capsid protein. The capsid protein is enveloped by a lipid bilayer, which is connected to the capsid through the tegument. The complete viral particle with dsDNA, capsid, and tegument is known as virion. This comparatively larger genome of HSV1 and HSV2 contains almost 74 ORFs but the number of protein-coding genes can be up to 84 with 94 putative ORFs. Various proteins are encoded by these genes that are involved in the capsid, tegument, and envelop formation as well as the replication and infectivity of virus.

The complex genomes of HSV1 and HSV2 can be classified into two regions known as long unique region (UL) and short unique region (US). Among 74 ORFs, 12 genes reside on US while 56 genes belong to UL. The virus utilized the RNA polymerase II of infected host for its transcription. The protein expression can be categorized into three groups. The earliest group encodes for the proteins that are needed for the functioning of second and third groups. Immediately after the expression of early group proteins, the second group starts expression, which is mainly involved in the infectivity of virus. In the end, the third group expresses, which mainly encode for viral particles.

The assembly of herpes virus complex takes place in two distinct compartments, nucleus and cytoplasm. The conserved proteins of virus play a major role in nuclear egress, viral genome packaging, and capsid formation, whereas mostly no conserved proteins are involved in secondary egress and envelopment. This indicates that from evolutionary point of view, the nuclear association of virus is more ancient as compared to the viral association to cytosol. This mode of association makes them ancestrally more related to large dsDNA bacteriophages. Many cellular as well as viral proteins were identified to be the constituents of mature viral particles, and upon analyzation of purified extracellular viral particles, they were subjected to mass spectrometric analysis. The mature virion particle of HSV1, which is an alphaherpesvirus and serves as a model for herpes virus studies, contains 23 tegument viral proteins, 13 viral glycoproteins, 8 capsid and capsid-associated proteins, and up to 49 host protein virion. The most abundant and most frequently found proteins in herpes virus complex are annexin, beta-tubulin, ezrin/moesin, enolase, heat shock protein 90, heat shock protein 70, and actin. In addition to proteins, several cellular and viral RNAs have also been found in purified herpes virus complex.

Infection cycle of HSV

HSV undergoes two types of life cycle, lytic and latent. The replication of HSV happens within 15 h of infection. All the herpes viruses have the ability to enter latency, a dormancy phase where the virus enters sensory neurons, which are nondividing cells. During the latency phase, infectious progeny does not produce and only limited viral genes can transcribe. Upon facing the particular stimulus (heat, stress, ultraviolet light, menstruation, fever, hormonal changes, and physical trauma to the neuron fever), the dormant infected cells enter the lytic phase that result in the production of infectious viral particles.

In the lytic cycle, due to infection of HSV in the epithelial cells located in the mucosa, cell death occurs. HSV1 mostly invades ocular and oral epithelial cells while HSV2 target the genital areas, but both strains can cause infection in any area of the host body. In order to infect epithelial cells, the virus comes in contact with the cellular envelop of the host, resulting in the entry into the host. The virus invades the cell by fusing its envelop with host cell membrane. For this fusion, the virus possesses four envelop glycoproteins, that is, glycoprotein D (gD), a complex of glycoprotein H and glycoprotein L (gHL), and glycoprotein B (gB). Viral-induced membrane fusion occurs in three phases. During phase I, two membranes come into close proximity by binding the viral glycoprotein to the host cellular surface receptor. gD is recognized by cell surface receptors, so it is necessary for this phase. Phase II results in the initial mixing of virus with host membrane leaflets and the formation of a hemifusion intermediate pore. gD and gHL are required to be present in phase II. In phase III, stable fusion pore form and expand. gB assists in full fusion and pore formation. All these four proteins are prerequisites for HSV to be infectious; no cellular invasion could occur without any one of the glycoproteins.

After the completion of membrane fusion, the viral tegument and capsid are transported to the infected cell nucleus for virus replication. Both viral tegument and capsid remains attached until viral DNA enters into the host cell nucleus and makes their way by using host cytoskeleton proteins. Once the capsid reaches the host nucleus, capsid injects viral DNA in the nucleus. Capsid portal comprising 12 units of capsid protein UL6 facilitates the viral DNA injection in the host nucleus. This capsid portal is assembled on one of the capsid's verticals.

Replication of HSV

All the viruses belonging to three families of Herpesvirales follow almost the same major steps in viral replication. Evidence comes from the conservation of most of the genes involved in replication machinery. The basic steps involved in viral replication are reviewed below.

Suppression of cellular proteins of host is the first thing required for replication of viral DNA. This phenomenon is known as virion-associated host shutoff (VHS). VHS, a tegument protein coded by UL41 gene, controls this process. VHS protein is located in the cytoplasm and performs multiple functions like shutoff of host protein synthesis and host mRNA degradation. At the same time, viral DNA gets entry into the host nucleus. VHS appears in the host cell immediately after the infection and so it is called immediate-early protein. This protein does not differentiate between host and viral mRNA and start degrading viral mRNA. So VHS also interacts with the host alpha-trans inducing factor that downregulates its activity. After viral entry in the nucleus, it undergoes a recombination event and becomes circular, which is necessary for virus replication, and this circularization does not need any protein. This circularization results in origin-dependent replication.

Within the HSV genome, there are three sites reported as origin of replication, oirL, and two copies of oriS. OriL is located between two genes named UL29 and UL30, while oriS is located on either end of the US. Not all three regions are required at one time for replication; deletion of any ori results in viral replication. One theory is that some ori points may be required during lytic mode and other becomes activated in the latency mode of replication.

Table 18.1 Herpes virus genes and encoded proteins

Name of gene	Encoded protein	Name of gene	Encoded protein
UL5, UL8, and UL52	DNA helicase-primase	UL6, UL18, UL19, UL35, UL38	Capsid proteins
UL30	DNA polymerase	UL26.5	Scaffolding proteins
UL42	Helper protein for UL30	UL26	Maturation protease
UL9	Origin binding protein	UL25	Capsid-associated protein
UL29	ssDNA binding protein	UL31 and UL34	Movement protein

There are a number of viral genes that are necessary for origin-specific genome replication of HSV. Genes and their encoded products are listed in Table 18.1.

As soon as the viral particle comes in contact with the plasma membrane, it attaches itself with the membrane. Attachment is followed by immediate penetration of virion inside the cytosol where the virion capsid interacts with the cellular microtubules and cytoskeleton. This interaction aids the efficient delivery of viral particle from the plasma membrane to nuclear membrane. At this stage, the capsid opens up and releases the viral genome inside the host's nucleus and itself remains outside the nucleus.

Now, viral genome is exposed to the nuclear machinery where it harbors many processes, majorly the viral genome replication and transcription of viral genes. As a result of transcription, capsid proteins are generated with parallel production of replicated viral genome. Then the viral genome is encapsidated and ready for transport out of the nucleus, which starts with the fusion of encapsidated virus to the inner nuclear membrane.

Virus gets packed into a capsid using a complex of five capsid proteins. Scaffolding protein and maturation protease helps in the assembly of virus capsid, which then undergoes three steps of capsid formation. First of all, a partial capsid is formed, resulting in the formation of a closed spherical intermediate, and finally a closed icosahedral capsid is formed. The closed, spherical capsid is an open form of the virus than the icosahedral capsid but is an unstable conformation. UL25, a conserved protein, is part of the viral tegument and promotes envelop formation as the tegument joins the capsid with the envelop. Now, this capsid contains the viral DNA.

After nucleocapsid formation, these fused primary virions move to the perinuclear space and are transported out of the nucleus by budding through the outer nuclear membrane. Microfilament protein actin and inner nuclear membrane viral proteins (UL31 and UL34) help in the movement of the virus within the nucleus.

The primary virion matures when it comes in contact with the trans-Golgi network where the secondary envelopment of intercytosolic capsids is formed. This secondary envelopment is made up of viral glycoproteins that result in enveloped virion within a cellular vesicle. Now, this mature enveloped virion is ready to be transported out of the cell. It is moved to the inner surface of the plasma membrane where it fuses with the membrane and is exported out of the cell.

Vaccine and antiviral drugs development

The last 40 years was an era of antiviral drug development of high therapeutic value in treating life-threatening diseases, including HSV disease. This was possible due to technical advances such as cultivation of viruses in the laboratory, molecular biology studies, and identification of viral enzymes.

In case of HSV, so far no vaccine is available to eradicate HSV infection. However, antiviral drugs which have been developed are only helpful in reducing the pain during the initial episodes of the disease, resolving the symptoms a day or two later and subsequently relieving the discomfort slightly in the later stages of the disease. Moreover, treatment would only be helpful if started earlier, soon after the appearance of first sign of blisters and discomfort. Frequent attacks can also be reduced by taking antiviral drugs daily, indefinitely. Aciclovir, an artificially synthesized acyclic purine-nucleoside analog, is considered as the standard therapy against HSV infections and has been greatly helpful to control symptoms. Precursor drugs, famciclovir (converted to penciclovir) and valaciclovir (converted to aciclovir), have also been licensed and have better oral bio-availability than penciclovir and acyclovir, respectively.

Mucous membrane HSV infections

There are three ways to treat genital or oral HSV infection, that is, topical, oral, or intravenous aciclovir. Tropical therapy is the least effective and intravenous aciclovir is the most effective among the three to treat initial genital herpes infection, while a dose of 200 mg oral aciclovir for five times a day is recommended. Neither oral nor intravenous treatment is effective during acute infection of HSV for reducing the frequency of recurrences. Oral suspension of aciclovir can treat gingivostomatitis in children. During recurrent herpes infections, oral intake of aciclovir will only shorten virus shedding and healing time by 1 day (6 days treated vs. 7 days untreated) and it is possible when treatment has started within 24 h of onset of disease. Single dose of valaciclovir while two doses of famciclovir must be given. So, oral aciclovir, famciclovir, and valaciclovir would be effective in frequent recurrences. There is 80% reduction in frequency of recurrences by daily uptake of aciclovir and it prevents recurrences in 25%–30% patients. Symptomless shedding of virus is still possible; hence, person-to-person transmission is still possible. To reduce the risk of genital herpes transmission among heterosexual, HSV2–discordant couples, once-daily suppressive therapy with valaciclovir can be effective.

Intravenous or oral aciclovir therapy benefits immunocompromised hosts. Aciclovir prophylaxis can reduce the symptomatic HSV infection up to 70%, especially in those having chemotherapy and transplantation. Similar is true for famciclovir and valaciclovir. However, dose regimens range from 200 mg (thrice a day) to 800 mg (twice a day). Patients resistant to aciclovir would also be resistant to famciclovir or penciclovir or both.

Other HSV infections such as hepatitis, herpetic, pulmonary infection, proctitis, esophagitis, herpetic whitlow, and erythema multiform have been treated successfully by using aciclovir. HSV eye infections can also be treated effectively with aciclovir.

Molecular basis for antiviral drug development

Up till now, three drugs have been developed, that is, aciclovir, valaciclovir, and famciclovir. These drugs act through the function of two herpes virus enzymes and it is a linked process. The first herpes virus enzyme is thymidine kinase (TK), which phosphorylates aciclovir to monophosphate (in case of HSV) and then to diphosphate. Subsequent action of cellular enzymes makes them triphosphate. The second enzyme is the DNA polymerase encoded by the virus, which is the target inhibited by aciclovir triphosphate by a mechanism known as obligate chain termination.

There is another mechanism by which aciclovir triphosphate inhibits the activity of viral polymerase, which is competitive inhibition of the natural guanosine triphosphates for the DNA polymerase. It must be clear during the development of aciclovir drug that the levels of aciclovir triphosphate generated are sufficient only for chain termination.

The activity of the other two drugs is different as inhibition of HSV polymerase is not obligate chain termination because a pseudo 32 hydroxyl is present in the molecule to carry on polymerization. So, in this case, polymerase activity is terminated by altering the conformation of the replication complex that prevents DNA chain extension.

Toxicity of antiviral drugs

There are some adverse effects observed following the treatment with aciclovir, valaciclovir, and famciclovir drugs. A large dose of aciclovir could result in renal dysfunction by rapid intravenous infusion, but is reversible and usually rare. Risk of nephrotoxicity can be reduced by adequate hydration and slow infusion of aciclovir. Oral aciclovir therapy, even at doses of 800 mg five times daily, does not result in renal dysfunction and even prolonged treatment with aciclovir does not cause toxicity. Central nervous system disturbance is usually linked with intravenous acyclovir, which include disorientation, agitation, tremors, hallucination, and myoclonus.

Viral resistance to antiviral drugs

HSV can develop resistance against aciclovir by creating mutations in the viral gene that encodes the TK gene. TK is responsible for the activity of aciclovir as it phosphorylates; hence, selection of mutants with a TK gene is unable to phosphorylate aciclovir and the generation of TK-deficient mutants brings resistance in HSV to aciclovir. Second, alternate DNA polymerase has been detected in some HSV that cannot be targeted by aciclovir. Aciclovir-resistant HSV isolates have been recognized as the cause of pneumonia, esophagitis, encephalitis, and mucocutaneous infections in immune-compromised patients.

HSV vaccines

Prophylactic vaccination might be difficult to develop because HSV recurs in individuals with cell-mediated and humoral immunity. However, for prevention and therapy, two subunit vaccines have been studied. Results from a trial in which glycoproteins (gB and gD) were the vaccine subunits did not show any prophylactic or therapeutic effect. Another gD vaccine study showed positive results but men cannot be protected. Alternatively, a non-infectious HSV or naked DNA or a disabled infectious single-cycle vaccine can be used as vaccine. But the development of such virus constructs requires special care as infectious viruses are capable of inducing severe mortality if it infects the brain.

Molecular biology of RNA viruses

RNA viruses are known as nature's swiftest evolvers. Their high rates of mutation and replication have enabled them to hijack their DNA hosts machinery for their own purposes. Single-stranded RNA (ssRNA) viruses are one of the most studied types of viruses.

Negative-stranded RNA viruses are those that cannot directly make proteins. Instead, they are first transcribed into positive-sense RNA, which serves as a template for negative-sense RNAs.

Positive ssRNA viruses are those that are directly accessed by host translational machinery to form proteins. They replicate in the cytoplasm. Positive RNA viruses comprise over one-third of known viruses. They are divided into many superfamilies based on their distinct RNA replication mechanism. However, it shares enough similarities to discuss positive RNA viruses' replication as a class.

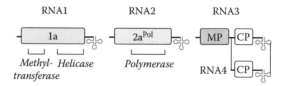

Figure 18.2 Schematic diagram of the BMV genome, showing genomic RNA1 (3.2 kb), RNA2 (2.9 kb), and RNA3 (2.1 kb), plus subgenomic RNA4.

Brome mosaic virus as a model virus

Brome mosaic virus (BMV) is a positive-strand RNA virus, which belongs to the family Begomoviridae. It is a member of alphavirus-like superfamily of human, animal, and plant RNA viruses. This virus infects grasses, including cereals. BMV is a model virus to study viral RNA replication, gene expression, recombination, and virus–host interactions due to some of its characteristics such as segmented genome, high RNA synthesis, ability to replicate in yeast, etc. BMV is a small icosahedral virus, which comprises a tripartite genome. It has three genomic RNAs and one subgenomic mRNA encapsidated by three nonenveloped capsids. RNA 1 encodes the replication factor and RNA the 1a protein. It has N-proximal capping domain and C terminal NTPase/RNA helicase-like domain, which are separated by a short proline-rich spacer. RNA2 encodes 2a polymerase, which has a central polymerase domain and an N terminal domain that interact with the helicase-like domain of 1a protein during infection. 3a protein is encoded from RNA3 for cell-to-cell movement in plants while subgenomic RNA4 encodes the coat protein (Figure 18.2).

The first eukaryotic viral RNA-dependent RNA polymerase was also isolated from BMV, which facilitated RNA studies. BMV was also the subject of earlier translation studies, providing the first definition of eukaryotic ribosome binding site. The sequence analysis of various RNA viruses, including BMV showed many conserved multiple domains in their RNA replication proteins. It gave the concept of virus superfamilies.

It was the first plant virus whose infectious transcripts were designed from cloned viral cDNA. This allowed the recombinant manipulative techniques to study other RNA viruses *in vivo* as well as *in vitro*. Plant and yeast studies show that BMV form membrane-bounded organelles for RNA replication. Multiple features of the assembly, structure, and function of the replication complexes and the tRNA-like3′ ends of BMV genomic RNAs suggest that positive-strand RNA viruses, retroviruses, and double-stranded RNA (dsRNA) viruses share a common ancestor.

Host range of BMV

BMV infects different monocots belonging to the Poaceae family. These include grasses and cereals. It causes brown streaks in *Zea mays*, *Bromus inermis*, and *Hordeum vulgare*. It can also cause infection in *Nicotiana benthamiana* and many *Chenopodium* species. Necrosis or chlorotic lesions are produced in BMV-infected chenopodium plants, while in *N. benthamiana*, infection is asymptomatic.

Infection cycle of BMV

There are some common features of the replication of positive-sense RNA viruses (Figure 18.3). These are the coordinate use of viral positive-sense genomic RNA as a

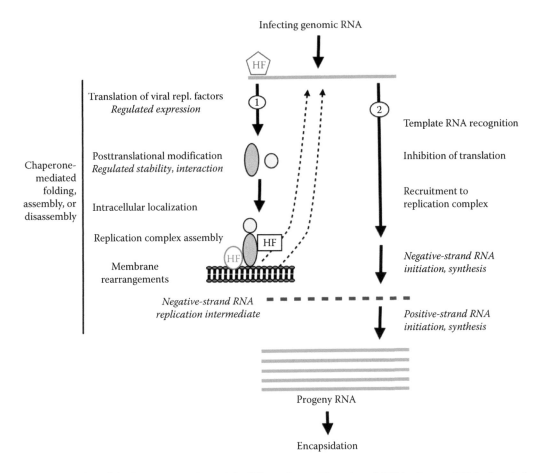

Figure 18.3 Simplified representation of the life cycle positive-strand RNA viruses. HF is denoted for host factors.

template for both translation and replication, assembly of replication complexes on intracellular membranes compartments and many folds production of positive- over negative-strand RNAs.

During the infection cycle, genomic RNA is first translated into viral proteins. Viral replication requires the close association of replication-associated proteins and host factors in virus-induced intracellular membranes. The formation of a replication complex is likely to depend on interactions among viral RNA, proteins, and host lipid membranes.

All BMV RNAs have 3′ UTR with three RNA pseudoknots along with an almost 200 nt sequence, which can fold into tRNA-like structures (TLS). These TLS are found to play an important role in BMV replication. As the formation of minus-strand RNA requires the recruitment of replication machinery, the 3′ recognition sites for the proper synthesis of RNA genome, TLS play a key role in this entire regulatory network.

Replicase enzyme recognizes the specific stem loop C present in TLS. *In vitro* studies demonstrated that SLC alone can bind with replicase but it cannot direct the synthesis of the minus strand unless it has the last 6–8 nt of the 3′ TLS. 1a and 2a proteins of BMV are involved in the replication of RNA 3 and transcription of RNA 4. BMV capsid protein studies also show its participation in the regulation of viral minus-strand RNA synthesis. It may play a role in the recognition of promoter for minus-strand RNA synthesis.

Then RNA-dependent RNA polymerase directs the formation of negative-strand RNA. Now this negative strand acts as a template for positive-strand RNA. Generally, a large excess of positive-sense strands is produced than negative-sense strands. The fact that positive-sense RNA strand serves as a template for both replication and translation suggests that both viral replication and translation are coordinated in viral infection cycle. A number of viral nonstructural proteins are phosphoproteins. Phosphorylation is important for reversible posttranslational modifications of proteins, therefore playing a major role in the regulation of many host cellular processes.

Successful infection of BMV requires the efficient spread of virus within the host, which is mediated by 3a protein. BMV can spread in the nearby cells with the help of 3a proteins but for cell-to-cell movement, it requires a capsid protein. 3a protein of the virus also provides it host specificity and symptom development. The capsid protein is also involved in the systemic movement of the virus.

BMV RNA 1 and RNA 2 encapsidate separately and RNA 3 and subgenomic RNA 4 are likely to be encapsidated in the same particle. There is a region present in RNA 1 that shows a high affinity for capsid protein. It may signal the virus to stop replication and guide the virion assembly.

Biology of BMV replication in yeast

A crucial aspect of RNA viral replication is that they use cell functions. Through this, they have evolved the capacity to hijack host mechanisms and proteins for their function. They directly or indirectly influence the host metabolic pathways ranging from transcription to cytoplasmic signaling. BMV RNA replication depends on 1a and 2a pol proteins and *cis* acting RNA signals both in plants and yeast. Negative-strand RNA intermediate synthesis requires the 3′ tRNA sequence as it has a replicase binding site. RNA synthesis takes place by *de novo* priming at the tRNA signature sequence (5′ … CCA3′) at the penultimate cytidylate end. The *de novo* property to initiate RNA synthesis at or near the 3′ terminus of a template is unique to RdRps. There is stem loop structure within the 3′ tRNA region, which is required to bind with specific replicase (Figure 18.4).

Replication of positive RNA strand is initiated from cytidylate as the penultimate nucleotide at the 3′ end of the negative strand. There is also a sequence rich in adenine and uracil nucleotides near initiation sites, which promote efficient RNA synthesis. Genomic RNA 3 and subgenomic RNA 4 are synthesized from negative-strand RNA 3. Subgenomic RNA 4 synthesis requires the interaction of replicase with promoter sequence, directly upstream of the RNA 4 initiation cytidylate.

In yeast, 1a causes 60–80 nm vesicular invaginations in the outer membrane of the endoplasmic reticulum in the absence of other viral factors, which is referred as "spherules." These spherules are surrounded by a lipid bilayer and vesicles are connected with cytoplasm through a neck-like opening inside them. This opening may serve as a channel for the export of progeny RNAs and ribonucleotide import. In each spherule, there are hundreds of 1as and only a few number of 2a proteins. It shows that 1a interacts with genomic RNA and then it recruits 2a for the RNA replication in ER spherules. Moreover, a variable level of interaction between 1a and 2a pol change the spherules in large, multilayer ER membrane stacks, which also promote RNA synthesis efficiently as spherules.

1a proteins do not have a transmembrane domain and resides on the cytoplasmic side of ER. It has an α helix, called helix A, which is required for the 1a-induced recruitment of viral RNAs and 2a protein, spherule formation and membrane rearrangements mediated

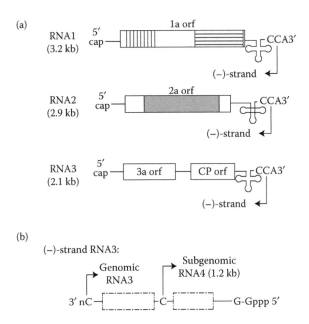

Figure 18.4 The genomic organization of BMV. (a) The RNAs are capped with a 7-methyl-guanylate at the 5′ end and possess a tRNA-like structure at the 3′ end. The initiation sites of minus-strand RNA synthesis from the penultimate cytidylate are indicated with an arrow. The protein-coding sequences are indicated in boxes. The position of the domain in protein 1a that is highly conserved in capping-associated proteins is denoted with vertical stripes. The region denoted with horizontal stripes contains helicase-like motifs. The gray region in 2a contains the central RdRp-like sequences. (b) Minus-strand RNA3 is used to promote synthesis of genomic RNA3 and subgenomic RNA4 (ORF, open reading frame).

by 1a protein. 1a–2a pol interactions occur before spherule formation and it is suggested that 1a–2a pol interactions and 1a-induced spherule formation are sequential but perhaps antagonistic. Study showed that 1a capping and helicase domain interact intramolecularly but not at intermolecular levels. There are at least three types of interactions present within 1a polymerase, which are intramolecular interactions between the capping and helicase-like domains, intermolecular interactions between capping domains, and independent intermolecular interactions between helicase-like domains.

Host factors involved in BMV replication

There are more than 100 genes that are involved in BMV replication either negatively or positively. DED1 is the host protein, which encodes an RNA helicase involved in the translation initiation of yeast mRNAs. It was found that mutations in DED1 suppress the BMV replication. LSM 1 is also another host gene that is required by the virus for the recruitment of 1a protein. LSM 1 gene encodes for the protein having sm motif through which it interacts with other sm motif-containing proteins involved in host RNA turnover and other cellular processes. Subunits of host translation initiation factors (eIF3) associated with viral RdRp complexes were also found in BMV-infected cells. Newly synthesized RNA and 1a and 2a polymerase are found to completely colocalize in ER membranes near the nucleus. Mutations in yeast YDJ1 gene also block the BMV replication. YDJ1 encodes chaperones that are localized in the ER membrane.

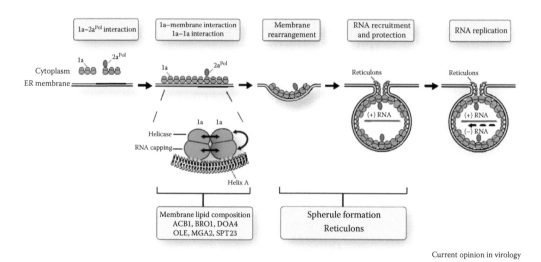

Current opinion in virology

Figure 18.5 Diagram showing the interactions and host factors involved in BMV RNA replication assembly complexes.

Cellular membranes are very important in the replication of positive RNA-strand viruses because RNA replication is performed in vesicular invaginations of endoplasmic reticulum membranes. Thus, lipids are the crucial part of these membranes and genes involved in lipid metabolism are important host factors involved in BMV replication. OLE1-encoded Λ9 fatty acid desaturase converts saturated fatty acids into unsaturated fatty acid (UFA). A mutation in OLE1-encoded Λ9 fatty acid desaturase causes 20-fold reduction in BMV replication. This is due to the decrease in UFA level as it affects the membrane composition in the vicinity of spherules. Many genes such as DOA4, MGA2, BRO1, and SPT23 are involved in OLE1 expression; hence, they also control BMV RNA replication in yeast (Figure 18.5).

There are reticulin proteins that are present in ER membranes. These are membrane-shaping proteins, which are responsible for partitioning and stabilization of highly curved ER membranes. This protein's functioning is due to its oligomerization ability. They reside in peripheral ER membrane. The virus protein 1a interacts with reticulins and then they relocalize from peripheral ER membrane into replication compartments. Deletion of reticulins progressively stops the spherule formation and ultimately viral replication. It was found that reticulins are involved in nuclear pore formation, so certainly they are involved in spherule formation and neck maintenance.

Viral 1a protein potentially interacts with other cellular pathways contributing to virus replication, like the factors that regulate membrane synthesis/composition, involved in trafficking, and membrane remodeling remains to be explored. Additionally, determining the liposome and proteome contents of purified replication compartments should help identify in more detail the viral and cellular components of these compartments.

Biology of retroviruses

Positive-strand RNA viruses comprise one-third of all known viruses. They have positive-sense messenger RNA genome encapsidated in their virions, which is directly translated into viral proteins.

Retroviruses are one of the main classes of RNA viruses. Retroviruses are positive-sense ssRNA viruses that replicate through a DNA intermediate. All the retroviruses encode two proteins, that is, *gag* and *pol*. gag is multidomain structural protein and *pol* has reverse transcriptase (RT) activity. All retroviruses replicate by reverse transcription of their DNA intermediate. Instead of using an RNA template for proteins, they use DNA intermediate for minus-strand RNA synthesis. They use different host factors for their cell entry, replication, and other major steps in their infection cycle. Human immunodeficiency virus (HIV) is a well-known retrovirus, which escapes from the human immune response by using multiple strategies.

Human immunodeficiency virus

HIV belongs to the family Retroviridae, subfamily Lentivirinae, and genus *Lentivirus*. It is a lentivirus because it is a slow virus. It takes a while for the virus to replicate enough to cause symptoms in the host. There are three species of *Lentivitrus*.

Human immunodeficiency virus (HIV): It causes acquired immune deficiency syndrome (AIDS)
Feline immunodeficiency virus (FIV)
Simian immunodeficiency virus (SIV)

HIV infection is the result of zoonotic infection of simian immunodeficiency syndrome from African primates.

Life cycle of HIV

HIV is an obligate intracellular parasite; it cannot replicate outside human cells. The virus enters the cell by binding with receptors present in the host cell membrane (Figure 18.6). Generally, the viral envelope protein gp120 recognizes two different host–cell-surface receptors. The first coreceptor is CDC4, which is mainly present on T lymphocytes and macrophages. The presence of CDC4 protein on different lymphocytes and monocytes/macrophages make them susceptible for HIV, although recent studies also showed the efficient virus entry in cells not expressing CDC4 protein.

The second coreceptor is a range of proteins from a class of seven transmembrane receptors. Among these, CCR5 (CC chemokine receptor 5) and CXCR4 (CXC chemokine receptor 4) are most important. There are hundreds of other related proteins in the class of seven transmembrane receptors, which are shown to be binding with HIV in *in vitro* studies. However, their function is still unclear *in vivo*.

The binding of gp120 with CDC4 and other coreceptor is followed by the conformational change in gp41. It causes the insertion of N-terminal fusion peptide region in the target host cell. The cell membrane is fused and viral contents enter the cytoplasm. At this moment, entry is dependent on the interaction of the C-terminal and the N-terminal of gp41 ectodomain (Figure 18.6).

All strains of HIV bind with gp120 but their affinity varies for CCR5 and CXCR4. The binding ability and tropism of the virus depends on the structure of the gp120 protein. Specific pattern of the sequence of variable (v) and other regions of gp120 affect the CDC4 binding and coreceptor affinity. Generally, viral strains that bind with CCR5 infect macrophages and T cell. They have less viral titer. But the strains that bind with CXCR4, infect only T cells. They are more active and produce high viral titer.

After entry, the RNA genome is first transcribed into ssDNA, which is further transcribed into dsDNA with the help of RT, copackaged in the virus particle. There are series

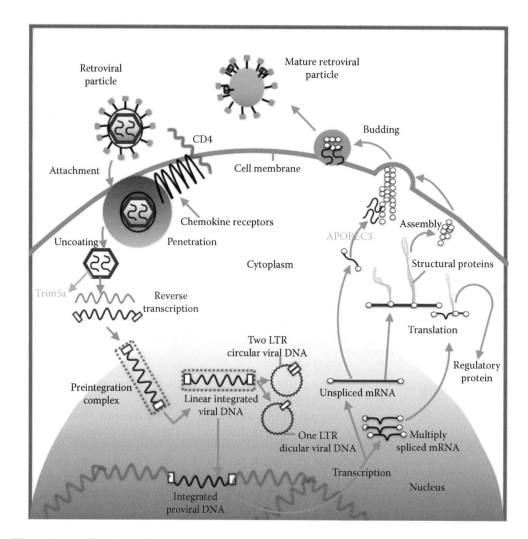

Figure 18.6 Life cycle of HIV showing entry, intermediate DNA formation, and integration in host genome, transcription and translation, packaging, and release of mature virus particles.

of steps involved in which viral long-terminal repeat (LTR) and host enzymes interacts during *de novo* priming of viral RNA, DNA, and removal of transcribed RNA strand. The dsDNA forms a complex with viral proteins (matrix, integrase, and Vpr) and host membrane and is then transported to the nucleus. Here, this dsDNA is either incorporated randomly in cell genome with the help of viral integrase or it forms stable DNA circles. The integrated form of DNA is called provirus. This proviral DNA is replicated along with normal host DNA and may persist in this form for a long period. It passes through multiple rounds of mitotic cell division.

For viral transcription, the 5′ of LTR acts as a primer in the presence of specific host transcription factors (promoter-specific transcription factor-SP1, and nuclear factor-kappa beta) and viral protein Tat. The newly transcribed RNA is either spliced to translate the viral proteins or it is exported from the nucleus as in unspliced form for packaging in newly formed virions. The spliced RNA is exported from the nucleus with the help of Rev protein. Viral proteins do not perform only multiple functions in the viral replication but they also interact with various host pathways to suppress immune response.

Vpr alters host–cell transcription and arrests infected cells at the G2/M phase of cell division. Nef cause downregulation of the CD4 receptor and MHC class I molecules. Vif counteracts cytidine deaminases enzymes that are present especially in macrophages and T cells. These are naturally occurring host defense mechanisms against retroviruses. These proteins include APOBEC3G and APOBEC3F and are degraded by HIV. vpu increases the degradation of CD4 in the endoplasmic reticulum and Vif is necessary for subsequent efficient infectivity of the newly produced viral particles.

Immature viral polyproteins are processed by protease to form functional proteins. They are assembled with unspliced HIV RNA to form new virions. gag, the main structural polypeptide, encodes the majority of these proteins while it itself uses the cellular proteins to go into the plasma membrane where it assembles with new viral particles. At this time, the p6 protein present at the C-terminal of gag interacts with tumor suppressor gene 101 (TSG101). It is known that this interaction is important for the release of newly formed virus particles. Initially, the unspliced form of RNA is assembled with immature polyproteins. After budding from the plasma membranes, these proteins are processed into their functional forms and rearranged into mature particles. vpu protein helps in the release of viral particle during the last stages of replication. It interacts with host factor tetherin. Tetherin is a host membrane-associated protein that hinders the release of viral particles. Without vpu, virus particles cannot be released.

Recent pandemics of HIV

There are two types of HIV:

HIV-1: It is the most common type of HIV that is prevalent in the world. It is severe than HIV-2.

HIV-2: It is confined to western central Africa and southern and western India. It is a less virulent type of HIV, which produces low viral titer, slow progression of the disease, and less transmissible. The difference in both types is due to difference in their zoonotic origin. HIV-1 is much like SIV isolated from chimpanzees and HIV-2 is similar with those from sooty mangabey.

There are four groups in HIV-1 type that are M, N, O, and P. M, N, and O are transmitted from chimpanzee and P from gorilla. Group N, O, and P are confined to western Africa while M is responsible for world HIV pandemic. It started 100 years ago and now it has nine subtypes. These are A, B, C, D, F, G, H, J, and K. Subtype B is more prevalent in Western Europe, America, and Australia. Subtype C is common in Africa and India and 48% of HIV-1 cases worldwide in 2007. In 2012, there were about 35.3 million people living with HIV-1 at a global scale. Among them, sub-Saharan Africa has the highest burden (70.8%) of HIV (Figure 18.7).

WHO and HIV

According to WHO, there are almost 35.0% people already infected with HIV and 1.2 million newly affected people worldwide at the end of 2013. Among them sub-Saharan Africa is the most affected region with 24.7 million people living with HIV at the end of 2013.

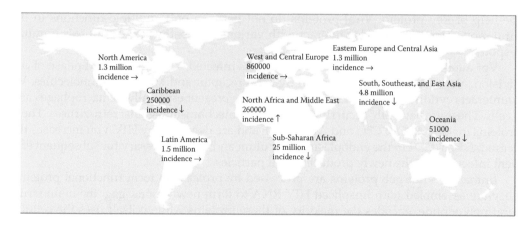

Figure 18.7 Estimated number of people living with HIV in 2012 and trends in the incidence of new infections from 2001 to 2012 by global region. (Data from UNAIDS 2013 report.)

WHO is working with countries to implement the Global Health Sector Strategy on HIV/AIDS for 2011–2015. WHO has six operational objectives for 2014–2015 to support countries effectively to achieve the global HIV targets. These are the following:

- Strategic use of antiretrovirals (ARVs) for HIV treatment and prevention
- Elimination of HIV in children and expanding access to pediatric treatment
- An improved health sector response to HIV among key populations
- Innovative approaches in HIV prevention, diagnosis, treatment, and care
- Strategic information for effective scale-up
- Stronger links between HIV and related health outcomes

WHO is a cosponsor of the Joint United Nations Programme on AIDS (UNAIDS). In UNAIDS, WHO leads activities on HIV treatment and care, and HIV and tuberculosis coinfection. It also coordinates with UNICEF to work on the elimination of mother-to-child transmission of HIV.

Genome organization of HIV

HIV has single-stranded, positive-sense RNA genome of about 9.7 kb in size, encapsidated by cone-shaped capsid (see Figure 18.9 later in the chapter). There are two RNA strands, each encoding for nine viral genes. The HIV structure is represented in Figure 18.8. The virus capsid has almost 2000 copies of viral p24 protein. It is further surrounded by an envelope, which comprises a lipid bilayer formed by an intracellular membrane of host cell when a new viral particle is released from the infected cell.

Each envelope subunit consists of two noncovalently linked viral proteins. One is glycoprotein (gp) 120, outer envelope protein, and the other is transmembrane, gp41 protein, which anchors glycoproteins to the surface of a virion. Host major histocompatibility complex (MHC) antigens and actin are present along with viral envelope proteins, embedded in a viral envelope. The envelope protein is relatively the most variable component of HIV because gp120 is structurally two regions; one is more constant (C) and the other is a highly variable (V) region. This variability of the envelope protein may be a product

Mature HIV-I virion

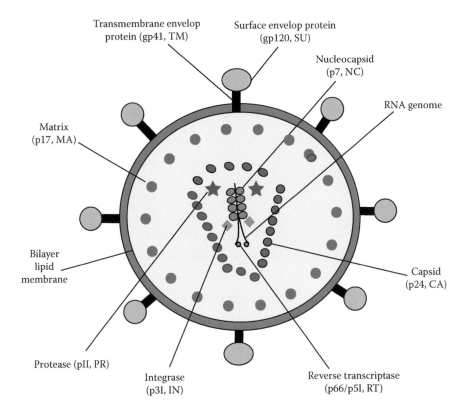

Figure 18.8 The structure of HIV showing different structural and nonstructural proteins of HIV.

of virus functionality because any change in amino acid changes the coreceptor binding. It also provides complex antigenic diversity.

The gene organization of HIV is highly efficient. It uses all three reading frames of nucleic acid sequence allowing overlapping genes. These nine genes encode different types of proteins, which can be broadly divided into categories, that is, structural, catalytic, regulatory, and accessory proteins (Table 18.2) (Figure 18.9). The RNA sequence of HIV also has its intrinsic properties along with the formation of functional proteins. Like the Rev, responsive element interacts with Rev protein and RNA transcripts are exported outside the nucleus of the cell. The LTR region in the integrated DNA provirus also acts as a transcription promoter. It also contains regions essential for reverse transcription, integration into the host–cell genome, and genomic RNA dimerization.

Replication and mutation in HIV

The mutation rate of HIV is very high due to very high rate of replication, the errors made during replication, and higher viral load. Mutation rate is high as RT is error prone; it converts virus RNA genome into DNA. So errors made by RT during replication allow the virus to escape from immune response. This high mutation rate of virus is responsible for immunological escape mutants and drug-resistant mutants.

Table 18.2 Major genes of HIV and their gene products

Protein class	Gene	Gene product
Structural	gag	p17 (matrix)
	env	p24 (capsid)
		p7 (nucleocapsid)
		gp120
		gp41
Catalytic	pol	Protease
		Reverse transcriptase
		Integrase
Regulatory	tat	tat
	rev	rev
Accessory	vpu	vpu
vif	vif	vif
	vpr	vpr
	nef	nef

In viral replication, three types of polymerases are involved, host DNA polymerase, and host RNA polymerase II and viral RT. DNA polymerase has good fidelity than RNA *pol* II and RT due to its proofreading and editing machinery. Mostly infected cells die before dividing, so *pol* II and RT are responsible for making errors during viral replication. However, previous studies show that there is a major contribution of RT in the mutation rate. The majority of errors arising during replication are missense mutations. There is some mutation in polymerase active site due to which RT cannot discriminate between correct dNTP and other dNTPs. Missense mutations are predominant, comprising 75% of the total, and 85% of these missense mutations are transitions. It is observed that RT tends to substitute A in place of other nucleotides and these mutations occur at hot spot regions.

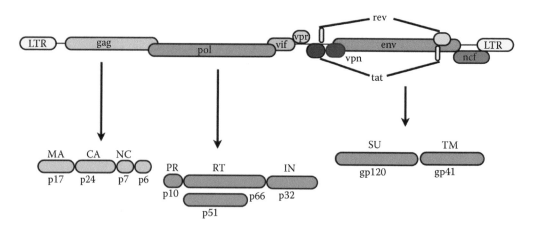

Figure 18.9 The genome organization of HIV. There are nine overlapped genes of HIV in single-stranded positive-sense RNA genome, encoding structural, enzymatic, and movement proteins. LTR is long-terminal repeat region.

These missense mutations in RT results in the alteration of nucleotide-binding pockets, which confers the resistance against nucleoside/nucleotide analogs and anti-HIV drugs. It also provides the virus an escape from host immune system.

Vaccine and antiviral drugs development

Antiretroviral therapy

Combination antiretroviral therapy (ART) regimens were developed in the late 1990s. These techniques helped to suppress viral replication and transformed HIV from fatal illness to manageable chronic disease. According to World Health Organization (WHO), studies showed that there is 44% reduction in disease severity in people who use ART immediately after diagnosis. Currently, there are almost 25 licensed drugs available that suppress viral replication at various steps in viral infection. ARTs are less dosed than initial protease inhibitor-based drugs, more effective and are less toxic. Standard ART regimens use a combination of two nucleoside RT inhibitors (emtricitabine or lamivudine together with one of abacavir, zidovudine, or tenofovir) with a nonnucleoside RT inhibitor, integrase, or protease inhibitor. To avoid resistance or intolerance, many nucleoside RT inhibitor-sparing regimens can be used together.

Usually within 3 months following antiretroviral therapy, the plasma viral load decreases below the detection level of many commercial assays. But the recovery of CD4 T cells is variable in different individuals. Adjunctive interleukin significantly increases CD4 T-cell counts; however, it might be involved in expanding the pool of CD4 T cells, latently infected with HIV.

Vaccine development

There are major challenges in developing the HIV vaccine due to higher mutation rate of HIV, uncertainty of how to create protective immunity, and difficulty in making immunogenic antigens. Generally, an induction of CD4 T cells response is a key to make vaccine against HIV. Many clinical trials of HIV vaccine are done but in only one trial RV144, conducted in Thailand, could vaccine protection be achieved with a 31% reduction in HIV acquisition. Hopefully, correlation of this trial together with new vector approaches that may improve the T-cell responses and identify targets for broadly neutralizing antibodies will result in the development of more effective vaccines.

Bibliography

Abram, M.E., A.L. Ferris, K. Das, O. Quinones, W. Shao, S. Tuske, W.G. Alvord, E. Arnold, and S.H. Hughesa. 2014. Mutations in HIV-1 reverse transcriptase affect the errors made in a single cycle of viral replication. *J. Virol.* 88(13):7589–7601.

Ahlquist, P., A.O. Noueiry, W.M. Lee, D.B. Kushner, and B.T. Dye. 2003. MINIREVIEW: Host factors in positive-strand RNA virus genome replication. *J. Virol.* 77(15):8181–8186.

Anna, J. and I. Jupin. 2007. Regulation of positive-strand RNA virus replication: The emerging role of phosphorylation. *Virus Res.* 129:73–79.

Bert, L.S. and P.J.W. Sean. 2013. Methods to study RNA virus molecular biology. *N.I.H.* 59(2):165–166.

Bresnahan, W.A. and T. Shenk. 2000. A subset of viral transcripts packaged within human cytomegalovirus particles. *Science* 288(5475):2373–2376.

Briddon, R.W. and J. Stanley. 2006. Subviral agents associated with plant single-stranded DNA viruses. *Virology* 344:198–210.

Davison, A.J. and M.D. Davison. 1995. Identification of structural proteins of channel catfish virus by mass spectrometry. *Virology* 206(2):1035–1043.

Davison, A.J., B.L. Trus, N. Cheng, A.C. Steven, M.S. Watson, C. Cunningham, R.-M. Le Deuff, and T. Renault. 2005. A novel class of herpes virus with bivalve hosts. *J. Gen. Virol.* 86(1):41–53.

Diaz, A. and X. Wang. 2014. Bromovirus-induced remodeling of host membranes during viral RNA replication. *Curr. Opin. Virol.* 9:104–110.

Greijer, A.E., C.A. Dekkers, and J.M. Middeldorp. 2000. Human cytomegalovirus virions differentially incorporate viral and host cell RNA during the assembly process. *J. Virol.* 74(19):9078–9082.

Kao, C.C., P. Ni, M. Hema, X. Huang, and B. Dragnea. 2011. The coat protein leads the way: An update on basic and applied studies with the Brome mosaic virus coat protein. *Mol. Plant Pathol.* 12(4):403–412.

Kao, C.C. and K. Sivakumaran. 2000. Brome mosaic virus, good for an RNA virologist's basic needs. *Mol. Plant Pathol.* 1(2):91–97.

Loret, S., G. Guay, and R. Lippé. 2008. Comprehensive characterization of extracellular herpes simplex virus type 1 virions. *J. Virol.* 82(17):8605–8618.

Maartens, G., C. Celum, and S.R. Lewin. 2014. HIV infection: Epidemiology, pathogenesis, treatment, and prevention. *Lancet* 384:258–271.

Maxwell, K.L. and L. Frappier. 2007. Viral proteomics. *Microbiol. Mol. Biol. Rev.* 71(2):398–411.

McGeoch, D.J., F.J. Rixon, and A.J. Davison. 2006. Topics in herpes virus genomics and evolution. *Virus Res.* 117(1):90–104.

Mettenleiter, T.C., B.G. Klupp, and H. Granzow. 2006. Herpes virus assembly: A tale of two membranes. *Curr. Opin. Microbiol.* 9(4):423–429.

Mettenleiter, T.C., B.G. Klupp, and H. Granzow. 2009. Herpes virus assembly: An update. *Virus Res.* 143(2):222–234.

Miranda, S.-X. and R. Oelrichs. 2009. *Basic HIV Virology*. HIV Management in Australasia a guide for clinical care, pp. 9–18.

Mubin, M., R.W. Briddon, and S. Mansoor. 2009. Diverse and recombinant DNA betasatellites are associated with a begomovirus disease complex of *Digera arvensis*, a weed host. *Virus Res.* 142:208–212.

Nawaz-ul-Rehman, M.S. and C.M. Fauquet. 2009. Evolution of geminiviruses and their satellites. *FEBS Lett.* 583:1825–1832.

Nawaz-ul-Rehman, M.S., S. Mansoor, R.W. Briddon, and C.M. Fauquet. 2009. Maintenance of an Old World betasatellite by a New World helper begomovirus and possible rapid adaptation of the betasatellite. *J. Virol.* 83:9347–9355.

Raoa, A.L.N. and C.C. Kao. 2015. The brome mosaic virus untranslated sequence regulates RNA replication, recombination, and virion assembly. *Virus Res.* 206:46–52.

Robert, W.L., S.B. Larson, and A. McPherson. 2002. The crystallographic structure of brome mosaic virus. *J. Mol. Biol.* 317:95–108.

Rojas, M.R., C. Hagen, W.J. Lucas, and R.L. Gilbertson. 2005. Exploiting chinks in the plant's armor: Evolution and emergence of geminiviruses. *Ann. Rev. Phytopathol.* 43:361–394.

Scholthof, K.-B.G., S. Adkins, H. Czosnek, P. Palukaitis, E. Jacquot, T. Hohn, B. Hohn et al. 2011. Top 10 plant viruses in molecular plant pathology. *Mol. Plant Pathol.* 12(9):938–954.

Schwartz, M., J. Chen, M. Janda, M. Sullivan, J. Boon, and P. Ahlquist. 2002. A positive-strand RNA virus replication complex parallels form and function of retrovirus capsids. *Mol. Cell.* 9:505–514.

Terhune, S.S., J. Schröer, and T. Shenk. 2004. RNAs are packaged into human cytomegalovirus virions in proportion to their intracellular concentration. *J. Virol.* 78(19):10390–10398.

Worobey, M. and E.C. Holmes. 1999. Evolutionary aspects of recombination in RNA viruses. *J. Gen. Virol.* 80:2535–2543.

chapter nineteen

Viral biotechnology
Production perspective

Kinza Waqar, Hafeez Ullah, and Alvina Gul

Contents

Abstract

Biotechnology deals with the manipulation of genetic material to achieve the desired outputs. Viruses have been popular agents used in biotechnological research and applications. Viruses are compact entities that have the ability to carry the genetic material to the cell and thus facilitate the safe transport and integration of genetic material to the target cell. Therefore viruses are used in various biotechnology applications such as production of vaccines, production of therapeutic protein, and improvement of crops with the help of plants viruses and also in cancer research. Bacteriophage (a special class of viruses) is used in various industrial biotechnological processes. Similarly viruses like particles (VLP) have interesting applications in research, gene therapy and vaccine production because of their non-infectious nature. Overall viruses are potent agents of research and applications of biotechnology.

Keywords: Biotechnology, Viruses, Oncolytic viruses, Virus like particles, Epitopes, Viral biotechnology.

Introduction

Viral biotechnology is the use of viruses and virus derived products in biotechnology applications. In dealing with any of the machinery, all the hardware depends on the blueprints and the strategies on which the machinery is developed and harmonized; similarly, living beings are programmed machines and fingerprints exist in the form of genetic material. Any change in the fingerprint will alter the morphology and physiology of the organism. In this aspect, we may say that viruses are the natural workshop for organismal repair. It is up to us how we use these viruses to harvest benefit. The use of viral entities for the benefit of human beings facilitates us to know the function of a specific genetic triplet or any combination of amino acids.

Viral biotechnology is an overlapping discipline of genomics, proteomics, virology, and immunology. In viral biotechnology, genomics deals with all the genetic material of an organism; whether it is a viral or nonviral entity. Proteomics is the protein expression in the source (virus) or the sink (nonvirus). Virology helps us to gather information on the viruses that may antagonize each other, which will lead to reduction or enhancement of an effect or defect. Immunology is the chemical analysis of viruses and their impact on the immunity of the organism, especially human beings.

This technology not only uses the virus and viral-like particles, but also helps to prepare and synthesize such chemicals by total synthesis pathways such that we may be able to form some of the genes and proteins in the laboratory and get them expressional and effective in the living system, and it is not necessary that the newly synthesized material is from either nucleic acid or protein category; it may also be a conjugate molecule. Thus, we can mimic some part of a virus or the whole virus.

Each gene is product specific. What are the products and effects of each gene? This can be understood by generating a minimal medium. Since each virus prepares only a short number of proteins, the efficacy of the minimal medium and targeting the specificity of particular genes is easy to analyze.

Eukaryotes have complex structures which enables them to survive in diverse chemical environment. Each outcome involves a series of chemical reactions. In such an environment, any change in any of the steps will not yield the final product. We are conversant with one-gene–one-protein hypothesis. Any genetic disorder whose genes exist at more than one loci can be easily explained through viral biotechnology.

In viral biotechnology, the use of viral particles, viral-like particles, and viral products has led to better and positive results directly or indirectly.

In the ever-changing environment of the biosphere, evolution is an inevitable process. Viral biotechnology provides us the possible varieties of a combination of genetic material, which enables us to deal with more diverse forms of a particular disease as well as a variety of diseases.

Structure of viruses

Viruses are small infectious parasites. These can grow in a natural environment and are also sometimes host specific. Viruses have two parts: core and cladding. The core is a single- or double-stranded polynucleotide chain (the nucleic acid, either deoxyribonucleic acid [DNA] or ribonucleic acid [RNA]), whereas the cladding is a polymeric protein, the capsid. Along with DNA, the RNA can also be double stranded in the viruses. Such a virus that contains the double-stranded RNA is called the retrovirus or the oncovirus. The nucleic acid may be a single molecule or it may have multiple units. The former is termed as the monopartite, whereas the latter is termed as the multipartite. On the basis of RNA, there are two structural categories (except single stranded and double stranded) of viruses: sense and antisense viruses. Those in which the RNA strand directly acts as the messenger RNA (mRNA) are termed as the sense or positive sense, whereas those in which a complementary strand from the viral RNA acts as the mRNA are termed as the antisense or the negative sense.

The viral genome is not a very long chain; therefore, it has limited capacity to synthesize a specific number of proteins. These proteins are very few in number and this is also a reason for the viruses to exist in a small number of morphological forms.

Unlike eukaryotes, viruses lack their own biosynthetic machinery, therefore, the multiplication of the virus is carried out inside the host cell, which qualifies the viruses as parasites.

The capsid and the core are collectively termed as the nucleocapsid. The capsid bears a characteristic number of protein monomers. These monomers are called the capsomers or the protomers. The classification of viruses is done mainly based on the number of these capsomers.

The viruses attach themselves to the host at the specific sites of the cell membrane for the safe inlet of the viral genome. This quality is evident in viruses that use only their genetic material to spread infections.

All the entities that form the structure of a virus are collectively termed as the virion.

From an external appearance, the viruses may be filamentous, helical, spherical, pleomorphic, or with edges and hadrons (Gelderblom, 1996).

Viruses as expression systems for protein expression

Protein synthesis requires the genetic material as an essential part. Each monomer of the polynucleotide chain of the nucleic acid contributes to one-third of the amino acid. Thus, three of the nucleotides in the form of a specific sequence contribute to a specific amino acid. A fragment of the nucleic acid (gene) results in the formation of a specific protein. In eukaryotes, there is a bulk of nucleic acid that is hard to specify for the specific type of protein whereas in viruses, the nucleotide chain is comparatively much smaller. In such a small sequence, it is easy to extract a specific preexisting combination for a specific protein.

A large amount of different types of proteins can be achieved in a short interval of time through viruses as compared with the eukaryotic expression system, particularly for the proteins that are intracellular in nature. The protein that can be synthesized through this method may also be posttransnationally modified. The viral expressions that are being availed for the protein expression in primates are adenovirus, lentivirus, and Semliki Forest virus. The adenovirus and the lentivirus are currently in use for better gene expression. The Semliki Forest virus is very important in the formation of membrane proteins.

A prominent feature of the viruses in protein expression is that of sense and antisense RNA-stranded viruses. This feature makes the translational process easy, fast, and much reliable.

Another type of virus that is important in gene expression is the baculovirus. These are rod-shaped viruses found mostly in insects. These are double-stranded DNA viruses that develop the polyhedrin protein around themselves. This protein is not essential for the propagation of the viruses; therefore, some heterologous proteins resulted in the presence of some promoters. The suitable host for the final products of the baculovirus is the *Escherichia coli*.

Protein expression is carried out through the following steps.

Nucleotide fragment (the gene)

Each virus exhibits a specific protein product that either results in pathogenicity or some other viral infections and functions. Since the viruses are parasites and express them only in the host cells, the required proteins for the viral coating is programmed through these fragments. The very same reason is exploited for the synthesis of different proteins.

Clone of the fragment

The proteins are synthesized from the mRNA. In viruses, this mRNA is termed as the sense strand. Sometimes, this strand is directly available as the genome of the virus and sometimes the complementary strand or the DNA strand is available to form the sense strand. The formation of the sense strand from the antisense or the double-stranded DNA helix is called cloning. In eukaryotes, only a specific required region is transcribed whereas a full clone has to be synthesized for the whole viral genome.

Expression constructs

After entering the targeted or the selected host and clone, the viral genome starts forming the required structural parts. These parts definitely include the proteins. If the whole steps are defined till these lines are under controlled and programmed supervision, then the protein thus obtained is truncated and tagged and its expression in the host is also being monitored.

Expression screening/optimization

If we are already conversant with the final product from the specific sequence of the viral genome, then we optimize the outcome by varying different factors.

Best construct

If we are not conversant with the final product, then we analyze the entire possible products and search for the best construct.

Transients/stable pool/clonal selection

The mammalian cells do have a specific antagonistic mechanism not only for physiological activity but also for the chemical one. All the viral products may not be stable in such an environment; so, only selected products are stable in this pool of chemicals. On the basis of a complete analysis and interaction of chemicals, a cell-based study is done for the particular virus and the particular host.

Scale-up and purification

Each chemical (protein and by-products) is then quantified and purified with respect to each cell type and function.

Although we confront the whole process of the protein expression from the viruses, to reach a specific protein product for the specific pharmaceutical use, we will have to prepare a specific medium. Moreover, we do not know all the viral genome products; maybe the toxicity produced from some of the products proves fatal to the host cell or the alteration in the chemical structure of the resulting protein is carried out before the final extraction.

To avoid toxicity and related problems the following measures are to be taken

One of the ways is the use of a tightly regulated promoter system that will result in a less number of possible by-products. A low dose of the plasmid is regulated so that the yield of the products may not harm to a greater extent before the compilation of the analysis. Specific type of cultures can be used to refer a semiminimal medium. Such media are termed as the batch-fed culture. Some of the reagents and chemicals also play a great role in maintaining the buffering ratio to provide the least toxicity. Since the viruses are host specific or cell specific, the growing of such viruses in the cell or medium which does not have any antagonistic behavior against the viral products will yield some better information about the medium to grow a targeted protein from that particular virus.

Use of plants in viral biotechnology

Another solution to express proteins and avoid toxicity is through the use of plant viruses. The plant viruses would verily cause the production of the by-products; however, the extraction of specific pharmaceutical proteins is much less arduous and less tedious. For the purpose, the viruses are categorized into two parts: "independent virus" and "minimal virus."

The independent viruses are just studies with respect to their activity. These are allowed to infect the host plant cells and then the activity is noted in the phloem cells where the product of these viruses is transferred. Minimal viruses are those viruses whose genomes are modified. This modification is made by knowing the complete code of the viral genome and then inserting some foreign genes in the genome so that the activity of the genome is well understood.

The study of protein expression through plant viruses is particularly important so that the plants are not affected in the manner as the animals are. Moreover, the cryoscopic extraction of protein from plants is far convenient than that from the animals (Saida et al., 2006).

Use of viral vectors

1. Viruses have been used for vaccine development
2. Viruses are used in the preparation of subunit vaccines
3. Viral vectors are used in gene therapy

Viral biology helps us to use the viruses in inducing immunity at the cellular and humoral level (Draper et al., 2010).

The viral genome has the capacity to mutate like how the human and other organisms evolute with the passage of time. Since the viruses are not technically living and since these need a biological system, diseases can be considered as the outcome of the parallel growth of viral and nonviral entities.

Different strategies may be adopted to develop a vaccine; these may be weekend, attenuated, killed, some subunits, or conjugated. Different types of viral vaccines and their uses for a particular disease are given hereunder.

Use of viral epitopes to boost immunity

Epitope is the locus on the surface of an antigen molecule at which an antibody attaches itself. It is also known as an antigenic determinant. This determinant is part of a large molecule and some chemical receptors detect its presence in the target or host cells. While using viruses as a vaccine, we observe that there are certain points on the virus that let the host's defense mechanism to activate. These points are some polypeptides or some glycan to which the antibodies from the host cells attach and thus activate the immunity of the cell or the organism. These points are called the epitopes. Although the attenuated viruses are weakened, sometimes they may cause disease due to some mutation in them. Therefore, using these attenuated viruses as a vaccine is risky. Since the epitopes control the activation of the defense system, instead of using the whole virus as a vaccine, only epitopes are used and this proves more effective than vaccines which use complete virus structures. The epitopes increase the number of antibodies, thus activating the defense mechanism, and no infectious or oncolytic viral particles are associated to the administration of the vaccine, as a result the immunity of the host boosts up. We observe two types of epitopes: T-cell epitopes and B-cell epitopes. The former is linked with the major histocompatibility complex (MHC), whereas the latter is linked with the antibodies or B cells. The T-cell epitopes are recognized by the linear amino acid sequence of the string whereas the B-cell epitopes are recognized by the three-dimensional (3D) structure of the epitope that is presented to the antibodies for interface (Chen et al., 2011).

The following are some of the diseases for which the epitopes have been either detected or synthesized through their recognition from the attachment to the viral surface.

Foot-and-mouth virus (Zhang et al., 2010): Mapping and identification of these epitopes are mainly done in animals such as mice (Herd et al., 2006) and pigs (Díaz et al., 2009).

The use of virus-like particles

Complex molecules that can multiply inside the living organism, like the viruses do, are called the virus-like particles (VLPs). However, the necessary condition for the VPLs is that these are void of nucleic acids.

If a nucleic acid is removed from a virus, it will also be a VPL, and researches made till now mainly surround these hollow viral proteins. These are extracted through different cell lines, for example, mammalian, yeast, or plant cell lines (Santi et al., 2006). A new VPL called "lipoparticle" has been synthesized. It is not derived from the viral source (Jones et al., 2008; Willis et al., 2008).

Since VPLs mainly involve the cover of viruses, it will contain epitopes as well. These epitopes can serve as the immunity booster for the target cells (Akahata et al., 2010). A VPL is a better vaccine against influenza than the attenuated viruses (Landry et al., 2010). As VPLs have tropism similar to the natural virus and show the same mechanism of uptake by the cells; they may also be used for gene transfer in genetic engineering (Petry et al., 2003).

The VPL "lipoparticle" is used for the detection of antibodies and to produce immunogens (Willis et al., 2008). It has been observed that the biological systems do bear a mechanism to inhibit VPLs being piled up (Chromy et al., 2003).

Oncolytic viruses and cancer clinical trials

Oncolytic viruses are those naturally occurring or genetically engineered attenuated viruses that have the ability to invade and destroy malignant and tumorous cells. This concept was introduced in the twentieth century (Kelly and Russell, 2007). In the early decades of the century, the focus was mainly on oncotropic viruses, which could easily get access to the inner environment of the cell but technically are unable to complete their life cycle inside the cell as a result of which are unable to induce cytotoxic effects, for example, canine parvovirus, canarypox virus, and baculovirus (Griffith et al., 2001; Mäkelä et al., 2006). In 1950, some naturally occurring viruses were passed through clinical trials to test for oncolytic behavior, but using these viruses could prove harmful to the cells as these viruses could cause other diseases to the cell (Southam et al., 1956; Huebner et al., 1956). It was in 1990 that the trials conducted to evaluate the use of viruses as anticancerous particles proved fruitful (Stanford et al., 2010; Quetglas et al., 2012).

As soon as an oncolytic virus gets activated in the malignant cells, it switches on the mechanism of the host cell and as a result the cells undergo lysis. This lysis produces membrane-bound vesicles, which exhibit the oncolytic virus. Thus, the virus evades the immune system of the organism and is safely transmitted to the other malignant cells (the neighboring cells) (Roulston et al., 1999; Hay and Kannourakis, 2002). Some of the viruses have also evolved their capsid in a protein so that they can escape from the cellular immune system of the host and let the viral activation move on (Johnson and Huber, 2002). To make the activity of an oncolytic virus more effective, specific targets are made. This targeting is done in various strategies. The strategies applied till now are transductional targeting (involves changes in the capsid or the envelope) (Muik et al., 2011; Ayala-Breton et al., 2012), transcriptional targeting (insertion and expression of some extra genes essential for the cell cycle) (Post and Van Meir, 2003; Passer et al., 2010), translational targeting (use of some essential translational viral genes) (Yang et al., 2009), posttranslational targeting (use of destabilization domains) (Banaszynski et al., 2008; Glass et al., 2009), oncogenetic targeting (attenuating the viruses by deleting some virulent genes) (Coffey et al., 1998; Stojdl et al., 2003), and microRNA targeting (to insert tissue-specific microRNA-binding elements to find an escape from productive infections) (Sugio et al., 2011) (Figure 19.1).

Oncolytic viruses due to their ability to active cell lysis in the cancerous cell cause elimination of the cancerous cells while they do not harm the healthy cells. The viruses replicate only in cancerous cells and on cell lysis they further bind to other available cancerous cells and thus cause reduction in the number of these cells.

The viruses that are used as an oncolytic virus are adenovirus (Cerullo et al., 2012), herpes simplex virus (Li and Zhang, 2010), and poxvirus (Perkus et al., 1995).

Adenoviruses are used against glioma (Chiocca et al., 2004), sarcoma (Galanis et al., 2005), head and neck cancer (Nemunaitis et al., 2001), pancreatic cancer (Hecht et al., 2003), prostate carcinoma (DeWeese et al., 2001), colorectal cancer (Reid et al., 2005), bladder carcinoma (Burke et al., 2012), ovarian cancer (Kimball et al., 2010), and multiple solid tumors (Nemunaitis et al., 2007). Attenuated herpes simplex viruses are used against hepatic metastasis (Fong et al., 2009), pancreatic carcinoma (Nakao et al., 2011), melanoma (Kaufman et al., 2010), head and neck cancer (Mace et al., 2008), breast carcinoma (Kimata

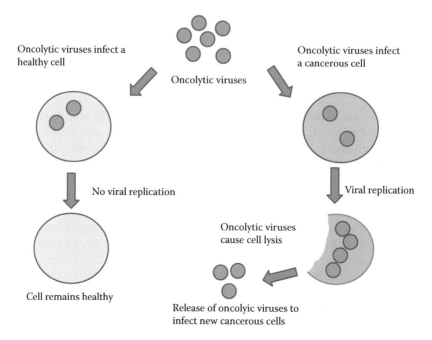

Figure 19.1 Mechanism of action of oncolytic viruses.

et al., 2006), glioma (Harrow et al., 2004), and other solid malignancies (Hu et al., 2006). Vaccinia virus is used against melanoma and hepatic cellular carcinoma (Heo et al., 2013); reovirus against melanoma, glioma, and other solid tumors (Vidal et al., 2008); parvovirus against glioblastoma multiforme (Geletneky et al., 2012); and Newcastle disease virus against glioma and other solid tumors (Freeman et al., 2006). In virotherapy against tumors and cancer, the main issue is safe delivery of the virus to the target place, although much has been done in this aspect but a lot more is required to improve the strategies (Tsuji et al., 2013).

Bacteriophage and phage therapy

The use of phage viruses against the bacterial diseases is called phage therapy. If this control is outside the organism (Zuzanna et al., 2012), only on bacterial colonies, then it is termed as biological control (Wilhite et al., 2001). In most of the bacterial diseases, the bacteria do not respond to the antibiotic as they develop resistance against the antibiotic. This resistance is developed due to the formation of polysaccharide cysts around the bacterial cell wall. This proves to be a hindrance for the treatment of the disease. Thus, a method was required to deal with this problem. Since all the viruses are host specific, the phages will also be specific for the specific strain and if we come to know of a bacteriophage for a particular disease-causing bacterium, we will be able to manage the cure by the use of that phage virus on a victim. Bacteriophages were discovered in 1915 by Frederick Twort and the discovery was confirmed in 1917 by Felix d'Herelle (Shasha et al., 2004). Soon after the discovery, the phage therapy was introduced around 1920 (Thiel, 2004; Parfitt, 2005). Although there were chances that the bacteria could develop resistance against the viruses, viruses can also develop to harm and thus can still infect the developed bacterial walls. An extra benefit for phage therapy is not only that the phage causes the lysis of disease-causing bacterium, but also the viral envelope proteins may act as an antibiotic (Courchesne et al., 2009; O'Flaherty et al., 2009). The common problem is that if any disease-causing bacterium is not known with respect to any of the phage, a mixture of the phage will have to be prepared so that the lytic cycle may infect all the possible infections. Still the evolution of both the species, that is, bacteria as well as the phage inside the organism may result in further complexities (Abedon et al., 2011).

Future prospects

All the pure and applied sciences are meant for the benefit of man. Till now, more than 400 proteins have been discovered. The use of their interactional data and genomic expression will help us to combat not only the genetic diseases but also other chemical and pathological diseases.

Conclusion

The use of viruses in biotechnology has proved its vitality not only in controlling the diseases by an antagonistic pathway, but also reduces the threat of an epidemic through shots or vaccination. The same strain of the virus that is causing some disease may be helpful in controlling the other in a controlled manner. Not only the whole virus may help induce immunity but a single protein derived from the viral particle may also be helpful in boosting immunity. Through the use of this prokaryote, the eukaryotes can be made to utilize for the betterment of human health and environmental conditions.

References

Abedon ST, Kuhl SJ, Blasdel BG, and Kutter EM. 2011. Phage treatment of human infections. Phage treatment of human infections. *Bacteriophage.* 1(2): 66–85.

Akahata W, Yang ZY, Andersen H, Sun S, Holdway HA, Kong WP et al. 2010. A VLP vaccine for epidemic Chikungunya virus protects non-human primates against infection. *Nat Med.* 16(3): 334–8. doi: 10.1038/nm.2105. PMC 2834826. PMID 20111039.

Ayala-Breton C, Barber GN, Russell SJ, and Peng KW. 2012. Retargeting vesicular stomatitis virus using measles virus envelope glycoproteins. *Hum Gene Ther.* 23:484–91. doi: 10.1089/hum.2011.146.

Banaszynski LA, Sellmyer MA, Contag CH, Wandless TJ, and Thorne SH. 2008. Chemical control of protein stability and function in living mice. *Nat Med.* 14:1123–7. doi: 10.1038/nm.1754.

Burke JM, Lamm DL, Meng MV, Nemunaitis JJ, Stephenson JJ, Arseneau JC et al. 2012. A first in human phase 1 study of CG0070, a GM-CSF expressing oncolytic adenovirus, for the treatment of nonmuscle invasive bladder cancer. *J Urol.* 188:2391–7. doi: 10.1016/j.juro.2012.07.097.

Cerullo V, Vähä-Koskela M, and Hemminki A. 2012. Oncolytic adenoviruses: A potent form of tumor immunovirotherapy. *Oncoimmunology.* 1:979–81. doi: 10.4161/onci.20172.

Chen P, Rayner S, and Hu K-H. 2011. Advances of bioinformatics tools applied in virus epitopes prediction. *Virol Sin.* 26(1): 1–7.

Chiocca EA, Abbed KM, Tatter S, Louis DN, Hochberg FH, Barker F et al. 2004. A phase I open-label, dose-escalation, multi-institutional trial of injection with an E1B-attenuated adenovirus, ONYX-015, into the peritumoral region of recurrent malignant gliomas, in the adjuvant setting. *Mol Ther.* 10:958–66. doi: 10.1016/j.ymthe.2004.07.021.

Chromy LR, Pipas JM, and Garcea RL. 2003. Chaperone-mediated *in vitro* assembly of polyomavirus capsids. *Proc Natl Acad Sci USA.* 100(18): 10477–82. doi: 10.1073/pnas.1832245100. PMC 193586. PMID 12928495.

Coffey MC, Strong JE, Forsyth PA, and Lee PW. 1998. Reovirus therapy of tumors with activated Ras pathway. *Science.* 282:1332–4. doi: 10.1126/science.282.5392.1332.

Courchesne NM, Parisien A, and Lan CQ. 2009. Production and application of bacteriophage and bacteriophage-encoded lysins. *Recent Pat. Biotechnol.* 3:37–45.

Díaz I, Pujols J, Ganges L et al. 2009. *In silico* prediction and *ex vivo* evaluation of potential T-cell epitopes in glycoproteins 4 and 5 and nucleocapsid protein of genotype-I (European) of porcine reproductive and respiratory syndrome virus. *Vaccine.* 27(41): 5603–11.

DeWeese TL, van der Poel H, Li S, Mikhak B, Drew R, Goemann M et al. 2001. A phase I trial of CV706, a replication-competent, PSA selective oncolytic adenovirus, for the treatment of locally recurrent prostate cancer following radiation therapy. *Cancer Res.* 61:7464–72.

Draper SJ and Heeney JL. 2010. Viruses as vaccines vectors for infectious diseases and cancer. *Nat Rev Microbiol.* 8:62–73. doi: 10.1038/nrmicro2240.

Fong Y, Kim T, Bhargava A, Schwartz L, Brown K, Brody L et al. 2009. A herpes oncolytic virus can be delivered via the vasculature to produce biologic changes in human colorectal cancer. *Mol Ther.* 17:389–94. doi: 10.1038/mt.2008.240.

Freeman AI, Zakay-Rones Z, Gomori JM, Linetsky E, Rasooly L, Greenbaum E et al. 2006. Phase I/II trial of intravenous NDV-HUJ oncolytic virus in recurrent glioblastoma multiforme. *Mol Ther.* 13:221–8. doi: 10.1016/j.ymthe.2005.08.016.

Galanis E, Okuno SH, Nascimento AG, Lewis BD, Lee RA, Oliveira AM et al. 2005. Phase I–II trial of ONYX-015 in combination with MAP chemotherapy in patients with advanced sarcomas. *Gene Ther.* 12:437–45. doi: 10.1038/sj.gt.3302436.

Glass M, Busche A, Wagner K, Messerle M, and Borst EM. 2009. Conditional and reversible disruption of essential herpes virus proteins. *Nat Methods.* 6:577–9. doi: 10.1038/nmeth.1346.

Gelderblom HR. 1996. *Medical Microbiology*, 4th edition, Baron S, ed. University of Texas Medical Branch at Galveston, Galveston, TX. ISBN-10: 0-9631172-1-1.

Geletneky K, Huesing J, Rommelaere J, Schlehofer JR, Leuchs B, Dahm M et al. 2012. Phase I/IIa study of intratumoral/intracerebral or intravenous/intracerebral administration of parvovirus H-1 (ParvOryx) in patients with progressive primary or recurrent glioblastoma multiforme: ParvOryx01 protocol. *BMC Cancer.* 12:99. doi: 10.1186/1471-2407-12-99.

Griffith TS, Kawakita M, Tian J, Ritchey J, Tartaglia J, Sehgal I et al. 2001. Inhibition of murine prostate tumor growth and activation of immunoregulatory cells with recombinant canarypox viruses. *J Natl Cancer Inst.* 93:998–1007. doi: 10.1093/jnci/93.13.998.

Harrow S, Papanastassiou V, Harland J, Mabbs R, Petty R, Fraser M et al. 2004. HSV1716 injection into the brain adjacent to tumour following surgical resection of high-grade glioma: Safety data and long-term survival. *Gene Ther.* 11:1648–58. doi: 10.1038/sj.gt.3302289.

Hay S and Kannourakis G. 2002. A time to kill: Viral manipulation of the cell death program. *J Gen Virol.* 83:1547–64.

Hecht JR, Bedford R, Abbruzzese JL, Lahoti S, Reid TR, Soetikno RM et al. 2003. A phase I/II trial of intratumoral endoscopic ultrasound injection of ONYX-015 with intravenous gemcitabine in unresectable pancreatic carcinoma. *Clin Cancer Res.* 9:555–61.

Herd KA, Mahalingam S, Mackay IM et al. 2006. Cytotoxic T-lymphocyte epitope vaccination protects against human metapneumovirus infection and disease in mice. *J Virol.* 80(4): 2034–44.

Heo J, Reid T, Ruo L, Breitbach CJ, Rose S, Bloomston M et al. 2013. Randomized dose-finding clinical trial of oncolytic immunotherapeutic vaccinia JX-594 in liver cancer. *Nat Med.* 19:329–36. doi: 10.1038/nm.3089.

Hu JC, Coffin RS, Davis CJ, Graham NJ, Groves N, Guest PJ et al. 2006. A phase I study of OncoVEXGM-CSF, a second-generation oncolytic herpes simplex virus expressing granulocyte macrophage colony-stimulating factor. *Clin Cancer Res.* 12:6737–47. doi: 10.1158/1078-0432.CCR-06-0759.

Huebner RJ, Rowe WP, Schatten WE, Smith RR, and Thomas LB. 1956. Studies on the use of viruses in the treatment of carcinoma of the cervix. *Cancer.* 9:1211–8. doi: 10.1002/1097-0142(195611/12)9:6<1211::AID-CNCR2820090624>3.0.CO;2-7.

Johnson DC and Huber MT. 2002. Directed egress of animal viruses promotes cell to cell spread. *J Virol.* 76:1–8.

Jones JW, Greene TA, Grygon CA, Doranz BJ, and Brown MP. 2008. Cell-free assay of G-protein-coupled receptors using fluorescence polarization. *J Biomol Screen.* 13(5): 424–9. doi: 10.1177/1087057108318332. PMID 18567842.

Kaufman HL, Kim DW, DeRaffele G, Mitcham J, Coffin RS and Kim-Schulze S. 2010. Local and distant immunity induced by intralesional vaccination with an oncolytic herpes virus encoding GM-CSF in patients with stage IIIc and IV melanoma. *Ann Surg Oncol.* 17:718–30. doi: 10.1245/s10434-009-0809-6.

Kelly E, Russell SJ. 2007. History of oncolytic viruses: Genesis to genetic engineering. *Mol Ther.* 15:651–9.

Kimball KJ, Preuss MA, Barnes MN, Wang M, Siegal GP, Wan W et al. 2010. A phase I study of a tropism-modified conditionally replicative adenovirus for recurrent malignant gynecologic diseases. *Clin Cancer Res.* 16:5277–87. doi: 10.1158/1078-0432.CCR-10-0791.

Kimata H, Imai T, Kikumori T, Teshigahara O, Nagasaka T, Goshima F et al. 2006. Pilot study of oncolytic viral therapy using mutant herpes simplex virus (HF10) against recurrent metastatic breast cancer. *Ann Surg Oncol.* 13:1078–84. doi: 10.1245/ASO.2006.08.035.

Landry N, Ward BJ, Trépanier S, Montomoli E, Dargis M, Lapini G, and Vézina LP. 2010. Preclinical and clinical development of plant-made virus-like particle vaccine against avian H5N1 influenza. *PLoS ONE.* 5(12): e15559. doi: 10.1371/journal.pone.0015559. PMC 3008737. PMID 21203523.

Li H and Zhang X. 2010. Oncolytic HSV as a vector in cancer immunotherapy. *Methods Mol Biol.* 651:279–90. doi: 10.1007/978-1-60761-786-0_16.

Mace AT, Ganly I, Soutar DS, and Brown SM. 2008. Potential for efficacy of the oncolytic herpes simplex virus 1716 in patients with oral squamous cell carcinoma. *Head Neck.* 30:1045–51. doi: 10.1002/hed.20840.

Mäkelä AR, Matilainen H, White DJ, Ruoslahti E, and Oker-Blom C. 2006. Enhanced baculovirus-mediated transduction of human cancer cells by tumor-homing peptides. *J Virol.* 80:6603–11. doi: 10.1128/JVI.00528-06.

Muik A, Kneiske I, Werbizki M, Wilflingseder D, Giroglou T, Ebert O et al. 2011. Pseudotyping vesicular stomatitis virus with lymphocytic choriomeningitis virus glycoproteins enhances infectivity for glioma cells and minimizes neurotropism. *J Virol.* 85:5679–84. doi: 10.1128/JVI.02511-10.

Nakao A, Kasuya H, Sahin TT, Nomura N, Kanzaki A, Misawa M et al. 2011. A phase I dose-escalation clinical trial of intraoperative direct intratumoral injection of HF10 oncolytic virus in non-resectable patients with advanced pancreatic cancer. *Cancer Gene Ther.* 18:167–75. doi: 10.1038/cgt.2010.65.

Nemunaitis J, Khuri F, Ganly I, Arseneau J, Posner M, Vokes E et al. 2001. Phase II trial of intratumoral administration of ONYX-015, a replication-selective adenovirus, in patients with refractory head and neck cancer. *J Clin Oncol.* 19:289–98.

Nemunaitis J, Senzer N, Sarmiento S, Zhang YA, Arzaga R, Sands B et al. 2007. A phase I trial of intravenous infusion of ONYX-015 and Enbrel in solid tumor patients. *Cancer Gene Ther.* 14:885–93. doi: 10.1038/sj.cgt.7701080.

O'Flaherty S, Ross RP, and Coffey A. 2009. Bacteriophage and their lysins for elimination of infectious bacteria. *FEMS Microbiol. Rev.* 33:801–19.

Parfitt T. 2005. Georgia: An unlikely stronghold for bacteriophage therapy. *Lancet.* 365(9478): 2166–7. doi: 10.1016/S0140-6736(05)66759-1. PMID 15986542.

Passer BJ, Cheema T, Zhou B, Wakimoto H, Zaupa C, Razmjoo M et al. 2010. Identification of the ENT1 antagonists dipyridamole and dilazep as amplifiers of oncolytic herpes simplex virus-1 replication. *Cancer Res.* 70:3890–5. doi: 10.1158/0008-5472.CAN-10-0155.

Perkus ME, Taylor J, Tartaglia J, Pincus S, Kauffman EB, Tine JA, and Paoletti E. 1995. Live attenuated vaccinia and other poxviruses as delivery systems: Public health issues. *Ann N Y Acad Sci.* 754:222–33. doi: 10.1111/j.1749-6632.1995.tb44454.x.

Petry H, Goldmann C, Ast O, and Lüke W. 2003. Use of virus like particle for gene transfer. *Curr Opin Mol Ther.* (5):524–8.

Post DE and Van Meir EG. 2003. A novel hypoxia-inducible factor (HIF) activated oncolytic adenovirus for cancer therapy. *Oncogene.* 22:2065–72. doi: 10.1038/sj.onc.1206464.

Quetglas JI, John LB, Kershaw MH, Alvarez-Vallina L, Melero I, Darcy PK et al. 2012. Virotherapy, gene transfer and immunostimulatory monoclonal antibodies. *Oncoimmunology.* 1:1344–54.

Reid TR, Freeman S, Post L, McCormick F, and Sze DY. 2005. Effects of Onyx-015 among metastatic colorectal cancer patients that have failed prior treatment with 5-FU/leucovorin. *Cancer Gene Ther.* 12:673–81. doi: 10.1038/sj.cgt.7700819.

Roulston A, Marcellus RC, and Branton PE. 1999. Viruses and apoptosis. *Annu Rev Microbiol.* 53:1–60.

Saida F, Uzan M, Odaert B, and Bontems F. 2006. Expression of highly toxic genes in *E. coli*: Special strategies and genetic tools. *Curr Protein Pept Sci.* 7:47–56.

Santi L, Huang Z, and Mason H. 2006. Virus like particles production in green plants. *Methods.* 40(1): 66–76. doi: 10.1016/j.ymeth.2006.05.020. PMC 2677071. PMID 16997715.

Shasha SM, Sharon N, and Inbar M. 2004. Bacteriophages as antibacterial agents. *Harefuah.* (in Hebrew) 143(2): 121–5, 166. PMID 15143702.

Stanford MM, Bell JC, and Vähä-Koskela MJ. 2010. Novel oncolytic viruses: Riding high on the next wave? *Cytokine Growth Factor Rev.* 21:177–83. doi: 10.1016/j.cytogfr.2010.02.012.

Stojdl DF, Lichty BD, TenOever BR, Paterson JM, Power AT, Knowles S et al. 2003. VSV strains with defects in their ability to shutdown innate immunity are potent systemic anti-cancer agents. *Cancer Cell.* 4:263–75. doi: 10.1016/S1535-6108(03)00241-1.

Southam CM, Hilleman MR, and Werner JH. 1956. Pathogenicity and oncolytic capacity of RI virus strain RI-67 in man. *J Lab Clin Med.* 47:573–82.

Sugio K, Sakurai F, Katayama K, Tashiro K, Matsui H, Kawabata K et al. 2011. Enhanced safety profiles of the telomerase-specific replication-competent adenovirus by incorporation of normal cell-specific microRNA-targeted sequences. *Clin Cancer Res.* 17:2807–18. doi: 10.1158/1078-0432. CCR-10-2008.

Thiel K. 2004. Old dogma, new tricks—21st century phage therapy. *Nat Biotechnol.* 22(1): 31–6. doi: 10.1038/nbt0104-31. PMID 14704699. Retrieved 2007-12-15.

Tsuji T, Nakamori M, Iwahashi M, Nakamura M, Ojima T, Iida T et al. 2013. An armed oncolytic herpes simplex virus expressing thrombospondin-1 has an enhanced *in vivo* antitumor effect against human gastric cancer. *Int J Cancer.* 132:485–94. doi: 10.1002/ijc.27681.

Willis S, Davidoff C, Schilling J, Wanless A, Doranz BJ, and Rucker J. 2008. Use of virus-like particles as quantitative probes of membrane protein interactions. *Biochemistry.* 47(27): 6988–90. doi: 10.1021/bi800540b. PMC 2741162. PMID 18553929.

Wilhite SE, Lumsden RD, and Straney DC. 2001. *Appl Environ Microbiol.* 67(11):5055–62. doi: 10.1128/ AEM.67.11.5055-5062.2001. PMCID: PMC93271.

Vidal L, Pandha HS, Yap TA, White CL, Twigger K, Vile RG et al. 2008. A phase I study of intravenous oncolyticreovirus type 3 Dearing in patients with advanced cancer. *Clin Cancer Res.* 14:7127–37. doi: 10.1158/1078-0432.CCR-08-0524.

Yang X, Chen E, Jiang H, Muszynski K, Harris RD, Giardina SL et al. 2009. Evaluation of IRES-mediated, cell-type-specific cytotoxicity of poliovirus using a colorimetric cell proliferation assay. *J Virol Methods.* 155:44–54. doi: 10.1016/j.jviromet.2008.09.020.

Zhang ZW, Zhang YG, Wang YL et al. 2010. Screening and identification of B cell epitopes of structural proteins of foot-and-mouth disease virus serotype Asia1. *Vet Microbiol.* 140(1–2): 25–33.

Zuzanna D-K, Grażyna M-S. 2012. Learning from bacteriophages—Advantages and limitations of phage and phage-encoded protein applications. *Curr Protein Pept Sci.* 13(8): 699–722. Published online December 2012. doi: 10.2174/138920312804871193. PMCID: PMC3594737.

chapter twenty

Cell-free biosystems

Manju Sharma and SM Paul Khurana

Contents

Abstract

Raising concerns to save our environment against climate change and to proceed towards the safe future for self and generations is the utmost motivation to look for technologies which are sustainable and cost effective. The world is in dire need of technology which could convert renewable substrates to a variety of products, viz. bioenergy (like alcohols, hydrogen, jet fuels) and biochemicals. Cost is major concern for these processes. Since long, microbial fermentation has been used as basic process to develop various products. Now, the development of cell-free biosystems for biomanufacturing (CFB2)—complex, cell-free systems that catalyze the conversion of renewable substrates to a variety of products—are promising choice over fermentation. As compared to whole cell entity, cell free biosystems are comprised of synthetic enzymatic pathways with a number of advantages, such as enhanced product yield, easy process control and regulation, speedy reaction rate, broad reaction conditions, tolerance toward toxic compounds and products. In this chapter, we have discussed the current status of cell free biosystems which are emerging as the most promising tool in present scenario.

Keywords: Biofuels, Cell free pathways, Bioenergy, *In vitro* synthetic biology, Enzymatic fuel cells.

Introduction

In this modern era, everyone is looking for greener processes that are safe, eco-friendly, and cost-efficient for the production of sustainably derived biochemicals and biofuels (Figure 20.1). The currently available methods for biotransformation are mainly based on genetic modification of microbes. But elucidation of complex metabolic pathways in these living organisms is needed to obtain desired products. Sometimes, using microbes

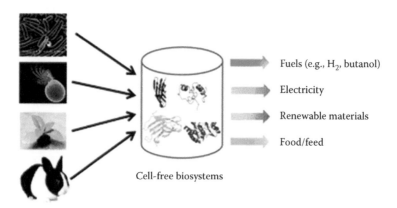

Fuels (e.g., H_2, butanol)

Electricity

Renewable materials

Food/feed

Cell-free biosystems

Figure 20.1 Living system capable to work as cell-free biosystem to convert important produce. (Adapted from Rollin JA, Tsz Kin Tamb, and Y-H Percival Zhang. 2013. *Green Chem.* 1708–1719.)

as system for various purposes can turn the different compound produced to be fatal to them. The variation in the processes of engineering and cellular objectives is difficult to cover and also poses significant scaling challenges. Furthermore, the ability to engineer biological systems up to expectation is limited. Having novel low-cost biomanufacturing podium cell-free biosystem is now a suitable option as compared to microbe-based fermentation used since ages. The accessibility without membranes and easy-to-control quality makes cell-free system foremost advantageous, resulting in higher product yields and quicker reaction rates with lesser interference of toxic compounds.

Since ages, the production of fermented food and beverages has been credited to natural microbial systems. A change began in the 1850s, followed by the use of live yeast cells for fermentation. This discovery by Louis Pasteur was the moment of inception of modern biotechnology. Moritz Traube's landmark work on acknowledging proteins as biocatalysts further led to the discovery of the first cell-free ethanol fermentation using yeast extract. The German chemist Buchner was awarded the Nobel Prize for enabling the biochemistry of fermentation study *in vitro* later in 1907. The research proceeded in this direction with additional fermentation products with the most notable being the production of acetone by *Clostridium acetobutylicum* during World War I, the production of penicillin for antibacterial use in World War II, and production of amino acids in the 1960s.

In the mid-twentieth century, the use of enzymes such as tannase, amylase, protease, and specifically glucose isomerase (GI) increased. Of these enzymes, GI is most extensively used in industries to prepare high-fructose corn syrup. With the advent of immobilization of enzymes, the efficiency of conversion enhanced several times since 1916 up until 1990 and is moving ahead with a bang.

In the beginning, at a time, single enzyme was immobilized to use in reaction. Later in the 1970s with progression more complex systems were observed with cofactor regeneration, living cells, and two enzymes. Consequently, a weight-based total turnover number (TTN_W) of more than 106 kg of product was obtained per kg of enzyme utilization, for more than 2 years before substitution (Zhang et al. 2010). Simultaneously, other enzyme technologies such as extensively used polymerase chain reaction, made possible by the finding of Taq polymerase, debittering of fruit juices, biodiesel production, interesterification of food fats and oils, pectin hydrolysis, and the creation of dustless detergent enzymes using techniques such as encapsulation were reviewed by Robert et al. (2013).

Based on biocatalysts used, biotransformation can be classified into microbial versus cell-free systems (Figure 20.2). The cell-free biosystems could become an unruly technology to microbial fermentation (MiF), particularly to develop high-impact low-value biocommodities, cell extracts, or purified enzymes. The cell-free biosystem is an *in vitro* system capable of using more than three catalytic components in a single reaction vessel, which differs from one, two, or three enzyme; biocatalysis can be treated as a straightforward cell-free system for biomanufacturing (CBF2) with a multienzymatic approach for industrial production (Guterl et al. 2012; You and Zhang 2012). Further, this can be divided into four classes, depending on an increasing order in complexity of reactions and biocatalysts: (1) single enzyme, (2) multienzyme in one pot, (3) cell-free fermentation, and (4) whole-cell fermentation.

Biotransformation can also be classified based on product price (Figure 20.3). The array of high-value products include taxol, protein drugs, antibodies, chiral compounds, antibiotics, etc.; their price varies from thousands to billions of dollars per kilogram. On the contrary, ethanol, hydrogen, lactic acid, and many others are biocommodities with low

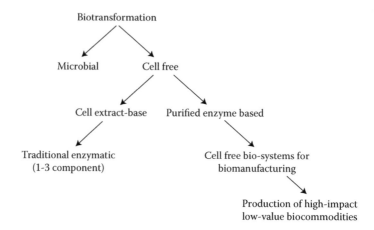

Figure 20.2 Classification of biotransformation based on biocatalyst. (Adapted from Rollin JA, Tsz Kin Tamb, and Y-H Percival Zhang. 2013. *Green Chem.* 1708–1719.)

selling price that do not exceed beyond $1 to several dollars per kilogram (Lynd et al. 1999; Zhang and Lynd 2008).

Sometimes, it is difficult to operate complicated biotransformations by implementing a single enzyme or a multienzyme mixture; therefore, whole-cell biocatalysts with or without genetic modifications could be used. For example, alcohol fermentation and cheese production mediated by wild-type microorganisms have been used for thousands of years. With the beginning of modern biotechnology, entire cells have been modified to produce numerous high-value products, from low-value products, such as butanol, fatty acid esters, ethanol, hydrogen, electricity, etc., to products such as enzymes, vitamins, antibiotics, vaccines, antibodies, taxol, artemisinin, glycoproteins, and lycopene.

Need for cell-free system

Progression of cell-free system has played an important role in understanding fundamental biology since 1897. For instance, the discovery of the genetic code by Nirenberg, which led to him receiving the Nobel Prize in 1968, disclosed how protein synthesis takes place in a cell. Since then other fundamental discoveries like understanding of eukaryotic

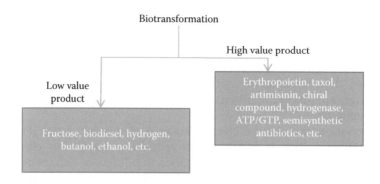

Figure 20.3 Product price-based classification of biotransformation.

translation has been unravel by cell-free system. The cell-free system has provided unlimited designing opportunities and understanding to control *in vivo* systems, which has led to the rapid development of engineering foundations for cell-free systems. The efforts have paved the way for the development of coactivated catalytic, programmed circuits, spatially organized pathways ensembles rational optimization of synthetic multienzyme pathways, and scalability from the microliter to the 100-liter level. Nowadays, cell-free systems have the answer to how nature's designs work. This also enables biosynthetic routes to sustainable fuels, concordant materials, and novel chemicals.

Cell-free biosystem biomanufacturing

The ability to self-replicate with continuous production of biocatalyst makes a microbial system different from the cell-free biosystems. Therefore, many strategies such as the use of heat-stable enzyme, immobilization, and protein engineering could be used to extend the working life of cell-free biosystems biomanufacturing (CFB2). Once the stability is achieved in enzyme production, the costs of biocatalyst will go down automatically. In addition to biocatalyst cost, several other factors such as pathway complexity (including ATP and cofactor requirements), relative importance of the reaction rate, and yield in the reaction along with maturity of current production technology matters a lot for the manufacture of a given biocommodity. Cell extracts or purified enzymes are the base for cell-free biosystems. Altogether it can utilize one to three enzymes, which makes it different from the traditional enzymatic methods. The cell-free biosystems are defined as *in vitro* biosystems of biomanufacturing comprising more than three enzymes in one vessel to facilitate the industrial production with the following advantages.

Quick reaction rates

In CFB2, the absence of a cellular membrane by and large permits more rapid reaction rates over microbial systems. The enzymatic fuel cells (EFCs) are more powerful in comparison with microbial fuel cells as they do not have cellular membranes to limit the mass transfer and high enzyme loading per volume or area. The reaction solution is therefore free of a large proportion of redundant biomolecules as CFB2 do not allow their accumulation (Osman et al. 2011).

Tolerance of toxic compounds and products

CFB2 systems are generally more tolerant to toxic compounds and products than microorganism. The absence of cellular membrane in these systems makes CFB2 better to tolerate alcohols with a carbon number greater than two. An innovative cell-free approach has been devised artificially by minimizing glycolytic reaction cascade at the single coenzyme level for converting glucose into ethanol and isobutanol. The solvent tolerance test has proved that the production of butanol using cell-free system is not affected in the presence of isobutonal up to 4% (v/v), whereas a little isobutanol concentration (1% v/v) hinders the production of microbial butanol due to membrane disintegration (Guterl et al. 2012).

Catalysis of nonnatural reactions

CFB2 system has a potential to use intracellular and extracellular enzymes to implement biotransformation, which is very difficult to pursue with the help of chemical catalyses or living microbes. For example, it is possible to convert β-1,4-glucosidic bond-linked cellulose to α-1,4-glucosidic bond-linked starch in one pot with the intervention of intracellular and extracellular enzymes (You et al. 2013). The cellulose to starch conversion could be achieved by the following enzymes, that is,

1. Cellulase cocktail—Required to convert cellulose into cellobiose by hydrolyzing the substrate partially
2. Cellobiose phosphorylase—Changes cellobiose into glucose-1-phosphate and glucose
3. Alpha-glucan phosphorylase (from potato)—Converts glucose-1-phosphate into starch

Maximize product yield

Operation of undesired chemical pathways along with significant desired pathway diminishes the product yield. Eliminating the synthesis/reproduction of cell mass, unwanted side pathways, as well as redesigning unnatural pathways to minimize or prevent the formation of by-products are the key factors to obtain a high yield with the intervention of CFB2.

Reaction equilibrium shift

The pathway energetics can move the equilibrium intermediates to the final product side by applying multistep biocatalytic transformations. Reactions having isoenergetic last step, causing the reaction to have dominated equilibrium constant of two products, but by changing this step to exergonic, the process could easily be shifted to the desired product.

Applications of CFB2

The system has three distinct applications:

- Cell-free protein synthesis (CFPS) makes the production of high-value proteins and antibodies easy.
- *In vitro* synthetic biology (ivSB) to produce priced drugs and fine chemicals.
- Synthetic enzymatic pathway biotransformation (SEPB) facilitates the production of low-value biocommodities in bulk, namely, amylose, alcohols, organic acids, jet fuels, hydrogen, etc.

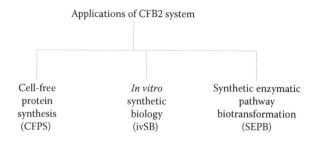

Cell-free protein synthesis

The protein expression system developed using living cells have their own limitations:

- Several expressed proteins are difficult to dissolve and aggregate in inclusion bodies.
- The presence of intercellular proteases of the host cells may digest the proteins.
- A number of proteins are difficult to be produced in living cells because of their toxicity.
- The amino acid metabolic system in the host cells can cause isotope dilution and diffusion for amino acids with selective stable isotope labeling.
- Sometimes, it seems difficult to grow organisms in deuterium-labeled media.
- Prokaryote cannot glycosylate proteins.

CFPS harnesses the synthetic power of biology, which includes

- Programming of ribosomal translational machinery of the cell to create macromolecular products
- Use of polymerase chain reaction (PCR), which are efficient cellular replication machinery to create a DNA amplifier
- Establishing a transformative technology of CFPS system has extensive applications in protein engineering, postgenomic research, and biopharmaceutical development (Hodgman and Jewett 2012; Kai et al. 2012; Swartz 2012).

Nirenberg and Matthaei in 1961 have demonstrated *in vitro* translation of protein using cell extracts, presently being used as an important tool in molecular biology. A CFPS continuous-flow system was developed using ultrafiltration followed by dialysis and inclusion of extract with optimization of the reaction conditions (Spirin et al. 1988). In addition, they succeeded in developing a site-directed stable isotope labeling method for a protein by using a CFPS system.

The PURE (protein synthesis using recombinant elements) is a recent approach to the CFPS, based on modular reconstitution of the translational machinery of the cell from affinity-purified protein components (Wang et al. 2012), including initiation factors (IF1, IF2, IF3), elongation factors (EF-Tu, EF-Ts, EF-G), ribosome recycling factors, release factors (RF1, RF2, RF3), 20 aminoacyl-tRNA synthetases, methionyl tRNA formyltransferase, and pyrophosphatase.

The PURE system has been commercialized (e.g., PURESYSTEM, Cosmo Bio, Tokyo, Japan; PURExpress, New England Biolabs, Beverly, MA, USA), and is available for laboratory research applications.

The following are the advantages of PURE systems:

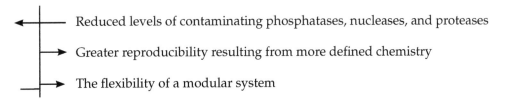

Reduced levels of contaminating phosphatases, nucleases, and proteases

Greater reproducibility resulting from more defined chemistry

The flexibility of a modular system

Being modular, the PURE system supports a variety of modifications for specialized applications as follows:

Printing protein arrays

CFPS is an important link between the established technology of DNA microarrays and protein arrays, allowing multiplexed, robotic processing to be used to print protein arrays from gene chips. Utilization of protein microarrays as a promising platform for high-throughput screening in postgenomic biomedical research allows the functional analysis of complete proteomes and provide a tool for vaccine development and personalized medicine (Berrade et al. 2011).

Ribosome display

For analyzing the protein functionality, ribosome analysis can be used as a platform for protein engineering, which relies on the stability of the ribosomal translation complex (comprising ribosome, mRNA, and nascent polypeptide) in the absence of release factors. Having a rational choice of omitting release factors and capability to accommodate incorporation of nonnatural amino acids using PURE cell-free system is well suited to this application (Watts and Forster 2012).

Isotopic labeling

The expression of proteins in a living cell or cell extract is always complicated by metabolic processes that dilute the isotope or scramble labeling patterns due to molecular transformations even after applying strategies to suppress side reactions. Labeling is very much important for biological nuclear magnetic resonance (NMR) and tracer experiments. The severely edited metabolic map in PURE CFPS may be expected to eliminate many of these problems, because the interfering enzymes and metabolites are not present.

Incorporation of nonnatural amino acids

Having advantage of expanding genetic code is perhaps the most significant development in ribosomal protein synthesis that adds new dimensions to protein engineering (Liu and Schultz 2010). The approach is conceptually simple; it needs a bioorthogonal cognate pair of transfer ribonucleic acid (tRNA) and aminoacyl-tRNA synthetase to suppress nonsense or frameshift mutations, putting in nonnatural amino acids into the growing polypeptide and thereby creating a chemical toolbox for protein engineering. By systematic evolution of alike tRNA/aminoacyl-tRNA sythetases, more than 50 distinct nonnatural amino acids have been incorporated into proteins in this way, addition of fluorescent tags or reactive groups contribute novel chemistry or functionality in the protein product (De Graaf et al. 2009; Hartman et al. 2007). Bioorthogonal chemistry has been successfully employed in protein functional studies to visualize protein expression, measure protein activity, identify protein interaction partners, track protein localization, and study turnover of proteins in living systems (Lim and Lin 2010).

In vitro synthetic biology

Contrary to *in vivo* synthetic biology, cell-free synthetic biology does not need supportive ancillary processes for cell growth and viability, but it offers a great deal to speed up the synthetic pathways optimization. There are two classes of synthetic biology: *in vivo* and *in vitro* (Meyer et al. 2007).

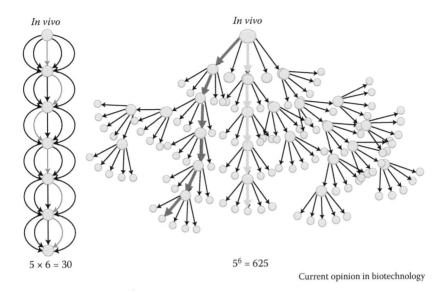

Figure 20.4 Comparison of complexity between *in vitro* and *in vivo* synthetic biological system. The red lines denote the pathways optimized to produce desired products in maximum amount. (Adapted from Zhang Y-HP, Jibin S, and Zhong JJ. 2010. *Curr Opin Biotechnol.* 21:1–7.)

Figure 20.4 is an example of the relatively simplest pathway having six cascade biochemical reactions with five choices at each step. *In vitro* systems would have 30 combinations as each enzymatic step can be changed with another enzyme, whereas *in vivo* system may have $5^6 = 625$ combinations where every step has linkage with others along with the regulation of gene transcription, translation, and stability of messenger ribonucleic acid (mRNA).

Yet, *in vitro* cell-free synthetic biology is in its early stage compared to *in vivo* living biological entity-based synthetic biology, but interpreted much better. Modifying a living system is difficult as compared to assembling a new system. *In vitro* system provides much flexibility to engineer pathways as it is free membranes and/or cell walls, has low degree of complexity, physiology, and cellular viability issues (Zhang et al. 2008).

Beyond biochemical analysis, the successful recapitulation of biological function *in vitro* has encouraged attempts to use cell-free systems for product synthesis. ivSB has given importance to establish basic easy-to-apply molecular toolkit for engineering purpose. It provides a platform to construct protein cascades, genetic circuits, compartmentalization, spatial organization, and minimal systems to get a bigger picture of life's chemistry.

Programming circuits

Using DNA as a language to write synthetic genetic programs that rewire and reprogram organisms is a driving force in synthetic biology. The generation of a synthetic genetic oscillator (Elowitz and Leibler 2000) and a synthetic genetic toggle switch (Gardner et al. 2000) is a landmark development in this field. The focus areas to make genetic circuits to work well are promoters and regulatory sites, used as circuit-like connectivity to execute logical operations, although the complexity of function implemented is still in a developing stage.

Other options of transcriptional control are nucleic acid switches and oscillators (Kim and Winfree 2011). They developed a three-node ring oscillator and a two-switch negative-feedback oscillator, an amplified negative-feedback oscillator, demonstrating high correlation with computational modeling. This model presents a mounting list of unbeaten applications of synthetic approaches to cell-free genetic wiring that include the following:

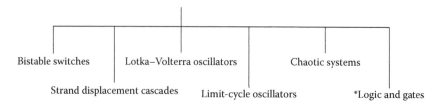

*The **AND gate** is a basic digital **logic gate** that implements logical conjunction—it behaves according to the truth table. (So A and B could be 0 or 1, respectively.)

Minimal cells

There are two approaches, namely, the top-down reduction approach, which works on minimizing cellular genomes to reduce complexity *in vivo* (Gibson et al. 2010) and another approach of genome reduction *in vivo*, the "bottom-up" integration of DNA/RNA/protein/membrane syntheses *in vitro*, which is also being worked out to build a minimal biological system that is self-replicating or autopoietic from individual biomolecular parts that function together. The synthesis of active membrane-bound proteins from a purified translation system in liposomes to convert *sn*-glycerol-3-phosphate to phosphatidic acid, an important precursor to membrane synthesis, is a footstep ahead to a self-sustaining system.

Compartmentalization

In vitro compartmentalization (IVC) has strongly contributed a lot in exploring enzymatic activity. Therefore, 63-fold higher record fastest hydrolases activity of bacterial phosphotriesterase was recorded using the IVC technique (Griffiths and Tawfik 2003).

The introduction of DNA hydrogels, also called "P-gels" as synthetic biocompatible compartments, also increased the transcription and translation yield. The "P-gels" can also confine enzyme systems to a specific geometry using lipid disks for membrane-bound protein expression.

Cell-free systems and spatial organization

For optimizing the synthetic networks to function at the maximum rate, there is a need to increase the enzyme's concentration, and their proximity and other mediating factors have been worked out. Synthetic biologists have devised numerous methods of specific geometric arrangement for enzyme immobilizing or tethering. The mRNA–protein blend molecules tethered to complementary DNA sequences arranged on microchips via mRNA to achieve surface tethering (Figure 20.5a). During the process, a slight change in DNA composition on the chip, can lead to building, optimizing and reconstructing the new pathways.

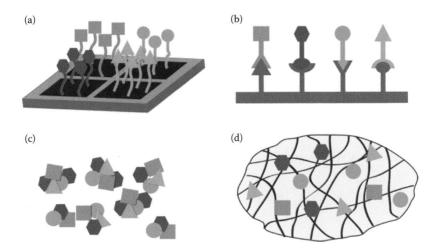

Figure 20.5 Spatial organization shows (a) mRNA display, (b) protein scaffolding, (c) CLEA particles, and (d) foam dispersion.

DNA origami nanostructures together with tethering enzyme pathways could be used to create on/off transcriptional switches and trapping products as potential cell-free reaction applications.

The promising method of recruiting enzymes spatially via protein scaffolding has significantly decreased substrate diffusion lengths to increase reaction rates many folds (Figure 20.5b). By this way, an observation was made on protein–protein interactions for tracing the enzyme pathway. The idea behind this was that they function as a metabolic channel and not as dispersed enzymes individually. The stoichiometric optimizations for efficient working are easy to achieve by making slight change in protein scaffold interaction domains.

In covalently linked enzyme aggregate (CLEA) particles (Figure 20.5c), enzyme ensembles are easily precipitated in the presence of ammonium sulfate and then cross-linked with glutaraldehyde. This process is toxic to cells but was found fit and safe for the synthesis of nucleotide analogs (Scism and Bachmann 2010).

Tween-20 and Ranaspumin-2 (Rsn-2) are detergents and surfactants that can be used as foam to spatially disperse enzymes and liposomes (Hyo-Jick and Carlo 2006) (Figure 20.5d). The simulation of photosynthetic system was built in tungara frog and nest foam are made from the Rsn-2 protein. Approximately, threefold increased production of glucose was observed. The Rsn-2 has exclusive property to generate a foam structure at low protein concentrations without affecting the cell membranes and enzyme activity.

Synthetic pathway biotransformation

The cell-free biocommodity production using synthetic enzymatic pathway biotransformation (SEPB/SyPaB) has differences based on the type of product formed from other techniques such as CFPS and ivSB). The principles of engineering (viz. design, standardization, and extraction) are used in synthetic biology and bring together other sciences (biology and chemistry) in order to shape up novel systems and functions of biological importance. The functional aspects of these designed devises are noticed much better than their natural counterparts. The science of synthetic biology is tapping the potential of simple and very important biological processes and increasingly reshaping complicated biological entities such as parts, devices, and systems.

SyPaB is composed of

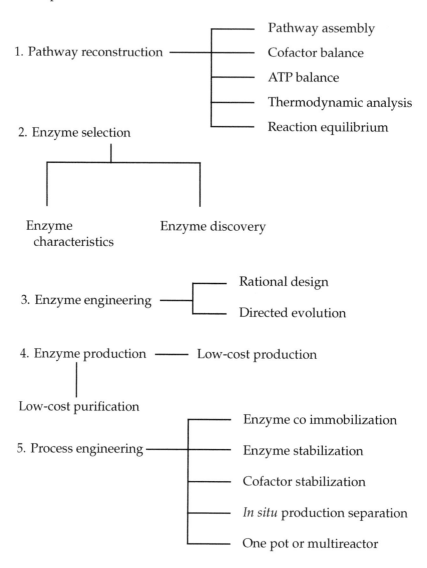

1. Pathway reconstruction
   - Pathway assembly
   - Cofactor balance
   - ATP balance
   - Thermodynamic analysis
   - Reaction equilibrium

2. Enzyme selection
   - Enzyme characteristics
   - Enzyme discovery

3. Enzyme engineering
   - Rational design
   - Directed evolution

4. Enzyme production — Low-cost production
   - Low-cost purification

5. Process engineering
   - Enzyme co immobilization
   - Enzyme stabilization
   - Cofactor stabilization
   - *In situ* production separation
   - One pot or multireactor

Designing of each step needs a meticulous input for reconstructing a cell-free synthetic enzymatic pathway from specialist of the particular area, which is considered as a core point of this process.

Necessary and judicious modifications in natural metabolic pathways are the base for designing synthetic pathway. The biochemical reactions need to be designed carefully as they have many ways to operate.

MICROBIAL SYSTEM

Many processes (viz. metabolism regulation, self-duplication, and formation of a desired product) in microbes are governed in the presence of thousands of proteins present in microbes.

versus

SyPaB

SyPaB contains only the enzymes responsible for the desired transformation without side pathways or cell duplication. For example, even lesser fraction of cellular enzymes (less than 20 enzymes) are sufficient for desired product generation from the substrate.

Table 20.1 Comparison between criteria of SyPaB and MiF

Feature	SyPaB	MiF
Product yield	Theoretic or high[a]	Low or modest
Product titer	High	Low or modest
Productivity	High	Low or modest
Process control	Easy	Modest or hard
Reaction conditions	Broad	Narrow
Product purity	High	Low

Source: Adapted from Zhang Y-HP and Jibin Sand Zhong JJ. 2010. *Curr Opin Biotechnol* 21:1–7.

[a] If the reactions are irreversible or products are removed *in situ*.

SyPaB system in many ways is preferred over MiF (Table 20.1) due to the following reasons:

1. Absence of cell membrane or wall; otherwise it could slow down the transportation of substrate/product.
2. Transport of substrate/product across the membrane does not require energy.
3. As cellular proteins and other biomacromolecules are not part of the SyPaB system; therefore, high concentration of biocatalysts can be used in the reactors.

Recently, the optimization of enzyme ratio and higher substrate concentration has enhanced the production rates of hydrogen by nearly 20-fold (Ye et al. 2009). Further, applying much higher substrate concentration, elevated temperatures, higher enzyme loading, metabolite channeling, higher specific enzyme activity, and optimization of the key enzyme ratio, the production rate could be increased by another 1000-fold (Zhang 2009).

Examples of SyPaB

The use of low-cost renewable carbohydrate ($0.18 per kg) as an option for the production of biocommodities will likely become a key future industrial feedstock (Zhang 2009). Biocommodities developed with utilization of renewable carbohydrates would decrease greenhouse gas emissions. This has better future prospects for safe environment and improved rural economy.

Production of biofuels

High-yield production of hydrogen

Presently, hydrogen is gaining importance as a promising option for future clean and green energy. Hydrogen can produce high energy and is utilized efficiently with minimum wastage. Its production cost is meager with utilization of abundant biomass ruling out the net carbon emissions during the process (Zhang 2009).

A pathway has been designed using synthetic enzymatic system for high-yield hydrogen production. The pathway includes four steps (Figure 20.6), which are catalyzed by various enzymes: (1) glucan phosphorylase, (2) phosphoglucomutase, (3) G-6-P dehydrogenase; 6PGDH, 6-phosphogluconate dehydrogenase; R5PI, phosphoribose isomerase; Ru5PE, ribulose 5-phosphate epimerase; TKL, transketolase; TAL, transaldolase; TIM, triose phosphate isomerase; ALD, aldolase; FBP, fructose-1,6-bisphosphatase; PGI, phosphoglucose isomerase, and (4) hydrogenase.

A theoretical yield of 12 H_2 per glucose unit is expected to be obtained when a continuous reactor is run.

The reaction conditions are selected based on numerous criteria:

1. Temperature and pH (all enzymes are active at 30–32°C and pH 7.5).
2. Buffers (the HEPES buffer is used because some buffers, e.g., Tris, inhibit the activities of some enzymes).
3. 4 mM phosphate and 2 mM $NADP^+$ are chosen as substrates and inhibitors (moderate concentrations are appreciated).
4. Enzymes activators (10 mM Mg^{2+} and 0.5 mM Mn^{2+}) are required to accelerate the reaction but caution to be taken using high concentration Mg^{2+} prone to precipitate free inorganic phosphate.
5. Appropriate coenzymes added to run the reaction smoothly (e.g., transketolase should be facilitated with coenzyme thiamine pyrophosphate).

Production of alcohols

Nowadays, it is a well-known fact that the biological production of ethanol and butanol is not a problem. The feasibility of ethanol production without live cells has been

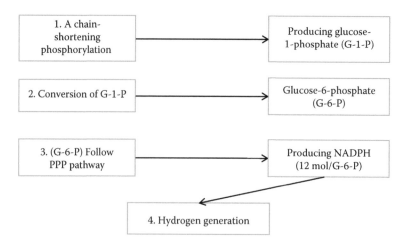

Figure 20.6 High-yield hydrogen generation via synthetic enzymatic pathway.

demonstrated using yeast extract which catalyzes the reaction (Buchner 1897). The complete conversion of glucose into ethanol is hampered due to adenosine triphosphate (ATP) imbalance (i.e., two ATP produced per glucose). The ATP balance can be achieved by adding ATPase, which facilitates the yeast glycolytic enzyme system to go for higher ethanol yield. Alternatively, it is feasible to replace costly ATPase with arsenate. The arsenate is structurally and chemically similar to inorganic phosphate and many enzymes that require phosphate will also use arsenate. Consequently, the cell-free system produces 99% yield of ethanol, that is, 180 g/L of glucose can be converted into 90 g/L of ethanol within 8 h. The lesser stability of arsenate organic compounds than phosphate analog compounds leads to spontaneous hydrolysis followed by rapid decomposition that results in dissipation of high-energy phosphate bonds (Nelson and Cox 2008).

Butanol is a liquid alcohol with four-carbon number, and has been receiving much attention as a better biofuel than ethanol because of its lower water absorption, a higher energy content, improved blending ability, and easy usage in internal combustion engines. However, butanol cannot give yield and titer as high as ethanol (Lee et al. 2008) with minimum risk of bacteriophage contamination (Zhang 2010) even though it has a higher density and is more hydrophobic. Scientists are looking at SyPaB as a transformative system to troubleshoot MiFs with wide applicability due to the following reasons:

1. Complete absence of by-product formation and cell mass synthesis proceeds toward high product yield.
2. Having no interference of cellular membranes; enzymes used can tolerate higher levels of organic solvents and enhance the chances of a high product titer.
3. Simple controlled process (i.e., one-step conversion as all three enzymes can work together).
4. Minimum risk of bacteriophage contamination.
5. Costly medium nutrients are not required.

The SyPaB system also needs removal of *in situ* product by prevaporation, liquid–liquid extraction, or gas stripping similar to MiF. These downstream operations can drive the overall reaction toward the increase of butanol formation.

Generation of electricity

EFCs have a capacity to store high energy as well as replenish themselves rapidly like fuel cells. To have a viable commercial enzymatic biofuel cell is still a dream because of large low-power output and poor enzyme stability, although numerous companies (e.g., Sony) are trying their best on EFCs (Sakai et al. 2009). Changing chemicals to generate electrical energy using EFCs prefer enzymes over chemical catalysts for conversion. They catalyze oxidation of fuels (sugars or alcohols or organic acids) at the anode and cathode, which shows the reduction of the oxidant (Figure 20.7). The enzyme cells are designed in a way that they can induce complete as well as partial extraction of chemical energy. A good number of EFCs are designed to induce incomplete oxidation of chemicals to extract

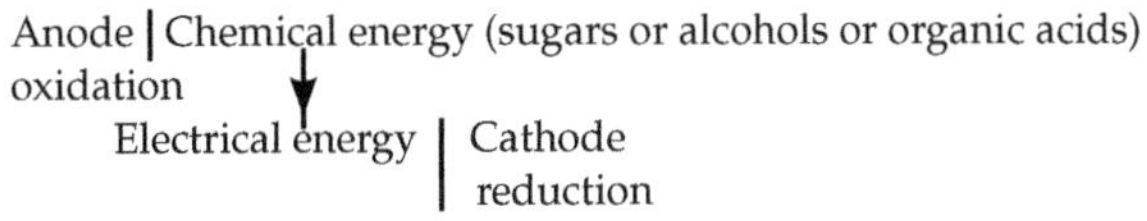

Anode | Chemical energy (sugars or alcohols or organic acids)
oxidation ↓
Electrical energy | Cathode
reduction

Figure 20.7 Conversion of chemical energy to electrical energy.

only a small portion of chemical energy from compounds and convert it to electricity. Carrying out a complete conversion of chemical energy-rich compounds such as sugars or complicated chemicals to electricity has fourfold advantages, namely, high-energy storage density, efficient utilization of high energy, high-power density, and little product inhibition.

Sokic-Lazic and Minteer (2008) tried to mimic the complete citric acid cycle on a carbon electrode and demonstrated immobilization up to 11 enzymes, including six dehydrogenases necessary to generate NAD(P)H and $FADH_2$. The rest five were nonenergy-producing enzymes required for the entire cycle, on the surface of an electrode. Densities of current increase when polymethylene green as a mediator immobilize more dehydrogenases on an anode by electron transfer. During this process, power density is increased by 8.71 times over a single enzyme (alcohol dehydrogenase)-based ethanol/air fuel cell. However, as per the design principles of SyPaB, the pathway demands little modification.

Limitations

Coactivation of multiple biochemical networks in a single integrated platform was thought to be the core idea while working with the cell-free systems. However, the cell-free systems have a major limitation of not being capable to do so. Despite the advantages of CFB2, several obstacles must be defeated before applying them to the industrial scale. The most critical limitations are

1. Produce low-cost enzyme for running reactions
2. Enzyme should have prolonged stability
3. Engineering of cofactor
4. Optimization and modeling of different pathways

Conclusions

Cell-free biotransformation system has been in a state of evolution and it could optimize many aspects as discussed in this chapter. CFPS has emerged as an important and effective alternative to both cell-based expression systems and solid-phase protein synthesis along with commercialization of PURE expression technology, and is generating new possibilities for applications in biotechnology, ranging from microscale to industrial scale and in diverse areas, including protein microarrays, biotherapeutics, and biomaterials.

Cell-free biosystems have an immense scope to be utilized in catalyzing complex conversions. Furthermore, this technology can be used for producing specialty chemicals, biopharmaceuticals, fine chemicals, and biocommodities more than MiF. Such technologies are still in a state of evolution, but the continual reduction in enzyme cost and improvements in enzyme engineering techniques will bring cell-free biosystems upfront for biomanufacturing. In future, this will become a widely popular technique for industrial bioconversion.

Key points

- Quick synthesis of cell-free protein is the biggest advantage of this technique to obtain an expressed phenotype (protein) from a genotype (gene). Being independent of host cells, *in vitro* synthesis of toxic proteins or proteins prone to proteolytic degradation and functional protein assays are possible in a few hours time.

- High-yield alcohol production could be achieved by keeping ATPase balance as imbalance prevents complete conversion of glucose.
- The arsenate is chemically and structurally similar to phosphate and so can replace costly ATPase.
- Hydrogen production from cheap biomass is a shortcut for producing low-cost hydrogen with net carbon emission.
- Several key enabling technologies such as enhanced speed of reaction and lower cost is a central theme to synthetic biology.
- DNA sequencing, modeling, fabrication of genes, behavior of synthetic genes, and correct monitoring and measuring gene behavior are essential tools in synthetic biology.

Institute and eminent scientists working on cell-free systems and their applications

Northwestern University: Has an exclusive laboratory for cell-free systems and synthetic biology, with objectives to build up cell-free biology as a competent technology for biomanufacturing of different products, namely, therapeutics to save life, advanced materials, and sustainable chemicals. Understanding biological systems concerned with protein synthesis and metabolism is the focal point to design, construct, and modify processes with assurance to advance new paradigms for synthetic biology.

Biological Systems Engineering Department, Virginia Tech, Blacksburg, Virginia, USA: In the department of Biological Systems Engineering (BSE), scientists have realized the synergistic force to amalgamate concepts of three basic sciences such as biology, chemistry, and physics, with engineering science and design principles, and to solve problems in biological systems.

Cell-Free Bioinnovations Inc., Blacksburg, Virginia, USA: It is realized the market of economically competitive produce biofuels (e.g., hydrogen and electricity), biomaterials, biochemicals, and fine chemicals. Having an agenda to develop cell-free enzymatic biosystems that can not only contribute to develop sustainable products for day-to-day use, but food and feed from nonfood biomass could also be obtained.

Gate Fuels Inc., Blacksburg, Virginia, USA: Gate Fuels Inc. has led the way to bring together enzyme engineering, metabolic engineering, and synthetic biology to realize their combined potential. It has directed its efforts to develop recombinant *Bacillus* strains capable of converting nonfood biomass into value-added chemical compounds, such as lactic acid. Cellulase engineering and the manipulation of *Bacillus* species together have contributed to the establishment of a consolidated bioprocessing (CBP) strain to accomplish a demand for low-cost lactic acid.

Professor Y.-H. Percival Zhang: An associate professor at Virginia Tech (USA). He aspires to revolutionize key challenges of the energy–water–food nexus with the help of innovative technology of biology system.

Dr. Tsz Kin Tam: A PhD in chemistry research with intentions to develop switchable biosensors and enzymatic biofuel cells controlled by external biochemical and physical signals.

Joseph A. Rollin: A PhD candidate in the Biological Systems Engineering Department at Virginia Tech, where his research focuses on the conversion of sugars to hydrogen, using a cell-free nonnatural synthetic enzymatic pathway.

References

Berrade L, Garcia AE, and Camarero JA. 2011. Protein microarrays: Novel developments and applications. *Pharm Res.* 28:1480–1499.

Buchner E. 1897. Alkoholische Gahrung ohne Hefezellen (Vorlaufige Mittheilung). *Ber Chem Ges.* 30:117–124.

De Graaf AJ, Kooijman M, Hennink WE, and Mastrobattista E. 2009. Nonnatural amino acids for site-specific protein conjugation. *Bioconjug Chem.* 20:1281–1295.

Elowitz MB and Leibler S. 2000. A synthetic oscillatory network of transcriptional regulators. *Nature* 403:335–338.

Gardner TS, Cantor CR, and Collins JJ. 2000. Construction of a genetic toggle switch in *Escherichia coli*. *Nature* 403:339–342.

Gibson DG, Glass JI, Lartigue C, Noskov VN, Chuang RY, Algire MA, Benders GA et al. 2010. Creation of a bacterial cell controlled by a chemically synthesized genome. *Science* 329:52–56.

Griffiths AD and Tawfik DS. 2003. Directed evolution of an extremely fast phosphotriesterase by *in vitro* compartmentalization. *EMBO J.* 22:24–35.

Guterl JK, Garbe D, Carsten J, Steffler F, Sommer B, Reiße S, Philipp A et al. 2012. Cell-free metabolic engineering: Production of chemicals by minimized reaction cascades. *ChemSusChem* 11:2165–2172.

Hartman MC, Josephson K, Lin CW, and Szostak JW. 2007. An expanded set of amino acid analogs for the ribosomal translation of unnatural peptides. *PLoS One* 2:e972.

Hodgman CE and Jewett MC. 2012. Cell-free synthetic biology: Thinking outside the cell. *Metab Eng.* 14:261–269.

Hyo-Jick C and Carlo DM. 2006. Biosynthesis within a bubble architecture. *Nanotechnology.* 17:2198.

Kai L, Roos C, Haberstock S, Proverbio D, Ma Y, Junge F, Karbyshev M, Dötsch V, and Bernhard F. 2012. Systems for the cell-free synthesis of proteins. *Methods Mol Biol.* 800:201–225.

Kim J and Winfree E. 2011. Synthetic *in vitro* transcriptional oscillators. *Mol Syst Biol.* 7:465.

Lee SY, Park JH, Jang SH, Nielsen LK, Kim J, and Jung KS. 2008. Fermentative butanol production by clostridia. *Biotechnol Bioeng.* 101:209–228.

Lim RKV and Lin Q. 2010. Bioorthogonal chemistry—A covalent strategy for the study of biological systems. *Sci China Chem.* Jan 1, 53(1):61–70.

Liu CC and Schultz PG. 2010. Adding new chemistries to the genetic code. *Annu Rev Biochem.* 79:413–444.

Lynd LR, Wyman CE, and Gerngross TU. 1999. Biocommodity engineering. *Biotechnol Prog.* 15:777–793.

Meyer A, Pellaux R, and Panke S. 2007. Bioengineering novel *in vitro* metabolic pathways using synthetic biology. *Curr Opin Microbiol.* 10:246–253.

Nelson DL and Cox MM. 2008. *Lehninger Principles of Biochemistry*, 5th edition. New York: WH Freeman.

Nirenberg MW and Matthaei JH. 1961. The dependence of cell-free protein synthesis in *E. coli* upon naturally occurring or synthetic polyribonucleotides. *Proc Natl Acad Sci USA* 47(10):1588–1602.

Osman M, Shah A, and Walsh F. 2011. Recent progress and continuing challenges in bio-fuel cells. Part I: Enzymatic cells. *Biosens Bioelectron.* 26:3087–3102.

Rollin JA, Tsz Kin Tamb, and Y-H Percival Zhang. 2013. New biotechnology paradigm: Cell-free biosystems for biomanufacturing. *Green Chem.* 15:1708–1719.

Sakai H, Nakagawa T, Tokita Y, Hatazawa T, Ikeda T, Tsujimura S, and Kano K. 2009. A high-power glucose/oxygen biofuel cell operating under quiescent conditions. *Energy Environ Sci.* 2:133–138.

Scism RA and Bachmann BO. 2010. Five-component cascade synthesis of nucleotide analogues in an engineered self-immobilized enzyme aggregate. *ChemBioChem.* 11:67–70.

Sokic-Lazic D and Minteer SD. 2008. Citric acid cycle biomimic on a carbon electrode. *Biosens Bioelectron.* 24(4):939–944.

Spirin AS, Baranov VI, Ryabova LA, Ovodov SY, and Alakhov YB. 1988. A continuous cell free translation system capable of producing polypeptides in high yield. *Science* 242(4882):1162–1164.

Swartz JR. 2012. Transforming biochemical engineering with cell-free biology. *AIChE J.* 58:5–13.

Wang HH, Huang PY, Xu G, Haas W, Marblestone A, Li J, Gygi S, Forster A, Jewett MC, and Church GM. 2012. Multiplexed *in vivo* His-tagging of enzyme pathways for *in vitro* single-pot multi-enzyme catalysis. *ACS Synth Biol.* 1:43–52.

Watts RE and Forster AC. 2012. Update on pure translation display with unnatural amino acid incorporation. *Methods Mol Biol.* 805:349–365.

Ye X, Wang Y, Hopkins RC, Adams MWW, Evans BR, Mielenz JR, and Zhang Y-HP. 2009. Spontaneous high-yield production of hydrogen from cellulosic materials and water catalyzed by enzyme cocktails. *ChemSusChem* 2(2):149–152.

You C and Zhang YHP. 2012. Cell-free biosystems for biomanufacturing. In: Zhong J-J, editor. *Advances in Biochemical Engineering/Biotechnology*: Springer, Berlin. 89–119.

You C, Chen H, Myung S, Sathitsuksanoh N, Ma H, Zhang X-Z, Li J, and Zhang YHP. 2013. Enzymatic transformation of nonfood biomass to starch. *Proc Natl Acad Sci USA* 110:7182–7187.

Zhang YH, Wang Y, and Ye X. 2008. Biofuels production by cell-free synthetic enzymatic technology. In: Columbus F, editor. *Biotechnology: Research, Technology and Applications.* Hauppauge, NY: Nova Science Publishers. 143–157.

Zhang Y-HP. 2009. A sweet out-of-the-box solution to the hydrogen economy: Is the sugar-powered car science fiction? *Energy Environ Sci.* 2(2):272–282.

Zhang Y-HP. 2010. Production of biocommodities and bioelectricity by cell-free synthetic enzymatic pathway biotransformations: Challenges and opportunities. *Biotechnol Bioeng.* 105:663–677.

Zhang Y-HP, Jibin S, and Zhong JJ. 2010. Biofuel production by *in vitro* synthetic enzymatic pathway biotransformation. *Curr Opin Biotechnol.* 21:1–7.

Zhang Y-HP and Lynd LR. 2008. New generation biomass conversion: Consolidated bioprocessing. In: Himmel ME, editor. *Biomass Recalcitrance: Deconstructing the Plant Cell Wall for Bioenergy.* Blackwell Publishing: Chichester, UK. 480–494.

chapter twenty-one

Magnetic nanoparticles with multifunctional water-soluble polymers for bioapplications

Muhammad Irfan Majeed, Muhammad Asif Hanif, Haq Nawaz, and Bien Tan

Contents

Abstract

Magnetic nanoparticles (MNPs) have obtained much scientific attention due to their unique magnetic properties such as superparamagnetism, low Curie temperature, high coercivity, and high magnetic susceptibility, which are being employed in various bioapplications such as magnetic separation and immobilization of biomolecules, resonance imaging (MRI), magnetic fluid hyperthermia, biolabeling, and in the development of drug delivery systems for controlled release of drugs and magnetic sensors. Coprecipitation and thermal decomposition are most commonly used procedures employed for

MNP preparation. Multifunctional water-soluble polymers (MWPs) can play an important role in the preparation and functionalization of MNPs, as they have the ability to control the size and morphology of the MNPs efficiently due to the presence of functional groups such as $-NH_2$, $-COOH$, $-SH$, $-OH$, etc. Moreover, multifunctional polymers make MNPs more stable, uniform, and water soluble in addition to giving them rich surface chemistry, thus opening up new ways for their modification and functionalization and increasing their scope of bioapplications. In this chapter, we have discussed various advantages of using MWP ligands for preparation and functionalization of MNPs for various bioapplications with recent reported examples.

Keywords: Magnetic, Nanoparticles, Water-soluble, Polymer ligands, Bioapplications.

Introduction

Inorganic nanoparticles

Inorganic nanoparticles (NPs) such as noble metal, metal oxide, magnetic and semiconductor NPs, or quantum dots (QDs) have found great prospects in various fields of science and engineering such as electronics, information technology, and chemical industry as well as in diagnostic and therapeutic biomedical research areas due to their unique optical, electrical, magnetic, and catalytic properties.[1–3] Owing to their extremely small diameters (usually in the range of 10–100 nm), NPs possess high surface area and surface free energy, which result in their aggregation unless protected by some secondary material which helps in keeping these NPs apart.[4–6]

The surface modification of the NPs is carried out either during NP synthesis or in a postsynthesis step using a variety of materials ranging from a simple coating of a relatively inert material, for example, silica[7,8] to use of organic molecules that are linked to the NP surface covalently or electrostatically.[9] Polymeric ligands (especially water-soluble) are indeed another highly efficient group of capping ligands that are used for surface modification of inorganic NPs due to their several advantages over nonpolymeric capping ligands.[4] Water-soluble polymer ligands not only provide steric and electrostatic hindrance to the NPs toward aggregation but also help to disperse the NPs in aqueous media, which is a prerequisite for bioapplications of NPs.[10,11]

Since the Turkevich-Frens preparation of colloidal gold by sodium citrate reduction method,[12] numerous physical and chemical processes have been reported for the preparation of metal/metal oxide NPs in aqueous media by introducing miscellaneous capping ligands for this purpose. However, most of these reports using nonpolymeric capping ligands have several disadvantages such as poor control over nucleation and growth process, which leads to NPs with large diameters and poor monodispersity, low solubility and stability resulting in aggregation, and undesirable chemical reactivity restricting their scope of applications in various fields such as quantized capacitance charging,[13] single electron transistors assembly,[14] thermal gradient imaging,[15] and biomedical research areas.[16–18]

Magnetic nanoparticles

Magnetic nanoparticles (MNPs) have always been a hot topic in NP research due to their unique properties such as superparamagnetism, low Curie temperature, high coercivity, and high magnetic susceptibility, and therefore they have numerous potential applications in ferrofluids, magnetoptics, catalysis, information storage, spintronics, as well as in several biomedical applications such as magnetic resonance imaging (MRI), magnetic fluid hyperthermia (MFH), magnetic separation of biomolecules, drug delivery, etc.[6,19–25] However, in order to fabricate some practical magnetic nanostructure materials or devices, it is very essential to choose the stable materials with high magnetization and magnetic susceptibility as well as versatile physicochemical properties especially for bioapplication.[26,27] Magnetic iron oxide NPs (MIONs) due to their advantages such as low toxicity and biocompatibility are considered to be the most favorable candidates for numerous bioapplications.[23,28] There are numerous methodologies developed for preparation of MIONs such as microemulsion processing, ultrasound and photo irradiation, laser pyrolysis, and hydrothermal routes; however, thermal decomposition and coprecipitation are two commonly used methods for MION synthesis having their own advantages and limitations.[29]

The thermal decomposition method involves the pyrolysis of organic iron salts at high temperatures in organic solvents and subsequent oxidation, which results in the preparation of highly monodisperse iron oxide NPs with high monodispersity, crystallization, and large saturation magnetization. During the thermal decomposition method, nucleation and growth processes remain separate and complex hydrolysis reactions are not involved, which offers better control over size and shape of MNPs.[30–32] However, these NPs are water insoluble and require postsynthesis ligand exchange procedures for bioapplications[33,34] due to which aggregation of NPs and loss in their magnetic properties occur.[35] Moreover, surface modification of the MNPs obtained by thermal decomposition in organic phase is an even harder process since phase transfer as well as surface modification require high use of surface coupling agents, which cannot be achieved without compromising magnetic and other properties of NPs.[34,36]

On the contrary, coprecipitation method yields MNPs that are directly dispersed in aqueous phase and are feasible for bioapplications. Unlike thermal decomposition method, it does not employ the use of high temperature and expensive organic metal precursors and solvents. Therefore, coprecipitation approach seems to be more efficient, cost effective, and environment friendly; however, there are some limitations of conventional coprecipitation method such as low monodispersity, crystallinity, and saturation magnetization due to low reaction temperature and less control over complex hydrolysis reactions of the iron precursors and difficulty in isolation of nucleation and growth steps.[37]

Moreover, since physicochemical properties of MNPs depend on their size and shape, the development of new protocols for the synthesis of nonaggregated NPs with a well-controlled mean size and a narrow size distribution is essential. A vital aspect in the NP synthesis procedure is its capability to keep NPs physically apart and prevent irreversible agglomeration. The different types of protective molecules or capping ligands are employed for this purpose, which bind to the NP surface and avoid their aggregation in aqueous solution.[38] Therefore, it is necessary to develop some easy and efficient method for water-soluble MNP preparation with better control over size and magnetic properties.

Preparation of MNPs with multifunctional water-soluble polymers as capping ligands

Multifunctional water-soluble polymers (MWPs) play an important role in the preparation of inorganic (metal, metal oxide, and semiconductor) NPs by providing several beneficial properties to them such as monodispersity, multiple functionality, better control over size and shape, excellent solubility and stability in aqueous media, and biocompatibility. Moreover, other physicochemical properties of these NPs, which depend on their size, shape, and surface chemistry such as catalytic chemical reactivity, optoelectronic, and magnetic properties can also be well tailored by using water-soluble polymers as capping ligands. Water-soluble polymer ligands are easy to synthesize, having long-chain flexible molecular structures decorated with desirable multiple functional groups in abundance, with different molecular weights, help in tailoring the properties of NPs according to the choice of applications as compared to rigid short molecule capping ligands with limited choices. Water-soluble polymers have been proven as excellent capping ligands for the preparation of inorganic NPs in aqueous phase.[10]

MWPs are also helpful in the preparation of MIONs. Intrinsic properties of these polymer ligands such as long-chain molecules and an abundance and variety of functional groups (such as −COOH, −SH, −OH, −NH$_2$, etc.) present on polymer chains can efficiently cap the NPs and control their size and distribution during synthesis. These functional groups present on the polymer chains can be linked to the NP surface through electrostatic or covalent interactions and can effectively arrest their growth as well as provide them excellent water solubility, chemical stability, and rich surface chemistry, which opens up ways for easy modification and functionalization of NPs for bioapplications. For example, there are several types of functional groups that can be exploited to anchor the polymers on the iron oxide NPs[39] as depicted in Figure 21.1.

In the aqueous coprecipitation process, polymer ligands can efficiently control the size and shape of the MIONs due to their long-chain molecules bearing a variety of functional groups such as −SH, −COOH, −OH, −NH$_2$, etc. in abundance. Therefore, nowadays, surface modification of MNPs with MWPs is the most effective method to prepare MNPs with small size, good monodispersity, controllable morphology, excellent biocompatibility, easy biofunctionalization ability, good environmental adaptability, and ability to overcome other disadvantages of using nonpolymeric capping ligands for NPs.[4,10,40–42]

For example, Li et al. reported the synthesis of multifunctional, highly water-soluble, and highly stable ultrasmall MIONs using a trithiol end-functionalized water-soluble polymer (PTMP-PMAA) as capping ligand through the coprecipitation method.[37] The molar ratio between the iron precursors and carboxylic acid groups present on the polymer was

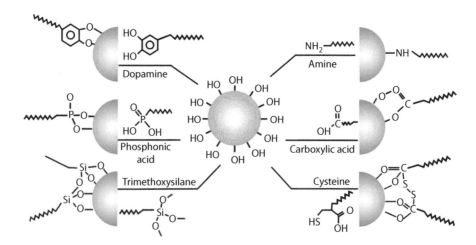

Figure 21.1 Functional groups usually employed to anchor polymers on MIONs.

described to play an important role in determining the size and size distribution of NPs by separation of NP nucleation and growth processes. Carboxylic acid functional groups of polymer ligand also provided high water solubility and stability to NPs over a wide pH range. The NPs were successfully functionalized in postsynthesis steps by utilizing the thiol and carboxylic acid groups present on their surface as shown in Figure 21.1. These NPs could find several applications such as in biolabeling, magnetic separation, and assays of biomolecules.[37]

Currently, monodisperse MNPs are widely applied in fields of electronic device fabrication and information storage, and are considered to be one of the best functional materials in biomedical research areas such as magnetically targeted drug delivery and release,[21,43-46] magnetic separation of cells and biomolecules,[47,48] magnetic MRI[49] and cancer diagnosis and treatment, etc.[50-52] For biomedical applications, the MNPs should have the following properties such as low toxicity, biocompatibility, good dispersibility in biological media, rich surface chemistry for easy biofunctionalization, and above all, excellent magnetic properties that make their detection and manipulation within the organism easy, as well as recyclability.[41]

MNPs have a unique superparamagnetism property when their diameter is smaller than a certain critical value (D_p). Owing to this superparamagnetic property, these MNPs show a magnetic behavior only under the influence of an external magnetic field but do not aggregate in its absence; as a result, the monodispersity and size of the MNPs mainly determine whether they are suitable or not for a certain bioapplication. However, for bioapplications, it is very essential for these MNPs to have certain properties such as small size, biocompatibility, water solubility, chemical stability, target recognition ability, high magnetic susceptibility and superparamagnetism, and rich surface chemistry for required functionalization in several cases.[26]

Properties of MNPs prepared with MWPs

Water-soluble polymers that are mainly used as capping ligands for the NPs include polyethylene glycol (PEG),[53-55] polyvinyl alcohol (PVA),[52,56,57] polyvinyl pyrrolidone (PVP),[58-60] dendritic polyamido amine (PAMAM),[61] polyethyleneimine (PEI),[62-64] mercaptan, and acrylic polymers such as pentaerythritol tetrakis (3-mercaptopropionate) polymethacrylic acid (PTTM-PMAA)[37,65] and dodecanethiol-polymethacrylic acid (DDT-PMAA)[11,26] and also some natural polymers such as starch, dextran, chitosan, and plant resins, biomacromolecules such as polynucleic acids, polypeptides, etc.[3,66-68] Modification of the NPs using MWPs as capping ligands mainly has the following advantages.

Controllability

Polymers play an important role during the preparation of MNPs in two ways; first, they can prevent coagulation of nuclei induced by van der Waals forces to obtain nearly monodisperse NPs; second, because of strong affinities of abundant functional groups present on the polymer backbone to the surface atoms of growing NPs, they can efficiently enfold NPs produced to prevent them from further grouping up into larger aggregates.[40] The size, monodispersity, and morphology of NPs can be precisely controlled by tuning the concentration and molecular weight of MWPs.[6,26,37,69]

Lu et al. demonstrated a one-step protocol for size and shape controlled synthesis of monodisperse and water-soluble Co NPs through the wet chemical reduction method using alkyl thioether end-functionalized polymer ligand poly(methacrylic acid) DDT-PMAA as capping ligand.[26] Through optimization of molecular weight and concentration of the DDT-PMAA, they were able to obtain 2–7.5-nm ultrasmall monodisperse spherical NPs as well as larger spherical Co NPs with 80 nm diameter. Similarly, polymer ligand was also helpful in anisotropic growth of NPs into 15 × 36 nm nanorods. These Co NPs and nanorods were superparamagnetic at room temperature while highly stable in the form of aqueous solution for up to several weeks. Owing to the abundant presence of carboxylic acid (−COOH) groups on their surface, they were demonstrated to be useful for numerous bioapplications through chemical functionalization and modification. Especially, the formation of the Co nanorods with the help of polymer ligand in aqueous phase can open up new ways for many exciting studies, where the physical properties (magnetic response, MR relaxivities, dipole–dipole interactions) of the NPs depend on their anisotropic shapes. It may also be interesting to study the effect of cellular uptake on the different shaped NPs obtained through the use of water-soluble polymer ligands.[70] The efficiency of these Co NPs as contrast in MRI agent was also demonstrated and they were found to be useful as negative contrast agents with similar efficiency than that of the iron oxide NPs.[71] Similarly, polyvinylpyrrolidone (PVP) played an important role in the preparation of nearly monodisperse hollow Fe_2O_3 nanoovals (NMHFNs), through a solution process followed by annealing, which showed soft ferromagnetic behavior at room temperature having scope in photocatalysis, gas sensors, and magnetic carriers.[25]

Stability

NPs because of their high surface area and large surface energy tend to coagulate.[72] Through mechanisms of electrostatic repulsion, steric hindrance, and vacancy stabilization effects, MWPs can change the physical and chemical properties of the NP surface to keep them stable for quite a long time even when ionic strength and pH of the solutions are changed[73] to opposite situation without any capping ligands. In many cases, MWPs also provide excellent stability to NPs in aqueous medium against chemical oxidation, which was otherwise not possible. For instance, Co NPs prepared with DDT-PMAA were stable in open environment for several days, which otherwise is not possible because of the highly oxidative nature of Co.[26] Superparamagnetic iron oxide NPs (SPIONs) due to their biocompatibility are often applied in various bioapplications; however, the quality and performance of these NPs significantly depend on several factors such as NP size, stability, thickness of dispersant layer, and control over their interfacial chemistry.[29]

Easy modification

Since MWPs usually bear a variety of functional groups present in abundance on the polymer backbone, it is very easy to make the NPs gain different physical and chemical properties through modification of these functional groups in several ways, for example, through host–guest interactions and direct chemical modification of these functional groups.[74] These designable and modifiable properties of NPs with multifunctional polymer ligands increase their scope of applications. Host–guest interactions refer to the process in which two or more chemical entities combine through weak electrostatic interactions or hydrogen bonding. Functional groups present at the NPs surface act as host molecules and can be combined with a variety of guest molecules selectively to form a supramolecule system required for a particular application.[35]

For example, alpha cyclo dextrin (α-CD) can act as a host of a typical supramolecule system. It forms host–guest systems with organic molecules and inorganic ions mainly through two ways: they form cage complexes through "endorecognition" (van der Waals force, dispersion force, hydrophobic interaction) and can form pipe-shaped surface interaction products through "exorecognition" (hydrogen bonding force).[75,76] Hexane solution of monodispersed ferric oxide NPs or silver NPs stabilized with oleic acid was mixed with aqueous solution of α-CD. The NPs could transfer from organic phase to aqueous phase because of complexation between α-CD with oleic acid (Figure 21.2).[77] Direction chemical modification of functional groups such as carboxyl group, amino group, hydroxyl group, etc. of polymer ligands present on the NP surface refers to the chemical modification of these functional groups through characteristic chemical reactions such as esterification, acylation, condensation, etc., which would lead to the change in the chemical properties of NPs.

Latham et al.[78] used trifluoroethylester-polyethyleneglycol-thiol polymer (TFEE-PEG-SH) as capping ligand, in order to prepare water-soluble gold and FePt alloy MNPs. TFEE-PEG-SH was very soluble in water and its trifluoro ether group was easy to leave so it could be easily replaced with other chemical moieties for further functionalization. It was found that different end group would influence the size of NPs and thus owned different properties after trifluoro ether group was replaced by butane, pyridine, and biotin.

The polymer ligands are usually attached to the NP surface either through a single or a combination of a variety of different interaction modes such as electrostatic forces, hydrophobic/hydrophilic interactions, and covalent bonding between surface atoms of NPs and functional groups present on the polymer backbone. We can realize the ligand exchange on the NP surface using the difference of interaction between ligands and NPs and can make the NPs suitable for a particular bioapplication.

In 2006, Choi et al.[79] prepared hydrophobic heterodimer FePt-Au NPs by chemical reduction of metal precursors using 1-hexadecylamine as capping ligand. Then, reckoning on the principle that the chemical bonding between S and Au was stronger than the electrostatic interaction between N and Au, they successfully exchanged the 1-hexadecylamine with a thiol (—SH) functionalized water-soluble polymer ligand amino dihydrolipoic acid esterified polyethylene glycol (DHLA-PEG-NH$_2$) thus rendering the hydrophobic heterodimer FePt-Au NPs soluble in water phase, which could be successively combined with antibody through biocoupling. The prepared composite could be used in biodetection as a bioprobe.

Self-assembly of NPs

Superparamagnetic NPs behave as a single magnetic domain and can be attracted to an applied magnetic field due to their very small size (10–30 nm); however, they do not retain a significant magnetism and coercivity when the applied magnetic field is removed and thus, are suitable for numerous applications where NPs agglomeration is

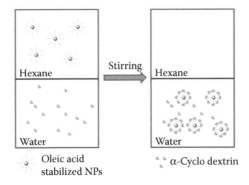

Figure 21.2 A schematic diagram of phase transfer of oleic acid-stabilized MNPs from organic (top layers) to aqueous phase (bottom layers) by surface modification using α-CD.

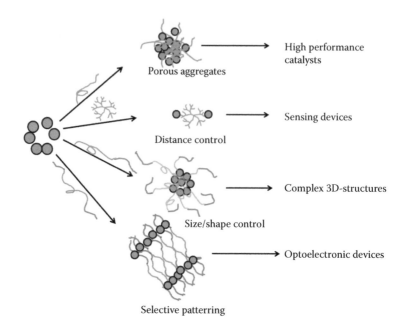

High performance
catalysts

Porous aggregates

Sensing devices

Distance control

Complex 3D-structures

Size/shape control

Optoelectronic devices

Selective patterring

Figure 21.3 Different approaches for self-assembly of MNPs using polymers.

undesirable. However, they have very low magnetization due to very small size and cannot be manipulated with moderate magnetic fields, which limit their applications. One way to overcome this problem is to increase their size, which can only be achieved at the expense of superparamagnetic–ferromagnetic transition resulting in poor dispersion and agglomeration of NPs. The other way of increasing magnetization while retaining the superparamagnetic properties is to self-assemble these NPs into large complex superparamagnetic structures by imbedding them in a nonmagnetic matrix such as polymer shell. Self-assembly of NPs into highly ordered structures, exhibiting novel collective properties different from the individual NPs, can be achieved through several different strategies.[80] MWPs can also be helpful for the chemical fabrication of highly ordered assemblies of superparamagnetic NPs[81] (as shown in Figure 21.3) with novel properties and applications in the fields of spintronics, catalysis, separation, photonic crystals, biomedicine, and biotechnology.[82]

Stimuli-responsive polymers for MNPs

Polymers that have the ability to respond to external stimuli such as pH, temperature, electric or magnetic field, light and/or chemical and biological stimuli, etc. are called stimuli-responsive polymers and are a cheaper and more easily tailored type of smart materials in comparison to metals or ceramics. Stimuli-responsive polymers have a wide range of applications that include sensors, drug delivery systems (DDS), gene transfection, and tissue engineering.[83]

The polymers that respond to a change in the temperature and pH are called thermo-responsive and pH-responsive polymers, respectively. Many groups have used such kind of stimuli-responsive polymers as coatings for the MNPs and developed novel pH- and thermo-responsive DDS. Magnetic thermo-responsive nanocarriers can be prepared by using some temperature-responsive polymers such as poly(N-isopropylacrylamide) (PNIPAM), poly(ethylene oxide)-poly(propylene oxide)-poly(ethylene oxide) (F127), etc.[84]

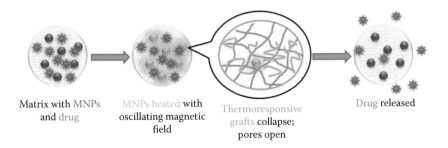

Matrix with MNPs and drug   MNPs heated with oscillating magnetic field   Thermoresponsive grafts collapse; pores open   Drug released

Figure 21.4 Depiction of a grafted polymer-based magnetothermal responsive delivery system.

Zakharchenko et al. demonstrated the preparation of interesting thermo-responsive polymer films consisting of a bilayer of poly(*N*-vinylcaprolactam) (PVCL) on top of PNIPAAM with encapsulated MNPs.[85] At temperatures more than lower critical solution temperature (LCST), the films were flat and allowed the adsorption of NPs, cells, and drugs onto their surface while on cooling it down, the films rolled up to entrap these entities, which could be released by increasing the temperature again. Moreover, the encapsulation of MNPs enabled the magnetic manipulation of the films by an external magnetic field.[85]

Zhang et al. transferred water-insoluble MIONs into aqueous phase by coating them with PNIPAAM and prospected their use in drug delivery and biological sensing.[86] Similarly, modification of inorganic NPs with thermo-responsive polymers for the development of magnetically controlled DDS has been shown by several groups.[87–89] In some cases, the magnetic field was used to heat the system and release the entrapped drugs from the thermo-responsive polymer coatings[90,91] as depicted in Figure 21.4.

Stimuli-responsive "intelligent" polymers that respond to the change in pH are called pH-responsive polymers such as poly(methacrylic acid), poly(acrylic acid), polyvinyl pyridine, poly(1-vinylimidazole) (PVIm), etc., and are promising for a variety of bioapplications such as DDS.[92–96] Kim et al. stabilized the iron oxide NPs with PEO-based water-soluble amphiphilic copolymers, which have pH-dependent phase transition behaviors and were synthesized through atom-transfer radical polymerization (ATRP) and reversible addition/fragmentation chain transfer (RAFT) radical polymerization of 1-vinylimidazole using a PEO-based RAFT agent and AIBN.[97] The magnetite NPs prepared with pH-responsive PEO-based water-soluble block copolymers showed high potential as a contrast agent for MRI.

Addition of a targeting agent to such multifunctional DDS improves their efficiency and reduces their toxicity to the healthy cells due to their high selectivity for cancer cells. Chen et al. demonstrated the preparation of a targeted DDS by encapsulating hydrophobic SPIONs with a novel biocompatible, biodegradable, and pH-responsive amphiphilic diblock polymer (HAMAFA-b-DBAM), chemotherapeutic agent doxorubicin (DOX), and conjugation with a targeting agent folic acid for cancer cells.[98] Folic acid is a well-known targeting agent for several types of cancer cells, which overexpress the folic acid receptors as compared with the normal healthy cells. Thus, folic acid bearing conjugates showed a significantly higher targeting capability and toxicity to cancer cells due to the selective release of drug in mildly acidic endosomal/lysosomal compartments and degradation of the pH-responsive bonds, which result in an improved imaging signal and cancer therapy.[98]

Chang et al. prepared novel water-soluble and pH-responsive anticancer drug nanocarriers by attachment of doxorubicin-functionalized polyamidoamine dendrimers

(PAMAM-DOX) to the SPIONs.[99] Magnetite NPs were stabilized by mPEG-G2.5 PAMAM dendrimers and the DOX was conjugated to the dendrimer segments of oleylamine-stabilized NPs through hydrazone bonds, which can easily undergo a cleavage at acidic pH conditions and thus create an opportunity for development of an ideal pH-responsive drug release system. The drug release profiles of this system showed the hydrolytic release of drug at pH 5.0, which is a lysosomal pH condition while magnetite NPs facilitated the cellular uptake of the drug in the tumor cells due to their enhanced permeability and retention (EPR) effect. These pH-responsive drug carriers were described to have a high potential for MRI and cancer diagnosis and therapy even without a specific targeting agent.[99]

Biocompatibility

MNPs have close interaction with the cells and can pass through biological membranes, which offer numerous opportunities for their applications in biomedicine due to their small dimensions, for example, serving as contrast agents for MRI, magnetic hyperthermia, and drug delivery. However, the most important need for these bioapplications is the biocompatibility of these inorganic NPs, that is, they should have none or least toxicity to the healthy cells, higher cellular uptake, as well as the ability to dodge the immune system.[100] MWPs have been proven to diminish or reduce the toxicity of MNPs by making a biocompatible polymer shell around the NPs. Several types of natural and synthetic MWPs such as PEG,[101] PVA,[102,103] PAA,[104] PMAA,[105] PEI,[62] PEO,[97] chitosan,[106] dextran,[107] starch,[108] etc. have been used for this purpose. Owing to their intrinsic hydrophilic and biocompatible nature, MWPs not only reduce the possible toxicity of the NPs but also prolong their blood circulation times by providing stealth from immune system, which can otherwise recognize NPs as foreign bodies and rapidly eliminate them from the blood stream by the macrophages and other cells of the reticuloendothelial system (RES).[109] It is commonly believed that NPs with neutral and hydrophilic surfaces have longer half-life,[110,111] which is achievable as such by coating of the NPs with hydrophilic polymers, which prevent opsonization by shielding the surface charges, increasing surface hydrophilicity, and sterically repulsing the blood components, and a good example of such polymers is polyethylene glycol (PEG),[112] which is proven to be nonimmunogenic, nontoxic, and protein resistant.[113] Polymer coating of NPs is usually achieved either *in situ*, that is, the preparation of NPs in the presence of polymers, or by postsynthesis polymer coating of NPs, and polymer coating may involve electrostatic interactions, chemical conjugation, or both.[100]

Iron oxide NPs have been extensively used for biomedical applications due to their unique magnetic properties, low toxicity, and biodegradability. They have been coated with different materials to control their surface chemistry such as with dextran,[107] citrate,[114] or aminosilane[115] and prevent their aggregation *in vivo* but those coatings were not found to be effective in dodging the RES after intravenous injection. Therefore, much attention is now focused on coating these iron oxide NPs with hydrophilic polymers.

For example, Wan et al. prepared Fe_3O_4 NPs through aqueous coprecipitation of iron precursors by a base in the presence of two different types of PEG-containing copolymers: (PEG-*b*-PGA) and (PGA-*g*-PEG) and thus polymer coating was achieved *in situ* during nucleation and growth of NPs.[116,117] The copolymers were attached to Fe_3O_4 NPs through coordination of PGA via its 1,2-diols to the Fe atoms on the NP surface (Figure 21.5), and NP dispersion stability was attributed to the extension of the PEG chains into the aqueous medium. Furthermore, by varying the graft density of copolymers, the size of the NPs was successfully controlled from 4 to 18 nm. Similarly, poly(oligoethylene glycol methacrylate-*co*-methacrylic acid) (P(OEGMA-*co*-MA)) was used to prepare 10–25-nm iron oxide

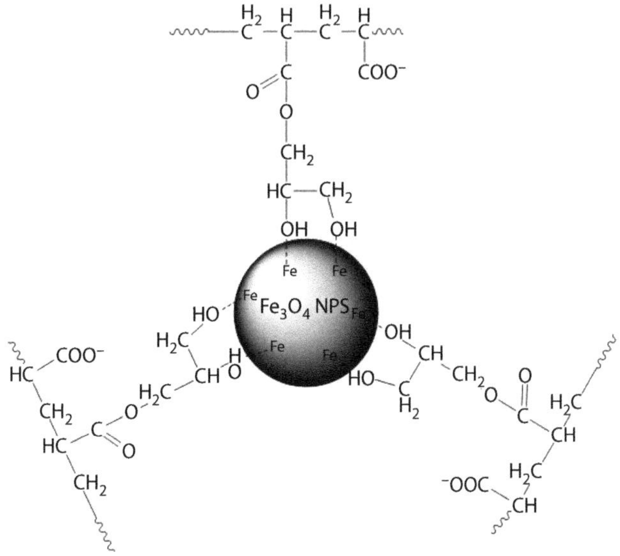

Figure 21.5 Schematic illustration of interaction between iron oxide surface and PGA.

NPs through the coprecipitation method.[118] These PEGylated NPs showed long-term colloidal stability in physiological buffer as well as long blood circulation time of 6 hours as obtained after an intravenous injection into rats.

More recently, Li et al. prepared Fe_3O_4 and $MnFe_2O_4$ NPs through the coprecipitation method in the presence of trithiol end-functionalized polymethacrylic acid.[20,105] Polymer ligand not only efficiently controlled the size and size distribution of NPs but also rendered them soluble in water. *In vitro* and *in vivo* experiments demonstrated these NPs as highly biocompatible and nontoxic with long blood circulation time and were used as MRI contrast agents.

Bioapplications of MNPs prepared with MWPs

Inorganic NPs have found a great deal of interest because of their unique properties and persistently forthcoming applications in biomedical research and clinical diagnosis. For example, gold and other noble metal NPs and semiconductor QDs are being used in biolabeling, cell tracking, cancer diagnosis, and hyperthermia treatment, and in the development of new and efficient biosensors and bioassays.[16,119,120] Similarly, MNPs are being employed as contrast agents in MRI, hyperthermia treatment of cancer cells, magnetic separations, magnetically assisted drug and gene delivery, etc.[18,27,121] However, there are some limitations in the use of these inorganic NPs in biomedicine such as toxicity, low stability in physiological environment, and short blood circulation time, which is because these NPs once injected into the body are considered as foreign particles by the immune system and are cleared out of the body within a short time.[100]

Fortunately, these problems can be overcome by the use of hydrophilic polymer shell around the NPs which could defy the reticuloendothelial system (RES) by pretending them to be water droplets.[100] Polymer shell also reduces the possible toxicity of the NPs by preventing them from undergoing any chemical reactions within the body. Furthermore, polymer shell around MNPs can serve as platform for the attachment of multiple functional

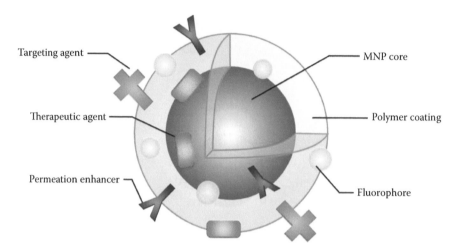

Figure 21.6 Schematic illustration of magnetic nanoparticle with hydrophilic polymer shell containing different ligands to make single multifunctional NP-based platform.

agents as depicted in Figure 21.6. Numerous reports regarding the use of biocompatible water-soluble polymers as capping ligands for preparation of MNPs[44] and their potential for biomedical applications will be briefly discussed in the following sections.

Bioimaging

Nowadays, there are several noninvasive optical bioimaging techniques, which include computed tomography (CT), positron emission tomography (PET), single photon emission CT (SPECT), ultrasound (US), optical imaging (OI), and MRI, each one having its own advantages and limitations over the others but they are all complementary.[122] The advances in the development of modern diagnostic techniques and tools in biomedical research are highly anticipated to the synthesis of nanoparticle-based imaging contrast agents. A remarkable progress has been made in the field of biomedical imaging in the past two decades due to considerable advances made in electronics, information technology, and more recently, nanotechnology. Several types of nanosized contrast agents such as noble metal NPs, MNPs, and QDs have been introduced for noninvasive bioimaging. These NPs have several advantages over old fluorescent dye-based contrast agents such as higher quantum yield, photostability, biocompatibility, easy functionalization, etc.; thus they have improved sensitivity and detection limits of the instruments.

MNPs are being effectively applied as contrast agents in MRI. Multimodal detection systems based on fluorescent and MNPs have further facilitated deep tissue imaging by combined optical and MRI techniques.[122] The fabrication of novel targeted luminescent and MNPs with MWPs would play a vital role in the development of contrast agents for diagnosis, imaging, and therapeutic technologies of the new era.[122]

Based on the principle of nuclear magnetic resonance, MRI has now become a powerful and essential technique in the field of biomedical research, clinical diagnosis, and healthcare and is frequently used to visualize internal structures of the body in great details with high resolution. The efficiency of MRI depends on several factors, including power and sensitivity of instruments, procedure adopted, and choice of contrast agents, which can improve the quality of an MRI scan by shortening the relaxation time of the hydrogen nuclei present in their neighborhood.[20] These contrast agents are generally classified into two categories; paramagnetic or T_1-contrast agents, which reduce T_1 (longitudinal or spin-lattice relaxation time) and superparamagnetic or T_2-contrast agents shortening the T_2 (transverse or spin–spin relaxation time) and T_2^*. Both of these T_1 and T_2 contrast agents have different mechanisms of action, which results in positive enhancement (i.e., brighter image) in T_1-weighted MRI and negative enhancement (i.e., darker image) in T_2-weighted MRI, respectively, and their efficiency depends on their capacity to alter the relaxation times.

At present, usual MRI contrast agents are gadolinium (Gd) and manganese (Mn) containing complexes or iron oxide NPs belonging to positive and negative contrast agent's families, respectively. Disadvantages of Gd- and Mn-based MRI contrast agents include low proton relaxation rate, short clear time by cells, and nonbiocompatibility. Furthermore, there is little known about toxicity of Gd and Mn to cells when they are chelated. MNPs are a new kind of MRI contrast agents, which belong to the main type of negative MRI contrast agents, and they can efficiently enter into specific cells, which is strongly affected by their size and surface properties (electric charge and chemical structure). Internalization of MNPs can be realized through fluid-phase endocytosis and receptor-mediated endocytosis. Efficient internalization of MNPs makes a single-cell imaging possible *in vivo* using MRI.[10] Multifunction MNPs prepared with water-soluble polymer ligands can be efficiently used in MRI for several reasons.

Parkes et al. demonstrated the application of Co NPs prepared with DDT-PMAA as contrast agent in MRI.[71] Two samples of Co NPs having different sizes were prepared by varying the concentration and molecular weight of polymer ligand. The relaxivities r_1 and r_2 of the two samples were determined and the effect of particle size and MRI parameters (i.e., temperature, magnetic field strength, and dispersing media) on the relaxivities (r_1 and r_2) was investigated. It was noted that temperature, magnetic field strength, and NPs dispersing media have little effect on the relaxivities of the Co NPs. However, the larger NPs had significantly higher r_1 than the smaller NPs, whereas r_2 value was slightly increased at 1.5 T. The higher r_1 value in larger Co NPs was attributed to the thicker hydrophilic polymer coating around the Co core and which may have influenced relaxation rate of the protons in it vicinity. The highest values for r_1 and r_2 of these Co NPs were obtained to be 7.4 and 88 $mM^{-1} S^{-1}$, respectively, measured at 1.5 T, which were described to be similar to those of iron oxide NPs with a larger size.

From the earlier example, it can be assumed that Co NPs prepared with water-soluble polymer ligands have the potential to be used as contrast agents in MRI, but in comparison to the iron oxide NPs, Co NPs may pose some risk of Co^{2+} toxicity to the subject. However, polymer ligands can still avoid this risk by making a dense covering shell around Co NPs, and preventing oxidation Co core and release of Co^{2+} ions into solution.[71] Furthermore, the hydrophilic polymer shell around NPs will also provide special properties such as preserving magnetic properties of NPs, increasing blood circulation time by providing stealth against body's immune system, and easy functionalization due to the rich surface chemistry of polymer shell for the attachment of targeting molecules on NPs.[41]

Li et al. prepared ultrasmall highly water-soluble magnetite NPs (3.3 ± 0.5 nm) through the high-temperature coprecipitation route using a multifunctional water-soluble polymer (PTMP-PMAA) as capping ligand.[37] Later on, these MNPs were successfully employed as dual-contrast agents in MRI.[20] The efficiency of these NPs as MRI contrast agents was estimated by calculating the longitudinal (r_1) and transverse (r_2) relaxivities *in vivo* as well as *in vitro* experiments. The results demonstrated that the longitudinal and transverse relaxivities of MNPs were two to six times higher than that of commercially available T_1-positive contrast agent Gd-DPTA and dual T_2-negative contrast agent

SHU-555C. In short, these MNPs showed a great potential as dual-contrast agents of MRI, especially as a substitute to Gd-based T_1-contrast agents both in terms of performance and biocompatibility.[20]

Magnetofection

Among other potential bioapplications, gene delivery or DNA transfection is an outstanding application in the biotechnology of MNPs. In biotechnology, gene delivery means introduction of some foreign genes into cells to supplement flawed genes (gene therapy) or to provide some extra biological characteristic functions to cells or organisms (genetic engineering). For gene delivery, the desired genes are required to be carried inside the host cells by a carrier or vector, which is often a viral vector or sometimes a synthetic nonviral gene delivery system. Nonviral vectors which may comprise liposomes, polymers, or inorganic NPs have benefits over viral vectors, such as easy synthesis and reduced risk of cytotoxicity and immunogenicity.[123]

In recent years, polymer-modified gold and MIONs with size range 10–500 nm have been used as carriers for gene delivery.[124,125] The particular gene of interest is immobilized on the surface of these NPs and then NPs are introduced into the host cells through different processes such as microinjection, gene gun, impalefection, hydrostatic pressure, electroporation, etc. However, these processes suffer from lack of efficiency. In order to increase the efficiency of the gene delivery process, the NPs are required to bear suitable functional groups for increased gene-loading capacity. In case of gene carrier based on MNPs, the efficiency of the process can be increased by manipulating the NPs with the help of an external magnetic field to the desired location[126,127] as illustrated in Figure 21.7.

Targeting gene or DNA can be immobilized on to the NP surface in several ways but it is more suitable to load through electrostatic interactions because it gives a good loading capacity and easy release of DNA once inside the cell. Since the DNA is negatively charged, the NPs should have a high positive charge density for efficient DNA loading through electrostatic interactions.[45,63] Polyethylene imines (PEIs) are believed to be important polymeric material because of its ability to make complex with DNA by strong

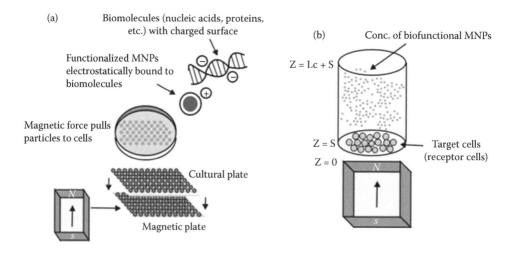

Figure 21.7 Process of magnetofection: (a) array of cell cultures positioned on the top of an array of cubical magnets and (b) the magnetic force pulls MNPs along with genes bound on their surface toward the cells.

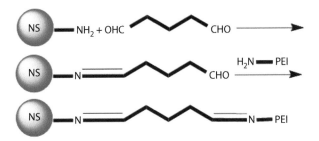

Figure 21.8 Schematic illustration of the coupling of PEI to the NH$_2$-NS NPs with glutaraldialde-hyde linkers.

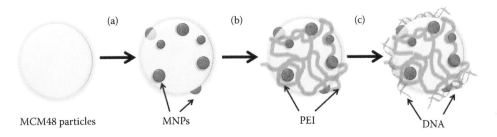

Figure 21.9 Particles for magnetofection: MNPs loaded on polymer particles (a) and subsequent loading of charged polymer (PEI) (b) and DNA molecules (c).

electrostatic interactions between its positively charged amine groups and the negatively charged phosphate backbone of DNA.[63]

McBain et al. reported the synthesis of an effective DNA delivery system from MNPs and PEI. PEI with small molecular weight (423–1800) was attached covalently to the composite iron oxide NPs with silica dextran coating size 250 nm as shown in Figure 21.8. This gave more firm attachment and provides an equal distribution of PEI even in small amounts across the NPs surface as compared to just electrostatic interactions between NPs and polymer. These NPs showed a high DNA loading capacity even at lower N:P ratio and good magnetic properties, which made them right candidates for a magnetic transfection of DNA (magnetofection) without any significant PEI toxicity.[63]

Yiu et al. used mesoporous silica MCM-48 and decorated them with MIONs and then their subsequent coating with positively charged PEI resulted into a new type of 300-nm-sized MNP-based gene delivery system (Figure 21.9). The DNA was attached to these final NP surface electrostatically and this system was reported to be 400 times more efficient than for DNA transfection compared with other commercial products.[128,129]

Magnetic hyperthermia

Magnetic hyperthermia is another important bioapplication in which MNPs are heated up locally with the help of an external oscillating magnetic field resulting in toxic amounts of heat to the tumor cells or it can be used to assist chemotherapy and radiotherapy treatments of tumors leading to more effective cancer cell destruction.[24] It has now been well established that superparamagnetic NPs have a high efficiency of absorbing the energy of an alternative magnetic field and converting it into thermal energy, and also that the tumor cells are more sensitive to an increase in temperature than healthy cells. These facts can

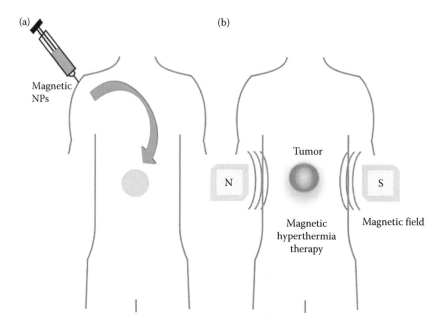

Figure 21.10 Diagram showing the intravenous injection of MNPs suspension in human body: (a) MNPs first travel to tumor and (b) then an oscillating external magnetic field induces hyperthermia.

be used *in vivo* to increase the temperature of tumor tissue locally and destroy the pathological cells by hyperthermia.[130,131] Magnetic hyperthermia has the advantage of fewer side effects compared with chemotherapy and radiotherapy due to its ease and capability of targeting cancerous tissues and thus is a promising technique for cancer treatment.[132]

In local MFH, MNPs are injected directly into the tumor site or into the blood stream from where they are driven and concentrated in the tumor site by some tumor-specific antibody targeting agent present on NPs. After that, the tumor is exposed to an alternating magnetic field causing the MNPs to oscillate and generate thermal energy through magnetic relaxation mechanisms. This process leads to a rise in local tissue temperature resulting in damage to the tumor cells as depicted in Figure 21.10. However, in all hyperthermia treatments, the temperature profile is the most important parameter to be controlled and an ideal temperature profile is that in which normal healthy cells remain at body temperature while the temperature of tumor tissue is raised and maintained around 45°C throughout the process.

Hyperthermia treatment can also be used for assistance of chemotherapy and radiotherapy as it increases the perfusion in the tumor tissue and local oxygen concentration, which are ideal conditions for absorption of drugs and γ-radiation to destroy the tumor cells. Thus, it might also be helpful to reduce the side effects of chemo- and radiotherapy by reducing their doses without compromising their therapeutic benefits. In order to obtain better results in magnetic hyperthermia, the MNPs should be less than 20 nm in diameter and homogeneous in size and shape as these parameters affect magnetic relaxation mechanisms, which are important for localized heating of tumor cells to between 42°C and 45°C.[133–135]

SPIONs due to their several other advantages such as biocompatibility, low toxicity, escapability from reticuloendothelial system (RES), and low protein adsorption are thought to be ideal candidates for magnetic hyperthermia therapy.[136,137] Like most of the

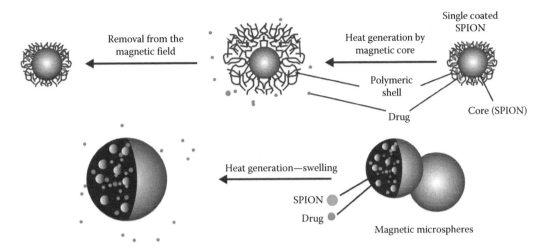

Figure 21.11 Schematic illustration of a multipurpose NP system active in hyperthermia and drug delivery processes.

in vivo bioapplication MNPs, in magnetic hyperthermia, superparamagnetic NPs are preferred because they do not aggregate in the absence of external magnetic field. When the size of a ferromagnetic NPs is reduced below a critical size, they get a single magnetic domain. They do not show any magnetization in the absence of external magnetic field, thus agglomeration of the NPs can be avoided within the capillaries.[138] pH- and thermo-sensitive polymers can be used to prepare the smart systems consisting of a magnetic core for heat production and stimuli-sensitive polymer shell to carry therapeutic and targeting agents, thus combining the benefits of chemotherapy and hyperthermia.[139] When such multifunctional NPs reach the targeted site, the magnetic core generates heat on the application of external alternating magnetic field, which causes the structural/conformational changes in the polymer shell of NPs, resulting in the release of loaded drug.[91,140,141] A schematic diagram of such a system is shown in Figure 21.11.[24] Similarly, some polymeric shells (e.g., PEG-based hydrogels) have the capability of cross linking, which can help in more controlled release of drug.[142,143]

Hayashi et al. prepared folic acid (FA) functionalized SPIONs with a natural polymer β-cyclodextrin (CD) for delivery of hydrophilic and lipophilic drugs. Upon application of a high-frequency magnetic field, drugs were released from the CD cavity on the particles and system could be switched on and off by on site performing drug delivery and hyperthermia at the same time.[144,145]

MWPs can play an important role in the preparation of superparamagnetic NPs for hyperthermia treatment of tumors *in vivo,* because of several advantages which they can provide to these NPs such as better control over size, shape, and stability. Multifunctional surface of these NPs provide opportunities for easy and simultaneous functionalization with different targeting, therapeutic, and labeling agents resulting in multifunctional nanoprobes. MWPs can serve as an excellent dispersant for these MNPs with better satiability in physiological environment, stealth from (RES), which can prolong their blood circulation time, and better specific deposition at tumor site. Moreover, thermal and pH-sensitive polymer shells of these NPs can further increase the efficiency of a targeted drug delivery. To achieve all of these benefits, more extensive research is required in this direction.

Magnetic separations

MNPs have also been used for magnetically controlled separation/isolation of significant amounts of hydrophilic and hydrophobic molecules from water. That is why they have several potential applications in biology, medicine, and biotechnology, including the separation of proteins, nucleic acids, and isolation of enzymes, cells, and pathogens from biological fluids such as human blood and cell lysates as depicted in Figure 21.12.[146] Therefore, MNPs prepared with MWPs due to their controlled size, superparamagnetic behavior, high magnetization, biocompatibility, and multifunctional surface chemistry have a various bioapplications such as in detection, diagnosis, separation, immunoassay, and therapy.

Chockalingam et al. reported the synthesis of collagen-coated gum arabic-modified MIONs (CCGAMIONs) and demonstrated their application for instant detection/isolation of pathogenic bacteria *Staphylococcus aureus* at ultralow concentrations, achieving a detection limit of 8 cfu/mL in 3 minutes.[147]

Majewski et al. described the synthesis of dual-responsive MNPs for nonviral gene delivery and cell separation.[125] Through ligand exchange process, the oleic acid stabilized SPIONs were transferred into water phase and functionalized with dopamine group and then 2-(dimethylamino)ethyl methacrylate (DMAEMA) was polymerized from the NPs surface by atom-transfer radical polymerization (ATRP) to prepare (γ-Fe$_2$O$_3$@PDMAEMA), which were highly water-soluble but also exhibited the pH- and temperature-dependent reversible aggregation determined by turbidimetry. The resulting NPs were nontoxic and showed two times higher magnetofection efficiency compared with conventional PEI-based transfection agents. Furthermore, after magnetofection, the transfected cells got magnetic properties and they were able to be selectively isolated by magnetic separation as shown in Figure 21.13.

Drug delivery

In modern drug therapy, NP-based controlled DDS have obtained a huge significance because of their intrinsic properties such as variety of materials, low toxicity, and improved drug efficacy as compared to the conventional DDS.[21] In the development of an efficient DDS, NPs can serve as a platform for attachment of different useful chemical moieties to make one multifunctional probe. For example, simultaneous attachment of fluorescent labelling, targeting and therapeutic agents to the surface of magnetic nanoparticles will result in a multifunctional nanoprobe. Physicochemical properties of NPs such as size, shape, high surface area, chemistry, and charge also play an important role in efficient

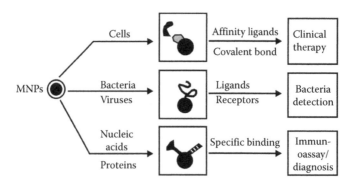

Figure 21.12 Potential applications of MNPs in bioseparation.

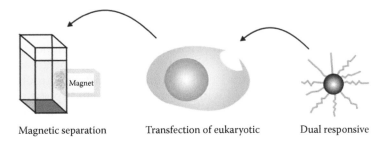

Magnetic separation Transfection of eukaryotic Dual responsive

Figure 21.13 Magnetic separation of magnetically transfected living cells.

delivery of therapeutic agents to the right place, at the right time, and in the right amount through their targeted delivery and sustained release.[148] There are several types of NPs that can be used in the development of NPs-based DDS[149] as shown in Figure 21.14.

Inorganic NPs prepared with water-soluble polymer ligands as capping agents are also being used in the development of novel and effective targeted DDS. Water-soluble polymer ligands would offer several advantages to NPs in this regard, such as biocompatible coating of the hydrophilic polymers around NPs could defy the body's immune system prolonging their blood circulation time. Hydrophilic polymer shell also provides NPs with stability, solubility, and rich surface chemistry for the attachment of targeting, fluorescent, moieties. A variety of hydrophobic/hydrophilic drugs can be loaded onto/into the polymer shell through electrostatic or covalent interactions. Polymer shell also provides stability to chemically sensitive drugs and prevents them from degradation, facilitates intracellular entry, and can serve as intracellular or intercellular drug depots.[149] Several types of nanomaterials are used as carriers in DDS as shown in Figure 21.14, but we will focus only on inorganic NPs with polymer coatings due to their advantages.

Polymer shell allows uploading of drug either through attachment or entrapment and provides good pharmacokinetic control. Several types of hydrophilic (natural and synthetic) polymers have been used for modification of inorganic NPs for the development of DDS such as starch,[150] chitosan,[151] gelatins,[152] dextran, plant resins,[153] etc., and synthetic polymers such as PEG,[154] polyacrylic acid (PAA),[155] PVA,[21] PEI,[64] etc.

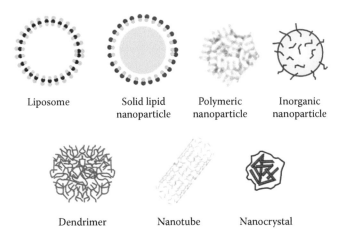

Liposome Solid lipid Polymeric Inorganic
    nanoparticle nanoparticle nanoparticle

Dendrimer Nanotube Nanocrystal

Figure 21.14 Various types of NPs used in biomedical research and drug delivery.

PEG is the most commonly used biologically inert polymer, which is used for modification of drugs and increases blood circulation time, biodistribution, and bioavailability of the drugs. PEG has also been extensively used for modification of inorganic NPs and is mostly covalent attached to the surface of NPs; moreover, the therapeutic agents and other functionalities can be appended to the PEG shell through functional groups.[100]

Presently, nanomaterials prepared with water-soluble monodisperse polymer are the main carriers used for a targeted drug delivery and controlled drug release,[156] because these small particles can enter into human organs and tissues that most of the large particles cannot. For example, NPs smaller than 50 nm can reach spleen, bone marrow, and tumor tissue through lymph or by penetrating through liver endothelium. Moreover, NPs can penetrate through many biological barriers to reach the site of interest. For instance, they can deliver drugs to brain by penetrating blood–brain barrier (BBB) and they can also enrich in lymph nodes through an oral administration.[149]

NPs modified with MWPs possess many advantages such as they can deliver drugs specifically to the affected site and can remain there for quite a long time. Drugs can be released controllably through diffusion and degradation of NPs. Biodegradable polymer nanomaterials can be hydrolyzed enzymatically or simply through pH change in a certain period of time and excreted through physiological pathways, avoiding side effects of the accumulating carrier materials, and so are more biocompatible, safer, and reliable. Biomolecules such as proteins, amino acids, and nucleic acids *in vivo* have stronger interaction with NPs and can replace water-soluble polymer ligands through ligand exchange[157] realizing the applications of these NPs in biolabeling of proteins, antibodies and polypeptides, biosensor of proteins, and pathogen separation.[158,159]

Thus, water-soluble polymer ligands can play an important role in the preparation of inorganic NPs for the development of a new and effective targeted DDS in future by high payload, functionalization with intelligent targeting moieties, drug bioavailability, biodistribution, and efficient evasion of the RES. Moreover, kinetic profiles of the drugs can be improved for their controlled and sustained release. Otherwise, the availability of the drug to the places like BBB[160] will not be possible.

Chertok et al. modified commercially available carboxylic acid-functionalized MIONs with low-molecular weight PEI, Mw = 1200.[64] The carboxylic acid groups on the surface of the MIONs were reacted with amine groups of the PEI through well-known 1-ethyl-3-(3-dimethylaminopropyl)-carbodiimide (EDC) coupling reaction. The presence of primary and secondary amine groups and positive charge on the PEI modified was varied with ninhydrin test and zeta potential measurements. The modified MNPs referred as GPEI were used as a model to demonstrate the preferential delivery, and accumulation of positively charged drug delivery carriers to the brain tumors. The results demonstrated that positively charged NPs have high affinity and selectivity to the tumor cells, which was also in accordance to the literature.[161,162]

Kayal et al. prepared 10-nm magnetite NPs through coprecipitation route, subsequently coated them with PVA and were loaded with an anticancer drug doxorubicin (DOX) through hydrogen boding between hydroxyl group ($-OH$) of PVA and amine group ($-NH_2$) of DOX as confirmed by Fourier transform infrared spectroscopy (FTIR). The *in vitro* drug loading and release analysis of these PVA-coated magnetite NPs showed that up to 45% of absorbed drug was released in 80 hours. The drug loading was increased with an increase in PVA contents of NPs, which was attributed to the increase in ($-OH$) groups of PVA present on NP surface. Adsorption of DOX was initially fast but then it slowed down and reached a saturation value. Similarly, drug release profiles showed an initial rapid drug release and it slowed down to 6 hours and with the increase in PVA contents, the drug release was slowed down due to higher interactions between polymer and drug adsorption and the release processes were controlled by Fickian diffusion process. These DOX-loaded PVA-coated magnetite NPs were demonstrated to have a potential for application as magnetic drug carriers for magnetically targeted drug delivery.[102]

Recently, we have reported the preparation of highly water-soluble MIONs through coprecipitation technique using dodecanethiol-polymethacrylic acid (DDT-PMAA),[163] which, due to its intrinsic properties, efficiently controls the size of the MNPs and also provides them superb water solubility, high stability against aggregation, and oxidation. Moreover, it provides MNPs with good biocompatibility and multifunctional surface having carboxylic acid and thioether groups. Polymer molecular weight and concentration were optimized to prepare MIONs with small size (4.6 ± 0.7 nm) and high magnetization (50 emu g^{-1}). MTT assays were performed with MNPs, which proved them to be highly biocompatible. A DDS based on these MNPs was designed by conjugating them with DOX, an anticancer drug. The drug was loaded on to the NPs in covalent and electrostatic modes and the efficiency of this system was determined with HepG2 cells through MTT assay. It was found that the activity of free drug and drug–NP conjugates was dose- and time-dependent and drug–NP conjugates showed higher toxicity than free drug.[163] Cell viability of HepG2 cells was initially dropped in case of free DOX, which was attributed to the higher drug uptake by cells through simple Fickian diffusion.[164–167] However, later on, free DOX showed a lower cytotoxicity than drug–NP conjugates, which was attributed to more efficient cellular uptake of drug–NP conjugates through endocytosis in comparison with the simple diffusion of free drug. Furthermore, it was found that anticancer efficiency of electrostatic drug–NP conjugates (DOX/MIONs) was more than covalent conjugates (DOX–MIONs) due to the fact that free DOX causes cell death by interacting with DNA and inhibiting its function inside the cell nucleus. It was suggested that in case of electrostatic drug–NP conjugates (DOX/MIONs), the DOX was released into the cytoplasm and penetrated to the nucleus whereby they chelated with DNA and caused cell death.[168] The earlier results were also verified with investigation of intracellular localization of all the three drugs (pure DOX, DOX–MIONs, and DOX/MIONs) in HepG2 cells by confocal laser scanning microscopy (CLSM) after 4 and 8 hours of incubation of HepG2 cells with drugs having equivalent concentration of DOX 25 µg/mL. The cells were washed with buffer and treated with DAPI to stain the nuclei (blue fluorescence) before microscopy. It is known that the DOX itself has the ability to stain the nuclei (red fluorescence). Figure 21.15 represents the confocal laser scanning microscopic images of the HepG2 cells after 4 and 8 hours of incubation with all the three drugs along with a control experiment.

It is obvious from Figure 21.15 that all the three drugs showed dissimilar distribution patterns in HepG2 cells. Particularly, cells incubated with free DOX showed a purple color inside the cell nuclei due to the colocalization of red fluorescence and blue fluorescence (merged together), suggesting that free DOX was mainly distributed in the nuclei of the cells. This is in consistence with the well-accepted cytotoxic mechanism of free DOX, which shows that cytotoxicity largely arises from its direct intercalation into DNA and consequently the inhibition of DNA replication and RNA biosynthesis as discussed earlier.[169]

In contrast, HepG2 cells that were treated with DOX–MIONs were homogeneously distributed in the cytoplasm and weak red fluorescence could be observed in cell nuclei, which suggests that DOX–MIONs was mainly distributed in the cytoplasm while little

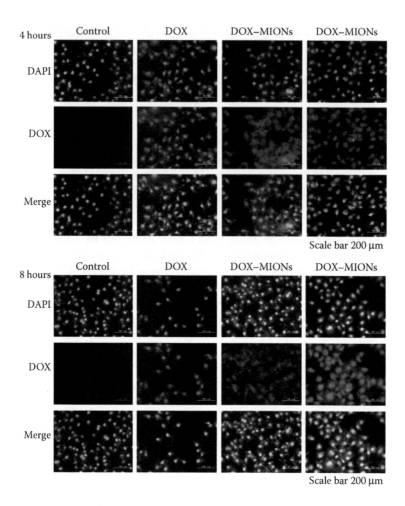

Figure 21.15 Laser confocal fluorescence microscopic images of HepG2 cells (stained with DAPI) after 4 and 8 hours of incubation with free DOX and DOX–MIOINs conjugates with equivalent DOX concentration (25 μg/mL).

DOX–MIONs penetrated the nuclei. This suggests that some alternative cytotoxicity mechanisms such as inhibition of mitochondrial function,[170] generation of free radicals,[171] and direct membrane effects[172] may have been adopted for DOX upon conjugation to MIONs.[164] However, in the case of DOX/MIONs, the main fraction of DOX can be observed inside the nuclei (red fluorescence), which is probably due to the pH-responsive release of DOX from DOX/MIONs in tumor cells.[173,174] Because DOX/MIONs was formed through the electrostatic interactions of DOX and MIONs, in the acid environment of tumor cells, the electrostatic interactions were weakened, which triggered the release of DOX from DOX/MIONs.[164] However, small amounts of DOX still remained in the cytoplasm, which can be attributed to unreleased DOX. Finally, it was concluded that the cytotoxic mechanism of DOX/MIONs was probably similar to that of free DOX. As a result, the binding mode of DOX to MIONs could influence intracellular distribution of DOX (Figure 21.16) and led to different cytotoxic efficiency. Conclusively, the drug-loading mode in a DDS has a pivotal role toward its ultimate therapeutic efficiency and is schematically represented in Figure 21.16 in this particular case study.

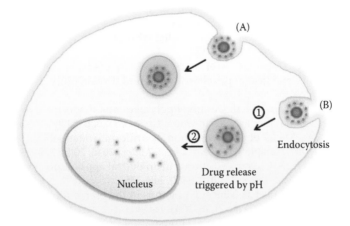

Figure 21.16 Schematic diagram of the effect of DOX loading method on its intracellular localization. DOX–MIONs is usually dispersed in the cytoplasm (A) and the main fraction of DOX present in the cell nuclei is released from DOX/MIONs (B).

Conclusion and prospects

Preparation of MNPs with MWPs offers several advantages such as better control over size and morphology, monodispersity, biocompatibility, multifunctional surface chemistry, easy functionalization, and modification ability, therefore these NPs have a large scope of applications in diverse areas of research such as in fabrication of optoelectronic devices, in the development of selective and efficient catalytic systems for industrial chemical processes, in preparation of early disease diagnosis and effective treatment systems, in formulation of vital drug delivery and controlled release systems, such as biocompatibility, excellent water solubility, long blood circulation time, etc.

In future, it would be of great importance to have a better understanding of the influences of MWP properties (molecular weight and molecular weight distribution, degree and branching, variety, and number of functional groups present on the polymer backbone) on the formation and stability of NPs and it will also help to integrate this idea toward the design and use of new polymer-based materials for NPs.

NP–polymer composites are another type of novel materials that have numerous applications in electronics, sensors, and catalysis. Similarly, core–shell NPs with combined unique properties of two or more different materials into one system can be prepared with the use of MWPs. Polymer ligands will also help to develop the NP assemblies with unique morphologies and definite interparticle spacing, which are otherwise difficult to obtain. Shortly, MWPs will continue to provide an endless account of possibilities for both fundamental and applied research.

References

1. Laurent S., Forge D., Port M. et al. Magnetic iron oxide nanoparticles: Synthesis, stabilization, vectorization, physicochemical characterizations, and biological applications. *Chem. Rev.*, **2008**, 108(6): 2064–2110.
2. Mornet S., Vasseur S., Grasset F. et al. Magnetic nanoparticle design for medical applications. *Prog. Solid State Chem.*, **2006**, 34(2–4): 237–247.

3. Kim E. H., Ahn Y., and Lee H. S. Biomedical applications of superparamagnetic iron oxide nanoparticles encapsulated within chitosan. *J. Alloys Compd.*, **2007**, 435: 633–636.
4. Rozenberg B. A. and Tenne R. Polymer assisted fabrication of nanoparticles and nanocomposites. *Prog. Polym. Sci.*, **2008**, 33(1): 40–112.
5. Seino S., Kusunose T., Sekino T. et al. Synthesis of gold/magnetic iron oxide composite nanoparticles for biomedical applications with good dispersibility. *J. Appl. Phys.*, **2006**, 99(8): 101–103.
6. Wang Z., Tan B., Hussain I. et al. Design of polymeric stabilizers for size-controlled synthesis of monodisperse gold nanoparticles in water. *Langmuir*, **2006**, 23(2): 885–895.
7. Niu D. C., Li Y. S., Ma Z. et al. Preparation of uniform, water-soluble, and multifunctional nanocomposites with tunable sizes. *Adv. Funct. Mater.*, **2010**, 20(5): 773–780.
8. Jang J. H. and Lim H. B. Characterization and analytical application of surface modified magnetic nanoparticles. *Microchem. J.*, **2010**, 94(2): 148–158.
9. Song X. G., Ji X. H., Jun L. et al. Molecular ligands in the preparation and surface modification of gold nanocrystals. *Prog. Chem.*, **2008**, 20(1): 11–18.
10. Huang X., Zhang H., Liang L. Y. et al. Preparation of nanoparticles with multi-functional water-soluble polymer ligands. *Prog. Chem.*, **2010**, 22(5): 953–961.
11. Hussain I., Graham S., Wang Z. et al. Size-controlled synthesis of near-monodisperse gold nanoparticles in the 1–4 nm range using polymeric stabilizers. *J. Am. Chem. Soc.*, **2005**, 127(47): 16398–16399.
12. Turkevich J., Stevenson P. C., and Hillier J. The formation of colloidal gold. *J. Phys. Chem.*, **1953**, 57(7): 670–673.
13. Chen S., Ingram R. S., Hostetler M. J. et al. Gold nanoelectrodes of varied size: Transition to molecule-like charging. *Science*, **1998**, 280(5372): 2098–2101.
14. Tsai L. C., Cheng I. C., Tu M. C. et al. Formation of single electron transistors using self-assembly of nanoparticle chains. *J. Nanopart. Res.*, **2010**, 12(8): 2859–2864.
15. Boyer D. Photothermal imaging of nanometer-sized metal particles among scatterers. *Science*, **2002**, 297(5584): 1160–1163.
16. Saha K., Agasti S. S., Kim C. et al. Gold nanoparticles in chemical and biological sensing. *Chem. Rev.*, **2012**, 112(5): 2739–2779.
17. Liu Z., Kiessling F., and Gätjens J. Advanced nanomaterials in multimodal imaging: Design, functionalization, and biomedical applications. *J. Nano. Mater.*, **2010**, 2010: 1–15.
18. Katz E. and Pita M. Biomedical applications of magnetic particles. *Fine Particles in Medicine and Pharmacy*. Egon Matijević (Ed.), Springer, USA; 2012:147–173.
19. Mateo C., Vázquez C., Buján N. M. C. et al. Synthesis and characterization of $CoFe_2O_4$–PVP nanocomposites. *J. Non-Cryst. Solids.*, **2008**, 354(47–51): 5236–5237.
20. Li Z., Yi P. W., Sun Q. et al. Ultrasmall water-soluble and biocompatible magnetic iron oxide nanoparticles as positive and negative dual contrast agents. *Adv. Funct. Mater.*, **2012**, 22(11): 2387–2393.
21. Zhou L., He B., and Zhang F. Facile one-pot synthesis of iron oxide nanoparticles cross-linked magnetic poly(vinyl alcohol) gel beads for drug delivery. *ACS Appl. Mater. Interfaces*, **2012**, 4(1): 192–199.
22. Nie L. B., Wang X. L., Li S. et al. Amplification of fluorescence detection of DNA based on magnetic separation. *Anal. Sci.*, **2009**, 25(11): 1327–1331.
23. Tai Y. L., Wang L., Fan G. Q. et al. Recent research progress on the preparation and application of magnetic nanospheres. *Polym. Int.*, **2011**, 60(7): 976–994.
24. Laurent S., Dutz S., Häfeli U. O. et al. Magnetic fluid hyperthermia: Focus on superparamagnetic iron oxide nanoparticles. *Adv. Colloid Interface Sci.*, **2011**, 166(1): 8–23.
25. Zhong J. Y. and Cao C. B. Nearly monodisperse hollow Fe_2O_3 nanoovals: Synthesis, magnetic property and applications in photocatalysis and gas sensors. *Sens. Actuators B*, **2010**, 145(2): 651–656.
26. Lu L. T., Tung L. D., Robinson I. et al. Size and shape control for water-soluble magnetic cobalt nanoparticles using polymer ligands. *J. Mater. Chem.*, **2008**, 18(21): 2453.
27. Huang S. H. and Juang R. S. Biochemical and biomedical applications of multifunctional magnetic nanoparticles: A review. *J. Nanopart. Res.*, **2011**, 13(10): 4411–4430.

28. Xie J., Huang J., Li X. et al. Iron oxide nanoparticle platform for biomedical applications. *Curr. Med. Chem.*, **2009**, 16(10): 1278–1294.

29. Lu A. H., Salabas E. L., and Schüth F. Magnetic nanoparticles: Synthesis, protection, functionalization, and application. *Angew. Chem. Int. Ed.*, **2007**, 46(8): 1222–1244.

30. Casula M. F., Jun Y., Zaziski D. J. et al. The concept of delayed nucleation in nanocrystal growth demonstrated for the case of iron oxide nanodisks. *J. Am. Chem. Soc.*, **2006**, 128(5): 1675–1682.

31. Sun S. and Zeng H. Size-controlled synthesis of magnetite nanoparticles. *J. Am. Chem. Soc.*, **2002**, 124(28): 8204–8205.

32. Park J., Lee E., Hwang N. M. et al. One-nanometer-scale size-controlled synthesis of monodisperse magnetic iron oxide nanoparticles. *Angew. Chem. Int. Ed.*, **2005**, 44(19): 2872–2877.

33. Gu H., Xu K., Xu C. et al. Biofunctional magnetic nanoparticles for protein separation and pathogen detection. *Chem. Commun.*, **2006**, 1(9): 941–949.

34. Huh Y. M., Jun Y., Song H. T. et al. *In vivo* magnetic resonance detection of cancer by using multifunctional magnetic nanocrystals. *J. Am. Chem. Soc.*, **2005**, 127(35): 12387–12391.

35. Bourlinos A. B., Bakandritsos A., Georgakilas V. et al. Surface modification of ultrafine magnetic iron oxide particles. *Chem. Mater.*, **2002**, 14(8): 3226–3228.

36. Woo K. and Hong J. Surface modification of hydrophobic iron oxide nanoparticles for clinical applications. *IEEE Trans. Magn.*, **2005**, 41(10): 4137–4139.

37. Li Z., Tan B., Allix M. et al. Direct coprecipitation route to monodisperse dual-functionalized magnetic iron oxide nanocrystals without size selection. *Small*, **2008**, 4(2): 231–239.

38. Couto G. G., Klein J. J., Schreiner W. H. et al. Nickel nanoparticles obtained by a modified polyol process: Synthesis, characterization, and magnetic properties. *J. Colloid Interface Sci.*, **2007**, 311(2): 461–468.

39. Boyer C., Whittaker M. R., Bulmus V. et al. Magnetic nanoparticles and targeted drug delivery. *NPG Asia Mater.*, **2010**, 2(1): 23–30.

40. Grubbs R. B. Roles of polymer ligands in nanoparticle stabilization. *Polym. Rev.*, **2007**, 47(2): 197–215.

41. Chanana M., Mao Z. W., and Wang D. Y. Using polymers to make up magnetic nanoparticles for biomedicine. *J. Biomed. Nanotechnol.*, **2009**, 5(6): 652–668.

42. Wu W., He Q., and Jiang C. Magnetic iron oxide nanoparticles: Synthesis and surface functionalization strategies. *Nanoscale Res. Lett.*, **2008**, 3(11): 397–415.

43. Chomoucka J., Drbohlavova J., Huska D. et al. Magnetic nanoparticles and targeted drug delivering. *Pharmacol. Res.*, **2010**, 62(2): 144–149.

44. Sun C., Lee J., and Zhang M. Magnetic nanoparticles in MR imaging and drug delivery. *Adv. Drug Deliv. Rev.*, **2008**, 60(11): 1252–1265.

45. McBain S. C., Yiu H. H. P., and Dobson J. Magnetic nanoparticles for gene and drug delivery. *Int. J. Nanomed.*, **2008**, 3(2): 169–180.

46. Durán J. D. G., Arias J. L., Gallardo V. et al. Magnetic colloids as drug vehicles. *J. Pharm. Sci.*, **2008**, 97(8): 2948–2983.

47. Horak D., Babic M., Mackova H. et al. Preparation and properties of magnetic nano- and microsized particles for biological and environmental separations. *J. Sep. Sci.*, **2007**, 30(11): 1751–1772.

48. Liu Y. L., Jia L., and Xing D. Recent advances in the preparation of magnetic microspheres and its application in bio-separation and concentration fields. *Chin. J. Anal. Chem.*, **2007**, 35(8): 1225–1232.

49. Na H. B., Song I. C., and Hyeon T. Inorganic nanoparticles for MRI contrast agents. *Adv. Mater.*, **2009**, 21(21): 2133–2148.

50. Fornara A., Johansson P., Petersson K. et al. Tailored magnetic nanoparticles for direct and sensitive detection of biomolecules in biological samples. *Nano Lett.*, **2008**, 8(10): 3423–3428.

51. Peng S., Wang C., Xie J. et al. Synthesis and stabilization of monodisperse Fe nanoparticles. *J. Am. Chem. Soc.*, **2006**, 128(33): 10676–10677.

52. Makhluf S. B., AbuMukh R., Rubinstein S. et al. Modified PVA–Fe$_3$O$_4$ nanoparticles as protein carriers into sperm cells. *Small*, **2008**, 4(9): 1453–1458.

53. Sun C. R., Du K., Fang C. et al. PEG-mediated synthesis of highly dispersive multifunctional superparamagnetic nanoparticles: Their physicochemical properties and function *in vivo*. *ACS Nano*, **2010**, 4(4): 2402–2410.

54. Hu F., Jia Q., Li Y. et al. Facile synthesis of ultrasmall PEGylated iron oxide nanoparticles for dual-contrast T_1- and T_2-weighted magnetic resonance imaging. *Nanotechnology*, **2011**, 22(24): 245604.

55. Wang C. H., Liu C. J., Wang C. L. et al. Optimizing the size and surface properties of polyethylene glycol (PEG)-gold nanoparticles by intense x-ray irradiation. *J. Phys. D: Appl. Phys.*, **2008**, 41(19): 195301.

56. Mahmoudi M., Simchi A., and Imani M. Cytotoxicity of uncoated and polyvinyl alcohol coated superparamagnetic iron oxide nanoparticles. *J. Phys. Chem. C*, **2009**, 113(22): 9573–9580.

57. Khanna P. K., Gokhale R. R., Subbarao V. et al. Synthesis and optical properties of CdS/PVA nanocomposites. *Mater. Chem. Phys.*, **2005**, 94(3): 454–459.

58. Arshi N., Ahmed F., Kumar S. et al. Comparative study of the Ag/PVP nanocomposites synthesized in water and in ethylene glycol. *Curr. Appl. Phys.*, **2011**, 11(1): 346–349.

59. Zhang Y., Liu J. Y., Ma S. et al. Synthesis of PVP-coated ultra-small Fe_3O_4 nanoparticles as a MRI contrast agent. *J. Mater. Sci.: Mater. Med.*, **2010**, 21(4): 1205–1210.

60. Lee H. Y., Lee S. H., Xu C. et al. Synthesis and characterization of PVP-coated large core iron oxide nanoparticles as an MRI contrast agent. *Nanotechnology*, **2008**, 19(16): 165101.

61. Ji M. L., Yang W. L., Ren Q. G. et al. Facile phase transfer of hydrophobic nanoparticles with poly(ethylene glycol) grafted hyperbranched poly(amido amine). *Nanotechnology*, **2009**, 20(7): 075101.

62. Wang X., Zhou L., Ma Y. et al. Control of aggregate size of polyethyleneimine-coated magnetic nanoparticles for magnetofection. *Nano Res.*, **2010**, 2(5): 365–372.

63. McBain S. C., Yiu H. H. P., El Haj A. et al. Polyethyleneimine functionalized iron oxide nanoparticles as agents for DNA delivery and transfection. *J. Mater. Chem.*, **2007**, 17(24): 2561–2565.

64. Chertok B., David A. E., and Yang V. C. Polyethyleneimine-modified iron oxide nanoparticles for brain tumor drug delivery using magnetic targeting and intra-carotid administration. *Biomaterials*, **2010**, 31(24): 6317–6324.

65. Schaeffer N., Tan B., Dickinson C. et al. Fluorescent or not? Size-dependent fluorescence switching for polymer-stabilized gold clusters in the 1.1–1.7 nm size range. *Chem. Commun.*, **2008**, 0(34): 3986–3988.

66. Fabris L., Antonello S., Armelao L. et al. Gold nanoclusters protected by conformationally constrained peptides. *J. Am. Chem. Soc.*, **2005**, 128(1): 326–336.

67. Quan C. Y., Chen J. X., Wang H. Y. et al. Core–shell nanosized assemblies mediated by the α–β cyclodextrin dimer with a tumor-triggered targeting property. *ACS Nano*, **2010**, 4(7): 4211–4219.

68. Levy R. Peptide-capped gold nanoparticles: Towards artificial proteins. *ChemBioChem*, **2006**, 7(8): 1141–1145.

69. Gonzaga F., Singh S., and Brook M. A. Biomimetic synthesis of gold nanocrystals using a reducing amphiphile. *Small*, **2008**, 4(9): 1390–1398.

70. Hao N., Li L., Zhang Q. et al. The shape effect of PEGylated mesoporous silica nanoparticles on cellular uptake pathway in hela cells. *Microporous Mesoporous Mater.*, **2012**, 162(0): 14–23.

71. Parkes L. M., Hodgson R., Lu L. T. et al. Cobalt nanoparticles as a novel magnetic resonance contrast agent relaxivities at 1.5 and 3 Tesla. *Contrast Media Mol. Imaging*, **2008**, 3(4): 150–156.

72. Rouhana L. L., Jaber J. A., and Schlenoff J. B. Aggregation-resistant water-soluble gold nanoparticles. *Langmuir*, **2007**, 23(26): 12799–12801.

73. Yang Q., Li W., Wei D. X. et al. Stabilization and its mechanisms of the metal nanoparticles/polymer systems. *Prog. Chem.*, **2006**, 18(2): 290–297.

74. Xin H., Hui Z., Liyun L. et al. Preparation of nanoparticles with multi-functional water-soluble polymer ligands. *Prog. Chem.*, **2010**, 22 (5): 953–961.

75. Kool E. T. Preorganization of DNA: Design principles for improving nucleic acid recognition by synthetic oligonucleotides. *Chem. Rev.*, **1997**, 97(5): 1473–1488.

76. Dorokhin D., Tomczak N., Han M. et al. Reversible phase transfer of (CdSe/ZnS) quantum dots between organic and aqueous solutions. *ACS Nano*, **2009**, 3(3): 661–667.

77. Wang Y., Wong J. F., Teng X. et al. Pulling nanoparticles into water: Phase transfer of oleic acid stabilized monodisperse nanoparticles into aqueous solutions of α-cyclodextrin. *Nano Lett.*, **2003**, 3(11): 1555–1559.

78. Latham A. H. and Williams M. E. Versatile routes toward functional, water-soluble nanoparticles via trifluoroethylester–PEG–thiol ligands. *Langmuir,* **2006**, 22(9): 4319–4326.
79. Choi J. S., Jun Y. W., Yeon S. I. et al. Biocompatible heterostructured nanoparticles for multimodal biological detection. *J. Am. Chem. Soc.,* **2006**, 128(50): 15982–15983.
80. Roy S., Vincent R., Tyler N. et al. Polymer-mediated self-assembly of nanoparticles. In *Dekker Encyclopedia of Nanoscience and Nanotechnology,* 2nd edition. Six volume set (print version). C. I. Contesc and K. Putyera (Eds), CRC Press, Florida; 2008:3443–3455.
81. Shenhar R., Norsten T. B., and Rotello V. M. Polymer-mediated nanoparticle assembly: Structural control and applications. *Adv. Mater.,* **2005**, 17(6): 657–669.
82. Bao N. Z. and Gupta A. Self-assembly of superparamagnetic nanoparticles. *J. Mater. Res.,* **2011**, 26(2): 111–121.
83. Ward M. A. and Georgiou T. K. Thermoresponsive polymers for biomedical applications. *Polymers,* **2011**, 3(3): 1215–1242.
84. Sun L., Huang C., Gong T. et al. A Biocompatible approach to surface modification: Biodegradable polymer functionalized superparamagnetic iron oxide nanoparticles. *Mater. Sci. Eng. C,* **2010**, 30(4): 583–589.
85. Zakharchenko S., Puretskiy N., Stoychev G. et al. Temperature controlled encapsulation and release using partially biodegradable thermo-magneto-sensitive self-rolling tubes. *Soft Matter,* **2010**, 6(12): 2633–2636.
86. Zhang S., Zhang L., He B. et al. Preparation and characterization of thermosensitive PNIPAA-coated iron oxide nanoparticles. *Nanotechnology,* **2008**, 19(32): 325608.
87. Liu C., Guo J., Yang W. et al. Magnetic mesoporous silica microspheres with thermo-sensitive polymer shell for controlled drug release. *J. Mater. Chem.,* **2009**, 19(27): 4764–4770.
88. Zhang J. and Misra R. D. K. Magnetic drug-targeting carrier encapsulated with thermosensitive smart polymer: Core–shell nanoparticle carrier and drug release response. *Acta Biomater.,* **2007**, 3(6): 838–850.
89. Yuan Q., Venkatasubramanian R., Hein S. et al. A stimulus-responsive magnetic nanoparticle drug carrier: Magnetite encapsulated by chitosan-grafted-copolymer. *Acta Biomater.,* **2008**, 4(4): 1024–1037.
90. Purushotham S. and Ramanujan R. V. Modeling the performance of magnetic nanoparticles in multimodal cancer therapy. *J. Appl. Phys.,* **2010**, 107(11): 114701.
91. Brazel C. Magnetothermally-responsive nanomaterials: Combining magnetic nanostructures and thermally-sensitive polymers for triggered drug release. *Pharm. Res.,* **2009**, 26(3): 644–656.
92. Morishita M., Goto T., Nakamura K. et al. Novel oral insulin delivery systems based on complexation polymer hydrogels: Single and multiple administration studies in Type 1 and 2 diabetic rats. *J. Controlled Release,* **2006**, 110(3): 587–594.
93. Feng Z., Wang Z., Gao C. et al. Template polymerization to fabricate hydrogen-bonded poly(acrylic acid)/poly(vinylpyrrolidone) hollow microcapsules with a pH-mediated swelling–deswelling property. *Chem. Mater.,* **2007**, 19(19): 4648–4657.
94. Serra L., Doménech J., and Peppas N. A. Drug transport mechanisms and release kinetics from molecularly designed poly(acrylic acid-g-ethylene glycol) hydrogels. *Biomaterials,* **2006**, 27(31): 5440–5451.
95. Tonge S. R. and Tighe B. J. Responsive hydrophobically associating polymers: A review of structure and properties. *Adv. Drug Deliv. Rev.,* **2001**, 53(1): 109–122.
96. Asayama S., Sekine T., Kawakami H. et al. Design of aminated poly(1-vinylimidazole) for a new pH-sensitive polycation to enhance cell-specific gene delivery. *Bioconjugate Chem.,* **2007**, 18(5): 1662–1667.
97. Kim K., Kim T. H., Choi J. H. et al. Synthesis of a pH-sensitive PEO-based block copolymer and its application for the stabilization of iron oxide nanoparticles. *Macromol. Chem. Phys.,* **2010**, 211(10): 1127–1136.
98. Chen D., Xia X., Gu H. et al. pH-responsive polymeric carrier encapsulated magnetic nanoparticles for cancer targeted imaging and delivery. *J. Mater. Chem.,* **2011**, 21(34): 12682–12690.
99. Chang Y. L., Meng X. L., Zhao Y. L. et al. Novel water-soluble and pH-responsive anticancer drug nanocarriers: Doxorubicin-PAMAM dendrimer conjugates attached to superparamagnetic iron oxide nanoparticles (IONPs). *J. Colloid Interface Sci.,* **2011**, 363(1): 403–409.

100. Neoh K. G. and Kang E. T. Functionalization of inorganic nanoparticles with polymers for stealth biomedical applications. *Polym. Chem.*, **2011**, 2(4): 747–759.

101. Gupta A. K. and Wells S. Surface-modified superparamagnetic nanoparticles for drug delivery prep, charac, and cytotoxicity studies. *IEEE Trans. Nanobiosci.*, **2004**, 3(1): 66–73.

102. Kayal S. and Ramanujan R. V. Doxorubicin loaded PVA coated iron oxide nanoparticles for targeted drug delivery. *Mater. Sci. Eng. C*, **2010**, 30(3): 484–490.

103. Cengelli F. Interaction of functionalized superparamagnetic iron oxide nanoparticles with brain structures. *J. Pharmacol. Exp. Ther.*, **2006**, 318(1): 108–116.

104. Elkhoury J. M., Caruntu D., Connor C. J. et al. Poly(allylamine) stabilized iron oxide magnetic nanoparticles. *J. Nanopart. Res.*, **2007**, 9(5): 959–964.

105. Li Z., Wang S. X., Sun Q. et al. Ultrasmall manganese ferrite nanoparticles as positive contrast agent for magnetic resonance imaging. *Adv. Healthcare Mater.*, **2013**, 2(7): 958–964.

106. Ge Y. Q., Zhang Y., He S. Y. et al. Fluorescence modified chitosan-coated magnetic nanoparticles for high-efficient cellular imaging. *Nanoscale Res. Lett.*, **2009**, 4(4): 287–295.

107. Hradil J., Pisarev A., Babič M. et al. Dextran-modified iron oxide nanoparticles. *China Particuol.*, **2007**, 5(2): 162–168.

108. Kim D. K., Mikhaylova M., Wang F. H. et al. Starch-coated superparamagnetic nanoparticles as MR contrast agents. *Chem. Mater.*, **2003**, 15(23): 4343–4351.

109. Owens D. E. and Peppas N. A. Opsonization, biodistribution and pharmacokinetics of polymeric nanoparticles. *Int. J. Pharm.*, **2006**, 307(1): 93–102.

110. Zahr A. S., Davis C. A., and Pishko M. V. Macrophage uptake of core–shell nanoparticles surface modified with poly(ethylene glycol). *Langmuir*, **2006**, 22(19): 8178–8185.

111. Vonarbourg A., Passirani C., Saulnier P. et al. Parameters influencing the stealthiness of solloidal drug delivery systems. *Biomaterials*, **2006**, 27(24): 4356–4373.

112. Moghimi S. M., Hunter A. C., and Murray J. C. Long-circulating and target-specific nanoparticles: Theory to practice. *Pharmacol. Rev.*, **2001**, 53(2): 283–318.

113. Zhang M., Desai T., and Ferrari M. Proteins and cells on PEG immobilized silicon surfaces. *Biomaterials,* **1998**, 19(10): 953–960.

114. Kotsmar C., Yoon K. Y., Yu H. et al. Stable citrate-coated iron oxide superparamagnetic nanoclusters at high salinity. *Ind. Eng. Chem. Res.*, **2010**, 49(24): 12435–12443.

115. Yamaura M., Camilo R. L., Sampaio L. C. et al. Preparation and characterization of (3-aminopropyl) triethoxysilane-coated magnetite nanoparticles. *J. Magn. Magn. Mater.*, **2004**, 279(2): 210–217.

116. Wan S., Zheng Y., Liu Y. et al. Fe_3O_4 nanoparticles coated with homopolymers of glycerol mono(meth)acrylate and their block copolymers. *J. Mater. Chem.*, **2005**, 15(33): 3424–3430.

117. Wan S., Huang J., Yan H. et al. Size-controlled preparation of magnetite nanoparticles in the presence of graft copolymers. *J. Mater. Chem.*, **2006**, 16(3): 298–303.

118. Lutz J. F., Stiller S., Hoth A. et al. One-pot synthesis of PEGylated ultrasmall iron-oxide nanoparticles and their *in vivo* evaluation as magnetic resonance imaging contrast agents. *Biomacromolecules*, **2006**, 7(11): 3132–3138.

119. Yeh Y. C., Creran B., and Rotello V. M. Gold nanoparticles: Preparation, properties, and applications in bionanotechnology. *Nanoscale*, **2012**, 4(6): 1871–1880.

120. Dykman L. and Khlebtsov N. Gold nanoparticles in biomedical applications: Recent advances and perspectives. *Chem. Soc. Rev.*, **2012**, 41(6): 2256–2282.

121. Petkar K. C., Chavhan S. S., Agatonovik-Kustrin S. et al. Nanostructured materials in drug and gene delivery: A review of the state of the art. *Crit. Rev. Ther. Drug.*, **2011**, 28(2): 101–164.

122. Sharma P., Brown S., Walter G. et al. Nanoparticles for bioimaging. *Adv. Colloid Interface Sci.*, **2006**, 123(126): 471–485.

123. Salem A. K., Searson P. C., and Leong K. W. Multifunctional nanorods for gene delivery. *Nat. Mater.*, **2003**, 2(10): 668–671.

124. Xiao B., Wan Y., Wang X. Y. et al. Synthesis and characterization of N-(2-hydroxy)propyl-3-trimethyl ammonium chitosan chloride for potential application in gene delivery. *Colloids Surf. B*, **2012**, 91(1): 168–174.

125. Majewski A. P., Schallon A., Jérôme V. et al. Dual-responsive magnetic core–shell nanoparticles for nonviral gene delivery and cell separation. *Biomacromolecules*, **2012**, 13(3): 857–866.

126. Berry C. C. Progress in functionalization of magnetic nanoparticles for applications in biomedicine. *J. Phys. D: Appl. Phys.*, **2009**, 42(22): 224003.
127. Furlani E. P. and Ng K. C. Nanoscale magnetic biotransport with application to magnetofection. *Phys. Rev. E*, **2008**, 77(6): 1–8.
128. Yiu H. H. P., McBain S. C., El Haj A. J. et al. A triple-layer design for polyethyleneimine-coated, nanostructured magnetic particles and their use in DNA binding and transfection. *Nanotechnology*, **2007**, 18(43): 435601.
129. Yiu H. H. P., McBain S. C., Lethbridge Z. A. D. et al. Preparation and characterization of poly-ethylenimine-coated Fe_3O_4-MCM-48 nanocomposite particles as a novel agent for magnet-assisted transfection. *J. Biomed. Mater. Res. A*, **2010**, 92(1): 386–392.
130. Jordan A., Scholz R., Wust P. et al. Endocytosis of dextran and silan-coated magnetite nanoparticles and the effect of intracellular hyperthermia on human mammary carcinoma cells *in vitro*. *J. Magn. Magn. Mater.*, **1999**, 194(3): 185–196.
131. Moroz P., Jones S. K., and Gray B. N. Magnetically mediated hyperthermia: Current status and future directions. *Int. J. Hyperther.*, **2002**, 18(4): 267–284.
132. Maier H. K., Ulrich F., Nestler D. et al. Efficacy and safety of intratumoral thermotherapy using magnetic iron oxide nanoparticles combined with external beam radiotherapy on patients with recurrent glioblastoma multiforme. *J. Neurooncol.*, **2011**, 103(2): 317–324.
133. Rosensweig R. E. Heating magnetic fluid with alternating magnetic field. *J. Magn. Magn. Mater.*, **2002**, 252(0): 370–374.
134. Pankhurst Q. A., Connolly J., Jones S. K. et al. Applications of magnetic nanoparticles in bio-medicine. *J. Phys. D: Appl. Phys.*, **2003**, 36(13): 167–181.
135. Andrade Â., Ferreira R., Fabris J. et al. Coating nanomagnetic particles for biomedical applications. *Biomed. Eng.*, **2011**, 2(5): 157–176.
136. Mahmoudi M., Milani A. S., and Stroeve P. Synthesis, surface architecture and biological response of superparamagnetic iron oxide nanoparticles for application in drug delivery: A review. *Int. J. Biomed. Nanosci. Nanotechnol.*, **2010**, 1(2): 164–201.
137. Mahmoudi M., Simchi A., and Imani M. Recent advances in surface engineering of superpara-magnetic iron oxide nanoparticles for biomedical applications. *JICS*, **2010**, 7(2): 1–27.
138. Mornet S., Vasseur S., Grasset F. et al. Magnetic nanoparticle design for medical diagnosis and therapy. *J. Mater. Chem.*, **2004**, 14(14): 2161–2175.
139. Dimitrov I., Trzebicka B., Müller A. H. E. et al. Thermosensitive water-soluble copolymers with doubly responsive reversibly interacting entities. *Prog. Polym. Sci.*, **2007**, 32(11): 1275–1343.
140. Brazel C. S. and Peppas N. A. Pulsatile local delivery of thrombolytic and antithrombotic agents using poly(*N*-isopropylacrylamide-*co*-methacrylic acid) hydrogels. *J. Controlled Release*, **1996**, 39(1): 57–64.
141. Chilkoti A., Dreher M. R., Meyer D. E. et al. Targeted drug delivery by thermally responsive polymers. *Adv. Drug Deliv. Rev.*, **2002**, 54(5): 613–630.
142. Mahmoudi M., Simchi A., Imani M. et al. Superparamagnetic iron oxide nanoparticles with rigid cross-linked polyethylene glycol fumarate coating for application in imaging and drug delivery. *J. Phys. Chem. C*, **2009**, 113(19): 8124–8131.
143. Meenach S. A., Anderson K. W., and Hilt J. Z. Synthesis and characterization of thermorespon-sive poly(ethylene glycol)-based hydrogels and their magnetic nanocomposites. *J. Polym. Sci. A: Polym. Chem.*, **2010**, 48(15): 3229–3235.
144. Hayashi K., Ono K., Suzuki H. et al. High-frequency, magnetic-field-responsive drug release from magnetic nanoparticle/organic hybrid based on hyperthermic effect. *ACS Appl. Mater. Interfaces*, **2010**, 2(7): 1903–1911.
145. Mertoglu M., Garnier S., Laschewsky A. et al. Stimuli responsive amphiphilic block copo-lymers for aqueous media synthesised via reversible addition fragmentation chain transfer polymerisation (RAFT). *Polymer*, **2005**, 46(18): 7726–7740.
146. Tai Y., Wang L., Yan G. et al. Recent research progress on the preparation and application of magnetic nanospheres. *Polym. Int.*, **2011**, 60(7): 976–994.
147. Chockalingam A., Babu H., Chittor R. et al. Gum arabic modified Fe_3O_4 nanoparticles cross linked with collagen for isolation of bacteria. *J. Biotechnol.*, **2010**, 8(1): 30–39.

148. Murakami T. and Tsuchida K. Recent advances in inorganic nanoparticle-based drug delivery systems. *Mini Rev. Med. Chem.*, **2008**, 8(2): 175–183.
149. Faraji A. H. and Wipf P. Nanoparticles in cellular drug delivery. *Bioorg. Med. Chem.*, **2009**, 17(8): 2950–2962.
150. Alexiou C., Schmid R. J., Jurgons R. et al. Targeting cancer cells: Magnetic nanoparticles as drug carriers. *Eur. Biophys. J. Biophy.*, **2006**, 35(5): 446–450.
151. Tang H. Y., Guo J., Sun Y. et al. Facile synthesis of pH sensitive polymer-coated mesoporous silica nanoparticles and their application in drug delivery. *Int. J. Pharm.*, **2011**, 421(2): 388–396.
152. Gaihre B., Khil M. S., and Kim H. Y. *In vitro* anticancer activity of doxorubicin-loaded gelatin-coated magnetic iron oxide nanoparticles. *J. Microencapsulation*, **2011**, 28(4): 286–293.
153. Xing R. M. and Liu S. H. Facile synthesis of fluorescent porous zinc sulfide nanospheres and their application for potential drug delivery and live cell imaging. *Nanoscale*, **2012**, 4(10): 3135–3140.
154. Manju S. and Sreenivasan K. Gold nanoparticles generated and stabilized by water soluble curcumin–polymer conjugate: Blood compatibility evaluation and targeted drug delivery onto cancer cells. *J. Colloid Interface Sci.*, **2012**, 368(1): 144–151.
155. Wang T. T., Zhang X. L., Pan Y. et al. Fabrication of doxorubicin functionalized gold nanorod probes for combined cancer imaging and drug delivery. *Dalton Trans.*, **2011**, 40(38): 9789–9794.
156. Partha R., Mitchell L. R., Lyon J. L. et al. Buckysomes: Fullerene-based nanocarriers for hydrophobic molecule delivery. *ACS Nano*, **2008**, 2(9): 1950–1958.
157. Woehrle G. H., Brown L. O., and Hutchison J. E. Thiol-functionalized, 1.5-nm gold nanoparticles through ligand exchange reactions: Scope and mechanism of ligand exchange. *J. Am. Chem. Soc.*, **2005**, 127(7): 2172–2183.
158. Elbakry A., Zaky A., Liebl R. et al. Layer-by-layer assembled gold nanoparticles for siRNA delivery. *Nano Lett.*, **2009**, 9(5): 2059–2064.
159. Kong T., Zeng J., Wang X. et al. Enhancement of radiation cytotoxicity in breast-cancer cells by localized attachment of gold nanoparticles. *Small*, **2008**, 4(9): 1537–1543.
160. Chertok B., Moffat B. A., David A. E. et al. Iron oxide nanoparticles as a drug delivery vehicle for MRI monitored magnetic targeting of brain tumors. *Biomaterials*, **2008**, 29(4): 487–496.
161. Thurston G., McLean J. W., Rizen M. et al. Cationic liposomes target angiogenic endothelial cells in tumors and chronic inflammation in mice. *J. Clin. Invest.*, **1998**, 101(7): 1401–1413.
162. Harush F. O., Debotton N., Benita S. et al. Targeting of nanoparticles to the clathrin-mediated endocytic pathway. *Biochem. Biophys. Res. Commun.*, **2007**, 353(1): 26–32.
163. Majeed M. I., Lu Q., Yan W. et al. Highly water-soluble magnetic iron oxide (Fe_3O_4) nanoparticles for drug delivery: Enhanced *in vitro* therapeutic efficacy of doxorubicin and MION conjugates. *J. Mater. Chem. B*, **2013**, 1(22): 2874–2884.
164. Qin Y., Sun L., Li X. et al. Highly water-dispersible TiO_2 nanoparticles for doxorubicin delivery: Effect of loading mode on therapeutic efficacy. *J. Mater. Chem.*, **2011**, 21(44): 18003–18010.
165. Arora H. C., Jensen M. P., Yuan Y. et al. Nanocarriers enhance doxorubicin uptake in drug-resistant ovarian cancer cells. *Cancer Res.*, **2012**, 72(3): 769–778.
166. Dalmark M. and Storm H. H. A Fickian diffusion transport process with features of transport catalysis: Doxorubicin transport in human red blood cells. *J. Gen. Physiol.*, **1981**, 78(4): 349–364.
167. Skovsgaard T. and Nissen N. I. Memberane transport of anthracyclines. *Pharmacol. Ther.*, **1982**, 18(3): 293–311.
168. Gu Y. J., Cheng J., Man C. W. et al. Gold-doxorubicin nanoconjugates for overcoming multidrug resistance. *Nanomed. Nanotechnol. Biol. Med.*, **2012**, 8(2): 204–211.
169. Momparler R. L., Karon M., Siegel S. E. et al. Effect of adriamycin on DNA, RNA, and protein synthesis in cell-free systems and intact cells. *Cancer Res.*, **1976**, 36(8): 2891–2895.
170. Nicolay K. and De Kruijff B. Effects of adriamycin on respiratory chain activities in mitochondria from rat liver, rat heart and bovine heart. Evidence for a preferential inhibition of Complex III and IV. *Biochim. Biophys. Acta Bioenergetics*, **1987**, 892(3): 320–330.
171. Takemura G. and Fujiwara H. Doxorubicin-induced cardiomyopathy: From the cardiotoxic mechanisms to management. *Prog. Cardiovasc. Dis.*, **2007**, 49(5): 330–352.

172. Minotti G., Menna P., Salvatorelli E. et al. Anthracyclines: Molecular advances and pharmacologic developments in antitumor activity and cardiotoxicity. *Pharmacol. Rev.,* **2004**, 56(2): 185–229.

173. Li Q., Wang X., Lu X. et al. The incorporation of daunorubicin in cancer cells through the use of titanium dioxide whiskers. *Biomaterials,* **2009**, 30(27): 4708–4715.

174. Nigam S., Barick K. C. and Bahadur D. Development of citrate-stabilized Fe_3O_4 nanoparticles: Conjugation and release of doxorubicin for therapeutic applications. *J. Magn. Magn. Mater.,* **2011**, 323(2): 237–243.

chapter twenty-two

Industrial biotechnology
Its applications in food and chemical industries

*Syed Ali Imran Bokhari, Muhammad Sarwar Khan,
Nosheen Rashid, and Muhammad Irfan Majeed*

Contents

Abstract

Industrial biotechnology, also known as "white biotechnology,"
demands the development of new biological technologies replacing chemical processes with biological processes that could convert
renewable raw materials into chemicals and bioenergy, which would
ultimately lead to sustainability. With the development of molecular biology and contemporary sciences, there is acceleration in the
understanding of a variety of processes, including gene regulation,
which is the basis of hyper-production of commercially viable products. Development of new techniques, such as enzyme engineering,
which involves directed evolution, and rational design and metabolic engineering of microorganisms and cells has paved the path
for the use of white biotechnological approaches in the chemical

industry. One of the striking recent achievements is the development of a technique known as multiplex automated genome engineering that can be used to engineer the entire genome of the bacteria may lead to the evolution of new bacterial strains for the production of proteins and synthetic products. Nevertheless, the biggest challenge for industrial biotechnology is to develop those processes that are cost-effective and more reliable at the industrial level. Although a number of processes that have been adopted by the industry are cost-effective, there is a huge demand for the establishment of the new processes that are reliable, rapid, and cost-effective at an industrial level.

Keywords: Industrial biotechnology, Fermentation, Metabolites, Enzymes.

Introduction

The word "biotechnology" was coined by Karoly Ereky in 1919. The history of biotechnology dates back to BC when brewing and bread-making processes began. It was discovered later in the 1800s that the processes were carried out by yeast. Industrial biotechnology can be divided into five different periods consisting of early period (till 1850s), understanding fermentation (1850s to 1900), development of microbiology as a new field (1900–1940), development of biochemistry and new discovery of bioproducts of industrial value (1940–1970), and advancements in molecular biology and emergence of recombinant DNA technology (1975 to date) (Buchholz and Collins, 2013).

Industrial or white biotechnology is the development and use of biosystems for sustainable and environment-friendly processing and the production of chemicals, materials, and energy. Industrial biotechnology is regarded as one of the main contributors in the field of industrial chemistry, especially green chemistry. Nowadays, a number of industrial chemicals, including food additives, vitamins, solvents, pharmaceuticals, bulk chemicals, and biofuels, are made using renewable resources such as agricultural crops and agricultural wastes. The benefits of industrial biotechnology for the current industry include the use of renewable resource, decrease in waste production and treatment costs, higher reaction rates, increase in conversion efficiency, lower energy consumption, and improvement in product purity. All of these benefits make industrial biotechnology very attractive to industrialists and research scientists.

Industrial microbes, their selection and improvement

Majority of the industrial microbiological products come from microbes; hence, selection of an industrial microbe and its strain improvement is an essential part of industrial biotechnology. The most used microbes are bacteria, fungi, and algae. Microbes are isolated using standard microbiological procedures and identified by morphological and molecular techniques.

The information about the microbes responsible for the production of particular metabolite can be obtained from the web and technical literature. The microbes can be purchased from microbial culture collections or it can be isolated from the local environment. While isolating the microbes from their environment, selection is the key to get the target organism. If the organism under consideration grows on a particular substrate, the enrichment technique can be employed. Other conditions, including pH of the culture media, temperature, and even the toxin, can also be used to select the organism of choice.

In order to make the process of metabolite production commercially viable, a number of strategies are being used. Microbes are normally not hyper-producer of any metabolite, which results in high cost of production at an industrial scale. For the improvement of industrial processes and making them economically feasible, the fermentation processes are to be improved. The optimization of metabolite production can be achieved by the combination of culture conditions and substrate of choice. This can help the industry to produce metabolites at the highest rate. Notably, there are issues with the production of the metabolite at the industrial level due to the reason that naturally occurring microbes have tight gene regulation system and do not produce a metabolite beyond a particular limit. This is why the strain improvement is considered to be one of the important research and development area in the field of the industrial biotechnology. It is important to mention that improving the strain may result in some problems or may have particular requirements, which were not presented by a wild strain. A mutant, which is auxotrophic for certain metabolite, may require that metabolite as a nutrient in media, or require high aeration rate or perhaps entirely different fermentation process. In essence, the mutant strain must be thoroughly examined for its potential cost affectivity before bringing it to the industrial process.

Conventional approach involves random mutagenesis for the improvement of microbes. In this technique, physical or chemical agents that can mutate DNA are used. The physical agents include x-rays, gamma rays, and UV light. The x-rays are produced from machines that are commercially available whereas a radioactive material is used for the production of gamma rays. The x-rays or gamma rays are also called ionization radiation as these radiation remove the outer-shell electron causing molecules to ionize forming the reactive radicals/free radicals, which can damage the DNA. Similarly, ultraviolet light (200–300 nm) can mutate the DNA for which the UV lamps are used, which emit 254 nm and are of 15 watt power. The process of the UV mutagenesis must be carefully performed as exposure to the visible light can reverse the mutation by photo reactivation process. The UV light causes dimerization of adjacent pyrimidines particularly thymine, although dimerization can also occur between two adjacent cytosine or two adjacent cytosine and thymine residues. There are three groups of chemical mutagens, including the mutagens that affect nonreplicating DNA such as nitrosoguanidine (MNNG/NTG), nitrous oxide, and ethyl methane sulfonate (EMS). The second type of chemical mutagens is the incorporation of base analogs during replication, for example, 5-bromouracil and 2-aminopurine. The third type of chemical mutagens is also known as intercalating agents and can lead to frameshift mutations, for example, acridine and ethidium bromide. The exposure time in either case is optimized to give 3 log kill.

The selection procedure after mutation is the key to obtain an industrially viable microbe. Mutation can bring changes in the DNA but searching desired changes among the undesirable is a tough task. Scientists devise the selection procedure by keeping in mind the need for improvement without affecting the vital functions of the microbe. The aims may include hyper-production of end products or producing the repressed mutant. To hyper-produce end product, the feedback mechanism is the target. The selection for microorganism in which feedback mechanism is disrupted can be done by using a toxic analog of end product or by producing an auxotrophic mutant for a particular metabolite. The repressed mutant is produced by using a toxic analog of metabolite causing repression.

With the advent of recombinant DNA technology, some impossible industrial products have become possible. A number of nonmicrobial products, especially pharmaceuticals, can be produced at industrial scale by cloning the desired gene in the microbes. There are several reports to substantiate the fact that a new metabolic pathway can be introduced in the microbes for the production of the nonprotein product. Another cutting-edge technology called as multiplex automated genome engineering (MAGE) (Wang et al., 2009) may prove to be very helpful in producing virtually any bioproduct in economical way at a large scale.

Fermentation technology

Fermentation technology is the backbone of industrial biotechnology. After selection and/or improvement of the microorganism for industrial production of required metabolite, fermentation technology is used. Fermentation process may include the design of the culture media for growth of microorganism and its formulation, sterilization of media and fermentor, inoculum production and its aseptic transfer to the fermentor, microbial growth (fermentation) under optimal conditions, and product extraction and purification.

There are three main types of fermentation process on the basis of media feed. A batch fermentation process can be explained as one in which all the nutrients are added at once and the production is carried out for a particular time period. As the fermentation process is completed, the product is harvested. In the second type, the nutrients are added and the product is harvested in a continuous way. A third type of fermentation process is called fed batch process in which a particular nutrient is fed in a fermentor in intermittent way and the product is harvested at the end of the process. Fermentation process may be of different types, including surface, solid state, or submerged. Submerged fermentation is the most common type of fermentation although surface and solid-state fermentations are also used, for example, in the production of citric acid and compost, respectively.

Growth pattern of an industrial microbe can be modeled to predict and compare the production of a particular metabolite by utilizing microbial growth and product production kinetics. For batch cultivation, various growth kinetic parameters such as specific growth rate (μ), cell yield coefficient ($Y_{X/S}$), and product formation parameters such as specific product yield ($Y_{P/X}$) and product yield ($Y_{P/S}$) can be calculated for different organisms and for mutants to compare the optimal utilization of substrate and maximal production of product at industrial level (Bokhari et al., 2008). The same concepts can be used for continuous or fed batch cultivation of microbes.

Different types of fermentor or bioreactor design are used for the process of fermentation, including stirred tank reactor, bubble column fermentor, airlift fermentor, fluidized bed, and packed bed column fermentor (Figure 22.1). Each design has its own importance.

Industrial biotechnology in food industry

Right from the very beginning, living organisms or their derivatives are used in food processing, for example, production of alcoholic beverages, cheese, yogurt, and bread. One of the main contributors in food industry is the enzymes. Enzymes act on their substrate and convert them into product. Such conversion is of indispensible value to food processing, for example, in cheese production, rennin is used, which breaks down casein, a milk protein. The production of enzymes that are used in the food industry is an important field of industrial biotechnology. Efforts are being made for the improvement of enzyme production techniques and enzyme quality using protein engineering and directed evolution techniques.

Starch industry

Starch is the primary source of energy for human beings. Starch consists of amylase and amylopectin. Industrial processing of starch had been mainly done by using acid

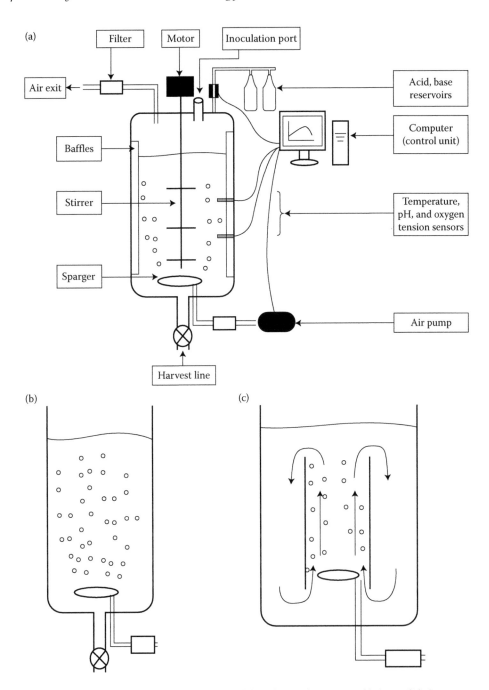

Figure 22.1   (a) A stirred tank reactor. (b) A bubble column fermentor. (c) An airlift fermentor.

hydrolysis; however, nowadays, α-amylase enzyme is used for this purpose. The enzymatic process is better than conventional methods such as hydrolysis, since less undesirable by-products are produced. Previously, the α-amylase enzyme was produced from *Bacillus amyloliquefaciens*. But as the liquefaction process of starch requires high temperature, a thermostable α-amylase from *Bacillus licheniformis* was later adopted.

Glucoamylase and pullulanase are used for conversion of starch into dextrose syrup. One of the important applications of dextrose syrup is its conversion into high-fructose corn syrup (HFCS), which is used in the production of candies, confectionaries, etc. The enzyme used to carry out this process is glucose isomerase, which is obtained from *Streptomyces* sp. (Bhosale et al., 1996).

A major breakthrough in the field of starch hydrolysis is the development of amylase, which is able to efficiently liquefy raw starch granules and is obtained from a bacterium *Anoxybacillus* sp. Amylase has a domain that binds to the starch and it is not present in traditional enzymes used for the process (Vikso-Nielsen et al., 2006). When a chimeric enzyme was obtained by adding a starch binding domain from *Aspergillus niger* to glu-coamylase of *Saccharomyces cerevisiae*, it was observed that such enzymes can also hydro-lyze insoluble starch (Latorre-Garcia et al., 2005).

Dairy industry

Humans have been using milk to produce yogurt, cheese, and butter from a long time. Processing of milk into various products is not a new technology and was developed in an empirical way. Nowadays, these processes have been improved and run at industrial scale where constant organoleptic quality of the product is the key to market. Lactic acid bacteria and various enzymes are used to produce a number of dairy products. Various species of *Lactobacillus*, *Lactococcus*, *Streptococcus*, and *Leuconostoc* are used in the produc-tion of various dairy products.

Cheese production is essentially a protein precipitation process. Casein is a major milk protein, which is located in micelles of milk. An enzyme found in the stomach of a calf called rennin hydrolyzes specifically peptide bond between phenylalanine (105) and methionine (106) resulting in the destabilization of micelles and subsequent aggregation of protein in curd formation (Dalgleish, 1999). The rennet can be obtained from calf stom-ach, microbial sources (chymosin from *Rhizomucor miehei*), and recombinant chymosin expressed and obtained from *Escherichia coli*, *A. niger*, etc.

Lipases are used in cheese-ripening process. Lipases release free fatty acid, which impart characteristic flavor in case of Italian and blue cheese (Aehle, 2004). Lipases for cheese ripening are obtained from *Aspergillus* sp. and *R. miehei*, which specifically removes fatty acid from triglycerides present at 1 and 3 positions.

Other enzymes are also used in dairy products, for example, lysozyme, which reduces the risk of growth of *Clostridium tyrobutyricum* and may cause late blowing in cheese (Cunningham et al., 1991). Lactase is used to produce lactose-free milk, which can be used for lactose-intolerant people or in ice cream production (Vesa et al., 2000). It is also used in the production of galacto-oligosaccharides, which act as prebiotic and gives numerous other health benefits (Park and Oh, 2010).

Bread industry

To make breads with increased bread volume and with improved bread texture, amy-lases and xylanases are used. Amylase obtained from *Aspergillus oryzae* is used in bread making, which not only reduces dough viscosity but also releases maltose, resulting in improved yeast fermentation. Xylanase degrades insoluble arabinoxylan, which disrupts the gas cells, resulting in poor dough raising. Therefore, xylanase helps in improving bread volume and crumb structure (Jiang et al., 2005).

Fruit processing

A wide variety of enzymes are used for fruit processing. The commonly used enzymes for processing of fruits are polygalacturonases, arabinanases, feruloyl esterases, arabino-furanosidases, exopolygalacturonases, and pectin methylesterases. Polygalacturonases are the main pectin (complex polysaccharide found in cell wall of the plants)-degrading enzyme. Commercially, *Aspergillus* spp. are used to produce these pectin-degrading enzymes (Jayani et al., 2005).

In enzyme-assisted maceration of fruits, the juice extraction is maximized by enzymatic cell wall degradation. Enzymes also help in reducing viscosity and volume of solid wastes generated in conventional food processing. Amylase is used in case of apple juice, which reduces haze problem and also helps in the production of clear juice (Kashyap et al., 2001).

Food additive production

Amino acid

Amino acids are produced in biotechnological processes and have numerous applications in chemical market (12%), food sector (32%), and animal feed sector (56%) (Leuchtenberger et al., 2005). L-Phenylalanine, L-aspartate, and L-glutamate have applications in food industry. *Corynebacterium glutamicum* is used at industrial scale for the production of L-glutamate, which is used as a flavor enhancer in the form of monosodium glutamate (Ikeda, 2003).

Yeast production

Yeast (*S. cerevisiae*) is utilized as an additive in the bread-making process. It is produced at industrial scale through an aerobic fermentation process. The most commonly used carbon source for the production of yeast is sugarcane molasses. Fed batch cultivation is usually the fermentation process of choice, which helps to avoid the production of ethanol through overflow metabolism. The yeast production process takes about 15 hours. The final yield at industrial scale is about 50–60 grams of cells per liter. After fermentation, downstream processing is done and the broth is removed by centrifugation. At this stage, the yeast can be marketed as cream yeast. To make compressed yeast, the broth is further removed by using plate and frame filter or rotary vacuum filter and extruded through extruder to give a block-like shape. To make it dry, the yeast suspension is dried using a fluidized bed dryer (Enfors, 2001).

Citric acid

Citric acid has numerous applications and one of them is the use of citric acid as an acidulant in the production of juices and beverages. Citric acid can also be used in nonfood products, for example, in cosmetics. Industrial production of citric acid is achieved by surface or submerged fermentation of *A. niger* (Okafor, 2007).

Chemical industry

Paper and pulp industry

Paper and pulp industry is one of the most suitable industries for the application of enzymes. One of the reasons is that the raw material used is mainly wood, which is degradable naturally, and nature possesses all the required enzymes to selectively and completely degrade all the components of wood. Studies on an enzyme application in paper and pulp industries began in the 1970s (Ma and Jiang, 2002). But at that time, it was not considered feasible because of high costs of enzyme production. However, with the increase in our understanding about the enzyme, the enzyme production and environmental issues made it one of the most attractive alternatives to chemical processes. By mid-1980s, enzyme usage was started in the paper and pulp industry as a part of bleaching process (Bajpai, 1999).

The main application of xylanase enzyme is in the paper and pulp industry. Use of xylanase enzyme for biobleaching of cooked pulp is an economical as well as environmentally safe procedure (Viikari et al., 1994). The two hypotheses are considered for xylanase action in the paper pulp biobleaching. First, it is considered that the enzyme degrades xylan, which is relocated or reprecipitated during the pulping process (Viikari et al., 1994). Degradation of this relocated xylan by enzyme will result in exposure of entrapped lignin to the bleaching compounds, resulting in more final brightness of cellulose pulp. The second hypothesis is based on the capability of lignin to form covalent bonds with cellulose and hemicelluloses. It is thought that some of the bonds between lignin and polysaccharides escape during the pulping process (Buchert et al., 1992). So, the enzyme cleaves xylan and the covalently linked lignin is released. The use of xylanase enzyme as a bleaching agent has reduced the usage of chlorine-based chemicals by 27% on the one hand and improved certain properties of paper on the other (Bokhari et al., 2010).

Bioplastic

Polyhydroxyl butyrate (PHB) and polyhydroxyl alkanoates (PHA) are the two main types of natural polymers used for the production of bioplastics. These polymers can be used as alternative for polypropylene or polyethylene because of similar thermoplastic properties. PHB is produced by a number of bacteria, including *Bacillus* sp., *Pseudomonas* sp., and *Cupriavidus necator*. It has also been recently produced with *E. coli* harboring PHB genes from *Azotobacter vinelandii* and *C. necator* (Centeno-Leija et al., 2014). PHB is also used in tissue repair, artificial organ development, and drug delivery (Chen and Wang, 2013). PHB is produced using fermentation process. For this purpose, batch, fed batch, or continuous fermentation processes are employed.

PHA production is reported from a number of bacterial species, including *Ralstonia* sp., *Alcaligenes* sp., and *Pseudomonas* sp. PHA is an environment-friendly degradable bioplastic, which, depending upon its formula, varies in flexibility and toughness (Verlinden et al., 2007). Both PHA and PHB are currently expensive than plastics derived from petrochemicals; therefore, they are only used in sophisticated applications such as in the field of medicine. Research is going on to make their production processes more feasible and economical so that synthetic plastic can be replaced by bioplastic.

Biofuel

Since fossil fuel is a nonrenewable resource and is depleting at a very high rate, the biggest challenge to scientists today is to search for renewable energy resources, which should be enough to meet the present and future energy demands. Biofuel may be the answer to this problem. Biofuel is the fuel produced from renewable resources and it includes bioethanol, biodiesel, biobutanol, etc. Bioethanol was actually introduced as an additive for petroleum fuels (Otero et al., 2007). In 2014, the ethanol industry produced 24.5 billions of gallons of ethanol fuel (http://ethanolrfa.org/pages/World-Fuel-Ethanol-Production). A great deal of research is being done on improving the efficiency of bioethanol production. One of the major problems in ethanol production is the stress on yeast, which ferments the feed stock. A number of techniques, including transposon-based mutant production and deletion mutation, were used to select genetically modified ethanol-resistant strain (Kim et al., 2011; Zheng et al., 2011). Global transcription machinery engineering technology can also help in developing ethanol-tolerant yeast (Yang et al., 2011).

Production of *n*-butanol using *Clostridium* sp. is also established (Ezeji et al., 2007). One of the challenges in producing biobutanol using *Clostridium* sp. was its complex physiology and difficulties in genetic engineering (Inui et al., 2008). Owing to these limitations, *E. coli* attained focus. As a result, a set of genes from *Clostridium* sp. involved in the production of butanol were cloned and after a series of genetic engineering experiments, a strain of *E. coli* was obtained with a yield of 552 mg of butanol per liter (Atsumi et al., 2008).

Biodiesel is produced by transesterification of fats and oils with methanol or ethanol. Animal fats or vegetable oil can be used for this purpose. Microbes such as yeast, fungi, bacteria, and microalgae also produce oils, which can be used for the production of biodiesel (Li et al., 2008). By cloning purvate decarboxylase from *Zymomonas* sp. and acyltransferase from *Acinetobacter* sp. in *E. coli*, fatty acid ethyl esters were produced (Kalscheuer et al., 2006). In transesterification of fatty acids, the use of lipase is a well-established process (Fukuda et al., 2009).

Pharmaceutical industry

Pharmaceutical industry is one of the important industry in which biotechnology is involved. In 2013, the pharmaceutical and biotechnology sector increased R&D spending by 2.4% and became at the top of R&D investing sector (www.ifpma.org). In the history of biotechnology, one of the major breakthroughs was the discovery of penicillin by Sir Alexander Fleming in 1928. It took about 14 years to make a penicillin production process commercial. Until 1942, the production of penicillin was improved by about 140,000 times by using sequential mutagenesis (Buchholz and Collins, 2013). By 2009, the total antibiotic sale was about $42 billion with an average growth of 4% per year (Hamad, 2010).

Antibiotics are one of the important secondary metabolites produced by microbes. The commercially available antibiotics are produced by fermentation, semisynthetic, and chemical processes depending on cost economics. Generally, microbial fermentation process is used to produce a basic active molecule, for example, 6-amino penicillinic acid, which is then modified through chemical processes, which may affect various properties of antibiotics, including pharmacokinetics, resulting in better antibiotic for medical and veterinary use. Most commonly produced antibiotics are β-lactam antibiotics, which form 65% of total antibiotic production (Elander, 2003).

Other important pharmaceutical products are recombinant proteins, which is produced by genetically modified microbes at industrial scale. The first recombinant protein was insulin approved by Food and Drug Administration (FDA) in 1982. Later, a number of pharmaceutically important proteins were cloned in microbes (*E. coli*, *Bacillus subtilis*, *S. cerevisiae*), insect cells (*Autographa californica*) and animal cells (Hamster kidney cells), and produced at industrial scale (Schmidt, 2007). The list of some recombinant proteins and their application is presented in Table 22.1.

Table 22.1 Some Examples of Proteins Produced by Recombinant DNA Technology
at Industrial Scale

Protein	Function	Application	Approval
Insulin	Hormone helps in maintaining blood glucose level	Treatment of diabetes	1982
Hepatitis B surface antigen (vaccine)	Immunogenic agent	Immunization against HBV	1986
Interferon	Signaling protein involved in stimulation of immune system	Cancer treatment, HCV treatment	1986
Blood clotting factor	Clotting of blood	Treatment of hemophilia A	1992
DNAase	Breakdown of DNA	Cystic fibrosis treatment	1994
Tissue plasminogen activator	Dissolves blood clot	Treatment of heart attack and stroke	1996

Note: For details, see Kresse (2001).

References

Aehle W. 2004. *Enzymes in Industry*, 2nd Edition. Chapter 5, Wiley—VCH Verlag GmbH, Weinheim.

Atsumi S, Cann AF, Connor MR, and Shen CR. 2008. Metabolic engineering of *Escherichia coli* for 1-butanol production. *Metabol. Eng.* 10, 305–311.

Bajpai P. 1999. Application of enzymes in the pulp and paper industry. *Biotechnol. Prog.* 15, 147–157.

Bhosale SH, Rao MB, and Deshpande VV. 1996. Molecular and industrial aspects of glucose isomerase. *Microbiol. Rev.* 60, 280–300.

Bokhari SAI, Latif F, and Rajoka MI. 2008. Kinetics of high-level of β-glucosidase production by a 2-deoxyglucose-resistant mutant of *Humicola lanuginosa* in submerged fermentation. *Braz. J. Microbiol.* 39, 724–733.

Bokhari SAI, Rajoka MI, Javed A, Shafiq ur Rehman, Ishtiaq ur Rehman, and Latif F. 2010. Novel thermodynamics of xylanase formation by a 2-deoxy-D-glucose resistant mutant of *Thermomyces lanuginosus* and its xylanase potential for biobleachability. *Bioresour. Technol.* 101, 2800–2808.

Buchholz K and Collins J. 2013. The roots—A short history of industrial microbiology and biotechnology. *Appl. Microbiol. Biotechnol.* 97, 3747–3762.

Buchert J, Ranua M, Kantelinem A, and Viikari L. 1992. The role of two *Trichoderma reesei* xylanases in the bleaching of kraft pulp. *Appl. Microbiol. Biotechnol.* 37, 825–829.

Centeno-Leija S, Huerta-Beristain G, Giles-Gomez M, Bolivar F, Gosset G, Martinez A. 2014. Improving poly-3-hydroxybutyrate production in Escherichia coli by combining the increase in the NADPH pool and acetyl-CoA availability. *Antonie Van Leeuwenhoek* 105, 687–696.

Chen G and Wang Y. 2013. Medical applications of biopolyesters polyhydroxyalkanoates. *Chin. J. Polym. Sci.* 31, 719–736.

Cunningham FF, Proctor VA, and Goetsch SJ. 1991. Egg-white lysozyme as a food preservative: An overview. *World's Poult. Sci. J.* 47, 141–163.

Dalgleish DG. 1999. The enzymatic coagulation of milk. *Cheese: Chemistry, Physics and Microbiology*, vol. 1, 2nd Edition. (Editor. Fox PF), Chapman and Hall, London. 1, 69–100.

Elander RP. 2003. Industrial production of β-lactam antibiotics. *Appl. Microbiol. Biotechnol.* 61, 385–392.

Enfors S. 2001. Baker's yeast. In *Basic Biotechnology*, 2nd Edition. (Editors Ratledge C and Kristiansen B.), Cambridge University Press, Cambridge, UK. 377–389.

Ezeji TC, Qureshi N, and Blaschek HP. 2007. Bioproduction of butanol from biomass: From genes to bioreactors. *Curr. Opin. Biotechnol.* 18, 220–227.

Fukuda H, Kondo A, and Tamalampudi S. 2009. Bioenergy: Sustainable fuels from biomass by yeast and fungal whole cell biocatalysts. *Biochem. Eng. J.* 44, 2–12.

Hamad B. 2010. The antibiotic market. *Nat. Drug. Discov.* 9, 675–676.

Ikeda M. 2003. Amino acid production processes. In *Advances in Biochemical Engineering/Biotechnology*, (Editors Scheper T, Faurie R, and Thommel J.), Springer, Berlin. 79, 1–35.

Inui M, Suda M, Kimura S, Yasuda K, Suzuki H, Toda H, Yamamoto S, Okino S, Suzuki N, and Yukawa H. 2008. Expression of *Clostridium acetobutylicum* butanol synthetic genes in *Escherichia coli*. *Appl. Microbiol. Biotechnol.* 77, 1305–1316.

Jayani RS, Saxena S, and Gupta R. 2005. Microbial pectinolytic enzymes: A review. *Process Biochem.* 40, 2931.

Jiang ZQ, Yang SQ, Tan SS, Li LT, and Li XT. 2005. Characterization of a xylanase from the newly isolated thermophilic *Thermomyces lanuginosus* CAU44 and its application in bread making. *Lett. Appl. Microbiol.* 41, 69–76.

Kalscheuer R, Stolting T, and Steinbuchel A. 2006. Microdiesel: *Escherichia coli* engineered for fuel production. *Microbiology* 152, 2529–2536.

Kashyap DR, Vohra PK, Chopra S, and Tewari R. 2001. Applications of pectinases in the commercial sector: A review. *Bioresour. Technol.* 77, 215–27.

Kim HS, Kim NR, Yang J, and Choi W. 2011. Identification of novel genes responsible for ethanol and/or thermotolerance by transposon mutagenesis in *Saccharomyces cerevisiae*. *Appl. Microbiol. Biotechnol.* 91, 1159–1172.

Kresse GB. 2001. Recombinant proteins of high value. In *Basic Biotechnology*, 2nd Edition. (Editors Ratledge C and Kristiansen B.) Cambridge University Press, Cambridge, UK. 437.

Latorre-Garcia L, Adam AC, Manzanares P, and Polaina J. 2005. Improving the amylolytic activity of *Saccharomyces cerevisiae* glucoamylase by the addition of a starch binding domain. *J. Biotechnol.* 118, 167–176.

Leuchtenberger W, Huthmacher K, and Drauz K. 2005. Biotechnological production of amino acids and derivatives: Current status and prospects. *Appl. Microbiol. Biotechnol.* 69, 1–8.

Li Q, Du W, and Liu D. 2008. Perspectives of microbial oils for biodiesel production. *Appl. Microbiol. Biotechnol.* 80, 749–756.

Ma HJ and Jiang C. 2002. Enzyme application in paper and pulp industry. *Prog. Paper Recycl.* 11, 36–47.

Okafor N. 2007. *Industrial Microbiology and Biotechnology*, Science Publisher, New Hampshire, USA. 368.

Otero J, Panagiotou G, and Olsson L. 2007. Fueling industrial biotechnology growth with bioethanol. *Adv. Biochem. Eng. Biotechnol.* 108, 1–40.

Park A and Oh D. 2010. Galacto-oligosaccharide production using microbial β-galactosidase: Current state and perspectives. *Appl. Microbiol. Biotechnol.* 85, 1279–1286.

Schmidt FR. 2007. From gene to product: The advantage of integrative biotechnology. In *Handbook of Pharmaceutical Biotechnology*, (Editor. Gad SC.), John Wiley and Sons Inc, New Jersey, USA.

Vesa HT, Marteau P, and Korpela R. 2000. Lactose intolerance. *J. Am. Coll. Nutr.* 19, 165S–167S.

Verlinden RAJ, Hill DJ, Kenward MA, Williams CD, and Radecka I. 2007. Bacterial synthesis of biodegradable polyhydroxyalkanoates. *J. Appl. Microbiol.* 102, 1437–1449.

Viikari L, Kantelinen A, Sundquist J, Linko M. 1994. Xylanases in bleaching, from an idea to the industry. *FEMS Microbiol. Rev.* 13, 335–350.

Vikso-Nielsen A, Andersen C, Hoff T, and Pedersen S. 2006. Development of new α—Amylases for raw starch hydrolysis. *Biocatal. Biotransform.* 24, 121–127.

Wang HH, Isaacs FJ, Carr PA, Sun ZZ, Xu G, Forest CR, and Church GM. 2009. Programming cells by multiplex genome engineering and accelerated evolution. *Nature* 460, 894–898.

Yang J, Baeb JY, Lee YM, Kwon H, Moon HY, Kang HA, Yee SB, Kim W, and Choi W. 2011. Construction of *Saccharomyces cerevisiae* strains with enhanced ethanol tolerance by mutagenesis of the TATA-binding protein gene and identification of novel genes associated with ethanol tolerance. *Biotechnol. Bioeng.* 108, 1776–1787.

Zheng DQ, Wu XC, Tao XL, Wang PM, Li P, and Chi XQ. 2011. Screening and construction of *Saccharomyces cerevisiae* strains with improved multi-tolerance and bioethanol fermentation performance. *Bioresour. Technol.* 102, 3020–3027.

chapter twenty-three

Environmental biotechnology
Approaches for ecosystem conservation

Vasavi Mohan, Mohammed Khaliq Mohiuddin, and Yog Raj Ahuja

Contents

Abstract

Natural bioremediation processes have helped the sustenance of diverse ecological habitats through biochemical intervention of plant and microbial systems. Nature has evolved processes such as bioremediation, biodegradation, mineralization and solubilization, and detoxification of different hazardous substances accumulating in the ecosystems over a period of time to non-toxic materials. Owing to industrialization and increased urban population, the ratio between wastage and natural decomposition has changed. This has necessitated the adoption of biotechnological approaches for the efficient management/utilization of waste. Pollutants arising from domestic, industrial, or agricultural processes affect mostly land and water resources, apart from greenhouse gases or other particulate pollutants in the air. These pollutants threaten the conservation of these ecosystems and the vast biodiversity contained in them. Environmental biotechnology is an integrated multidisciplinary approach which utilizes the huge biochemical potential of microorganisms and plants for environmental conservation. To achieve this goal, we should understand the normal ecosystem involving microbial and plant systems, their interacting pathways, and role of specific gene products in maintaining a given habitat. Advanced molecular systems from genomics to metabolomics have the potential to give the environmental biologist a complete understanding of microbial–plant–ecosystem interaction, in order to adapt effective remediation processes. Based on this knowledge, modern molecular tools could be used for engineering vectors to tackle waste management in an environmentally friendly manner. Bioremediation processes through a scientific approach, utilizing the action of microbes and plants, keeping in view the biotic and abiotic factors that influence the degradation rates, are now being researched worldwide for habitat preservation and biomonitoring.

Keywords: Bioremediation, Ecosystem, Conservation, Molecular, Environmental biotechnology.

Introduction

The adoption of biotechnological approaches for efficient management of wastes, and aiding natural bioremediation for the conservation of habitat, is one of the choicest areas of research in most developing countries. Pollutants arising from various sources as a result of urbanization are a constant threat to the vast biodiversity contained in an ecosystem (Figure 23.1 illustrates the causes for altered ecosystem). Along with development, we have inherited a legacy of toxic waste that has percolated through layers of our ecosystems; it lies with us now to rid the earth of these contaminants, restore natural processes, taking help from nature's own mechanisms.

The Chernobyl nuclear reactor accident in Ukraine is a classic example. In 1989, 3 years after the explosion, the International Atomic Energy Agency (IAEA) assessed the extent of damage around the power plant. Findings showed radioactive emissions and toxic metals, including iodine, cesium-137, strontium, and plutonium, that are potentially harmful to soil, plant, animal, and human health in general. Iodine, if inhaled/ingested, can accumulate in the thyroid gland, delivering high doses of radiation as it decays. Similarly, cesium-137 (radioactive) can deliver an internal dose of radiation if ingested in any form, before being eliminated metabolically (UNSEAR, 2008).

A wide range of pesticides used against farming pests and microbes directly affect the soil nitrogen balance, suppress nitrifying bacteria, interfere with ammonification in soils, and adversely affect crop productivity in the long run. Not all of these pesticides are biodegradable and their persistence in soils has proven deleterious effects on humans and livestock, owing to their entry into the food chain (Aktar MW et al., 2009). These are just a couple of examples in which human intervention has disturbed the ecosystems around us, which in turn affect humans themselves.

Restoring ecosystems to their natural composition and functioning takes place normally with the help of microbes, plants, etc. but with effective technological inputs, the

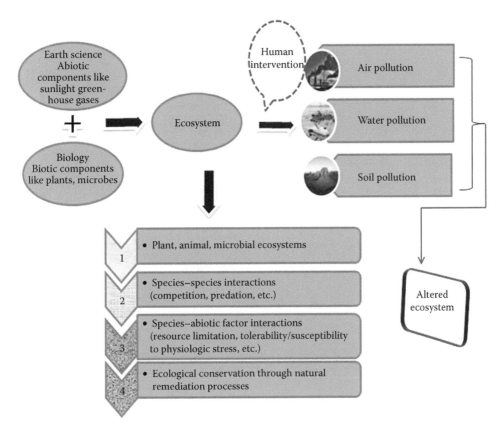

Figure 23.1 Ecosystem.

remediation process can be improved and speeded up. Courses in environmental biotechnology are now being promoted *on par* with other life sciences to help students understand the normal ecosystem involving microbial and plant systems and their interacting pathways, based on which the appropriate intervention strategies can be worked out through standardizing efficient molecular tools for biomonitoring and remediation.

Plant and microbial ecosystems

Ecology is an interdisciplinary field that includes biology (biotic component) and earth science (abiotic components such as sunlight, greenhouse gases, and earth) (Figure 23.2 illustrates the interactions between biotic and abiotic components). The structure and function of an ecosystem is controlled by the interaction between its biotic and abiotic components by mechanisms related to the cycling of matter and energy flow. The ecosystem is represented by the totality of interactions, which include species–species (e.g., competition and predation), species–abiotic factor interactions (e.g., resource limitation, responses, and physiological stress).

The effect of the ecosystem on plant growth depends on its biotic and abiotic factors such as nitrogen availability, levels of light, soil moisture and other nutrients, symbionts, associated plants, and animal communities. The science of ecology deals with interaction between the same species and with other organisms present in the environment. Its main

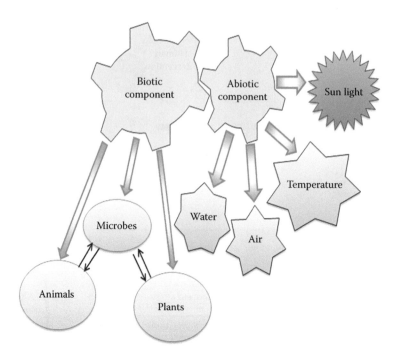

Figure 23.2 Interaction of biotic and abiotic components.

dynamic components include biomass, population, and diversity, with constant competition between and within these ecosystems. Biodiversity refers to different species, the
variety of gene pools present in diverse kinds of ecosystems (Zak DR et al., 2003).

Understanding the ecosystem starts from observing nature's different phenomena
such as sunlight being utilized by plants to convert it into a useful energy source. Studying
the interaction between plants and the atmospheric gases present in a given ecosystem
will be useful for assessing agricultural productivity and food quality. It should be remem-
bered that ecosystems are areas in which communities are self-sustaining. The main focus
of the environmental biotechnologist is to preserve the natural ecosystem by developing
methods to minimize the adverse impact of urbanization and human intervention on our
planet (Heggie L and Halliday KJ, 2005; Grünhage H and Haene HD, 2008; Uttara S et al.,
2012; Arabi U et al., 2013; Cruz-Cruz CA et al., 2013).

Microbes and environment

The natural environment (soil, water, and air) has a number of microbial communities
apart from the plants and animals that inhabit this planet. The study of different microbial
communities and their physiological features, chemical, and biological properties is the
main scope of environmental microbiology. The subject deals with the study of microbes
in the natural ecosystem and in man-made ecosystems (such as bioreactors) for tapping
the commercial potential of these organisms.

Planet Earth is covered by different ecosystems and different climatic conditions
in which microbial fauna are omnipresent. The known microbial species are less than
1% of the total actual microbes existent on earth. Microbes show their endless tolerance
to extremity by growing in unfavorable environments such as high temperatures, salt

concentrations, alkaline, and acidic conditions. Microbial communities have a special importance in biogeochemical cycles. Their effects on the environment can be beneficial or harmful or as perceived by human need, based on observation and manipulation using biotechnological tools. They participate in the carbon, nitrogen, and phosphorus cycles, in oxygen production, biomass control, and biodegrading dead matter in the environmental niche they occupy. Vitamin K_2 or menaquinone produced by some bacteria in the ileum takes care of the vitamin K requirements in humans and prevents coagulopathies (Conly JM and Stein K, 1992). In humans, there are over 400 different species of bacteria; some may live symbiotically like some gut bacteria that aid in digestion, protect from other harmful bacteria, and fungi.

Microbes associate with plants symbiotically through the root nodules or mycorrhizal association. In legumes, plants and nitrogen-fixing bacteria interact at the postembryonically developed root nodule for nitrogen fixing where the microbe identifies the specific host for an association (Harrison MJ, 1998; Silva FV et al., 2014). The symbiotic association of a rhizobium species present in the root nodule of *Phaseolus mungo* was found to be the probable reason for production of ascorbic acid by the organism in the root nodule (Ghosh S et al., 2008). Genes regulating symbiosis—nodC for nodulation and nifH for nitrogen fixation studied in bacteria isolated from the root nodules of *Vigna unguiculata* (Brazil rainforests) helped group the *Bradyrhizobium* strains together. Microbial pesticides are used to get rid of specific plant pests. Slow and controlled microbial degradation of organic matter leads to the formation of compost; further, mites, springtails, millipedes, centipedes, earthworms, and other soil fauna help in the breakdown and enrichment of these compost materials that are useful as fertilizer.

Revolutionary changes that have taken place in the scientific knowledge leading to advancement in technology have helped the study of microbes in the environment through the improved understanding of their composition and physiology. Several sewage treatment techniques use biofiltration, the use of bacteria that break down pollutants, for converting toxic organic waste to non-toxic form and then releasing it in the surroundings. The use of DNA-based technologies by molecular phylogeneticists for the construction of microbial phylogeny has made it easier for the identification of new microbes in an ecosystem. An example is the molecular clock technique that used nucleotide sequence substitutions in DNA to measure the timing of evolutionary changes, that is, to predict when in geological history the two species diverged.

Plants: Their role in ecosystem conservation

Plants are known as primary producers; they are vital components of the vast biological diversity, with an economic and cultural importance. They have an important role to play in maintaining environmental balance and ecosystem stability, in addition to providing habitat for animal life. The importance of plant ecology started with the role of plants in creating the oxygenated atmosphere of earth, an event that occurred some two billion years ago (Keddy PA, 2007). The biodiversity in plants can be attributed to their adaptability to different climatic conditions.

Plant communities prevail in different environmental conditions based on the physiology of the plant and the other components of a given ecosystem. Understanding the plant ecosystem is important to see the relationship between plant–water and seasonal balances, to show a significant change in productivity in agriculture and the conservation of the natural ecosystem by maintaining the plant–plant and plant–environment interactions, which will facilitate individual plants to grow as a part of a community. Plant community/ plant population dynamics depends on the surrounding ecosystem (biotic and abiotic); abiotic components include water and greenhouse gases and biotic components include different plant species and microbes living in the system. In addition, plant pathogens affect the natural composition of plant communities that are responsible for evolutionary changes, through the development of resistant genotype and thereby maintenance of plant species diversity.

Soil

Soil is a unique biological ecosystem with an abundant microflora and fauna. It is said that an average of one billion (1,000,000,000) microbes inhabit 1 gram of soil, which represents several thousands of species. Microbes can grow in diverse conditions such as aerobic, anaerobic, with or without energy, and light source. A small clump of soil has a gradient of oxygen from its edges to the center, each O_2 concentration making the perfect habitat for a certain kind of organism (Dance A, 2008). They can utilize the chemical components (by using biogeochemical cycles such as carbon, nitrogen, and phosphorus cycles) of the ecosystem and can create suitable conditions for the benefit of their own and other organisms (Figure 23.3 illustrates soil ecosystem).

Microbes in the soil live in a symbiotic relationship with the flora in a given ecosystem. The interaction between nutrient-rich plant root surroundings and microbes is known as the rhizosphere. The colonization of microbes in the rhizosphere may be for mutualism or parasitism. The ability to utilize the root exudates is dependent on the capability of microbes present in the rhizosphere for effective colonization on the root surface and to interact or compete with the other microbes (Whipps JM, 2001; Bais HP et al., 2006; Lambers H et al., 2009).

Contamination of soils by human activity and urbanization has caused toxic substances to enter plant systems by absorption from the soils, into the food chain via grazing cattle, such as cows and other livestock that feed on plants growing in these contaminated soils. These accumulated toxins in meat and milk products are eventually consumed

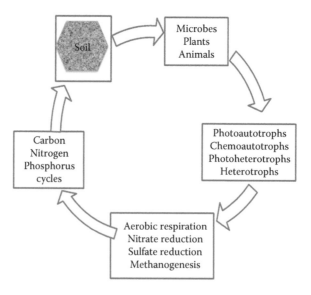

Figure 23.3 Soil ecosystem.

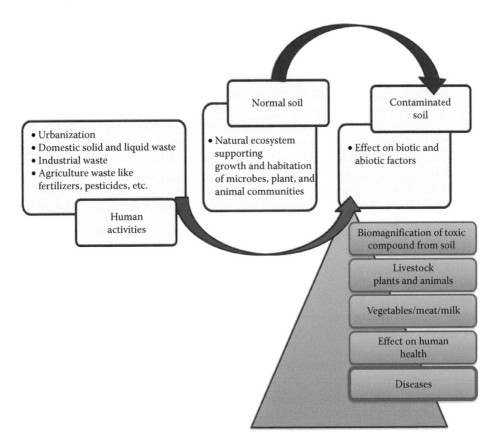

Figure 23.4 Assaulted soil ecosystem.

by humans. Mining of raw materials such as oil, coal, iron, gold ore, etc. for production, transportation to different parts from the place of mining and their use and improper disposal, accidental spills, etc. often contaminate the soil (Figure 23.4 illustrates assaulted soil ecosystem). This in turn affects the ecosystem by acting on the different micro- and macroflora. Sometimes, natural microbial flora withstand these adversities, and remediation of contaminated or polluted soils takes place. It seems that the trace metal contamination of soil from various sources will increase now with the growing rate of agricultural and industrial activities. The application of pesticides and fertilizers, in addition to sewage, has added to the load of trace elements in soil. A number of ecological standards such as NOEC (no observed ecological consequences), LOEC (lowest observed effect concentration), MPL (maximum permissible loading), PAA (permissible annual application), etc. are used to assess significant levels of trace element in soils. While plants play a significant role in cycling trace elements owing to their natural ability to adapt to variable soils, they also carry these contaminants in the process from soil, water, and air to the food chain.

The analysis of plants is a good indicator of the health of the soil/water in which they grow. Lichens have a wide geographical distribution and are long-lived. They allow high accumulation of pollutants throughout their thalli and can be used effectively for ecosystem biomonitoring work. Dr. Harry Harmens, Centre of Ecology and Hydrology, Bangor, suggests that mosses are more sensitive to atmospheric pollution as they directly prevail on air and moisture from the atmosphere. It is far cheaper to collect them for biomonitoring purposes than to sample air- and rain-borne pollutants from time to time.

The present advancement in environmental biotechnology and the presence of different approaches to aid the development of bioremediation techniques enhances the possibility of efficient and cost-effective bioremediation methods for soil treatment. Researchers are using Omics techniques to find unique biomarkers such as genes, enzymes, or metabolites of a bioremediation process. Advanced molecular systems from genomics to metabolomics have the potential to give the environmental biologist a complete understanding of the microbial–ecosystem interaction and to adapt remediation processes more effectively. For remediation approaches, microbes take a foreseat as compared to plants since they are ubiquitous and many of them can survive under extreme conditions as seen at contaminated niches in need of remediation. High biodiversity in the microbial system itself is an indicator of its adaptability to varied environments.

Water

Water is the most common solvent on the earth, out of which only 1% of earth water is available for drinking. The aquatic ecosystem is mainly divided into the fresh water ecosystem and marine ecosystem. Millions of micro- and macroorganisms are a part of an equally vast water ecosystem. The microbial population in a natural water body (aquatic ecosystem) is determined by the physical and chemical properties of that habitat (Figure 23.5 illustrates aquatic ecosystem).

Human habitation around water bodies and their subsequent use/misuse have caused different types of pollutants to disturb that ecological niche. Discharged domestic sewage and industrial effluents into nearby water bodies are the main sources of water pollution. Studies are needed to precisely characterize the concentration/populations of microbial flora to define their detection systems in drinking and non-potable water bodies, which will be useful in the conservation of the aquatic ecosystem (Figure 23.6 illustrates assaulted aquatic ecosystem).

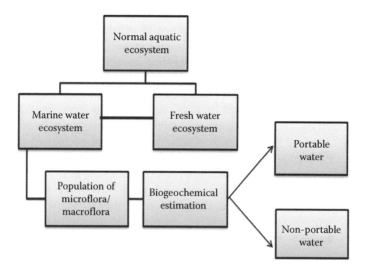

Figure 23.5 Normal aquatic ecosystem.

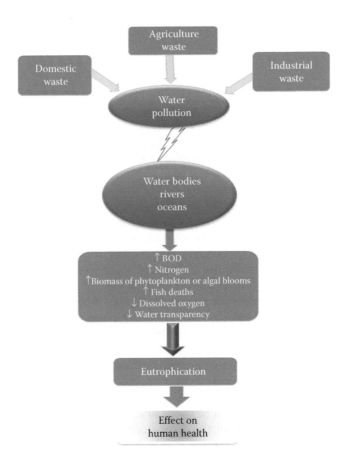

Figure 23.6 Assaulted aquatic ecosystem.

It is interesting to know that viruses that are harmless to animals and plants are important in regulating aquatic ecosystems—both fresh and saltwater. These viruses are essential players in the recycling of carbon in the marine environment. They infect the bacteria present in aquatic environments, and the organic substances released from these bacterial cells stimulate fresh algal as well as bacterial growth. The role of microbes in the fermentation industry to generate products such as bread, cheese, wine, beer, or soya sauce is another major application of food biotechnology tools. Apart from this, viruses have been used as vectors to replace a defective gene in the human genome as a part of gene therapy to produce healthy copies of the gene in question. A knowledge of the genetic diversity of these pathogens would help improve devising newer strategies for public health and better our understanding of sciences such as evolution, taxonomy, and pathogenicity (Sharma A, 2012).

Marine ecosystems offer a huge repository of microbes with a wide commercial potential; lagoons in French Polynesia are being tested for microbes that could produce compounds with pharmaceutical, cosmetic application, or even as an alternative for the usage of plastics. Microalgae like phytoplankton are an important part of the marine ecosystem and they regulate the amount of carbon in the atmosphere. Professor Gram's research from the Technical University of Denmark indicates that marine bacteria may be sources of bioactive compounds with anticancer properties, to treat antibiotic-resistant infections, or perhaps for biofuel production (Machado H et al., 2015).

Air

The area close to the earth's surface is divided into four interconnected regions called geospheres. The four spheres are lithosphere (stone), atmosphere (air), hydrosphere (water), and biosphere (life). Atmosphere, the outermost zone of air, is composed of 79% nitrogen, 21% oxygen, 1% argon, CO_2, and other gases. In addition to these, it also consists of particulate matter and water vapors. Air pollution can cause damage to ecological resources such as the ozone layer, water quality, soil, and animal and plant life. Nitrogen and sulfur deposition from air, etc. causes the acidification of soil and water bodies. The excess nitrogen in air and soil disrupt the N_2 cycle, thereby altering plant and microbial communities living around. In water bodies, an excess of nitrogen, other metals (mercury), and toxic compounds such as pesticides deposited from the atmosphere, bioaccumulate in the food chain, which can adversely affect behavioral, neurological, and reproductive aspects in aquatic and terrestrial livestock (Figure 23.7 illustrates causes and effects of pollution).

Air pollution has now become almost an endless by-product of industrialization (Figure 23.8 illustrates air pollution and biomagnification). A classic example would be industrial pollution in the nineteenth century, which necessitated the Alkali Act in 1862 and the Rivers Pollution Act in 1876; finally in 1952, it took the London smog to bring about the Clean Air Act in 1956 that reduced smoke levels and SO_2 from the atmosphere. Sadly, by then, the increase in the number of motor vehicles and the replacement of coal by oil for heating buildings led to an increase in the CO_2, NO_x, PAH, O_3, and peroxyacetyl nitrate (PAN) levels (Alloway BJ and Ayres DC, 1997).

Owing to industrialization, urbanization, and deforestation, the ratio of the important gases in air and balance within the ecosystems have been changed and disturbed, leading to increased temperatures and the greenhouse effect. The major interest of our society and/or the environmental biologists in particular should be to protect the environment and free it of/optimize the levels of greenhouse gases in order to avoid adverse

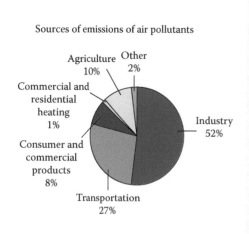

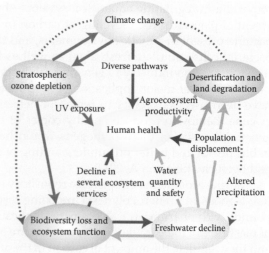

Figure 23.7 Causes and effects of pollution.

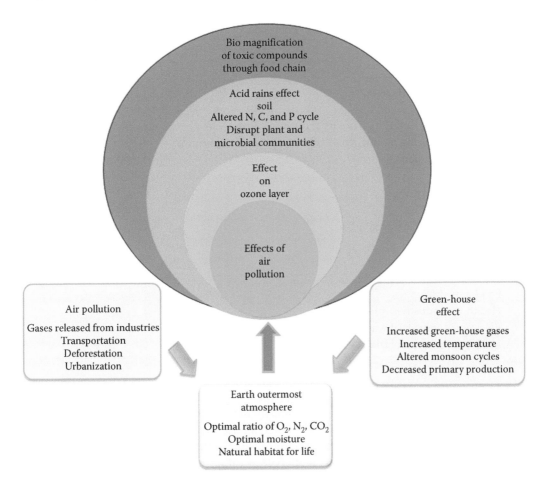

Figure 23.8 Pollutants from air to food chain.

effects. Maintenance of functions of different ecosystems may depend on the maintenance of high biodiversity on planet Earth through appropriate damage control and disaster management.

Environmental pollutants—Issues

Pollutants from human activities are the principal cause for the deteriorating quality of soil, water, air, and gene pool in existing life forms that inhabit those ecosystems, and eventually results in the disappearance of species that cannot cope with the persistence of this problem. Some organic pollutants such as hexachlorobenzene, aldrins, and polychlorinated biphenyls, which dissolve poorly in water are easily stored in fatty tissue and may be passed to infants through breast milk. A sizable proportion of pollutants do not have any immediate effects on health and are not easily detectable, but may show their impact later on after several years of constant exposure (Table 23.1 illustrates common pollutants of soil, water, and air). Apart from the lack of nutritious diet and physical activity owing to present-day lifestyles playing a part, a good deal of evidence points to many environmental pollutants acting as endocrine disruptors to precipitate metabolic disease

Table 23.1 Common pollutants of soil, water, and air

Pollutant/source	Soil	Water	Air
Oil/petroleum waste	√	√	√
Pesticides/chemicals	√	√	√
Fertilizers	√	√	√
Cadmium (Cd), lead (Pb), mercury (Hg), and arsenic (As)	√	√	√
Dust arising during mining (coal/gold ore/uranium/thorium)	√	√	√
Radiation pollution	√	√	√
Paint industry	√	√	√
Cement industry	√	√	√
Ceramic industries	√	√	√
Power plant/nuclear/coal plant	√	√	√
Automobile	√	√	√
Textile (coloring agents)	√	√	√
Paper/printing industry	√	√	√
Food industries	√	√	√
Healthcare industries (pharma)	√	√	√
Electronics, industrial waste (Cd, Pb, Hg, and As)	√	√	√
Leather tanning industrial waste	√	√	√

(Kirkley AG and Sargis RM, 2014). Studies have shown an alarming rate of increase in cases with metabolic disorders. Research has revealed a strong association between exposure to crystalline silica, organic solvents such as trichloroethylene and pesticides in soil to systemic lupus erythematosus in humans (Parks CG and Cooper GS, 2006). Crystalline silica acts as an immune adjuvant; it increases inflammatory responses and antibody production. It is a documented fact that environmental factors contribute to autoimmunity (Pollard KM and Kono DH, 2013). We breathe in about 3000 gallons of air each day. Polluted air can affect people of all age groups, more likely with immediate effects on the health of the elderly and children. Such air can visibly affect crop yield and commercial forests, apart from mixing into water and reaching soil in the surrounding areas. Caciari T et al. (2014) recruited 50 outdoor workers and 50 indoor workers to their study and tested them for plasma levels of free T3, T4, and thyrotropin stimulating hormone (TSH). TSH levels showed a significant rise in outdoor workers exposed to urban pollution compared to the controls. Heavy metal toxicity of milk from cattle and other foods may be responsible for the unprecedented increase in the cases of children showing low hemoglobin levels and iron deficiency anemia.

Other major problems with these environmental pollutants include global warming, global climate change, loss of diverse species due to their inability to combat the high concentration of pollutants, etc. As various governments have recognized this problem and have regulations against polluting activities, it also increases the cost factor to maintain a given contaminant under accepted limits. There is indeed a greater need, now more than ever, to understand our ecological systems and then facilitate/modify processes for cost-effective waste disposal. There is also a need for a local group/body that is prepared to manage a major catastrophic event, which may cause widespread pollution of a certain kind/kinds of pollutants (accidental radioactive leaks, huge oil spills, sewage leaks into domestic areas, etc.).

Challenges such as high persistence of hazardous and xenobiotic substances is a major challenge for environmental remediation even with the aid of biotechnological tools. The major focuses for environmental biotechnology is (1) to develop efficient methods for waste management by the cleaning of effluents and treating the solid/liquid waste, (2) to develop alternative and environmentally friendly processes, and (3) to produce more environment friendly products with the materials recovered after bioremediating an environment. The above goals are achieved by using microbes or plants (as biocatalysts) through environmentally friendly processes which give less by-products and the treatment or procedures can also run in a low-budget scenario. The final products may be made more biodegradable by the development of biodegradable plastic, paper bags, remolded/ processed paper, etc., production of new types of detergents by using microbial enzymes, detoxifying, leather treatment, and cotton industry usage of natural colors. As per the concept of cleaner production, the United Nations Environment Programme addresses all requirements for the production of a product but with the object of minimal short- and long-term risks to humans and the environment in general. Such concepts have an application worldwide and could be made imperative to industrial areas.

This chapter will be incomplete if we failed to mention that pollution also takes place due to the excessive usage of electronic gadgets and cell phones, and due to tower radiation, which have been showing damage to health and the environment around us, according to the research work done by some groups. These pollutants belong to that class of agents that had been causing damage over a long period of time and adequate steps to curb them are not taken till the damage is very pronounced. For the time being, it is good to at least monitor the effects of this radiation and report any form of damage that may be seen to humans/environment/livestock in general.

Bioremediation

There are various processes by which bioremediation is accomplished; aerobic, anaerobic, and by co-metabolic means. Aerobic bioremediation comprises of reactions involving microbes that use a carbon substrate as the electron donor and oxygen as the electron acceptor. This process is faster; however, a larger volume of waste is generated. It is more expensive than anaerobic bioremediation, which involves microbial activity in processes such as fermentation, methanogenesis, sulfate, and nitrate reducing conditions. Based on which basic contaminant is being worked upon, a subset of these mechanisms come into play. In anaerobic metabolism, nitrate, sulfate, carbon dioxide, oxidized materials, or organic compounds replace oxygen as the electron acceptor. In cometabolic bioremediation, organisms such as *Nitrosomonas, Nitrobacter,* and *Arthrobacter,* by aerobic methods, and *Clostridium, Streptomyces,* and other methanogens, by anaerobic processes, degrade contaminants such as polycyclic aromatic hydrocarbons (PAHs), polychlorinated biphenyls (PCBs), pesticides, etc. Here, the organism does benefit from any gain of energy or carbon by degrading the contaminant, instead, the contaminant is degraded by a side reaction (EPA, 2006). However, the primary step is to identify priority areas that need attention for remediation and then to identify major contaminants at that site in order to apply the most effective bioremediating tool.

Phytoremediation

Global industrialization has resulted in large quantities of toxic waste being released into surrounding environments; current remediation merely shifts these wastes to a designated site. Through the process of photosynthesis, plants use solar energy to convert the various chemical components of soil to less toxic forms or other biologically available content that gets deposited in parts above the ground. Hence, by the process of phytoremediation, soil contaminants are converted or transferred or stabilized to less toxic and useful forms. Newer methods such as genetic manipulation and plant transformation can be used to introduce novel traits in plants that impart improved tolerance toward pollutants, ability to degrade them, increased accumulation of these pollutants, and production of plants with higher biomass.

Dr. Kochian, an expert who has researched at the U.S. Plant, Soil, and Nutrition Laboratory at Ithaca, has studied plant responses when they are exposed to stress in the surrounding environment and has characterized the strategies used by plants to tolerate soil toxicity. He has also extensively worked on plant mineral nutritional aspects. Plants have been used to remediate toxic heavy metal content and radioisotopes from nearby soil, for example, cadmium, a known toxic heavy metal gets taken up by the roots and is deposited in edible plant portions. So now, a matter of grave concern for researchers worldwide is to device the strategies to prevent toxic metal contaminants from entering the food chain.

Some plant species known as metal hyperaccumulators extract contaminants from the soil and concentrate them in the easily harvested plant parts such as stems, shoots, and leaves. While acting like cleansers, their ability to tolerate/survive high levels of heavy metals such as zinc, cadmium, and nickel in soils, is an added beneficial aspect. One such promising hyperaccumulator is the plant *Thlaspi caerulescens* (alpine pennycress), a small, weedy member of the broccoli and cabbage family that thrives on soils rich in zinc and cadmium (Figure 23.9 refers to a representative picture of the plant). While the toxicity happens to other plants at a level of 1000 ppm of Zn or 20–50 ppm of Cd in its shoots, *T. caerulescens* can accumulate up to 30,000 ppm zinc and 1,500 ppm Cd in its shoots, while remaining nearly unaffected by symptoms of metal toxicity. The crops thus grown could be cut, baled, burnt to ashes, and sold as ore. Electricity generated by the burning could

Figure 23.9 *Thlaspi caerulescens.*

partially offset biomining costs. Ashes of alpine pennycress grown on a high-Zn soil in Pennsylvania yielded 30%–40% Zn, which is as high as high-grade ore.

According to research work undertaken at the DuPont Company, corn and *Zea mays* can gather very high levels of lead contaminants. *Z. mays*, a monocot in the Poaceae family, cultivated extensively next only to wheat and rice, yields highly consumed products such as corn meal, corn flour, cornflakes, cooking oil, beer, and animal feed. The risk of these metal contaminants reaching the human food chain through these plant products is a matter of concern that cannot be ignored. In tests with metals toxins introduced into artificial soil (a mixture of sand and vermiculite), *Brassica juncea* and *Brassica carinata*, members of the mustard family, appeared to be most effective in removing large quantities of chromium, lead, copper, and nickel metals. Phytotech, Inc., a NJ-based company, reported in 1996 that it had developed transgenic strains of sunflowers, *Helianthus* sp., that could remove 95% of surrounding toxic contaminants in a record short time of 24 hours. Subsequently, *Helianthus* was planted on a styrofoam raft and placed at one end of a contaminated pond near Chernobyl; it was recorded that in 12 days, the cesium concentrations within its roots was 8000 times that of the water, whereas the strontium concentrations were 2000 times that of the water.

The properties of the roots and shoots from seven different populations of *Medicago sativa* (alfalfa) were evaluated for their nickel-binding ability from aqueous solutions. While the experiments showed that the optimal pH for nickel to bind to the alfalfa plant was pH 5–6, binding-time-dependency studies indicated that 80% (approximately) of the nickel ions bound to the alfalfa plant tissues in less than 5 minutes, where 4.1 mg of nickel was found bound per gram of alfalfa biomass. Metal recovery experiments showed that it was possible to remove over 90% of the bound nickel from the alfalfa biomass.

Gum kondagogu (*Cochlospermum gossypium*—family Bixaceae) (Figure 23.10 refers to a representative picture of the plant) exudes a gummy substance that is nontoxic and has been used in food additives (Vinod VTP and Sashidhar RB, 2010). In the recent past, gum kondagogu has been shown to have properties of a biosorbent and removed toxic lead,

Figure 23.10 *Cochlospermum gossypium.*

cadmium, nickel, and chromium from aqueous systems. The gum has functional groups that impart to it an ability to bind metal, thus making it a natural material for bioremediation of toxic metal contaminants especially from industrial effluents.

Microbial remediation

Enriching bacteria by culturing and introducing them into a contaminated area or using grassroot remediators such as mycorrhizal fungi or vermiculture can rejuvenate the food web in the soil by introducing beneficial organisms such as bacteria and fungi into the contaminated site. According to Gayle's principle of microbial infallibility, for every organic compound present, there is a microbe or enzyme system that is capable of its degradation (Gayle, 1952). Hence, a good biodegradation strategy can be worked out by selecting and concocting those microbes that are capable of catabolizing specific contaminants (Tables 23.2 and 23.3). For this purpose, processes such as bioventing, biostimulation, bioaugmentation, and mineralization are helpful (Table 23.4).

Fungi have a very vital role to play in managing and healing damaged ecosystems. They are nature's tools for decomposing complex, carbon-based plant/tree material such as lignin and cellulose into life-supporting soils. Fungi can promote the breakdown of larger hydrocarbons allowing microorganisms and plants to carry on with further remediation processes. Saprophytic fungi break down hydrocarbons and pesticides by making use of their digestive enzymes. Mycoremediation is also effective for accumulation of heavy metal toxins and eventually concentrate them in the fungal fruiting body or the "mushroom."

Table 23.2 Biodegradation of soil contaminated with oil or petroleum by microbes

S. no.	Microbial species	Substrate	Utilized as	Reference
1	*Acinetobacter* sp.	*n*-Alkanes of chain length C10–C40	Sole source of carbon	Throne-Holst et al. (2007)
2	*Gordonia, Brevibacterium, Aeromicrobium, Dietzia, Burkholderia,* and *Mycobacterium*	Petroleum-contaminated soil	Degraders of hydrocarbons	Chaillan et al. (2004)
3	*Pseudomonas putida, Rhodococcus erythroplotis,* and *Bacillus thermoleovorans*	Hexadecane is aliphatic fraction of crude oil and major component of diesel	Hexadecane degradation	Chenier et al. (2003)
4	*Aspergillus niger*	Hexadecane	Hexadecane degradation	Stroud et al. (2008)
5	*Desulfatibacillum alkenivorans*	Estuarine sediment, which utilizes C13–C18 alkanes	Mesophilic sulfate reducer	
6	*Amorphoteca, Neosartorya, Talaromyces, Graphium, Candida, Yarrowia,* and *Pichia*	Oil-contaminated soil	Degraders of petroleum	Chaillan et al. (2004)

Table 23.3 Pesticide degrading microbes

S. no.	Microbial species	Substrate	Utilized as	Reference
1	*Pseudomonas* sp.	Dieldrin, organochlorine pesticides	Breakdown of dieldrin in the soil	Matsumura et al. (1968)
2	*Aerobacter aerogenes*	Metabolism of DDT	Anaerobic transformation of DDT by bacteria	Wedemeyer (1967)
3	*Bacillus, Pseudomonas, Arthrobacter,* and *Micrococcus*	Organochlorine pesticides	Organochlorine degradation	Langlois et al. (1970)
4	*Sphingobacterium* sp.	DDT	Biodegradation of DDT	Fang et al. (2010)
5	*Trichoderma* sp.	DDD degradation	DDD degradation	Ortega et al. (2011)
6	*Pseudomonas* sp., *Micrococcus* sp., *Bacillus* sp.	Aldrin	Degradation	Patil et al. (1970)
7	*Micrococcus* sp., *Arthrobacter* sp., *Bacillus* sp.	Endrin	Degradation	Patil et al. (1970)
8	Bacteria	Organophosphate pesticide	Degradation	Liu et al. (2003) and Hong et al. (2005)
9	*Sphingobium* sp.	JQL4-5 Fenpropathrin-degrading bacterium	Degradation	Yuanfan et al. (2010)

Using fungi for bioremediation requires a large amount of mycelia. Mycelium refers to the vegetative part of the fungi that has fine and dense branched network of white thread-like hyphae. These mycelia act as filters and release enzymes needed for breakdown of chemical substances. Mycoremediation installations need a lot of mycelia; these can be bought from mushroom farms, or cultivated in-house. Cultivating mushrooms requires some amount of training and gets better with experience—it is important to provide the right amounts of moisture, shade, air, and temperature to grow. Myceliated straw can be used to remediate environments with oil spills. Straw bales are inoculated with mushroom spawn, and placed in the path of a polluted or a contaminated water body for nearly

Table 23.4 Soil mineralizing microbes

S. no.	Microbial species	Action	Reference
1	*Azotobacter, Azospirillum, Rhizobium,* and *Bradyrhizobium*	Reduction of atmospheric nitrogen	Postgate (1982), National Research Council (1994), and Hubbell and Kidder (2009)
2	*Desulfovibrio vulgaris*	Bioremediation of toxic metal ions	Heidelberg et al. (2004)
3	*Bacillus subtilis, Pseudomonas putida,* and *Pseudomonas fluorescens*	Phosphate solubilization	Rodríguez et al. (1999)

complete remediation. In order to be efficient remediators, some fungal species need to be left undisturbed while others thrive on agitation. Different mushrooms can handle different temperature ranges and also different species have selected preferences for the contaminants they target. Different strains of those species can be more effective than others.

The survival and thriving of a *Pseudomonas fluorescens* species has been demonstrated using TNT for a sole source of N_2 and acetate as the carbon source. 10% of the transformed trinitrotoluene (TNT) was accounted for by high-performance liquid chromatography (HPLC) analysis; a significant amount of the TNT nitrogen present was converted to biomass. 2-Amino-4,6-dinitrotoluene was identified as a dead-end metabolite. Yet another example is the *Pseudomonas guezennei* identified by Pacific Biotech; this bacterium is known to secrete a natural polyester substance, polyhydroxyalkanoate, which could replace plastics for a biodegradable packaging material. These are but few successful examples of nature's ingredients that can help bioremediate a certain niche, however such solutions are also grossly underutilized due to the lack of characterization of contaminants present in a particular environment.

Advanced molecular methods for bioremediation

Molecular-based technologies can be used to investigate the environment (natural vs. affected) and provide potential solutions through engineering novel and effective microbes for bioremediation, development of environment friendly receptor-based pesticides or targeted pesticides, conservation of the environment through mineral recycling, waste utilization, pollution control, and bioremediation of toxicants in treatment of domestic and industrial waste. The contribution of developing molecular-based technologies in areas like food production and security, climate change, and ecosystem management is by the innovative application of methods described via technological inputs.

Metagenomics involves the study of genetic material, recovered directly from environmental samples and is used widely to categorize microbial communities. It is a recent science which helps study the diversity in these microbes where whole genome sequencing studies are extremely helpful in defining the species in a given ecosystem along with information on pathways for interaction between different species. To develop an understanding of the available fauna/flora from nature and manipulate it to apply to a particular environment of our interest, new technologies like the study of metagenomics (also known as ecogenomics) will help in studying the natural ecosystems in order to create man-made remediation and recovery systems. This science can help us understand how microbes cope with a certain hazardous pollutant; the information generated can be used in remediation by processes such as biostimulation or bioaugmentation.

Multiparallel analyses of plant, whole genome, exome, and proteins are key to today's functional genomics initiatives. Metabolite profiling for a comparative display of gene functioning provides deeper insight into complex regulatory processes and to determine phenotype directly. Using gas chromatography/mass spectrometry (GC/MS), distinct compounds from *Arabidopsis thaliana* leaf extracts were compared and each genotype possessed a distinct metabolic profile (Quanbeck SM et al., 2012). Bioengineered *Arabidopsis* plants expressing the selenocysteine methyltransferase (SMTA) gene from the *Astralagus bisulcatus*, a species effective in clearing selenium waste, contain eight times more selenium in their biomass when grown on selenite compared to nontransgenic controls. Comparing the gene expression profiles of *A. thaliana* and closely related *Arabidopsis hallerri*, which is tolerant to cadmium and hyperaccumulates zinc, has been found to be helpful in identifying genes crucial for metal tolerance.

Metal detoxification in plants requires three classes of proteins—metallothionein, phytochelatin, and glutathione. Genetic engineering of the genes responsible for the expression of these proteins has been used to successfully enhance the effectiveness of phytoremediation. Given the application of hyperaccumulators as good candidates in phytoremediation, cloning, and expression of relevant genes (heavy metal accumulating and xenobiotic-degrading) have resulted in improved remediation rate for application to large areas containing these environmental contaminants.

Huge hauls of explosives have been released into different land and marine habitats from time to time. In May 1999 issue of *Nature Biotechnology*, research groups from United Kingdom have reported that transgenic tobacco plants can help clean up explosive material. Genetic engineering applied to bioremediation processes demonstrated that the bacterial gene, NADPH-dependent nitroreductase, bioengineered into tobacco plants, improved their tolerance and helped them degrade high levels of TNT. Similarly, transgenic *Arabidopsis* plants, engineered to express the xplA gene from *Rhodococcus* bacteria, showed high resistance to RDX. However, for the wide range of organic and inorganic pollutants affecting soil, air, and water habitats, very few phytoremediation methods are available; a better understanding of molecular pathways that lead to degradation of pollutants by plant and microbe systems will help evolve appropriate waste management and biosafety strategies (Tan DX et al., 2007, Cherian S and Oliveira MM, 2005). Given all these benefits of bioengineering organisms for remediation, it is essential that realistic and applicable tools should be in place for analyzing the risks of any of these genetically engineered organisms being released into the environment. Using currently available biotechnological tools to utilize functional genomics approaches will enhance the potential of these methods to design ecosystem conservation strategies.

Environment and conservation

Ecosystems worldwide are being subject to rapid changes due to anthropogenic activities resulting in reduced adaptability of the species contained in them. This has led to deficiency in the natural resources that these ecosystems would otherwise provide for sustenance of human life. For this, it is also essential that efforts should be made to replace nonrenewable resources with renewable forms, for example, the production of cellulase and sugars, 2,3-butanediol and other value-added chemicals from weeds (Motwani M et al., 1993; Sharma SK et al., 2002). Changing ocean ecosystems will have an undesirable impact on marine microenvironments and raise oceanic acidity levels, which most organisms will not be able to thrive in. Creating ocean reserves, understanding, and preserving the microbiome of these areas by biotechnological intervention may be able to minimize damage and replace the lost microflora and fauna of marine habitats, thus restoring ecological balance (Figure 23.11 illustrates ecological conservation for the protection of humankind). While technology can improve and speed up the pace of remediation pertaining to an ecosystem, it is more than essential to identify the contaminating sources disrupting the natural ecological balance of a certain niche and take adequate measures to eliminate or minimize their inflow.

In order to reduce or reverse this anthropogenic environmental change, mechanisms are needed that will predict the effect of human actions on ecosystems. Harfoot et al. (2014) elaborated the first mechanistic general ecosystem model (GEM) of ecosystem structure and function, globally applicable to both land and aquatic environments. The model covers almost all the organisms in ecosystems and describes the biology and behavioral aspects of individual organisms, the interactions between them and those with the environment,

Figure 23.11 Ecological conservation for the protection of humankind.

modeling their fate and predicting the ecosystem structure and function in response to human pressures. These types of models hold promise to predict the ecological implications of human activities on our planet in the future.

In addition to these systems, media needs to play a powerful role in disseminating information on conservation and remediation methods, apart from portraying the larger picture of doom in the absence of these efforts. Regulations must make it mandatory for the industrial sector not to adopt passive ways of disposal, which just mean dilute and disperse a given contaminant, but take to reactive environmental strategies, which involve not only initiating measures for least harmful waste generated but also practicing on-site remediation and recovery to the maximum extent possible, which can be done only with the involvement of biotechnological aids. Such strategies and companies would attract the support of both local communities and authorities and give them a competitive advantage over others in the same field.

Scope of environmental biotechnology

Every industry involved in the production or manufacturing of some sort of product should have its own in-house remediation unit with appropriately trained experts to monitor the disposal and recovery processes from time to time. In the wake of a growing world population and the constant depletion of nonrenewable resources, it is vital that we look at alternate forms of catering to the energy needs of our community by promoting thrust areas such as biotechnology. Environmental biotechnology holds the key to offering feasible solutions on management of environmental wastes and promises to provide goods and services to humans even in a low-economic setup like ours. In India, active research is being pursued and promoted from various centers involved in waste treatment, recovery and utilization, and for replacing nonrenewable sources with renewable forms.

Uses in the field of agriculture such as large-scale production of biofertilizers and biopesticides help reduce the overuse of chemical fertilizers which are otherwise damaging the environment and ecosystem. Waste management is another field where environmental biotechnology has an important role to play as genetically modified organisms have increased ability to degrade biological and chemical wastes or have the ability to utilize organic feedstock to obtain useful products. Environmental biotechnology basically involves identifying/altering organisms acting on target pollutants to be involved in the remediation process, to either optimize or enhance the biological processes occurring there, in order to make them more efficient. Ultimately, environmental biotechnology aims at sustainable and feasible ways of controlling environmental pollution. Environmental biotechnology is the path leading to a cleaner, safer, and pollution-free planet.

Conclusion and future perspective

As with other technologies, environmental biotechnology is all set to provide sustained services in the field of remediation and preventing pollution, monitoring pollutants, increased food production through genetically modified crops, and replacing nonrenewable resources with renewable forms. Worldwide data indicate that India stands sixth in energy demand and accounts for 3.5% of the world's commercial energy consumption (OECD, 2011). Hydrogen, produced from natural gas and coal, is the "fuel of the future"; an alternative will be biohydrogen production from microorganisms, which integrates waste management for clean energy production using renewable biomass and sunlight. The use of nanobiotechnology and living microorganisms in bio-microelectronic devices has opened up newer areas for interdisciplinary understanding for scientists to offer revolutionary perspectives that have immediate application. Advances in allied areas such as metagenomics, genetic, metabolic engineering, and cell technology have vastly widened the potential for the applicability of biotechnology to all possible aspects of ecosystem preservation. In future, this science offers a new bioeconomy with a big market for commercialization of products, bioenergy production, and services offered by environmental biotechnology in developed countries and the developing world alike. The economic viability of innovative biotechnological products and their safety monitoring will also be an area of importance in future. As newer challenges continue to crop up, protecting the environment and managing all kinds of waste with biotechnological tools can continue to be successful only with the involvement of policy makers to incorporate these processes into industry.

Summary

Environmental biotechnology will be the frontier area for future research on ecological conservation through the utilization and production of renewable resources. Bioremediating ecosystems and preventing pollutants from traveling through the food chain to humans will need consistent and integrated approaches from ecologists, molecular biologists, policy makers, and industry. The efficiency of identifying pollutants from industrial, agricultural, and domestic sources, and then bioengineering plants and microbes for their degradation, can be enhanced by making use of various ecosystem models being developed. They can be used to predict the implications of anthropological activities in a given environment so that strategies can be put in place to counter the imbalance generated. Mankind and economy in general can benefit from the advances made in the allied sciences of biotechnology applicable to healthcare, bioremediation, the chemical industry, and food production. Habitat and resource preservation, mediated through biotechnological sciences, is the need of the present day and also for times to come.

Take home points

- Nature has sufficient resources to provide us with goods and services, provided we preserve its basic structure and functional components.
- As it is, there are natural remediation processes that come into play constantly to maintain equilibrium of a given ecosystem.
- Anthropogenic activities have challenged the pace at which these natural remediation systems are working.
- While development has partly or completely thrown off-gear the ecological balance in several niches, we have the biotechnological tools that can help us restore this balance, taking lessons from nature.
- As the pace of development cannot be slowed down, it is for eco-conservationalists and biotechnologists to play an active role in identifying and applying suitable remediation strategies to specific target pollutants/industries.
- Advanced biotechnological techniques are now available to understand the makeup of each ecosystem and the interactions between different organisms contained in them.
- Model systems are available and can be developed to study an environment and the effect of human activities on it.
- Biotechnology-based approaches should be rapidly adopted by policy makers and industries to meet solutions on waste management that are affordable even in a low-cost setup.
- Bioenergy production, use of renewable raw materials, bioplastics, bio-based chemicals, "designer" strains for biodegradation, biosafety monitored genetically modified crops, and biotechnology for the healthcare industry are frontier areas for research and development.

References

Aktar, M.W., Sengupta, D., and Chowdhury, A., Impact of pesticides use in agriculture: Their benefits and hazards. *Interdisc Toxicol*, 2(1), 2009:1–12.

Alloway, B.J. and Ayres, D.C., *Chemical Principles of Environmental Pollution*. Blackie Academic & Professional Glasgow London. Second Edition, CRC Press. 1997.

Arabi, U., Indian agriculture under global climate change: Need for a rethinking on viable adaptation and mitigation policies. *EXCEL Int J Mult Man Stud*, 3(3), 2013:103–121.

Bais, H.P., Weir, T.L., Perry L.G. et al., The role of root exudates in rhizosphere interactions with plants and other organisms. *Ann Rev Plant Biol*, 57, 2006:233–266.

Caciari, T., Casale, T., Capozzella, A. et al., Thyroid hormones in male workers exposed to urban stressors. *Ann Ig*, 26(2), 2014:167–175.

Chaillan, F., Le, Fleche. A., Bury, E. et al., Identification and biodegradation potential of tropical aerobic hydrocarbon-degrading microorganisms. *Res Microbiol*, 155(7), 2004: 587–595.

Chenier, M.R., Beaumier, D., Roy, R. et al., Impact of seasonal variations and nutrient inputs on nitrogen cycling and degradation of hexadecane by replicated river biofilms. *Appl Environ Microbiol*, 69, 2003:5170–5177.

Cherian, S. and Oliveira, M.M., Transgenic plants in phytoremediation: Recent advances and new possibilities. *Environ Sci Technol*, 15;39(24), 2005:9377–9390.

Conly, J.M. and Stein, K., The production of menaquinones (vitamin K2) by intestinal bacteria and their role in maintaining coagulation homeostasis. *Prog Food Nutr Sci*, 16(4), 1992:307–343.

Cruz-Cruz, C.A., González-Arnao M.T., and Engelmann, F., Biotechnology and conservation of plant biodiversity. *Resources*, 2(2), 2013:73–95.

Dance, A., Soil ecology: What lies beneath. *Nature*, 455, 2008:724–725.

EPA, Engineering Issue. 2006. *In Situ* and *Ex Situ* Biodegradation Technologies for Remediation of Contaminated Sites. EPA-625-R-06-015.

Fang, H., Dong, B., Yan, H. et al., Characterization of a bacterial strain capable of degrading DDT congeners and its use in bioremediation of contaminated Soil. *J Hazard Mater*, 184(1–3), 2010:281–289, ISSN 0304-3894.

Gayle, E.F., 1952. *The Chemical Activities of Bacteria*. Academic Press, London.

Ghosh, S., Maiti, T.K., and Basu, P.S., Bioproduction of ascorbic acid in root nodule and root of the legume pulse *Phaseolus mungo*. *Curr Microbiol*, 56(5), 2008:495–498.

Grünhage, L. and Haene, H.D., PLATIN (PLant-ATmosphere INteraction)—A model of biosphere/atmosphere exchange of latent and sensible heat, trace gases and fine-particle constituents. *Forestry Res*, 4(58), 2008:253–266.

Harfoot, M.B.J., Newbold, T., Tittensor, D.P. et al., Emergent global patterns of ecosystem structure and function from a mechanistic general ecosystem model. *PLoS Biol*, 12(4), 2014:e1001841.

Harrison, M.J., Development of the arbuscular mycorrhizal symbiosis. *Curr Opin Plant Biol*, 1, 1998:360–365.

Heggie, L. and Halliday, K.J., The highs and lows of plant life: Temperature and light interactions in development. *Int J Dev Biol*, 49(5–6), 2005:675–687.

Heidelberg, J.F., Seshadri, R., Haveman, S.A. et al., The genome sequence of the anaerobic, sulfate-reducing bacterium *Desulfovibrio vulgaris* Hildenborough. *Nat Biotechnol*, 22(5), 2004:554–559.

Hong, L., Zhang, J.J., Wang, S.J., Zhang, X.E., and Zhou, N.Y. Plasmid-borne catabolism of methyl parathion and p-nitrophenol in Pseudomonas sp. strain WBC-3. *Biochem Biophys Res Commun*, 334(4), 2005:1107–1114.

Hong, Y., Zhou, J., Hong, Q. et al., Characterization of a fenpropathrin-degrading strain and construction of a genetically engineered microorganism for simultaneous degradation of methyl parathion and fenpropathrin. *J Environ Manage*, 91(11), 2010:2295–2300.

Hubbell, D.H. and Kidder, G., *Biological Nitrogen Fixation*. University of Florida IFAS Extension Publication, Florida, SL16, 2009:1–4.

Keddy, PA., *Plants and Vegetation: Origins, Processes, Consequences*. Cambridge University Press, Cambridge, 2007.

Kirkley, A.G. and Sargis, R.M., Environmental endocrine disruption of energy metabolism and cardiovascular risk. *Curr Diab Rep*, 14, 2014:494.

Lambers, H., Mougel, C., Jaillard, B. et al., Plant-microbe-soil interactions in the rhizosphere: An evolutionary perspective. *Plant Soil*, 321, 2009:83–115.

Langlois, B.E., Collins, J.A., and Sides, K.G., Some factors affecting degradation of organochlorine pesticide by bacteria, *J Dairy Sci*, 53(12), 1970:1671–1675.

Liu, Z., Hong, Q., Xu, J.H. et al., Cloning, Analysis and fusion expression of methyl parathion hydrolase. *Acta Genet Sin*, 30(11), 2003:1020–1026.

Machado, H., Sonnenschein, E., Melchiorsen, J., and Gram, L., Genome mining reveals unlocked bioactive potential of marine Gram-negative bacteria. *BMC Genomics*, 16, 2015: ISSN: 1471-2164.

Matsumura, F., Boush, G.M., and Tai, A., Breakdown of dieldrin in the soil by a microorganism. *Nature*, 219(5157), 1968:965–967.

Motwani, M., Seth. R., Daginawala et al., Microbial-production of 2,3-butanediol from water hyacinth. *Bioresour Technol*, 44, 1993:187–195.

National Research Council, *Biological Nitrogen Fixation Research Challenges*. National Academy Press, Washington, DC, 1994.

OECD 2011, *Future Prospects for Industrial Biotechnology*. OECD Publishing, http://dx.doi.org/10.1787/9789264126633-en.

Ortega, N.O., Nitschke, M., and Mouad. A.M., Isolation of Brazilian marine fungi capable of growing on DDD pesticide. *Biodegradation*, 22, 2011:43–50.

Parks, C.G. and Cooper, G.S., Occupational exposures and risk of systemic lupus erythematosus: A review of the evidence and exposure assessment methods in population- and clinic-based studies. *Lupus*, 15(11), 2006:728–736.

Patil, K.C., Matsumura, F., and Boush, G.M., Degradation of endrin, aldrin, and DDT by soil microorganisms. *J Appl Microbiol*, 19(5), 1970:879–881, ISSN 1365-2672.

Pollard, K.M. and Kono, D.H., Requirements for innate immune pathways in environmentally induced autoimmunity. *BMC Med*, 11, 2013: 100. doi: 10.1186/1741-7015-11-100.

Postgate, J.R., *Fundamentals of Nitrogen Fixation*. Cambridge University Press, New York, NY, 1982, ISBN 10:0521284945/ISBN 13:9780521284943.

Quanbeck, S.M., Brachova1, L., and Campbell, A.A., Metabolomics as a hypothesis-generating functional genomics tool for the annotation of Arabidopsis thaliana genes of "unknown function". *Front Plant Sci*, 3, 2012: 15. doi: 10.3389/fpls.2012.00015.

Rodríguez, H. and Fraga, R., Phosphate solubilizing bacteria and their role in plant growth promotion. *Biotechnol Adv*, 17(4–5), 1999:319–339.

Sharma, A., Bhattacharya, A., Bora, C.R. et al., Diversity of enteropathogens in river Narmada and their environmental and health implications. *Microorganisms in Environmental Management*, Springer, London, pp. 35–60. 2012, ISBN 978-94-004-2228-6.

Sharma, S.K., Kalra, K.L., and Grewal, H.S., Enzymatic saccharification of pretreated sunflower stalks. *Biomass Bioenergy*, 23(3), 2002:237–243.

Silva, F.V., De Meyer, S.E., de Araujo, J.L.S. et al., *Bradyrhizobium manausense* sp. nov., isolated from effective nodules of *Vigna unguiculata* grown in Brazilian Amazon rainforest soils. *Int J Syst Evol Microbiol*, 64, 2014: 2358–2363.

Stroud, J.L., Paton, G.I., and Semple, K.T., Linking chemical extraction to microbial degradation of 14C-hexadecane in soil. *Environ Pollut*, 156(2), 2008:474–481.

Tan, D.X., Manchester, L.C., and Helton, P., Phytoremediative capacity of plants enriched with melatonin. *Plant Signal Behav*, 2(6), 2007:514–516.

Throne-Holst, M., Wentzel, A., and Ellingsen, T.E., Identification of novel genes involved in long-chain *n*-alkane degradation by *Acinetobacter* sp. strain DSM 17874. *Appl Environ Microbiol*, 73(10), 2007:3327–3332.

UNSEAR 2008 (United Nations Scientific Committee on the effects of Atomic Radiation). *Source and Effects of Ionizing Radiation*, 2008; (1) United Nations Publication, ISBN 978-92-142274-0.

Uttara, S., Bhuvandas, N., and Aggarwal, V., Impacts of urbanization on environment. *IJREAS*, 2(2), 2012: 1637–1645.

Vinod, V.T.P. and Sashidhar, R.B., Bioremediation of industrial toxic metals with gum kondagogu (Cochlospermum gossypium): A natural carbohydrate biopolymer. *Indian J Biotechnol*, 10, 2010:113–120.

Wedemeyer, G., Dechlorination of 1,1,1-trichloro-2,2-bis(*p*-chlorophenyl) ethane by *Aerobacter aerogene*. *J Appl Microbiol*, 15(3), 1967:569–574, ISSN 1365-2672.

Whipps, J.M., Microbial interactions and biocontrol in the rhizosphere. *J Exp Bot*, 52(suppl 1), 2001:487–511.

Yuanfan, H., Jin, Z., Qing, H. et al., Characterization of a fenpropathrin-degrading strain and construction of a genetically engineered microorganism for simultaneous degradation of methyl parathion and fenpropathrin. *J Environ Manage*, 91(11), 2010; 2295–2300, ISSN 0301-4797.

Zak, D.R., Holmes, W., White, D.C. et al., Plant diversity, soil microbial communities, and ecosystem function: Are there any links? *Ecology*, 84(8), 2003:2042–2050.

Web sources

1. Phytoremediation: Using Plants to Clean up Soils, http://www.ars.usda.gov/is/ar/archive/jun00/soil0600.htm.
2. Sunflowers Bloom in Tests to Remove Radioactive Metals from Soil and Water, http://www.herbmuseum.ca/content/industrial-hemp-bioremediation.
3. http://earthrepair.ca/resources/type/mycoremediation/.
4. http://www.e-herbar.net/main.php?g2_itemId=17812.
5. http://www.britannica.com/EBchecked/media/4914/Cochlospermum-gossypium.
6. http://clu-in.org/techfocus/default.focus/sec/Bioremediation/cat/Overview/.
7. Bennett, B., Repacholi, M., and Carr, Z., Health Effects of the Chernobyl Accident and Special Health Care Programmes, Report of the UN Chernobyl Forum Expert Group "Health." WHO Library Cataloguing-in-Publication Data, Geneva, 2006; ISBN 92 4 159417 9.

8. http://www.uvm.edu/~dwang/interact/intro.html.
9. http://rochels.weebly.com/uploads/1/0/7/6/10768381/chapter_2.7.pdf.

Further reading

1. Manivasagan, V., Kannadasan, T., and Sathya, G., Recent trends in environmental biotechnology. *Adv. Bio Tech*, 11(1), 2011:23–26.
2. OECD, *Future Prospects for Industrial Biotechnology*, OECD Publishing, Paris 2011. http://dx.doi.org/10.1787/9789264126633-en.
3. http://www.bioteach.ubc.ca/Journal/V02I01/bioremediation.pdf.
4. Bioremediation. *Methods and Protocols Series: Methods in Molecular Biology*, 599, Cummings, S.P. (Ed.) 2010. Humana Press, a part of Springer Science+Business media, LLC.

chapter twenty-four

Marine biotechnology
Focus on anticancer drugs

Amit Rastogi, Sameen Ruqia, and Alvina Gul

Contents

Abstract

Marine biotechnology is a multidisciplinary field of research which may be defined as the study of how various organisms in the marine ecosystem and the actions of the ocean can be used to provide services and valuable products to us. In such a manner, the ocean's sponges, mollusks, algae, microbes, sea cucumbers, mangroves, soft corals, ascidians, tunicates, etc. are the main source for the discovery and development of novel anticancer drugs which are not found in terrestrial counterparts. Some of the examples of these kinds of drugs are xinghaiamine A, fucoxanthin, bryostanin-1, kahalalide F, halichondrin B, philinopside A and philinopside E, sansalvamide A, and dolastanin. These marine-based drugs showed great anticancer activity *in vitro* via inhibiting various cancer cell lines as well as in different animal models. These marine-based anticancer drugs, novel and structurally unique present an alternative to conventional drugs to treat cancer, targeted several signaling pathways, mechanisms, and molecules such as topoisomerase I or II, MAPK, NFYB, HDAC, PKCs, CDKs, MetAPs, microtubule assemblies, actin, Bcl-2 proteins, v-ATPase, via blocking calcium ion/channels, caspase pathways, STAT-3, via inhibiting angiogenesis, metastasis, and several others pathways and molecules. Most of them are in clinical trial pipelines, some have been rejected at various stages of clinical trials due to their toxicity profile. A few of them are commercially available. This field of research is however in its initial phases and needs development.

Keywords: Marine drugs, Antitumor agents, Marine biotechnology, Anticancer CAM.

Introduction

Marine biotechnology is a multidisciplinary field of research which may be defined as the study of how various organisms in the marine ecosystem and the actions of the ocean can be used to provide services and valuable products to human beings which include aquacultures, cosmetics, pharmaceuticals, nutraceuticals, therapeutics, medical devices, drug delivery systems, industrial enzymes, wound dressing, biofuels, environmental products, biosensors, tissue engineering, herbicides, fungicides, pesticides, feed, and foods. Marine biotechnology is defined as the key to unlock the potential of the unique biodiversity of marine organisms and ecosystems (Marine Biotechnology, 2005; Thakur et al., 2006). Approximately 6 million annual deaths are caused by cancer (Rennie and Rusting, 1996). Cancer is a condition in which cells grow abnormally due to uncontrolled division. These cells are capable of invading other tissues too. The spread of cancer cells to other tissues and organs is mediated by the circulatory system (Weinberg, 1996).

Oceans cover a large area (over 70%) of the Earth's surface and possess more than 300,000 of all the described species from marine sources that shows a rich biodiversity (Jirge et al., 2010).

It is estimated that almost 18,000 different compounds have been extracted from marine sources and among these more than 150 have been found to possess potent cytotoxic activity against tumor cells (Kumar et al., 2011). Due to its immense biodiversity and natural product providing promise, marine biotechnology has gained the attention of many researchers from all over the world for the last 44 years. However, research in this field is still in its early phases, which means that many marine originated fauna and flora are yet to be exploited for novel bioactive compounds. In the last two decades, various natural products obtained from marine sources have been employed for pharmaceutical uses. Various marine sources have been reported to produce structurally unique and novel bioactive compounds which are antiviral, antifungal, antibacterial, antidepressant, antimalarial, anticancer, and antitumor agents. These compounds are considered to be unique due to their diverse structural and mechanical characteristics (Thakur et al., 2005, Diers et al., 2008, Jimenez et al., 2009).

There are many natural products obtained from marine sources which play an immense role in chemical defense of their respective organisms. These natural products may have significant influence on the whole ecological environment (Paul et al., 2007). One of the most important aspects of studying marine products is novel drug development. Now science has progressed so much that biosynthetic pathways can be manipulated in eukaryotic as well as prokaryotic systems. Manipulation of biosynthetic pathways is a major step toward the production of potential drugs. Many of the genes which are responsible for the biological production of novel drugs have not only been identified but also sequenced. The research on snails from the genus *Conus* has resulted in fruitful production of the first ever marine drug named Prialt. The drug is approved by the regulatory authorities in USA and is used for the treatment of chronic pain originating in the spinal cord (Molinski et al., 2009).

Pharmaceutical R&D and institutes in marine biotechnology research

At the present time, much medical and pharmacological research is being performed on world's oceans. Pharma MAR, which is a pharmaceutical sciences institute, is considered to be one of the leaders in developing drugs from marine sources. Another noteworthy institute for marine biotechnology is Aquapharma Biodiversity Ltd. It is a UK-based company which not only discovers but also commercializes pharmaceutical and nutraceutical products. Nerus Pharmaceutical, a leading marine biotechnology-based industry in the United States, has a strong research and development (R&D) division and has played a significant role in contributing to develop marine natural products as well as in the discovery of new pharmaceutical agents. Its main R&D focus is marine microbial-based novel drug discovery. Some other marine biotechnology-based prominent companies include Wyeth, Aventis, Novartis, and Neurex. In India, some selected institutes such as NIO, Goa; CDRI, Lucknow; Bose Institute, Kolkata; CIFE, Mumbai; RRL, Bhubaneswar; Annamalai University and Mumbai University are engaged in the development of marine-based drug discoveries from various sources (Thakur et al., 2005).

Research grant support history for marine biology and/or biotechnology

Marine biotechnology research is supported in the United States by the National Institutes of Health (NIH) and NOAA Sea Grant Program. The research conducted with the support of NIH focuses on the development and commercialization of novel and potent anticancer drugs originating from marine organisms.

NIH research support is more recent and is focused on the development of novel anticancer drugs from marine biodiversity. Recently, the National Science Foundation (NSF) got much attention for scientific research related to marine biodiversity and its potential to produce novel drugs (Fenical et al., 1996).

Conventional route to treat cancer and its side effects

Conventional cancer therapies include all the medical treatments that doctors use to treat cancer patients. These conventional treatments are researched and tested for their safety and effectiveness. Based on conventional cancer therapies, cancer can be treated in three different ways. These techniques include chemotherapy which is therapy through chemicals, radiation therapy in which nonmutagen radiations are used to kill cancer cells, and hormone therapy which can be used in special cases of cancer. These therapies can also be used in combination. Sometimes, hormone therapy may require removal of an organ, for instance in case of breast cancer (Malaker and Ahmad, 2013).

Side effects of conventional cancer therapies

Chemotherapy has adverse effects on hair, bone marrow, nails, and digestive system cells as it not only targets cancer cells but also affects other parts of the body. Radiation therapy in certain cases has been associated with an increase in the proliferation of cancer cells. There are certain side effects associated with hormone therapy. Some of these include formation of blood clots, gaining of weight, change in appetite pattern, tiredness, retention of body fluids, flashes, and nausea. Multiple drug resistance (MDR) can be considered as one of the limitations of chemotherapy. MDR is basically caused by overexpression of some integral membrane transporters which can decrease accumulation of drugs and can cause resistance to multiple drugs. The cells which are multidrug resistant are also resistant to cytotoxicity and hence are unaffected by chemotherapy (Malaker and Ahmad, 2013).

Complementary and alternative therapies

Complementary and alternative therapies are also known as CAM. These are the treatments that people use besides their conventional therapeutic regimens or in place of them. These therapies boost the physical and emotional health of the patient and help in relieving the symptoms of cancer and the side effects of conventional therapies but are unable to cure cancer (White 2001, Cancer Support, Conventional, Complementary and Alternative Therapies, 2012).

Why? Development of marine-based therapeutics

More than 60% of commercially available drugs for cancer treatment are of natural origin. Around 2500 metabolites with antiproliferative activity have been reported during the last decade. Marine resources for anticancer drug discovery are used due to several reasons. These reasons include the inherent activity of marine peptides which are largely unexplored, ability to revolve against challenges, structural diversity, and reduced risks of adverse side effects. Marine habitat-derived bioactive compounds are structurally unique, novel, potent cytotoxic, even at nanomolar scale concentration, and most of them have the novel mechanism of action to treat cancer with few side effects, while some of them are successful in treating cancer with an unknown mechanism of action (Folmer et al., 2010, Gate et al., 2012, Malaker and Ahmad, 2013).

There are many disadvantages of conventional drug therapies. For this reason, marine biospheres and biodiversity can contribute significantly to provide novel and unique anticancer drugs. Marine-derived drugs are structurally unique and represent an alternative therapy to conventional drugs. Some of them are in the market and others are in various phases of clinical trials.

Antitumor compounds from marine sponge

Sponges (also called primitive filter-feeders) are multicellular organisms belonging to the phylum Porifera (meaning "pore bearers"). This phylum is considered a gold mine due to its diversity in producing novel secondary metabolites. Scientists have discovered more than 5000 sponge species but it is estimated that there are approximately 8000 marine sponges on the planet. Marine sponges are a very potent source of novel anticancer, antiviral, anti-inflammatory, immunosuppressive, neurosuppressive, antimalarial, antifouling, antitumor, and cardiovascular agents. Much research has been performed in which sponges have been used to treat sunstroke, testicular tumors, infectious wounds, dropsy, stomach aches, and bone fractures. It is due to this fact that drug discovery from marine sources is based on marine sponges, since the discovery of sponge nucleosides, including spongothymidine in *Cryptotethya crypta*. These nucleosides act as a base for the production of Ara-C or simply cytosine arabinoside which is the first ever potent marine-derived anticancer agent. Ara-A, an antiviral drug, was also produced. Ara-A is also known as Vidarabine which is potently active against varicella zoster and herpes simplex. Cytarabine also known as Cytosar-U or Depocyt is another chemotherapeutic agent. This agent can be used for the treatment of cancers and tumors such as leukemia and non-Hodgkin lymphoma (Wang et al., 1997, Sipkema et al., 2005, Essack et al., 2011, Perdicaris et al., 2013).

The first natural steroid 6-hydroxyamine, isolated from *Cinachyrella* sp., is shown to inhibit the aromatase enzyme. This enzyme plays an important role in the estrogen pathway and is involved in the final catalysis step of estrone and testosterone to estradiol. Estrogens (estrone and estradiol) formation involves the active participation of the aromatase enzyme. It has been reported that estrogen conversion occurs in blood as well as in breast tumor cells (Sipkema et al., 2005).

Marine sponges are involved in the production of various anticancer alkaloids such as pyridoacridine alkaloids (pentacyclic alkaloids, hexacyclic alkaloids, heptacyclic alkaloids, and octacyclic alkaloids), indole alkaloids (bisindole alkaloids, peptidoindoles, β-carbolines, and trisindole alkaloids), pyrrole alkaloids (bromopyrrole alkaloids, pyrroloquinones, pyrroloquinoline alkaloids, and pyrroloacridine), pyridine alkaloids, isoquinoline alkaloids, guanidine alkaloids, aminoimidazole alkaloids, and steroidal alkaloids. These alkaloids have been shown to possess great cytotoxic activity, and antitumor and antiproliferative activity against various cancer cell lines. These alkaloids possess multiple mechanisms for anticancer therapy such as cytotoxicity, and for the production of reactive oxygen species which are characteristic features of pyridoacridine alkaloids (Kumar et al., 2011, Lee and Su, 2011).

Antitumor compound from marine actinomycetes

Actinomycetes are economically and biotechnologically the most valuable prokaryotes on the planet and are characterized by a complex life cycle. Distribution of novel marine actinomycetes has been reported everywhere in oceanic environments and habitats. Some novel marine actinomycetes such as *Aplysina aerophoba* and *Theonella swinhoei* are found associated with Mediterranean sponges while some actinomycetes *Rhopaloeides odorabile*, *Candidaspongia flabellata*, and *Pseudoceratina clavata* are associated with Great Barrier Reef sponges. Some marine actinomycetes *Dietzia maris*, *Kocuria erythromyxa*. *Rhodococcus erythropolis* has been reported to be isolated from sediment core, at a depth of 1225 m off Hokkaido. However, distribution of marine actinomycetes in the ocean is largely unexplored. Marine actinomycetes face oceanic environmental challenges and adapt themselves to survive in extremely harsh environmental conditions which make them producers of novel antibacterial, antifungal, antimalarial, anti-inflammatory, and anticancer drugs. Marine actinomycetes are known to inhibit ornithine decarboxylase (ODC), which is a directed transcriptional target of oncogene *myc* and is overexpressed in various tumor cells. ODCs may be an important target for the chemoprevention of cancer. *Saliniketals* from marine actinomycetes are known to inhibit ODC (Lam, 2006, Olano et al., 2009). Some examples of marine actinomycetes-based novel anticancer drugs are given below.

Recently, discovered novel pyridinium containing cytotoxic antibiotic *1-(10-aminodecyl) pyridinium* is a white waxy solid with molecular formula $C_{15}H_{27}N_2$. It is isolated from the bioactive actinomycete stain DVR D4 which is obtained from a marine sediment sample from the Visakhapatnam coast of the Bay of Bengal, India. The stain DVR D4 is identified as *Amycolatopsisalba* on the basis of biochemical properties and 16S rDNA analysis. The compound is found to be a potent antibacterial agent against Gram-negative as well as Gram-positive bacteria and potent cytotoxic against breast (MCF-7), cervix (HeLa), and brain (U87MG) cancer cell lines *in vitro*. Maximum inhibitions of 60.46%, 39.64%, and 41.85% for breast (MCF-7), cervix (HeLa), and brain (U87MG) cancer cell lines, respectively, are observed at 1000 µg/mL concentration of this compound (Dasari et al., 2012).

Resistoflavine (1), a novel qunione-related cytoxic agent, obtained from marine-derived actinomycete *Streptomyces chibaensis* AUBN$_1$/7 is isolated from sediment samples from the Bay of Bengal, India. The physicochemical properties of the compound indicate that it is a pale yellow-colored powder (solid) with molecular formula $C_{22}H_{16}O_7$. The compound shows *in vitro* potent cytotoxic activity against HMO2 (gastric adenocarcinoma) and HePG2 (hepatic carcinoma) cancer cell lines with LC$_{50}$ value 0.013 and 0.016 µg/mL. This compound also shows weak antibacterial activity against Gram-negative and Gram-positive bacteria (Gorajana et al., 2007).

Other novel marine actinomycetes-based anticancer compounds are *chandrananimycins, 3,6-disubstituted indoles*, and *diazepinomicin (ECO-4601)* isolated by *Actinomadura* sp., *Streptomyces* sp., and *Micromonospora* sp., respectively (Lam et al., 2006).

Lagunamide C is isolated from marine cyanobacterium *Lyngbya majuscula*. Lagunamide C is a potent cytotoxic and antimalarial cyclodepsipeptide. Lagunamide C is able to exhibit potent cytotoxic activity at nanomolar level against P388, A549, PC3, HCT8, and SK-OV cancer cell lines with IC$_{50}$ value 2.4, 2.6, 2.1, and 4.5 nM, respectively (Tan et al., 2008, Tripathi et al., 2011, Duffy et al., 2012).

Anticancer drug from mollusks

Mollusks are invertebrate animals belonging to the phylum *Mollusca*. Around 85,000 species of mollusks are recognized so far. Mollusks are the largest marine phylum comprising about 23% of all known and named marine organisms and are involved in the production of many pharmaceutically active compounds. *Sea hare*, a shelled organism, produces bioactive metabolites used in the treatment of cancerous tumors.

Keenamide A (a cytotoxic cyclic hexapeptide) is isolated from the notaspidean mollusk *Pleurobranchus forskalii*, and shows antitumor activity via unknown mechanisms. *Keenamide A* shows significant activity against P388, A549, MEL-20, and HT-29 tumor cell lines (Wesson et al., 1996).

Dolastanins are cytotoxic cyclic and linear peptides obtained from sea hare, *Dolabella auricularia*, a mollusk which is found in the Indian Ocean. The linear pentapeptide *Dolastatin* 10 was discovered by researchers at Arizona State University from *Dolabella auricularia*, but now it can be isolated from cyanobacteria *Symploca* sp. *VP642* as well. Dolastanin 10 shows an outstanding inhibitory effect against several forms of skin cancers in laboratory studies. The most active antineoplastic substance dolastanin 10 disrupts microtubule spindles and inhibits microtubule assembly. At subnanomolar concentration, dolastatin 10 is cytotoxic against lymphocytic leukemia cells. Dolastanin 10 induces apoptosis via Bcl-2 phosphorylation and increases p53 expression in lymphoma cell lines. More recently, other dolastanins (dolastanin H and isodolastanin H) have been isolated and are shown to possess high cytotoxicity (Blunden, 2001, Pandey et al., 2013, Varshney et al., 2013). In phase I clinical trials, dolastanin 10 reports good tolerability and presented myelosuppression at the dose-limiting toxicity. Bone marrow toxicity at the initial clinical trials level as well as peripheral sensory neuropathies, pain, swelling, and erythema at the injection site are some side effects caused by Dolastanin 10 (Folmer et al., 2010, Suarez-Jimenez et al., 2012).

Dolastanin 15 is an antineoplastic agent which interferes with microtubule assembly and causes mitotic arrest. Dolastanin 15 is found to be effective in inhibiting cell growth of L1210 murine leukemic cells (Varshney and Singh, 2013). The dolastanin 15 analog, *LU103793*, failed in phase II clinical trials because this compound did not show significant antitumor activity. Hence, clinical development of this compound has been discontinued (Le Tourneau et al., 2007).

A synthetic dolastatin analog, *ILX-651 (tasidotin)*, has successfully completed phase I clinical trials and is currently undergoing phase II anticancer clinical trials. *Genzyme Corporation*, Cambridge, Massachusetts has rights for its clinical trial and for development of an anticancer drug from this compound (Folmer et al., 2010).

TZT-1027, a newly synthesized dolastanin 10 derivative, has antitumor activity against a variety of transplantable tumors in mice. It is found to be effective against P388 leukemia and B16 melanoma (Kobayashi et al., 1997).

ES-285 (PM 95118, IL0111), also known as spisulosine, obtained from *Spisula polynyma*, a marine mollusk, is an analog of the phospholipid sphingosine. ES-285 completed preclinical trials with promising antineoplastic potential against a wide variety of malignancies, ranging from hematological malignancies to solid tumors. It induces programmed cell death *in vitro*. This compound has undergone clinical validation in phase I trials. In phase I clinical trials, the maximum tolerated dose (*MTD*) was optimized with a patient having an advanced solid tumor and found to be 200 mg/m^2 (Schöffski et al., 2011).

Antitumor compounds from tunicates (ascidians)

One of the largest and most diverse phyla is that of ascidians. They are also known as Urochordata. These organisms comprise almost 3000 species, most of which have marine habitats. They live in shallow water as well as the deep sea. Ascidians are also known as tunicates. This name has been coined by Lamarck. They are named tunicates because they originated in a polysaccharide-containing tunic which envelopes the animals and forms their skeleton (Shenkar and Swalla, 2011). Tunicates are known to produce complex antitumor compounds; examples of such kinds of antitumor compounds are *Didemnin B*, *Aplidine*, and *ET-743*.

Didemnins are cyclic depsipeptide compounds isolated from tunicates of the genus *Trididemnum solidum*, which are collected from the Caribbean Sea. More than nine didemnins (didemnin A–E, G, X, and Y) were isolated from these tunicates. Among them, *didemnin B* is known to have the most potent antitumor and antiproliferative activity for cancerous cell lines. Didemnin B acts at GDP-binding protein elongation factor. This compound is also known to exert antitumor activity via protein synthesis inhibition and also has antiviral and immunosuppressive activities (Le Tourneau et al., 2007, Sarfaraj et al., 2012, Malakar and Ahmad et al., 2013).

Ecteinascidins are obtained from the Caribbean tunicate *Ecteinascidia turbinata*, which are known to exhibit significant antitumor activity against murine and human tumor cell lines. Ecteinascidin-743 (*trabectedin*; Yondelis®) is a tetrahydroisoquinolene alkaloid isolated from the same tunicate. *Ecteinascidin-743* is identified to have a broad-spectrum antitumor activity (Thakur et al., 2005). This compound acts by selective alkylation of guanine residues in DNA minor groove as well as by telomerase dysfunction of tubulin.

This compound alters transcription factors and proteins (Schwartsmann et al., 2001, Le Tourneau et al., 2007, Gate et al., 2012, Newman and Cragg, 2014).

Synthetic diazonamide A shows high cytotoxicity against ovarian carcinoma 1A9, breast carcinoma MCS7, and taxol-resistant 1A9/PTX10 cell lines. Diazonamide A is a potent inhibitor of tubulin assembly equivalent to dolastanin (into microtubules), causing cells to accumulate at the G2/M phase of cell cycle. It does not inhibit binding of vinblastin, dolastanin 10, or GTP exchange with tubulin and also does not stabilize colchicine binding to tubulin. Diazonamide A inhibits tubulin assembly by binding to a unique site on the tubulin dimer and/or binds to a "peptide site" but only at the end of growing tubes (Lindquist et al., 1991, Nicolaou et al., 2002, Cruz-Monserrate et al., 2003).

A number of alkaloids namely *kuanoniamines A–D* have been isolated along with known *shermilamine B* from a Micronesian tunicate and its prosobranch mollusk predator *Chelyonotus simperi*. All these compounds show cytotoxicity against KB cells (Carroll and Scheuer, 1990).

Antitumor compounds from marine algae

Marine algae are also called the "photosynthetic machinery" of the marine environment. They are considered to be the first photosynthetic cellular plants which have 50% contribution to global photosynthesis and describe about 90% of total marine flora (Dhargalkar and Neelam, 2005). Marine algae play a significant role in contributing to marine species diversity (Rinehart et al., 1981). Hopkins Marine Station is reported to constitute more than 5000 known species of green algae, over 4000 species of red algae, and 1500 species of brown algae which are exclusively found in marine habitats (Mostafa et al., 2005). In the Western Atlantic region, over 600 algal species have been reported in the Caribbean, and around 400 species reported on the north-eastern tropical coast of Brazil, whereas the western coast of Africa contain at least 300 algal species. The Indo-West Pacific is reported as a "hotspot" of algal diversity. In the Philippines, over 900 species of marine algae have been reported which constitutes 250 species of green algae, 506 of red, and 154 of brown algae. The Eastern Pacific region is reported to be the poorest region to hold algal species with no more than 300 algal species (Figueiredo and Costeira, 2009).

Marine algae have been used in China to prepare traditional and folk medicines for over 2000 years. Uses of marine algae on a large scale have also been reported in Ancient Egyptian and Indian ayurvedic medicines. For the first time in the 1960s, algae were used in Western medicines to treat breast cancer (Varshney et al., 2013).

Marine algae has been reported to produce *novel halogenated metabolites* via diverse and unique biosynthesis pathways. Such compounds have various biological activities such as antifungal, antibacterial, antiviral, anti-inflammatory, antiproliferative, cytotoxicity, antifouling, and several other significant biological activities. These halogenated compounds have been reported for the first time in *Carpodetus serratus*, which represents the largest group of algal antifungal compounds and are involved in the growth inhibition of a marine pathogenic fungus *Lindra thalassiae*. The production of many secondary metabolites via marine algae are now thought to be a result of the defense mechanism developed by these organisms due to their adaptation to survive in extreme marine conditions with varying amounts of space, light, and nutrients which has shaped their physiology to produce these kinds of unique secondary metabolites. Chemical ecology may be another reason for the production of such kinds of bioactive secondary metabolites from marine algae. Macroalgae from tropical areas has been reported to have been screened heavily for the discovery of novel halogenated compounds with such kinds of biological activities. The ecological interaction between algae and herbivorous inverte-brates may be another reason for such kinds of biologically active secondary metabolites (Cabrita et al., 2010).

Marine algae vary dramatically in size and can be divided into unicellular organisms such as microalgae such as blue-green algae or cyanobacteria and multicellular organ-isms referred to as macroalgae, showing low-level thallus differentiation. Macroalgae are visible and are organized in filamentous or parenchymatous thalli. Larger algae that have a very complex differentiated thallus are named *seaweeds* and when they are attached by holdfast with upright stems and fronds as tall as trees, they are known as *kelps* (*Macrocystis*, a kelp reaches 65 m height). Around 150,000 known seaweed species have been reported worldwide from which only 2% species have been studied and identi-fied till date (Gomez et al., 2010).

Microalgae include diatoms, dinoflagellates, and cyanobacteria. These are collectively known as phytoplanktons and have been reported to produce novel anticancer agents (Folmer et al., 2010). *Nostoc, Lyngbya, Symploca,* and *Calothrix* are the main blue-green gen-era studied for this purpose (Zheng et al., 2010). Dinoflagellates are marine unicellular protozoans. Dinoflagellates are known to produce some of the largest and most complex polyketides (Bold and Wynne, 1978, Schwartsmann et al., 2001, Kadam and Prabhasankar, 2010).

Hierridin B is isolated from picocyanobacterium *Cyanobium* sp. LEGE 06113 which has been obtained from the Atlantic coast of Portugal. Previously, this compound was isolated from filamentous epiphytic cyanobacterium *Phormidium ectocarpi* SAG 60.90 where it was known to possess antiplasmodial activity. The structure of Hierridin B was confirmed by NMR and MS analyses. In a concentration–response assay, up to 100 µg m/L concentration of Hierridin B is found to be cytotoxic against colon adenocarcinoma HT-29 cells with IC_{50} value 100.2 µM (Leão et al., 2013).

Coibamide A, symplostanin 1, and *caulerpenyne* are other marine algae-based anticancer compounds. Coibamide A and symplostanin 1 are isolated from the marine cyanobacte-rium *Leptolyngbya* sp. and *Symploca hydnoides*, respectively, while caulerpenyne is isolated from the green algae *Caulerpa taxifolia* (Varshney et al., 2013).

Another marine famed carotenoid *fucoxanthin* with molecular formula $C_{42}H_{58}O_6$ has been reported to be isolated from numerous classes of microalgae such as bacillariophytes, bolidophytes, chrysophytes, silicoflagellates, pinguiophytes, and also from brown mac-roalgae. This compound and its deacetylated metabolite *fucoxanthinol* has been reported to have antiproliferative and anticancer activity *in vitro* in various cancer cell lines and in var-ious animal models. This compound has been reported to have an anticarcinogenic effect which results in apoptosis, cell cycle arrest, antiangiogenesis, and metastasis suppression effect of this compound via triggering up- and downregulation of various signaling path-ways and molecules such as Bcl-2 proteins, caspase pathways, MAPK and GADD45, NFYB, CYP3A4 enzymes, and several other molecules which are involved in anticancer activities (Kumar et al., 2013).

Anticancer compounds from mangroves and salt marsh plants

Anticancer compound *xylogranatins A–D* have been isolated from mangrove *Xylocarpus granatum*. Other famed mangrove anticancer compounds are limonoids *granaxylocarpins A and B*. Granaxylocarpins A and B which are found to be cytotoxic against P-388 leukemia cells with IC_{50} value <10 µM. *Naphtoquinones 3-chlorodeoxylapachol* and *stenocarpoquinone B* are other famed mangrove cytotoxic compounds isolated from *Avicennia germinans* and

Avicennia marina, respectively. Marine marsh grass, cord-grass, or grasswort have also been reported to produce anticancer compounds such as *Juncusol* from marsh grass *Juncus roemerianus*, isorhamnetin 3-O-β-Dd-*glucopyranoside* from grasswort *Salicornia herbacea* (Folmer et al., 2010).

Antitumor compounds from marine bacteria and fungi

Marine bacteria constitute ~10% of the living biomass carbon of the biosphere. The distribution of bacteria in seawater occurs through a series of biological and physicochemical factors, and there are mainly two fundamental causes of this distribution, the amount of available decaying organic matter, and the density of planktonic organisms in water because planktonic organisms are the main source of food for bacteria. Marine environments provide low nutrition, high salinity, and high pressure for microorganisms to survive. The immense chemical and biochemical diversity of marine microorganisms makes them a rich and effective source of novel drugs. Extremely complex marine environments make microorganisms different from terrestrial microorganisms (Thakur et al., 2005, Jimenez et al., 2009, Olano et al., 2009, Gonzales-Parraga et al., 2011).

The association of bacteria with marine algal species has been reported for their higher growth. Moreover these bacteria may be partially involved in the production of specific vitamins which are needed for the higher growth of these algae. Bacteria associated with seaweeds have also been found to show anticancer activity (Gomez et al., 2010). Interactions of bacteria and seaweed have also been reported to influence the secretion of bioactive compounds and such kinds of interaction may be positive such as metabiosis and symbiosis, or negative such as parasitism, predation, and competition. Marine bacteria are known to serve as food for diverse groups of organisms due to their relative incapability to escape from predators (Schlegel and Jannasch, 2006). Marine bacteria have also been reported to associate with a sea worm of *Polychaetes* species to show strong antimicrobial activity (Fenical, 1993, Sunjaiy-Shankar et al., 2010).

A new class of anticancer agents includes the *vascular disrupting agents*. These are gaining attention from all around the world due to their low toxicity profile, which specifically targets established tumor blood vessels. In rodent cancer models, vascular disrupting agents are responsible for massive hemorrhagic necrosis in tumors, because these agents cause rapid shutdown of blood flow in established solid tumors within a few minutes. Vascular disrupting agents principally target the vascular endothelial growth factor (VEGF) pathway to inhibit tumor angiogenesis. These agents can broadly be classified into two major groups: microtubule binding agents and flavonoids. A great example of vascular disrupting agents from marine fungus *Aspergillus* sp. is *plinabulin* (NPI-2358) which is a synthetic analog of phenylahistin (NPI-2350). It has been found active against advanced solid tumors with low toxicity profile but some side effects such as nausea, vomiting, chest pain, tumor pain, back pain, dehydration or electrolyte disturbances, mental confusion, and myocardial infarction have been reported. Plinabulin, the structurally novel compound, interacts with the interfacial region of α- and β-tubulin, which partially overlaps with the colchicine binding site. It has also been reported that Plinabulin shows antitumor activity in animal models alone and it is reported to show synergistic effect with chemotherapy agents, including taxanes, without worsening, and even improving chemotherapy tolerability in some cases (Mita et al., 2010).

Marine-derived fungus *Scopulariopsis brevicaulis* is able to produce two novel cyclodepsipeptides *scopularide A and B*. Both are able to inhibit pancreatic and colon tumor cell lines while their mechanism of inhibition is still unknown (Yu et al., 2008). *Sansalvamide A* is structurally unique cyclic depsipeptide that acts to inhibit topoisomerase I. *Sansalvamide A* is obtained from the fungus of genus *Fusarium* which is a host on the marine plant *Halodule wrightii* found on Little San Salvador Island, the Bahamas (Heiferman et al., 2010).

Other marine fungal-derived anticancer compounds are *6-methoxyspirotryprostatin B, 18-oxotryprostatin A*, and *14-hydroxyterezine D* and are isolated from an ethyl acetate extract of a marine-derived fungal strain *Aspergillus sydowi* PFW1-13. All these compounds are cytotoxic against A-549 and HL-60 cell lines and are found to show weak cytotoxicity against both cell lines (Zhang et al., 2008).

Anticancer compound from bryozoans

Bioactive compounds obtained from bryozoans are known to have anticancer effects. Bryostatin was discovered in marine bryozoans *Bugula neritina*. These bryostatins are produced by bacterial symbiont *Candidatus Endobugula sertula*, which is present through all life stages of *Bugula neritina*. This symbiosis provides an antipredatory defense for its host. Possibly, this symbiosis produces defensive compounds other than bryostatins. It is possible that bryostatins give a competition advantage to *B. neritina* because it is a relatively dominant and cosmopolitan species. *Bugula neritina* is not only the source of bryostanin. Bryostatins are also discovered in snail *Polycera atra* from the group Nudibranchia. Bryostatin 10 is most abundant in larvae, probably for its antipredatory effects, but it is difficult to determine the biochemical mechanisms by which bryostatins deter predators. Bryostanin 1 is a macrocyclic lactone isolated and identified by researchers at Arizona State University. Bryostatin 1 is a potential drug for treatment of leukemia, lymphomas, melanomas, solid tumors, traumatic brain injury, depression, Alzheimer disease, and other CNS disorders and can activate innate immunity. However, it has also been observed that bryostatin 1 causes intense muscle hyperalgesia. Bryostatin-1 is shown to be a potent activator of protein kinases C (PKCs), and has antagonistic effects on tumor-promoting phorbol esters. Furthermore, bryostatin-1 inhibits the production of components of matrix metalloproteinases family, downregulates multidrug-resistance 1 (MDR1) gene expression, modulates bcl-2 and p53 gene expression, and induces apoptosis. *Merle 23*, a derivative of bryostatin 1, differs from it only at four positions. It behaves like a phorbol ester in U-937 human leukemia cells, while bryostatin 1 antagonizes many phorbol ester responses in cells (Le Tourneau et al., 2007, Sinko et al., 2012).

Conclusion

As described above, marine biodiversity of flora and fauna is able to provide a number of anticancer drugs, which are unique in their structure, mode of action, and are not found in terrestrial counterparts. The extreme harsh conditions of the marine environment make them provide such kinds of drugs. However, there is an urgent need to develop new culture techniques for these kinds of microbes, which are difficult to cultivate under ordinary laboratory conditions. The development of new screening strategies to find new anticancer drugs from these microbes is also in demand. Moreover, there is also an urgent need to develop new fermentation techniques to scale up the production of drugs from these microbes. Molecular biology techniques and genetic engineering are currently being used to optimize production of these drugs. Anticancer drugs which have an unknown mechanism of action should be identified soon. Investigation of new marine flora and fauna and their symbiont identification may also lead the way to discover new anticancer drugs.

References

Blunden G, 2001, Biologically active compounds from marine organisms, *Phytotherapy Research*, 15(2), 89–94.

Bold HC and Wynne MJ, 1978, *Introduction to the Algae: Structure and Reproduction*. Englewood Cliffs, New Jersey, USA, pp. 706.

Cabrita MT et al., 2010, Halogenated compounds from marine algae, *Marine Drugs*, 8, 2301–2317.

Cancer Support, Conventional, Complementary and Alternative Therapies, 2012, Available online: http://www.macmillan.org.uk/Cancerinformation/Cancertreatment/Complementarythera pies/Complementaryalternativetherapies.aspx.

Carroll AR and Scheuer PJ, 1990, Kuanoniamines A, B, C, and D: Pentacyclic alkaloids from a tunicate and its prosobranch mollusk predator *Chelyonotus simperi*, *Journal of Organic Chemistry*, 5514, 4426–4431.

Cruz-Monserrate Z, Vervoort HC, Bai R, Newman DJ, Howell SB, Los G, Mullaney JT, Williams MD, Pettit GR, Fenical W, and Hamel E, 2003, Diazonamide A and a synthetic structural analog: Disruptive effects on mitosis and cellular microtubules and analysis of their interactions with tubulin, *Molecular Pharmacology*, 63, 1273–1280.

Dasari VRRK, Muthyala MKK, Nikku MY, and Donthireddy SRR, 2012, Novel Pyridinium compound from marine actinomycete, *Amycolatopsis alba* var. nov. DVR D4 showing antimicrobial and cytotoxic activities *in vitro*, *Microbiological Research*, 167(6), 346–351.

Dhargalkar VK and Pereira N, 2005, Seaweed: Promising plant of the millennium, *Science and Culture*, 71, 60–66.

Diers JA, Ivey KD, El-Alfy A, Shaikh J, Wang J, Kochanowska AJ, Stoker JF, Hamann MT, and Matsumoto RR, 2008, Identification of antidepressant drug leads through the evaluation of marine natural products with neuropsychiatric pharmacophores, *Pharmacology Biochemistry and Behavior*, 89(1), 46–53.

Duffy R, Wade C, and Chang R, 2012, Discovery of anticancer drugs from antimalarial natural products: A MEDLINE literature review, *Drug Discovery Today*, 17, 942–953.

Essack et al., 2011, Recently confirmed apoptosis-inducing lead compounds isolated from marine sponge of potential relevance in cancer treatment, *Marine Drugs*, 9, 1580–1606.

Fenical W, 1993, Chemical studies of marine bacteria: Developing a new resource, *Chemical Reviews*, 93(5), 1673–1683.

Fenical W et al., 1996, Marine biodiversity and the marine cabinet: The status of new drugs from marine organisms, *Oceanography*, 9(1), 23–27.

Figueiredo MAO and Costeira PZ, 2009, Marine algae and plants, *Tropical Biology and Conservation Management*, EOLSS, IV, 190–202.

Folmer F, Jaspars M, Dicato M, and Diederich M, 2010, Photosynthetic marine organisms as a source of anticancer compounds, *Phytochemistry Reviews*, 9, 557–579.

Gate SS, Jathin S, and Bandawane DD, 2012, Application of marine products in pharmacy, *Journal of Biotechnology and Biotherapeutics*, 2(9), 37–44.

Gomez et al., 2010, Antibacterial and anticancer activity of seaweeds and bacteria associated with their surface, *Revista de Biologia Marina y Oceanografia*, 45, 267–275.

Gonzales-Parraga et al., 2011, Marine micro-organisms: The world also changes, Science Against Microbial Pathogens: Communicating Current Research and Technological Advances, 1281–1292.

Gorajana A, Venkatesan M, Vinjamuri S, Kurada BVVSN, Peela S, Jangam P, Poluri E, and Zeeck A, 2007, Resistoflavine, cytotoxic compound from a marine actinomycete, *Streptomyces chibaensis* $AUBN_1/7$, *Microbiological Research*, 162(4), 322–327.

Heiferman MJ, Salabat MR, Ujiki MB, Strouch MJ, Cheon EC, Silverman RB, and Bentrem DJ, 2010, Sansalvamide induces pancreatic cancer growth arrest through changes in the cell cycle, *Anticancer Research*, 30(1), 73–78.

http://www.nsf.gov/funding/pgm_list.jsp?org=bio.

Jimenez JT, Sturdíkova M, and Studik E, 2009, Natural products of marine origin and their perspectives in the discovery of new anti-cancer drugs, *Acta Chimica Slovaca*, 2(2), 63–74.

Jirge SS et al., 2010, Marine: The ultimate source of bioactives and drugs metabolites, *International Journal of Research in Ayurveda and Pharmacy*, 1(1), 55–62.

Kadam SU and Prabhasankar P, 2010, Marine foods as functional ingredients in bakery and pasta products, *Food Research International* 43, 1975–1980.

Kobayashi M et al., 1997, Antitumor activity of TZT-1027, A novel dolastanin 10 derivative, *Japanese Journal of Cancer Research*, 88, 316–327.

Kumar D et al., 2011, Marine natural alkaloids as anticancer agents, *Opportunity, Challenge and Scope of Natural Products in Medicinal Chemistry*, 213–268.

Kumar SR et al., 2013, Fucoxanthin: A marine carotenoid exerting anti-cancer effects by affecting multiple mechanisms, *Marine Drugs*, 11, 5130–5147.

Lam KS, 2006, Discovery of novel metabolites from marine actinomycetes, *Current Opinion in Microbiology*, 9(3), 245–251.

Le Tourneau C, Raymond E, and Faivre S, 2007, Aplidine: A paradigm of how to handle the activity and toxicity of a novel marine anticancer poison, *Current Pharmaceutical Design*, 13(33), 1–13.

Leão PN et al., 2013, Antitumor activity of hierridin B, a cyanobacterial secondary metabolite found in both filamentous and unicellular marine strains. *PLoS ONE*, 8(7): e69562.

Lee NL and Su JH, 2011, Tetrahydrofuran cembranoids from the cultured soft coral *Lobophytum crassum*, *Marine Drugs*, 9, 2526–2536.

Lindquist N, Fenical W, Van Duyne GD, and Jon Clardy, 1991, Isolation and structure determination of diazonamides A and B, unusual cytotoxic metabolites from the marine ascidian Diazona chinensis, *Journal of the American Chemical Society*, 113(6), 2303–2304.

Malakar et al., 2013, Therapeutic potency of anti-cancer peptides derived from marine organism, *International Journal of Engineering and Applied Sciences*, 2(3), 82–94.

Marine biotechnology, 2005, National Marine Research Strategy, National Marine Biotechnology Programme, Available online [http://www.marine.ie/home/research/SeaChange/National MarineBiotechnology/Marine+Biotechnology.htm].

Mita MM, Spear MA, Yee LK, Mita AC, Heath EI, Papadopoulos KP, Federico KC et al., 2010, Phase 1 first-in-human trial of the vascular disrupting agent plinabulin (NPI-2358) in patients with solid tumors or lymphomas. *Clinical Cancer Research*, 16(23), 5892–5899.

Molinski TF, Dalisay DS, Lievens SL, and Saludes JP, 2009, Drug development from marine natural products, *Nature Review Drug Discovery*, 8, 69–85.

Mostafa MH et al., 2005, Active biological materials inhibiting tumor initiation extracted from marine algae, *Egyptian Journal of Aquatic Research*, 31(1), 146–155.

Newman DJ and Cragg GM, 2014, Marine-sourced anti-cancer and cancer pain control agents in clinical and late preclinical development, *Marine Drugs*, 12, 255–278.

Nicolaou et al., 2002, Total synthesis of diazonamide A, *Angewandte Chemie International Edition*, 41, 3495.

Olano C et al., 2009, Anti-tumor compounds from marine actinomycetes, *Marine Drugs*, 7, 210–248.

Pandey PK et al., 2013, Potent anti-cancer compounds from the ocean, *International Journal of Biodiversity and Conservation*, 5(8), 455–460.

Paul VJ, Arthur K, Ritson-Williams R, Ross C, and Kotty S, 2007, Chemical defenses: From compounds to communities, *Biological Bulletin* 213, 226–251.

Perdicaris S et al., 2013, Bioactive natural substances from marine sponges: New developments and prospects for future pharmaceuticals, *Natural Produts Chemistry and Research*, 1(3), 1–8.

Rennie J and Rusting R, 1996, Making headway against cancer, *Scientific American*, 275, 56.

Rinehart KL et al., 1981, Marine natural products as sources of antiviral, antimicrobial, and antineoplastic agents, *Pure and Applied Chemistry* 53, 759–871.

Sarfaraj HM et al., 2012, Marine natural products: A lead for anti-cancer, *Indian Journal of Geo-Marine Sciences*, 41(1), 27–39.

Schlegel HG and Jannasch HW, 2006, Prokaryotes in their habitats. In: Balows A, Truper HG, Dworken M, Harder W, Scheifer KH (eds) *The Prokaryotes*. Springer Verlag, New York, pp. 137–184.

Schöffski P et al., 2011, Spisulosine (ES-285) given as a weekly three-hour intravenous infusion: Results of a phase I dose-escalating study in patients with advanced solid malignancies, *Cancer Chemotherapy and Pharmacology*, 68(6), 1397–1403.

Schwartsmann G, Brondani da Rocha A, Berlinck RGS, and Jimeno J, 2001, Marine organisms as a source of new anticancer agents, *Lancet Oncology*, 2, 221–225.

Shenkar N and Swalla BJ, 2011, Global diversity of ascidiacea. *PLoS ONE*, 6(6), e20657. doi: 10.1371/journal.pone.0020657.

Sills AK, Williams JI, Tyler BM, Epstein DS, Sipos EP, Davis JD, McLane MP et al., 1998, Squalamine inhibits angiogenesis and solid tumor growth in vivo and perturbs embryonic vasculature, *Cancer Research*, 58(13), 2784–2792.

Sinko J, Rajchard J, Balounova Z, and Fikotova L, 2012, Biologically active substances from water invertebrates: A review, *Veterinarni Medicina*, 57(4), 177–184.

Sipkema D et al., 2005, Marine sponges as pharmacy, *Marine Biotechnology*, 7, 142–162.

Suarez-Jimenez G, Burgos-Hernandez A, and Ezquerra-Brauer J, 2012, Bioactive peptides and depsipeptides with anticancer potential: Sources from marine animals, *Marine Drugs*, 10(5), 963–986.

Sunjaiy-Shankar CV, Jeba-Malar AH, and Punitha SMJ, 2010, Antimicrobial activity of marine bacteria associated with *Polychaetes*, *Bioresearch Bulletin*, 1, 24–28.

Tan LT, Chang YY, and Ashootosh T, 2008, Besarhanamides A and B from the marine cyanobacterium *Lyngbya majuscula*, *Phytochemistry*, 69, 2067–2069.

Thakur NL et al., 2005, Marine natural products in drug discovery, *Natural Product Radiance*, 4(6), 471–477.

Thakur NL et al., 2006, Marine biotechnology: An overview, *Indian Journal of Biotechnology*, 5, 263–268.

Tripathi A, Puddick J, Prinsep MR, Rottmann M, Chan KP, Chen DY, and Tan LT, 2011, Lagunamide C, a cytotoxic cyclodepsipeptide from the marine cyanobacterium *Lyngbya majuscula*, *Phytochemistry*, 72, 2369–2375.

Varshney A and Singh V, 2013, Effects of algal compounds on cancer cell lines, *Journal of Experimental Biology and Agricultural Sciences*, 1(5), 337–352.

Varshney A et al., 2013, Effects of algal compounds on cancer cell lines, *Journal of Experimental Biology and Agricultural Sciences*, 1(9), 337–352.

Wang WS et al., 1997, High-dose cytarabine and mitoxantrone as salvage therapy for refractory non-Hodgkin's lymphoma, *Japanese Journal of Clinical Oncology*, 27(3), 154–157.

Weinberg RA, 1996, How cancer arises, *Scientific American*, 275, 62–70.

Wesson KJ and Hamann MT, 1996, Keenamide A, A bioactive cyclic peptide from the marine mollusk *Pleurobranchus forskalii*, *Journal of Natural Products*, 59(6), 629–631.

White JD, 2001, Complementary, alternative, and unproven methods of cancer treatment. In: DeVita VT Jr, Hellman S, and Rosenberg SA (eds) *Cancer: Principles and Practice of Oncology*, 6th ed. Lippincott Williams & Wilkins, Philadelphia, PA, pp. 3147–57.

Yu Z, Lang G, Kajahn I, Schmaljohann R, and Imhoff JF, 2008, Scopularides A and B, cyclodepsipeptides from a marine sponge-derived fungus, *Scopulariopsis brevicaulis, Journal of Natural Products*, 71(6), 1052–1054.

Zhang M, Wang WL, Fang YC, Zhu TJ, Gu QQ, and Zhu WM, 2008, Cytotoxic alkaloids and antibiotic nordammarane triterpenoids from the marine-derived fungus *Aspergillus sydowi, Journal of Natural Products*, 71(6), 985–989.

Zhang Y, Zhu T, Fang Y, Liu H, Gu Q, and Zhu W, 2007, Carbonarones A and B, new bioactive g-pyrone and a-pyridone derivatives from the marine-derived fungus Aspergillus carbonarius. *The Journal of Antibiotics*, 60, 153–157.

Zheng YL, Lu XL, Lin J, Chen HM, Yan XJ, Wang F, and Xu WF, 2010, Direct effects of fascaplysin on human umbilical vein endothelial cells attributing the anti-angiogenesis activity, *Biomedicine and Pharmacotherapy*, 64, 527–533.

chapter twenty-five

Engineering genomes for biofuels

Niaz Ahmad, Muhammad Aamer Mehmood,
Steven J. Burgess, and Muhammad Sarwar Khan

Contents

Abstract

Global warming, limited oil supplies, energy security, and economic concerns are among the factors driving the search for alternatives to fossil fuels. Biofuels, including bioethanol, biodiesel, and biohydrogen, can be produced renewably, and as such are considered an essential component of strategies to minimize dependency on fossil fuels and reduce greenhouse gas (GHG) emissions. In this chapter, we provide a survey of the various systems used for the production of biofuels, including higher plant-derived bioethanol and biodiesel, the use of microorganisms for the production of biofuels, and solar-driven hydrogen production. We go on to show how genetic engineering has been used to improve the yields of various biofuels and discuss future perspectives.

Keywords: Bioethanol, Biodiesel, Biohydrogen, Metabolic engineering, Plant biomass, Algae.

Introduction

The 1970s oil crisis raised the price of crude oil from $3 to $39.40 per barrel, triggering the search for alternative fuel sources. It was during this era that researchers began to explore both the possibility of utilizing plant and algal biomass as a source of hydrocarbons, and the feasibility of developing artificial photosynthetic systems for the production of fuels and electricity (Tachibana et al., 2012; Bonke et al., 2015). These early efforts were hampered by technical limitations, and as restrictions on oil were eased, funding was cut. Recently, interest in biofuels has been revived due to concerns about limited oil supplies and raising CO_2 emissions at a time when increasing industrialization and a rising global population is creating an increased demand for energy, with a projected increase in consumption of ~50% by 2050, the vast majority of which will originate from developing countries.

Much of the carbon released during the production and combustion of biofuels is sequestered during the growth of biomass, and subsequently the use of biofuels results in significant reductions in CO_2 emissions compared to fossil fuels. Technical advances have made it possible to convert plant and algal biomass into liquid or gaseous transportation fuels, and these strategies can be grouped into two different categories depending on the nature of biomass used. First-generation biofuels are obtained from sugars and vegetable oils extracted from food crops such as corn and sugarcane, whereas second-generation biofuels are produced from other than nonfood plant biomass such as lignocellulose. Third-generation biofuels broadly refer to new and hybrid processing technologies for the conversion of organic biomass such as algae into biofuels. However, the techniques for the second- and third-generation biofuels are largely in their infancy, making these fuels costly. Extensive research work has been undertaken and is being directed to develop alternative fuels at commercial scale with competitive production costs. In this chapter, we look at research employing genetic engineering to develop methods and techniques for the production of low-cost second- and third-generation biofuels.

Engineering plants for biofuel production

Need for plant-based biofuels

Components of plant biomass, including starch, cellulose, and hemicellulose, can be converted to fermentable sugars to produce bioethanol, and their oils can be turned into biodiesel.

Bioethanol is currently made from the fermentation of starch and sucrose obtained from major food crops such as corn and sugarcane (Ruth, 2008), resultantly increased demand has raised economic, environmental, and social concerns (FAO, 2007). More agricultural land for fuel production means less for other crops, and could result in an ecological imbalance and increase in food prices (Schnoor, 2006). Resultantly, lignocellulosic biomass has emerged as an attractive alternative feedstock as it does not compete with the food crops and is available in large quantities worldwide. Ethanol produced from lignocellulose has the potential to reduce greenhouse gas (GHG) emissions up to 65% compared to corn-ethanol (Montenegro, 2006; Adler et al., 2007; Morales et al., 2015).

Subsequently, many countries have committed to produce ethanol from lignocellulosic plant biomass. According to the Biofuels Digest, 64 countries worldwide, including 13 in America, 12 in the Asia-Pacific, 11 in Africa and the Indian Ocean, and 2 non-EU countries in Europe have approved targets or mandates for biofuels (Lane, 2015). For example, in 2006, the US Government announced plans to replace 30% of fossil fuels with that of lignocellulosic ethanol, and approved setting a mandate of 36 billion gallons of renewable fuel by 2022, of which 16 billion gallons is to be from cellulosic ethanol, and the European Parliament has specified via the Renewable Energy Directive that biofuels must represent 10% of all fuels consumed within the EU by 2020. Private companies are also taking interest in renewable biofuel research, with BP providing $500 million over a period of 10 years to develop next-generation biofuels from plant biomass (Mannan, 2010), and almost all the leading oil and gas companies, including Shell, Chevron, and Valero, have shown interest in second-generation renewable biofuels (Kanes et al., 2010).

Lignocellulosic plant biomass

The plant cell wall is the prime source of lignocellulosic biomass. Based on composition, plant cell walls are grouped into two types, type I and type II. Type-I cell walls are present in dicots and contain equal amounts of glucan and xyloglucan. Type-II cell walls are found in monocots, and compared to type-I cell walls, they have reduced amounts of xyloglucans and pectin, contain glucuronoarabinoxylan cross-links and lack some structural proteins (Rubin, 2008).

The majority of carbohydrates present in the primary cell wall are cellulose, hemicelluloses, and pectin (Pauly and Keegstra, 2008). Cellulose microfibrils are cross-linked with hemicelluloses and are embedded in a matrix of pectin, which acts to cement the cellulose–hemicellulose network (Ding and Himmel, 2006). Pectin comprises up to 35% of dry matter in dicotyledonous plants and has a major role in the synthesis of primary walls (Xiao and Anderson, 2013).

The major components of plant secondary cell walls are cellulose (35%–50%), hemicellulose (20%–35%), and lignin (10%–25%) (Rubin, 2008). Cellulose is made up of glucose molecules connected by glycosidic bonds, hemicellulose is a mixture of 5- and 6-carbon molecules, and lignin is a polymer of phenylpropanoid. Lignin polymers are attached with cellulosic and hemicellulosic fibrils, providing structural support during growth and resistance to insects and pathogens (Vanholme et al., 2008). So far, cellulose is the only polysaccharide that has been utilized for the production of bioethanol due to the availability of degrading enzymes. But like cellulose, hemicellulose and pectin can also be used as feedstocks.

The process of producing lignocellulosic bioethanol involves (1) removal of lignin by either heating or treating biomass with a strong acid to expose celluloses and hemicelluloses, (2) enzymatic hydrolysis of the lignin-free plant biomass to convert it into 5- or 6-carbon simple sugars, and (3) microbial fermentation of released sugars to produce ethanol (Guo et al., 2015) (Figure 25.1). The conversion of cellulosic material into soluble

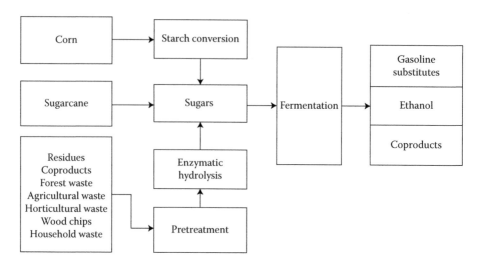

Figure 25.1 Production of ethanol and other gasoline-equivalent fuels from plant materials. To produce ethanol from plant materials, plant biomass is harvested, compacted, and transported to biofuel plants to extract sugars. Lignocellulosic plant biomass undergoes additional steps pretreatment or a delignification process usually through heat or acid to remove lignin, and to expose cellulosic fibrils for enzymatic digestion for recovering sugars. These sugars are then fermented to ethanol, and gasoline substitutes and coproducts are recovered. The nature of plant biomass used differentiates resultant biofuels into first-generation biofuels if food crops are used and second-generation biofuels if nonfood plant biomass is used. The costs associated with the pretreatment and enzymatic hydrolysis makes lignocellulosic ethanol more costly than corn- or sugarcane-ethanol as sugars can be extracted easily from sugarcane and corn-starch.

sugars is carried out using three different classes of cellulolytic enzymes (Tomme et al., 1995). These include endoglucanases, exoglucanases, and β-glucosidase (Ziegler et al., 2000; Ziegelhoffer et al., 2001). Hydrolysis of the cell wall begins with the random cleavage of crystalline cellulose by endoglucanases, producing cellulose chain ends, which are then bound by an exoglucanase that cleaves off cellobiose units. These chains are finally broken down into glucose monomers by β-glucosidases (Sticklen, 2006).

The biggest challenge for the biofuel industry is to produce ethanol from lignocellulosic biomass at a price which is competitive with ethanol derived from corn or sugarcane. Several factors currently impact upon the economic viability of this process; foremost of which are the cellulolytic digestion of plant biomass into fermentable sugars, as a result of the high price of cellulases produced in bioreactors from microbial systems, and the cost associated with the pretreatment of lignocellulosic feedstock for the removal of lignin. The preremoval of lignin by acid and/or heat pretreatments mean cellulases must be active at low pH and high temperatures. Enzymes for commercial use are therefore isolated from thermophilic and acidophilic organisms, and their characterization for industrial applications is an active area of research (Bauer et al., 2006). Culturing extremophiles is often a challenging task, meaning commercial cellulases are usually heterologously expressed.

These two factors mean the costs of ethanol derived from lignocellulose are 2–3-fold higher than ethanol from corn (Lynd et al., 2008). These are the areas where plant genetic engineering can be helpful in bringing down the production costs to a competitive level.

Cost-effective production of cell wall-degrading enzymes in plants

Plants offer an attractive route for the inexpensive production of cell wall-degrading enzymes (CWDEs). Transgenic plants expressing such enzymes can be grown in large quantities with minimum inputs using existing infrastructures. Further, using plants for the production of cellulases means enzymes can be used without costly purification. A number of cellulolytic enzymes, including endoglucanases, exoglucanase, xylanase, acetyl xylan esterase, beta glucosidase, and lipase, have been expressed in tobacco (Jiang et al., 2011; Llop-Tous et al., 2011), *Arabidopsis* (Borkhardt et al., 2010), sugarcane (Harrison et al., 2011), maize (Hood et al., 2007; Mei et al., 2009), and rice (Oraby et al., 2007) using nuclear transformation. Most of these enzymes were expressed at low levels, and the use of constitutive promoters and adjusting the codon-usage of the coding sequences has been recommended to obtain higher concentrations (Jung et al., 2012b).

Research on heterologous production of CWDEs suggests that the subcellular localization of cellulases helps in proper folding of the enzymes and increases stability (Jung et al., 2014). Additionally, it is important to limit cellulase expression in the cytoplasm of leaf cells, and to make sure they are not produced in floral parts such as flowers and seeds to avoid unwanted phenotypes (Jung et al., 2012b). Resultantly, transformation of chloroplasts is viewed as an effective strategy for the production of CWDEs (McCann and Carpita, 2008; Verma et al., 2010). Chloroplasts provide a gene containment system arising from plastid maternal inheritance in flowering plants, and engineering its genome, known as the plastome, usually results in high levels of expression of recombinant proteins due to its high copy number (Bock and Khan, 2004; Ahmad and Mukhtar, 2013; Bock, 2014).

A number of attempts have been made to express CWDEs in higher plant chloroplasts and their accumulation and activities were assessed with a view to commercial exploitation. For example, Yu et al. (2007) expressed two *Thermobifida fusca* thermostable enzymes, endoglucanase (CelA) and exocellobiohydrolase (CelB) in tobacco chloroplasts, which accumulated to a level of 2% and 4% of total soluble proteins (TSP), respectively (Yu et al., 2007), and both the chloroplast-made enzymes were found to be effective in hydrolyzing crystalline cellulosic material. Subsequently, attempts were made to increase the expression levels, by N-terminal fusion of 14 amino acids-coding sequences dubbed as the downstream box (DB) resulting in a 5-fold increase in the accumulation of CelA in plant leaves (Gray et al., 2009).

Another study expressed two fungal xylanases, Xyn10A or Xyn11B, from *Aspergillus niger* and a bacterial xylanase, XynA, from *Clostridium cellulovorans* in tobacco chloroplasts to determine the factors driving higher expression of cellulases in higher plant chloroplasts. The study found that judicious use of *cis*-elements such as 5′-untranslated regions, ribosomal binding sites (RBS), DB sequences, and the overall design of the construct were crucial in determining the expression level of these enzymes in higher plant chloroplasts (Kolotilin et al., 2013).

In another study investigating the low-cost production of multiple cellulolytic enzymes in a single host, bacterial enzymes, including β-glucosidase, xyloglucanase, exoglucanase, and endoglucanase from *T. fusca* were expressed in tobacco chloroplasts, and were found to efficiently degrade cellulosic and hemicellulosic plant material, whilst accumulating at 5%–40% of TSP (Petersen and Bock, 2011).

Verma et al. (2010) expressed a cocktail of bacterial and fungal enzymes, including endoglucanases, exoglucanase, pectate lyases, cutinase, swollenin, xylanase, acetyl xylan esterase, beta glucosidase, and lipase in *Escherichia coli* and tobacco chloroplasts to compare their relative production costs in these systems. The study found that the cost of endoglucanase produced in chloroplasts was 3100-fold lower, and that of pectate lyase was 1057- or 1480-fold lower than commercially available enzymes. In addition, the chloroplast-made enzymes were found to be more active compared to those made in *E. coli*, releasing up to 3625% more fermentable sugars from plant biomass.

These results show that plant-based expression systems offer an inexpensive and efficient method of producing CWDEs. Ideally, it would be preferable to express them in energy crops such as miscanthus and switchgrass but chloroplast transformation is not yet established in these crops, and more work is needed to determine the minimum expression levels that are needed for a complete hydrolysis of lignocellulosic plant biomass without external supplies.

Overcoming the recalcitrance of plant biomass

Current processes to remove lignins have been developed empirically due to limited knowledge about their chemistry and biogenesis. Large-scale genomics studies have identified genes involved in cellulose and hemicellulose biosynthesis as well as those which influence plant height, stem thickness, and branching number (Kalluri et al., 2007; Busov et al., 2008) which should help direct efforts at altering plant biomass for increased yields of cellulose.

Similarly, a number of genes controlling the biosynthesis of lignin has been identified and investigated (see Poovaiah et al., 2014, for review). This has led to the development of strategies to modify lignin content and to change the lignin profile to reduce pretreatment costs. For example, downregulation of three cytochrome P450 "early pathway" enzymes, cinnamate 4-hydroxylase (C4H), coumaroyl shikimate 3-hydroxylase/coumarate 3-hydroxylase (C3H), and coniferaldehyde 5-hydroxylase/ferulate 5-hydroxylase (F5H) using RNAi significantly reduced lignin content of alfalfa (Reddy et al., 2005). This study showed that downregulation of early pathway enzymes has a more dramatic effect than the "late" pathway enzymes such as caffeic acid 3-*O*-methyltransferase (COMT) and cinnamyl alcohol dehydrogenase (CAD). Similarly, downregulation of the expression of a number of lignin–biosynthesis enzymes such as CAD and COMT in poplar (Pilate et al., 2002), cinnamoyl CoA reductase (CCR) in *Populus* species (Van Acker et al., 2014), CAD and C3H in alfalfa (Ralph et al., 2006), COMT, *p*-hydroxycinnamoyl-CoA:quinate/shikimate *p*-hydroxycinnamoyltransferase (HCT), C3H, C4H, caffeoyl-CoA *O*-methyltransferase (CCo-AOMT) and F5H in alfalfa (Chen and Dixon, 2007), COMT in sugarcane (Jung et al., 2012a, 2013), COMT in switchgrass (Baxter et al., 2014), and overexpression of *Panicum virgatum* (switchgrass) transcription factor, *Pv*MYB4 (Shen et al., 2012), suggest that altering lignin profile by genetic means could be a potential route to lowering pretreatment costs as well as improving sugar recovery.

Another strategy could be to divert carbon from lignin synthesis to cellulose formation, as evidenced by the downregulation of 4-coumarate CoA ligase (4CL), which resulted in a decrease in lignin content (45%) and increase in cellulose content (15%) in poplar (Hu et al., 1999). Similar findings were observed in another study where downregulating the expression of two enzymes of the monolignol pathway, CCR and CAD, resulted in 50% reduction in lignin content, which was compensated for increase in xylose and glucose associated with the cell wall, resulting in a normal phenotype (Chabannes et al., 2001). However, alteration of lignin often results in vascular collapse (Li et al., 2010), meaning it must be applied in a tissue-specific manner.

An alternative way of making cellulose and hemicellulose more accessible to hydrolytic enzymes is to introduce "loosening" agents such as expansins (Cosgrove, 2000), swollenins (Saloheimo et al., 2002), and glycanases (Park et al., 2004) or making the lignin backbone more labile for chemical treatments (Wilkerson et al., 2014). The exact mechanism of how they work is not fully known; however, they have been shown to possess the potential to increase sugar recovery from plant biomass (Pauly and Keegstra, 2008).

Various strategies for the direct conversion of lignocellulosic biomass into ethanol without undergoing costly pretreatment are also being developed. For example, Chung et al. (2014) engineered a thermophilic, anaerobic, cellulolytic bacterium *Caldicellulosiruptor bescii* by deleting a gene coding for lactate dehydrogenase and expressing a dual-functional acetaldehyde/alcohol dehydrogenase of *Clostridium thermocellum*. The resulting changes in *C. bescii* allowed an efficient conversion of switchgrass biomass into ethanol without any costly pretreatment. The initial results suggest that combining all processes from lignin removal to ethanol biosynthesis in a single step is possible, although further optimization is needed to increase ethanol yields.

Biodiesel production from plants

Biodiesel does not require any major engine modifications and can be used directly or in blends with conventional sources. It is produced from transesterification of oil and fats with monohydric short-chain alcohols, usually methanol or ethanol (Guo et al., 2015). Transesterification is a three-step process (Figure 25.2) by which triglycerides present in vegetable or animal oils are converted into fatty acid alkyl esters (FAAE)—collectively known as biodiesel—and glycerol, a waste product, in the presence of a catalyst at elevated temperatures (Moser, 2011). Biodiesel is predominantly made from edible oil feedstocks such as rapeseed, sunflower, coconut, palm oil, and soybean, but jatropha, caster, polanga, recycled edible oil, and animal fats are also used. The enzymatic conversion of oil into biodiesel is less energy intensive compared to chemical conversion.

Considerable advances have been made in optimizing procedures for producing biodiesel from oils (plants and animals), determining feedstock qualities as well as identification of high-oil-yielding feedstocks (Moser, 2011). The key to sustainable biodiesel production lies in the continuous supply of oils and lipids without competing with food production (Guo et al., 2015). The rapid increase in demand and the limited supply of oils have spurred research into cost-effective production of fatty acids using microorganisms such as *E. coli* and algae. Metabolic engineering to produce advanced biofuels holds great potential to deliver next-generation biofuels (see the next section for details).

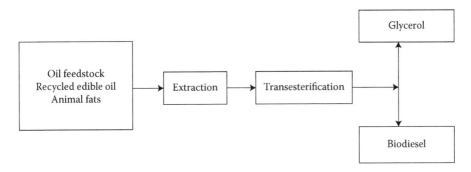

Figure 25.2 Production of biodiesel from vegetable oils and animal fats. Flowchart showing different steps in the production of biodiesel from vegetable oils and animal fats.

Engineering microbial genomes for biofuel production

Synthetic biology and metabolic engineering facilitate the production of next-generation or "advanced" biofuels (Singh et al., 2014) as a means of overcoming the limitations of traditional biofuel production processes. The vast amount of genetic resources, coupled with detailed knowledge about the biochemistry and cell biology of *E. coli*, has made this bacterium a model system for the production of advanced biofuels (Tee et al., 2014).

Although rewiring the biofuel-producing pathways in microbes, particularly *E. coli* and *Saccharomyces cerevisiae*, seems practical, fine-tuning of gene expression mechanisms is required to maximize yields. Further, hosts that have a higher tolerance toward toxic biofuels such as butanol, with capability to grow on nonsugar substrates would be a promising alternative (Singh et al., 2014). This section deals with advances in the field of metabolic engineering to produce next-generation biofuels, highlights issues, and assesses the potential of advanced biofuels to meet commercial requirements.

Engineering microbial genomes for bioalcohol production

Alcohols

Higher-chain alcohols (C_4–C_5) can be used as replacements for gasoline or they can be used as blends depending upon their miscibility and energy content. For instance, butanol (a C_4 alcohol) has 84% of the energy content of gasoline and is completely miscible with it. Resultantly, genetic modification of microorganisms has been attempted for sustainable production of these compounds. For example, *Clostridium* has been engineered to produce butanol from different feedstocks such as glucose (Lynd et al., 2002), corn flour (Jeffries and Jin, 2004), glycerol (Dien et al., 2003), and syngas (mixture of H and CO) (Ingram et al., 1987). However, under the stringent anaerobic conditions required to produce these alcohols, yields were low, either due to the bacterium's slow growth, intolerance to butanol, or a limited understanding of the metabolism (Ohta et al., 1991; Zhang et al., 1995). This challenge can be addressed by the introduction of this butanol pathway into *E. coli*, and to those strains such as *S. cerevisiae* and *Lactobacillus brevis*, which are able to tolerate high alcohol concentrations (Ho et al., 1998; Hanai et al., 2007; Zhang et al., 2011). Other than these, *Pseudomonas putida* and *Bacillus subtilis* have managed to tolerate the toxic levels of alcohols using efflux pumps and by changing the cell wall composition, respectively (Jones and Woods, 1986). For example, *E. coli* has been engineered to produce various isomeric forms of higher (C_4–C_5) alcohols from glucose and fatty acids with high yield and specificity (Atsumi et al., 2008a; Steinbusch et al., 2008; Mattam and Yazdani, 2013; Singh et al., 2014). Similarly, the cyanobacterium *Synechococcus elongatus* can produce butanol using atmospheric CO_2 via photosynthesis (Jojima et al., 2008). Microbes have been engineered to produce butanol at up to 20–30 g L^{-1} by increasing NADH, acetyl-CoA, and ATP pools (Lee et al., 2008; Shen et al., 2011).

Pentanol (a C_5 alcohol) is another potential biofuel molecule but it cannot currently be produced in sufficient quantities for commercial applications. Expressing the *nud*F gene from *B. subtilis* in *E. coli* resulted in yields of 2-methyl-1-butanol at up to 110 mg L^{-1} (Withers et al., 2007) but still it is too low to meet commercial levels. Introduction of the *2-keto-acid* or *Ehrlich pathway* has also been pursued as a means of producing higher-carbon alcohols (Yan and Liao, 2009) and involves the stepwise decarboxylation of keto acids (amino acid intermediates) into aldehydes, followed by their reduction to alcohols. The Ehrlich pathway is indigenous to some yeast species, which produce fusel alcohols as by-products during fermentation. This pathway can also be used to produce amino-acid-based alcohols, such as *n*-propanol, isobutanol, and *n*-butanol from isoleucine, valine, and norvaline, respectively (Yan and Liao, 2009). Out of these, isobutanol is believed to be suitable for commercialization due to its higher octane number than ethanol as a result of its branched structure. Introduction of the Ehrlich pathway by expressing a 2-keto-acid decarboxylase and alcohol dehydrogenase in *E. coli* resulted in the overproduction of isobutanol (Atsumi et al., 2008b). Further manipulations, including overexpression of 2-ketoisovalerate genes, knockdown of competing pathways, and substitution of the native acetolactate synthase enzymes, increased yields up to 22 g L^{-1} (Atsumi et al., 2008b). Building on this approach, the Ehrlich pathway has also been engineered into *Corynebacterium glutamicum* (Smith et al., 2010), *S. cerevisiae* (Chen et al., 2011), *C. acetobutylicum*, and *S. elongatus* (Atsumi et al., 2009) to metabolize protein-rich substrates such as algal hydrolysate, which is a potentially abundant feedstock (Huo et al., 2011).

In another study, the engineered *E. coli* was shown to convert short-chain fatty acids (C$_3$–C$_7$) into their corresponding alcohols with a remarkable conversion efficiency of almost 100% (Mattam and Yazdani, 2013).

The isopentenols (isoprenol and prenol) are also believed to be alternative to traditional alcohol-based fuels. However, they cannot be produced in large quantities by natural host organisms due to their cytotoxic nature. The mevalonate-dependent (MVA) pathway was introduced into *E. coli* to synthesize isoprenol and prenol from isopentenyl pyrophosphate (IPP) and dimethylallyl pyrophosphate (DMAPP) intermediates but yields were low (Zheng et al., 2013). To address the sensitivity of *E. coli* to isopentenol, transcriptomic analysis was employed to identify genes involved in isopentenol tolerance (Foo et al., 2014) identifying 40 differentially expressed genes as potential candidates. Further analysis identified two transporter genes *Met*R and *Mdl*B associated with the strain growth, and their expression increased isopentenol tolerance by 55% and 12%, respectively. A positive correlation was observed between the intracellular accumulation of MdlB and isopentenol production indicating that by coupling introduction of alcohol-producing pathways with a relevant transporter, it is possible to alleviate the cytotoxic effects of fuel production.

Saccharomyces cerevisiae: A natural host

Saccharomyces cerevisiae is the most popular fermentation yeast used for the scale up of bioconversion of sugars to alcohols, but it is unable to metabolize nonglucose sugars (pentose, xylose, and arabinose). Bioethanol production using LC hydrolyzates requires robustness and a strain capable of D-xylose fermentation. Aiming at this, 13 genes, including *Xyl*A (from *Clostridium phytofermentans*) which encodes D-xylose isomerase (XI), and the genes encoding enzymes of the pentose phosphate pathway (PPP) from *C. phytofermentans* were inserted into the genome of the *S. cerevisiae* strain ethanol red. The genetically modified strain exhibited a higher D-xylose consumption rate of 1.1 g g^{-1} DW h^{-1} resulting in an average increase in ethanol yields of 32% due to improved utilization of D-xylose (Demeke et al., 2013). The integration of pentose metabolic pathways for efficient bioconversion of pentose to biofuels in *S. cerevisiae* has also been explored, with strains engineered to produce 2-methyl-1-butanol (Cann and Liao, 2008).

Engineering microbial genomes for biodiesel production

Fatty acids

Fatty acids exist in the form of triglycerides and are synthesized by fatty acid synthase (FAS). Long-chain fatty acids (FAs) may be freed from the acyl carrier proteins (ACPs) by thioesterase through hydrolysis. Free fatty acids (FFAs) are a direct precursor to many high-value compounds, including biofuels (Youngquist et al., 2013). However, the FAs cannot be directly used as transportation fuels due to the presence of ionic carboxyl groups, which need to be converted into nonionic (hydrophobic) molecules such as fatty alcohols, hydrocarbons, fatty acid ethyl esters (FAEEs), and fatty acid methyl esters (FAMEs) before use. Heterologous expression of bacterial or plant thioesterase in *E. coli* has been shown to produce FAs with chain length ranging from C_8 to C_{18} (Lennen et al., 2010; Liu et al., 2010; Steen et al., 2010). A 3–5-fold increase in FA yields from *E. coli* was achieved when ACPs and acyl-ACP-TEs from *Streptococcus pyogenes* (strain MGAS10270) were expressed in *E. coli* K-12 W3110 and MG1655. Similarly, the content of $C_{18:1}$ and $C_{18:2}$ FFAs could be improved 8–42-fold in comparison to the control (Lee et al., 2013). A fatty alcohol titer of 60 mg L^{-1} has been achieved in *E. coli* through heterologous expression of an acyl-CoA synthase and acyl-CoA reductase or an acyl-ACP reductase from different genome sources (Reiser and Somerville, 1997; Steen et al., 2010; Tan et al., 2011). FAMEs and FAEEs (biodiesel) have also been heterologously produced at the titer of 1.5 g L^{-1} (Zhang et al., 2012) via conversion of FAs to acyl-CoA, followed by esterification using an ester synthase (Kalscheuer et al., 2006; Steen et al., 2010).

Medium-chain fatty acids (MCFAs; C_4–C_{12}) can also be used for biofuel synthesis; however, MCFA production is hampered by a rapid synthesis of their acyl-ACP precursors. To address this limitation, a ketoacyl synthase was introduced into *E. coli*, along with MCFA synthesizing genes, to degrade the ACP precursors, subsequently slowing down the acyl-ACP extension and diverting phospholipid production to MCFA synthesis (Torella et al., 2013). The degradation of the proteins which induce the synthesis of acyl-ACP precursors was believed to redirect metabolic flux toward the synthesis of MCFAs. The demonstration of this strategy in *E. coli* suggests that it could be useful in future applications aimed at diverting the flux away from or toward the desired product, as well as for the modification of the chain length.

Monounsaturated fatty acids (MUFAs) are also believed to be potential biodiesel substrates in terms of low-temperature fluidity and oxidative stability (Cao et al., 2014). Biodiesel produced from plant or microbial lipids always contain polyunsaturated and saturated fatty acid alkyl esters, which limit its applications. It is therefore desirable to increase both the yields of MUFA and their relative proportion in lipids. In an effort to address that question, *E. coli* (BL21 DE3) was engineered to produce free MUFAs using the fatty acyl-ACP thioesterase and fatty acid desaturase from *Arabidopsis thaliana*. To boost yields, the endogenous *fadD* gene (encoding acyl-CoA synthetase) was disrupted to block fatty acid catabolism while the native acetyl-CoA carboxylase (ACCase) was overexpressed to boost fatty acid biosynthesis. The engineered strain produced 82.6 mg L^{-1} (FFAs) when grown on glucose reaching about 3.3% of maximum theoretical yield. Metabolite analysis showed that two types of the MUFAs, palmitoleate and *cis*-vaccenate, constituted more than 75% of the free fatty acid. Further increases in yields were achieved by a fed-batch fermentation method to a titer of 1.27 g L^{-1} without affecting overall fatty acid composition (Cao et al., 2014).

In a study, a "push-and-pull" strategy was used to enhance lipid production in oleaginous yeast *Yarrowia lipolytica* where *ACC1* and *DGA1* were overexpressed, influencing the first and the last steps of the lipid biosynthesis pathway. The genetically engineered strain accumulated up to 62% of its dry cell mass as lipids through *de novo* synthesis (Tai and Stephanopoulos, 2013). Similarly, the native lipogenesis pathway of *Y. lipolytica* was engineered to directly convert lipids into biodiesel. Trilevel metabolic control produced cells that could accumulate lipids up to 90% of their cellular content and titer surpassed 25 g L^{-1} lipids, representing a 60-fold increase in comparison with the parental strain (Blazeck et al., 2014).

Methyl ketones

Methyl ketones ranging from C_{11} to C_{15} have a favorable cetane number (a parameter that reflects the combustion quality of fuel in compression engines) for use as transportation fuels. Synthesis of methyl ketones in *E. coli* has been reported with an average titer of 380 mg L^{-1} (Goh et al., 2012), and occurs in three steps: (i) β-oxidation of fatty acids to β-ketoacyl-CoAs, (ii) hydrolysis to β-keto-FAs (fatty acids or fatty alcohols) by a thioesterase, and (iii) decarboxylation to produce methyl ketones.

Microalgae

Interest in utilizing algae for biodiesel production began in the wake of the 1970s oil crisis as part of NREL's "Aquatic Species Program" (Sheehan et al., 1998). Due to a fall in crude oil prices, research stalled in the 1990s, but in recent years, interest has been revived as microalgae are believed to be up to 300 times more productive than traditional energy crops on an area basis (Chisti, 2007; Rincon et al., 2014). However, currently, the commercialization of microalgae-based biofuels is hindered by the high costs incurred during large-scale cultivation, harvesting of cells, drying, and oil extraction. Research is therefore focused on the isolation of fast-growing algal strains with hyper lipid accumulation (Keasling, 2010; Scott et al., 2010), manipulation of fatty acid biosynthesis (Georgianna and Mayfield, 2012; Peralta-Yahya et al., 2012), and development of low-cost cultivation and extraction processes (Posten, 2009; Gill et al., 2013; Rincon et al., 2014).

Increasing lipid yield

Enhancing lipid accumulation is a direct means of reducing input costs and subsequently much effort has gone into understanding how lipids are synthesized. Commonly lipids are accumulated when nutrients become limiting (for instance, nitrogen), typically coinciding with the onset of the stationary phase. Subsequently, lipid synthesis pathways are specifically regulated to balance cell growth and lipid accumulation (Beopoulos et al., 2009, 2011). However, starch is believed to be the major storage product during nutrient stress and is preferentially accumulated, with lipid production dependent on external carbon supply (Fan et al., 2012; Ramanan et al., 2013). Resultantly, application of an "acetate boost" to nitrogen-starved cultures has been utilized to significantly increase lipid yields from *Chlamydomonas reinhardtii* (Goodson et al., 2011) and blocking starch synthesis has been shown to help redirect carbon toward lipid production, with increased yields reported from starchless mutants of *C. reinhardtii* (Wang et al., 2010) and *Chlorella* (Ramazanov and Ramazanov, 2006).

Large-scale functional genomics studies, including the reverse as well as forward genetics, have substantially increased our understanding of the key genes involved in the synthesis of algal lipids (Gonzalez-Ballester et al., 2010; Miller et al., 2010; Castruita et al., 2011; Boyle et al., 2012; Schmollinger et al., 2014), and have helped to identify 86 candidate transcription factors involved in TAG accumulation, making them potential targets for increasing yields. Microalgae have two ways to accumulate lipids intracellularly: (1) *ex novo* incorporation of exogenous lipids or (2) *de novo* lipid synthesis via pathways located in either the chloroplast or endoplasmic reticulum (ER) (Fan et al., 2011; Goodson et al., 2011).

It was shown that the targeted knockdown of a multifunctional enzyme (lipase/phospholipase/acyltransferase) enhanced lipid accumulation in *Thalassiosira pseudonana* up to 4.4-fold as compared to the wild-type strain without compromising growth (Trentacoste et al., 2013). The CrDGAT2a,b, and c overexpression did not increase lipid content (La Russa et al., 2012), requiring a better understanding of the metabolic network (Chang et al., 2011; reviewed by Koussa et al., 2014).

Similarly, the *Streptomyces coelicolor* TAG biosynthesis pathway was successfully reconstructed in *E. coli* strain by expressing *SCO958*, *lppβ* genes, genes encoding the *S. coelicolor* acetyl-CoA carboxylase complex, and mutation in *fadE* which encodes for the acyl-CoA dehydrogenase and catalyzes the first step of the β-oxidation cycle in *E. coli* and a deletion in the *dgkA* gene. The engineered strain could accumulate TAG up to 4.85% of cell dry weight (722.1 mg TAG L^{-1}) in batch cultivation (Comba et al., 2014). This study has revealed that availability of diacylglycerol (DAG) is one of the critical factors to achieve higher yields of the storage compound. Therefore, it is crucial to enhance our understanding about the regulatory mechanism of carbon flow to develop a competitive process for neutral lipid production in *E. coli*.

Increasing biomass production

In addition to improving lipid yields, efforts have been directed at increasing algal growth rates and culture densities, which has been aided by the detailed annotation of microalgal genomes (Georgianna and Mayfield, 2012). It has been observed that the highest growth rates are often associated with the use of heterotrophic culture conditions. For instance, the growth productivity of *Chlorella vulgaris* was reported to be 120 g L^{-1} compared to its control under heterotrophic conditions (Doucha and Livansky, 2012). Under these conditions, high cell densities are maintained through the secretion of molecules that facilitate cellular communication and inhibit the growth of species other than microalgae. Therefore, higher growth rates may also address contamination problems. Both *C. vulgaris* and *C. reinhardtii* were shown to produce such molecules that improve algal growth and inhibit contaminating bacteria (Teplitski et al., 2004; Rajamani et al., 2008). Therefore, studying quorum-sensing genes has the potential to enhance growth rates as well as to eradicate contamination risks.

Efforts have also been directed at improving solar to biomass conversion efficiencies. In direct sunlight, the photosynthetic apparatus of algae are quickly saturated leading to the wasteful dissipation of excess energy (Ort et al., 2011). Resultantly, both forward (Melis et al., 1998; Polle et al., 2002; Melis, 2009; Kirst et al., 2012; Cazzaniga et al., 2014) and reverse (Beckmann et al., 2009; Oey et al., 2013) genetic approaches have been used to identify mutants of *C. reinhardtii*, *Dunaliella salina*, and the fast-growing *Chlorella sorokiniana* with reduced light-harvesting antennae that show improved photon use efficiencies and increased biomass accumulation at high culture densities.

Cyanobacteria and red algae possess photosynthetic pigments arranged in a membrane-associated structure known as the phycobilisome (PC), composed of pigmented phycobiliproteins and linker proteins (Grossman et al., 1993; Adir, 2005). Productivity was increased in phycocyanin-deficient *Synechocystis* PCC 6714 at high light intensities (Nakajima and Ueda, 1999) and a reduction of the number of PC particles caused reduced photoinhibition and increased biomass accumulation under high light and carbon limitation (Lea-Smith et al., 2014), but gains are dependent on culturing conditions (Page et al., 2012), with 2–3 PC particles described as optimal (Kwon et al., 2013; Lea-Smith et al., 2014) for biomass production.

These latter findings demonstrate the difficulties in transferring efficiency gains from the laboratory into an industrial context, and caution that approaches aimed solely at reducing photoprotective mechanisms required to dissipate excess light energy may be overly simplistic. Efforts have begun to address this issue with studies in cyanobacteria (Nogales et al., 2012) and algae (Mettler et al., 2014) assessing the impact of light and carbon availability. These studies point to the critical importance of alternative electron transport processes in balancing of ATP/NADPH ratios for optimal growth, which are dependent on light intensity, carbon availability (Nogales et al., 2012), and light regime intensities (Mettler et al., 2014).

Engineering microbial genomes for petroleum-replica biofuels

Current biofuels (bioalcohols and biodiesels) require extensive downstream processing, and the majority of them are not fully compatible with modern internal combustion engines. Therefore, biofuels (hydrocarbons or their equivalents) that are chemically "identical" to the fossil fuels, known as *petroleum-replica* fuels (Figure 25.3), have a significant advantage over their traditional counterparts (Howard et al., 2013).

Isoprenoids-based biofuels

Isoprenoids are the compounds that may also be exploited as advanced biofuels because of the branches and rings present in their hydrocarbon chain (Lee et al., 2008; Peralta-Yahya and Keasling, 2010). Their branches can withstand pressure-induced free radical generation, avoiding premature ignition, and hence improve the octane number (fuel quality parameter). Isoprenoid-based biofuels may be generated from isopentenyl diphosphate and dimethylallyl diphosphate (Kuzuyama, 2002), using either the mevalonate

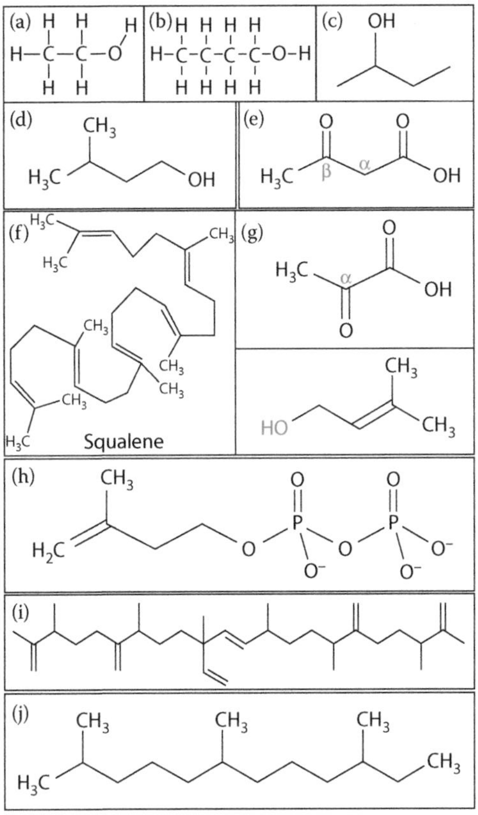

Figure 25.3 Chemical structures of different biofuel molecules. (a) Ethanol, (b) 1-butanol, (c) 2-butanol, (d) 2-methyl-1-butanol, (e) acetoacetic acid (a keto acid), (f) squalene (a terpenoid), (g) pyruvic acid, (h) isopentenyl pyrophosphate (a terpenoid), (i) botryococcene (a terpenoid), and (j) farnesane (biofuel close to commercialization).

or deoxyxylulose-5-phosphate pathway. The C_{15} isoprenoid molecules bisabolene and farnesane have cetane numbers of 58 and 52, respectively, which sit within the standard spectrum (40–60) for diesel. Plants are the natural source of isoprenoids albeit in a low quantity, the alga *Botryococcus braunii* can produce isoprenoids in higher quantities but its slow growth also makes it unsuitable for commercial applications (Banerjee et al., 2002).

Among isoprenoid-derived biofuels, farnesane may be the first to be commercialized (Rude and Schirmer, 2009; Renninger and Mcphee, 2010). A California (USA)-based company, Amyris, is using *S. cerevisiae* PE-2 to produce farnesene which is hydrogenated to produce farnesane, which is being evaluated for its performance in engines as an advanced biofuel (Renninger and Mcphee, 2010; Ubersax and Platt, 2010). Bisabolene is another potential biofuel. For high-level production of bisabolene, different plant isoprenoid synthases were expressed in *S. cerevisiae* and *E. coli* to produce up to 900 mg L^{-1} (Peralta-Yahya et al., 2011), this may possibly be further improved by using the bisabolene synthase from *Abies grandis* which is the most efficient version so far identified (McAndrew et al., 2011). The terpene biosynthetic pathway has also been engineered in *Streptomyces venezuelae* to overproduce bisabolene (Phelan et al., 2014).

Polyketide-derived hydrocarbons

The polyketide synthesis pathway produces hydrocarbons with varied structures and has the potential to create petroleum-replica fuels but is the least studied pathway for biofuel production. The synthesis of polyketides involves the decarboxylation of malonyl-CoA by the multidomain polyketide synthases (PKS) (Pfeifer and Khosla, 2001). PKSs have been widely studied over the last three decades, generating enough knowledge to engineer them for biofuel production. Modular type-I PKS provide best control over the product because of their multidomain structures (Menzella et al., 2005; Yuzawa et al., 2012). Each PKS contains a set of catalytic domains specialized for one particular round of polyketide chain extension. This mechanism provides an opportunity for engineering a modular type-I PKS in several ways: (1) by changing the precursor used for polyketide synthesis via manipulating the acyltransferase domain (Jacobsen et al., 1997; Yuzawa et al., 2012), (2) by modifying the structure of each extension unit by domain mutation (Yuzawa et al., 2012), and (3) by modifying the chain length of polyketides through module deletion, substitution, or *de novo* design (Menzella et al., 2005).

Pinene dimers

Pinene is another potential biofuel precursor, which is a 10-carbon ($C_{10}H_{16}$) bicyclic monoterpene and can be used as aviation fuel. Aviation fuel substitutes require high-energy contents which may be attained with the presence of ring structures, similar to those found in pinene dimers (Harvey et al., 2009). It is found in the resins of many conifers and nonconiferous plants. It is also produced by many insects as a communication molecule and can be chemically converted into isopentanol and pinene dimers (Harvey et al., 2009). The pinene synthesis pathway has been introduced into an *E. coli* strain which was able to utilize cellulose and xylose to produce up to 1.5 mg L^{-1} of pinene using switchgrass hydrolyzate as a carbon source (Bokinsky et al., 2011).

Chemical structures of different biofuel molecules are shown in Figure 25.3.

Biofuel-tolerating cells

The study of bacterial transporter proteins belonging to the resistance-nodulation-cell division family (RND) show that overexpression of transporters can improve the production of limonene 1.5-fold in *E. coli* compared to a control (Dunlop, 2011). However, due to strucural complexity and large size (Nikaido and Takatsuka, 2009), the RND-based secretion system does not seem practical in other hosts such as yeast and algae (Doshi et al., 2013) and they are present only in Gram-negative bacteria and cyanobacteria. Alternatively, transporters that belong to the ATP-binding cassette (ABC) protein family are ubiquitously found in all five kingdoms. ABC exporters are also able to export a range of extremely "greasy" molecules (Pohl et al., 2005) and have been utilized for the secretion of biofuel molecules (zeaxanthin and canthaxanthin) from *E. coli* (Doshi et al., 2013). Similarly, these exporters may be tested for their ability to secret other biofuel molecules, including nonpolar hydro-carbons, squalene, and botryococcenes for an easy recovery from the culture medium, but the most suitable transporters need to be evaluated.

Hydrogen production

As its combustion results in the production of water vapor, hydrogen produced from renewable energy sources represents an attractive alternative to conventional fossil fuels (for a review of production technologies, see Holladay et al., 2009). H_2 can be produced biologically by anaerobic fermentation of biomass (for review, see Merlin Christy et al., 2014) or by photolysis of water, the latter of which is performed by algae and cyanobacteria. In algae hydrogen production is catalysed by an [FeFe]-hydrogenase (Ghirardi et al., 2007; Burgess et al., 2011; Quintana et al., 2011; Torzillo and Seibert, 2013; Dubini and Ghirardi, 2014), whereas cyanobacterial species use either a [NiFe]-hydrogenase or a nitrogenase, which in the case of the later H_2 is produced as a bi-product of ammonium fixation, or in the absence of nitrogen (Tamagnini et al., 2002; Dutta et al., 2005). In this section, we will focus on solar hydrogen production.

Algal hydrogen production

Algal H_2 production is catalyzed by an [FeFe]-hydrogenase located in the chloroplast (Terashima et al., 2010), with the electrons required to reduce protons to molecular H_2 derived from the photosynthetic electron transport chain (PET) via transfer from the fer-redoxin PetF (Winkler et al., 2009).

The ability of algae to produce H_2 has long been known (Gaffron, 1939), but was consid-ered unsuitable for commercial applications due to the extreme oxygen sensitivity of [Fe]-hydrogenase assembly and catalysis (Stripp et al., 2009; Swanson et al., 2015). To address this challenge, Melis et al. developed a process for the temporal separation of growth and H_2 production—by depriving cultures of sulfur, photosynthetic activity can be reduced below the rate of respiration, thereby driving cultures into a hypoxic state and enabling sustained H_2 production over a period of days (Melis et al., 2000).

During sulfur deprivation, the reductant for H_2 production comes from a variety of sources (as reviewed in Hemschemeier et al., 2013), including (i) a light-dependent path-way: with electrons derived from photosystem II activity, (ii) an indirect pathway: whereby electrons from the breakdown of starch (Chochois et al., 2010) enter the photosynthetic electron transport chain through the activity of a type-II NADH dehydrogenase (NDA2; Jans et al., 2008; Desplats et al., 2009), and (iii) from the direct transfer of electrons to fer-redoxin by pyruvate ferredoxin:oxidoreductase (PFOR), which can utilize either pyruvate or oxaloacetate as substrates, releasing CO_2 (Noth et al., 2013; van Lis et al., 2013).

Nonphotochemical quenching, cyclic electron transport, and activation of fermenta-tion pathways are all thought to act as competing electron sinks during algal H_2 produc-tion, and have therefore been suggested as targets for manipulation to increase yields (Timmins et al., 2009; Hemschemeier and Happe, 2011; Nguyen et al., 2011). Preliminary transcriptomic (Mus et al., 2007; Toepel et al., 2013; Yang et al., 2013), proteomic (Terashima et al., 2010), and metabolomics (Timmins et al., 2009; Doebbe et al., 2010; Subramanian et al., 2014) analyses helped to define the fermentative pathways activated during anoxia (see Catalanotti et al., 2013 for review). Mutants have been generated effecting pyruvate formate lyase (Philipps et al., 2011; Burgess et al., 2012; Catalanotti et al., 2012), alcohol dehydrogenase 1 (Magneschi et al., 2012), acetate kinase (Yang et al., 2014), and [FeFe]-hydrogenases HYDA1 and HYDA2 (Meuser et al., 2012) and the impact upon fermentative metabolism assessed. PFL1-KO has been shown to increase H_2 yields in the dark in certain genetic backgrounds (Philipps et al., 2011), but not others (Catalanotti et al., 2012), and while helping to further refine metabolic models, no significant gains in light-driven H_2 production have yet been observed through metabolic engineering.

Direct fusion of HYDA1 to PETF was able to increase H_2 production rates by reducing competition with other pathways (primarily the activity of ferredoxin NAPDH:oxidoreductase; FNR) (Yacoby et al., 2011), and knockdown of FNR resulted in a 2.5-fold increase in H_2 yields during sulfur depletion (Sun et al., 2013). However, as FNR is vital for providing NADPH for CO_2 fixation, affecting its activity is likely to have detrimental effects on algal growth under photoautotrophic conditions. A subtler approach was suggested by Rumpel et al. (2014) who showed that it is possible to redirect electrons away from FNR to HYDA1 by making amino acid substitutions to both PETF and FNR, resulting in up to 5-fold increases in hydrogenase activity in a reconstituted assay.

Overexpression of NDA2 was able to increase the reductant for H_2 production via the indirect pathway, resulting in increased starch accumulation and H_2 production (Baltz et al., 2014). However, the largest increased yields have been achieved through disrupting cyclic electron transport, including the knockout line of proton gradient regulator 5 like protein 1 (PGRL1) (Tolleter et al., 2011) and state transition mutant *stm6* (Kruse et al., 2005), defective in the mitochondrial transcription factor MOC1.

The complexity of factors influencing algal hydrogen production is demonstrated by the investigation of a series D1 mutants with increased H_2 production ability, the precise reasons for the improved yields variously included increased starch accumulation, prolonged H_2 production, and increased photoinhibition (Torzillo et al., 2009; Faraloni and Torzillo, 2010; Scoma et al., 2012).

Cyanobacterial hydrogen production

Hydrogen is also produced by diazotrophic cyanobacteria as a by-product of nitrogenase activity (as reviewed by Dutta et al., 2005). The nitrogenase is highly oxygen sensitive (Gallon, 1981) and resultantly nitrogen-fixing cyanobacteria have evolved either temporal or spatial separation of nitrogen fixation and photosynthesis. Filamentous species develop specialized cells known as heterocysts, which possess a modified cell wall that acts as a barrier to O_2 diffusion (Halimatul et al., 2014) helping to maintain low oxygen levels and enabling nitrogenase activity (Wolk et al., 2004), whereas unicellular species photosynthesize during the day, accumulating starch reserves that are used to drive processes at night that require anoxic conditions (Bandyopadhyay et al., 2011).

The nitrogenase is often accompanied by either a bidirectional [NiFe]-hydrogenase or a unidirectional uptake hydrogenase encoded by large and small subunits referred to as *hupL* and *hupS*, respectively (Oxelfelt et al., 1998). These enzymes act to quickly metabolize the H_2 produced (Oxelfelt et al., 1998) and resultantly knockout of the uptake hydrogenase has become a key approach for increasing yields from filamentous cyanobacteria, with mutants in *hupL* or *hupS* resulting in 2–7-fold improvements (Sveshnikov et al., 1997; Happe et al., 2000; Masukawa et al., 2002; Yoshino et al., 2007; Khetkorn et al., 2012).

The unicellular *Cyanothece* sp. strain ATCC 51142 has been reported to produce high levels of H_2 under aerobic conditions with argon purging: 373 µmol mg^{-1} chlorophyll *a* (Chl a) h^{-1} (Bandyopadhyay et al., 2010), with the majority of hydrogen produced via the nitrogenase, and the reductant coming from photosystem I activity and mitochondrial respiration (Min and Sherman, 2010). These high rates have yet to be repeated (Skizim et al., 2012), but if confirmed would represent a significant increase over maximum rates reported from filamentous cyanobacteria (167.6 µmol of H_2 mg of Chl h^{-1}) which were achieved using a mutant strain of *Anabaena variabilis* sp. ATCC 29413 (Dutta et al., 2005) affected in hydrogenase maturation (Shestakov and Mikheeva, 2013).

In contrast to filamentous cyanobacteria, where H_2 production is spatially separated from photosynthesis, disruption of the *Cyanothece* uptake hydrogenase abolished nitrogenase activity, with the authors suggesting that *hupL* plays a vital role in protecting the nitrogenase from O_2 inhibition (Zhang et al., 2014). Basal levels of hydrogen production in *Cyanothece* are improved by the external supply of glycerol (Bandyopadhyay et al., 2010; Feng et al., 2010) and a number of additional strategies have been suggested for improving yields (Skizim et al., 2012), foremost of which is shifting nitrogenase activity away from nitrogen fixation toward the production of hydrogen, which can be achieved either by single nucleotide substitutions (Barney et al., 2004), or disrupting production of homocitrate—which binds to the catalytic site and is required for optimal nitrogenase activity (Masukawa et al., 2007).

Some nondiazotrophic cyanobacteria also produce H_2 via the activity of a bidirectional [NiFe]-hydrogenase, but yields are generally lower (Dutta et al., 2005). Attempts at manipulating metabolism to increase internal NADPH pools increase H_2 yields by up to 3-fold (Kumaraswamy et al., 2013), and disruption of competing fermentative pathways is an area for further improvement—as demonstrated in the *ldhA* mutant of *Synechococcus* sp. strain PCC 7002, which resulted in up to 5-fold increase in yields, with targeting of formate and ethanol producing pathways still possible (McNeely et al., 2010).

Conclusion

Energy has become an essential parameter for production and development processes, whereas conventional sources of energy are gradually diminishing, and escalating fuel prices are alarming for many oil-importing countries such as Pakistan. The situation warrants an increased search and research for alternate sources of energy. Contextual information on systems, including higher plants and microorganisms for the production of bioethanol, biodiesel, and solar-driven hydrogen production, has been discussed. Contrary to microorganisms, cost-effective biofuel production through plant genome is at its early development, conceivably because of the above-mentioned differences of the systems. Hence, developing routine transformation protocols for plants that could be used as greater source of biomass will remain a challenge. Efficient and reproducible transformation of prokaryotic organelles within the plant cell to express CWDEs cost effectively and efficiently is only one aspect that can be developed, yet it warrants combining with developing an *in vitro* cell culture protocol that allows efficient recovery and purification of homoplasmic shoots without compromising the regeneration with successive subcultures.

Acknowledgments

We thank HEC (Pakistan) and BBSRC (UK) for funding our work. We apologize to all those colleagues whose work could not be discussed here due to space restrictions.

References

Adir, N. 2005. Elucidation of the molecular structures of components of the phycobilisome: Reconstructing a giant. *Photosynthesis Research*. 85:15–32.

Adler, P.R., S.J. Del Grosso, and W.J. Parton. 2007. Life-cycle assessment of net greenhouse-gas flux for bioenergy cropping systems. *Ecological Applications*. 17:675–691.

Ahmad, N. and Z. Mukhtar. 2013. Green factories: Plastids for the production of foreign proteins at high levels. *Gene Therapy and Molecular Biology*. 15:14–29.

Atsumi, S., A.F. Cann, M.R. Connor, C.R. Shen, K.M. Smith, M.P. Brynildsen, K.J. Chou, T. Hanai, and J.C. Liao. 2008a. Metabolic engineering of *Escherichia coli* for 1-butanol production. *Metabolic Engineering*. 10:305–311.

Atsumi, S., T. Hanai, and J.C. Liao. 2008b. Non-fermentative pathways for synthesis of branched-chain higher alcohols as biofuels. *Nature*. 451:86–89.

Atsumi, S., W. Higashide, and J.C. Liao. 2009. Direct photosynthetic recycling of carbon dioxide to isobutyraldehyde. *Nature Biotechnology*. 27:1177–1180.

Baltz, A., K.-V. Dang, A. Beyly, P. Auroy, P. Richaud, L. Cournac, and G. Peltier. 2014. Plastidial expression of type II NAD(P)H dehydrogenase increases the reducing state of plastoquinones and hydrogen photoproduction rate by the indirect pathway in *Chlamydomonas reinhardtii*. *Plant Physiology*. 165:1344–1352.

Bandyopadhyay, A., T. Elvitigala, E. Welsh, J. Stöckel, M. Liberton, H. Min, L.A. Sherman, and H.B. Pakrasi. 2011. Novel metabolic attributes of the genus *Cyanothece*, comprising a group of unicellular nitrogen-fixing cyanobacteria. *mBio*. 2:e00214–00211.

Bandyopadhyay, A., J. Stöckel, H. Min, L.A. Sherman, and H.B. Pakrasi. 2010. High rates of photobiological H_2 production by a cyanobacterium under aerobic conditions. *Nature Communications*. 1:139.

Banerjee, A., R. Sharma, Y. Chisti, and U. Banerjee. 2002. *Botryococcus braunii*: A renewable source of hydrocarbons and other chemicals. *Critical Reviews in Biotechnology*. 22:245–279.

Barney, B.M., R.Y. Igarashi, P.C. Dos Santos, D.R. Dean, and L.C. Seefeldt. 2004. Substrate interaction at an iron-sulfur face of the FeMo-cofactor during nitrogenase catalysis. *Journal of Biological Chemistry*. 279:53621–53624.

Bauer, S., P. Vasu, S. Persson, A.J. Mort, and C.R. Somerville. 2006. Development and application of a suite of polysaccharide-degrading enzymes for analyzing plant cell walls. *Proceedings of the National Academy of Sciences*. 103:11417–11422.

Baxter, H.L., M. Mazarei, N. Labbe, L.M. Kline, Q. Cheng, M.T. Windham, D.G. Mann, C. Fu, A. Ziebell, and R.W. Sykes. 2014. Two-year field analysis of reduced recalcitrance transgenic switchgrass. *Plant Biotechnology Journal*. 12:914–924.

Beckmann, J., F. Lehr, G. Finazzi, B. Hankamer, C. Posten, L. Wobbe, and O. Kruse. 2009. Improvement of light to biomass conversion by de-regulation of light-harvesting protein translation in *Chlamydomonas reinhardtii*. *Journal of Biotechnology*. 142:70–77.

Beopoulos, A., J. Cescut, R. Haddouche, J.-L. Uribelarrea, C. Molina-Jouve, and J.-M. Nicaud. 2009. *Yarrowia lipolytica* as a model for bio-oil production. *Progress in Lipid Research*. 48:375–387.

Beopoulos, A., J.-M. Nicaud, and C. Gaillardin. 2011. An overview of lipid metabolism in yeasts and its impact on biotechnological processes. *Applied Microbiology and Biotechnology*. 90:1193–1206.

Blazeck, J., A. Hill, L. Liu, R. Knight, J. Miller, A. Pan, P. Otoupal, and H.S. Alper. 2014. Harnessing *Yarrowia lipolytica* lipogenesis to create a platform for lipid and biofuel production. *Nature Communications*. 5: doi: 10.1038/ncomms4131.

Bock, R. 2014. Engineering chloroplasts for high-level foreign protein expression. In *Chloroplast Biotechnology*. Pal Maliga (Ed.), Springer, New York. 93–106.

Bock, R. and M.S. Khan. 2004. Taming plastids for a green future. *Trends in Biotechnology*. 22:311–318.

Bokinsky, G., P.P. Peralta-Yahya, A. George, B.M. Holmes, E.J. Steen, J. Dietrich, T.S. Lee, D. Tullman-Ercek, C.A. Voigt, and B.A. Simmons. 2011. Synthesis of three advanced biofuels from ionic liquid-pretreated switchgrass using engineered *Escherichia coli*. *Proceedings of the National Academy of Sciences*. 108:19949–19954.

Bonke, S.A., M. Wiechen, D.R. MacFarlane, and L. Spiccia. 2015. Renewable fuels from concentrated solar power: Towards practical artificial photosynthesis. *Energy & Environmental Science.* 8:2791–2796.

Borkhardt, B., J. Harholt, P. Ulvskov, B.K. Ahring, B. Jørgensen, and H. Brinch-Pedersen. 2010. Autohydrolysis of plant xylans by apoplastic expression of thermophilic bacterial endo-xylanases. *Plant Biotechnology Journal.* 8:363–374.

Boyle, N.R., M.D. Page, B. Liu, I.K. Blaby, D. Casero, J. Kropat, S.J. Cokus et al. 2012. Three acyl-transferases and nitrogen-responsive regulator are implicated in nitrogen starvation-induced triacylglycerol accumulation in *Chlamydomonas. Journal of Biological Chemistry.* 287:15811–15825.

Burgess, S.J., B. Tamburic, F. Zemichael, K. Hellgardt, and P.J. Nixon. 2011. Solar-driven hydrogen production in green algae. *Advances in Applied Microbiology.* 75:71–110.

Burgess, S.J., G. Tredwell, A. Molnàr, J.G. Bundy, and P.J. Nixon. 2012. Artificial microRNA-mediated knockdown of pyruvate formate lyase (PFL1) provides evidence for an active 3-hydroxybutyrate production pathway in the green alga *Chlamydomonas reinhardtii. Journal of Biotechnology.* 162:57–66.

Busov, V.B., A.M. Brunner, and S.H. Strauss. 2008. Genes for control of plant stature and form. *New Phytologist.* 177:589–607.

Cann, A.F. and J.C. Liao. 2008. Production of 2-methyl-1-butanol in engineered *Escherichia coli. Applied Microbiology and Biotechnology.* 81:89–98.

Cao, Y., W. Liu, X. Xu, H. Zhang, J. Wang, and M. Xian. 2014. Production of free monounsaturated fatty acids by metabolically engineered *Escherichia coli. Biotechnology for Biofuels.* 7:59.

Castruita, M., D. Casero, S.J. Karpowicz, J. Kropat, A. Vieler, S.I. Hsieh, W. Yan, S. Cokus, J.A. Loo, and C. Benning. 2011. Systems biology approach in *Chlamydomonas* reveals connections between copper nutrition and multiple metabolic steps. *The Plant Cell Online.* 23:1273–1292.

Catalanotti, C., A. Dubini, V. Subramanian, W. Yang, L. Magneschi, F. Mus, M. Seibert, M.C. Posewitz, and A.R. Grossman. 2012. Altered fermentative metabolism in *Chlamydomonas reinhardtii* mutants lacking pyruvate formate lyase and both pyruvate formate lyase and alcohol dehydrogenase. *The Plant Cell Online.* 24:692–707.

Catalanotti, C., W. Yang, M.C. Posewitz, and A.R. Grossman. 2013. Fermentation metabolism and its evolution in algae. *Frontiers in Plant Science.* 4:150.

Cazzaniga, S., L. Dall'Osto, J. Szaub, L. Scibilia, M. Ballottari, S. Purton, and R. Bassi. 2014. Domestication of the green alga *Chlorella sorokiniana*: Reduction of antenna size improves light-use efficiency in a photobioreactor. *Biotechnology for Biofuels.* 7:157.

Chabannes, M., A. Barakate, C. Lapierre, J.M. Marita, J. Ralph, M. Pean, S. Danoun, C. Halpin, J. Grima-Pettenati, and A.M. Boudet. 2001. Strong decrease in lignin content without significant alteration of plant development is induced by simultaneous down-regulation of cinnamoyl CoA reductase (CCR) and cinnamyl alcohol dehydrogenase (CAD) in tobacco plants. *The Plant Journal.* 28:257–270.

Chang, R.L., L. Ghamsari, A. Manichaikul, E.F.Y. Hom, S. Balaji, W. Fu, Y. Shen et al. 2011. Metabolic network reconstruction of *Chlamydomonas* offers insight into light-driven algal metabolism. *Molecular Systems Biology.* 7:518.

Chen, F. and R.A. Dixon. 2007. Lignin modification improves fermentable sugar yields for biofuel production. *Nature Biotechnology.* 25:759–761.

Chen, X., K.F. Nielsen, I. Borodina, M.C. Kielland-Brandt, and K. Karhumaa. 2011. Increased isobu-tanol production in *Saccharomyces cerevisiae* by overexpression of genes in valine metabolism. *Biotechnology for Biofuels.* 4:2089–2090.

Chisti, Y. 2007. Biodiesel from microalgae. *Biotechnology Advances.* 25:294–306.

Chochois, V., L. Constans, D. Dauvillée, A. Beyly, M. Solivérès, S. Ball, G. Peltier, and L. Cournac. 2010. Relationships between PSII-independent hydrogen bioproduction and starch metabolism as evidenced from isolation of starch catabolism mutants in the green alga *Chlamydomonas reinhardtii. International Journal of Hydrogen Energy.* 35:10731–10740.

Chung, D., M. Cha, A.M. Guss, and J. Westpheling. 2014. Direct conversion of plant biomass to ethanol by engineered *Caldicellulosiruptor bescii. Proceedings of the National Academy of Sciences.* 11:8931–8936.

Comba, S., M. Sabatini, S. Menendez-Bravo, A. Arabolaza, and H. Gramajo. 2014. Engineering a *Streptomyces coelicolor* biosynthesis pathway in *Escherichia coli* for high yield triglyceride production. *Biotechnology for Biofuels*. 7:172.

Cosgrove, D.J. 2000. Loosening of plant cell walls by expansins. *Nature*. 407:321–326.

Demeke, M.M., H. Dietz, Y. Li, M.R. Foulqui-Moreno, S. Mutturi, S. Deprez, T. Den Abt, B.M. Bonini, G. Liden, and F. Dumortier. 2013. Development of a D-xylose fermenting and inhibitor tolerant industrial *Saccharomyces cerevisiae* strain with high performance in lignocellulose hydrolysates using metabolic and evolutionary engineering. *Biotechnology for Biofuels*. 6:89.

Desplats, C., F. Mus, S. Cuiné, E. Billon, L. Cournac, and G. Peltier. 2009. Characterization of Nda2, a plastoquinone-reducing type II NAD (P) H dehydrogenase in *Chlamydomonas* chloroplasts. *Journal of Biological Chemistry*. 284:4148–4157.

Dien, B., M. Cotta, and T. Jeffries. 2003. Bacteria engineered for fuel ethanol production: Current status. *Applied Microbiology and Biotechnology*. 63:258–266.

Ding, S.-Y. and M.E. Himmel. 2006. The maize primary cell wall microfibril: A new model derived from direct visualization. *Journal of Agricultural Food and Chemistry*. 54:597–606.

Doebbe, A., M. Keck, M. La Russa, J.H. Mussgnug, B. Hankamer, E. Tekçe, K. Niehaus, and O. Kruse. 2010. The interplay of proton, electron, and metabolite supply for photosynthetic H_2 production in *Chlamydomonas reinhardtii*. *Journal of Biological Chemistry*. 285:30247–30260.

Doshi, R., T. Nguyen, and G. Chang. 2013. Transporter-mediated biofuel secretion. *Proceedings of the National Academy of Sciences*. 110:7642–7647.

Doucha, J. and K. Livansky. 2012. Production of high-density *Chlorella* culture grown in fermenters. *Journal of Applied Phycology*. 24:35–43.

Dubini, A. and M. Ghirardi. 2014. Engineering photosynthetic organisms for the production of bio-hydrogen. *Photosynthesis Research*. 1–13.

Dunlop, M.J. 2011. Engineering microbes for tolerance to next-generation biofuels. *Biotechnology for Biofuels*. 4:32.

Dutta, D., D. De, S. Chaudhuri, and S.K. Bhattacharya. 2005. Hydrogen production by cyanobacteria. *Microbial Cell Factories*. 4:36.

Fan, J., C. Andre, and C. Xu. 2011. A chloroplast pathway for the *de novo* biosynthesis of triacylglycerol in *Chlamydomonas reinhardtii*. *FEBS Letters*. 585:1985–1991.

Fan, J., C. Yan, C. Andre, J. Shanklin, J. Schwender, and C. Xu. 2012. Oil accumulation is controlled by carbon precursor supply for fatty acid synthesis in *Chlamydomonas reinhardtii*. *Plant Cell Physiology*. 53:1380–1390.

FAO. 2007. A review of the current state of bioenergy development in G8 + 5 countries. Italy, Rome: United Nation's Food and Agricultural Organization.

Faraloni, C. and G. Torzillo. 2010. Phenotypic characterization and hydrogen production in *Chlamydomonas reinhardtii* QB-binding D1-protein mutants under sulfur starvation: Changes in Chl fluorescence and pigment composition. *Journal of Phycology*. 46:788–799.

Feng, X., A. Bandyopadhyay, B. Berla, L. Page, B. Wu, H.B. Pakrasi, and Y.J. Tang. 2010. Mixotrophic and photoheterotrophic metabolism in *Cyanothece* sp. ATCC 51142 under continuous light. *Microbiology*. 156:2566–2574.

Foo, J.L., H.M. Jensen, R.H. Dahl, K. George, J.D. Keasling, T.S. Lee, S. Leong, and A. Mukhopadhyay. 2014. Improving microbial biogasoline production in *Escherichia coli* using tolerance engineering. *mBio*. 5:e01932–01914.

Gaffron. 1939. Reduction of carbon dioxide with molecular hydrogen in green algae. *Nature*. 143:204–205.

Gallon, J.R. 1981. The oxygen sensitivity of nitrogenase: A problem for biochemists and microorganisms. *Trends in Biochemical Sciences*. 6:19–23.

Georgianna, D.R. and S.P. Mayfield. 2012. Exploiting diversity and synthetic biology for the production of algal biofuels. *Nature*. 488:329–335.

Ghirardi, M.L., M.C. Posewitz, P.-C. Maness, A. Dubini, J. Yu, and M. Seibert. 2007. Hydrogenases and hydrogen photoproduction in oxygenic photosynthetic organisms. *Annual Review of Plant Biology*. 58:71–91.

Gill, S.S., M.A. Mehmood, U. Rashid, M. Ibrahim, A. Saqib, and M.R. Tabassum. 2013. Waste-water treatment coupled with biodiesel production using microalgae: A bio-refinery approach. *Pakistan Journal of Life and Social Sciences*. 11:179–189.

Goh, E.-B., E.E. Baidoo, J.D. Keasling, and H.R. Beller. 2012. Engineering of bacterial methyl ketone synthesis for biofuels. *Applied and Environmental Microbiology.* 78:70–80.

Gonzalez-Ballester, D., D. Casero, S. Cokus, M. Pellegrini, S.S. Merchant, and A.R. Grossman. 2010. RNA-seq analysis of sulfur-deprived *Chlamydomonas* cells reveals aspects of acclimation critical for cell survival. *The Plant Cell Online.* 22:2058–2084.

Goodson, C., R. Roth, Z.T. Wang, and U. Goodenough. 2011. Structural correlates of cytoplasmic and chloroplast lipid body synthesis in *Chlamydomonas reinhardtii* and stimulation of lipid body production with acetate boost. *Eukaryotic Cell.* 10:1592–1606.

Gray, B.N., B.A. Ahner, and M.R. Hanson. 2009. High-level bacterial cellulase accumulation in chloroplast-transformed tobacco mediated by downstream box fusions. *Biotechnology and Bioengineering.* 102:1045–1054.

Grossman, A.R., M.R. Schaefer, G.G. Chiang, and J.L. Collier. 1993. The phycobilisome, a light-harvesting complex responsive to environmental conditions. *Microbiological Reviews.* 57:725–749.

Guo, M., W. Song, and J. Buhain. 2015. Bioenergy and biofuels: History, status, and perspective. *Renewable and Sustainable Energy Reviews.* 42:712–725.

Halimatul, H.S.M., S. Ehira, and K. Awai. 2014. Fatty alcohols can complement functions of heterocyst specific glycolipids in *Anabaena* sp. PCC 7120. *Biochemical and Biophysical Research Communications.* 450:178–183.

Hanai, T., S. Atsumi, and J. Liao. 2007. Engineered synthetic pathway for isopropanol production in *Escherichia coli. Applied and Environmental Microbiology.* 73:7814–7818.

Happe, T., K. Schütz, and H. Böhme. 2000. Transcriptional and mutational analysis of the uptake hydrogenase of the filamentous cyanobacterium *Anabaena variabilis* ATCC 29413. *Journal of Bacteriology.* 182:1624–1631.

Harrison, M.D., J. Geijskes, H.D. Coleman, K. Shand, M. Kinkema, A. Palupe, R. Hassall, M. Sainz, R. Lloyd, and S. Miles. 2011. Accumulation of recombinant cellobiohydrolase and endoglucanase in the leaves of mature transgenic sugar cane. *Plant Biotechnology Journal.* 9:884–896.

Harvey, B.G., M.E. Wright, and R.L. Quintana. 2009. High-density renewable fuels based on the selective dimerization of pinenes. *Energy & Fuels.* 24:267–273.

Hemschemeier, A., D. Casero, B. Liu, C. Benning, M. Pellegrini, T. Happe, and S.S. Merchant. 2013. Copper response regulator1–dependent and –independent responses of the *Chlamydomonas reinhardtii* transcriptome to dark anoxia. *The Plant Cell Online.* 25:3186–3211.

Hemschemeier, A. and T. Happe. 2011. Alternative photosynthetic electron transport pathways during anaerobiosis in the green alga *Chlamydomonas reinhardtii. Biochimica et Biophysica Acta (BBA)—Bioenergetics.* 1807:919–926.

Ho, N.W., Z. Chen, and A.P. Brainard. 1998. Genetically engineered *Saccharomyces* yeast capable of effective cofermentation of glucose and xylose. *Applied and Environmental Microbiology.* 64:1852–1859.

Holladay, J.D., J. Hu, D.L. King, and Y. Wang. 2009. An overview of hydrogen production technologies. *Catalysis Today.* 139:244–260.

Hood, E.E., R. Love, J. Lane, J. Bray, R. Clough, K. Pappu, C. Drees, K.R. Hood, S. Yoon, and A. Ahmad. 2007. Subcellular targeting is a key condition for high-level accumulation of cellulase protein in transgenic maize seed. *Plant Biotechnology Journal.* 5:709–719.

Howard, T.P., S. Middelhaufe, K. Moore, C. Edner, D.M. Kolak, G.N. Taylor, D.A. Parker, R. Lee, N. Smirnoff, and S.J. Aves. 2013. Synthesis of customized petroleum-replica fuel molecules by targeted modification of free fatty acid pools in *Escherichia coli. Proceedings of the National Academy of Sciences.* 110:7636–7641.

Hu, W.-J., S.A. Harding, J. Lung, J.L. Popko, J. Ralph, D.D. Stokke, C.-J. Tsai, and V.L. Chiang. 1999. Repression of lignin biosynthesis promotes cellulose accumulation and growth in transgenic trees. *Nature Biotechnology.* 17:808–812.

Huo, Y.-X., K.M. Cho, J.G.L. Rivera, E. Monte, C.R. Shen, Y. Yan, and J.C. Liao. 2011. Conversion of proteins into biofuels by engineering nitrogen flux. *Nature Biotechnology.* 29:346–351.

Ingram, L., T. Conway, D. Clark, G. Sewell, and J. Preston. 1987. Genetic engineering of ethanol production in *Escherichia coli. Applied and Environmental Microbiology.* 53:2420–2425.

Jacobsen, J.R., C.R. Hutchinson, D.E. Cane, and C. Khosla. 1997. Precursor-directed biosynthesis of erythromycin analogs by an engineered polyketide synthase. *Science*. 277:367–369.

Jans, F., E. Mignolet, P.A. Houyoux, P. Cardol, B. Ghysels, S. Cuiné, L. Cournac, G. Peltier, C. Remacle, and F. Franck. 2008. A type II NAD (P) H dehydrogenase mediates light-independent plastoquinone reduction in the chloroplast of *Chlamydomonas*. *Proceedings of the National Academy of Sciences*. 105:20546–20551.

Jeffries, T. and Y.-S. Jin. 2004. Metabolic engineering for improved fermentation of pentoses by yeasts. *Applied Microbiology and Biotechnology*. 63:495–509.

Jiang, X.-R., X.-Y. Zhou, W.-Y. Jiang, X.-R. Gao, and W.-L. Li. 2011. Expressions of thermostable bacterial cellulases in tobacco plant. *Biotechnology Letters*. 33:1797–1803.

Jojima, T., M. Inui, and H. Yukawa. 2008. Production of isopropanol by metabolically engineered *Escherichia coli*. *Applied Microbiology and Biotechnology*. 77:1219–1224.

Jones, D.T. and D.R. Woods. 1986. Acetone-butanol fermentation revisited. *Microbiological Reviews*. 50:484–524.

Jung, J.H., W.M. Fouad, W. Vermerris, M. Gallo, and F. Altpeter. 2012a. RNAi suppression of lignin biosynthesis in sugarcane reduces recalcitrance for biofuel production from lignocellulosic biomass. *Plant Biotechnology Journal*. 10:1067–1076.

Jung, J.H., W. Vermerris, M. Gallo, J.R. Fedenko, J.E. Erickson, and F. Altpeter. 2013. RNA interference suppression of lignin biosynthesis increases fermentable sugar yields for biofuel production from field-grown sugarcane. *Plant Biotechnology Journal*. 11:709–716.

Jung, S.K., B.E. Lindenmuth, K.A. McDonald, H. Hwang, M.Q. Bui, B.W. Falk, S.L. Uratsu, M.L. Phu, and A.M. Dandekar. 2014. *Agrobacterium tumefaciens* mediated transient expression of plant cell wall-degrading enzymes in detached sunflower leaves. *Biotechnology Progress*. 30:905–915.

Jung, S.K., V. Parisutham, S.H. Jeong, and S.K. Lee. 2012b. Heterologous expression of plant cell wall degrading enzymes for effective production of cellulosic biofuels. *Journal of Biomedicine and Biotechnology*. 2012:1–10.

Kalluri, U.C., S.P. DiFazio, A.M. Brunner, and G.A. Tuskan. 2007. Genome-wide analysis of Aux/IAA and ARF gene families in *Populus trichocarpa*. *BMC Plant Biology*. 7:59–73.

Kalscheuer, R., T. Stolting, and A. Steinbuchel. 2006. Microdiesel: *Escherichia coli* engineered for fuel production. *Microbiology*. 152:2529–2536.

Kanes, S., D. Forster, and L. Wilkinson. 2010. Biofuels Outlook. *Equity Research Industry Report*, Scotia Capital, Toronto.

Keasling, J.D. 2010. Manufacturing molecules through metabolic engineering. *Science*. 330:1355–1358.

Khetkorn, W., P. Lindblad, and A. Incharoensakdi. 2012. Inactivation of uptake hydrogenase leads to enhanced and sustained hydrogen production with high nitrogenase activity under high light exposure in the cyanobacterium *Anabaena siamensis* TISTR 8012. *Journal of Biological Engineering*. 6:19.

Kirst, H., J.G. Garcia-Cerdan, A. Zurbriggen, T. Ruehle, and A. Melis. 2012. Truncated photosystem chlorophyll antenna size in the green microalga *Chlamydomonas reinhardtii* upon deletion of the *TLA3-CpSRP43* Gene. *Plant Physiology*. 160:2251–2260.

Kolotilin, I., A. Kaldis, E.O. Pereira, S. Laberge, and R. Menassa. 2013. Optimization of transplastomic production of hemicellulases in tobacco: Effects of expression cassette configuration and tobacco cultivar used as production platform on recombinant protein yields. *Biotechnology for Biofuels*. 6:65–80.

Koussa, J., A. Chaiboonchoe, and K. Salehi-Ashtiani. 2014. Computational approaches for microalgal biofuel optimization: A review. *BioMed Research International*. 2014:649453.

Kruse, O., J. Rupprecht, K.-P. Bader, S. Thomas-Hall, P.M. Schenk, G. Finazzi, and B. Hankamer. 2005. Improved photobiological H_2 production in engineered green algal cells. *Journal of Biological Chemistry*. 280:34170–34177.

Kumaraswamy, G.K., T. Guerra, X. Qian, S. Zhang, D.A. Bryant, and G.C. Dismukes. 2013. Reprogramming the glycolytic pathway for increased hydrogen production in cyanobacteria: Metabolic engineering of NAD+-dependent GAPDH. *Energy & Environmental Science*. 6:3722–3731.

Kuzuyama, T. 2002. Mevalonate and nonmevalonate pathways for the biosynthesis of isoprene units. *Bioscience, Biotechnology, and Biochemistry*. 66:1619–1627.

Kwon, J.-H., G. Bernát, H. Wagner, M. Rögner, and S. Rexroth. 2013. Reduced light-harvesting antenna: Consequences on cyanobacterial metabolism and photosynthetic productivity. *Algal Research*. 2:188–195.

La Russa, M., C. Bogen, A. Uhmeyer, A. Doebbe, E. Filippone, O. Kruse, and J.H. Mussgnug. 2012. Functional analysis of three type-2 DGAT homologue genes for triacylglycerol production in the green microalga *Chlamydomonas reinhardtii*. *Journal of Biotechnology*. 162:13–20.

Lane, J. 2015. Biofuels mandates around the world: 2015. *Biofuels Digest*: http://www.biofuelsdigest.com/bdigest/2014/2012/2031/biofuels-mandates-around-the-world-2015/.

Lea-Smith, D.J., P. Bombelli, J.S. Dennis, S.A. Scott, A.G. Smith, and C.J. Howe. 2014. Phycobilisome-deficient strains of *Synechocystis* sp. PCC 6803 have reduced size and require carbon-limiting conditions to exhibit enhanced productivity. *Plant Physiology*. 165:705–714.

Lee, S., S. Park, and J. Lee. 2013. Improvement of free fatty acid production in *Escherichia coli* using codon-optimized *Streptococcus pyogenes* acyl-ACP thioesterase. *Bioprocess and Biosystems Engineering*. 36:1519–1525.

Lee, S.K., H. Chou, T.S. Ham, T.S. Lee, and J.D. Keasling. 2008. Metabolic engineering of microorganisms for biofuels production: From bugs to synthetic biology to fuels. *Current Opinion in Biotechnology*. 19:556–563.

Lennen, R.M., D.J. Braden, R.M. West, J.A. Dumesic, and B.F. Pfleger. 2010. A process for microbial hydrocarbon synthesis: Overproduction of fatty acids in *Escherichia coli* and catalytic conversion to alkanes. *Biotechnology and Bioengineering*. 106:193–202.

Li, X., N.D. Bonawitz, J.-K. Weng, and C. Chapple. 2010. The growth reduction associated with repressed lignin biosynthesis in *Arabidopsis thaliana* is independent of flavonoids. *The Plant Cell Online*. 22:1620–1632.

Liu, T., H. Vora, and C. Khosla. 2010. Quantitative analysis and engineering of fatty acid biosynthesis in *Escherichia coli*. *Metabolic Engineering*. 12:378–386.

Llop-Tous, I., M. Ortiz, M. Torrent, and M.D. Ludevid. 2011. The expression of a xylanase targeted to ER-protein bodies provides a simple strategy to produce active insoluble enzyme polymers in tobacco plants. *PloS ONE*. 6:e19474.

Lynd, L.R., M.S. Laser, D. Bransby, B.E. Dale, B. Davison, R. Hamilton, M. Himmel, M. Keller, J.D. McMillan, J. Sheehan, and C.E. Wyman. 2008. How biotech can transform biofuels. *Nature Biotechnology*. 26:169–172.

Lynd, L.R., P.J. Weimer, W.H. Van Zyl, and I.S. Pretorius. 2002. Microbial cellulose utilization: Fundamentals and biotechnology. *Microbiology and Molecular Biology Reviews*. 66:506–577.

Magneschi, L., C. Catalanotti, V. Subramanian, A. Dubini, W. Yang, F. Mus, M.C. Posewitz, M. Seibert, P. Perata, and A.R. Grossman. 2012. A mutant in the *ADH1* gene of *Chlamydomonas reinhardtii* elicits metabolic restructuring during anaerobiosis. *Plant Physiology*. 158:1293–1305.

Mannan, R. 2010. Intellectual property landscape and patenting opportunity in biofuels. *Journal of Commercial Biotechnology*. 16:33–46.

Masukawa, H., K. Inoue, and H. Sakurai. 2007. Effects of disruption of homocitrate synthase genes on *Nostoc* sp. Strain PCC 7120 photobiological hydrogen production and nitrogenase. *Applied and Environmental Microbiology*. 73:7562–7570.

Masukawa, H., M. Mochimaru, and H. Sakurai. 2002. Disruption of the uptake hydrogenase gene, but not of the bidirectional hydrogenase gene, leads to enhanced photobiological hydrogen production by the nitrogen-fixing cyanobacterium *Anabaena* sp. PCC 7120. *Applied Microbiology and Biotechnology*. 58:618–624.

Mattam, A.J. and S.S. Yazdani. 2013. Engineering *E. coli* strain for conversion of short chain fatty acids to bioalcohols. *Biotechnology for Biofuels*. 6:128.

McAndrew, R.P., P.P. Peralta-Yahya, A. DeGiovanni, J.H. Pereira, M.Z. Hadi, J.D. Keasling, and P.D. Adams. 2011. Structure of a three-domain sesquiterpene synthase: A prospective target for advanced biofuels production. *Structure*. 19:1876–1884.

McCann, M.C. and N.C. Carpita. 2008. Designing the deconstruction of plant cell walls. *Current Opinion in Plant Biology*. 11:314–320.

McNeely, K., Y. Xu, N. Bennette, D.A. Bryant, and G.C. Dismukes. 2010. Redirecting reductant flux into hydrogen production via metabolic engineering of fermentative carbon metabolism in a cyanobacterium. *Applied and Environmental Microbiology*. 76:5032–5038.

Mei, C., S.H. Park, R. Sabzikar, C. Ransom, C. Qi, and M. Sticklen. 2009. Green tissue-specific production of a microbial endo-cellulase in maize (*Zea mays* L.) endoplasmic-reticulum and
mitochondria converts cellulose into fermentable sugars. *Journal of Chemical Technology and
Biotechnology*. 84:689–695.

Melis, A. 2009. Solar energy conversion efficiencies in photosynthesis: Minimizing the chlorophyll
antennae to maximize efficiency. *Plant Science*. 177:272–280.

Melis, A., J. Neidhardt, and J. Benemann. 1998. *Dunaliella salina* (Chlorophyta) with small chlorophyll antenna sizes exhibit higher photosynthetic productivities and photon use efficiencies
than normally pigmented cells. *Journal of Applied Phycology*. 10:515–525.

Melis, A., L. Zhang, M. Forestier, M.L. Ghirardi, and M. Seibert. 2000. Sustained photobiological
hydrogen gas production upon reversible inactivation of oxygen evolution in the green alga
Chlamydomonas reinhardtii. *Plant Physiology*. 122:127–136.

Menzella, H.G., R. Reid, J.R. Carney, S.S. Chandran, S.J. Reisinger, K.G. Patel, D.A. Hopwood, and
D.V. Santi. 2005. Combinatorial polyketide biosynthesis by *de novo* design and rearrangement
of modular polyketide synthase genes. *Nature Biotechnology*. 23:1171–1176.

Merlin Christy, P., L.R. Gopinath, and D. Divya. 2014. A review on anaerobic decomposition and
enhancement of biogas production through enzymes and microorganisms. *Renewable and
Sustainable Energy Reviews*. 34:167–173.

Mettler, T., T. Mühlhaus, D. Hemme, M.-A. Schöttler, J. Rupprecht, A. Idoine, D. Veyel et al. 2014.
Systems analysis of the response of photosynthesis, metabolism, and growth to an increase
in irradiance in the photosynthetic model organism *Chlamydomonas reinhardtii*. *The Plant Cell
Online*. 26:2310–2350.

Meuser, J.E., S. D'Adamo, R.E. Jinkerson, F. Mus, W. Yang, M.L. Ghirardi, M. Seibert, A.R. Grossman,
and M.C. Posewitz. 2012. Genetic disruption of both *Chlamydomonas reinhardtii* [FeFe]-
hydrogenases: Insight into the role of HYDA2 in H_2 production. *Biochemical and Biophysical
Research Communications*. 417:704–709.

Miller, R., G. Wu, R.R. Deshpande, A. Vieler, K. Gartner, X. Li, E.R. Moellering, S. Zauner, A.J.
Cornish, and B. Liu. 2010. Changes in transcript abundance in *Chlamydomonas reinhardtii* following nitrogen deprivation predict diversion of metabolism. *Plant Physiology*. 154:1737–1752.

Min, H. and L.A. Sherman. 2010. Hydrogen production by the unicellular, diazotrophic cyanobacterium *Cyanothece* sp. Strain ATCC 51142 under conditions of continuous light. *Applied and
Environmental Microbiology*. 76:4293–4301.

Montenegro, M. 2006. The big three. *Grist Environmental News*. http://grist.org/news. *Retrieved on
2008-12-10*.

Morales, M., J. Quintero, R. Conejeros, and G. Aroca. 2015. Life cycle assessment of lignocellulosic bioethanol: Environmental impacts and energy balance. *Renewable & Sustainable Energy
Reviews*. 42:1349–1361.

Moser, B.R. 2011. Biodiesel production, properties, and feedstocks. In *Biofuels*. Dwight Tomes,
Prakash Lakshmanan, David Songstad (Eds), Springer, New York. 285–347.

Mus, F., A. Dubini, M. Seibert, M.C. Posewitz, and A.R. Grossman. 2007. Anaerobic acclimation in
Chlamydomonas reinhardtii: Anoxic gene expression, hydrogenase induction, and metabolic
pathways. *Journal of Biological Chemistry*. 282:25475–25486.

Nakajima, Y. and R. Ueda. 1999. Improvement of microalgal photosynthetic productivity by reducing the content of light harvesting pigment. *Journal of Applied Phycology*. 11:195–201.

Nguyen, A.V., J. Toepel, S. Burgess, A. Uhmeyer, O. Blifernez, A. Doebbe, B. Hankamer, P. Nixon,
L. Wobbe, and O. Kruse. 2011. Time-course global expression profiles of *Chlamydomonas reinhardtii* during photo-biological H_2 production. *PLoS ONE*. 6:e29364.

Nikaido, H. and Y. Takatsuka. 2009. Mechanisms of RND multidrug efflux pumps. *Biochimica et
Biophysica Acta (BBA)—Proteins and Proteomics*. 1794:769–781.

Nogales, J., S. Gudmundsson, E.M. Knight, B.O. Palsson, and I. Thiele. 2012. Detailing the opti-
mality of photosynthesis in cyanobacteria through systems biology analysis. *Proceedings of the
National Academy of Sciences*. 109:2678–2683.

Noth, J., D. Krawietz, A. Hemschemeier, and T. Happe. 2013. Pyruvate: Ferredoxin oxidoreductase
is coupled to light-independent hydrogen production in *Chlamydomonas reinhardtii*. *Journal of
Biological Chemistry*. 288:4368–4377.

Oey, M., I.L. Ross, E. Stephens, J. Steinbeck, J. Wolf, K.A. Radzun, J. Kügler, A.K. Ringsmuth, O. Kruse, and B. Hankamer. 2013. RNAi knock-down of LHCBM1, 2 and 3 increases photosynthetic $H_{(2)}$ production efficiency of the green alga *Chlamydomonas reinhardtii*. *PloS ONE*. 8:e61375.

Ohta, K., D. Beall, J. Mejia, K. Shanmugam, and L. Ingram. 1991. Metabolic engineering of *Klebsiella oxytoca* M5A1 for ethanol production from xylose and glucose. *Applied and Environmental Microbiology*. 57:2810–2815.

Oraby, H., B. Venkatesh, B. Dale, R. Ahmad, C. Ransom, J. Oehmke, and M. Sticklen. 2007. Enhanced conversion of plant biomass into glucose using transgenic rice-produced endoglucanase for cellulosic ethanol. *Transgenic Research*. 16:739–749.

Ort, D.R., X. Zhu, and A. Melis. 2011. Optimizing antenna size to maximize photosynthetic efficiency. *Plant Physiology*. 155:79–85.

Oxelfelt, F., P. Tamagnini, and P. Lindblad. 1998. Hydrogen uptake in *Nostoc* sp. Strain PCC 73102. Cloning and characterization of a hupSL homologue. *Archives of Microbiology*. 169:267–274.

Page, L.E., M. Liberton, and H.B. Pakrasi. 2012. Reduction of photoautotrophic productivity in the cyanobacterium *Synechocystis* sp. Strain PCC 6803 by phycobilisome antenna truncation. *Applied and Environmental Microbiology*. 78:6349–6351.

Park, Y.W., K.i. Baba, Y. Furuta, I. Iida, K. Sameshima, M. Arai, and T. Hayashi. 2004. Enhancement of growth and cellulose accumulation by overexpression of xyloglucanase in poplar. *FEBS Letters*. 564:183–187.

Pauly, M. and K. Keegstra. 2008. Cell-wall carbohydrates and their modification as a resource for biofuels. *The Plant Journal*. 54:559–568.

Peralta-Yahya, P.P. and J.D. Keasling. 2010. Advanced biofuel production in microbes. *Biotechnology Journal*. 5:147–162.

Peralta-Yahya, P.P., M. Ouellet, R. Chan, A. Mukhopadhyay, J.D. Keasling, and T.S. Lee. 2011. Identification and microbial production of a terpene-based advanced biofuel. *Nature Communications*. 2:483.

Peralta-Yahya, P.P., F. Zhang, S.B. Del Cardayre, and J.D. Keasling. 2012. Microbial engineering for the production of advanced biofuels. *Nature*. 488:320–328.

Petersen, K. and R. Bock. 2011. High-level expression of a suite of thermostable cell wall-degrading enzymes from the chloroplast genome. *Plant Molecular Biology*. 76:311–321.

Pfeifer, B.A. and C. Khosla. 2001. Biosynthesis of polyketides in heterologous hosts. *Microbiology and Molecular Biology Reviews*. 65:106–118.

Phelan, R.M., O.N. Sekurova, J.D. Keasling, and S.B. Zotchev. 2014. Engineering terpene biosynthesis in *Streptomyces* for production of the advanced biofuel precursor bisabolene. *ACS Synthetic Biology*: doi: 10.1021/sb5002517.

Philipps, G., D. Krawietz, A. Hemschemeier, and T. Happe. 2011. A pyruvate formate lyase-deficient *Chlamydomonas reinhardtii* strain provides evidence for a link between fermentation and hydrogen production in green algae. *The Plant Journal*. 66:330–340.

Pilate, G., E. Guiney, K. Holt, M. Petit-Conil, C. Lapierre, J.-C. Leplé, B. Pollet, I. Mila, E.A. Webster, and H.G. Marstorp. 2002. Field and pulping performances of transgenic trees with altered lignification. *Nature Biotechnology*. 20:607–612.

Pohl, A., P.F. Devaux, and A. Herrmann. 2005. Function of prokaryotic and eukaryotic ABC proteins in lipid transport. *Biochimica et Biophysica Acta (BBA)—Molecular and Cell Biology of Lipids*. 1733:29–52.

Polle, J.E.W., S. Kanakagiri, E. Jin, T. Masuda, and A. Melis. 2002. Truncated chlorophyll antenna size of the photosystems—A practical method to improve microalgal productivity and hydrogen production in mass culture. *International Journal of Hydrogen Energy*. 27:1257–1264.

Poovaiah, C.R., M. Nageswara-Rao, J.R. Soneji, H.L. Baxter, and C.N. Stewart. 2014. Altered lignin biosynthesis using biotechnology to improve lignocellulosic biofuel feedstocks. *Plant Biotechnology Journal*. 12:1163–1173.

Posten, C. 2009. Design principles of photo-bioreactors for cultivation of microalgae. *Engineering in Life Sciences*. 9:165–177.

Quintana, N., F. Van der Kooy, M.D. Van de Rhee, G.P. Voshol, and R. Verpoorte. 2011. Renewable energy from cyanobacteria: Energy production optimization by metabolic pathway engineering. *Applied Microbiology and Biotechnology*. 91:471–490.

Rajamani, S., W.D. Bauer, J.B. Robinson, J.M. Farrow III, E.C. Pesci, M. Teplitski, M. Gao, R.T. Sayre, and D.A. Phillips. 2008. The vitamin riboflavin and its derivative lumichrome activate the LasR bacterial quorum-sensing receptor. *Molecular Plant–Microbe Interactions*. 21:1184–1192.

Ralph, J., T. Akiyama, H. Kim, F. Lu, P.F. Schatz, J.M. Marita, S.A. Ralph, M.S. Reddy, F. Chen, and R.A. Dixon. 2006. Effects of coumarate 3-hydroxylase down-regulation on lignin structure. *Journal of Biological Chemistry*. 281:8843–8853.

Ramanan, R., B.H. Kim, D.H. Cho, S.R. Ko, H.M. Oh, and H.S. Kim. 2013. Lipid droplet synthesis is limited by acetate availability in starchless mutant of Chlamydomonas reinhardtii. *FEBS Letters*. 587:370–377.

Ramazanov, A. and Z. Ramazanov. 2006. Isolation and characterization of a starchless mutant of *Chlorella pyrenoidosa* STL-PI with a high growth rate, and high protein and polyunsaturated fatty acid content. *Phycological Research*. 54:255–259.

Reddy, M.S., F. Chen, G. Shadle, L. Jackson, H. Aljoe, and R.A. Dixon. 2005. Targeted down-regulation of cytochrome P450 enzymes for forage quality improvement in alfalfa (*Medicago sativa* L.). *Proceedings of the National Academy of Sciences*. 102:16573–16578.

Reiser, S. and C. Somerville. 1997. Isolation of mutants of *Acinetobacter calcoaceticus* deficient in wax ester synthesis and complementation of one mutation with a gene encoding a fatty acyl coenzyme A reductase. *Journal of Bacteriology*. 179:2969–2975.

Renninger, N.S. and D.J. Mcphee. 2010. Fuel compositions comprising farnesane and farnesane derivatives and method of making and using same. US Patent 7,846,222.

Rincon, L., J. Jaramillo, and C. Cardona. 2014. Comparison of feedstocks and technologies for biodiesel production: An environmental and techno-economic evaluation. *Renewable Energy*. 69:479–487.

Rubin, E.M. 2008. Genomics of cellulosic biofuels. *Nature*. 454:841–845.

Rude, M.A. and A. Schirmer. 2009. New microbial fuels: A biotech perspective. *Current Opinion in Microbiology*. 12:274–281.

Rumpel, S., J.F. Siebel, C. Fares, J. Duan, E. Reijerse, T. Happe, W. Lubitz, and M. Winkler. 2014. Enhancing hydrogen production of microalgae by redirecting electrons from photosystem I to hydrogenase. *Energy & Environmental Science*. 7:3296–3301.

Ruth, L. 2008. Bio or bust? The economic and ecological cost of biofuels. *EMBO Reports*. 9:130–133.

Saloheimo, M., M. Paloheimo, S. Hakola, J. Pere, B. Swanson, E. Nyyssönen, A. Bhatia, M. Ward, and M. Penttilä. 2002. Swollenin, a *Trichoderma reesei* protein with sequence similarity to the plant expansins, exhibits disruption activity on cellulosic materials. *European Journal of Biochemistry*. 269:4202–4211.

Schmollinger, S., T. Mühlhaus, N.R. Boyle, I.K. Blaby, D. Casero, T. Mettler, J.L. Moseley et al. 2014. Nitrogen-sparing mechanisms in *Chlamydomonas* affect the transcriptome, the proteome, and photosynthetic metabolism. *The Plant Cell Online*. 26:1410–1435.

Schnoor, J.L. 2006. Biofuels and the environment. *Environmental Science and Technology*. 40:4042.

Scoma, A., D. Krawietz, C. Faraloni, L. Giannelli, T. Happe, and G. Torzillo. 2012. Sustained H$_2$ production in a *Chlamydomonas reinhardtii* D1 protein mutant. *Journal of Biotechnology*. 157:613–619.

Scott, S.A., M.P. Davey, J.S. Dennis, I. Horst, C.J. Howe, D.J. Lea-Smith, and A.G. Smith. 2010. Biodiesel from algae: Challenges and prospects. *Current Opinion in Biotechnology*. 21:277–286.

Sheehan, J., T. Dunahay, J. Benemann, and P. Roessler. 1998. *A Look Back at the U.S. Department of Energy's Aquatic Species Program: Biodiesel from Algae*. Golden, National Renewable Energy Laboratory. Report no. NREL/TP-580-24190. Colorado.

Shen, C.R., E.I. Lan, Y. Dekishima, A. Baez, K.M. Cho, and J.C. Liao. 2011. Driving forces enable high-titer anaerobic 1-butanol synthesis in *Escherichia coli*. *Applied and Environmental Microbiology*. 77:2905–2915.

Shen, H., X. He, C.R. Poovaiah, W.A. Wuddineh, J. Ma, D.G. Mann, H. Wang, L. Jackson, Y. Tang, and C. Neal Stewart. 2012. Functional characterization of the switchgrass (*Panicum virgatum*) R2R3-MYB transcription factor *Pv*MYB4 for improvement of lignocellulosic feedstocks. *New Phytologist*. 193:121–136.

Shestakov, S. and A.M. L. Mikheeva, N. Ravin and K. Skryabin. 2013. Genomic analysis of *Anabaena variabilis* mutants PK17 and PK84 that are characterised by high production of molecular hydrogen. *Advances in Microbiology*. 3:350–365.

Singh, V., I. Mani, D.K. Chaudhary, and P.K. Dhar. 2014. Metabolic engineering of biosynthetic pathway for production of renewable biofuels. *Applied Biochemistry and Biotechnology.* 172:1158–1171.

Skizim, N.J., G.M. Ananyev, A. Krishnan, and G.C. Dismukes. 2012. Metabolic pathways for photobiological hydrogen production by nitrogenase- and hydrogenase-containing unicellular cyanobacteria *Cyanothece. Journal of Biological Chemistry.* 287:2777–2786.

Smith, K.M., K.-M. Cho, and J.C. Liao. 2010. Engineering *Corynebacterium glutamicum* for isobutanol production. *Applied Microbiology and Biotechnology.* 87:1045–1055.

Steen, E.J., Y. Kang, G. Bokinsky, Z. Hu, A. Schirmer, A. McClure, S.B. Del Cardayre, and J.D. Keasling. 2010. Microbial production of fatty-acid-derived fuels and chemicals from plant biomass. *Nature.* 463:559–562.

Steinbusch, K.J., H.V. Hamelers, and C.J. Buisman. 2008. Alcohol production through volatile fatty acids reduction with hydrogen as electron donor by mixed cultures. *Water Research.* 42:4059–4066.

Sticklen, M. 2006. Plant genetic engineering to improve biomass characteristics for biofuels. *Current Opinion in Biotechnology.* 17:315–319.

Stripp, S.T., G. Goldet, C. Brandmayr, O. Sanganas, K.A. Vincent, M. Haumann, F.A. Armstrong, and T. Happe. 2009. How oxygen attacks [Fe–Fe] hydrogenases from photosynthetic organisms. *Proceedings of the National Academy of Sciences.* 106:17331–17336.

Subramanian, V., A. Dubini, D.P. Astling, L.M.L. Laurens, W.M. Old, A.R. Grossman, M.C. Posewitz, and M. Seibert. 2014. Profiling *Chlamydomonas* metabolism under dark, anoxic H_2-producing conditions using a combined proteomic, transcriptomic, and metabolomic approach. *Journal of Proteome Research.* 13:5431–5451.

Sun, Y., M. Chen, H. Yang, J. Zhang, T. Kuang, and F. Huang. 2013. Enhanced H_2 photoproduction by down-regulation of ferredoxin-NADP$^+$ reductase (FNR) in the green alga *Chlamydomonas reinhardtii. International Journal of Hydrogen Energy.* 38:16029–16037.

Sveshnikov, D.A., N.V. Sveshnikova, K.K. Rao, and D.O. Hall. 1997. Hydrogen metabolism of mutant forms of *Anabaena variabilis* in continuous cultures and under nutritional stress. *FEMS Microbiology Letters.* 147:297–301.

Swanson, K.D., M.W. Ratzloff, D.W. Mulder, J.H. Artz, S. Ghose, A. Hoffman, S. White et al. 2015. [FeFe]-hydrogenase oxygen inactivation is initiated at the H cluster 2Fe subcluster. *Journal of the American Chemical Society.* 137:1809–1816.

Tachibana, Y., L. Vayssieres, and J.R. Durrant. 2012. Artificial photosynthesis for solar water-splitting. *Nature Photonics.* 6:511–518.

Tai, M. and G. Stephanopoulos. 2013. Engineering the push and pull of lipid biosynthesis in oleaginous yeast *Yarrowia lipolytica* for biofuel production. *Metabolic Engineering.* 15:1–9.

Tamagnini, P., R. Axelsson, P. Lindberg, F. Oxelfelt, R. Wünschiers, and P. Lindblad. 2002. Hydrogenases and hydrogen metabolism of cyanobacteria. *Microbiology and Molecular Biology Reviews.* 66:1–20.

Tan, X., L. Yao, Q. Gao, W. Wang, F. Qi, and X. Lu. 2011. Photosynthesis driven conversion of carbon dioxide to fatty alcohols and hydrocarbons in cyanobacteria. *Metabolic Engineering.* 13:169–176.

Tee, T.W., A. Chowdhury, C.D. Maranas, and J.V. Shanks. 2014. Systems metabolic engineering design: Fatty acid production as an emerging case study. *Biotechnology and Bioengineering.* 111:849–857.

Teplitski, M., H. Chen, S. Rajamani, M. Gao, M. Merighi, R.T. Sayre, J.B. Robinson, B.G. Rolfe, and W.D. Bauer. 2004. *Chlamydomonas reinhardtii* secretes compounds that mimic bacterial signals and interfere with quorum sensing regulation in bacteria. *Plant Physiology.* 134:137–146.

Terashima, M., M. Specht, B. Naumann, and M. Hippler. 2010. Characterizing the anaerobic response of *Chlamydomonas reinhardtii* by quantitative proteomics. *Molecular & Cellular Proteomics: MCP.* 9:1514–1532.

Timmins, M., W. Zhou, L. Lim, S.R. Thomas-Hall, A. Doebbe, O. Kruse, B. Hankamer, U.C. Marx, S.M. Smith, and P.M. Schenk. 2009. The metabolome of *Chlamydomonas reinhardtii* following induction of anaerobic H_2 production by sulphur deprivation. *Journal of Biological Chemistry.* 284:23415–23425.

Toepel, J., M. Illmer-Kephalides, S. Jaenicke, J. Straube, P. May, A. Goesmann, and O. Kruse. 2013. New insights into *Chlamydomonas reinhardtii* hydrogen production processes by combined microarray/RNA-seq transcriptomics. *Plant Biotechnology Journal.* 11:717–733.

Tolleter, D., B. Ghysels, J. Alric, D. Petroutsos, I. Tolstygina, D. Krawietz, T. Happe et al. 2011. Control of hydrogen photoproduction by the proton gradient generated by cyclic electron flow in *Chlamydomonas reinhardtii*. *The Plant Cell Online*. 23:2619–2630.

Tomme, P., R.A.J. Warren, N.R. Gilkes, and R.K. Poole. 1995. Cellulose hydrolysis by bacteria and fungi. *In Advances in Microbial Physiology*. Vol. 37. Robert K. Poole (Ed.), Academic Press, London. 1–81.

Torella, J.P., T.J. Ford, S.N. Kim, A.M. Chen, J.C. Way, and P.A. Silver. 2013. Tailored fatty acid synthesis via dynamic control of fatty acid elongation. *Proceedings of the National Academy of Sciences*. 110:11290–11295.

Torzillo, G., A. Scoma, C. Faraloni, A. Ena, and U. Johanningmeier. 2009. Increased hydrogen photoproduction by means of a sulfur-deprived *Chlamydomonas reinhardtii* D1 protein mutant. *International Journal of Hydrogen Energy*. 34:4529–4536.

Torzillo, G. and M. Seibert. 2013. Hydrogen Production by *Chlamydomonas reinhardtii*. In *Handbook of Microalgal Culture*. John Wiley & Sons, Ltd. 417–432.

Trentacoste, E.M., R.P. Shrestha, S.R. Smith, C. Gle, A.C. Hartmann, M. Hildebrand, and W.H. Gerwick. 2013. Metabolic engineering of lipid catabolism increases microalgal lipid accumulation without compromising growth. *Proceedings of the National Academy of Sciences*. 110:19748–19753.

Ubersax, J.A. and D.M. Platt. 2010. Genetically modified microbes producing isoprenoids. US Patents 12/791,596.

Van Acker, R., J.-C. Leplé, D. Aerts, V. Storme, G. Goeminne, B. Ivens, F. Légée, C. Lapierre, K. Piens, and M.C. Van Montagu. 2014. Improved saccharification and ethanol yield from field-grown transgenic poplar deficient in cinnamoyl-CoA reductase. *Proceedings of the National Academy of Sciences*. 111:845–850.

van Lis, R., C. Baffert, Y. Couté, W. Nitschke, and A. Atteia. 2013. *Chlamydomonas reinhardtii* chloroplasts contain a homodimeric pyruvate:ferredoxin oxidoreductase that functions with FDX1. *Plant Physiology*. 161:57–71.

Vanholme, R., K. Morreel, J. Ralph, and W. Boerjan. 2008. Lignin engineering. *Current Opinion in Plant Biology*. 11:278–285.

Verma, D., A. Kanagaraj, S. Jin, N.D. Singh, P.E. Kolattukudy, and H. Daniell. 2010. Chloroplast-derived enzyme cocktails hydrolyse lignocellulosic biomass and release fermentable sugars. *Plant Biotechnology Journal*. 8:332–350.

Wang, Z.T., N. Ullrich, S. Joo, S. Waffenschmidt, and U. Goodenough. 2010. Algal lipid bodies: Stress induction, purification, and biochemical characterization in wild-type and starchless *Chlamydomonas reinhardtii*. *Eukaryotic Cell*. 8:1856–1868.

Wilkerson, C., S. Mansfield, F. Lu, S. Withers, J.-Y. Park, S. Karlen, E. Gonzales-Vigil, D. Padmakshan, F. Unda, and J. Rencoret. 2014. Monolignol ferulate transferase introduces chemically labile linkages into the lignin backbone. *Science*. 344:90–93.

Winkler, M., S. Kuhlgert, M. Hippler, and T. Happe. 2009. Characterization of the key step for light-driven hydrogen evolution in green algae. *Journal of Biological Chemistry*. 284:36620–36627.

Withers, S.T., S.S. Gottlieb, B. Lieu, J.D. Newman, and J.D. Keasling. 2007. Identification of isopentenol biosynthetic genes from *Bacillus subtilis* by a screening method based on isoprenoid precursor toxicity. *Applied and Environmental Microbiology*. 73:6277–6283.

Wolk, C.P., A. Ernst, and J. Elhai. 2004. Heterocyst metabolism and development. In *The Molecular Biology of Cyanobacteria SE—27*. Vol. 1. D. Bryant, editor. Springer, Netherlands. 769–823.

Xiao, C. and C.T. Anderson. 2013. Roles of pectin in biomass yield and processing for biofuels. *Frontiers in Plant Science*. 4: 61–67. doi: 10.3389/fpls.2013.00067.

Yacoby, I., S. Pochekailov, H. Toporik, M.L. Ghirardi, P.W. King, and S.G. Zhang. 2011. Photosynthetic electron partitioning between [FeFe]-hydrogenase and ferredoxin:NADP$^{(+)}$-oxidoreductase (FNR) enzymes in vitro. *Proceedings of the National Academy of Sciences*. 108:9396–9401.

Yan, Y. and J.C. Liao. 2009. Engineering metabolic systems for production of advanced fuels. *Journal of Industrial Microbiology & Biotechnology*. 36:471–479.

Yang, S., M. Guarnieri, S. Smolinski, M. Ghirardi, and P. Pienkos. 2013. *De novo* transcriptomic analysis of hydrogen production in the green alga *Chlamydomonas moewusii* through RNA-Seq. *Biotechnology for Biofuels*. 6:1–17.

Yang, W., C. Catalanotti, S. D'Adamo, T.M. Wittkopp, C.J. Ingram-Smith, L. Mackinder, T.E. Miller et al. 2014. Alternative acetate production pathways in *Chlamydomonas reinhardtii* during dark anoxia and the dominant role of chloroplasts in fermentative acetate production. *The Plant Cell Online*. 26:4499–44518.

Yoshino, F., H. Ikeda, H. Masukawa, and H. Sakurai. 2007. High photobiological hydrogen production activity of a *Nostoc* sp. PCC 7422 uptake hydrogenase-deficient mutant with high nitrogenase activity. *Marine Biotechnology*. 9:101–112.

Youngquist, J.T., J.P. Rose, and B.F. Pfleger. 2013. Free fatty acid production in *Escherichia coli* under phosphate-limited conditions. *Applied Microbiology and Biotechnology*. 97:5149–5159.

Yu, L.X., B.N. Gray, C.J. Rutzke, L.P. Walker, D.B. Wilson, and M.R. Hanson. 2007. Expression of thermostable microbial cellulases in the chloroplasts of nicotine-free tobacco. *Journal of Biotechnology*. 131:362–369.

Yuzawa, S., W. Kim, L. Katz, and J.D. Keasling. 2012. Heterologous production of polyketides by modular type I polyketide synthases in *Escherichia coli. Current Opinion in Biotechnology*. 23:727–735.

Zhang, F., J.M. Carothers, and J.D. Keasling. 2012. Design of a dynamic sensor-regulator system for production of chemicals and fuels derived from fatty acids. *Nature Biotechnology*. 30:354–359.

Zhang, F., S. Rodriguez, and J.D. Keasling. 2011. Metabolic engineering of microbial pathways for advanced biofuels production. *Current Opinion in Biotechnology*. 22:775–783.

Zhang, M., C. Eddy, K. Deanda, M. Finkelstein, and S. Picataggio. 1995. Metabolic engineering of a pentose metabolism pathway in ethanologenic *Zymomonas mobilis. Science*. 267:240–243.

Zhang, X., D.M. Sherman, and L.A. Sherman. 2014. The uptake hydrogenase in the unicellular diazotrophic cyanobacterium *Cyanothece* sp. Strain PCC 7822 protects nitrogenase from oxygen toxicity. *Journal of Bacteriology*. 196:840–849.

Zheng, Y., Q. Liu, L. Li, W. Qin, J. Yang, H. Zhang, X. Jiang, T. Cheng, W. Liu, and X. Xu. 2013. Metabolic engineering of *Escherichia coli* for high-specificity production of isoprenol and prenol as next generation of biofuels. *Biotechnology for Biofuels*. 6:57.

Ziegelhoffer, T., J.A. Raasch, and S. Austin-Phillips. 2001. Dramatic effects of truncation and subcellular targeting on the accumulation of recombinant microbial cellulase in tobacco. *Molecular Breeding*. 8:147–158.

Ziegler, M.T., S.R. Thomas, and K.J. Danna. 2000. Accumulation of a thermostable endo-1,4-β-D-glucanase in the apoplast of *Arabidopsis thaliana* leaves. *Molecular Breeding*. 6:37–46.

Index

T - #0155 - 111024 - C652 - 254/178/30 - PB - 9780367872472 - Gloss Lamination